Lecture Notes in Artificial Intelligence 6321

Edited by R. Goebel, J. Siekmann, and W. Wahlster

Subseries of Lecture Notes in Computer Science

José Luis Balcázar
Francesco Bonchi Aristides Gionis
Michèle Sebag (Eds.)

Machine Learning and Knowledge Discovery in Databases

European Conference, ECML PKDD 2010
Barcelona, Spain, September 20-24, 2010
Proceedings, Part I

Springer

Series Editors

Randy Goebel, University of Alberta, Edmonton, Canada
Jörg Siekmann, University of Saarland, Saarbrücken, Germany
Wolfgang Wahlster, DFKI and University of Saarland, Saarbrücken, Germany

Volume Editors

José Luis Balcázar
Universidad de Cantabria
Departamento de Matemáticas, Estadística y Computación
Avenida de los Castros, s/n, 39071 Santander, Spain
E-mail: joseluis.balcazar@unican.es

Francesco Bonchi
Aristides Gionis
Yahoo! Research Barcelona
Avinguda Diagonal 177, 08018 Barcelona, Spain
E-mail: {bonchi, gionis}@yahoo-inc.corp

Michèle Sebag
TAO, CNRS-INRIA-LRI, Université Paris-Sud
91405, Orsay, France
E-mail: sebag@lri.fr

Cover illustration: Decoration detail at the Park Güell, designed by Antoni Gaudí, and one of the landmarks of modernist art in Barcelona. Licence Creative Commons, Jon Robson.

Library of Congress Control Number: 2010934301

CR Subject Classification (1998): I.2, H.3, H.4, H.2.8, J.1, H.5

LNCS Sublibrary: SL 7 – Artificial Intelligence

ISSN 0302-9743
ISBN-10 3-642-15879-X Springer Berlin Heidelberg New York
ISBN-13 978-3-642-15879-7 Springer Berlin Heidelberg New York

springer.com

Printed in Germany

Typesetting: Camera-ready by author, data conversion by Scientific Publishing Services, Chennai, India
Printed on acid-free paper 06/3180

Preface

The European Conference on Machine Learning and Principles and Practice of Knowledge Discovery in Databases, ECML PKDD 2010, was held in Barcelona, September 20–24, 2010, consolidating the long junction between the European Conference on Machine Learning (of which the first instance as European workshop dates back to 1986) and Principles and Practice of Knowledge Discovery in Data Bases (of which the first instance dates back to 1997). Since the two conferences were first collocated in 2001, both machine learning and data mining communities have realized how each discipline benefits from the advances, and participates to defining the challenges, of the sister discipline. Accordingly, a single ECML PKDD Steering Committee gathering senior members of both communities was appointed in 2008.

In 2010, as in previous years, ECML PKDD lasted from Monday to Friday. It involved six plenary invited talks, by Christos Faloutsos, Jiawei Han, Hod Lipson, Leslie Pack Kaelbling, Tomaso Poggio, and Jürgen Schmidhuber, respectively. Monday and Friday were devoted to workshops and tutorials, organized and selected by Colin de la Higuera and Gemma Garriga. Continuing from ECML PKDD 2009, an industrial session managed by Taneli Mielikainen and Hugo Zaragoza welcomed distinguished speakers from the ML and DM industry: Rakesh Agrawal, Mayank Bawa, Ignasi Belda, Michael Berthold, José Luis Flórez, Thore Graepel, and Alejandro Jaimes. The conference also featured a discovery challenge, organized by András Benczúr, Carlos Castillo, Zoltán Gyöngyi, and Julien Masanès.

From Tuesday to Thursday, 120 papers selected among 658 submitted full papers were presented in the technical parallel sessions. The selection process was handled by 28 area chairs and the 282 members of the Program Committee; additional 298 reviewers were recruited. While the selection process was made particularly intense due to the record number of submissions, we heartily thank all area chairs, members of the Program Committee, and additional reviewers for their commitment and hard work during the short reviewing period. The conference also featured a demo track, managed by Ulf Brefeld and Xavier Carreras; 12 demos out of 24 submitted ones were selected, attesting to the high impact technologies based on the ML and DM body of research.

Following an earlier tradition, seven ML and seven DM papers were distinguished by the program chairs on the basis of their exceptional scientific quality and high impact on the field, and they were directly published in the Machine Learning Journal and the Data Mining and Knowledge Discovery Journal, respectively. Among these papers, some were selected by the Best Paper Chair Hiroshi Motoda, and received the Best Paper Awards and Best Student Paper Awards in Machine Learning and in Data Mining, sponsored by Springer.

A topic widely explored from both ML and DM perspectives was graphs, with motivations ranging from molecular chemistry to social networks. The point of matching or clustering graphs was examined in connection with tractability and domain knowledge, where the latter could be acquired through common patterns, or formulated through spectral clustering. The study of social networks focused on how they develop, overlap, propagate information (and how information propagation can be hindered). Link prediction and exploitation in static or dynamic, possibly heterogeneous, graphs, was motivated by applications in information retrieval and collaborative filtering, and in connection with random walks.

Frequent itemset approaches were hybridized with constraint programming or statistical tools to efficiently explore the search space, deal with numerical attributes, or extract locally optimal patterns. Compressed representations and measures of robustness were proposed to optimize association rules. Formal concept analysis, with applications to pharmacovigilance or Web ontologies, was considered in connection with version spaces.

Bayesian learning features new geometric interpretations of prior knowledge and efficient approaches for independence testing. Generative approaches were motivated by applications in sequential, spatio-temporal or relational domains, or multi-variate signals with high dimensionality. Ensemble learning was used to support clustering and biclustering; the post-processing of random forests was also investigated.

In statistical relational learning and structure identification, with motivating applications in bio-informatics, neuro-imagery, spatio-temporal domains, and traffic forecasting, the stress was put on new learning criteria; gradient approaches, structural constraints, and/or feature selection were used to support computationally effective algorithms.

(Multiple) kernel learning and related approaches, challenged by applications in image retrieval, robotics, or bio-informatics, revisited the learning criteria and regularization terms, the processing of the kernel matrix, and the exploration of the kernel space. Dimensionality reduction, embeddings, and distance were investigated, notably in connection with image and document retrieval.

Reinforcement learning focussed on ever more scalable and tractable approaches through smart state or policy representations, a more efficient use of the available samples, and/or Bayesian approaches.

Specific settings such as ranking, multi-task learning, semi-supervised learning, and game-theoretic approaches were investigated, with some innovative applications to astrophysics, relation extraction, and multi-agent systems. New bounds were proved within the active, multi-label, and weighted ensemble learning frameworks.

A few papers aimed at efficient algorithms or computing environments, e.g., related to linear algebra, cutting plane algorithms, or graphical processing units, were proposed (with available source code in some cases). Numerical stability was also investigated in connection with sparse learning.

Among the applications presented were review mining, software debugging/ process modeling from traces, and audio mining.

To conclude this rapid tour of the scientific program, our special thanks go to the local chairs Ricard Gavaldà, Elena Torres, and Estefania Ricart, the Web and registration chair Albert Bifet, the sponsorship chair Debora Denato, and the many volunteers that eagerly contributed to make ECML PKDD 2010 a memorable event.

Our last and warmest thanks go to all invited speakers and other speakers, to all tutorial, workshop, demo, industrial, discovery, best paper, and local chairs, to the area chairs and all reviewers, to all attendees — and overall, to the authors who chose to submit their work to the ECML PKDD conference, and thus enabled us to build up this memorable scientific event.

July 2010

José L Balcázar
Francesco Bonchi
Aristides Gionis
Michèle Sebag

Organization

Program Chairs

José L Balcázar
Universidad de Cantabria and
Universitat Politècnica de Catalunya, Spain
http://personales.unican.es/balcazarjl/

Francesco Bonchi
Yahoo! Research
Barcelona, Spain
http://research.yahoo.com

Aristides Gionis
Yahoo! Research
Barcelona, Spain
http://research.yahoo.com

Michèle Sebag
CNRS
Université Paris Sud, Orsay Cedex, France
http://www.lri.fr/ sebag/

Local Organization Chairs

Ricard Gavaldà	Universitat Politècnica de Catalunya
Estefania Ricart	Barcelona Media
Elena Torres	Barcelona Media

Organization Team

Ulf Brefeld	Yahoo! Research
Eugenia Fuenmayor	Barcelona Media
Mia Padullés	Yahoo! Research
Natalia Pou	Barcelona Media

Workshop and Tutorial Chairs

Gemma C. Garriga	University of Paris 6
Colin de la Higuera	University of Nantes

Best Papers Chair

Hiroshi Motoda AFOSR/AOARD and Osaka University

Industrial Track Chairs

Taneli Mielikainen Nokia
Hugo Zaragoza Yahoo! Research

Demo Chairs

Ulf Brefeld Yahoo! Research
Xavier Carreras Universitat Politècnica de Catalunya

Discovery Challenge Chairs

András Benczúr Hungarian Academy of Sciences
Carlos Castillo Yahoo! Research
Zoltán Gyöngyi Google
Julien Masanès European Internet Archive

Sponsorship Chair

Debora Donato Yahoo! Labs

Web and Registration Chair

Albert Bifet University of Waikato

Publicity Chair

Ricard Gavaldà Universitat Politècnica de Catalunya

Steering Committee

Wray Buntine
Walter Daelemans
Bart Goethals
Marko Grobelnik
Katharina Morik
Joost N. Kok
Stan Matwin
Dunja Mladenic
John Shawe-Taylor
Andrzej Skowron

Area Chairs

Samy Bengio
Bettina Berendt
Paolo Boldi
Wray Buntine
Toon Calders
Luc de Raedt
Carlotta Domeniconi
Martin Ester
Paolo Frasconi
Joao Gama
Ricard Gavaldà
Joydeep Ghosh
Fosca Giannotti
Tu-Bao Ho
George Karypis
Laks V.S. Lakshmanan
Katharina Morik
Jan Peters
Kai Puolamäki
Yucel Saygin
Bruno Scherrer
Arno Siebes
Soeren Sonnenburg
Alexander Smola
Einoshin Suzuki
Evimaria Terzi
Michalis Vazirgiannis
Zhi-Hua Zhou

Program Committee

Osman Abul
Gagan Agrawal
Erick Alphonse
Carlos Alzate
Massih Amini
Aris Anagnostopoulos
Annalisa Appice
Thierry Artières
Sitaram Asur
Jean-Yves Audibert
Maria-Florina Balcan
Peter Bartlett
Younes Bennani
Paul Bennett
Michele Berlingerio
Michael Berthold
Albert Bifet
Hendrik Blockeel
Mario Boley
Antoine Bordes
Gloria Bordogna
Christian Borgelt
Karsten Borgwardt
Henrik Boström
Marco Botta
Guillaume Bouchard
Jean-Francois Boulicaut
Ulf Brefeld
Laurent Brehelin
Bjoern Bringmann
Carla Brodley
Rui Camacho
Stéphane Canu
Olivier Cappé
Carlos Castillo
Jorge Castro
Ciro Cattuto
Nicolò Cesa-Bianchi
Nitesh Chawla
Sanjay Chawla
David Cheung
Sylvia Chiappa
Boris Chidlovski
Flavio Chierichetti
Philipp Cimiano
Alexander Clark
Christopher Clifton
Antoine Cornuéjols
Fabrizio Costa
Bruno Crémilleux
James Cussens
Alfredo Cuzzocrea

Florence d'Alché-Buc
Claudia d'Amato
Gautam Das
Jeroen De Knijf
Colin de la Higuera
Krzysztof Dembczynski
Ayhan Demiriz
Francois Denis
Christos Dimitrakakis
Josep Domingo Ferrer
Debora Donato
Dejing Dou
Gérard Dreyfus
Kurt Driessens
John Duchi
Pierre Dupont
Saso Dzeroski
Charles Elkan
Damien Ernst
Floriana Esposito
Fazel Famili
Nicola Fanizzi
Ad Feelders
Alan Fern
Daan Fierens
Peter Flach
George Forman
Vojtech Franc
Eibe Frank
Dayne Freitag
Elisa Fromont
Patrick Gallinari
Auroop Ganguly
Fred Garcia
Gemma Garriga
Thomas Gärtner
Eric Gaussier
Floris Geerts
Matthieu Geist
Claudio Gentile
Mohammad Ghavamzadeh
Gourab Ghoshal
Chris Giannella
Attilio Giordana
Mark Girolami
Shantanu Godbole
Bart Goethals
Sally Goldman
Henrik Grosskreutz
Dimitrios Gunopulos
Amaury Habrard
Eyke Hüllermeier
Nikolaus Hansen
Iris Hendrickx
Melanie Hilario
Alexander Hinneburg
Kouichi Hirata
Frank Hoeppner
Jaakko Hollmen
Tamas Horvath
Andreas Hotho
Alex Jaimes
Szymon Jaroszewicz
Daxin Jiang
Felix Jungermann
Frederic Jurie
Alexandros Kalousis
Panagiotis Karras
Samuel Kaski
Dimitar Kazakov
Sathiya Keerthi
Jens Keilwagen
Roni Khardon
Angelika Kimmig
Ross King
Marius Kloft
Arno Knobbe
Levente Kocsis
Jukka Kohonen
Solmaz Kolahi
George Kollios
Igor Kononenko
Nick Koudas
Stefan Kramer
Andreas Krause
Vipin Kumar
Pedro Larrañaga
Mark Last
Longin Jan Latecki
Silvio Lattanzi

Anne Laurent
Nada Lavrac
Alessandro Lazaric
Philippe Leray
Jure Leskovec
Carson Leung
Chih-Jen Lin
Jessica Lin
Huan Liu
Kun Liu
Alneu Lopes
Ramón López de Màntaras
Eneldo Loza Mencía
Claudio Lucchese
Elliot Ludvig
Dario Malchiodi
Donato Malerba
Bradley Malin
Giuseppe Manco
Shie Mannor
Stan Matwin
Michael May
Thorsten Meinl
Prem Melville
Rosa Meo
Pauli Miettinen
Lily Mihalkova
Dunja Mladenic
Ali Mohammad-Djafari
Fabian Morchen
Alessandro Moschitti
Ion Muslea
Mirco Nanni
Amedeo Napoli
Claire Nedellec
Frank Nielsen
Siegfried Nijssen
Richard Nock
Sebastian Nowozin
Alexandros Ntoulas
Andreas Nuernberger
Arlindo Oliveira
Balaji Padmanabhan
George Paliouras
Themis Palpanas
Apostolos Papadopoulos
Andrea Passerini
Jason Pazis
Mykola Pechenizkiy
Dmitry Pechyony
Dino Pedreschi
Jian Pei
Jose Peña
Ruggero Pensa
Marc Plantevit
Enric Plaza
Doina Precup
Ariadna Quattoni
Predrag Radivojac
Davood Rafiei
Chedy Raissi
Alain Rakotomamonjy
Liva Ralaivola
Naren Ramakrishnan
Jan Ramon
Chotirat Ratanamahatana
Elisa Ricci
Bertrand Rivet
Philippe Rolet
Marco Rosa
Fabrice Rossi
Juho Rousu
Céline Rouveirol
Cynthia Rudin
Salvatore Ruggieri
Stefan Rüping
Massimo Santini
Lars Schmidt-Thieme
Marc Schoenauer
Marc Sebban
Nicu Sebe
Giovanni Semeraro
Benyah Shaparenko
Jude Shavlik
Fabrizio Silvestri
Dan Simovici
Carlos Soares
Diego Sona
Alessandro Sperduti
Myra Spiliopoulou

Gerd Stumme
Jiang Su
Masashi Sugiyama
Johan Suykens
Domenico Talia
Pang-Ning Tan
Tamir Tassa
Nikolaj Tatti
Yee Whye Teh
Maguelonne Teisseire
Olivier Teytaud
Jo-Anne Ting
Michalis Titsias
Hannu Toivonen
Ryota Tomioka
Marc Tommasi
Hanghang Tong
Luis Torgo
Fabien Torre
Marc Toussaint
Volker Tresp
Koji Tsuda
Alexey Tsymbal
Franco Turini
Antti Ukkonen
Matthijs van Leeuwen
Martijn van Otterlo
Maarten van Someren
Celine Vens
Jean-Philippe Vert
Ricardo Vilalta
Christel Vrain
Jilles Vreeken
Christian Walder
Louis Wehenkel
Markus Weimer
Dong Xin
Dit-Yan Yeung
Cong Yu
Philip Yu
Chun-Nam Yue
Francois Yvon
Bianca Zadrozny
Carlo Zaniolo
Gerson Zaverucha
Filip Zelezny
Albrecht Zimmermann

Additional Reviewers

Mohammad Ali Abbasi
Zubin Abraham
Yong-Yeol Ahn
Fabio Aiolli
Dima Alberg
Salem Alelyani
Aneeth Anand
Sunil Aryal
Arthur Asuncion
Gowtham Atluri
Martin Atzmueller
Paolo Avesani
Pranjal Awasthi
Hanane Azzag
Miriam Baglioni
Raphael Bailly
Jaume Baixeries
Jorn Bakker
Georgios Balkanas
Nicola Barbieri
Teresa M.A. Basile
Luca Bechetti
Dominik Benz
Maxime Berar
Juliana Bernardes
Aurélie Boisbunon
Shyam Boriah
Zoran Bosnic
Robert Bossy
Lydia Boudjeloud
Dominique Bouthinon
Janez Brank
Sandra Bringay
Fabian Buchwald
Krisztian Buza
Matthias Böck

José Caldas
Gabriele Capannini
Annalina Caputo
Franco Alberto Cardillo
Xavier Carreras
Giovanni Cavallanti
Michelangelo Ceci
Eugenio Cesario
Pirooz Chubak
Anna Ciampi
Ronan Collobert
Carmela Comito
Gianni Costa
Bertrand Cuissart
Boris Cule
Giovanni Da San Martino
Marco de Gemmis
Kurt De Grave
Gerben de Vries
Jean Decoster
Julien Delporte
Christian Desrosiers
Sanjoy Dey
Nicola Di Mauro
Joshua V. Dillon
Huyen Do
Stephan Doerfel
Brett Drury
Timo Duchrow
Wouter Duivesteijn
Alain Dutech
Ilenia Epifani
Ahmet Erhan Nergiz
Rémi Eyraud
Philippe Ezequel
Jean Baptiste Faddoul
Fabio Fassetti
Bruno Feres de Souza
Remi Flamary
Alex Freitas
Natalja Friesen
Gabriel P.C. Fung
Barbara Furletti
Zeno Gantner
Steven Ganzert
Huiji Gao
Ashish Garg
Aurelien Garivier
Gilles Gasso
Elisabeth Georgii
Edouard Gilbert
Tobias Girschick
Miha Grcar
Warren Greiff
Valerio Grossi
Nistor Grozavu
Massimo Guarascio
Tias Guns
Vibhor Gupta
Rohit Gupta
Tushar Gupta
Nico Görnitz
Hirotaka Hachiya
Steve Hanneke
Andreas Hapfelmeier
Daniel Hsu
Xian-Sheng Hua
Yi Huang
Romain Hérault
Leo Iaquinta
Dino Ienco
Elena Ikonomovska
Stéphanie Jacquemont
Jean-Christophe Janodet
Frederik Janssen
Baptiste Jeudy
Chao Ji
Goo Jun
U Kang
Anuj Karpatne
Jaya Kawale
Ashraf M. Kibriya
Kee-Eung Kim
Akisato Kimura
Arto Klami
Suzan Koknar-Tezel
Xiangnan Kong
Arne Koopman
Mikko Korpela
Wojciech Kotlowski

Alexis Kotsifakos
Petra Kralj Novak
Tetsuji Kuboyama
Matjaz Kukar
Sanjiv Kumar
Shashank Kumar
Pascale Kuntz
Ondrej Kuzelka
Benjamin Labbe
Mathieu Lajoie
Hugo Larochelle
Agnieszka Lawrynowicz
Gregor Leban
Mustapha Lebbah
John Lee
Sau Dan Lee
Gayle Leen
Florian Lemmerich
Biao Li
Ming Li
Rui Li
Tiancheng Li
Yong Li
Yuan Li
Wang Liang
Ryan Lichtenwalter
Haishan Liu
Jun Liu
Lei Liu
Xu-Ying Liu
Corrado Loglisci
Pasquale Lops
Chuan Lu
Ana Luisa Duboc
Panagis Magdalinos
Sebastien Mahler
Michael Mampaey
Prakash Mandayam
Alain-Pierre Manine
Patrick Marty
Jeremie Mary
André Mas
Elio Masciari
Emanuel Matos
Andreas Maunz
John McCrae
Marvin Meeng
Wannes Meert
Joao Mendes-Moreira
Aditya Menon
Peter Mika
Folke Mitzlaff
Anna Monreale
Tetsuro Morimura
Ryoko Morioka
Babak Mougouie
Barzan Mozafari
Igor Mozetic
Cataldo Musto
Alexandros Nanopoulos
Fedelucio Narducci
Maximilian Nickel
Inna Novalija
Benjamin Oatley
Marcia Oliveira
Emauele Olivetti
Santiago Ontañón
Francesco Orabona
Laurent Orseau
Riccardo Ortale
Aomar Osmani
Aline Paes
Sang-Hyeun Park
Juuso Parkkinen
Ioannis Partalas
Pekka Parviainen
Krishnan Pillaipakkamnatt
Fabio Pinelli
Cristiano Pitangui
Barbara Poblete
Vid Podpecan
Luigi Pontieri
Philippe Preux
Han Qin
Troy Raeder
Subramanian Ramanathan
Huzefa Rangwala
Guillaume Raschia
Konrad Rieck
François Rioult

Ettore Ritacco
Mathieu Roche
Christophe Rodrigues
Philippe Rolet
Andrea Romei
Jan Rupnik
Delia Rusu
Ulrich Rückert
Hiroshi Sakamoto
Vitor Santos Costa
Kengo Sato
Saket Saurabh
Francois Scharffe
Leander Schietgat
Jana Schmidt
Constanze Schmitt
Christoph Scholz
Dan Schrider
Madeleine Seeland
Or Sheffet
Noam Shental
Xiaoxiao Shi
Naoki Shibayama
Nobuyuki Shimizu
Kilho Shin
Kaushik Sinha
Arnaud Soulet
Michal Sramka
Florian Steinke
Guillaume Stempfel
Liwen Sun
Umar Syed
Gabor Szabo
Yasuo Tabei
Nima Taghipour
Hana Tai
Frédéric Tantini
Katerina Tashkova
Christine Task
Alexandre Termier
Lam Thoang Hoang
Xilan Tian
Xinmei Tian
Gabriele Tolomei
Aneta Trajanov
Roberto Trasarti
Abhishek Tripathi
Paolo Trunfio
Ivor Tsang
Theja Tulabandhula
Boudewijn van Dongen
Stijn Vanderlooy
Joaquin Vanschoren
Philippe Veber
Sriharsha Veeramachaneni
Sebastian Ventura
Alessia Visconti
Jun Wang
Xufei Wang
Osamu Watanabe
Lorenz Weizsäcker
Tomas Werner
Jörg Wicker
Derry Wijaya
Daya Wimalasuriya
Adam Woznica
Fuxiao Xin
Zenglin Xu
Makoto Yamada
Liu Yang
Xingwei Yang
Zhirong Yang
Florian Yger
Reza Bosagh Zadeh
Reza Zafarani
Amelia Zafra
Farida Zehraoui
Kai Zeng
Bernard Zenko
De-Chuan Zhan
Min-Ling Zhang
Indre Zliobaite

Sponsors

We wish to express our gratitude to the sponsors of ECML PKDD 2010 for their essential contribution to the conference: the French National Institute for Research in Computer Science and Control (INRIA), the Pascal2 European Network of Excellence, Nokia, Yahoo! Labs, Google, KNIME, Aster data, Microsoft Research, HP, MODAP (Mobility, Data Mining, and Privacy) a Coordination Action type project funded by EU, FET OPEN, the Data Mining and Knowledge Discovery Journal, the Machine Learning Journal, LRI (Laboratoire de Recherche en Informatique, Université Paris-Sud -CNRS), ARES (Advanced Research on Information Security and Privacy) a national Spanish project, the UNESCO Chair in Data Privacy, Xerox, Universitat Politecnica de Catalunya, IDESCAT (Institut d'Estadistica de Catalunya), and the Ministerio de Ciencia e Innovacion (Spanish government).

Table of Contents – Part I

Invited Talks (Abstracts)

Regular Papers

Table of Contents – Part II

Regular Papers

Table of Contents – Part III

Regular Papers

Demo Papers

Mining Billion-Node Graphs: Patterns, Generators and Tools

Christos Faloutsos

Computer Science Department,
Carnegie Mellon University
christos@cs.cmu.edu

What do graphs look like? How do they evolve over time? How to handle a graph with a billion nodes? We present a comprehensive list of static and temporal laws, and some recent observations on real graphs (like, e.g., "eigenSpokes"). For generators, we describe some recent ones, which naturally match all of the known properties of real graphs. Finally, for tools, we present "oddBall" for discovering anomalies and patterns, as well as an overview of the PEGASUS system which is designed for handling Billion-node graphs, running on top of the "hadoop" system.

J.L. Balcázar et al. (Eds.): ECML PKDD 2010, Part I, LNAI 6321, p. 1, 2010.

Structure Is Informative: On Mining Structured Information Networks

Jiawei Han

Department of Computer Science,
University of Illinois at Urbana-Champaign
hanj@cs.uiuc.edu

Many objects in the real world are interconnected, forming complex information networks. There have been a lot of studies on mining homogeneous information networks where objects and links are either treated as of the same type, such as friends linking with friends, or treated indiscriminatively, without structural or type distinction. However, real-world objects and links often belong to distinct types, such as students, professors, courses, departments, teach and advise in a university network, and such typed networks form structured, heterogeneous information networks.

We explore methodologies on mining such structured information networks and introduce several interesting new mining methodologies, including integrated ranking and clustering, classification, role discovery, data integration, data validation, and similarity search. We show that structured information networks are informative, and link analysis on such networks becomes powerful at uncovering critical knowledge hidden in large networks.

J.L. Balcázar et al. (Eds.): ECML PKDD 2010, Part I, LNAI 6321, p. 2, 2010.

Intelligent Interaction with the Real World

Leslie Pack Kaelbling

Computer Science and Artificial Intelligence Laboratory (CSAIL)
Massachusetts Institute of Technology
lpk@csail.mit.edu

Since the inception of the field if AI, one of the visions of artificial intelligence has been robust, intelligent, general-purpose robots that interact with the real world. We have made useful progress in that direction, but there is still a long way to go. I will characterize one view of how we might achieve this goal, describe some intermediate results, and characterize important technical and methodological problems that must be solved to make that vision real.

J.L. Balcázar et al. (Eds.): ECML PKDD 2010, Part I, LNAI 6321, p. 3, 2010.

Mining Experimental Data for Dynamical Invariants - From Cognitive Robotics to Computational Biology

Hod Lipson

Mechanical & Aerospace Engineering and Computing & Information Science
Cornell University
hod.lipson@cornell.edu

For centuries, scientists have attempted to identify and document analytical laws that underlie physical phenomena in nature. Despite the prevalence of computing power, the process of finding natural laws and their corresponding equations has resisted automation. A key challenge to finding analytic relations automatically is defining algorithmically what makes a correlation in observed data important and insightful. By seeking dynamical invariants, we go from finding just predictive models to finding deeper conservation laws. We demonstrated this approach by automatically searching motion-tracking data captured from various physical systems, ranging from simple harmonic oscillators to chaotic double-pendula. Without any prior knowledge about physics, kinematics, or geometry, the algorithm discovered Hamiltonians, Lagrangians, and other laws of geometric and momentum conservation. The discovery rate accelerated as laws found for simpler systems were used to bootstrap explanations for more complex systems, gradually uncovering the "alphabet" used to describe those systems. Applications to modeling physical and biological systems will be shown.

J.L. Balcázar et al. (Eds.): ECML PKDD 2010, Part I, LNAI 6321, p. 4, 2010.

Hierarchical Learning Machines and Neuroscience of Visual Cortex

Tomaso Poggio

Center for Biological & Computational Learning
McGovern Institute
Computer Science and Artificial Intelligence Laboratory
Department of Brain and Cognitive Sciences
Massachusetts Institute of Technology
tp@csail.mit.edu

Learning is the gateway to understanding intelligence and to reproducing it in machines. A classical example of learning algorithms is provided by regularization in Reproducing Kernel Hilbert Spaces. The corresponding architecture however is different from the deep hierarchies found in the brain. I will sketch a new attempt (with S. Smale) to develop a mathematics for hierarchical kernel machines centered around the notion of a recursively defined derived kernel and directly suggested by the neuroscience of the visual cortex.

Relevant papers can be downloaded from
http://cbcl.mit.edu/publications/index-pubs.html

J.L. Balcázar et al. (Eds.): ECML PKDD 2010, Part I, LNAI 6321, p. 5, 2010.

Formal Theory of Fun and Creativity

Jürgen Schmidhuber

IDSIA & USI & SUPSI
Switzerland
juergen@idsia.ch

To build a creative agent that never stops generating non-trivial & novel & surprising data, we need two learning modules: (1) an adaptive predictor or compressor or model of the growing data history as the agent is interacting with its environment, and (2) a general reinforcement learner. The LEARNING PROGRESS of (1) is the FUN or intrinsic reward of (2). That is, (2) is motivated to invent interesting things that (1) does not yet know but can easily learn. To maximize expected reward, in the absence of external reward (2) will create more and more complex behaviors that yield temporarily surprising (but eventually boring) patterns that make (1) quickly improve. We discuss how this principle explains science & art & music & humor, and how to scale up previous toy implementations of the theory since 1991, using recent powerful methods for (1) prediction and (2) reinforcement learning.

J.L. Balcázar et al. (Eds.): ECML PKDD 2010, Part I, LNAI 6321, p. 6, 2010.

Porting Decision Tree Algorithms to Multicore Using FastFlow

Marco Aldinucci[1], Salvatore Ruggieri[2], and Massimo Torquati[2]

[1] Computer Science Department, University of Torino, Italy
aldinuc@di.unito.it
[2] Computer Science Department, University of Pisa, Italy
{ruggieri,torquati}@di.unipi.it

Abstract. The whole computer hardware industry embraced multicores. For these machines, the extreme optimisation of sequential algorithms is no longer sufficient to squeeze the real machine power, which can be only exploited via thread-level parallelism. Decision tree algorithms exhibit natural concurrency that makes them suitable to be parallelised. This paper presents an approach for *easy-yet-efficient* porting of an implementation of the C4.5 algorithm on multicores. The parallel porting requires minimal changes to the original sequential code, and it is able to exploit up to 7× speedup on an Intel dual-quad core machine.

Keywords: parallel classification, C4.5, multicores, structured parallel programming, streaming.

1 Introduction

Computing hardware has evolved to sustain an insatiable demand for high-end performances along two basic ways. On the one hand, the increase of clock frequency and the exploitation of instruction-level parallelism boosted the computing power of the single processor. On the other hand, many processors have been arranged in multi-processors, multi-computers, and networks of geographically distributed machines. This latter solution exhibits a superior peak performance, but it incurs in significant software development costs. In the last two decades, the parallel computing research community aimed at designing languages and tools to support the seamless porting of applications and the tuning of performances [3,13,21,22]. These languages, apart from few exceptions that also focus on code portability [13,22], require a redesign of the application logic in an explicitly parallel language or model.

Up to now, clock speed and algorithmic improvements have exhibited a better performance/cost trade-off than application redesign, being the possibility to preserve the existing code its most important component. Data mining is not an exception in this regard. By surveying the papers in the main scientific conferences and journals, there is a diminishing number of proposals for parallel implementations of data mining algorithms in the last few years. After all, only a small percentage of data analysis projects can afford the cost of buying (and

J.L. Balcázar et al. (Eds.): ECML PKDD 2010, Part I, LNAI 6321, pp. 7–23, 2010.

maintaining) a parallel machine and a data mining software capable of exploiting it. In most cases, data reduction techniques (such as sampling, aggregation, feature selection) can mitigate the problem while waiting the advancement in memory and computational power of low-cost workstations.

Nowadays, however, this vision should be reinterpreted. After years of continual improvement of single core chips trying to increase instruction-level parallelism, hardware manufacturers realised that the effort required for further improvements is no longer worth the benefits eventually achieved. Microprocessor vendors have shifted their attention to thread-level parallelism by designing chips with multiple internal cores, known as Multicore or Chip Multiprocessors [19]. However,this process does not always translate into greater CPU performance: multicore are small-scale but full-fledged parallel machines and they retain many of their usage problems. In particular, sequential code will get no performance benefits from them. A workstation equipped with a quad-core CPU but running sequential code is wasting 3/4 of its computational power. Developers, including data miners, are then facing the challenge of achieving a trade-off between performance and human productivity (total cost and time to solution) in developing and porting applications to multicore. Parallel software engineering engaged this challenge trying to design tools, in the form of high-level sequential language extensions and coding patterns, aiming at simplifying the porting of sequential codes while guaranteeing the efficient exploitation of concurrency [2,3,13,22].

This paper focuses on achieving this trade-off on a case study by adopting a methodology for the *easy-yet-efficient* porting of an implementation of the C4.5 decision tree induction algorithm [15] onto multicore machines. We consider the YaDT (Yet another Decision Tree builder) [17] implementation of C4.5, which is a from-scratch and efficient C++ version of the well-known Quinlan's entropy-based algorithm. YaDT is the result of several data structure re-design and algorithmic improvements over Efficient C4.5 [16], which is in turn is a patch to the original C4.5 implementation improving its performance mainly for the calculation of the entropy of continuous attributes. In this respect, we believe that YaDT is a quite paradigmatic example of sequential, already existing, complex code of scientific and commercial interest. In addition, YaDT is an example of extreme algorithmic sequential optimisation, which makes it unpractical to design further optimisations. Nevertheless, the potential for improvements is vast, and it resides in the idle core CPUs on the user's machine.

Our approach for parallelising YaDT is based on the FastFlow programming framework [1], a recent proposal for parallel programming over multicore platforms that provides a variety of facilities for writing efficient lock-free parallel patterns, including pipeline parallelism, task parallelism and Divide&Conquer (D&C) computations. Besides technical features, FastFlow offers an important methodological approach that will lead us to parallelise YaDT with minimal changes to the original sequential code, yet achieving up to 7× boost in performance on a Intel dual-quad core. MIPS, FLOPS and speedup have not to be the only metrics in software development. Human productivity, total cost and time to solution are equally, if not more, important.

The rest of the paper is organised as follows. In Sect. 2, the FastFlow programming environment is introduced. We recall in Sect. 3 the C4.5 decision tree construction algorithm, including the main optimisations that lead to YaDT. Then the parallelisation of YaDT is presented in detail in Sect. 4, followed by experimental evaluation and discussion in Sect. 5. Finally, we report related works in Sect. 6, and summarise the contribution of the paper in the conclusions.

2 The FastFlow Parallel Programming Environment

FastFlow is a parallel programming framework aiming to *simplify* the development of *efficient* applications for multicore platforms, being these applications either brand new or ports of existing legacy codes. The key vision underneath FastFlow is that effortless development and efficiency can be both achieved by raising the level of abstraction in application design, thus providing designers with a suitable set of parallel programming patterns that can be compiled onto efficient networks of parallel activities on the target platforms. To fill the abstraction gap, as shown in Fig. 1, FastFlow is conceptually designed as a stack of layers that progressively abstract the shared memory parallelism at the level of cores up to the definition of useful programming constructs and patterns.

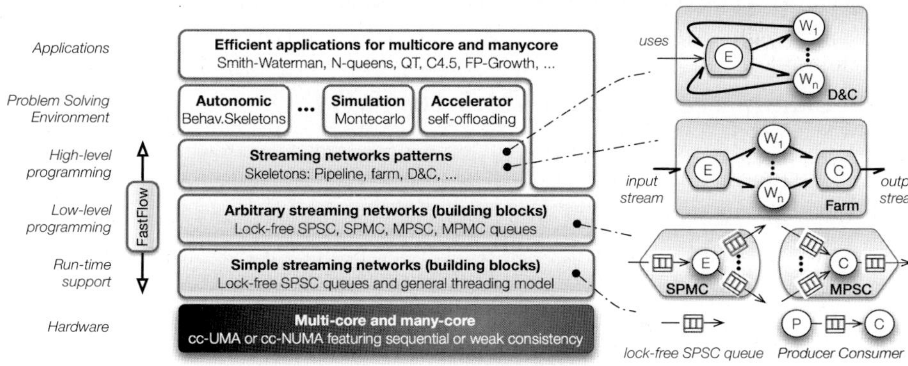

Fig. 1. FastFlow layered architecture with pattern examples

At the lowest tier of the FastFlow system we have the architectures that it targets: cache-coherent multiprocessors, and in particular commodity homogeneous multicore (e.g. Intel core, AMD K10, etc.).

The second tier provides mechanisms to define simple streaming networks whose *run-time support* is implemented through correct and efficient lock-free Single-Producer-Single-Consumer (SPSC) queues. This kind of queues do not requires any lock or memory barrier,[1] and thus they constitute a solid ground for a low-latency synchronisation mechanism for multicore. These synchronisations, which are asynchronous and non-blocking, do not induce any additional

[1] For Total Store Order processors, such as Intel core, AMD 10.

cache invalidation as it happens in mutual exclusion primitives (e.g. locks and interlocked operations), and thus do not add any extra overhead.

The third tier generalises one-to-one to one-to-many (SPMC), many-to-one (MPSC), and many-to-many (MPMC) synchronisations and data flows, which are implemented using only SPSC queues and arbiter threads. This abstraction is designed in such a way that arbitrary networks of activities can be expressed while maintaining the high efficiency of synchronisations.

The next layer up, i.e., *high-level programming*, provides a programming framework based on parallelism exploitation patterns (a.k.a. *skeletons* [5]). They are usually categorised in three main classes: Task, Data, and Stream Parallelism. FastFlow specifically focuses on Stream Parallelism, and in particular provides: *farm*, *farm-with-feedback* (i.e. Divide&Conquer), *pipeline*, and their arbitrary nesting and composition. These high-level skeletons are actually factories for parametric patterns of concurrent activities, which can be instantiated with sequential code (within white circles in Fig. 1) or other skeletons, then cross-optimised and compiled together with lower FastFlow tiers. The skeleton disciplines concurrency exploitation within the generated parallel code: the programmer is not required to explicitly interweave the business code with concurrency related primitives.

We refer to [1] for implementation details. FastFlow is open source available at `http://sourceforge.net/projects/mc-fastflow/` under LGPLv3 license.

3 Decision Trees: From C4.5 to YaDT

A decision tree is a classifier induced by supervised learning from a relation $\mathcal{T}$ called the *training set*. Tuples in $\mathcal{T}$ are called *cases*. An attribute C of the relation is called the *class*, while the remaining ones $A_1, \ldots, A_m$ are called the *predictive attributes*. The domain of an attribute $dom(A_i)$ can be discrete, namely a finite set of values, or continuous, namely the set of real numbers. Also, the special value *unknown* is allowed in $dom(A_i)$ to denote unspecified or unknown values. The domain of the class $dom(C) = \{c_1, \ldots, c_{NC}\}$ is discrete and it does not include the unknown value.

A *decision tree* is a tree data structure consisting of *decision nodes* and *leaves*. A leaf specifies a class value. A decision node specifies a *test* over one of the predictive attributes, which is called the attribute *selected* at the node. For each possible outcome of the test, a child node is present. A test on a discrete attribute A has h possible outcomes $A = d_1, \ldots, A = d_h$, where $d_1, \ldots d_h$ are the known values in $dom(A)$. A test on a continuous attribute has 2 possible outcomes, $A \leq t$ and $A > t$, where t is a *threshold* value determined at the node.

3.1 The C4.5 Tree-Induction Algorithm

The C4.5 decision tree induction algorithm [15] is a constant reference in the development and analysis of novel proposals of classification models [12]. The core[2] algorithm constructs the decision tree top-down. Each node is *associated* with a set of weighted cases, where weights are used to take into account unknown attribute values. At the beginning, only the root is present, with associated the whole training set $\mathcal{T}$. At each node a D&C algorithm is adopted to select an attribute for splitting. We refer the reader to the method `node::split` in Fig. 2 from the YaDT implementation of the algorithm.

Let T be the set of cases associated at the node. For every $c \in dom(C)$, the weighted frequency $freq(c, T)$ of cases in T whose class is c is computed (§2.2 – throughout the paper, we use the §M.N to reference line N from the pseudo-code in Fig. M). If all cases in T belong to the same class or the number of cases in T is less than a certain value then the node is set to a leaf (§2.3-4). If T contains cases belonging to two or more classes, then the *information gain* of each attribute at the node is calculated (§2.6-7). Since the information gain of a discrete attribute selected in an ancestor node is necessarily 0, the number of attributes to be considered at a node is variable (denoted by `getNoAtts` in §2.6).

For a discrete attribute A, the information gain of splitting T into subsets $T_1, \dots, T_h$, one for each known value of A, is calculated [3]. For A continuous, cases in T with known value for A are first ordered w.r.t. such an attribute. Let $v_1, \dots, v_k$ be the ordered values of A for cases in T. Consider for $i \in [1, k-1]$ the value $v = (v_i + v_{i+1})/2$ and the splitting of T into cases T_1^v whose value for the attribute A is lower or equal than v, and cases T_2^v whose value is greater than v. For each value v, the information gain $gain_v$ is computed by considering the splitting above. The value v' for which $gain_{v'}$ is maximum is set to be the *local threshold* and the information gain for the attribute A is defined as $gain_{v'}$.

The attribute A with the highest information gain is selected for the test at the node (§2.8). When A is continuous, the *threshold* of the split is computed (§2.9-10) as the greatest value of A in the *whole* training set $\mathcal{T}$ that is below the local threshold. Finally, let us consider the generation of the child nodes (§2.12-14). When the selected attribute A is discrete, a child node for each known value from $dom(A)$ is created, and cases in T are partitioned over the child nodes on the basis of the value of attribute A. When A is continuous two child nodes are created, and cases from T with known value of A are partitioned accordingly to the boolean result of the test $A \leq t$, where t is the threshold of the split. Cases in T whose value for attribute A is unknown are added to the set of cases of every child, but their weights are rebalanced.

3.2 From C4.5 to YaDT

The original Quinlan's implementation of C4.5 maintains the training set as an array of cases. Each case is an array of attribute values. The decision tree is grown depth-first. The computation of information gain takes $O(r)$ operations

[2] In this paper, we concentrate on the *growth* phase of the algorithm. The subsequent *prune* phase is computationally less expensive.

[3] As follows: $gain(T, T_1, \dots, T_h) = info(T) - \sum_{i=1}^{h} \frac{|T_i|}{|T|} \times info(T_i)$, where $info(S) = -\sum_{j=1}^{NC} \frac{freq(c_j,S)}{|S|} \times log_2(\frac{freq(c_j,S)}{|S|})$ is the entropy function.

```
    void node::split () {
2.2   computeFrequencies();
      if (onlyOneClass() || fewCases())
2.4     set_as_leaf ();
      else {
2.6     for(int i=0;i<getNoAtts();++i)
          gain[i]= gainCalculation(i);
2.8     int best = argmax(gain);
        if (attr[best].isContinuous())
2.10      findThreshold(best);
        ns=attr[best].nSplits();
2.12    for(int i=0;i<ns;++i)
          childs.push_back(
2.14          new node(selectCases(best,i)));
      }
2.16 }
```

Fig. 2. The original YaDT node splitting procedure

```
     bool node::splitPre() {
3.2    computeFrequencies();
       if (onlyOneClass() || fewCases()) {
3.4      set_as_leaf ();
         return true;
3.6    }
       return false;
3.8  } void node::splitAtt(i) {
       gain[i]= gainCalculation(i);
3.10 } void node::splitPost() {
       int best = argmax(gain);
3.12   if (attr[best].isContinuous())
         findThreshold(best);
3.14   ns=attr[best].nSplits();
       for(int i=0;i<ns;++i)
3.16     childs.push_back(
             new node(selectCases(best,i)));
3.18 }
```

Fig. 3. Partitioning of the `node:: split` method into three steps

for discrete attributes, where $r = |T|$ is the number of cases at the node; and $O(r\, log\, r)$ operations for continuous attributes, where sorting is the predominant task. Finally, searching for the threshold of the selected continuous attribute (§2.10) requires $O(|\mathcal{T}|)$ operations, where $\mathcal{T}$ is the whole training set. This linear search prevents the implementation being truly a D&C computation.

Efficient C4.5 (EC4.5) [16] is a patch software improving the efficiency of C4.5 in a number of ways. Continuous attribute values in a case are stored as indexes to the pre-sorted elements of the attribute domain. This allows for adopting a binary search of the threshold in the set of domain values at §2.10, with a computational cost of $O(log\, d)$ operations where $d = max_i|dom(A_i)|$. At each node, EC4.5 calculates the information gain of continuous attributes by choosing the best among three strategies accordingly to an analytic comparison of their efficiency: the first strategy adopts *quicksort*; the second one adopts *counting sort*, which exploits the fact that in lower nodes of the tree continuous attributes ranges tend to be narrow; the third strategy calculates the local threshold using a main-memory version of the RainForest [7] algorithm, without any sorting.

YaDT [17] is a from scratch C++ implementation of C4.5. It inherits the optimisations of EC4.5, and adds further ones, such as searching the local threshold for continuous attributes by splitting at *boundary* values (Fayyad and Irani method). Concerning data structures, the training set is now stored by columns, since most of the computations scan data by attribute values. Most importantly, the object oriented design of YaDT allows for encapsulating the basic operations on nodes into a C++ class, with the advantage that the growing strategy of the decision tree can now be a parameter (depth first, breadth first, or any other top-down growth). By default, YaDT adopts a breadth first growth – which has a less demanding memory occupation. Its pseudo-code is shown in Fig. 4 as method `tree::build`. Experiments from [16,17] show that YaDT reaches up to 10× improvement over C4.5 with only 1/3 of its memory occupation.

```
void tree::build() {
  queue<node *> q;
  node *root = new node( allCases );
  q.push(root);
  while( !q.empty() ) {
    node *n = q.front();
    q.pop();
    n->split();
    for(int i=0;i<n->nChilds();++i)
      q.push( n->getChild(i) );
  }
}
```

Fig. 4. YaDT tree growing procedure

```
    void tree::build_ff() {
5.2   node *root = new node( allCases );
      E=new ff_emitter(root,PAR_DEGREE);
5.4   std::vector<ff_worker*> w;
      for(int i=0;i<PAR_DEGREE;++i)
5.6     w.push_back( new ff_worker());
      ff_farm<ws_scheduler>
5.8       farm(PAR_DEGREE*QSIZE);
      farm.add_workers(w);
5.10  farm.add_emitter(E);
      farm.wrap_around();
5.12  farm.run_and_wait_end();
    }
```

Fig. 5. YaDT-FF D&C setup

```
void * ff_emitter::svc(void * task) {
  if (task == NULL) {
    task=new ff_task(root,BUILD_NODE);
    int r = root->getNoCases();
    setWeight(task, r);
    return task;
  }
  node *n = task->getNode();
  nChilds = n->nChilds();
  if (noMoreTasks() && !nChilds)
    return NULL;
  for(int i=0; i < nChilds; i++) {
    node *child = n->getChild(i);
    ctask=new ff_task(child,BUILD_NODE);
    int r = child->getNoCases();
    setWeight(ctask, r);
    ff_send_out(ctask);
  }
  return FF_GO_ON;
}

void * ff_worker::svc(void * task) {
  node *n = task->getNode();
  n->split();
  return task;
}
```

Fig. 6. Emitter and Worker definition for the NP strategy

4 Parallelising YaDT

We propose a parallelisation of YaDT, called YaDT-FF, obtained by stream parallelism. Each decision node is considered a task that generates a set of sub-tasks; these tasks are arranged in a stream that flows across a *farm-with-feedback* skeleton which implements the D&C paradigm. The FastFlow D&C schema is shown in the top-right corner of Fig. 1. Tasks in the stream are scheduled by an *emitter* thread towards a number of *worker* threads, which process them in parallel and independently, and return the resulting tasks back to the emitter. For the parallelisation of YaDT, we adopt a two-phases strategy: first, we accelerate the `tree::build` method (see Fig. 4) by exploiting task parallelism among node processing, and we call this strategy *Nodes Parallelisation* (NP); then, we add the parallelisation of the `node::split` method (see Fig. 2) by exploiting parallelism also among attributes processing, and we call such a strategy *Nodes & Attributes Parallelisation* (NAP). The two strategies share the same basic setup method, `tree::build_ff` shown in Fig. 5, which creates an emitter object (§5.2-3) and an array of worker objects (§5.4-6). The size of the array, `PAR_DEGREE`, is the parallelism degree of the farm. The root node of the decision tree is passed to the constructor of the emitter object, so that the stream can be initiated from it. The overall farm parallelisation is managed by the FastFlow layer through a `ff_farm` object, which creates feedback channels between the emitter and the

```
void * ff_emitter :: svc(void * task) {
 if (task == NULL ) {
  if (root->splitPre()) return NULL;
  int r = root->getNoCases();
  int c = root->getNoAtts();
  for(int i=0;i<c;++i) {
   task=new ff_task(root,BUILD_ATT);
   task->att = i;
   setWeight(task, r);
   ff_send_out(task);
  }
  root->attTasks = c;
  return FF_GO_ON;
 }
 node *n = task->getNode();
 if (task->isBuildAtt()) {
  if (--n->attTasks>0)
   return FF_GO_ON;
  n->splitPost();
 }
 nChilds = n->Childs();
 if (noMoreTasks() && !nChilds)
  return NULL;
 for(int i=0; i < nChilds; i++) {
  node *child = n->getChild(i);
  int r = child->getNoCases();
  int c = child->getNoAtts();
  if (!buildAttTest(r,c)) {
   ctask=new ff_task(child,BUILD_NODE);
   setWeight(ctask, r);
   ff_send_out(ctask);
  } else {
   if (child->splitPre()) continue;
   for(int j=0;j<c;++j) {
    ctask=new ff_task(child,BUILD_ATT);
    ctask->att = j;
    setWeight(ctask, r);
    ff_send_out(ctask);
   }
   child->attTasks = c;
  }
 return FF_GO_ON;
}

void * ff_worker :: svc(void * task) {
 node *n = task->getNode();
 if (task->isBuildAtt())
  n->splitAtt(task->att);
 else
  n->split();
 return task;
}
```

Fig. 7. Emitter and Worker definition for the NAP strategy

workers (§5.7-11). Parameters of `ff_farm` include: the size `QSIZE` of each worker input queue, and the scheduling policy (`ws_scheduler`), which is based on tasks weights. Basically, such a policy assigns a new task to the worker with the lowest total weight of tasks in its own input FIFO queue. The emitter class `ff_emitter` and the worker class `ff_worker` define the behaviour of the farm parallelisation through the class method `svc` (short name for *service*) that is called by the FastFlow run-time to process input tasks. Different parallelisation strategies can be defined by changing only these two methods. The implementation of the NP and the NAP strategies are shown in Fig. 6 and Fig. 7 respectively.

NP strategy (Fig. 6). At start-up the `ff_emitter::svc` method is called by the FastFlow run-time with a NULL parameter (§6.2). In this case, a task for processing the root node is built, and its weight is set to the number of cases at the root (§6.3-5). Upon receiving in input a task coming from a worker, the emitter checks the termination conditions (§6.10), and then produces in output the sub-tasks corresponding to the children of the node (§6.12-18). The `ff_send_out` method of the FastFlow runtime allows for queueing tasks without returning from the method. Finally, the `FF_GO_ON` tag in the return statement (§6.19) tells the run-time that the computation is not finished (this is stated by returning NULL), namely further tasks must be waited for from the input channel. The `ff_worker::svc` method for a generic worker (§6.22-25) merely calls the node splitting algorithm `node::split`, and then it immediately returns the computed task back to the emitter. The overall coding is extremely simple and intuitive – almost a rewriting of the original `tree::build` method. Moreover, it is quite generalisable to any top-down tree-growing algorithm with greedy choice of the splitting at each node. The weighted scheduling policy is the most specific part; in particular, for the use of weights that are linear in the number of cases at the node. This is motivated by the experimental results of [16, Fig. 1], which show how the YaDT implementation of `node::split` exhibits a low-variance elapsed time per case for the vast majority of nodes.

NAP strategy (Fig. 7). The NAP strategy builds over NP. For a given decision node, the emitter follows a D&C parallelisation over its children, as in the case of the NP strategy. In addition, for each child node, the emitter may decide to parallelise the calculation of the information gains in the `node::split` method (§2.6-7). In such a case, the stopping criterion at §2.3 must be evaluated prior to the parallelisation, and the creation of the child nodes must occur after all the information gains are computed. This leads to partitioning the code of `node::split` into three methods, as shown in Fig. 3.

For the root node, attribute parallelisation is always the case (§7.3-10). A task with label `BUILD_ATT` is constructed for each attribute, with the field `att` recording the attribute identifier (the index `i`). Tasks are weighted and queued. The information about how many tasks are still to be completed is maintained in the `attTasks` field of the decision node – such a field is added to the original `node` class. Upon receiving in input a task coming from a worker, the emitter checks whether it concerns the processing of an attribute (§7.16). If this is the case (§7.17-20), the `attTasks` counter is decremented until the last attribute task arrives, and then the `node::splitPost` method is called to evaluate the best split. At this point, the emitter is given a processed node (either from a worker, or as the result of the `node::splitPost` call). Unless the termination conditions occur (§7.22), the emitter proceeds with outputing tasks. The `buildAttTest` at §7.28 controls for each child node whether to generate a single node processing task, or one attribute processing task for each attribute at the child node. In the former case (§7.29-31), we proceed as in the NP strategy; in the latter case (§7.33-38), we proceed as for the root node[4]. Based on the task label, the `ff_worker::svc` method for a generic worker (§7.46-51) merely calls the node splitting procedure or the information gain calculation for the involved attribute.

Let us discuss in detail two relevant issues. Let r be the number of cases and c the number of attributes at the node.

The first issue concerns task weights. Node processing tasks are weighted with r (§7.30), as in the NP strategy. Although attribute processing tasks have a finer grain, which suggests a lower weight, there exists a synchronisation point – all attribute tasks must be processed before the emitter can generate tasks for the child nodes. By giving a lower weight, we run the risk that all attribute tasks are assigned to the most unloaded worker, thus obtaining a sequential execution of the attribute tasks. For these reasons, attribute processing tasks are weighted with r as well (§7.9,§7.37).

[4] Notice that tasks for node processing are labelled with `BUILD_NODE`, while tasks for attribute processing are labelled with `BUILD_ATT`.

The second issue concerns the test `buildAttTest`, which decides whether to perform nodes or attributes parallelisation. We have designed and experimented three cost models. Attribute parallelisation is chosen respectively when:

- ($\alpha < r$) the number of cases is above some hand-tuned threshold value α;
- ($|\mathcal{T}| < c\, r\, log\, r$) the average grain of node processing (quicksort is $r\, log\, r$ on average) is higher than a threshold that is dependent on the training set. Intuitively, the threshold should be such that the test is satisfied at the root node, which is the coarser-grained task, and for nodes whose size is similar. Since the average grain of processing an attribute at the root is $|\mathcal{T}|\, log\, |\mathcal{T}|$, we fix the threshold to a lower bound for such a value, namely to $|\mathcal{T}|$;
- ($|\mathcal{T}| < c\, r^2$) the worst-case grain of node processing (quicksort is r^2) is higher than a threshold that is dependent on the training set. As in the previous case, the threshold is set to $|\mathcal{T}|$. The higher value cr^2, however, leads to selecting attributes processing more often than the previous case, with the result of task over-provisioning.

All tests are monotonic in the number r of cases at the node. Hence, if the nodes parallelisation is chosen for a node, then it will be chosen for all of its descendants. As we will see in Sec. 5, the third cost model shows the best performance.

5 Performance Evaluation

In this section we show the performances obtained by YaDT-FF. The datasets used in the tests and their characteristics are reported in Table 1. They are publicly available from the UCI KDD archive, apart from *SyD10M9A* which is synthetically generated using function 5 of the QUEST data generator. All presented experimental results are taken performing 5 runs, excluding the higher and the lower value obtained and computing the average of the remaining ones.

Experimental framework. All experiments were executed on two different Intel workstation architectures: *Nehalem)* a dual quad-core Xeon E5520 Nehalem (16 HyperThreads) @2.26GHz with 8MB L3 cache and 24 GBytes of main memory with Linux x86_64. *Harpertown)* a dual quad-core Xeon E5420 Harpertown @2.5GHz 6MB L2 cache and 8 GBytes of main memory, with Linux x86_64. They are a quite standard representative of current and immediately preceding

Table 1. Training sets used in experiments, and size of the induced decision tree

			No. of attributes			Tree	
$\mathcal{T}$ name	$\vert\mathcal{T}\vert$	NC	discr.	contin.	total	size	depth
Census PUMS	299,285	2	33	7	40	122,306	31
U.S. Census	2,458,285	5	67	0	67	125,621	44
KDD Cup 99	4,898,431	23	7	34	41	2,810	29
Forest Cover	581,012	7	44	10	54	41,775	62
SyD10M9A	10,000,000	2	3	6	9	169,108	22

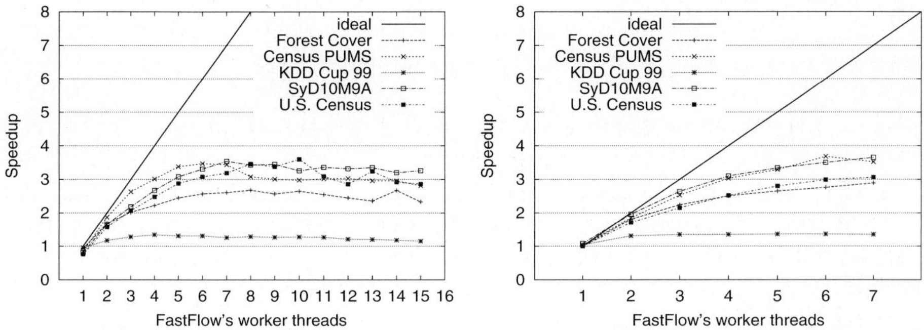

Fig. 8. NP strategy speedup. Nehalem box (left), Harpertown box (right).

generation of (low-cost) server boxes. The Nehalem-based machine exploits Simultaneous MultiThreading (SMT, a.k.a. HyperThreading) with 2 contexts per core and the novel Quickpath interconnect equipped with a distributed cache coherency protocol. SMT technology makes a single physical processor appear as two logical processors for the operating system, but all execution resources are shared between the two contexts: caches of all levels, execution units, etc.

Performance. Let us start considering the *NP strategy*, i.e., the parallelisation of nodes processing. The obtained speedup is shown in Fig. 8. The maximum speedup is similar on both architectures, and quite variable from a dataset to another; it ranges from 1.34 to 3.54 (with an efficiency of 45%). As one would expect, exploiting inter-nodes parallelism alone is not enough to reach a close to optimal speedup, because a large fraction of the computing time is spent in the coarse-grained nodes (those in the higher levels of the tree), thus lacking parallelism. This phenomenon has been already observed in previous work on the parallelisation of decision tree construction over distributed memory architectures [9]. These systems, however, suffer from load balancing problems, which we will handle later on, and high costs of communications, which in shared memory architectures do not occur. Summarising, although the NP strategy yields a modest speedup, it is worth noting that the effort required to port the sequential code was minimal.

The *NAP strategy* aims at increasing the available parallelism by exploiting concurrency also in the computation of the information gain of attributes. This is particularly effective for nodes with many cases and/or attributes, because it reduces the sequential fraction of the execution. As presented in Sec. 4, the emitter relies on a *cost model* in order to decide whether to adopt attributes parallelisation. We have tested the three cost models discussed in Sec. 4. Fig. 12 shows that the test $|\mathcal{T}| < cr^2$ provides the best performance for almost all datasets. This is justified by the fact that the test exhibits an higher task over-provisioning if compared to the test $|\mathcal{T}| < cr\,log\,r$, and it is dataset-tailored if compared to $\alpha < r$. In all of the remaining experiments, we use that model.

The speedup of YaDT-FF with the NAP strategy is shown in Fig. 9. It ranges from 4 to 7.5 (with an efficiency of 93%). The speedup gain over the NP

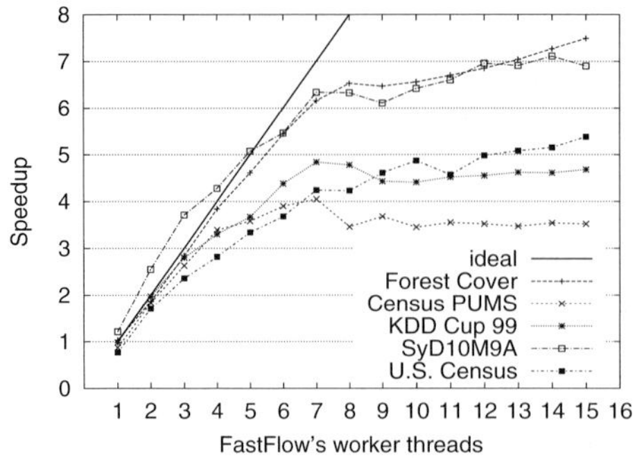

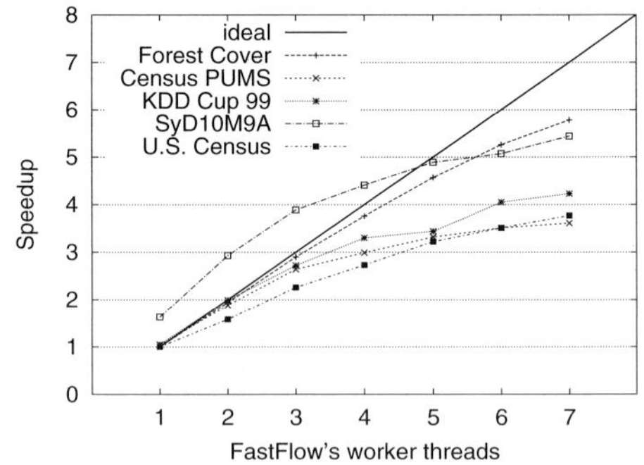

Fig. 9. NAP strategy speedup. Nehalem box (left), Harpertown box (right).

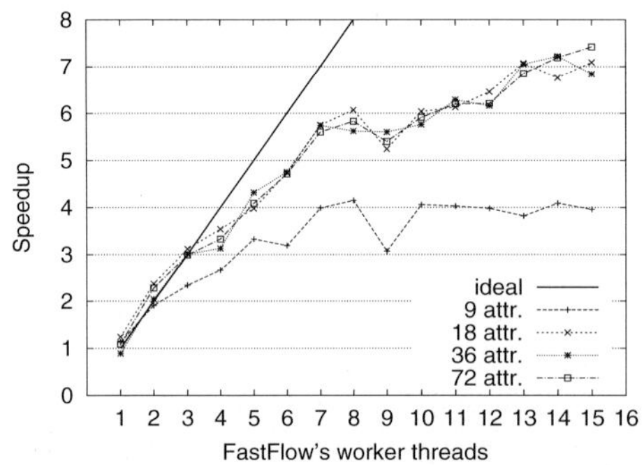

Fig. 10. Speedup vs no. of attributes for $1M$ sample cases from *SyD10M9A*

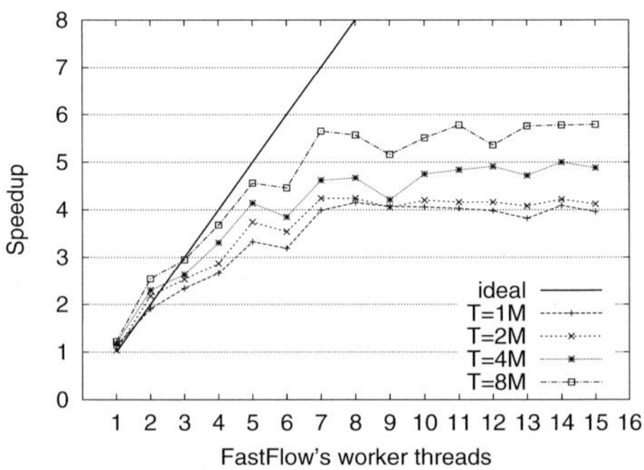

Fig. 11. Speedup vs no. of sample cases (T) from *SyD10M9A*

strategy is remarkable. Only for the *Census PUMS* dataset, the smallest dataset as for number of cases, the speedup gain is just +12% over NP. Notice that the *SyD10M9A* dataset apparently benefits from a super-linear speedup. Actually, this happens because the speedup is plotted against the number of farm workers. Hence, the fraction of work done by the emitter thread is not considered, yet not negligible as shown in Fig. 14.

YaDT-FF also exhibits a good scalability with respect to both the number of attributes (Fig. 10) and to the number of cases (Fig. 11) in the training set. The plots refer to subsets of the *SyD10M9A* dataset possibly joined with randomly distributed additional attributes. In the former case, the maximum speedup (7×) is reached as soon as the number of attributes doubles the available hardware parallelism (18 attributes for 8 cores). In the latter case, the achieved speedup increases with the number of cases in the training set.

Load-balancing. The parallelisation of decision tree construction algorithms may suffer from load balancing issues due to the difficulties in predicting the time needed for processing a node or a sub-tree. This is exacerbated in the parallelisation of the original C4.5 implementation, because of the linear search of the threshold (§2.10). Fig. 14 shows that load balancing is not a critical issue for YaDT-FF with the NAP strategy. We motivate this by two main reasons: 1) the NAP strategy produces a significant over-provisioning of tasks with respect to the number of cores; these tasks continuously flow (in a cycle) from the emitter to the workers and they are subject to online scheduling within the emitter; 2) FastFlow communications are asynchronous and exhibit very low overhead [1]. This makes it possible to sustain all the workers with tasks to be processed for the entire run. This also reduces the dependence of the achieved speedup from the effectiveness of the scheduling policy. Nevertheless, such dependence exists.

Fig. 13 shows results for three different scheduling policies: 1) Dynamic Round-Robin (DRR); 2) On-Demand (OD); 3) Weighted Scheduling (WS). The DRR policy schedules a task to a worker in a round-robin fashion, skipping workers with full input queue (with size set to 4096). The OD policy is a fully online scheduling, i.e., a DDR policy where each worker has an input queue of size 1. The WS policy is a user-defined scheduling that can be set up by assigning

	Total Execution Time (sec.)		
$\mathcal{T}$ name	$\lvert\mathcal{T}\rvert < cr^2$	$\alpha < r$	$\lvert\mathcal{T}\rvert < cr \log r$
Census PUMS	**0.85**	0.85	0.91
U.S. Census	**3.28**	3.51	3.35
KDD Cup 99	**3.76**	3.80	3.77
Forest Cover	**2.64**	2.66	2.73
SyD10M9A	16.90	**16.68**	18.16

Effectiveness of *buildAttTest(c,r)* for different attributes parallelisation cost models. $\lvert\mathcal{T}\rvert$ = no. of cases in the training set, c = no. of attributes at the node, r = no. of cases at the node, and $\alpha = 1000$. Bold figures highlight the best results.

Fig. 12. Attributes parallelisation tests (Nehalem, 7 worker threads)

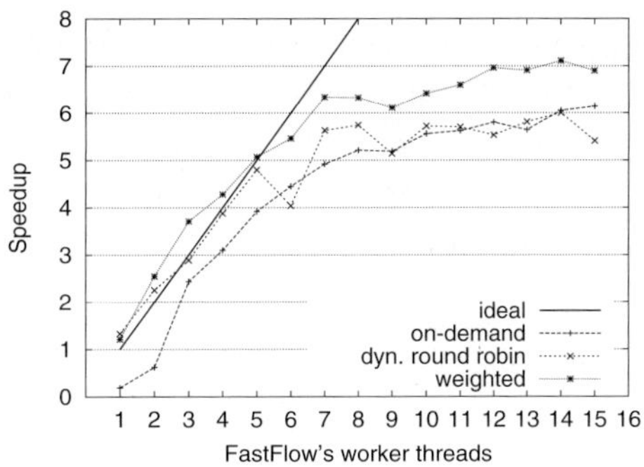

Fig. 13. Speedup of different scheduling policies over *SyD10M9A*

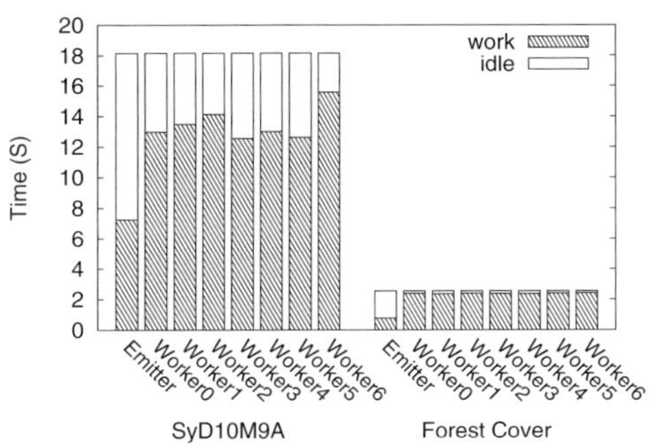

Fig. 14. YaDT-FF execution breakdown (Nehalem, 7 worker threads)

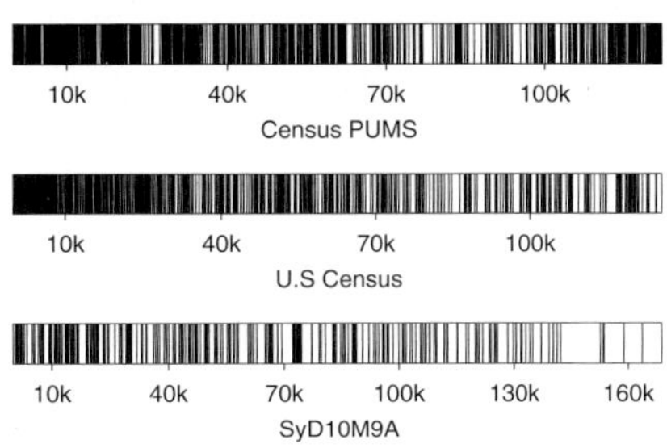

Fig. 15. Nodes (white) vs attributes (black) parallelisation choices

Table 2. YaDT vs YaDT-FF on a Nehalem quad-core (E= Emitter, W=Worker)

		1E+1W	1E+2W	1E+3W	
$\mathcal{T}$ name	Seq.Time (S)		Time (S)		Max Boost
Census PUMS	4.46	4.3	2.37	1.69	2.64×
U.S. Census	17.67	17.97	11.17	7.8	2.26×
KDD Cup 99	18.11	17.26	9.12	6.67	2.71×
Forest Cover	16.99	16.97	8.74	5.86	2.90×
SyD10M9A	103.21	93.95	52.34	39.37	2.62×

weights to tasks through calls to the `setWeight` method. YaDT-FF adopts a WS policy, with the weight of a task set to the number r of cases at the node.

It is immediate to observe from Fig. 13 that all the scheduling policies are fairly efficient. WS exhibits superior performance because it is tailored over the YaDT-FF algorithm; it actually behaves as a quite efficient online scheduling. Finally, we show in Fig. 15 how often nodes parallelisation has been chosen by the emitter against the attributes parallelisation (we recall that the test $|\mathcal{T}| < cr^2$ was fixed). Black stripes lines in the figure denote attributes parallelisation choices whereas white stripes denote nodes parallelisation ones. As expected, the former occurs more often when processing the top part of the decision tree (from left to the right, in the figure).

Simultaneous MultiThreading. We briefly evaluate the benefits achieved using the Nehalem HyperThreaded box. SMT is essentially a memory latency hiding technique that is effective when different threads in a core exhibit a shared working set that induces high cache hit rate. However, even in non-ideal conditions, SMT is able to moderately increase instructions per clock-cycle count, hence the overall performance, by partially hiding costly memory operations with threads execution. The comparison between the two graphs in Fig. 9 shows that several datasets benefit of (about) 30% improvement due to SMT; some others, such as *Census PUMS* and *KDD Cup*, show only a modest benefit (about 12%). These figures match the expected benefit for this kind of architectures [19]. As future work, we believe that the effectiveness of SMT can be further improved by devising a *cache-aware* weighted scheduling policy.

6 Related Work

Over the last decade, parallel computing aimed at addressing three main classes of data mining issues: 1) solve inherently distributed problems, e.g., mining of datasets that are bound to specific sites due to privacy issues; 2) manage larger datasets by exploiting the aggregate memories of different machines; 3) decrease the processing time of mining algorithms. In many cases, the latter two issues have been jointly addressed by trying to bring in-core datasets that are out-of-core on a single machine. Such an approach, which often requires the redesign of the algorithms or the introduction of new scalable data structures, is loosing interest with the ever-increasing availability of main memory space. Our work distinguishes from this approach, even if it clearly belongs to the third class.

Considering the parallelisation methodology, related works can be categorised as follow: 1) exploiting attributes parallelism by partitioning the training set by columns and then adopting data parallelism [11,18]; 2) exploiting nodes parallelism by independently building different nodes or sub-trees adopting task parallelism [6]; 3) combining the two above in various fashions [20,23]. Several works focus on distributed-memory machines, including SPRINT [18], ScalParC [11], pCLOUDS [20], and the approach of [6]. The use of scalable data structures and of efficient load balancing techniques, trying to minimise costly data redistribution operations, are the most important factors to obtain good performance. As an example, the earliest SPRINT parallel algorithm adopts scalable SLIQ data structure for representing the dataset, but it suffers from communication bottlenecks addressed in the successor system ScalParC. pCLOUDS [20] combines both the data parallel and the task parallel approaches. It exploits data parallelism for large decision nodes, then it switches to a task parallel approach as soon as the nodes become small enough. The proposal of [6] categorises tasks in three different classes: large, intermediate and small ones. Large tasks process a decision node. Intermediate tasks process a sub-tree up to a given number of nodes. Small tasks sequentially process the whole sub-tree of a node. YaDT-FF takes inspiration from the two latter works and distinguish from them for: 1) it does not need the redesign of the sequential algorithm but rather an *easy-yet-efficient* porting of the existing code; 2) it targets multicore rather than distributed memory machines; 3) it adopts an effective cost model for deciding whether to parallelise on nodes or on attributes. Few works target data mining systems on multicore [4,8,10,14], but none specifically decision tree algorithms.

7 Conclusions

Nowadays, and for foreseeable future, the performance improvement of a single core will no longer satisfy the ever increasing computing power demand. For this, computer hardware industry shifted to multicore, and thus the extreme optimisation of sequential algorithms is not longer sufficient to squeeze the real machine power. Software designers are then required to develop and to port applications on multicore. In this paper, we have presented the case study of decision tree algorithms, porting YaDT using the FastFlow parallel programming framework. The strength of our approach consists in the minimal change of the original code with, at the same time, a non-trivial parallelisation strategy (nodes and attributes parallelism plus weighted problem-aware load balancing) and notable speedup. Eventually, we want to stress the results in the case of a low cost quad-core architecture that may be currently present in the desktop PC of any data analyst. Table 2 shows that the parallelisation of YaDT boosts up to 2.9×, with no additional cost to buy a specific parallel hardware.

References

1. Aldinucci, M., Meneghin, M., Torquati, M.: Efficient Smith-Waterman on multi-core with FastFlow. In: Proc. of the Euromicro Conf. on Parallel, Distributed and Network-based Processing (PDP), pp. 195–199. IEEE, Pisa (2010)
2. Asanovic, K., Bodik, R., Demmel, J., Keaveny, T., Keutzer, K., Kubiatowicz, J., Morgan, N., Patterson, D., Sen, K., Wawrzynek, J., Wessel, D., Yelick, K.: A view of the parallel computing landscape. CACM 52(10), 56–67 (2009)
3. Blumofe, R.D., Joerg, C.F., Kuszmaul, B.C., Leiserson, C.E., Randall, K.H., Zhou, Y.: Cilk: An efficient multithreaded runtime system. Journal of Parallel and Distributed Computing 37(1), 55–69 (1996)
4. Buehrer, G.T.: Scalable mining on emerging architectures. Phd thesis, Columbus, OH, USA (2008)
5. Cole, M.: Bringing skeletons out of the closet: A pragmatic manifesto for skeletal parallel programming. Parallel Computing 30(3), 389–406 (2004)
6. Coppola, M., Vanneschi, M.: High-performance data mining with skeleton-based structured parallel programming. Parallel Computing 28(5), 793–813 (2002)
7. Gehrke, J.E., Ramakrishnan, R., Ganti, V.: RainForest — A framework for fast decision tree construction of large datasets. Data Mining and Knowledge Discovery 4(2/4), 127–162 (2000)
8. Ghoting, A., Buehrer, G., Parthasarathy, S., Kim, D., Nguyen, A., Chen, Y.K., Dubey, P.: Cache-conscious frequent pattern mining on a modern processor. In: Proc. of the Intl. Conf. on Very Large Data Bases (VLDB), pp. 577–588 (2005)
9. Han, E., Srivastava, A., Kumar, V.: Parallel formulation of inductive classification parallel algorithm. Tech. rep., Department Computer and Information Science, University of Minnesota (1996)
10. Jin, R., Yang, G., Agrawal, G.: Shared memory parallelization of data mining algorithms: Techniques, programming interface, and performance. IEEE Transactions on Knowledge and Data Engineering 17, 71–89 (2005)
11. Joshi, M., Karypis, G., Kumar, V.: ScalParC: A new scalable and efficient parallel classification algorithm for mining large datasets. In: Proc. of IPPS/SPDP, pp. 573–579. IEEE, Los Alamitos (1998)
12. Lim, T., Loh, W., Shih, Y.: A comparison of prediction accuracy, complexity, and training time of thirthy-tree old and new classification algorithms. Machine Learning Journal 40, 203–228 (2000)
13. Park, I., Voss, M.J., Kim, S.W., Eigenmann, R.: Parallel programming environment for OpenMP. Scientific Programming 9, 143–161 (2001)
14. Pisharath, J., Zambreno, J., Ozisikyilmaz, B., Choudhary, A.: Accelerating data mining workloads: Current approaches and future challenges in system architecture design. In: Proc. of Workshop on High Performance and Distributed Mining (2006)
15. Quinlan, J.R.: C4.5: Programs for Machine Learning. Morgan Kaufmann, San Mateo (1993)
16. Ruggieri, S.: Efficient C4.5. IEEE Transactions on Knowledge and Data Engineering 14, 438–444 (2002)
17. Ruggieri, S.: YaDT: Yet another Decision tree Builder. In: 16th IEEE Int. Conf. on Tools with Artificial Intelligence (ICTAI), pp. 260–265. IEEE, Los Alamitos (2004)

18. Shafer, J.C., Agrawal, R., Mehta, M.: SPRINT: A scalable parallel classifier for data mining. In: Proc. of the Intl. Conf. on Very Large Data Bases (VLDB), pp. 544–555 (1996)
19. Sodan, A.C., Machina, J., Deshmeh, A., Macnaughton, K., Esbaugh, B.: Parallelism via multithreaded and multicore CPUs. IEEE Computer 43(3), 24–32 (2010)
20. Sreenivas, M.K., Alsabti, K., Ranka, S.: Parallel out-of-core divide-and-conquer techniques with application to classification trees. In: Proc. of IPPS/SPDP, pp. 555–562. IEEE, Los Alamitos (1999)
21. Thies, W., Karczmarek, M., Amarasinghe, S.P.: StreamIt: A language for streaming applications. In: Horspool, R.N. (ed.) CC 2002. LNCS, vol. 2304, pp. 179–196. Springer, Heidelberg (2002)
22. Vanneschi, M.: The programming model of ASSIST, an environment for parallel and distributed portable applications. Parallel Computing 28(12), 1709–1732 (2002)
23. Zaki, M., Ho, C.T., Agrawal, R.: Parallel classification for data mining on shared-memory multiprocessors. In: Proc. of the Intl. Conf. on Data Engineering (ICDE), pp. 198–205. IEEE, Los Alamitos (1999)

On Classifying Drifting Concepts in P2P Networks

Hock Hee Ang, Vivekanand Gopalkrishnan, Wee Keong Ng, and Steven Hoi

Nanyang Technological University, Singapore

Abstract. Concept drift is a common challenge for many real-world data mining and knowledge discovery applications. Most of the existing studies for concept drift are based on centralized settings, and are often hard to adapt in a distributed computing environment. In this paper, we investigate a new research problem, P2P concept drift detection, which aims to effectively classify drifting concepts in P2P networks. We propose a novel P2P learning framework for concept drift classification, which includes both reactive and proactive approaches to classify the drifting concepts in a distributed manner. Our empirical study shows that the proposed technique is able to effectively detect the drifting concepts and improve the classification performance.

Keywords: Concept drift, classification, peer-to-peer (P2P) networks, distributed classification.

1 Introduction

Recent years have witnessed a surge of emerging research for data mining and machine learning in peer-to-peer (P2P) environments [1,2,3]. This includes distributed classification in P2P networks, referred to "P2P classification", which aims to exploit the resources of all peers to collaboratively learn an accurate classification model that is representative of the entire network's data. P2P classification is a rapidly growing research topic in data mining due to its rich applications, such as user preference mining, recommendation systems, automated document organization, etc. In general, it has many open challenges, such as massive number of peers, arbitrarily connected and dynamic peers, and so on [4]. With consideration of the properties of P2P environments, an ideal P2P classification scheme should typically be: anytime (produce an answer at any time), asynchronous (peer dependencies are low), decentralized, highly scalable, tolerant of peer failures (failures should not result in catastrophic outcome) and privacy preserving (private data should not be revealed).

Besides the above mentioned challenges, another critical challenge for P2P classification is *concept drift* [2], which is an important problem in many data mining and machine learning applications. Concept drift refers to the learning problem where the target concept to be predicted, changes over time in some unforeseen behaviors. It is commonly found in many dynamic environments, such

J.L. Balcázar et al. (Eds.): ECML PKDD 2010, Part I, LNAI 6321, pp. 24–39, 2010.

as data streams, P2P systems, etc. Real-world examples include network intrusion detection, spam detection, fraud detection, epidemiological, and climate or demographic data, etc. It is crucial to address the concept drift problem in P2P classification, as the inability to adapt swiftly to the drifted concept often results in significant loss of classification accuracy.

Although concept drift has been actively studied, concept drift in a P2P environment has some fundamental difference from that of a typical centralized setting. Typical scenarios only model concept drift from a single source of data. In P2P networks, each peer can be viewed as an independent data source. Hence, P2P concept drift affects different peers in diverse ways, such as varying degree or varying time occurrence.

An ideal classification algorithm that deals with the problem of concept drift in a P2P setting should possess the aforementioned desirable properties for learning in a P2P setting, and also be able to adapt swiftly to the changes in concepts, without adversely affecting the peers. Although there has been much work on P2P classification [1,2,3], most do not address the problem of concept drift, and those that do simply assume the concept drift scenario of a centralized setting; i.e., all peers are affected in the same manner at the same time.

Most of the existing approaches that deal with concept drifts [5,6,7,8,9,10,11] are based on a centralized setting and cannot be easily adapted for a P2P environment. While the ensemble-based solution seems to be a viable option, existing works do not consider the problem of distributed concept drifts that may occur in a P2P environment. They are also not designed to be efficient for the demanding environment of P2P networks.

To address the above challenges, this paper presents a novel concept drift adaptation framework for performing distributed classification in P2P environments, and makes the following contributions:

- To the best of our knowledge, we are the first to formally examine the problem of concept drift for distributed classification in P2P environments. We illustrate the difference from the conventional centralized single-source concept drifts, and investigate their effects on the P2P classification problem.
- We propose a novel P2P classification framework to address the concept drift issue in a P2P environment, which includes both reactive and proactive adaptation approaches. Our framework is based on the ensemble paradigm, which mines meta data of different peers to achieve reactive and proactive concept drift adaptation.
- We theoretically and empirically justify the efficacy of our approach. In addition, we demonstrate that two aspects are crucial to the mutual benefits of peers, i.e., (1) the sharing of knowledge, and (2) the judicious choice on the usage of relevant knowledge.

The rest of this paper is organized as follows. We introduce the problem of concept drift and related work in Section 2. Section 3 presents our framework for learning with drifting concepts in P2P networks. The proposed framework is empirically examined in Section 4 and Section 5 concludes this paper.

2 Background and Related Work

2.1 Background

We first introduce the setup of classification in a P2P environment. Let us denote by $\mathbf{D} = [(x_{t,1}, \ldots, x_{t,d})_{t=1}^{\ell}]$ a training data set, and $\mathbf{y} = [y_1, \ldots, y_\ell]^{\mathrm{T}}$ the corresponding class labels, where ℓ denotes the total number of training data instances, d denotes the dimensionality of data points, $y_t \in \mathcal{Y}$, and $\mathcal{Y}$ denotes the class label space; e.g., $\mathcal{Y} = \{+1, -1\}$ for binary classification. Typically, each training instance $(\mathbf{x}, y)$ is drawn from some unknown but i.i.d. distribution $P(\mathbf{x}, y)$. Suppose there are N peers in the P2P network, each sample is a partition $\mathbf{D}_i$ of $\mathbf{D}$ where $\ell = \sum_i \ell_i$. The goal of P2P classification is to collaboratively learn a global prediction function $f : X \to Y$ from the training data of all peers, which maps a d-dimensional vector $\mathbf{x} = (x_1, \ldots, x_d)^T \in X$ to a corresponding class label $y \in Y$ of a target concept.

Consider a non-stationary environment where the target concept may change. We are given a series of training datasets $\{D^1, \ldots, D^t\}$, where t is the current time period, and D^t was drawn from some unknown probability distribution $P_t(\mathbf{x}, y)$. Given such a scenario, *concept drift* is defined as the change of the underlying unknown probability distribution, i.e., $P_{t-1}(\mathbf{x}, y) \neq P_t(\mathbf{x}, y)$, which has occurred from time period $t - 1$ to t. This obsoletes the models that were built on the old training data, and thus causes their prediction accuracy to drop. In addition, as each peer is *sub-sampling* from the changing unknown probability distribution, different peers may be affected differently by the same change in concept; e.g., due to a delayed concept drift, varying degree of drift, etc. Hence, an ideal classification solution for a concept drifting P2P network should possess the desired properties stated in the beginning of the paper, and should be able to promptly adapt to the concept drift without introducing adverse effects.

2.2 Related Work

Classification in P2P networks has been recently addressed by several studies [1,2,3]. While most of them were proposed to learn the concepts in an incremental manner, only the P2P decision tree [2] has been partially demonstrated on the concept drifting data. However, since the concept drift problem is not their focus, their experiments only assume some simple scenarios where concept drift occurs simultaneously for all peers and only the suddenly changing concept drift problem was examined.

In literature, the issue of concept drift has been actively studied in the context of data stream classification [5,6,7,8,9,10,11]. These approaches can be broadly divided into *single models* and *ensemble-based* approaches.

The category of *single model* based studies includes CVFDT [6], which is based on the VFDT that incrementally builds a decision tree based on the incoming data. New subtrees are constructed when the old subtrees become outdated and the latter is replaced when the former achieves better accuracy. The incremental aspect of the P2P decision tree [2] bears some resemblance to this approach. Another work by Xu *et al.* [11] is also based on single model, but their work focuses on learning from multiple streams, examining how the streams can be combined for handling the concept drift in a centralized manner. These approaches are however not suitable for P2P environments due to the nature of their centralized settings, which often bring many difficulties and require much effort for any extension of such work.

Another work whose setting is quite similar to ours was done by Chen *et al.* [5]. They explored the web mining task from multiple distributed data streams. At each distributed site, a local Bayesian Network (BN) model is learned and a subset of the relevant observations are sent to a centralized site. The centralized site then uses these partial observations to construct a BN model and link up with BN models of the distributed sites to obtain a collective BN model. However, their drawback is that a centralized site is required, which is often not available in a P2P environment.

Recently, ensemble approaches have been widely proposed for solving the concept drift problem [7,8,9,10,12]. In general, models are constructed on every new chunk of data and different model selection or weighting schemes are explored. One early ensemble based solution for learning drifting concepts was proposed by Street and Kim [8]. They proposed a simple solution to construct a classifier on every chunk of data, which the results are combined with majority voting. When the ensemble is completed, old models are replaced if their performance on the latest data chunk is lower than that of the new models. In the work by Wang *et al.* [10], models are also constructed on every new data chunk and the models are weighted according to their performance. In addition, pruning based on cost-sensitive and prediction confidence were explored to improve the efficacy.

More recently, ensemble solutions based on dynamic weighting have also been proposed. The dynamic weighted majority approach [7] maintains an ensemble of classifiers and creates new ones when global predictions are incorrect. Similarly, the weights of individual models are reduced when their predictions are incorrect. After every evaluation, the weights are normalized to prevent over-emphasizing on the latest models. When the weights of the models fall below a given threshold, the models are removed, but the last classifier is excluded. Tsymbal *et al.* [9] proposed three dynamic weighting schemes; viz., (1) dynamic selection—only best model is selected, (2) dynamic voting—weighted majority based on accuracy, and (3) dynamic voting with selection—only the top half of the models are selected for weighted majority voting. In addition, they adopted an instance-based k-nearest neighbour weighting scheme for the models. For each instance, its k nearest neighbour is selected and used for weighting the models. They empirically showed better performance than a simple weighted voting approach.

It is important to note that all the above approaches are essentially *reactive*, i.e., they only adapt to the drift after it happened. Efforts have also been made on exploring *proactive* solutions, which try to predict the change in concept before it happens. In the RePro [12] system, models are re-constructed when the stored historical models cannot correctly predict the new coming data. In addition, models are only stored when they are conceptually different from those existing

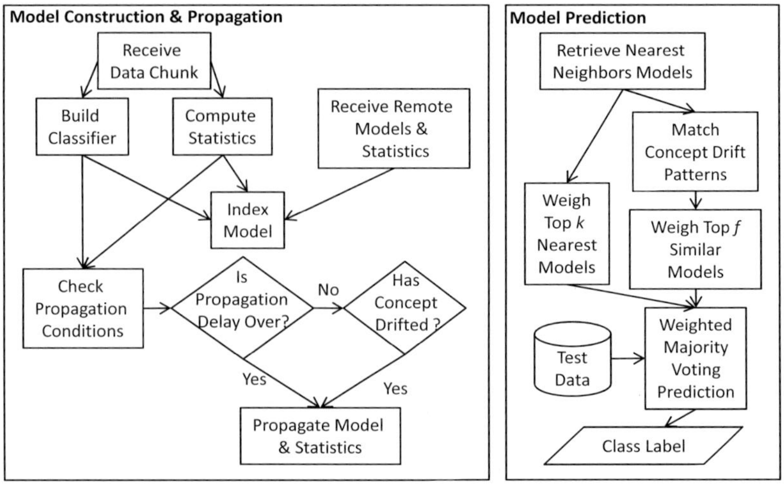

Fig. 1. Flowchart of the RePCoDE Framework

ones. In their study, the reactive approach chooses the best model that matches the latest data, while the proactive approach constructs a Markov Chain using historical data to predict the next possible concept given that the concept has occurred. The RePro system first uses the proactive approach to find suitable models and then falls back to the reactive approach when none is found.

While the existing ensemble solutions seem promising, they are unsuitable or insufficient for being deployed in the P2P environment due to several reasons. First, these approaches are based on centralized settings and only monitor a single stream. Due to the nature of P2P networks and the problem of distributed concept drifting, each peer should be treated as a separate data (sub) stream. In addition, the information of relationship between peers are also not utilized. They also do not carefully address the properties of the P2P network such as the massive size of the network and communication cost. Hence, they are not optimized for solving the concept drift challenge in such scenarios.

3 RePCoDE Framework

3.1 Overview

In this section, we propose a *Re*active and *P*roactive *Co*ncept *D*rift detection *E*nsemble (RePCoDE) framework for learning drifting concepts in P2P networks.

We first start by stating some basic assumptions for learning with concept drifts in a P2P environment. First, we assume that the data of all peers are drawn from the same unknown underlying probability distribution where the concept changes from time to time. However, concepts may not always drift instantaneously in all peers (some delays can occur). Also, we assume that for all peers, data arrive sequentially, but are grouped together and processed in chunks, each consisting of

Algorithm 1. RePCoDE Framework.

input: number of nearest neighbour voters k, number of proactive voters f, length of past statistics sequence b, proactive/reactive ratio λ, max propagation delay w, drift detection threshold T;

Last propagated model $M_L = \varnothing$;
current data chunk $\mathbf{D}_t = \emptyset$;
current model $M_t = \varnothing$;
current data statistics $S_t = \varnothing$;
propagation delay count $P_{delay} = w$;
while *stream not end* **do**
    if *new data chunk* $\boldsymbol{D}_t$ *arrives* **then**
        M_t = construct classification model based on $\mathbf{D}_t$;
        S_t = compute statistics on $\mathbf{D}_t$;
        M_L = check propagation conditions using Algorithm 2;
    if *new remote peer's model* M_r *and statistics* S_r *arrives* **then**
        Index model M_r using S_r;
    if *new test dataset* $\boldsymbol{D}_{test}$ *arrives* **then**
        predicted class labels $\mathbf{y}$ = predict class labels of $\mathbf{D}_{test}$ using Algorithm 3;

n data instances. In addition, we assume the time taken to gather each chunk is a single time period/step. Figure 1 gives an overview of the processes in our proposed solution. We briefly describe the idea of our approach below.

First, each peer monitors its own data stream and builds a classification model for every chunk of incoming labeled data. Next, it computes statistics of the data and uses them for indexing the corresponding classifier. If concept drift has occurred since last propagation or propagation waiting time reaches zero, one then propagates its current model together with data statistics to other peers.

Further, upon receiving the models and data statistics from other peers, each peer indexes the models using the corresponding data statistics. Contrary to most existing studies, our approach requires the propagation of data statistics in order to achieve the *proactive* adaptation. Due to the possibly large number of models indexed, we first choose the most relevant models and then weight them according to their performance on the most current data chunk. In addition, we also try to match the local data stream with that of other peers to select models that might better represent future data.

Finally, the class labels of unseen coming data are predicted using these selected models based on the weighted majority voting. The pseudo code of our algorithm is provided in Algorithm 1. For the remainder of this section, we first present detailed descriptions of the main phases of our proposed framework; viz., (1) training phase and (2) prediction phase, and then analyze time complexity and communication cost of our approach.

3.2 Training Phase

In a continuous manner, each peer independently gathers data and their corresponding labels until the size of the specified chunk n is reached. Each peer then constructs a local classification model based on the latest data chunk. As RePCoDE is an ensemble based approach, it is possible to use any type of classification algorithms such as decision tree, neural networks, Support Vector Machine, etc, for the model construction. However, due to the high frequency and short gap of data arrival, the algorithm used must have very low time complexity. In addition, as the models may need to be propagated at a later stage, we also need to consider the size of the resultant classification model. Hence, to meet the above two criteria, we choose the state-of-the-art linear SVM classification algorithm [13] in our experiments. A linear SVM model, consisting of only a single data vector, can be built in linear time complexity.

In addition to model construction, *statistics* of the data chunks are also computed. These statistics are concise representation of the data chunks, which are computed for the following reasons: (1) to facilitate the searching of models built using *similar* data, and (2) to reduce the communication cost required for propagation together with their respective models. Although there are many possible methods (such as Gaussian Mixture Model, clustering, etc.) to compute the statistics, we simply consider the linear SVM model which represents the decision hyperplane due to several reasons: (1) to streamline the model construction and statistics computation process, and (2) its low time and space complexity, and (3) its good ability of representing the training data in a separating manner.

Using the data statistics, the local models are then indexed locally. The index method used here has to be distance-aware, such as locality sensitive hashing (LSH) [14]. The purpose is to speed up the process of model retrieval and filtering. As the number of models can be infinite, we limit the size of the index m to control the space complexity of the framework. In this work, we adopt a simple yet computationally efficient replacement strategy for handling index overflow—to simply discard the oldest models, i.e., the newest model replaces the oldest in the index. For efficiency and suitability, LSH [14] is adopted for indexing in our implementation.

In addition to the indexing of local models, propagation conditions are also validated. Here, we propose two criteria for propagation: (1) presence of concept drift, and (2) propagation waiting delay. The first criterion uses the detection of concept drift to decide whether the local models should be propagated to other peers. Here, concept drift is examined by measuring the difference of the classification accuracy / prediction outputs between the last propagated model and the latest model from the latest data. Given a concept drift threshold $T \in [0, 1]$, the latest model and its data statistics are propagated to other peers if the difference is greater than T. By comparing with the last propagated model, we are able to detect both the sudden and gradual concept drifts. In the presence of gradual concept drifts, the difference in accuracy will slowly build up as more models are built and will eventually trigger the model propagation. Needless to say, for sudden concept drifts, the difference in accuracy will immediately satisfy the propagation criteria. Moreover, as a backup plan to insure against failure of concept drift detection, we impose a propagation waiting delay condition. By assuming the absence of any model propagation for w time steps, the

Algorithm 2. Check propagation conditions.

input : Current data chunk $\mathbf{D}_t$, Current model M_t, data statistics S_t, last propagated model M_L, propagation delay count P_{delay}, max propagation delay w, drift detection threshold T;

output: Last propagated model M_L, propagation delay count P_{delay};

if $P_{delay} > 0$ **then**
    $y_{current}$ = predict $\mathbf{D}_t$ using M_t;
    y_{last} = predict $\mathbf{D}_t$ using M_L;
    if $y_{last} - y_{current} > T$ **then**
        propagate M_t and S_t;
        $P_{delay} = w$;
    else $P_{delay} = P_{delay} - 1$
else
    propagate M_t and S_t;
    $P_{delay} = w$;

latest model and statistics are propagated to the network, regardless of whether concept drift has been detected or not. The propagation waiting delay condition ensures regular updates of peers' data trends and provides additional models to improve accuracy performance of the ensemble solution. An outline is provided in Algorithm 2.

As for the remote models, upon receiving the models and statistics from other peers, each peer indexes the remote models using the corresponding data statistics in the same manner similar to the processing of local models.

3.3 Prediction Phase

As mentioned earlier, our proposed framework adapts to concept drift using both reactive and proactive techniques. Hence, the models used for prediction are selected based on two criteria: (1) distance between the model's statistics and the statistics of the latest (local) data chunk (reactive) and (2) similarity between the sequence of statistics of the models and the sequence of statistics of the local peer's data, i.e., pattern matching of concept drift trends (proactive). This process is outlined in Algorithm 3.

First, a peer retrieves the statistics of the latest data. Then the statistics are used to retrieve $max(k, f * 2)$ models from the index, where k is the number of *nearest-neighbour* voters and f is the number of *proactive* voters. The models retrieved have statistics that are the nearest (out of all indexed models) in terms of Euclidean distance to the statistics of the latest data chunk. Out of all the retrieved models, we select only the top k models for the reactive prediction component. Here, the basic assumption is that if the models have statistics similar to the most current data, they are more likely to be trained on similar data and hence, achieve better accuracy for the most current data. In addition, the accuracy of each of the chosen models is validated using the most current labeled data chunk, which is then set as the weights of the model for weighted majority voting. Finally, the sum of all weights of selected models is normalized

Algorithm 3. Prediction.

input : test dataset $\mathbf{D}_{test}$, number of nearest neighbour voters k, number of proactive voters f, length of past statistics sequence b, current data chunk $\mathbf{D}_t$, current data statistics S_t, proactive/reactive ratio λ;
output: Predicted Labels $\mathbf{y}$;

```
model set M_set = ∅, reactive weights W_reactive = ∅, reactive model set
  M_reactive = ∅, Similarity Score W_sim = ∅, proactive weights W_proactive = ∅,
  proactive model set M_proactive = ∅, local past data statistics sequence S_local = ∅;
M_set = retrieve from index max(k, 2f) models nearest to S_t, in ascending order;
  M_reactive, W_reactive = find top k most accurate models and their weights from
  M_set based on D_t;
S_local = retrieve b past data statistics of local peer, e.g., {S_{t-b}, S_{t-b+1}, ..., S_t};
for i = 1 to 2f do
    S_temp = retrieve b past data statistics of M_set[i];
    W_sim[i] = compute sequence similarity between S_local and S_temp);
M_proactive, W_proactive = find top f most similar models and their weights from
  M_set, W_sim;
y = predict labels of D_test using models in M_reactive and M_proactive based on
  weighted majority voting (W_reactive, W_proactive) with ratio λ;
```

to 1 such that the weights of all models becomes a distribution. Note here that this approach is similar to previous works [7,9], which is a reactive technique as it is only based on what has happened; i.e., based on the latest labeled data chunk which is already observed. Hence, adaptation to the concept drift only starts after the drift is known/detected. However, the sharing of the different peers' models will allow the ensemble to achieve much better adaptation to concept drifts compared to only using local models. An ensemble solution also performs better when more (relevant, generalization error less than 50%) models are used.

It is obvious that using the reactive approach, there is still be an initial drop in accuracy when concept drift first starts. However, if we assume that the same concept drift occurs at different time steps for different peers, then it may be possible to learn from the concept drift patterns of peers' whose concept drift has already occurred. Hence, we propose the following proactive technique to select models that may be representative of the future data. First, we backtrack b time steps and retrieve the statistics of the data up until the most current time step. We define this as a sequence of the local data statistics. Next, for each model retrieved that is not constructed locally, we first check the existence of a *later* model received from the same peer. Then, in a similar manner, we backtrack to b *earlier* models received from the same peer and create a sequence of remote data statistics. Then, we compute the similarity of the two sequences to obtain a similarity score. In our implementation, we used the dynamic time warping (DTW) algorithm to perform similarity matching [15]. Once the similarity matching has been performed for all models retrieved from the index, we select f *later* models whose sequence of statistics is the most similar to the local sequence. These later models are termed as *proactive* models as they are deemed to be indicative of future data. Here, we assume that peers experiences concept drifts in the same patterns and the sequence similarity matching searches for peers who had similar concept drifts experience. The similarity score obtained is used to weigh the importance of the *proactive* models. Similar to the reactive approach, we also normalize the sum of the weights of all proactive models such that the weights of all models becomes a distribution.

As the proactive and reactive approaches select models and assign weights (importance of classifiers) based on different criteria, they result in different ensembles of classifiers with different voting weights. This leads us to one last question—how do we combine the reactive and proactive approaches? In this paper, we allow users to set a ratio parameter $\lambda \in [0, 1]$ that determines their relative importance. With $\lambda = 0$, only the reactive approach is used, and on the contrary, with $\lambda = 1$, only the proactive approach is used. Finally, the class labels of unlabelled data are obtained by performing weighted majority voting based on both the selected reactive and proactive models, where the balance between the two approach is determined by λ.

3.4 Complexity Analysis

For the model construction phase, we analyse the time complexity with respect to only a single data chunk since data are possibly infinite. For each model construction (Linear SVM), the time complexity is $O(\log(1/\epsilon)nd)$ for an ϵ-accurate solution where n is the size of the data chunk [13]. For computation of data statistics, no additional cost is incurred. For indexing the model based on LSH, suppose we have L hash tables and τ is the cost of computing one function, then the cost is $O(L\tau)$ for each model. Hence, the total cost for model construction is $O(\log(1/\epsilon)nd + L\tau)$.

The time complexity to predict a dataset D_{test} of size n_t is as follows (c.f., Algorithm 3). First, we retrieve $\max(k, 2f)$ models from the LSH index, which cost $O(dm^{1/c^2})$, where m is the size of the index and c is the approximation factor. To evaluate the accuracy of the models, the cost is $O(\max(k, 2f)nd)$. Next, the past data statistics are retrieved and sequence similarity computed, which cost $O(2fb)$ and $O(2fdb^2)$ respectively. The cost of retrieving the top k and f models is $O(k)$ and $O(f)$ respectively. Finally, the cost of predicting with $k + f$ models is $O((k + f)dn_t)$. Hence, the total cost for prediction is $O(dm^{1/c^2} + \max(k, 2f)nd + 2fb + fdb^2 + k + f + (k + f)dn_t)$.

3.5 Communication Cost

The only communication cost incurred is the cost of propagating the data statistics and models. Hence, the cost of propagation for each model (including its data statistics) is $O(d)$, as the Linear SVM model consist of only a single vector and since the Linear SVM model is used as the data statistics to represent the data, no additional cost is incurred. However, note that factors such as concept drift detection threshold and propagation waiting delay can affect the frequency of model propagation and hence increase the communication cost.

4 Experimental Results

We conduct extensive experiments for evaluation under varying distributed concept drift scenarios in order to examine how the sharing of knowledge among peers can improve concept drift adaptation and how our proposed approach performs better than other baseline approaches.

4.1 Experimental Setup

For evaluating the competing approaches, we assume a feedback type of concept drift system [16,17,18] where data arrive first followed by their class labels, which are made known after some time. This method of evaluation allows us to compare the accuracy of the competing approaches in the presence of concept drifts and also how quickly they can adapt to the drifts.

The P2P concept drift problems are simulated using the moving hyperplane synthetic data generator [6], which has been widely used in concept drift works, for simulating gradual concept drifts environments. The moving hyperplane generator – a hyperplane in d-dimensional space is expressed by the equation $\sum_{i=1}^{d} a_i x_i = a_0$, where a_i is the weight of the attribute x_i. Data instances satisfying $\sum_{i=1}^{d} a_i x_i \geq a_0$ are labelled as positive and negative otherwise. Concept drifts are achieved by changing the weights of the attributes and the direction of change [6]. To simulate varying occurrence of concept drifts in peers, they are split into equal groups g and the drift occurrence of each sequential group is delayed by S time steps. For instance, suppose there are 5 groups, and concept drift occurs at time step t_1 for group 1. Then concept drift will occur at time step $t_1 + S$ for group 2 and $t_1 + 2S$ for group 3 and so on. Note that every peer will draw a sample from the same concept (same a weights) although the master data stream is constantly drifting.

Experiments are conducted over 50 time steps, each with a data chunk of size n. Error is measured using the prequential methods, i.e., each data chunk is first used to test the classifier, and then used as the training data. Results presented are the average of all peers, over 10 independent runs. We used the LIBLINEAR [13] linear SVM package as the base classifier for all approaches and the LSHKIT [14] LSH library for the index in RePCoDE. All algorithms are implemented in C++. Default parameters are as follows: number of peers $N = 100$, proactive voters $f = 20$, nearest neighbour voters $k = 20$, proactive/reactive ratio $\lambda = 0.5$, propagation waiting delay $w = 4$, number of dimensions for moving hyperplane $d = 50$, number of drifting dimensions $= 30$, weight change per attribute for moving hyperplane $= 0.05$, probability of change in direction for the weights $= 10\%$. All parameters for LIBLINEAR are as default, except for $\epsilon = 0.01$. The number of peer groups g is 5 and the number of time shifts S is 4.

We compare RePCoDE to the following approaches — (1) single local classifier (Single) — the classifier used for prediction is trained on the most current local data chunk , (2) weighted ensemble of most recent local classifiers (Local) — the most recent local models are weighted and used for prediction, (3) weighted

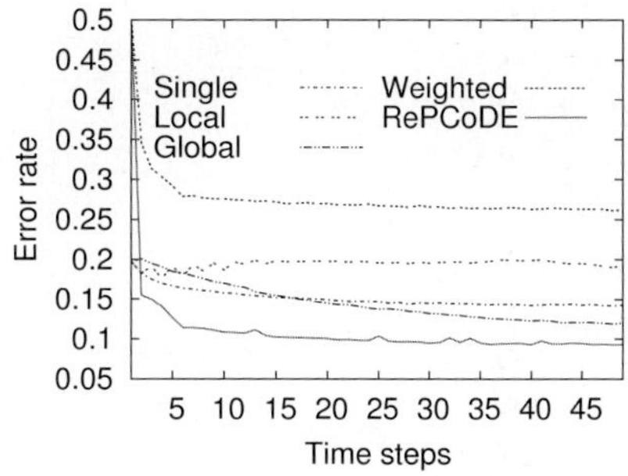

Fig. 2. Error on moving hyperplane dataset

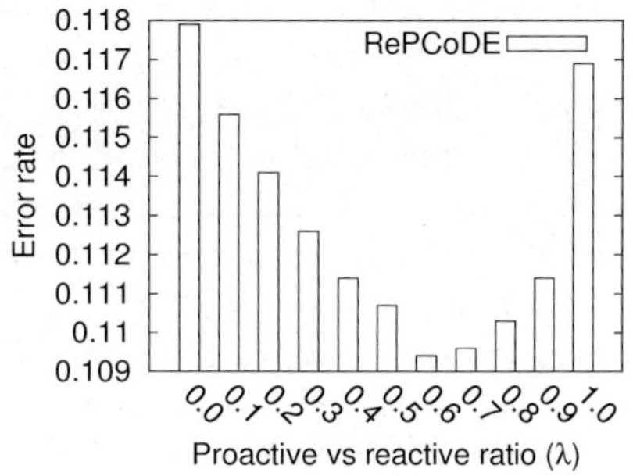

Fig. 3. Effect of proactive/reactive ratio λ on accuracy

ensemble of most recent local & remote classifiers (Global) — the most recent local and remote models are weighted and used for prediction, and (4) weighted ensemble of most accurate local & remote classifiers (Weighted) [8] — all models are weighted using the latest data chunk and the top most accurate models are used for prediction. The number of models used for Local, Global and Weighted is equal to the total number of models (nearest neighbour and proactive) used for RePCoDE. The different approaches are compared based on average error rate of every peer, average model propagated per time step (e.g., 0.5 average number of models sent means that every peer will propagate 1 model every 2 time steps) and computation time cost incurred.

4.2 Comparison

Here, we compare the error rates of the various approaches on a gradually drifting concept in a P2P network and presented the results in Figure 2. Observe that except in the beginning where other peers' knowledge is not yet available, RePCoDE achieves lower error for the rest of the experiment, followed by Global, Single, Local and Weighted. As Global is based on the most recent models of all peers, it is likely to include models of peers who have already experienced concept drift, and models of those who may be delayed even more than itself. Hence it has a higher error rate compared to RePCoDE. One possible reason Local has higher error rates than Single could be attributed to its inclusion of the outdated models. The only unexpected result is the Weighted approach, since the models chosen are the best on the current data chunk. This implies that the best model at current time step may not be the best for the next time step and hence the need for proactive approaches.

4.3 Parameter Sensitivity

Here, we varied various parameters, from both the approach and the environment to see how they affect the approaches.

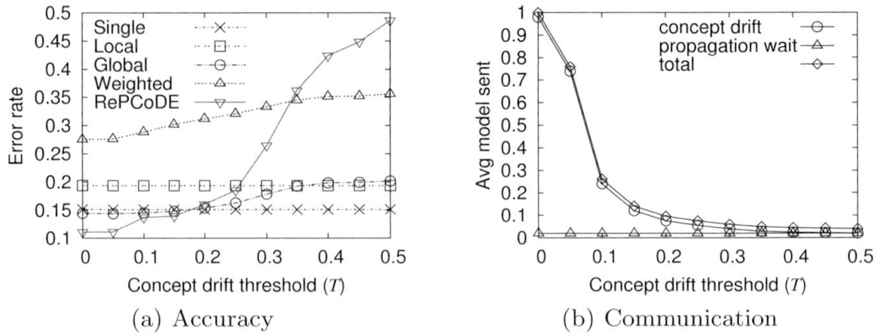

Fig. 4. Effect of drift detection threshold T on accuracy and communication costs

Reactive/Proactive Ratio λ. This experiment examines the effect of the Reactive/Proactive Ratio on the error rate. The Reactive/Proactive ratio λ varies from 0 to 1 and results are presented in Figure 3. Observe that neither of the extreme values perform well, meaning that using only the reactive or proactive approach is not sufficient. One has to combine both approaches to achieve low error rates. Empirically, a value that is somewhere in the middle is a good choice.

Drift Detection Threshold T. To understand how the concept drift detection affects RePCoDE, we set the propagation waiting delay to a large number such that it will not be activated, and varied the drift detection threshold T from 0 to 0.5. The results on accuracy and communication cost are presented in Figure 4. We see that as T increases, the error rate of all approaches except for Single and Local increase, with ReCoDE being affected the most, and the communication cost also decreases. This is because less concept drifts are detected, hence reducing the models propagated, affecting both communication and accuracy — a typical accuracy vs. cost trade off problem. As ReCoDE depends on the concept drift trends of other peers for the proactive approach to work well, with the failure to detect concept drifts, accuracy will be greatly affected (unless ratio is adjusted). However, note the lower error rates when concept drifts are properly detected. In addition, note that Single and Local are not affected as they are only based on the local models.

Propagation waiting delay w. Here, we study how the propagation waiting delay w can help ensure RePCoDE maintain high accuracy in the event of failure to detect concept drifts. We set the drift detection threshold T to 1 (i.e., no drift will be detected) and varied w from 2 to 10. The error rate and total communication cost are presented in Table 1. Results show that as w reduces, RePCoDE's error rate decreases, but the communication cost increases.

Number of nearest neighbour voters k. The results in Figure 5 show the error rate and time cost with respect to the number of nearest neighbour voters k varying from 5 to 50. The results show that a small number of k is sufficient to achieve satisfactory error rate, as error rate does not decrease further when

Table 1. Effect of propagation waiting delay w on accuracy and communication costs

Propagation waiting delay w	2	4	6	8	10
Error Rate	0.1805	0.2452	0.3092	0.3736	0.4368
Total Comm. Cost	0.3400	0.2000	0.1400	0.1200	0.1000

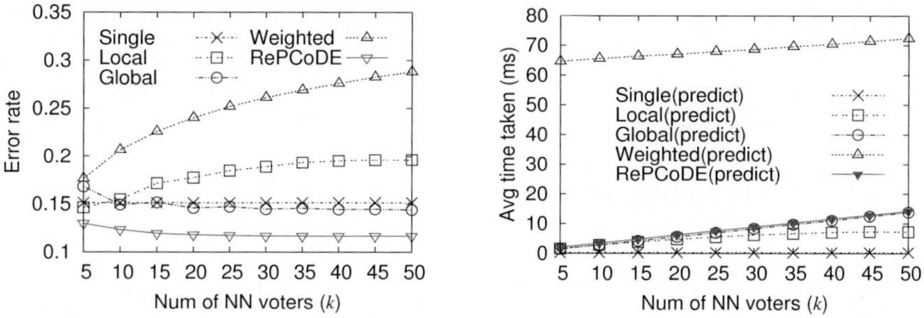

(a) Accuracy (b) Time

Fig. 5. Effect of number of nearest neighbour voters k on accuracy and time costs

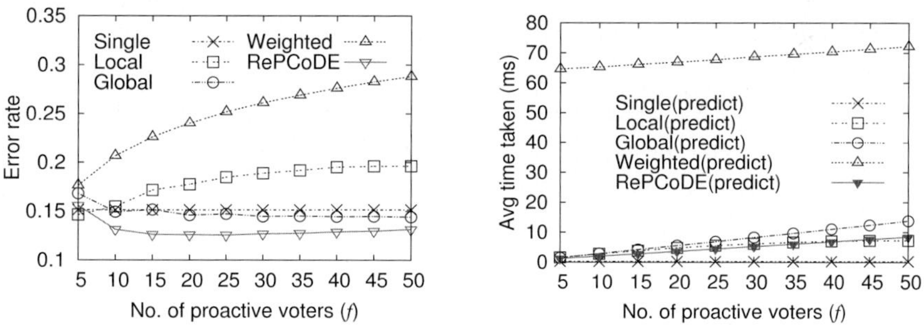

(a) Accuracy (b) Time

Fig. 6. Effect of number of proactive voters f on accuracy and time costs

k exceeds 25. In addition, as k increases, time cost increases linearly and with a small coefficient.

Number of proactive models f. The results in Figure 6 shows the error rate and time cost, as the number of proactive voters f is varied from 5 to 50. The results show that a small number of f is required. Initially, error rate decreases sharply with the increase in f but starts to gradually increase as f increase further. This implies that only a smaller number of proactive voters are able to predict the concept drift of the current peer. Similar to the nearest neighbour voters k, time cost for proactive voting increases linearly and with a small coefficient as f increases.

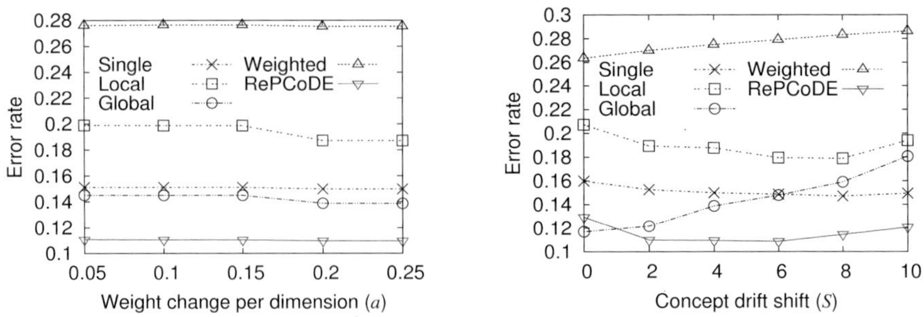

(a) Accuracy (b) Accuracy

Fig. 7. Effects of weight change and concept drift shift on accuracy

Amount of weight change a**.** This is to study how the increase in weight change for the drifting concept affects error rate (c.f., Figure 7(a)). The amount of weight change per dimension varies from 0.05 to 0.25. The results show that error rates of all approaches decrease with respect to the increase of weight change. The effects on ReCoDE, Single and Weighted are more gradual, while Local and Global are more sensitive. This coincides with the results presented in [10], where the larger weight increase can give more importance to certain dimensions making the problem easier to learn.

Number of time shifts S**.** This experiment examines how the delay in concept drift among peers affects error rates(c.f., Figure 7(b)). The number of time shift/delay S for different groups of peers is varied from 0 to 10. Observe that the error rate of Global and Weighted increases as S increases. This is because they are using the models of all peers, and as S increases, the mismatch in concepts among peers increases causing the increase in error. On the other hand, Single and Local are insensitive to the time shift as they are based only on local models. The error rate of RePCoDE only starts to gradually increase as S exceeds 6 and the increment is less than that of Global and Weighted, while achieving the lowest error rate. This demonstrates that RePCoDE is able to handle the problem of delayed occurrence of concept drift in P2P environments.

5 Conclusion

This paper studied the concept drift problem for distributed classification in P2P environments. We proposed a novel *Re*active and *P*roactive *Co*ncept *D*rift detection *E*nsemble (RePCoDE) framework, which is both efficient and accurate to overcome the concept drift issue for P2P classification. Experimental results showed that RePCoDE performs better than existing approaches that often can hardly handle concept drift in P2P environments. However, in order to achieve high accuracy, RePCoDE has to accurately detect concept drifts, which is dependent on the static drift detection threshold, and is prone to higher communication cost. In the event of poor concept drift detection, classification accuracy could be affected due to the inaccurate proactive voters and the static combination of proactive and reactive voters. In future work, we plan to study dynamic thresholds to reduce communication cost and ensure high accuracy at all times, and the dynamic combination of proactive and reactive voters.

References

1. Ang, H.H., Gopalkrishnan, V., Ng, W.K., Hoi, S.C.H.: Communication-efficient classification in P2P networks. In: Buntine, W., Grobelnik, M., Mladenić, D., Shawe-Taylor, J. (eds.) ECML/PKDD 2009. LNCS, vol. 5782, pp. 83–98. Springer, Heidelberg (2009)
2. Bhaduri, K., Wolff, R., Giannella, C., Kargupta, H.: Distributed decision-tree induction in peer-to-peer systems. Statistical Analysis and Data Mining 1(2), 85–103 (2008)
3. Luo, P., Xiong, H., Lü, K., Shi, Z.: Distributed classification in peer-to-peer networks. In: ACM SIGKDD, pp. 968–976 (2007)
4. Datta, S., Bhaduri, K., Giannella, C., Wolff, R., Kargupta, H.: Distributed data mining in peer-to-peer networks. Internet Computing 10(4), 18–26 (2006)
5. Chen, R., Sivakumar, K., Kargupta, H.: Distributed web mining using bayesian networks from multiple data streams. In: ICDM, pp. 75–82 (2001)
6. Hulten, G., Spencer, L., Domingos, P.: Mining time-changing data streams. In: KDD, pp. 97–106 (2001)
7. Kolter, J.Z., Maloof, M.A.: Dynamic weighted majority: An ensemble method for drifting concepts. Journal of Machine Learning Research 8, 2755–2790 (2007)
8. Street, W.N., Kim, Y.: A streaming ensemble algorithm (sea) for large-scale classification. In: KDD, pp. 377–382 (2001)
9. Tsymbal, A., Pechenizkiy, M., Cunningham, P., Puuronen, S.: Dynamic integration of classifiers for handling concept drift. Information Fusion 9(1), 56–68 (2008)
10. Wang, H., Fan, W., Yu, P.S., Han, J.: Mining concept-drifting data streams using ensemble classifiers. In: KDD, pp. 226–235 (2003)
11. Xu, Y., Wang, K., Fu, A.W.C., She, R., Pei, J.: Classification spanning correlated data streams. In: CIKM, pp. 132–141 (2006)
12. Yang, Y., Wu, X., Zhu, X.: Mining in anticipation for concept change: Proactive-reactive prediction in data streams. Data Mining and Knowledge Discovery 13(3), 261–289 (2006)
13. Hsieh, C.J., Chang, K.W., Lin, C.J., Keerthi, S.S., Sundararajan, S.: A dual coordinate descent method for large-scale linear SVM. In: ICML, pp. 408–415. ACM, New York (2008)
14. Dong, W., Wang, Z., Josephson, W., Charikar, M., Li, K.: Modeling lsh for performance tuning. In: CIKM, pp. 669–678 (2008)
15. Lemire, D.: Faster retrieval with a two-pass dynamic-time-warping lower bound. Pattern Recognition 42(9), 2169–2180 (2009)
16. Kubat, M.: A machine learning-based approach to load balancing in computer networks. Cybernetics and Systems 23(3-4), 389–400 (1992)
17. Widmer, G., Kubat, M.: Effective learning in dynamic environments by explicit context tracking. In: Brazdil, P.B. (ed.) ECML 1993. LNCS, vol. 667, pp. 227–243. Springer, Heidelberg (1993)
18. Kelly, M.G., Hand, D.J., Adams, N.M.: The impact of changing populations on classifier performance. In: KDD, pp. 367–371 (1999)

A Unified Approach to Active Dual Supervision for Labeling Features and Examples

Josh Attenberg[1], Prem Melville[2], and Foster Provost[3]

[1] Polytechnic Institute of NYU, Brooklyn, NY 11201
josh@cis.poly.edu
[2] IBM Research, Yorktown Heights, NY 10598
pmelvil@us.ibm.com
[3] NYU Stern School of Business, New York, NY 10012
fprovost@stern.nyu.edu

Abstract. When faced with the task of building accurate classifiers, active learning is often a beneficial tool for minimizing the requisite costs of human annotation. Traditional active learning schemes query a human for labels on intelligently chosen examples. However, human effort can also be expended in collecting alternative forms of annotation. For example, one may attempt to learn a text classifier by labeling words associated with a class, instead of, or in addition to, documents. Learning from two different kinds of supervision adds a challenging dimension to the problem of active learning. In this paper, we present a unified approach to such active dual supervision: determining which feature or example a classifier is most likely to benefit from having labeled. Empirical results confirm that appropriately querying for both example and feature labels significantly reduces overall human effort—beyond what is possible through traditional one-dimensional active learning.

1 Introduction

Active learning has been often used to reduce the amount of supervision required for effective learning. Traditionally, active learning research has focused on querying an oracle for labels on potentially informative examples. However, labeling effort may be better spent on providing alternative forms of supervision. Consider, for example, the task of *sentiment detection*, where given a piece of text as input, the desired output is a label that indicates whether this text expresses a positive or negative opinion. This problem can be cast as a typical binary text classification task, where a learner is trained on a set of documents that have been labeled based on the sentiment expressed in them. Alternatively, one could provide *labeled features*: for example, in the domain of movie reviews, words that evoke positive sentiment (e.g., "mesmerizing", "thrilling", etc.) may be labeled positive, while words that evoke negative sentiment (e.g., "boring", "disappointing", etc.) may be labeled negative. Through this kind of annotation a human conveys prior linguistic experience with a word by a sentiment label that reflects the emotion that the word evokes. The general setting of learning from both labels on examples and features is referred to as *dual supervision*.

J.L. Balcázar et al. (Eds.): ECML PKDD 2010, Part I, LNAI 6321, pp. 40–55, 2010.

This setting arises more broadly in tasks where, in addition to labeled documents, it is possible to provide domain knowledge in the form of words or phrases [26] or more sophisticated linguistic features that associate strongly with a class. Recent work [5,23,12] has demonstrated that feature supervision can greatly reduce the number of labels required to build high-quality classifiers. In general, example and feature supervision are complementary, rather than redundant.

This difference in information naturally leads to the problem of *active dual supervision*, or, how best to query a human resource to collect document labels *and* feature labels simultaneously, with the objective of building the highest quality model at the lowest cost. Much of the literature on active learning has focused on example-only annotation for classification problems. Less attention has been devoted to simultaneously acquiring alternative forms of supervisory domain knowledge. An exception, Sindhwani et al. apply classical uncertainty and experimental design-based active learning schemes to select labels for examples and features separately [24]. The present paper makes the following significant improvements over this prior work:

- Sindhwani et al. [24], at each iteration, randomly acquire a label for either an example or feature, and then probe the corresponding active learner. Here, we propose a holistic approach to active dual supervision based on an Expected Utility (estimated risk minimization) framework—within which, by optimizing the trade-offs between the costs and benefits of the different types of acquisitions, we deterministically select the most informative examples or features for labeling.
- We provide an instantiation of this framework for a recently introduced generative approach to dual supervision, instead of the graph-based dual supervision models used by Sindhwani et al. This generative approach, Pooling Multinomials [12], is comparable in performance to graph-based approaches and does not rely on unlabeled data. This is important for the present work. The Pooling Multinomials approach used here can be trained rapidly in an online fashion, rendering the otherwise computationally complex Expected Utility framework tractable.

Empirical results show that not only are we effective at actively selecting features for labeling, but that our unified approach to active dual supervision is better than the active learning of either instances or features in isolation.[1]

2 Dual Supervision

Most work in supervised learning has focused on learning from examples, each represented by a set of feature values and a class label. In dual supervision we consider an additional aspect: labels of features, which convey prior knowledge on associations of features to particular classes. This paper focuses solely on text

[1] A preliminary version of this work appeared in [15].

classification and all features represent term-frequencies of words; therefore, we use *feature* and *word* interchangeably.

While the active learning schemes explored in this paper are broadly applicable to any learner that can support dual supervision, we choose to focus on active learning for the Pooling Multinomials classifier [12] described below.

2.1 Pooling Multinomials

We introduce the Pooling Multinomials classifier as an approach to incorporate prior lexical knowledge into supervised learning for improved text classification. In the context of sentiment analysis, such lexical knowledge is available as the prior sentiment-polarity of words, while for classification, this knowledge comes from a human's term/class associations. Pooling Multinomials classifies unlabeled examples just as in multinomial Naïve Bayes classification, by predicting the class with the maximum likelihood, given by $argmax_{c_j} P(c_j) \prod_i P(w_i|c_j)$; where $P(c_j)$ is the prior probability of class c_j, and $P(w_i|c_j)$ is the probability of word w_i appearing in a document of class c_j. In the absence of background knowledge about the class distribution, we estimate the class priors $P(c_j)$ solely from the training data. However, unlike regular Naïve Bayes, the conditional probabilities $P(w_i|c_j)$ are computed using both labeled examples and labeled features. Given two models built using labeled examples and labeled features, the multinomial parameters of such models can be aggregated through a convex combination, $P(w_i|c_j) = \alpha P_e(w_i|c_j) + (1 - \alpha) P_f(w_i|c_j)$; where $P_e(w_i|c_j)$ and $P_f(w_i|c_j)$ represent the probability assigned by using the example labels and feature labels respectively, and α is the weight for combining these distributions. The weight indicates a level of confidence in each source of information, and Melville et al. [12] explore ways of automatically selecting this weight. However, in order to avoid confusion of our results with the choice of weight-selection mechanism, here we make the simplifying assumption that the two experts based on instance and feature labels are equally valuable, and as such set α to 0.5. The derivation and details of these models are not directly relevant to this paper, but can be found in [12].

Note that, though Pooling Multinomials is based on a Naïve Bayes generative model, it is a state-of-the-art approach for text classification. Our empirical results show that Pooling Multinomials outperforms linear SVMs, a popular technique for performing text classification that lacks a mechanism for handling feature labels. Melville et al. demonstrated that Pooling Multinomials performs better than alternative approaches to incorporating labeled features [12].

2.2 Experimental Setup

We conduct experiments on four binary text classification data set. The movies data set (2,000 examples, 5,000 features), introduced by Pang et al. [16] poses the task of classifying sentiment in movie reviews as positive or negative. *Politics* (107 examples, 1500 features) is based on posts from political blogs that were labeled as expressing positive or negative sentiments towards presidential candidates [12]. The *Baseball* (1,988 examples, 1,500 features) and *Science* (20,000 examples, 1,500 features) data sets are drawn from the 20-newsgroups[2] text collection where the task is to assign messages into the newsgroup in which they appeared. When performing analysis on these data sets, we use a bag-of-words representation, where documents are represented by term frequencies of the most frequent terms across all documents. All subsequent experiments present results averaged over 10-folds of cross validation.

2.3 Learning from Example vs. Feature Labels

Dual supervision makes it possible to learn from labeled examples and labeled features simultaneously. As in most supervised learning tasks, one would expect more labeled data of either form to lead to more accurate models. In this section we explore the influence of increased number of instance labels and feature labels independently, and also in tandem.

As with any active learning research, in order to study the effect of increasing number of labels we simulate a human oracle labeling data. In the case of examples this is straightforward, since all examples in these data sets have labels. However, in the case of features, we do not have a gold-standard set of feature labels. Ideally, we should have a human expert in the loop labeling each feature selected. However, such a manual process is not feasible for large scale, repeatable experiments. In order to simulate human responses to queries for feature labels, we construct a *feature oracle* in the following manner (as done in [5,24]). The information gain of words with respect to the known true class labels in the data set is computed using binary feature representations. Next, out of all available terms representing the data, the top $\sim \frac{1}{5}$ as ranked by information gain are assigned a label. This label is the class in which the word appears more frequently, corrected by the differences in base rate. The oracle returns a "don't know" response for the remaining words. As a result, this oracle simulates a human domain expert who is able to recognize and label the relevant task-specific words while being unable to assign a label to non-polar terms.

To demonstrate the basic value of dual supervision, Fig. 1 compares three schemes: Instances-then-features, Features-then-instances, and Passive Interleaving on the Movies data set. All three begin with a base set of training data, including labels for 10 randomly selected instances and 10 randomly selected features. As the name suggests, *Instances-then-features*, provides labels for randomly selected instances until all instances have been labeled, and then switches to labeling features. Similarly, *Features-then-instances* acquires labels for randomly selected features first and then switches to getting instance labels. In *Passive Interleaving* we probabilistically switch between issuing queries for randomly chosen instance and feature labels. In particular, at each step we choose to query for an instance with probability 0.36, otherwise we query for a feature label. The instance-query rate of 0.36 is selected based on the ratio of available instances (1,800) to available features (5,000) in the Movies set. For the learning curves presented in Fig. 1, the x-axis corresponds to the number of queries issued. As discussed earlier, in the case of features, the oracle may respond to a

[2] `http://archive.ics.uci.edu/ml/`

query with a class label or may issue a "don't know" response, indicating that no label is available. As such, the number of feature-queries on the x-axis does not correspond to the number of actual known feature labels. We would expect that on average 1 in 5 feature-label queries prompts a response from the feature oracle that results in a known feature label being provided.

At the end of the learning curves, each method has labels for all available instances and features; and as such, the last points of all three curves are identical. The results show that fixing the number of labeled features, and increasing the number of labeled instances steadily improves classification accuracy. This is what one would expect from traditional supervised learning curves. More interestingly, the results also indicate that we can fix the number of instances, and improve accuracy by labeling more features. Finally, results on Passive Interleaving show that though both feature labels and example labels are beneficial by themselves, dual supervision which exploits the interaction of examples and features does in fact benefit from acquiring both types of labels concurrently.

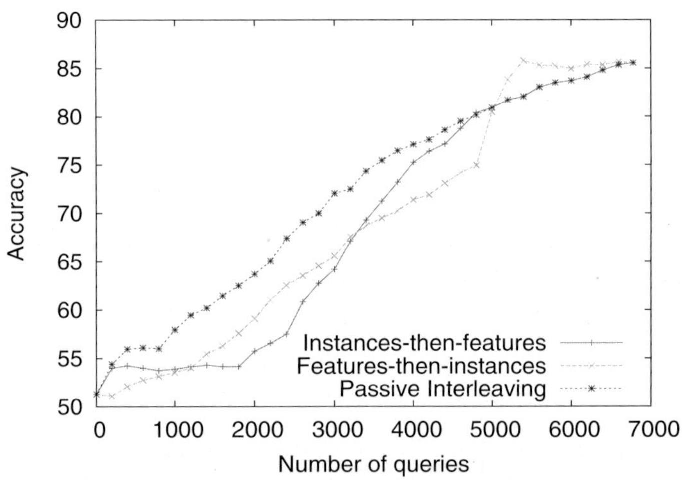

Fig. 1. Comparing the effect of instance and feature label acquisition in dual supervision

For all results above, we are selecting instances and/or features to be labeled uniformly at random. Based on previous work in active learning one would expect that we can select instances to be labeled more efficiently, by having the learner decide which instances it is most likely to benefit from. The results in this section suggests that actively selecting features to be labeled may also be beneficial. Furthermore, the Passive Interleaving results suggest that an ideal active dual supervision scheme would actively select both instances and features for labeling. We begin by exploring active learning for feature labels in the next section, and then consider the simultaneous selection of instances and features in Sec. 4.

3 Acquiring Feature Labels

Traditional active learning has primarily focused on selecting unlabeled *instances* to be labeled. The dual-supervision setting adds an additional aspect to active learning where labels may be acquired for features as well. In this section we focus on the task of active learning applied only to feature-label acquisition.

3.1 Feature Uncertainty vs. Certainty

In the traditional active learning setting, Uncertainty Sampling has earned a reputation as an effective and intuitive technique for selecting instances to get labeled [8]. In this approach, labels are requested for instances for which the current model gives the highest degree of uncertainty—which for binary classification with 0/1 loss are those instances nearest to the current classification boundary. Despite its simplicity, Uncertainty Sampling is often quite effective in practice, and has therefore become a standard for comparison for active learning research. This raises the question of whether one can apply the same principle to feature-label acquisition: select unlabeled features that the current model is most uncertain about.

Much like instance uncertainty, feature uncertainty can be measured in different ways, depending on the underlying method used for dual supervision. Since Pooling Multinomials builds a multinomial Naïve Bayes model, we can directly use the model's conditional probabilities of each feature f given a class. For ease of exposition we refer to the two classes in binary classification as *postive* (+) and *negative* (-), without loss of generality. Given the probabilities of f belonging to the positive and negative class, $P(f|+)$ and $P(f|-)$, we compute the uncertainty for f using the absolute value of the log-odds ratio, i.e.,

$$abs\left(\log\left(\frac{P(f|+)}{P(f|-)}\right)\right) \tag{1}$$

The smaller this value, the more uncertain the model is about the feature's class association. In every iteration of active learning we can select the features with the lowest certainty scores. We refer to this approach as *Feature Uncertainty*.

Though Uncertainty Sampling for features seems like an appealing notion, it may not lead to better models. If a classifier is uncertain about a feature, it may have insufficient information about this feature and may indeed benefit from learning its label. However, it is also quite likely that a feature has a low certainty score because it does not carry much discriminative information about the classes. In the context of sentiment detection, one would expect that neutral/non-polar words will appear to be uncertain words. For example, words such as "the" which are unlikely to help in discriminating between classes, are also likely to be considered the most uncertain. As we shortly report, on the *Movies* dataset, Feature Uncertainty ends up wasting queries on such words ending up with performance inferior to random feature queries. What works significantly better is an alternative strategy that acquires labels for features in *descending* order of the score in Eq 1. We refer to this approach as *Feature*

Certainty. A similar approach provided the best results in previous work [24]. We further improve on these results by the method described below.

3.2 Expected Feature Utility

The intuition underlying the feature certainty heuristic is that it serves to confirm or correct the orientation of model probabilities on different words during the active learning process. One can argue that feature certainty is also suboptimal in that queries may be wasted simply confirming confident predictions, which is of limited utility to the model. An alternative to using a certainty-based heuristic, is to directly estimate the expected value of acquiring each feature label. Such Expected Utility (Estimated Risk Minimization) approaches have been applied successfully to traditional active learning [19], and to active feature-value acquisition [14]. In this section we describe how this Expected Utility framework can be adapted for feature-label acquisition.

At every step of active learning for features, the next feature selected for labeling is the one that will result in the highest estimated improvement in classifier performance. Since the true labels of the unlabeled features are unknown prior to acquisition, it is necessary to estimate the potential impact of every feature query for all possible outcomes.[3] Hence, the decision-theoretic optimal policy is to ask for feature labels which, once incorporated into the data, will result in the highest increase in classification performance in *expectation.*

If f_j is the label of the j-th feature, and q_j is the query for this feature's label, then the Expected Utility of a feature query q_j can be computed as:

$$EU(q_j) = \sum_{k=1}^{K} P(f_j = c_k)\mathcal{U}(f_j = c_k) \tag{2}$$

Where $P(f_j = c_k)$ is the probability that f_j will be labeled with class c_k, and $\mathcal{U}(f_j = c_k)$ is the utility to the model of knowing that f_j has the label c_k. In practice, the true values of these two quantities are unknown, and the main challenge of any Expected Utility approach is to accurately estimate these quantities from the data currently available.

A direct way to estimate the utility of a feature label is to measure expected classification accuracy. However, small changes in the probabilistic model that result from acquiring a single additional feature label may not be reflected by a change in accuracy. Therefore, we use a finer-grained measure of classifier performance, Log Gain, which is computed as follows. For a model induced from a training set T, let $\hat{P}(c_k|x_i)$ be the probability estimated by the model that instance x_i belongs to class c_k; and $\mathbb{I}$ is an indicator function such that $\mathbb{I}(c_k, x_i) = 1$ if c_k is the correct class for x_i and $\mathbb{I}(c_k, x_i) = 0$, otherwise. Log Gain is then defined as:

[3] In the case of binary classification, the possible outcomes are a *positive* or *negative* label for a queried feature.

$$LG(x_i) = -\sum_{k=1}^{K} \mathbb{I}(c_k) \log \hat{P}(c_k|x_i) \tag{3}$$

Then the utility of a classifier, $\mathcal{U}$, can be measured by summing the Log Gain for all instances in the training set T. A lower value of Log Gain indicates a better classifier performance. We present an empirical comparison of different utility measures in Sec. 5.

In Eq. 2, apart from the measure of utility, we also do not know the true probability distribution of labels for the feature under consideration. This too can be estimated from the training data, by seeing how frequently the word appears in documents of each class. In the instance-based multinomial Naïve Bayes model we already collect these statistics in order to determine the conditional probability of a class given a word, i.e. $P(f_j|c_k)$. We can use these probabilities to get an estimate of the feature label distribution, $\hat{P}(f_j = c_k) = \frac{P(f_j|c_k)}{\sum_{k=1}^{K} P(f_j|c_k)}$.

Given the estimated values of the feature-label distribution and the utility of a particular feature query outcome, we can now estimate the Expected Utility of each unknown feature, selecting the features with the highest Expected Utility for labeling.

Though theoretically appealing, this approach can be computationally intensive if Expected Utility estimation is performed on all unknown features. In the worst case this requires building and evaluating models for each possible outcome of each unlabeled feature. In a setting with m features and K classes, this approach requires training $O(mK)$ classifiers. However, the complexity of the approach can be significantly alleviated by only applying Expected Utility evaluation to a sub-sample of all unlabeled features. Given the large number of features with no true class labels, selecting a sample of available features uniformly at random may be sub-optimal. Instead we select a sample of features based on Feature Certainty. In particular we select the top 100 unknown features that the current model is most certain about, and identify the features in this pool with the highest Expected Utility. We refer to this approach as *Feature Utility*. We use Feature Certainty to sub-sample the available feature queries, since this approach is more likely to select features for which the label is known by the oracle.

3.3 Active Learning with Feature Labels

We ran experiments comparing the three different active learning approaches described above on the Movies data set. Here we begin with a model trained on a random selection of 10 labeled features and 100 labeled instances.

The experiments in this section focus only on the selection of *features* to be labeled— in each iteration of active learning we select the next 10 feature-label queries, based on Feature Uncertainty, Feature Certainty, or Feature Utility. As a baseline, we also compare to the performance of a model that selects features uniformly at random. Our results are presented in Fig. 2.

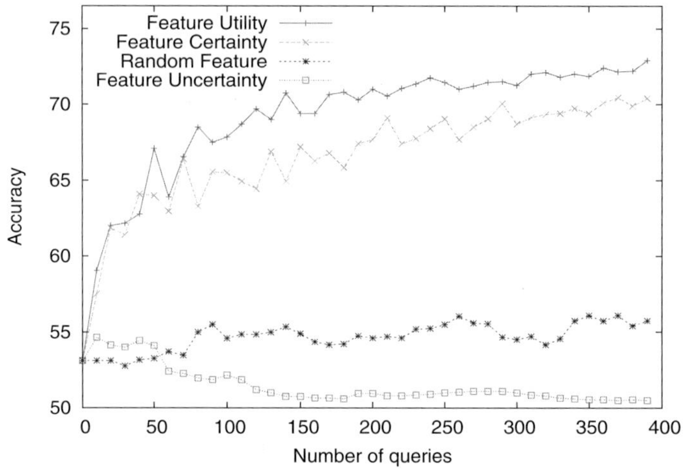

Fig. 2. Comparing different active learning approaches for acquiring feature labels

The results show that Feature Uncertainty performs worse than random sampling; however, the converse approach of Feature Certainty does remarkably well. This is in line with out discussion above in Sec. 3.1 The results for Feature Utility show that estimating the expected impact of potential labels for features does in fact perform much better than feature certainty. The results confirm that despite our crude estimations in Eq. 2, Feature Utility is an effective approach to active learning of feature labels. Furthermore, we demonstrate that by applying the approach to only a small sub-sample of certain features, we are able to make this method computationally feasible to use in practice. Increasing the size of the sample of candidate feature queries is likely to improve performance, at the cost of increased time in selecting queries.

4 Active Dual Supervision

In the previous section we demonstrated that actively selecting informative features to be labeled performs significantly better than random selection. This conclusion is congruent with the rich body of work showing the benefits over random selction of actively selecting *instances* for labeling. Furthermore, we have demonstrated in Sec. 2 that randomly selecting both feature and instance labels in tandem is better than either in isolation. An ideal active scheme should be able to assess if an instance or feature would be more beneficial at each step, and select the most informative instance or feature for labeling.

Fortunately, the Expected Utility method is very flexible, capable of addressing both types of acquisition within a single framework. Since the measure of utility is independent of the type of supervision and only dependent on the resulting classifier, we can estimate the expected utility of different forms of acquisitions in the same manner. For instance, Saar-Tsechansky et al. [20] use such an approach to estimate the utility of acquiring class labels and feature values (not labels), within one unified framework. A similar technique is utilized here, yielding a holistic approach to active dual supervision, where the Expected Utility of an instance or feature label query, q, can be computed as

$$EU(q) = \sum_{k=1}^{K} P(q = c_k) \frac{\mathcal{U}(q = c_k)}{\omega_q} \tag{4}$$

where ω_q is the cost of the query q, $P(q = c_k)$ is the probability of the instance or feature queried being labeled as class c_k, and utility $\mathcal{U}$ can be computed as in Eq. 3. By evaluating instances and features in the same units, and by measuring utility per unit cost of acquisition, such a framework facilitates explicit optimization of the trade-offs between the costs and benefits of the different types of acquisitions. For the sake of simplicity, we assume equal costs of acquisitions in this paper, i.e., $\omega_q = 1$. But in principle this framework can be used even if the cost of acquiring a feature label is different from the cost of acquiring an instance label. We refer to this combined instance and feature acquisition approach simply as Expected Utility. As before, we speed up query selection by first sub-sampling 100 features based on certainty and 100 instances at random, and then evaluate the Expected Utility on this candidate set.

Experiments are performed as before, comparing Expected Utility to Feature Utility and Passive Interleaving. Recall that Passive Interleaving corresponds to probabilistically interleaving queries for randomly chosen, not actively chosen, examples and features. For completeness, we also compare with an instance-only active learning method. Namely, we use Uncertainty Sampling [8], which has been shown to be a computationally efficient and effective approach in the literature. In particular, we select unlabeled examples to be labeled in order of decreasing uncertainty, measured in terms of the margin, as done in [13]. The margin on an unlabeled example is defined as the absolute difference between the class probabilities predicted by the classifier for the given example, i.e., $|P(+|x) - P(-|x)|$. We refer to the selection of instances based on this uncertainty as Instance Uncertainty, in order to distinguish it from Feature Uncertainty.

We compare the performance of any two methods, A and B, by computing the percentage reduction in classification error rate obtained by A over B at each acquisition phase and report the average reduction over all acquisition phases. We refer to this average as the *average percentage error reduction*. The reduction in error obtained with policy A over the error of policy B is considered to be

Table 1. Error reduction(%) of active learning approaches compared to Passive Interleaving

Data Set	Instance Uncertainty	Feature Utility	Expected Utility
Movies	**9.90**	**31.18**	**36.52**
Science	−3.88	**13.05**	**25.24**
Baseball	−101.98	−2.59	**39.61**
Politics	−7.04	−7.16	1.48

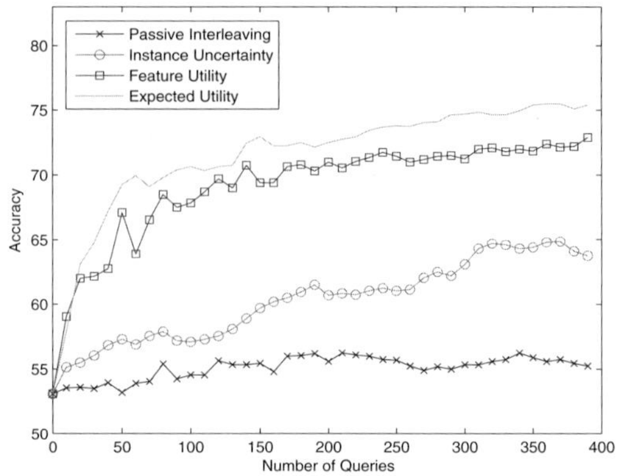

(a) Movies

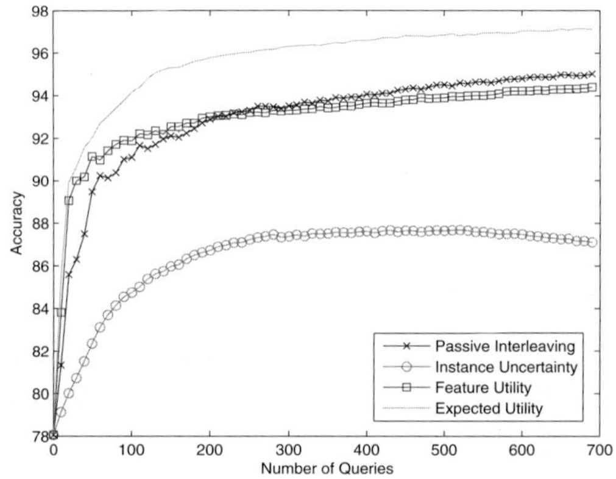

(b) Baseball

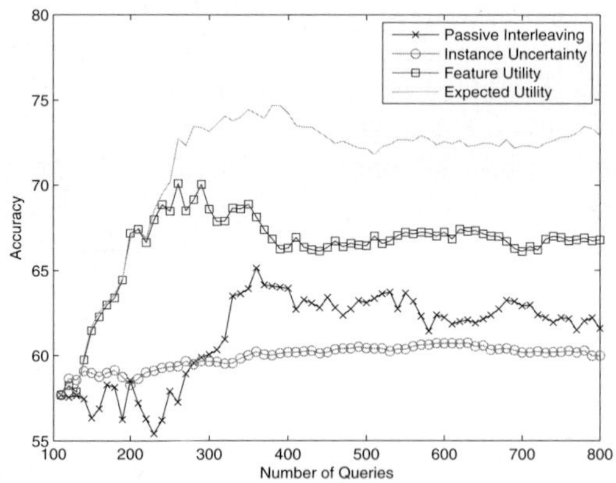

(c) Science

Fig. 3. Comparing Expected Utility to alternative label acquisition strategies significant if the errors produced by policy A are lower than the corresponding errors (i.e., at the same acquisition phase) produced by policy B, according to a paired t-test ($p < 0.05$) across all the acquisition phases [13]. In our experiments, we compare all active methods using Passive Interleaving as our baseline (B). Our results are summarized in Table 1, where statistically significant improvements over Passive Interleaving are shown in bold. We also present learning curves on three datasets in Fig. 3.

We observe that actively selecting instances or features for labeling is better than randomly selecting either. In general, effectively selecting features' labels, via Feature Utility, does even better than actively selecting only instances. However, in some cases, the advantage of actively selecting only one type of supervision is out-weighed by randomly selecting both instances and features in tandem. This phenomenon can be seen for *Baseball* in Fig. 3(b), where Uncertainty Sampling is clearly less effective than Passive Interleaving; even Feature Utility's initial advantage is lost by randomly selecting some instance labels in Passive Interleaving. However, by estimating the benefit of each type of acquisition at each step, the holistic Expected Utility approach is able to outperform active learning on instances and features in isolation, as well as randomly interleaving instance-labels with feature-labels. The savings from using Expected Utility for active dual supervision can be quite substantial. For instance, this approach achieves an accuracy of 75% on *Movies* with only 350 queries, while Passive Interleaving requires 10 times the number of queries to reach the same performance. The average reduction in error using Expected Utility over Passive Interleaving ranges from a 1.5% on *Politics* to 40% on *Baseball.*

5 Choice of Utility Measure

Up to now, in order to evaluate the utility of a potential label, we have measured the average log gain (Eq. 3) calculated on the training set by a classifier trained with the inclusion of the associated labeling. However, there are alternative ways to measure utility. One obvious choice is to measure utility based on classification accuracy on the training set. Both log gain and accuracy are "supervised" utilities measures, meaning that they are computed on examples for which we have labels. In contrast, in previous work in the traditional instance labeling setting, Roy and McCallum [19] use two "unsupervised" measures, computed on the pool of unlabeled examples. Their motivating objective is different from our desire to estimate directly the expected improvement in generalization performance. Instead, they try to "select those examples which maximizes [sic] the sharpness of the learner's posterior belief about the unlabeled examples" [19]. Namely, they use entropy and $(1 - \arg\max_{c_k} \hat{P}(c_k|x))$, where $\hat{P}(c_k|x)$ is the resulting classifier's predicted probability for class c_k . We will refer to the latter measure as maximum posterior. Lower values of these measures correspond to higher utilities in their setting.

We ran experiments as before for Feature Utility and the combined Expected Utility on the Movies data set, comparing five different measures of utility: log gain, accuracy, and entropy estimated on the training set, as well entropy and maximum posterior on the unlabeled pool. The results for Expected Utility are presented in Fig. 4. The performance of entropy and maximum posterior on the unlabeled pool, and accuracy are indistinguishable in the figure, and do not perform as well as the other measures. The results on Feature Utility (not shown) show a similar trend except that accuracy performs better than entropy and maximum posterior on the unlabeled pool, while still doing worse than log gain. By incorporating the predicted class probabilities, log gain is able to capture small changes in the classifier that may lead to an improvement that may not be reflected by a change in classification accuracy. Hence it tends to perform better than measuring utility based on accuracy. The unsupervised measures used by Roy and McCallum do not perform as well as they are focused on selecting examples that on average will make the predictions of the model the most certain on the not-yet-labeled examples. These results empirically support the use of log gain in the Expected Utility framework. For a deeper theoretical discussion of this measure see [20].

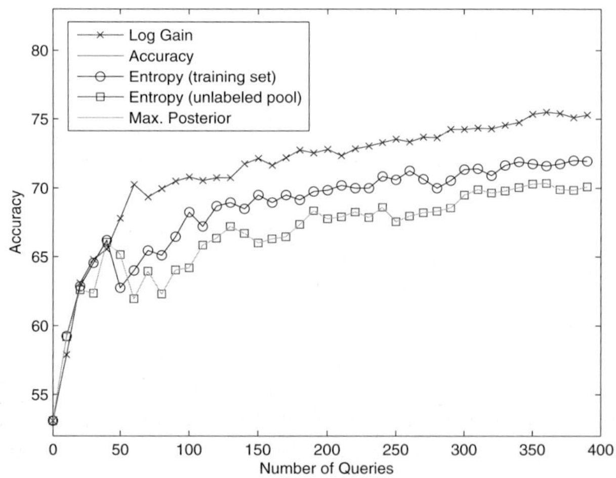

Fig. 4. Comparing different measures of utility. Accuracy, entropy (unlabeled) and max. posterior overlap.

6 Related Work

Active learning in the context of dual supervision models is a new area of research with very little prior work, to the best of our knowledge. Most prior work in active learning has focused on pooled-based techniques, where examples from an unlabeled pool are selected for labeling [3]. In contrast, active feature-value acquisition [14] and *budgeted learning* [11] focus on estimating the value of acquiring missing features, but do not deal with the task of learning from feature *labels*. Raghavan and Allan [17] and Raghavan et al. [18] study the problem of *tandem learning* where they combine uncertainty sampling for instances along with co-occurence based interactive feature selection. Godbole et al. [7] propose notions of feature uncertainty and incorporate the acquired feature labels, into learning by creating one-term mini-documents. Druck et al. [6] perform active learning via feature labeling using several uncertainty reduction heuristics. Sindhwani et al. [24] also study the problem of active dual supervision, applied to a graph-based dual supervision method. They explore various heuristic approaches to active learning for instances and features separately. In order to interleave selections from both instances and features, they randomly probe an active instance learner or an active feature learner for the next query. In contrast, we take a holistic approach to active dual supervision, where by estimating the potential value of features and instances on the same scale, we select the type of acquisition that is most likely to benefit our classifier. In contemporaneous work, Attenberg et al. [1] propose active dual supervision as one possible solution to the cold start problem often faced by active learners in settings with high class skew. Additionally, they propose tasking human domain experts with seeking and finding useful feature values directly, as opposed to the query/respose approach seen here.

While in principle the Expected Utility-based techniques presented here could be applied to any technique for dual supervision, Pooling Multinomials is well suited for our approach, since it can be implemented to be update-able (without retraining), making the computations in Expected Utility extremely efficient. It is a challenge to efficiently implement Expected Utility for the graph-based method used by Sindhwani et al., and this is a promising direction for future work.

Learning from labeled examples and features via dual supervision is itself a new area of research. Sindhwani et al. [22] use a kernel-based framework to build dual supervision into co-clustering models. Sindhwani and Melville [23] apply similar ideas for graph-based sentiment analysis. Note that, dual supervision should not be confused with Co-Training [2], in which the description of examples can be divided into two distinct views i.e. disjoint feature sets. There have also been previous attempts at using only feature supervision, mostly along with unlabeled documents. Much of this work [21,25,10,4] has focused on using labeled features to generate *pseudo-labeled examples* that are then used with well-known models. In contrast, Druck et al. [5] constrain the outputs of a multinomial logistic regression model to match certain reference distributions associated with labeled features. In a similar vein, Liang et al. [9] learn from labeled examples and constrains on model predictions.

7 Conclusions and Future Work

This paper presents a unified framework for active dual supervision, where the relative benefit of each type of acquisition is assessed based on the expected improvement of the resulting classifier.We demonstrated that not only is combining example and feature labels beneficial for modeling, but that actively selecting the most informative examples and features for labeling can significantly reduce the burden of labeling such data. For simplicity, we did not consider the different costs of acquiring labels. Presumably labeling a feature versus labeling an instance could incur very different costswhich could be monetary costs or time taken for each annotation. The general Expected Utility framework we present can directly handle such cost-benefit trade-offs, and empirically validating this is an avenue for future work. Furthermore, the mixing of multinomials based on labeled features and labeled examples exerts a strong influence on the probability estimates produced, and therefore the choices made in active learning. Another direction for future work is the investigation of the mixing parameter, α, and its influence on active dual supervision.

Human oracles may be able to provide a much richer set of background information than can be expressed via individual token polarities. For instance, "yeah" and "right" may both be terms denoting a positive sentiment polarity, while "yeah, right" may be used with sarcasm with a negative connotation[4]. Extending models to incorporate higher order information from oracles is another good direction for future work.

The Expected Utility framework we propose is general, in that it can be applied to any learning algorithm that supports dual supervision. However, it is a significant challenge to devise methods to efficiently estimate the terms in Eq. 4 that are appropriate to the learner of choice. As shown in Sec. 5, even the choice of measure of utility is not obvious, and can make a significant difference in results. As such, adapting this framework to other learners, such as the graph based approach in [24], is a challenging, but promising direction to explore.

Acknowledgments

We thank Vikas Sindhwani for invaluable help in preparing this document.

References

1. Attenberg, J., Melville, P., Provost, F.: Guided feature labeling for budget-sensitive learning under extreme class imbalance. In: BL-ICML 2010: Workshop on Budgeted Learning (2010)
2. Blum, A., Mitchell, T.: Combining labeled and unlabeled data with co-training. In: COLT (1998)
3. Cohn, D., Atlas, L., Ladner, R.: Improving generalization with active learning. Machine Learning 15(2), 201–221 (1994)
4. Dayanik, A., Lewis, D., Madigan, D., Menkov, V., Genkin, A.: Constructing informative prior distributions from domain knowledge in text classification. In: SIGIR (2006)
5. Druck, G., Mann, G., McCallum, A.: Learning from labeled features using generalized expectation criteria. In: SIGIR (2008)
6. Druck, G., Settles, B., McCallum, A.: Active learning by labeling features. In: EMNLP 2009, pp. 81–90. Association for Computational Linguistics (2009)
7. Godbole, S., Harpale, A., Sarawagi, S., Chakrabarti, S.: Document classification through interactive supervision of document and term labels. In: Boulicaut, J.-F., Esposito, F., Giannotti, F., Pedreschi, D. (eds.) PKDD 2004. LNCS (LNAI), vol. 3202, pp. 185–196. Springer, Heidelberg (2004)

[4] We would like to thank an anonymous reviewer for this suggestion given to a preliminary version of this paper

8. Lewis, D.D., Catlett, J.: Heterogeneous uncertainty sampling for supervised learning. In: Proc. of 11th Intl. Conf. on Machine Learning (ICML 1994) (July 1994)
9. Liang, P., Jordan, M.I., Klein, D.: Learning from measurements in exponential families. In: ICML (2009)
10. Liu, B., Li, X., Lee, W.S., Yu, P.: Text classification by labeling words. In: AAAI (2004)
11. Lizotte, D., Madani, O., Greiner, R.: Budgeted learning of naive-Bayes classifiers. In: UAI (2003)
12. Melville, P., Gryc, W., Lawrence, R.: Sentiment analysis of blogs by combining lexical knowledge with text classification. In: KDD (2009)
13. Melville, P., Mooney, R.J.: Diverse ensembles for active learning. In: Proc. of 21st Intl. Conf. on Machine Learning, ICML 2004 (2004)
14. Melville, P., Saar-Tsechansky, M., Provost, F., Mooney, R.: An expected utility approach to active feature-value acquisition. In: ICDM (2005)
15. Melville, P., Sindhwani, V.: Active dual supervision: Reducing the cost of annotating examples and features. In: NAACL HLT 2009 (2009)
16. Pang, B., Lee, L., Vaithyanathan, S.: Thumbs up? sentiment classification using machine learning techniques. In: EMNLP (2002)
17. Raghavan, H., Madani, O., Jones, R.: An interactive algorithm for asking and incorporating feature feedback into support vector machines. In: SIGIR (2007)
18. Raghavan, H., Madani, O., Jones, R.: Active learning with feedback on features and instances. J. Mach. Learn. Res. 7 (2006)
19. Roy, N., McCallum, A.: Toward optimal active learning through sampling estimation of error reduction. In: ICML (2001)
20. Saar-Tsechansky, M., Melville, P., Provost, F.: Active feature-value acquisition. In: Management Science (2009)
21. Schapire, R.E., Rochery, M., Rahim, M.G., Gupta, N.: Incorporating prior knowledge into boosting. In: ICML (2002)
22. Sindhwani, V., Hu, J., Mojsilovic, A.: Regularized co-clustering with dual supervision. In: NIPS (2008)
23. Sindhwani, V., Melville, P.: Document-word co-regularization for semi-supervised sentiment analysis. In: ICDM (2008)
24. Sindhwani, V., Melville, P., Lawrence, R.: Uncertainty sampling and transductive experimental design for active dual supervision. In: ICML (2009)
25. Wu, X., Srihari, R.: Incorporating prior knowledge with weighted margin support vector machines. In: KDD (2004)
26. Zaidan, O.F., Eisner, J.: Modeling annotators: A generative approach to learning from annotator rationales. In: EMNLP (2008)

Vector Field Learning via Spectral Filtering

Luca Baldassarre[1], Lorenzo Rosasco[2,3], Annalisa Barla[1], and Alessandro Verri[1]

[1] Università degli Studi di Genova - DISI
Via Dodecaneso 35, Genova, Italy
{baldassarre,barla,verri}@disi.unige.it
[2] Istituto Italiano di Tecnologia,
Via Morego, 30 16163 Genova, Italy
[3] CBCL, Massachusetts Institute of Technology
Cambridge, MA 02139 - USA
lrosasco@mit.edu

Abstract. In this paper we present and study a new class of regularized kernel methods for learning vector fields, which are based on filtering the spectrum of the kernel matrix. These methods include Tikhonov regularization as a special case, as well as interesting alternatives such as vector valued extensions of L2-Boosting. Our theoretical and experimental analysis shows that spectral filters that yield iterative algorithms, such as L2-Boosting, are much faster than Tikhonov regularization and attain the same prediction performances. Finite sample bounds for the different filters can be derived in a common framework and highlight different theoretical properties of the methods. The theory of vector valued reproducing kernel Hilbert space is a key tool in our study.

Keywords: Vector-valued Functions; Multi-task; Regularization; Spectral Filtering; Kernels.

1 Introduction

In this paper we study theoretical and computational properties of a class of kernel methods for learning a vector valued function. These methods are based on filtering the spectrum of the kernel matrix rather than empirical risk minimization. The idea of using kernel methods for vector field learning has been considered in [1] where the framework of vector valued reproducing kernel Hilbert spaces was adopted and the representer theorem for Tikhonov regularization was generalized to the vector valued setting. Our work can be seen as an extension of the work in [1] aimed in particular at: 1) investigating the application of spectral filtering schemes [2] to learning vector fields; 2) establishing consistency and finite sample bounds for Tikhonov regularization as well as for all spectral filters in the setting of vector valued learning. One of the main outcomes of our study is that iterative algorithms based on spectral filtering outperform Tikhonov regularization from the computational perspective, while preserving the good prediction performances.

J.L. Balcázar et al. (Eds.): ECML PKDD 2010, Part I, LNAI 6321, pp. 56–71, 2010.

Classical supervised learning focuses on the problem of estimating functions with scalar outputs: a real number in regression and one between two possible labels in binary classification. The starting point of our investigation is the observation that in many practical problems it is convenient to model the object of interest as a function with multiple outputs. In machine learning, this problem typically goes under the name of multi-task or multi-output learning and has recently attracted a certain attention. It is interesting to recognize at least two classes of problems with multiple output functions. The first class, that we might call multi-task learning, corresponds to the situation in which we have to solve several standard scalar learning problems (each with its own training set) that we assume to be related, so that we can expect to obtain a better solution if we attempt to solve them simultaneously. Application in user recommendation systems can be taken as an example. The second class of problems corresponds to learning vector valued functions. This situation is better described as a supervised learning problem where the outputs are vector valued and we have a single training set. For example, a practical problem is that of estimating the velocity field of an incompressible fluid from scattered spatial measurements.

The two problems are clearly related. Indeed, we can view tasks as components of a vector valued function or equivalently learning each component of a vector valued function as one of many scalar tasks. Nonetheless, there are also some differences that make the two problems different both from a practical and a theoretical point of view. For example, in multi-task learning the input points for each task can be represented by different features and the sample size might vary from one task to the other. In particular, each task can be sampled in a different way so that, in some situations, by assuming that the tasks are highly correlated, we can essentially augment the number of effective points available for each individual task. This effect does not occur while learning vector fields where each component is sampled at the same input points. Since the sampling procedures are different, the error analyses for multi-task and vector valued learning are also different. The latter case is closer to the scalar setting, whereas in the multi-task case the situation is more complex: one might have different cardinalities for the various tasks or be interested to evaluate the individual performance of each task.

In this paper, we focus primarily on vector field learning as a natural extension of the classical scalar setting. In particular, some of the theoretical results are specific to vector valued functions, but many of the computational ideas we discuss apply to general multi-task problems. We propose a new class of algorithms to learn multi-output functions, called *spectral filters*, that cannot be described in terms of penalized empirical risk minimization. Each algorithm performs a different filtering of the spectrum of the kernel matrix, designed to suppress contributions corresponding to small eigenvalues. These algorithms are motivated by the results connecting learning theory and regularization of ill-posed problems [3] and the relations between stability and generalization [4,5]. We provide a detailed computational analysis that takes into account the specific form of the kernel as well the regularization parameter choice step, from which it is clear that the various methods have different computational properties. The bound on the excess risk further illustrates these differences. Our experiments confirm that iterative spectral filters, that can be seen as extensions of L2 boosting [6] are often preferable to Tikhonov regularization.

The plan of the paper follows. After discussing previous work in the next subsection, in Sect.2 we recall some basic concepts. in Sect.3 we present the class of algorithms under study, while in Sect.4 we review examples of kernels. The finite sample bound on the excess risk and computational issues are discussed in Sect.5. The experimental analysis is conducted in Sect.6 and we conclude in Sect.7 proposing some future work.

1.1 Previous Work

Several recent works considered multi-output learning, especially multi-task, and proposed a variety of approaches. Starting from the work of [7], related ideas have been developed in the context of regularization methods [8], Gaussian processes [9,10]. The specific problem of learning a vector valued function has received less attention in machine learning. In statistics we mention the Curds & Whey method [11], Reduced Rank Regression [12], Filtered Canonical y-variate Regression [13] and Partial Least Squares [14]. Estimating vector fields is common in the context of geophysics and goes under the name of co-kriging [15]. Some attempts to extend machine learning algorithms from the scalar to the vector setting have been made [16,17]. A study of vector valued learning with kernel methods is started in [1], where regularized least squares are analyzed from the computational point of view. The error analysis of vector valued Tikhonov regularization is given in [18]. To the best of our knowledge the application of spectral filtering techniques to vector field learning has not yet been studied.

2 Basic Concepts

We start by presenting the setup of the problem, as well as the basic notions behind the theory of vector valued reproducing kernels.

Supervised Learning. The problem of supervised learning amounts to inferring an unknown functional relation given a finite *training set* of input-output pairs $\mathbf{z} = \{(x_i, y_i)\}_{i=1}^n$ that are assumed to be identically and independently distributed according to a fixed, but unknown probability measure $\rho(x, y) = \rho_X(x)\rho(y|x)$ on $\mathcal{Z} = \mathcal{X} \times \mathcal{Y}$. Here we are interested in vector valued learning where $\mathcal{Y} \subseteq \mathbb{R}^T$. A learning algorithm is a map from a training set $\mathbf{z}$ to an estimator $f_{\mathbf{z}} : \mathcal{X} \to \mathcal{Y}$. A good estimator should generalize to future examples and, if we choose the square loss, this translates into the requirement of having small *expected risk*

$$\mathcal{E}(f) = \int_{\mathcal{X}\times\mathcal{Y}} \|y - f(x)\|_T^2 d\rho(x, y) \ ,$$

where $\|\cdot\|_T$ denotes the euclidean norm in $\mathbb{R}^T$. The ideal estimator is the minimizer of the expected risk, that is the regression function $f_\rho(x) = \int_{\mathcal{Y}} y\rho(y|x)$, but cannot be directly calculated since ρ is unknown. Further, the search for a solution is often restricted to some space of hypotheses $\mathcal{H}$. In this case the best attainable error is $\mathcal{E}(f_{\mathcal{H}}) = \inf_{f \in \mathcal{H}} \mathcal{E}(f)$. The quality of an estimator can then be assessed considering the distribution of the *excess risk*, $\mathcal{E}(f_{\mathbf{z}}) - \mathcal{E}(f_{\mathcal{H}})$, and in particular we say that an estimator is consistent if

$$\lim_{n\to\infty} \mathrm{P}\left[\mathcal{E}(f_{\mathbf{z}}) - \mathcal{E}(f_{\mathcal{H}}) \geq \varepsilon\right] = 0$$

for all positive ε. A more quantitative result is given by finite sample bounds,

$$\mathrm{P}\left[\mathcal{E}(f_{\mathbf{z}}) - \mathcal{E}(f_{\mathcal{H}}) \geq \varepsilon(\eta, n)\right] \leq 1 - \eta\,, \quad 0 < \eta \leq 1\,.$$

In the following we'll be interested into hypotheses space defined by a kernel.

Vector Valued RKHS. The development of the theory of RKHS in the vector case is essentially the same as in the scalar case and we refer to [1,19] for further details and references. We consider functions having values in some euclidean space $\mathcal{Y} \subseteq \mathbb{R}^T$ with scalar product (norm) $\langle\cdot,\cdot\rangle_T$, ($\|\cdot\|_T$). A RKH space is a Hilbert space of functions $f : \mathcal{X} \to \mathbb{R}^T$, with scalar product (norm) denoted by $\langle\cdot,\cdot\rangle_\Gamma$ ($\|\cdot\|_\Gamma$), defined by a matrix valued kernel $\Gamma : \mathcal{X} \times \mathcal{X} \to \mathcal{B}(\mathbb{R}^T)$, where $\mathcal{B}(\mathbb{R}^T)$ is the space of $T \times T$ positive semi-definite matrices.
The kernel Γ has the following reproducing property: for all $c \in \mathbb{R}^T$ and $x \in \mathcal{X}$

$$\langle f(x), c\rangle_T = \langle f, \Gamma(\cdot, x)c\rangle_\Gamma\,. \tag{1}$$

We assume throughout that $\sup_{x\in\mathcal{X}} ||\Gamma(x,x)|| = \kappa < \infty$, where $||\cdot||$ is the operator norm, which implies $\|f(x)\|_T \leq \kappa\|f\|_\Gamma$. Similarly to the scalar case, it can be shown that for any reproducing kernel Γ, a unique RKHS can be defined.

3 Learning Vector Fields with Spectral Filtering

In this section we present the new class of algorithms that we study in this paper. Towards this end it is instructive to preliminarily recall the main features of Tikhonov regularization for scalar and vector problems.

3.1 Tikhonov Regularization

In the scalar case, Tikhonov regularization [20,21] in a RKHS $\mathcal{H}$, with kernel K, corresponds to the minimization problem

$$\min_{f\in\mathcal{H}}\{\frac{1}{n}\sum_{i=1}^{n}(y_i - f(x_i))^2 + \lambda\|f\|_{\mathcal{H}}^2\}\,.$$

Its solution is given by

$$f_{\mathbf{z}}^{\lambda}(\cdot) = \sum_{i=1}^{n} K(x_i,\cdot)c_i \qquad \text{with} \qquad (\mathbf{K} + \lambda n I)\mathbf{c} = \mathbf{y}\,, \tag{2}$$

with $\mathbf{K}_{ij} = K(x_i, x_j)$, $\mathbf{y} = (y_1, \ldots, y_n)$, $c_i \in \mathbb{R}$ and $\mathbf{c} = (c_1, \ldots, c_n)$. The final estimator $f_\mathbf{z}$ is determined by a parameter choice $\lambda_n = \lambda(n, \mathbf{z})$, so that $f_\mathbf{z} = f_\mathbf{z}^\lambda$.

In the case of vector valued output, i.e. $\mathcal{Y} \subseteq \mathbb{R}^T$, the simplest idea is to consider a naïve extension of Tikhonov regularization, reducing the problem to learning each component independently. Namely, the solution is assumed to belong to $\mathcal{H} = \mathcal{H}^1 \times \mathcal{H}^2 \cdots \times \mathcal{H}^T$, where the spaces $\mathcal{H}^1, \mathcal{H}^2, \ldots, \mathcal{H}^T$ are RKHS with norms $\|\cdot\|_{\mathcal{H}^1}, \ldots, \|\cdot\|_{\mathcal{H}^T}$. Then $f = (f^1, \ldots, f^T)$ and $\|f\|_\Gamma^2 = \sum_{j=1}^T \|f^j\|_{\mathcal{H}^j}^2$ and Tikhonov regularization amounts to solving the following problem

$$\min_{f^1 \in \mathcal{H}^1, \ldots, f^T \in \mathcal{H}^T} \left\{ \frac{1}{n} \sum_{i=1}^n \sum_{j=1}^T (y_i^j - f^j(x_i))^2 + \lambda \sum_{j=1}^T \|f^j\|_{\mathcal{H}^j}^2 \right\}, \tag{3}$$

which is equivalent to solve T *independent* scalar problems. Within the framework of vector valued RKHSs, the choice above corresponds to a diagonal matrix valued kernel of the form $\Gamma(x, x') = diag(K_1(x, x'), \ldots, K_T(x, x'))$.

Recently, a regularization scheme of the form (3) has been studied in [1] for general matrix valued kernels. In this case there is *no straightforward* decomposition of the problem and one of the main results in [1] shows that the regularized solution can be written as

$$f_\mathbf{z}^\lambda(\cdot) = \sum_{i=1}^n \Gamma(\cdot, x_i) c_i \qquad \text{with} \qquad (\mathbf{\Gamma} + \lambda n I)\mathbf{C} = \mathbf{Y}, \tag{4}$$

where $c_i \in \mathbb{R}^T$, $\mathbf{C} = (c_1, \ldots, c_n)$, $\mathbf{Y} = (y_1, \ldots, y_n)$ and the kernel matrix $\mathbf{\Gamma}$ is a $n \times n$ block matrix, where each block is a $T \times T$ scalar matrix, so that $\mathbf{\Gamma}$ is a $nT \times nT$ scalar matrix.

The above discussion highlights a first interesting observation. Assuming the components of the vector field to be independent results in a block diagonal kernel matrix. Conversely, working with a non-diagonal matrix, hence with a general kernel, it is possible to exploit the functional relations that exists among the components of the vector field.

3.2 Regularization via Spectral Filtering

We present the class of regularized kernel methods under study, referring to [2,22] for the scalar case. We call these methods *spectral filters* because they achieve a stable, hence generalizing, solution by *filtering out* the unstable components of the kernel matrix, that is the directions corresponding to small eigenvalues. The interpretation of regularization as a way to restore stability is classical in ill-posed inverse problems, where many algorithms besides Tikhonov regularization are used [23]. The connection between learning and regularization theory of ill-posed problems [3] motivates considering spectral filtering techniques.

Adding a penalty to the empirical risk has a stabilizing effect from a numerical point of view, since it transforms the problem from $\mathbf{\Gamma C} = \mathbf{Y}$ to $(\mathbf{\Gamma} + \lambda n\mathbf{I})\mathbf{C} = \mathbf{Y}$. Hence, the penalty reduces the instability due to the eigenvectors corresponding to the small eigenvalues of the kernel matrix.

The idea of spectral filtering algorithms is that other regularized matrices $g_\lambda(\mathbf{\Gamma})$ besides $(\mathbf{\Gamma} + \lambda n\mathbf{I})^{-1}$ can be defined. Each algorithm corresponds to a specific filter function and in general there is no natural interpretation in terms of penalized empirical risk minimization. The matrix valued function $g_\lambda(\mathbf{\Gamma})$ is described by a scalar function g_λ using spectral calculus. More precisely, if $\mathbf{\Gamma} = \mathbf{USU}^*$ is the eigendecomposition of $\mathbf{\Gamma}$ with $\mathbf{S} = diag(\sigma_1, \ldots, \sigma_n)$, then $g_\lambda(\mathbf{S}) = diag(g_\lambda(\sigma_1), \ldots, g_\lambda(\sigma_n))$ and $g_\lambda(\mathbf{\Gamma}) = \mathbf{U}g_\lambda(\mathbf{S})\mathbf{U}^*$. For example, in the case of Tikhonov regularization $g_\lambda(\sigma) = \frac{1}{\sigma+n\lambda}$.

Suitable choices of filter functions g_λ define estimators of the form (4) with coefficients given by

$$\mathbf{C} = g_\lambda(\mathbf{\Gamma})\mathbf{Y} \ . \tag{5}$$

From the computational perspective, a key point that we show in the following is that many filter functions allow to compute the coefficients $\mathbf{C}$ without explicitly computing the eigen-decomposition of $\mathbf{\Gamma}$.

Clearly not all filter functions are admissible. Roughly speaking, as λ decreases an admissible filter function $g_\lambda(\mathbf{\Gamma})$ should approximate $\mathbf{\Gamma}^{-1}$ and its condition number should increase.

Remark 1. Note that in the scalar case, manipulations of the kernel matrix have been extensively used to define (and learn) new kernels to be used in Tikhonov regularization [24]. In the approach we present here, rather than defining a new kernel, each spectral filter g_λ defines an *algorithm* which is not based on empirical risk minimization.

3.3 Examples of Spectral Regularization Algorithms

We proceed giving several examples of spectral filtering algorithms.

L2 Boosting. In the scalar setting this method has been interpreted as a way to combine weak classifiers corresponding to splines functions at the training set points [6] and is called Landweber iteration in inverse problems literature [23]. The method can also be seen as the gradient descent minimization of the empirical risk on the whole RKHS, with no further constraint. Regularization is achieved by early stopping of the iterative procedure, hence the regularization parameter is the number of iterations.
The coefficients (5) can be found by setting $\mathbf{C}^0 = 0$ and considering for $i = 1, \ldots, t$ the following iteration

$$\mathbf{C}^i = \mathbf{C}^{i-1} + \eta(\mathbf{Y} - \mathbf{\Gamma}\mathbf{C}^{i-1}) \ ,$$

where the step size η can be chosen to make the iterations converge to the minimizer of the empirical risk, see below. If we use (4) to write the empirical risk as $\|\mathbf{\Gamma}\mathbf{C} - \mathbf{Y}\|^2$, it is easy to see that this is simply gradient descent. Further, it is can be shown by induction that the solution at the $t-$th iteration is given by

$$\mathbf{C}^t = \eta \sum_{i=0}^{t-1} (I - \eta\mathbf{\Gamma})^i \mathbf{Y} \ ,$$

and it follows that the filter function is $G_\lambda(\sigma) = \eta \sum_{i=0}^{t-1}(I - \eta\sigma)^i$. Interestingly, this filter function has another interpretation that can be seen recalling that, given a matrix A, $\|A\| \in (0,1)$ ($\|\cdot\|$ is the operator norm), $\sum_{i=0}^{\infty} A^i = \frac{1}{\mathbf{I}-A}$. If we replace A with $I - \eta\mathbf{\Gamma}$ and $\|I - \eta\mathbf{\Gamma}\| < 1$, we get $\mathbf{\Gamma}^{-1} = \eta \sum_{i=0}^{\infty}(I - \eta\mathbf{\Gamma})^i$. Therefore, the filter function of L2 Boosting corresponds to the truncated power series expansion of $\mathbf{\Gamma}^{-1}$. The last reasoning also shows a possible way to choose the step-size. In fact, choosing $\eta = 1/\sigma_{max}$, where σ_{max} is the maximum eigenvalue of the kernel matrix $\mathbf{\Gamma}$, we are guaranteed that $\|I - \eta\mathbf{\Gamma}\| < 1$.

Accelerated L2 Boosting. This method is also called the ν-method and it is particularly interesting since it is significantly faster than L2 boosting. Usually, it can find the same solution in only $\sqrt{t}$ steps. The coefficients are found by setting $\mathbf{C}^0 = 0$, $\omega_1 = (4\nu + 2)/(4\nu + 1)$, $\mathbf{C}^1 = \mathbf{C}^0 + \frac{\omega_1}{n}(\mathbf{Y} - \mathbf{\Gamma}\mathbf{C}^0)$ and considering for $i = 2, \ldots, t$ the iteration given by

$$\mathbf{C}^i = \mathbf{C}^{i-1} + u_i(\mathbf{C}^{i-1} - \mathbf{C}^{i-2}) + \frac{\omega_i}{n}(\mathbf{Y} - \mathbf{\Gamma}\mathbf{C}^{i-1}) .$$

The derivation of the filter function is considerably more complicated and is given in [23], where the parameters ν, ω_i and u_i are also defined. The filter function is shown to be $G_t(\sigma) = p_t(\sigma)$ with p_t a polynomial of degree $t - 1$. This method can be proved to be faster than L2 boosting since the ν-method can find in $\sqrt{t}$ steps the same solution found by L2 boosting after t iterations. regularization parameter is the *square root* of the iteration number rather than the iteration number itself.

Iterated Tikhonov. This method is a combination of Tikhonov regularization and L2 boosting where we set $\mathbf{C}^0 = 0$ and consider for $i = 1, \ldots, t$ the iteration $(\mathbf{\Gamma} + n\lambda I)\mathbf{C}^i = \mathbf{Y} + n\lambda\mathbf{C}^{i-1}$. The filter function is:

$$G_\lambda(\sigma) = \frac{(\sigma + n\lambda)^t - (n\lambda)^t}{\sigma(\sigma + n\lambda)^t} .$$

This methods is motivated by the desire to circumvent some of the limitations of Tikhonov regularization, namely a saturation effect that prevents exploiting the smoothness of the target function beyond a given critical value– see [23,2] for further details.

Truncated Singular Values Decomposition. This method is akin to a projection onto the first principal components in a vector valued setting. The number of components depends on the regularization parameter. The filter function is defined as $G_\lambda(\sigma) = 1/\sigma$ if $\sigma \geq \lambda/n$ and 0 otherwise.

4 Matrix Valued Kernels

In this section, we briefly review some matrix valued kernels for vector fields learning - see [25,26,27]. In particular we discuss a class of kernels that leads to a faster implementation of spectral filtering algorithms. Before doing this we give an example of a general kernel that we will consider in our experimental section.

Divergence free and curl free fields. These kernels have been used in [28] for the problem of reconstructing divergence-free or curl-free vector fields and they can be used only for vector fields whose input and the output spaces have the same dimensions. The divergence-free kernel is

$$\Gamma_{df}(x,x') = \frac{1}{\sigma^2} e^{-\frac{||x-y||^2}{2\sigma^2}} A_{x,x'} \tag{6}$$

where

$$A_{x,x'} = \frac{(x-x')(x-x')^T}{\sigma^2} + \left((T-1) - \frac{||x-x'||^2}{\sigma^2}\right)\mathbf{I}$$

and the curl-free is

$$\Gamma_{cf}(x,x') = \frac{1}{\sigma^2} e^{-\frac{||x-x'||^2}{2\sigma^2}} \left(\mathbf{I} - \frac{(x-x')(x-x')^T}{\sigma^2}\right). \tag{7}$$

It is possible to consider a convex linear combination of these two kernels for learning any vector field and for reconstructing its divergence-free and curl-free parts separately (see the experiments in Sect.6).

4.1 Design of Decomposable Kernels.

A general class of kernels consists of kernels of the form

$$\Gamma(x,x') = K(x,x')A \tag{8}$$

where K is a scalar kernel and A a positive semidefinite $T \times T$ matrix that encodes how the outputs are related. This class of kernels allows to decouple the role played by the input and output spaces. As we show in Sect.5.1, it is possible to derive more efficient learning schemes using these kernels. The role of the matrix A can be understood by linking it to a regularizer on the components of the vector field.

Proposition 1. *Let Γ be a product kernel of the form in (8). Then the norm of any function in the corresponding RKHS can be written as*

$$||f||_{\Gamma}^2 = \sum_{\ell,q=1}^{T} A_{\ell q}^{\dagger} < f^{\ell}, f^{q} >_K , \tag{9}$$

where $A^{\dagger}$ is the pseudoinverse of A.

Proof. A function in the RKHS defined by the matrix valued kernel $\Gamma = KA$ can be written as $f(x) = \sum_i \Gamma(x,x_i)c_i = \sum_i K(x,x_i)Ac_i$ with $c_i \in \mathbb{R}^T$, so that the ℓ-th component is $f^{\ell}(x) = \sum_i \sum_{t=1}^{T} K(x,x_i)A_{\ell t}c_i^t$, where $c_i^t \in \mathbb{R}$ is the t-th component of c_i. Therefore, each f^{ℓ} belongs to $\mathcal{H}_K$ and it is a linear combination of the coefficients $\{c_i\}_{i=1}^n$ that depends on the ℓ-th row of the matrix A.

The norm of f is $||f||_{\Gamma}^2 = \sum_{i,j} \sum_{\ell,q=1}^{T} K(x_i,x_j)c_i^{\ell}A_{\ell q}c_j^q$ and the scalar products between its components are given by $\langle f^{\ell}, f^q\rangle_K = \sum_{i,j} \sum_{t,s=1}^{T} K(x_i,x_j)A_{\ell t}c_i^t A_{qs}c_j^s$. Combining these expressions, it is straightforward to obtain (9). □

The above result shows how to design a kernel by defining a penalty on the components of the vector field.

Common similarity. This kernel forces the components of the vector field to be similar to their average, $\Gamma_\omega(x, x') = K(x, x')(\omega\mathbf{1} + (1 - \omega)\mathbf{I})$. In fact, it is straightforward to show that the corresponding regularizer is

$$A_\omega \sum_{\ell=1}^{T} ||f^\ell||_K^2 + B_\omega \sum_{\ell=1}^{T} ||f^\ell - \frac{1}{T}\sum_{q=1}^{T} f^q||_K^2 , \tag{10}$$

where A_ω and B_ω are coefficients that depend on ω.

Graph regularization. A regularizer that forces stronger or weaker similarity between the components [29] is defined as

$$\frac{1}{2}\sum_{\ell,q=1}^{T} ||f^\ell - f^q||_K^2 M_{\ell q} + \sum_{\ell=1}^{T} ||f^\ell||_K^2 M_{\ell\ell} , \tag{11}$$

where M is a $T \times T$ positive weight matrix. The corresponding kernel is $\Gamma = KL^\dagger$, where $L = D - M$ and $D_{\ell q} = \delta_{\ell q}\left(\sum_{h=1}^{T} M_{\ell h} + M_{\ell q}\right)$.

Output components clustering. This regularizer is based on the idea of grouping the components into r clusters and enforcing the components in each cluster to be similar to their average [25]

$$J(f) = \epsilon_1 \sum_{c=1}^{r} \sum_{l \in I(c)} ||f^l - \overline{f}_c||_K^2 + \epsilon_2 \sum_{c=1}^{r} m_c ||\overline{f}_c||_K^2 , \tag{12}$$

where $\overline{f}_c$ is the mean of the components in cluster c and $I(c)$ is the index set of the components that belong to cluster c. Simple calculations show that the corresponding kernel is $\Gamma = KG^\dagger$, where $G_{lq} = \epsilon_1\delta_{lq} + (\epsilon_2 - \epsilon_1)M_{lq}$ and $M_{lq} = \frac{1}{m_c}$ if components l and q belong to the same cluster c of cardinality m_c, $M_{lq} = 0$ otherwise.

5 Computational and Sample Complexity

We present a computational complexity analysis of the spectral filters taking into account the choice of kernel. We show that for a specific class of matrix valued kernels it is possible to greatly reduce the computational complexity of the algorithms. Finally, we give the bound on the excess risk that leads to consistency.

5.1 Faster Implementation for Decomposable Kernels

The main point we make in this section is that, for kernels of the form $\Gamma = KA$, we can use the eigen-sytem of the matrix A to define a new coordinate system where the problem can be solved in a computational faster way. The outcome of this analysis is that the vector field learning problem can be reduced to solving T scalar regression problems.

If we denote with $u_1, \ldots, u_T$ the eigenvectors of A, we can write the vector $\mathbf{C} = (c_1, \ldots, c_n)$, with $c_i \in \mathbb{R}^T$, as $\mathbf{C} = \sum_{j=1}^T \tilde{c}^j \otimes u_j$, where $\otimes$ is the tensor product and $\tilde{c}^j = (\langle c_1, u_j \rangle_T, \ldots, \langle c_n, u_j \rangle_T)$. Similarly $\mathbf{Y} = \sum_{j=1}^T \tilde{y}^j \otimes u_j$, with $\tilde{y}^j = (\langle y_1, u_j \rangle_T, \ldots, \langle y_n, u_j \rangle_T)$. The above transformations are simply rotations in the output space. Moreover the kernel matrix $\mathbf{\Gamma}$ is given by the tensor product of the $n \times n$ scalar kernel matrix $\mathbf{K}$ and A, that is $\mathbf{\Gamma} = \mathbf{K} \otimes A$.
If we denote with λ_i, v_i $(i = 1, \ldots, n)$, the eigenvalues and eigenvectors of $\mathbf{K}$ and with σ_j $(j = 1, \ldots, T)$ the eigenvalues of A, we have the following result.

Proposition 2. *The solution of the vector valued problem* $\mathbf{C} = g_\lambda(\mathbf{\Gamma})\mathbf{Y}$ *can be obtained by solving* T *scalar problems*

$$\tilde{c}^j = g_\lambda(\sigma_j \mathbf{K})\tilde{y}^j \;, \qquad j = 1, \ldots, T \;. \tag{13}$$

Proof. Substituting the expressions for $\mathbf{C}$ and $\mathbf{Y}$ into $\mathbf{C} = g_\lambda(\mathbf{\Gamma})\mathbf{Y}$, we obtain

$$\sum_{j=1}^T \tilde{c}^j \otimes u_j = \sum_{j=1}^T g_\lambda(\mathbf{K} \otimes A)\tilde{y}^j \otimes u_j \;.$$

Working in the eigen-system $v_i \otimes u_j$ $(i = 1, \ldots, n$ and $j = 1, \ldots, T)$ of the matrix $\mathbf{K} \otimes A$ and recalling that the spectral filters operate on the eigenvalues of the kernel matrix, we have

$$\sum_{j=1}^T \tilde{c}^j \otimes u_j = \sum_{j=1}^T \sum_{i=1}^n g_\lambda(\lambda_i \sigma_j)\langle \tilde{y}^j, v_i \rangle v_i \otimes u_j = \sum_{j=1}^T g_\lambda(\sigma_j \mathbf{K})\tilde{y}^j \otimes u_j \;.$$

Since the eigenvectors u_j are orthonormal, the two sides of the equation must be equal term by term. It follows that $\tilde{c}^j = g_\lambda(\sigma_j \mathbf{K})\tilde{y}^j$ for $j = 1, \ldots, T$. □

The above equation shows that in the new coordinate system $\{u_1, \ldots, u_T\}$, we have to solve T essentially independent problems. Indeed, after rotating the outputs (and the coefficients) the only coupling is the rescaling of each kernel matrix by σ_j. For example, in the case of Tikhonov regularization, the j-th component is found solving $\tilde{c}^j = (\sigma_j \mathbf{K} + \lambda n \mathbf{I})^{-1}\tilde{y}^j = (\mathbf{K} + \frac{\lambda}{\sigma_j} n \mathbf{I})^{-1} \frac{\tilde{y}^j}{\sigma_j}$ and we see that the scaling term is changing the scale of the regularization parameter and of the outputs. The above calculation shows that all kernels of this form allow for a simple implementation at the price of the eigen-decomposition of the matrix A. Also, it shows that the coupling among the different tasks can be seen as a rotation and rescaling of the output points.

5.2 Regularization Path and Computational Complexity

Here we discuss the complexity of the whole *regularization path* for Tikhonov regularization and accelerated L2 boosting, since this algorithm turns out to be among the fastest. The regularization path is the set of solutions corresponding to many parameter values.

On one hand, when using Tikhonov regularization, for each value of the regularization parameter the solution is found by inverting a $nT \times nT$ matrix. On the other, most iterative methods require only matrix vector multiplications, and each step corresponds to a solution for a value of the regularization parameter, so that at step N we have computed the entire regularization path up to N. Therefore, in general, if we consider N parameter values we will have $O(N(nT)^3)$ time complexity for Tikhonov regularization and $O(N(nT)^2)$ for iterative methods.

In the special case of kernels of the form $\Gamma = KA$, we can diagonalize the matrix A and then work in a new coordinate system where the kernel matrix is block diagonal and all the blocks are the same, up to a rescaling. In this case the complexity of the vector field algorithm is essentially the same of T scalar problems – $O(TNn^3)$ for Tikhonov and $O(TNn^2)$ for iterative methods – plus the cost of computing the eigen-decomposition of A, which is $O(T^3)$.

5.3 Sample Complexity

Our main theoretical result is a finite sample bound on the excess risk for all algorithms based on spectral filtering. This result immediately leads to consistency and can be proven in a unified framework. Each algorithm is characterized by specific constants that might change from one algorithm to the other and we refer to [22] for their computation. In particular, here we underline to role played by the one of such constants, namely the qualification number $\overline{r}$, which controls the best achievable learning rate of the corresponding algorithm, as is illustrated in the following theorem. We need to assume that the input space is a separable metric space (not necessarily compact) and that the output space is a bounded set in $\mathbb{R}^T$, that is $sup_{y \in \mathcal{Y}} ||y||_T = M < \infty$. For the sake of simplicity we also assume that a minimizer of the expected risk on $\mathcal{H}$ exists and denote it with $f_{\mathcal{H}}$. Let us also define the integral operator $T_\Gamma f(x) = \int_{\mathcal{X}} \Gamma(x, x') f(x') \rho(x')$.

Theorem 1. *Assume* $||(T_\Gamma)^{-\nu} f_{\mathcal{H}}||_\Gamma \leq R$, *where* $\nu = r - \frac{1}{2}$. *If*

$$\frac{1}{2} \leq r \leq \overline{r} \qquad \text{and} \qquad \lambda_n = C n^{-\frac{1}{2r+1}} \log \frac{4}{\eta},$$

then, for $f_{\mathbf{z}} = f_{\mathbf{z}}^{\lambda_n}$, *we have, with probability* $1 - \eta$,

$$\mathcal{E}(f_{\mathbf{z}}) - \mathcal{E}(f_{\mathcal{H}}) \leq C' n^{-\frac{2r}{2r+1}} \log^2 \frac{4}{\eta}, \tag{14}$$

where C, C' *are constants that depend on* R *and* r, *but not on* n.

The index r describes the simplicity of the field estimation problem and $r > 1/2$ implies that $f_{\mathcal{H}}$ exists. The simpler the problem (r large), the faster the learning rate. The qualification number $\overline{r}$ represents how much each specific filter can exploit the simplicity of the learning problem. Tikhonov regularization, for instance, has a qualification number $\overline{r} = 1$, that yields a rate proportional to

$n^{-\frac{2}{3}}$, while for some iterative algorithms, such as the ν-method, the qualification number $\bar{r}$ can be arbitrarily large, yielding a bound arbitrarily close to n^{-1}. Note that the result above directly leads to consistency, since the limit for $n \to \infty$ is zero and, even when the expected risk does not achieve a minimum in $\mathcal{H}$, one can still show that there is a parameter choice ensuring convergence to $\inf_{f \in \mathcal{H}} \mathcal{E}(f)$. If the kernel is universal [30,27,31], then universal consistency [32] is ensured. In particular, note that the results in [31] allow to work on a non compact domain, at least if the kernel is well-behaved. Due to space reasons, we omit the proof which will be included in a longer version of this paper.

6 Empirical Analysis

We present a synthetic 2-dimensional vector field estimation problem in order to illustrate the benefits of using matrix valued kernels and the computational advantages of iterative spectral filters with respect to Tikhonov regularization. We consider the setting of [28] and first compare our vector valued regression approach with estimating each component of the field independently, in both cases using the ν-method, which is the fastest algorithm when the matrix valued kernel is not of the form $\Gamma = KA$. Finally, we compare the computation time of Tikhonov regularization and the ν-method to show that the latter is significantly faster and scales better with the number of training points.

The vector field is generated from a scalar field defined by the sum of 5 gaussians centered at $(0, 0)$, $(1, 0)$, $(0, 1)$, $(-1, 0)$ and $(0, -1)$ respectively. The covariances are all set to be $0.45\mathbf{I}$, where $\mathbf{I}$ is the 2×2 identity matrix. We compute the gradient of the scalar field and its perpendicula field. We then consider a convex combination of these two vector fields, controlled by a parameter γ. Examples of the resulting fields for $\gamma = 0$ and $\gamma = 1$ are shown in Fig.1.

Firstly, we consider the noiseless case. The vector field is constructed specifying a value of the parameter γ. The field is then computed on a 70×70 grid over the square $[-2, 2] \times [-2, 2]$. The models are trained on a uniform random

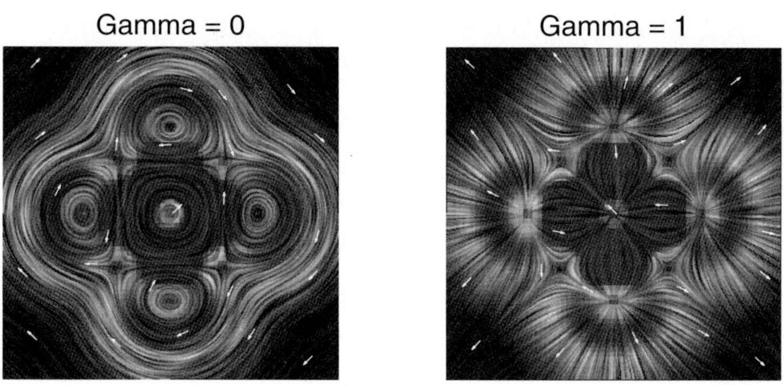

Fig. 1. Visualization of the 2-dimensional vector field for $\gamma = 0$, resulting in a divergence-free field, and for $\gamma = 1$ that yields a curl-free field

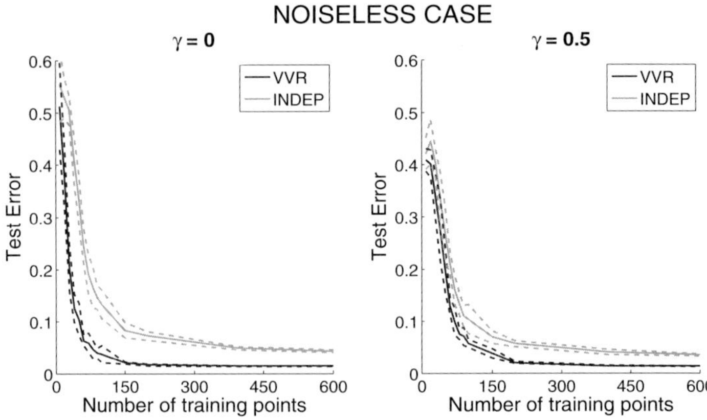

Fig. 2. Noiseless case. Test errors for the proposed vector valued approach (VVR) and for learning each component of the field independently (INDEP) as a function of the number of training points used for learning. Solid lines represent average test error, while dotted lines show the average test error plus/minus one standard deviation.

sample of points from this grid and their predictions on the whole grid (except the training points) compared to the correct field. The number of training examples is varied from 10 to 600. For each cardinality of the training set, the training and prediction process is repeated 10 times with a different randomization of the training points. We use a convex combination of the divergence-free (6) and curl-free (7) kernels, controlled by the parameter $\tilde{\gamma}$. We adopt a 5-fold cross validation to select the optimal number of iterations for the ν-method and the parameter $\tilde{\gamma}$. The width, σ, of these kernels was set to be 0.8.

We use an angular measure of error to compare two fields [33]. If $v_o = (v_o^1, v_o^2)$ and $v_e = (v_e^1, v_e^2)$ are the original and estimated fields, we consider the transformation $v \to \tilde{v} = \frac{1}{||(v^1,v^2,1)||}(v^1, v^2, 1)$. The error measure is then $err = arccos(\tilde{v}_e \cdot \tilde{v}_o)$. This error measure was derived by interpreting the vector field as a velocity field and it is convenient because it handles large and small signals without the amplification inherent in a relative measure of vector differences.

The results for the noiseless case are reported in Fig.2, which clearly shows the advantage of using a vector valued approach with the combination of curl-free and divergence-free kernels. We present only the results for the field generated with $\gamma = 0$ and $\gamma = 0.5$ since for the remaining fields the errors are set within these two examples. The prediction errors of the proposed approach via the ν-method are always lower than the errors obtained by regressing on each component independently, even when the training set is very large. The average value of the estimated parameter $\tilde{\gamma}$, converges to the true value of γ as the number of training points increases (result not shown for brevity), indicating that it is possible for the model to learn the field decomposition in an automatic way.

We then consider the case with normal noise whose standard deviation is independent from the signal and is chosen to be 0.3. We follow the same experimental

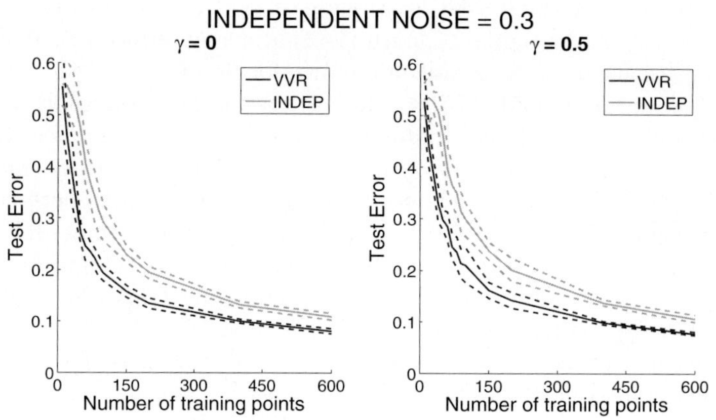

Fig. 3. Independent noise of standard deviation 0.3. Test errors for the proposed vector valued approach (VVR) and for learning each component of the field independently (INDEP) as a function of the number of training points used for learning. Solid lines represent average test error, while dotted lines show the average test error plus/minus one standard deviation.

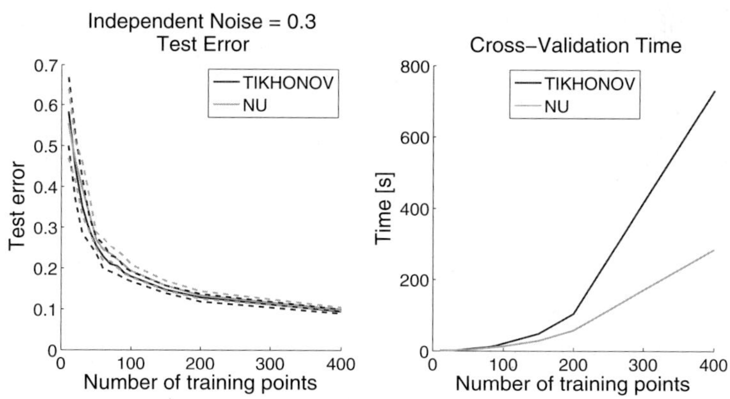

Fig. 4. (Left) Independent noise of standard deviation 0.3. Test errors for the proposed vector valued approach solved with Tikhonov regularization or the ν-method. (Right) Computation time for the whole regularization path for the ν-method and for Tikhonov regularization. The experiment was performed on Matlab on a desktop PC with AMD Athlon X2 64 3.2GHz and 2GB RAM.

protocol adopted for the noiseless case. The results are reported in Fig.3 and indicate that also in the presence of noise the proposed approach consistently outperforms regressing on each component independently. The advantage is stronger when fewer training points are available, but it is still present even at higher training set cardinalities. Again, the estimated value for the parameter $\hat{\gamma}$ well approximates the true value used for the creation of the vector field, indicating that it is possible for the model to learn the field decomposition in an automatic way also in presence of noise (result not shown for brevity).

In Fig.4 we compare Tikhonov regularization and the ν-method on the field generated with $\gamma = 0$. In the left panel are reported the test errors as a function of the number of training examples in the case with independent normal noise of standard deviation 0.3. We clearly see that the two algorithms perform equivalently. In the right plot are shown the cross validation learning times for Tikhonov regularization and the ν-method. For both methods, both the parameter $\hat{\gamma}$ and the regularization parameter have been estimated. For Tikhonov regularization 50 values of the regularization parameter, between 10^-5 and 10^-2, were assessed, while for the ν-method the iterations up to 500 were evaluated. The experimental results confirm our complexity analysis: the ν-method is significantly faster than Tikhonov regularization, while preserving its good generalization performance.

7 Conclusions

In this paper we considered the problem of learning vector valued functions and proposed a class of regularized kernel methods based on spectral filtering of the kernel matrix. Tikhonov regularization and L2 boosting are examples of methods falling in our framework. The complexity and empirical analysis showed the advantages of iterative algorithms with respect to the more standard Tikhonov regularization. A similar conclusion can be drawn comparing the learning rates of the spectral filters.

Acknowledgements. We would like to thank Ernesto De Vito for many useful discussions. This work has been partially supported by the EU Integrated Project Health-e-Child IST-2004-027749.

References

1. Micchelli, C.A., Pontil, M.: On learning vector–valued functions. Neural Computation 17 (2005)
2. Lo Gerfo, L., Rosasco, L., Odone, F., De Vito, E., Verri, A.: Spectral algorithms for supervised learning. Neural Computation 20 (2008)
3. De Vito, E., Rosasco, L., Caponnetto, A., De Giovannini, U., Odone, F.: Learning from examples as an inverse problem. JMLR 6 (2005)
4. Poggio, T., Rifkin, R., Mukherjee, S., Niyogi, P.: General conditions for predictivity in learning theory. Nature 428 (2004)
5. Bousquet, O., Elisseeff, A.: Stability and generalization. JMLR 2 (2002)
6. Bühlmann, P., Yu, B.: Boosting with the l_2-loss: Regression and classification. JASA 98 (2002)
7. Caruana, R.: Multitask learning. Mach. Learn. 28 (1997)
8. Argyriou, A., Maurer, A., Pontil, M.: An algorithm for transfer learning in a heterogeneous environment. In: Daelemans, W., Goethals, B., Morik, K. (eds.) ECML PKDD 2008, Part I. LNCS (LNAI), vol. 5211, pp. 71–85. Springer, Heidelberg (2008)

9. Boyle, P., Frean, M.: Dependent gaussian processes. In: NIPS (2005)
10. Chai, K., Williams, C., Klanke, S., Vijayakumar, S.: Multi-task gaussian process learning of robot inverse dynamics. In: NIPS (2009)
11. Breiman, L., Friedman, J.H.: Predicting multivariate responses in multiple linear regression. J. R. Statist. Soc. B 59(1) (1997)
12. Izenman, A.: Reduced-rank regression for the multivariate linear model. J. Multiv. Anal. 5 (1975)
13. van der Merwe, A., Zidek, J.V.: Multivariate regression analysis and canonical variates. Can. J. Stat. 8 (1980)
14. Wold, S., Ruhe, H., Wold, H., Dunn III, W.: The collinearity problem in linear regression. The partial least squares (pls) approach to generalizes inverses. SIAM J. Sci. Comput. 5 (1984)
15. Stein, M.L.: Interpolation of spatial data. Springer, Heidelberg (1999)
16. Brudnak, M.: Vector-valued support vector regression. In: IJCNN (2006)
17. Vazquez, E., Walter, E.: Multi output support vector regression. In: SYSID (2003)
18. Caponnetto, A., De Vito, E.: Optimal rates for regularized least-squares algorithm. Found. Comp. Math. (2006)
19. Carmeli, C., De Vito, E., Toigo, A.: Vector valued reproducing kernel Hilbert spaces of integrable functions and Mercer theorem. JAA 4 (2006)
20. Tikhonov, A.N., Arsenin, V.Y.: Solutions of Ill-posed Problems. John Wiley, Chichester (1977)
21. Evgeniou, T., Pontil, M., Poggio, T.: Regularization networks and support vector machines. Advances in Computational Mathematics 13(1), 1–50 (2000)
22. Bauer, F., Pereverzev, S., Rosasco, L.: On regularization algorithms in learning theory. J. Complex. 23(1) (2007)
23. Engl, H.W., Hanke, M., Neubauer, A.: Regularization of inverse problems. Kluwer, Dordrecht (1996)
24. Smola, A., Kondor, R.: Kernels and regularization on graphs. In: Schölkopf, B., Warmuth, M.K. (eds.) COLT/Kernel 2003. LNCS (LNAI), vol. 2777, pp. 144–158. Springer, Heidelberg (2003)
25. Evgeniou, T., Micchelli, C.A., Pontil, M.: Learning multiple tasks with kernel methods. JMLR 6 (2005)
26. Sheldon, D.: Graphical multi-task learning. Technical report, Cornell University (2008) (preprint)
27. Caponnetto, A., Micchelli, C.A., Pontil, M., Ying, Y.: Universal kernels for multi-task learning. JMLR 9 (2008)
28. Macêdo, I., Castro, R.: Learning divergence-free and curl-free vector fields with matrix-valued kernels. Technical report, Instituto Nacional de Matematica Pura e Aplicada (2008)
29. Micchelli, C.A., Pontil, M.: Kernels for multi-task learning. In: NIPS (2004)
30. Steinwart, I.: On the influence of the kernel on the consistency of support vector machines. JMLR 2 (2002)
31. Carmeli, C., De Vito, E., Toigo, A., Umanitá, V.: Vector valued reproducing kernel hilbert spaces and universality. JAA 4 (2010)
32. Devroye, L., Györfi, L., Lugosi, G.: A Probabilistic Theory of Pattern Recognition. Springer, Heidelberg (1996)
33. Barron, J., Fleet, D., Beauchemin, S.: Performance of optical flow techniques. IJCV 12(1) (1994)

Weighted Symbols-Based Edit Distance for String-Structured Image Classification*

Cécile Barat, Christophe Ducottet, Elisa Fromont, Anne-Claire Legrand, and Marc Sebban

Université de Lyon, F-42023, Saint-Etienne, France
CNRS, UMR 5516, Laboratoire Hubert Curien, 42023, Saint-Etienne, France
Université de Saint-Etienne, Jean Monnet, F-42023, Saint-Etienne, France

Abstract. As an alternative to vector representations, a recent trend in image classification suggests to integrate additional structural information in the description of images in order to enhance classification accuracy. Rather than being represented in a p-dimensional space, images can typically be encoded in the form of strings, trees or graphs and are usually compared either by computing suited metrics such as the (string or tree)-edit distance, or by testing subgraph isomorphism. In this paper, we propose a new way for representing images in the form of strings whose symbols are weighted according to a TF-IDF-based weighting scheme, inspired from information retrieval. To be able to handle such real-valued weights, we first introduce a new weighted string edit distance that keeps the properties of a distance. In particular, we prove that the triangle inequality is preserved which allows the computation of the edit distance in quadratic time by dynamic programming. We show on an image classification task that our new weighted edit distance not only significantly outperforms the standard edit distance but also seems very competitive in comparison with standard histogram distances-based approaches.

1 Introduction

Classification of images is of considerable interest in many image processing and computer vision applications. A common approach to represent the image content is to use histograms of color, texture and edge direction features (1; 2). Although they are computationally efficient, such histograms only use global information and so provide a crude representation of the image content. The current trend in image classification is towards the use of the *bag-of-visual-words* model that comes from the *bag-of-words* representation of text documents (3). This model requires four basic stages: (i) keypoints detection (ii) description, (iii) codebook creation and (iv) image representation. Keypoints refer to small regions of interest in the image. They can be sampled densely (4), randomly (5) or extracted with various detectors (6). Once extracted, keypoints are characterized

* This work is part of the ongoing ANR SATTIC 07-1_184534 research project and the Pascal2 Network of Excellence.

J.L. Balcázar et al. (Eds.): ECML PKDD 2010, Part I, LNAI 6321, pp. 72–86, 2010.

using a local descriptor, the most widely used being SIFT (7). A visual codebook is then learned over the collection of descriptors of a training set typically by using the k-means algorithm (8). Each cluster gives a visual word and each image can then be mapped into this new space of visual words leading to a *bag-of-visual-words*. Each word can be weighted either according to its frequency in the image, or using more sophisticated techniques such as mutual-information-based binarization (9) or TF-IDF-based weighting (Term Frequency-Inverse Document Frequency) (10; 11). Whatever the weighting scheme, two images are usually compared in a classification task by computing either a dot product or a given distance (*e.g.* L_1, L_2, L_∞) between their corresponding weighted feature vectors in the considered vector space.

Although working in a vector space brings many advantages, it does not allow the integration of additional structural information or topological relationships between the objects of the images. To overcome this drawback, an alternative to the vector representation-based approaches consists in representing images in the form of structured data such as strings, trees, or graphs. For instance, in order to code the topological relationship of so-called iconic objects in an image, the 2D string representation (12; 13; 14) uses the relative location of the visual words in the original 2D-space to build a graph of similarities. The comparison between two images coded in the form of strings is then achieved by searching for the largest common subsequence that satisfies a clique in that graph. In order to represent binary objects, Freeman (15) codes the boundaries of such objects in the form of sequences of symbols. Extending this principle in (16), Daliri proposes to map each contour of the objects into a string whose components are pairs of symbols, the first one representing the angle computed between the contour point and its neighbors and the other describing the normalized distance from the center of mass.

In the previous two approaches, the comparison between two images is achieved by computing the edit distance (17) between the corresponding string representations. Let us recall that the standard edit distance between two structured data $\mathbf{x}$ and $\mathbf{y}$ is defined as the less costly set of edits needed to transform $\mathbf{x}$ into $\mathbf{y}$, with the allowable edit operations being insertion, deletion, or substitution of symbols. If a usual way to use the edit distance is to assign a unit cost to each of the edit operations, many efforts have been made during the past few years to automatically learn more performing edit costs. In this context, a recent line of research has investigated the ways to model those costs in the form of the parameters of probabilistic state machines, such as pair-Hidden Markov models, stochastic transducers, or probabilistic automata (18; 19; 20; 21). The resulting *stochastic* edit distance (usually learned using an EM-based approach) can either be used in more performing neighborhood-based classifiers, or wrapped into *edit kernels* for improving Support Vector Machines (22).

While all the previously cited works mainly dealt with the improvement of weighted edit distances where the weights concern the *edit operations* between two symbols, we can note that no effort has been really made to develop new weighted edit distances where the weights are assigned to the *symbols* themselves.

And yet, such distances would allow us to take advantage of both vector and structured approaches. The objective of this paper is to fill this gap by presenting a new weighted symbols-based edit distance (Section 2). By allowing each symbol to be real-valued, this opens the door to new image representations. In this context, we propose a new way to encode images in the form of strings whose symbols are real-valued according to a TF-IDF weighting scheme (Section 3). A series of experiments is carried out in Section 4 that shows that our new weighted edit distance not only significantly outperforms the standard edit distance but also seems very competitive in comparison with standard histogram distances-based approaches. Before concluding, we propose an original way to automatically learn the costs of the edit operations, by taking into account the number of edges separating the visual words in a minimum spanning tree built on the visual codebook. Plugged in our weighted edit distance, we show that these edit costs allow us to improve the classification accuracy.

2 Weighted Edit Distance

In its basic form, the Levenshtein distance (or edit distance (17)) between two strings $\mathbf{x}(T)$ and $\mathbf{y}(V)$ of length T and V is defined as the minimum number of edits needed to transform $\mathbf{x}(T)$ into $\mathbf{y}(V)$, with the allowable edit operations being insertion, deletion, or substitution of a single character. Using the dynamic programming Algorithm 1, the edit distance $D(T,V)$ is computable in $\mathcal{O}(T \times V)$ and boils down to filling in a $(T+1) \times (V+1)$ matrix.

Rather than simply counting the number of required edit operations to change $\mathbf{x}(T)$ into $\mathbf{y}(V)$, the additive term 1 in the expressions d_1, d_2 and d_3 in Algorithm 1 can be replaced by the value of an edit cost function $c(x_r, y_k)$ that takes into account the nature of the symbols $x_r, y_k \in \Sigma \cup \{\lambda\}$ involved in the edit operation, where Σ is the alphabet and λ the empty symbol. In this case, the edit distance between $\mathbf{x}(T)$ and $\mathbf{y}(V)$ becomes the minimum cost of all sequences of edit operations which transform $\mathbf{x}(T)$ into $\mathbf{y}(V)$. As mentioned in (17), $D(T,V)$ remains a metric if the edit cost function $c(x_r, y_k)$ satisfies the two properties of *positive definiteness* and *symmetry*. It is computable in $\mathcal{O}(T \times V)$ if the *triangle inequality* is also fulfilled.

Rather than allowing the use of weights on each edit operation (as usually done in the literature), we propose in this section to authorize the management of weighted symbols during the calculation of the edit distance. By this way, an edit operation becomes a transformation of a weighed symbol into another one. This enables us to take into account the TF-IDF of each symbol of a string-structured image as a weight, and to compute the edit distance between two images represented with such strings of weighted symbols. We propose in the following an edit cost function that is able to manage at once two symbols and their corresponding weights. Henceforth, a string $\mathbf{x}(T)$ will be composed of T weighted symbols $\mathbf{x_1} \ldots \mathbf{x_T}$ where $\forall i = 1..T, \mathbf{x_i} = (x_i, w_{x_i})$ is made of a symbol $x_i \in \Sigma$ and a weight $w_{x_i} \in \mathbb{R}^*_+$.

```
Input: Two strings x(T) and y(V)
Output: Edit Distance D(T, V) between x(T) and y(V)
D(0, 0) ← 0;
for r=1 to T do
    D(r, 0) ← D(r − 1, 0) + 1;
end
for k=1 to V do
    D(0, k) ← D(0, k − 1) + 1;
end
for r=1 to T do
    for k=1 to V do
        if (x_r = y_k) then
            D(r, k) = D(r − 1, k − 1);
        end
        else
            d_1 ← D(r − 1, k − 1) + 1;
            d_2 ← D(r − 1, k) + 1;
            d_3 ← D(r, k − 1) + 1;
            D(r, k) ← min(d_1, d_2, d_3);
        end
    end
end
Return D(T, V);
```

Algorithm 1. Edit distance algorithm that returns the number of edit operations required to change a string $\mathbf{x}(T)$ into another $\mathbf{y}(V)$.

Definition 1. *The edit cost function* $c : ((\Sigma \times \mathbb{R}_+^*) \cup (\{\lambda\} \times \{0\})) \times ((\Sigma \times \mathbb{R}_+^*) \cup (\{\lambda\} \times \{0\})) \rightarrow \mathbb{R}_+$ *is defined as follows*[1]*:*
Let the symbols a, b *and the positive reals* n, m *be the components of two weighted symbols* $\mathbf{a} = (a, n)$ *and* $\mathbf{b} = (b, m)$

$$c(\mathbf{a}, \mathbf{b}) = \begin{cases} \max(n, m) \text{ if } a \neq b \\ |n - m| \text{ otherwise} \end{cases}$$

Plugging this function in an edit distance algorithm, we obtain the Algorithm 2. The underlying idea of the function c is graphically described in Figure 1. The edit cost between two weighted symbols (a, n) and (b, m) is close to the one computed between two strings where a and b are "virtually" repeated n and m times respectively. An alternative to our cost function would have consisted in actually repeating each symbol according to its corresponding weight. Despite the fact that this would quadratically increase (in the average of the weights) the algorithmic complexity of the edit distance calculation, this would lead to a loss of information by discarding the decimal part of the weight. We will show this behavior in the experimental part of this paper.

[1] Note that the weight of the empty string λ is always equal to 0 such that $\boldsymbol{\lambda} = (\lambda, 0)$.

Input: Two strings $\mathbf{x}(T)$ and $\mathbf{y}(V)$
Output: Edit Distance $D(T, V)$ between $\mathbf{x}(T)$ and $\mathbf{y}(V)$

1 $D(0,0) \leftarrow 0$;
2 **for** *r=1 to T* **do**
3     $D(r,0) \leftarrow D(r-1,0) + c(\mathbf{x_r}, \lambda)$;
4 **end**
5 **for** *k=1 to V* **do**
6     $D(0,k) \leftarrow D(0,k-1) + c(\lambda, \mathbf{y_k})$;
7 **end**
8 **for** *r=1 to T* **do**
9   **for** *k=1 to V* **do**
10         $d_1 \leftarrow D(r-1,k-1) + c(\mathbf{x_r}, \mathbf{y_k})$;
11         $d_2 \leftarrow D(r-1,k) + c(\mathbf{x_r}, \lambda)$;
12         $d_3 \leftarrow D(r,k-1) + c(\lambda, \mathbf{y_k})$;
13         $D(r,k) \leftarrow min(d_1, d_2, d_3)$;
14   **end**
15 **end**
16 Return $D(T, V)$;

Algorithm 2. Weighted edit distance algorithm that returns the minimum cost required to change a weighted string $\mathbf{x}(T)$ into another $\mathbf{y}(V)$.

Proposition 1. *Given the edit cost function c of Definition 1, Algorithm 2 is a generalization of Algorithm 1.*

Proof. Algorithm 2 generalizes Algorithm 1 if, for two unweighted strings, they both return the same edit distance. Two symbols $\mathbf{a} = (a, n)$ and $\mathbf{b} = (b, m)$ are unweighted if $n = m = 1$. In this case, $c(\mathbf{a}, \mathbf{b})$ returns

- either $|n - m| = 0$ if $a = b$ (and so lines 10 and 11 of Algorithm 1 and line 10 of Algorithm 2 are the same),
- or $\max(n, m) = 1$ if $a \neq b$ (and so lines 14, 15 and 16 of Algorithm 1 are the same as lines 10, 11 and 12 of Algorithm 2).

Therefore, the two algorithms return the same edit distance for two unweighted strings. □

Proposition 2. *Given the edit cost function c, Algorithm 2 returns a true distance function between* $\mathbf{x}(T)$ *and* $\mathbf{y}(V)$ *computable in* $\mathcal{O}(|T| \times |V|)$.

Proof. The edit distance computed from an edit cost function c is a metric if c fulfills the following two conditions, $\forall \mathbf{x}, \mathbf{y}, \mathbf{z} \in (\Sigma \times \mathbb{R}_+^*) \cup (\{\lambda\} \times \{0\})$:

1. $c(\mathbf{x}, \mathbf{y}) = 0$ if and only if $\mathbf{x} = \mathbf{y}$ (associated with the fact that $w_x \in \mathbb{R}_+^*$, this defines the *positive definiteness*). This is true because $c(\mathbf{x}, \mathbf{y}) = 0$ only if $|w_x - w_y| = 0$, *i.e.* when $x = y$ and $w_x = w_y$, so when $\mathbf{x} = \mathbf{y}$. On the other hand, this can not occur when $\mathbf{x} \neq \mathbf{y}$ because, except for λ, the weight of a symbol belongs to $\mathbb{R}_+^*$, so $max(w_x, w_y)$ cannot be equal to 0 in this case.

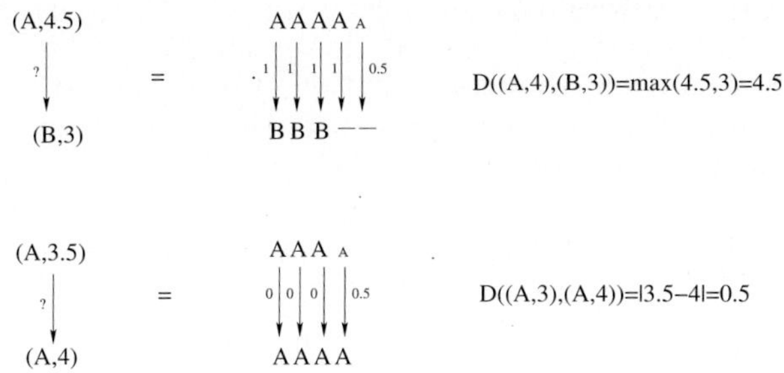

Fig. 1. Intuitive idea of the calculation of the edit distance between two weighted symbols

2. $c(\mathbf{x}, \mathbf{y}) = c(\mathbf{y}, \mathbf{x})$ (*symmetry*). This is always true because the two functions $max(w_x, w_y)$ and $|w_x - w_y|$ fulfill the symmetry condition.

Moreover, the edit distance is computable in $\mathcal{O}(T \times V)$ by the Algorithm 2 if c also satisfies the triangle inequality $c(\mathbf{x}, \mathbf{y}) \leq c(\mathbf{x}, \mathbf{z}) + c(\mathbf{z}, \mathbf{y})$ (see (17)). Since the output of c depends on the nature of the input symbols, let us study the different possible configurations:

1. If the symbols x, y, z are the same, the function c always returns $|n - m|$, that is the Manhattan function. Therefore, the property holds because the Manhattan function is a metric (d_1).
2. If the symbols x, y, z are different, the function c always returns $max(n, m)$, that is a specific case of the Tchebychev function. Therefore, the property holds because the Tchebychev function is a metric (d_∞).
3. If $x = y$ and $x \neq z$, one must satisfy $max(w_x, w_z) + max(w_z, w_y) \geq |w_x - w_y|$. This holds because $max(w_x, w_z) + max(w_z, w_y) \geq max(w_x, w_y)$ (cf case 2) and $max(w_x, w_y) \geq |w_x - w_y|$, $\forall w_x, w_y \in \mathbb{R}^*_+$.
4. If $y = z$ and $x \neq y$, one must satisfy $max(w_x, w_z) + |w_z - w_y| \geq max(w_x, w_y)$. 6 subcases must be studied:
   - If $w_z \in [w_x, w_y]$ we must prove that $w_z + w_y - w_z \geq w_y \Rightarrow w_y \geq w_y$, that is always true.
   - If $w_z \leq w_x \leq w_y$ we must prove that $w_x + (w_y - w_z) \geq w_y$, that is always true because $w_x - w_z \geq 0$.
   - If $w_x \leq w_y \leq w_z$ we must prove that $w_z + w_z - w_y \geq w_y$. Since $w_z + w_z - w_y = 2w_z - w_y \geq 2w_z - w_z = w_z \geq w_y$, the property holds.
   - If $w_z \in [w_y, w_x]$ we must prove that $w_x + (w_z - w_y) \geq w_x$, that is always true because $w_z - w_y \geq 0$.
   - If $w_z \leq w_y \leq w_x$ we must prove that $w_x + (w_y - w_z) \geq w_x$, that is always true because $w_y - w_z \geq 0$.
   - If $w_y \leq w_x \leq w_z$ we must prove that $w_z + w_z - w_y \geq w_x$. Since $w_z + w_z - w_y = 2w_z - w_y \geq 2w_z - w_z = w_z \geq w_x$, the property holds.
5. If $x = z$ and $x \neq y$, one must satisfy $|w_x - w_z| + max(w_z, w_y) \geq max(w_x, w_y)$. By symmetry, this case is proved as for case 4 □

3 Image Representation as a String of Weighted Symbols

As seen in the introduction, the representation of an image as a *bag-of-visual-words* has become the reference method in the field of image classification, inspired by the *bag-of-words* representation in text. Given a vocabulary defined by $\mathcal{V} = \{v_1, \ldots, v_j, \ldots, v_{|\mathcal{V}|}\}$, an image a_i is represented as a vector of weights $\mathbf{u_i} = (u_{i,1}, \ldots, u_{i,j}, \ldots, u_{i,|\mathcal{V}|})$. Each component $u_{i,j}$ of $\mathbf{u_i}$ gives the weight of the visual word v_j in the image a_i.

Unlike a text document, an image does not have a natural visual vocabulary. In the *bag-of-visual-words* approach, visual words represent small patches with a characteristic shape, texture or color that often appear in a reference image collection. Consequently, a visual vocabulary depends on the reference image collection and on the three following elements: the patch detector, the local descriptor and the patch classification method. The way of calculating $u_{i,j}$ weights also plays an important role in the image representation. The natural weighting is the number of occurrences of a visual word v_j in the image. This representation as an occurrence histogram is not the most efficient one and other strategies have been proposed (9). We choose to use a TF-IDF weighting scheme which has been shown to be relevant for image classification (10; 11). The method consists in multiplying a term $tf_{i,j}$ giving the representativeness of word v_j in image a_i to a term idf_j giving the discriminative power of this word in the reference collection. Thus, the weight $u_{i,j} = tf_{i,j} idf_j$ is high if the word appears frequently in the image and rarely in the reference collection.

Based on this vector *bag-of-visual-words* representation, we propose a new representation of an image as a string of weighted symbols. We first expose this new representation. Then we present a way to implement it.

3.1 The Proposed Representation

The principle of our representation is to consider the visual words as symbols whose weight is equal to the TF-IDF weight used in the vector representation. The question is how to structure the symbols as an ordered string.

Let us first note that visual words with zero TF-IDF weight are not taken into account in the string construction. Indeed, this happens when the term is not present in the image or when it appears in most of the documents which means it is not discriminative. This is consistent with the weighted distance previously defined which admits only one zero weighted symbol (the empty symbol λ). Then, the string associated with an image will have a length T equal to the number of visual words appearing at least once in this image.

Let us come to the question of symbol order which provides structural information and so can enhance the image representation. For that purpose, we must choose an order that takes into account visual words characteristics and possibly their distribution in the image. This is a difficult choice because of the high variability of the visual content of an image: two images of similar visual content can have different visual words spatial distributions. To circumvent this difficulty, we propose to use only the global characteristics of the visual vocabulary. For example, we can use the principle that if some visual words (*i.e.* symbols) are similar, it may be interesting to allow substitutions between them in the string. Such substitutions may occur if the corresponding symbols are close in the strings. Thus, the chosen order must group similar visual words. We finally propose to sort the symbols in descending order of their IDF discriminative power. The most discriminative symbols (those that are rare in the collection) will be at the beginning of the string while the less discriminative ones (those that are most common in the collection) will be at the end. This choice is based on the (quite likely) assumption that two similar symbols have a similar IDF discriminative power. Note that such an order is global to all strings since the discriminative power is defined from the entire collection.

Thus, the string $\mathbf{x_1} \dots \mathbf{x_r} \dots \mathbf{x_T}$ associated with a given image a_i is composed of T symbols $\mathbf{x_r} = (x_r, w_{x_r})$ with:

$$\begin{cases} j = ord_{idf}(r) \\ x_r = v_j \\ w_{x_r} = u_{i,j} \end{cases} \quad (1)$$

where $ord_{idf}(r)$ represents the position j of symbol x_r in the original vector $\mathbf{u_i}$.

A summary of the proposed representation is given in Figure 2.

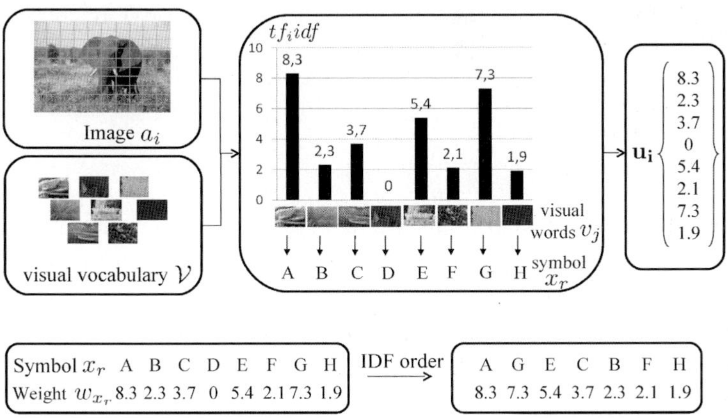

Fig. 2. Image representation as a string of weighted symbols

3.2 A Practical Implementation

In practice, to obtain the string associated with an image, we must choose the three different elements required to build the visual vocabulary (the patch detector, the local descriptor and the patch classification method) plus the TF-IDF weighting. Of course these choices depend on the nature of the images and the task to achieve. The four chosen elements are presented below.

The patch detector. We extract patches sampled in a 10×10 cells regular grid so as to extract 100 patches per image. This dense sampling was found to be more efficient than salient point detection for image classification, since salient points do not cover all the image spatial domain and lead to an incomplete description (4; 9). Moreover, dense sampling may produce a very large number of features.

The local descriptor. We choose to describe patches using a color descriptor. We transform the RGB components of the patch into three normalized components defined as $\frac{R}{R+G+B}$, $\frac{G}{R+G+B}$ and $\frac{R+G+B}{3\times 255}$. This color space presents two main advantages. First, it makes the first two variables independent of the third one representing the luminance. Second, it is very easy to compute. From the normalized components of a patch, we compute 6 features equal to the mean and the standard deviation of the three values.

The patch classification method. We learn a visual vocabulary $\mathcal{V}$ applying a k-means algorithm over all the computed patches on training images. We get k clusters of features whose centers represent k visual words, k being the size of the visual vocabulary. Local patches of images are mapped to their closest visual words using the euclidean distance. In our experiments, we extracted roughly 100000 patches and worked with different vocabulary sizes.

The TF-IDF weighting. Several formulations exist to calculate $tf_{i,j}$ and idf_j, but the okapi one proposed by Robertson *et al.*(23) is often reported superior to others in classification problems. We apply a modified version implemented in the lemur software[2] proposed by (24):

$$tf_{i,j} = \frac{k_1 n_{i,j}}{n_{i,j} + k_2(1 - b + b\frac{|a_i|}{a_{avg}})}$$

where $n_{i,j}$ is the occurrence of the word v_j in the image a_i, $|a_i|$ the number of visual words used to represent image a_i, a_{avg} the average number of visual words per image in the collection A. k_1, k_2 and b are three constants.

$$idf_j = \log \frac{|A| - |\{a_i | v_j \in a_i\}| + 0.5}{|\{a_i | v_j \in a_i\}| + 0.5}.$$

A main disadvantage of this formula is that it can be possibly negative, which has been discussed in (25). This happens when a term appears in more than half of the documents. Thus, we choose to floor the IDF values to 0.

4 Experiments

4.1 Experimental Protocol

To assess the relevance of the new edit distance computed with our TF-IDF-based weighting scheme, we carried out a series of experiments in image classification. The reference image database is the SIMPLIcity collection[3] (26), containing

[2] `http://www.lemurproject.com`

[3] `http://wang.ist.psu.edu/~jwang/test1.zip`

Fig. 3. Examples extracted from the SIMPLIcity collection

1000 images extracted from the COREL database. Each image (384×256 pixels) belongs to one of 10 meaningful categories: *African people, beaches, buildings, buses, dinosaurs, elephants, flowers, food, horses and mountains* (see Figure 3). Note that the different categories are equally distributed, each of them composed of 100 images.

Classification was performed using the 1-Nearest Neighbor (1-NN) rule. We conducted a 10-fold cross-validation to estimate the classifier accuracy. In order to evaluate our weighted edit distance (WED), we divided the experiments into two parts. The first series of experiments aims to analyze how the WED behaves compared to the standard edit distance (ED) and how weighting the symbols contributes to improve the image classification results. The second set of experiments compares the WED with the common metrics usually applied on feature vectors, *i.e.* the normalized dot product and different Minkowski distances such as the L_1, L_2, and L_∞. In this case, images are no more represented in the form of strings but rather by (unordered) vectors whose components are the TF-IDF weights of the visual vocabulary. To assess the impact of the visual vocabulary size, we carried out the experiments with an increasing number of visual words, from 20 to 80.

4.2 Weighted Edit Distance versus Standard Edit Distance

To compare the WED with the standard ED, we performed two series of experiments. First, the ED was applied on the same strings of symbols as for WED without considering their corresponding weights. Figure 4 clearly shows that our WED outperforms the classic ED. Whatever the alphabet size, the difference is statistically significant using a Student paired-t test (note that the highest p-value is equal to 0.02 for an alphabet size of 60 symbols). These results are not so surprising since by discarding the weights, we removed a part of the information about the symbols. Even if the standard ED is not able to deal with such real-valued weights, a fairer comparison would consist in converting each weighted symbol into a substring where the considered symbol would be repeated a number of times corresponding to its weight. But since weights are real valued, we set the number of repetitions to the integer value of the corresponding weight. As we can see on Figure 4, once again, WED performs better than the ED with repetitions.

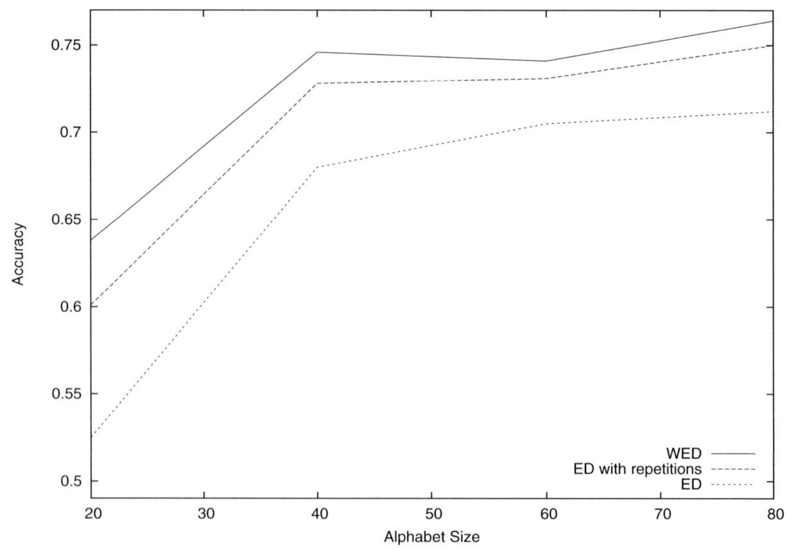

Fig. 4. Comparison between WED and the standard edit distance, computed either omitting the real weights (ED) or by repeating the symbols according to their corresponding weights (ED with repetitions).

Moreover, it is important to note that repeating the symbols generates larger string sizes that leads to a dramatic increase of the algorithmic complexity of the edit distance calculation. Indeed, the complexity becomes $\mathcal{O}(\overline{\lfloor w_x \rfloor} \times \overline{\lfloor w_y \rfloor} \times T \times V)$, where T and V still denote the lengths of the original strings $\mathbf{x}(T)$ and $\mathbf{y}(V)$, $\overline{\lfloor w_x \rfloor}$ and $\overline{\lfloor w_y \rfloor}$ being the average of the integer values of the weights.

4.3 Weighted Edit Distance versus Vector Space Metrics

Figure 5 compares the performance of the WED with that of classical vector space metrics, *i.e.* Minkowski distances (L_1, L_2, L_∞) and the normalized dot product. Several remarks can be made. First, WED is very efficient compared to the other metrics, particularly for small vocabularies. Indeed, it provides the best global behavior by returning the highest classification accuracy for vocabularies smaller than 60, and remaining competitive after. We can also note that for small vocabulary sizes WED significantly outperforms the normalized dot product, which is most often used to measure similarities between bags-of-words. Second, we can observe that the L_1 returns the closest results to WED ones. This behavior can be easily explained by the fact that when symbols are ordered (that is the case in our string-based representation), the L_1 is equivalent to the edit distance if the substitutions between two different symbols are not allowed. Therefore, the difference between the two curves WED and L_1 directly comes from the possibility to substitute symbols in the structured strings.

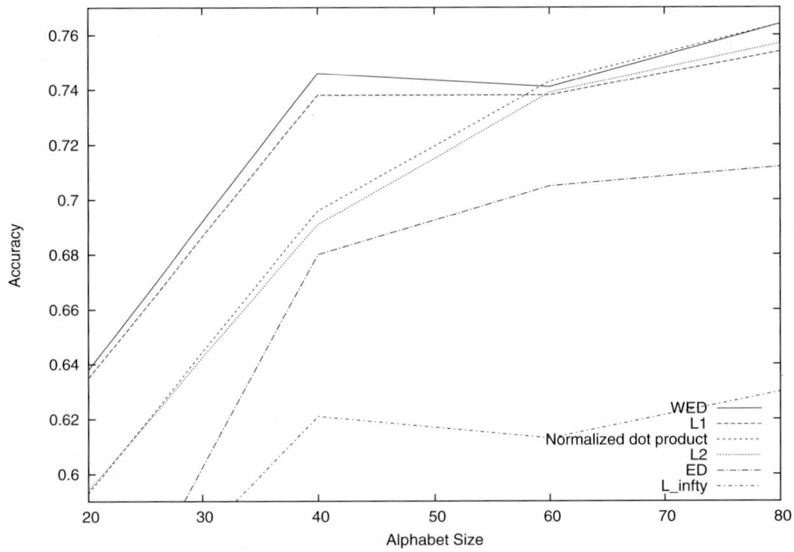

Fig. 5. Comparison between WED, the normalized dot product and some Minkowski metrics

4.4 Plugging Learned Edit Costs in the WED

So far, we mainly concentrated our efforts to take into account real-valued symbols in a new weighted edit distance. In this context, we paid little attention to the cost of the edit operations (*insertion*, *deletion* and *substitution*) that were set to 1 for all the previous experiments. However, as we explained in the introduction of this paper, there exists a huge literature about how to learn those edit costs often in the form of the parameters of stochastic state machines (18; 19; 20; 21; 27). The main drawback of these probabilistic approaches is that the returned parameters "only" satisfy constraints of statistical distribution, and so often do not fulfill the properties of a distance (the symmetry and the triangle inequality are often not ensured). Therefore, the resulting models provide a learned stochastic *edit similarity* rather than a true *edit distance*. To keep a distance function and also to allow the algorithm to compute the WED in quadratic time, we propose in the following a simple solution to determine efficient edit costs.

As explained in Section 3, the visual symbols used for the string comparison correspond to the centers of the clusters computed by a k-means algorithm. These centers are described by some coordinates in a 6-dimensional color space. Intuitively, it means that two centers are close in this space if their corresponding colors are close. When comparing (with an edit distance) two strings, it seems more relevant to favor substitutions of such symbols. This means that the substitution cost should vary according to the distance between one symbol and another.

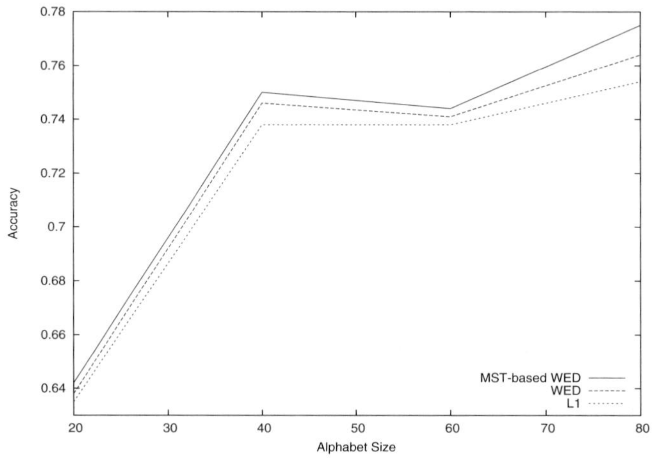

Fig. 6. Effect of the MST-based substitution costs on the performance of WED

To represent these distances, we decided to compute a Minimum Spanning Tree (MST) between the symbols (*i.e.* the center of the clusters) in the 6D space. We considered the complete simple undirected graph $G = (\mathcal{V}, E, \gamma)$ where $\mathcal{V}$ is the set of visual words (vertices), E is the set of all edges in G and $\gamma : e \in E \to \mathbb{R}$ is a function which assigns a weight to each edge of the graph. We considered here that $\gamma(v_1, v_2) = L_2(v_1, v_2)$ where $L_2(v_1, v_2)$ is the Euclidean distance between the visual words v_1 and v_2. Let us recall that a spanning tree of G is a subgraph of G (a tree) which connects all the vertices together. A *minimum* spanning tree is a spanning tree with weight (the sum of the weights of all the edges in the tree) less than or equal to the weight of every other spanning tree.

To set the cost of the substitution between two given symbols x_r and y_k, noted $\sigma(x_r, y_k)$, we used the number of edges between x_r and y_k in the MST. The cost of the insertion and deletion operations is left to 1. To be integrated into the edit distance calculation, $\sigma(x_r, y_k)$ is just multiplied to the edit cost function $c(\mathbf{x_r}, \mathbf{y_k})$ in line 10 of Algorithm 2. To ensure that this extended version of WED remains a true distance computable in quadratic time, we must prove that the substitution cost function σ provided by the MST is also a distance.

Proposition 3. *The substitution cost function* σ *is a metric.*

Proof. Let us prove the three properties of a distance:

- Since the MST is an undirected graph, $\sigma(x, y) = \sigma(y, x)$, and so the *symmetry* property holds.
- If $x \neq y$, by construction the number of edge in the MST between x and y is at least equal to 1. On the other hand, if $x = y$ then $\sigma(x, y) = 0$. Therefore, the *positive definiteness* holds.
- Since $\sigma(x, y)$ is associated to a path of the MST, this path is minimal. Therefore, $\sigma(x, y) \leq \sigma(x, z) + \sigma(z, y)$ and so the *triangle inequality* holds. □

The results of the classification task using the cost function σ in our WED are given in Figure 6. The figure shows that using these substitution costs instead of the naive ones used in the previous section allows us to always improve the classification accuracy and to increase the difference with the L_1 distance, in favor of WED.

5 Conclusion

In this paper, we have presented a new string edit distance in which symbols are allowed to be weighted. This metric opens the door to a better use of *bag-of-visual-words*-based image representations which are often constrained to be compared by vector space metrics. By organizing the visual words in the form of a string, it allows us to take into account additional structural information in the description of images. In this paper, we showed that our new edit distance is very competitive with the state of the art vector space metrics on an image classification task when the symbols are weighted by a TF-IDF measure inspired from information retrieval. We also proposed an extension of our approach by automatically determining the substitution costs from a minimum spannig tree built on the alphabet of *visual-words*. Even if we used our weighted distance with a nearest neighbor classifier, note that it can be easily integrated in *edit kernels* for which there exists a huge literature (see (22) for instance). Moreover, we claim that this distance could be applied to other fields (for example in molecular biology) where one might want to ease or make more difficult some operations on some specific parts of the string or of more complex structured data (*e.g.* trees or graphs). To be able to manage these more general cases, our approach should be extended to the tree or graph edit distances.

References

[1] Chapelle, O., Haffner, P., Vapnik, V.: SVMs for histogram-based image classification. IEEE transactions on Neural Networks 10(5), 1055 (1999)
[2] Vailaya, A., Figueiredo, M., Jain, A., Zhang, H.: Image classification for content-based indexing. IEEE Transactions on Image Processing 10(1), 117–130 (2001)
[3] Salton, G., Wong, A., Yang, C.S.: A vector space model for automatic indexing. Commun. ACM 18(11), 613–620 (1975)
[4] Jurie, F., Triggs, B.: Creating efficient codebooks for visual recognition. In: ICCV 2005: Proceedings of the Tenth IEEE International Conference on Computer Vision (ICCV 2005), Washington, DC, USA, vol. 1, pp. 604–610. IEEE Computer Society, Los Alamitos (2005)
[5] Vidal-Naquet, M., Ullman, S.: Object recognition with informative features and linear classification. In: ICCV 2003: Proceedings of the Ninth IEEE International Conference on Computer Vision, Washington, DC, USA, p. 281. IEEE Computer Society, Los Alamitos (2003)
[6] Mikolajczyk, K., Schmid, C.: Scale & affine invariant interest point detectors. Int. J. Comput. Vision 60(1), 63–86 (2004)
[7] Lowe, D.G.: Distinctive image features from scale-invariant keypoints. Int. J. Comput. Vision 60(2), 91–110 (2004)
[8] Bisgin, H.: Parallel clustering algorithms with application to climatology. Technical report, Informatics Institute, Istanbul Technical University, Turkey (2008)

[9] Nowak, E., Jurie, F., Triggs, B.: Sampling strategies for bag-of-features image classification. In: Leonardis, A., Bischof, H., Pinz, A. (eds.) ECCV 2006. LNCS, vol. 3954, pp. 490–503. Springer, Heidelberg (2006)

[10] Moulin, C., Barat, C., Géry, M., Ducottet, C., Largeron, C.: UJM at ImageCLEFwiki 2008. In: Peters, C., Deselaers, T., Ferro, N., Gonzalo, J., Jones, G.J.F., Kurimo, M., Mandl, T., Peñas, A., Petras, V. (eds.) Evaluating Systems for Multilingual and Multimodal Information Access. LNCS, vol. 5706, pp. 779–786. Springer, Heidelberg (2009)

[11] Yang, J., Jiang, Y.G., Hauptmann, A.G., Ngo, C.W.: Evaluating bag-of-visual-words representations in scene classification. In: MIR 2007: Proceedings of the international workshop on Workshop on multimedia information retrieval, pp. 197–206. ACM, New York (2007)

[12] Chang, S.K., Shi, Q.Y., Yan, C.W.: Iconic indexing by 2-d strings. IEEE Trans. Pattern Anal. Mach. Intell. 9(3), 413–428 (1987)

[13] Punitha, P., Guru, D.S.: Symbolic image indexing and retrieval by spatial similarity: An approach based on b-tree. Pattern Recogn. 41(6), 2068–2085 (2008)

[14] Hsieh, S.M., Hsu, C.C.: Retrieval of images by spatial and object similarities. Inf. Process. Manage. 44(3), 1214–1233 (2008)

[15] Freeman, H.: Computer processing of line-drawing images. ACM Comput. Surv. 6(1), 57–97 (1974)

[16] Daliri, M.R., Torre, V.: Robust symbolic representation for shape recognition and retrieval. Pattern Recognition 41, 1782–1798 (2008)

[17] Wagner, R.A., Fischer, M.J.: The string-to-string correction problem. J. ACM 21(1), 168–173 (1974)

[18] Durbin, R., Eddy, S., Krogh, A., Mitchison, G.: Biological sequence analysis. Cambridge University Press, Cambridge (1998)

[19] Ristad, E.S., Yianilos, P.N.: Learning string-edit distance. IEEE Trans. Pattern Anal. Mach. Intell. 20(5), 522–532 (1998)

[20] McCallum, A., Bellare, K., Pereira, P.: A conditional random field for disciminatively-trained finite-state string edit distance. In: Proceedings of the 21st Conference on Uncertainty in Artificial Intelligence (UAI 2005), pp. 388–400 (2005)

[21] Oncina, J., Sebban, M.: Learning stochastic edit distance: application in handwritten character recognition. Pattern Recogn. 39(9), 1575–1587 (2006)

[22] Bellet, A., Bernard, M., Murgue, T., Sebban, M.: Learning state machine-based string edit kernels. Pattern Recogn. 43(6), 2330–2339 (2010)

[23] Robertson, S.E., Walker, S., Hancock-Beaulieu, M., Gull, A., Lau, M.: Okapi at trec-3. In: Text REtrieval Conference, pp. 21–30 (1994)

[24] Zhai, C.: Notes on the lemur tfidf model. Technical report, Carnegie Mellon University (2001)

[25] Robertson, S.E., Walker, S.: On relevance weights with little relevance information. In: SIGIR 1997: Proceedings of the 20th annual international ACM SIGIR conference on Research and development in information retrieval, pp. 16–24. ACM, New York (1997)

[26] Wang, J.Z., Li, J., Wiederhold, G.: Simplicity: Semantics-sensitive integrated matching for picture libraries. In: Laurini, R. (ed.) VISUAL 2000. LNCS, vol. 1929, pp. 360–371. Springer, Heidelberg (2000)

[27] Boyer, L., Esposito, Y., Habrard, A., Oncina, J., Sebban, J.: Sedil: Software for edit distance learning. In: Daelemans, W., Goethals, B., Morik, K. (eds.) ECML PKDD 2008, Part II. LNCS (LNAI), vol. 5212, pp. 672–677. Springer, Heidelberg (2008)

A Concise Representation of Association Rules Using Minimal Predictive Rules

Iyad Batal and Milos Hauskrecht

Department of Computer Science
University of Pittsburgh
{iyad,milos}@cs.pitt.edu

Abstract. Association rule mining is an important branch of data mining research that aims to extract important relations from data. In this paper, we develop a new framework for mining association rules based on minimal predictive rules (MPR). Our objective is to minimize the number of rules in order to reduce the information overhead, while preserving and concisely describing the important underlying patterns. We develop an algorithm to efficiently mine these MPRs. Our experiments on several synthetic and UCI datasets demonstrate the advantage of our framework by returning smaller and more concise rule sets than the other existing association rule mining methods.

1 Introduction

The huge amounts of data collected today provide us with an opportunity to better understand the behavior and structure of many natural and man-made systems. However, the understanding of these systems may not be possible without automated tools that enable us to extract the important patterns in the data and present them in a concise and easy to understand form.

Rule induction methods represent a very important class of knowledge discovery tools. The advantage of these methods is that they represent the knowledge in terms of if-then rules that are easy to interpret by humans. This can facilitate the process of discovery and utilization of new practical findings. As an example, consider a knowledge discovery problem in medicine. Assume a rule mining algorithm identifies a subpopulation of patients that respond better to a certain treatment than the rest of the patients. If the rule clearly and concisely defines this subpopulation, it can speed up the validation process of this finding and its future utilization in patient-management.

Association rule mining is a very popular data mining technique to extract rules from the data. The original framework [1] has been extended to mine patterns from various domains [28,4,16]. The key strength of association rule mining is that it searches the space of rules completely by examining all patterns that occur frequently in the data. However, the main disadvantage is that the number of association rules it finds is often very large. Moreover, many rules are redundant because they can be naturally explained by other rules. This may hinder the discovery process and the interpretability of the results. The objective of this work is to filter out these redundant rules and provide the user with a small set of rules that are sufficient to capture the essential underlying structure of the data.

J.L. Balcázar et al. (Eds.): ECML PKDD 2010, Part I, LNAI 6321, pp. 87–102, 2010.

In this work, we focus on the problem of mining association rules that target a specific class variable of interest. We call these rules the *class association rules*. To achieve our goal, we first introduce the concept of the *minimal predictive rules* (MPR) set that assures a good coverage of the important patterns with very small number of rules. After that, we propose an algorithm for mining these rules. Briefly, our method builds the MPR set by examining more general rules first and gradually testing and adding more specific rules to the set. The algorithm relies on a statistical significance test to ensure that every rule in the result is significantly better predictor than any of its generalizations.

2 Methodology

In this section, we first define basic terminology used throughout the paper. After that, we present an example illustrating the challenges of rule mining and the limitations of existing methods. Next, we propose the minimal predictive rules (MPR) framework to address them. Finally, we present an algorithm for mining the MPRs.

2.1 Definitions

Our work focuses on mining relational databases, where each record is described by a fixed number of attributes. We assume that all attributes have discrete values (numeric attributes must be discretized [14]). We call an attribute value pair an *item* and a conjunction of items a *pattern* (sometimes patterns are also called *itemsets*). If a pattern contains k items, we call it a *k-pattern* (an item is a 1-pattern). We say that pattern P' is a subpattern of pattern P if $P' \subset P$ (P is a superpattern of P'). A rule is defined as R: $A \Rightarrow c$, where A is a pattern and c is the class label that R predicts. We say that rule R': $A' \Rightarrow c'$ is a subrule of rule R: $A \Rightarrow c$ if $c' = c$ and $A' \subset A$.

A pattern P can be viewed as defining a subpopulation of the instances (records) that satisfy P. Hence, we sometimes refer to pattern P as group P. If P' is a subpattern of P, then P' is a supergroup of P. Note that the empty pattern Φ defines the entire population. The support of pattern P, denoted as $sup(P)$, is the ratio of the number of records that contain P to the total number of records: $sup(P) \approx \Pr(P)$. The confidence of rule R: $A \Rightarrow c$ is the posterior probability of class c in group A: $conf(R) = sup(A \cup c)/sup(A) \approx \Pr(c|A)$. Note that confidence of the empty rule is the prior probability of the class: $conf(\Phi \Rightarrow c) \approx \Pr(c)$.

2.2 Example

Assume our objective is to identify populations which are at high risk of developing coronary heart disease (CHD). Assume that our dataset contains 200 instances and that the CHD prior is Pr(CHD)=30%. We want to evaluate the following 3 rules:

R1: Family history=yes $\Rightarrow$ CHD
[*sup*=50%, *conf*=60%]

R2: Family history=yes $\wedge$ Race=Caucasian $\Rightarrow$ CHD
[*sup*=20%, *conf*=55%]

R3: Family history=yes $\wedge$ Race=African American $\Rightarrow$ CHD
[*sup*=20%, *conf*=65%]

From the above rules, we can see that a positive family history is probably an important risk factor for CHD because the confidence of R1 (60%) is two times higher than CHD prior (30%). However, the problem is that we expect many rules that contain a positive family history in their antecedents to have a high confidence as well. So how can we know which of these rules are truly important for describing the CHD condition?

The original association rules framework [1] outputs all the frequent rules that have a higher confidence than a minimum confidence threshold (*min_conf*). For instance, if we set *min_conf*=50%, all of three rules will be returned to the user.

In order to filter out some of the uninteresting associations, the original support-confidence framework is sometimes augmented with a correlation measure. Commonly, a χ^2 test is used to assure that there is a significant positive correlation between the condition of the rule and its consequent [8,24,21,18]. However, because the posteriors of all three rules are much higher than the prior, we expect all of them to be statistically significant! Moreover, these rules will be considered interesting using most existing interestingness measures [15]. The main problem with this approach is that it evaluates each rule individually without considering the relations between the rules. For example, if we are given rule R2 by itself, we may think it is an important rule. However, by looking at all three rules, we can see that R2 should not be reported because it is more specific than R1 (applies to a smaller population) and has a lower confidence.

To filter out such redundant rules, [6] defined the confidence improvement constraint:

$$imp(A \Rightarrow c) = conf(A \Rightarrow c) - \max_{A' \subset A}\{conf(A' \Rightarrow c)\} > min\_imp$$

In practice, it is not clear how to specify this *min_imp* parameter. So the common convention is to set to zero ([18,17]). This means that we only report the rules that have a higher confidence than all of their subrules. If we applied the confidence improvement constraint to our working example, rule R2 will be removed and rule R3 will be retained. However, R3 may also be unimportant and its observed improvement in the confidence can be due to chance rather than actual causality. In fact, there is a high chance that a refinement of a rule, even by adding random items, leads to a better confidence. We will see later in the analysis in section 2.4 and in the experimental evaluation that the confidence improvement constraint can still output many spurious rules. So should we report rule R3? To answer this question, we define the *minimal predictive rules* concept.

2.3 Minimal Predictive Rules (MPR)

Definition 1. *A rule R: A⇒c is a minimal predictive rule (MPR) if and only if* R *predicts class* c *significantly better than all its sub-rules.*

This definition implies that every item in the condition (A) is an important contributor to the predictive ability of the rule. We call these rules *minimal* because removing any non-empty combination of items from the condition would cause a significant drop in the predictability of the rule. An MPR can be viewed as defining a "surprising" subpopulation, where the posterior probability of class c in the group A ($\Pr(c|A)$) is unexpected and cannot be explained by any convex combinations of A's subpatterns.

The MPR significance test: In order to check if a rule is significant with respect to its subrules, we use the binomial distribution as follows: Assume we are interested in testing the significance of rule R: $A \Rightarrow c$. Assume that group A contains N instances, out of which N_c instances belong to class c. Assume P_c represents the highest confidence achieved by any subrule of R: $P_c = \max_{A' \subset A} \Pr(c|A')$. The null hypothesis presumes that N_c is generated from N according to the binomial distribution with probability P_c. The alternative hypothesis presumes that the true underlying probability that generated N_c is significantly higher than P_c. Hence, we perform a *one sided* significance test (we are interested only in increases in the proportion of c) and calculate a p-value as follows:

$$p = Pr_{binomial}(x \geq N_c|N, P_c)$$

If this p-value is significant (smaller than a significance level α), we conclude that R significantly improves the predictability of c over all its subrules, hence is an MPR.

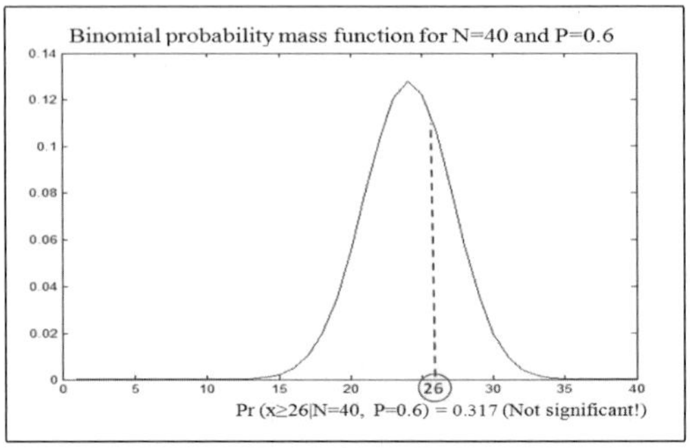

Fig. 1. The MPR significance test for rule R3

Example: Going back to our CHD example, rule R3 covers 40 instances, out of which 26 have CHD. For R3 to be an MPR, it should be significantly more predictive than all its simplifications, including rule R1. By applying the MPR significance test we get: $Pr_{binomial}(x \geq 26|40, 0.6) = 0.317$. As illustrated in Figure 1, we can see that R3 is not an MPR at significance level $\alpha = 0.05$. On the other hand, if we use the same binomial significance test to evaluate each rule individually against the CHD prior (by always setting P_c=Pr(CHD)), the p-values we get for R1, R2 and R3 are respectively, 5.13e-10, 8.54e-4 and 5.10e-6, meaning that all three rules are (very) significant!.

2.4 Spurious Patterns and Redundant Rules

In this section, we discuss and analyze the serious problem of redundancy in association rules. This problem is due to the manner in which large numbers of spurious rules are formed by adding irrelevant items to the antecedent of other rules. We show the deficiencies of the current approaches to dealing with this problem. Finally, we show how MPR can overcome the problem.

Consider a Bayesian belief network example in Figure 2, where we have a causal relation between variable X_1 and the class variable C. Assume that item X_1=1 is highly predictive of class c_1, so that Pr(C=c_1 | X_1=1) is significantly larger than Pr(C=c_1). Assume we have another variable X_2 that is completely independent of C: $X_2 \perp\!\!\!\perp C$. If we add any instantiation v_2 of variable X_2 to item X_1=1, the posterior distribution of

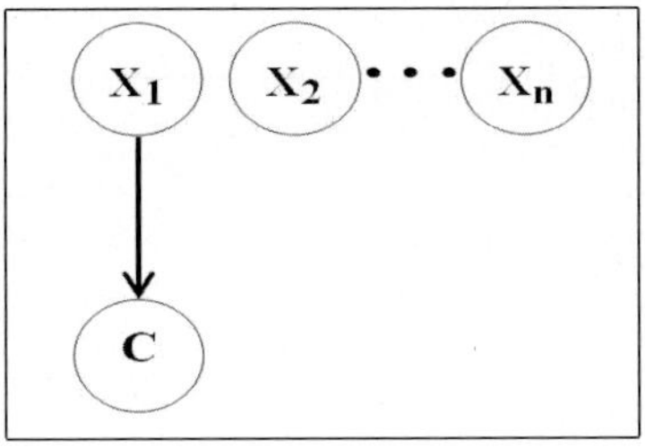

Fig. 2. Illustrating the redundancy problem in association rules

c_1 is Pr(C=c_1 | X_1=1 ∧ X_2=v_2) ≈ Pr(C=c_1 | X_1=1), i.e., *conf*(X_1=1 ∧ X_2=v_2 ⇒ c_1) ≈ *conf*(X_1=1 ⇒ c_1).

More generally, if we have many irrelevant variables such that $X_i \perp\!\!\!\perp C$: $i \in \{2..n\}$, the network structure implies that Pr(C=c_1 | X_1=1 ∧ X_2=v_2 ... ∧ X_n=v_n) ≈ Pr(C=c_1 | X_1=1) for every possible instantiation X_i=v_i. Clearly, the number of such spurious rules can become huge, which can easily overwhelm the user and prevent him from understanding the real patterns and causalities in the data.

Even by requiring the complex rules to have a higher confidence [6,17,18] or lift score [9] than their simplifications, the problem still exists and many of these redundant rules can easily satisfy this constraint. In fact, if X_i is a binary variable and *conf*(X_1=1 ∧ X_i=0 ⇒ c_1) < *conf*(X_1=1 ⇒ c_1), then we know for sure that *conf*(X_1=1 ∧ X_i=1 ⇒ c_1) > *conf*(X_1=1 ⇒ c_1). The same situation happens if we use the lift score instead of the confidence! Post-pruning the rules individually based on their correlations with the class (e.g. using the χ^2 test [8,24,21,18]) or based on their difference from the prior (e.g. using our binomial test) is not going to not help in this case.

Our frameworks tackles this problem because every item in an MPR should significantly contribute to improving the predictability of the rule. This means that if rule R: X_{q_1}=v_{q_1} ... ∧ X_{q_k}=v_{q_k}⇒C=c is an MPR, then there should exist a path from each variable X_{q_i} to the class C that is not blocked (d-separated) by the other variables in the rule: X_{q_i} not $\perp\!\!\!\perp$ $C \mid \{ X_{q_1}, X_{q_{i-1}}, ..., X_{q_{i+1}}, X_{q_k} \}$. Therefore, redundant rules are likely to be filtered out.

2.5 The Algorithm

In this section we explain our algorithm for mining the MPR set. The algorithm is outlined in Figure 3. Briefly, the algorithm explores the space by performing a level wise Apriori-like search. At each level (l), we first remove the candidate *l-patterns* that do not pass the minimum support threshold (line 6). Then we extract all MPRs from these frequent *l-patterns* (line 7). Finally, we generate the candidates for the next level (line 8).

Extracting MPRs. The process of testing if a rule $P \Rightarrow c$ is an MPR is not trivial because the definition requires us to check the rule against all its proper subrules. This would require to check 2^l subpatterns if P has length l. Our algorithm avoids this inefficiency by caching the statistics needed to perform the check within the *(l-1)-subpatterns* from the previous level. This part of the algorithm is outlined in Figure 4.

Algorithm 1: Mine all *MPRs*

```
Input: dataset: D, minimum support: min_sup
Output: minimal predictive rules: MPR
     // global data structure
01:  MPR=Φ, tbl_max_conf=hashtable()
     // the prior distribution of the classes
02:  tbl_max_conf[h(Φ)]=calculate_class_distribution(Φ, D)
03:  Cand=generate_1_patterns()
04:  l = 1
05:  while (Cand ≠ Φ)
         // remove candidates that are not frequent
06:      FP[l]=prune_infrequent(Cand, D, min_sup)
         // find all MPRs at level l
07:      extract_MPR(FP[l], D)
         // generate candidate (l+1)_patterns
08:      Cand=generate_candidates(FP[l])
09:      l = l + 1
10:  MPR=FDR_correction(MPR)
11:  return MPR
```

Fig. 3. The algorithm for mining MPRs from a dataset

To explain the method, it is useful to envision the progress of the algorithm as gradually building a lattice structure level by level, starting from the empty pattern Φ. An example lattice is shown in Figure 5. Every frequent *l-pattern* *P* is a node in the lattice with *l* children: one child for each of its *(l-1)-subpatterns*. The key idea of our algorithm is to store in each node P the maximum confidence score for every class that can be obtained in the sublattice with top *P* (including P itself): $max\_conf_P[c]$=max(Pr($c \mid P'$)): $\forall\, P' \subseteq P$. These *max_conf* values are computed from the bottom up as algorithm progresses. Initially, max_conf_Φ for the empty pattern is set to be the prior distribution of the class variable. In order to compute max_conf_P for pattern P, we first compute $conf_P$ (line 2), the distribution of the class variable in group P: $conf_P[c]$=Pr($c \mid P$). Then we use the *max_conf* values of P's direct children to compute $max\_conf\_children_P$ (line 3) so that $max\_conf\_children_P[c]$=max(Pr($c \mid P'$)): $\forall\ P' \subset P$. Finally, we compute max_conf_P by taking the element-wise maximum of two arrays: $conf_P$ and $max\_conf\_children_P$ (line 4).

After assigning the *max_conf* value for pattern P, we want to check if P forms an MPR. So for each class label c, we perform the MPR significance test to check if P predicts c significantly better than $max\_conf\_children_P[c]$. If the test is positive, we add the rule $P \Rightarrow c$ to the set of MPRs (line 8).

Please note that in our pseudo-code, we do not explicitly build the lattice. Instead, we use a hash table *tbl_max_conf* (Figure 3: line 1) to provide direct access to the *max_conf* values, so that *tbl_max_conf*[$h(P)$]=max_conf_P, where h is a hash function. Also note that none of the functions (*calculate_class_distribution*, *is_MPP* and *lossless_pruning*) requires another scan on the data because we can collect the class specific counts during the same scan that counts the pattern.

```
Algorithm 2: extract_MPR (FP[l], D)
//add pattern P ∈ FP[l] to MPR (a global variable) if P is
significantly more predictive than all its subpatterns
for each P ∈ FP[l]
    conf=calculate_class_distribution(P, D)
    max_conf_children=max {tbl_max_conf[h(S_{l-1})]} : S_{l-1}⊂P
    max_conf=max{ conf, max_conf_children }
    tbl_max_conf[h(P)]=max_conf
    for each class label c
        if ( is_MPR(P, c, max_conf_children, D) )
            MPR=MPR ∪ (P ⇒ c)
    lossless_pruning(P, max_conf, D, FP[l])

Function is_MPR(P, c, max_conf_children, D)
  //return true if P predicts c significantly better than all its subpatterns
  N=count(P, D)
  N_c=count(P ∪ c, D)
  p_value=Pr_binomial(x ≥ N_c | N, max_conf_children[c])
  if(p_value < α)
     return true
  return false

Function lossless_pruning(P, max_conf, D, FP[l])
  //prune P if it cannot produce any MPR
  for each class label c
     N_c=count(P ∪ c, D)
     p_value=Pr_binomial(x ≥ N_c | N_c, max_conf[c])
     //exit if P can potentially produce an MPR for any class c
     if(p_value < α)
        return ;
  //Prune P
  remove(P, FP[l])
```

Fig. 4. The algorithm for extracting MPRs from the frequent patterns at level l

Figure 5 illustrates the algorithm using a small lattice on a dataset that contains 200 instances from class c_1 and 300 instances from class c_2. For each pattern P (node), we show the number of instances in each class, the distribution of the classes (*conf*) and the maximum confidence from P's sublattice (*max_conf*). Let us look for example at pattern $I_1 \wedge I_2$. This pattern is predictive of class c_2: $conf(I_1 \wedge I_2 \Rightarrow c_2) = 0.75$. However, this rule is not an MPR since it does not significantly improve the predictability of c_2 over the subrule $I_2 \Rightarrow c_2$: $Pr_{binomial}(x \geq 75|100, 0.7) = 0.16$ (not significant at α=0.05). The MPRs from this example are: $I_1 \Rightarrow c_1$, $I_2 \Rightarrow c_2$ and $I_1 \wedge I_3 \Rightarrow c_1$.

Mining at low support. It is well know that the performance of the Apriori algorithm highly depends on the minimum support (*min_sup*) parameter. However, setting this parameter is not straightforward because the optimal *min_sup* can vary greatly across different datasets. In fact, [11,9] argued that it is sometimes of interest to mine low

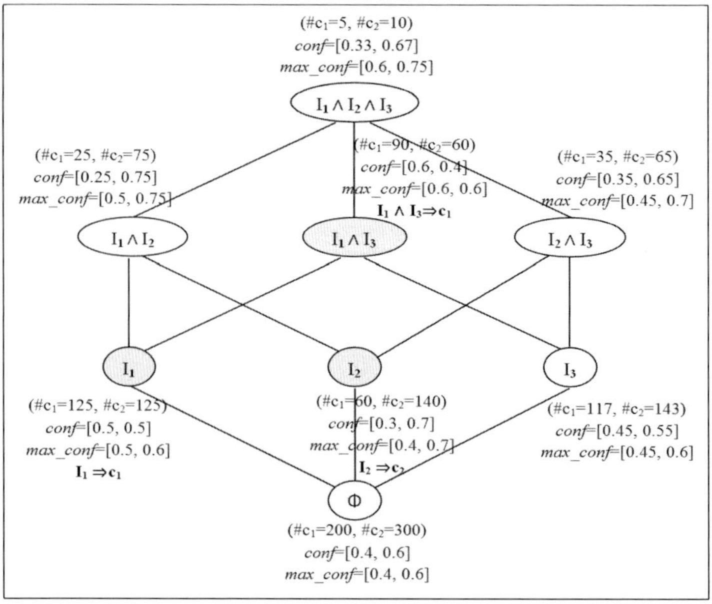

Fig. 5. An illustrative example showing the lattice associated with frequent pattern $I_1 \wedge I_2 \wedge I_3$. The MPRs are shaded.

support patterns. Therefore, in order not to miss any important pattern, the user may choose a low *min_sup* value. However, low *min_sup* raises two important concerns. First, the algorithm may return a huge number of rules, most which are very complex. Second, the algorithm may take very long time to finish.

In the following we argue that the MPR framework and our algorithm address both of these concerns. First, by requiring each MPR to be significantly better than all its subrules, the algorithm is biased towards choosing simple rules over more complex rules. Moreover, the MPR significance test incorporates the pattern's support because the chance of passing the test is lower for low support patterns. This acts as if the *min_sup* filter was built into the statistical significance test. Hence, very low support patterns are likely to be filtered out unless they are extremely predictive (with a very surprising distribution).

Second, the MPR significance test can help us to prune the search space. This early pruning is implemented by the *lossless_pruning* function in Figure 4. The idea is that we can prune pattern *P* if we guarantee that *P* cannot produce any MPR. However, because the pruning is applied while generating the patterns, we do not know what subgroups P will generate further in the lattice. To overcome this, let us define the *optimal subgroup* $P^*_{c_i}$ in group P with respect to class c_i to be the subgroup that contains all the instances from c_i and none of the instances from any other classes. Clearly, P cannot generate any subgroup better than $P^*_{c_i}$ for predicting class c_i. Now, we prune P if for every class c_i, $P^*_{c_i}$ is not significant with respect to the best confidence so far $max\_conf[c_i]$. Please note that this pruning technique does not miss any MPR because the *lossless_pruning* test is anti-monotonic.

As an example, consider pattern $P = I_1 \wedge I_2 \wedge I_3$ in Figure 5. This pattern contains 15 examples, 5 from class c_1 and 10 from class c_2. Both $P^*_{c_1}$ and $P^*_{c_2}$ are not significant with respect to the current best predictions 0.6 and 0.75 (respectively). Therefore, there is no need to further explore P's subgroups and the entire sublattice can be safely pruned.

Correcting for multiple testing. When multiple rules are tested in parallel for significance, it is possible to learn a number of false rules by chance alone. This is a common problem for all techniques that rely on statistical tests. For example, assume we have 10,000 random items (independent of the class variable). If we test the significance of each of them with a significance level α=0.05, then we expect about 500 items to be significant just by chance!

One approach to deal with the multiple hypothesis testing problem is to adjust the significance level at which each rule is tested. The most common way is the Bonferroni correction [25], which divides the significance level (α) by the number of tests performed. This approach is not suitable for rule mining because the number of rules tested is usually very large, resulting in an extremely low α and hence very few rules discovered. The other more recent approach, directly controls the false discovery rate (FDR) [7]: the expectation of the proportion of false discoveries in the result. We adopt the FDR correction method because it is less stringent and more powerful (has a lower Type II error) than the Bonferroni correction. We apply FDR as a post-processing step (Figure 3: line 10). It takes as input all potential MPRs with their p-values and outputs a subset of MPRs that satisfy the FDR criteria.

3 Experiments

In this section we present our experimental evaluation, first on synthetic datasets with known underlying patterns, and after that on several UCI classification datasets [2]. The experiments compare the performance of MPR against the following methods:

- ***complete:*** The set of all rules that cover more than *min_sup* examples in the data. We filter out useless rules by only including the ones that positively predict the class label (with a lift score [15] higher than one).
- ***closed:*** A subset of *complete* that corresponds to non-redundant rules based on the concept of closed frequent patterns [3].
- ***corr_chi:*** A subset of *complete* that contains the rules with significant positive correlations between the condition and conclusion according to the χ^2 test [8,24,21].
- ***prior_binom:*** A subset of *complete* that contains the rules that are significantly different from the prior according to the binomial statistical test (section 2.3).
- ***prior_FDR:*** A subset of *prior_binom* that is post-processed using the false discovery rate (FDR) technique [7] to correct for the multiple testing problem.
- ***conf_imp:*** A subset of *complete* that satisfies the confidence improvement constraint [6,18,17]: each rule should have a higher confidence than all its subrules.

For all methods, we set the minimum support threshold (*min_sup*) to 10% the number of records in the dataset (unless otherwise stated). For the methods that use a statistical test: *corr_chi*, *prior_binom*, *prior_FDR* and *MPR*, we use the conventional significance level $\alpha = 0.05$.

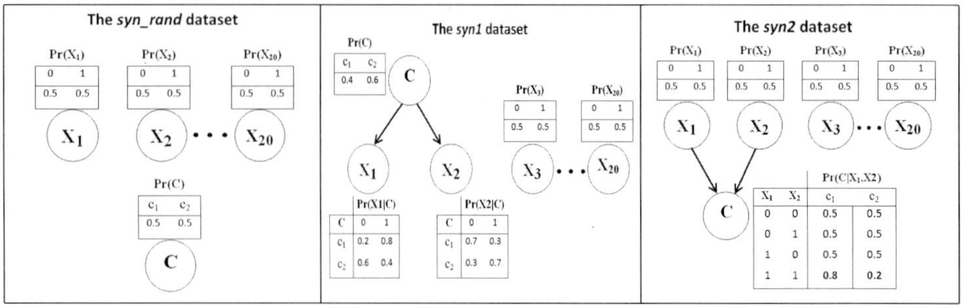

Fig. 6. Three Bayesian belief networks used to generate synthetic datasets: *syn_rand*, *syn1* and *syn2* (from left to right)

3.1 Experiments on Synthetic Data

The experiments on synthetic data (generated from pre-defined patterns) allow us to judge more objectively the quality of the algorithms' outputs by comparing the original and recovered patterns.

Data description. The synthetic data were generated from the three Bayesian belief networks (BBNs) in Figure 6. Each network consists of 20 random binary variables $\{X_1, ..., X_{20}\}$ and two class labels c_1 and c_2. The three networks are:

- ***syn_rand:*** In this network, all the attribute variables X_i and the class variable C are independent (*syn_rand* does not contain any pattern).
- ***syn1:*** In this network, X_1=1 is more frequent in class c_1: Pr(X_1=1 | C=c_1)=0.8 > Pr(X_1=1 | C=c_2)=0.4 and item X_2=1 is more frequent in class c_2: Pr(X_2=1 | C=c_2)=0.7 > Pr(X_2=1 | C=c_1)=0.3. Besides, the distribution of the class variable C is more biased towards c_2: Pr(C=c_2)=0.6. The attributes $\{X_3, ..., X_{20}\}$ are independent of each other and the class variable C.
- ***syn2:*** The main pattern in *syn2* is a conjunction of X_1=1 and X_2=1 that predicts class c_1: Pr(C=c_1 | X_1=1 and X_2=1)=0.8. For all other values of X_1 and X_2, the two classes (c_1 and c_2) are equally likely. The attributes $\{X_3, ..., X_{20}\}$ are independent of each other and the class variable C.

The datasets we analyze consist of 1000 examples randomly generated from these three networks.

Results: Table 1 summarizes the number of rules that each method produces on the three synthetic datasets. First notice that the number of rules in *complete* and *closed* are the same for all these datasets (since very few correlations exist). Also notice that *corr_chi* and *prior_binom* have similar results. This shows that the choice of the statistical test does not matter much and that the real problem with these approaches is that they evaluate each rule separately without considering the nested structure of the rules.

Let us look first at the results on the *syn_rand* dataset, which we know does not contain any pattern. MPR does not output any rule which is what we expect given that

Table 1. The number of rules for the different methods on the synthetic datasets

Dataset	*complete*	*closed*	*corr_chi*	*prior_binom*	*prior_FDR*	*conf_imp*	*MPR*
syn_rand	9,763	9,763	516	651	0	4,748	**0**
syn1	9,643	9,643	2,989	3,121	2,632	4,254	**6**
syn2	9,868	9,868	1,999	2,225	422	3,890	**5**

the class is independent of all attributes. On the other hand, *conf_imp* returns a huge number of rules. In fact, *conf_imp* returns almost half the total number of associations. This result agrees with our analysis in section 2.4. *prior_FDR* correctly removes all false positives from *prior_binom* on this random data.

Now consider the *syn1* dataset. Our algorithm extracts the following 6 MPRs: {X1=1 ⇒ c1, X1=0 ⇒ c2, X2=1 ⇒ c2, X2=0 ⇒ c1, X1=0 ∧ X2=1 ⇒ c2, X1=1 ∧ X2=0 ⇒ c1}[1]. In comparison, all other methods extract a vast number of rules. For example, even by using the FDR correction, the number of rules is 2,632 rules!

Table 2. The syn2 dataset: on the left is the set of all MPRs, in the middle is the top 5 *prior_binom* rules (out of 2,225 rules) and on the right is the top 5 *conf_imp* rules (out of 3,890 rules)

MPR	*prior_binom*	*conf_imp*
X1=1 ∧ X2=1 ⇒ c1 [*sup*=26.5%, *conf*=81.1%]	X1=1 ∧ X2=1 ⇒ c1 [*sup*=26.5%, *conf*=81.1%]	X1=1 ∧ X2=1 ∧ X8=0 ⇒ c1 [*sup*=12.3%, *conf*=85.4%]
X2=1 ⇒ c1 [*sup*=50.1%, *conf*=66.7%]	X1=1 ∧ X2=1 ∧ X14=0 ⇒ c1 [*sup*=14.5%, *conf*=84.8%]	X1=1 ∧ X2=1 ∧ X14=0 ⇒ c1 [*sup*=14.5%, *conf*=84.8%]
X1=1 ⇒ c1 [*sup*=51.2%, *conf*=63.9%]	X1=1 ∧ X2=1 ∧ X13=1 ⇒ c1 [*sup*=13.4%, *conf*=84.3%]	X1=1 ∧ X2=1 ∧ X13=1 ⇒ c1 [*sup*=13.4%, *conf*=84.3%]
X2=0 ⇒ c2 [*sup*=49.9%, *conf*=51.5%]	X1=1 ∧ X2=1 ∧ X9=1 ⇒ c1 [*sup*=14.1%, *conf*=83.7%]	X1=1 ∧ X2=1 ∧ X9=1 ⇒ c1 [*sup*=14.1%, *conf*=83.7%]
X1=0 ⇒ c2 [*sup*=48.8%, *conf*=49.0%]	X1=1 ∧ X2=1 ∧ X8=0 ⇒ c1 [*sup*=12.3%, *conf*=85.4%]	X1=1 ∧ X2=1 ∧ X18=0 ⇒ c1 [*sup*=13.9%, *conf*=83.5%]

Finally, let us look more closely at the results on the *syn2* dataset. Table 2 shows all MPRs (left), the top 5 ranked rules for *prior_binom* (same for *prior_FDR*) according to the p-values (center) and the top 5 ranked rules for *conf_imp* according to the confidence (right). Notice that *prior_binom* correctly ranks the real pattern (X_1=1 ∧ X_2=1) at the top. However, the following rules are redundant. For *conf_imp*, the real pattern is buried inside many spurious rules. This example clearly shows the deficiencies of these methods in concisely representing the actual patterns. On the contrary, by investigating the small number of MPRs, we can easily recover the structure of the underlying BBN.

3.2 Experiments on UCI Datasets

Data description. To further test the different methods, we use 9 public datasets from the UCI Machine Learning repository [2]. We discretize the numeric attributes into

[1] The last 2 rules combine the evidence of the two main patterns, hence improve the predictability. For example, Pr(c1|x1=1 ∧ x2=0) > Pr(c1|x1=1).

Table 3. UCI Datasets characteristics

dataset	# attributes	# items	# records	# classes
Adults	14	154	32,561	2
Heart disease	13	33	303	2
Lymphography	18	57	142	2
Pima diabetes	8	19	768	2
Breast cancer	9	41	286	2
Dermatology	12	47	366	6
Wine	13	39	178	3
Glass	10	22	214	2
Credit	15	69	690	2

Table 4. The number of rules for the different algorithms on several UCI datasets

Dataset	*complete*	*closed*	*corr_chi*	*prior_binom*	*prior_FDR*	*conf_imp*	*MPR*
Adults	2,710	2,042	2,616	2,620	2,619	374	**152**
Heart disease	5,475	5,075	4,820	4,825	4,784	1,021	**79**
Lymphography	31,594	5,840	15,740	15,032	11,627	978	**24**
Pima diabetes	466	448	345	350	337	144	**36**
Breast cancer	420	379	122	124	44	158	**10**
Dermatology	5,350	3,727	3,820	3,717	3,544	2,076	**96**
Wine	1,140	1,057	975	971	968	520	**116**
Glass	2,327	1,141	2,318	2,311	2,311	97	**20**
Credit	8,504	3,271	6,885	6,964	6,839	926	**49**
Average	6,443	2,553	4,182	4,102	3,675	699	**65**

intervals by minimizing the entropy based on the minimum description length principle [14] (supervised discretization). Table 3 shows the main characteristics of the datasets. The number of items in column 3 is the number of all distinct attribute value pairs.

Results. Table 4 shows the number of rules for each method on the 9 UCI datasets. First notice that evaluating the rules individually based on their statistical significance (*corr_chi*, *prior_binom* and *prior_FDR*) does not help much in reducing the number of rules (even by using the FDR correction). It is clear from the table that the number of MPRs is much smaller than the number of rules in the other approaches. On average, MPRs are about two orders of magnitude smaller than *complete* and about one order of magnitude smaller than *conf_imp*.

Now we need a way to check if this small set of MPRs can adequately describe the datasets. However, we do not know what are the real patterns for these datasets. Hence, to evaluate the usefulness of the rules, we use the classification accuracy of the corresponding rule based classifier. We define two simple classifiers:

- Weighted confidence classification (*w_conf*): To classify instance x, we weight each rule that satisfy x by its confidence and we choose the class that maximizes this weighted sum:
$$w\_conf(x) = \arg\max_{c_i}\{ \sum_{x \subseteq A} conf(A \Rightarrow c_i) \}$$

– Highest confidence classification (*h_conf*): We classify instance x according to the highest confidence rule that satisfy x (this method is used in [20]):

$$h\_conf(x) = \arg\max_{c_i}\{ \max_{x \subseteq A} conf(A \Rightarrow c_i) \}$$

We compare the classification accuracies obtained by using all association rules, using *conf_imp* rules (this approach was used in [17] for classification), and using MPRs[2]. All of the reported results are obtained using 5-folds cross validation. Remember that the lift score for all rules is bigger than one (the condition and consequent of each rule are positively correlated). Hence, *w_conf* consults only predictive rules.

Table 5. The classification accuracies (%) for *complete*, *conf_imp* and MPR using two rule classification techniques: weighted confidence classification (*w_conf*) and highest confidence classification (*h_conf*)

	complete		*conf_imp*		MPR	
Dataset	w_conf	h_conf	w_conf	h_conf	w_conf	h_conf
Adults	77.3	75.9	80.6	75.9	80.8	75.9
Heart disease	80.9	80.5	80.2	80.5	82.2	81.5
Lymphography	81.2	83.6	71.8	83.6	86.2	85.9
Pima diabetes	71.4	74.4	72.8	74.4	71.8	72.9
Breast cancer	73.8	72.0	74.1	72.0	72.4	73.4
Dermatology	68.0	69.4	58.2	69.4	63.4	65.6
Wine	88.8	93.3	86.4	93.3	88.2	92.8
Glass	94.4	100	93.5	100	95.8	100
Credit	80.4	85.4	76.7	85.4	77.8	85.4
Average	79.6	81.6	77.1	81.6	79.8	81.5

From Table 5, we can see that MPR does not sacrifice the classification accuracy. On average, all approaches produce comparable results for both *w_conf* and *h_conf*. An important benefit of using the compact set of MPRs for classification is that the classification time is very fast. For example, consider the Lymphography dataset. Instead of consulting 31,594 rules to classify each instance, MPR summarizes the classifier in only 24 rules. It is interesting to see that these 24 rules outperform the complete set of rules for both *w_conf* and *h_conf*.

Please note that we are not claiming that our approach can outperform the state-of-the-art frequent pattern-based classifiers [13,10]. Our objective is just to show that even though MPRs provide a huge compression of the association rules, they can still capture the essential underlaying patterns in the data by providing comparable classification performance to using all association rules.

Mining at low support. In this section, we study the performance and the output of the different methods under different support thresholds. Due to the space limitation, we only show the results on the *Heart disease* dataset.

[2] *corr_chi*, *prior_binom* and *prior_FDR* gave similar classification results as *complete*, hence we excluded them from this table to save space.

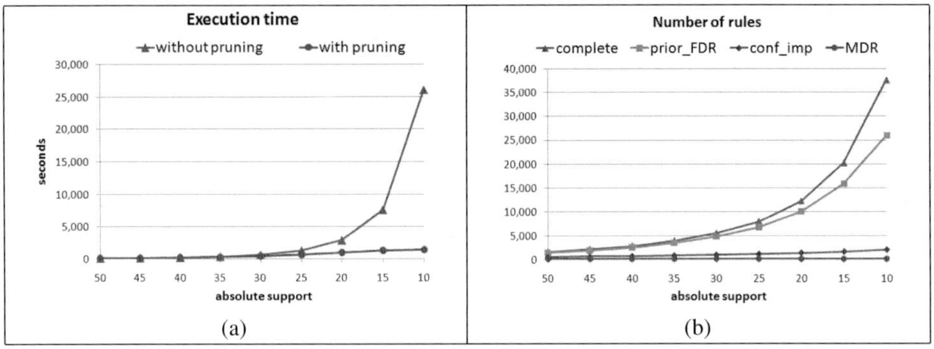

Fig. 7. The execution times (a) and the number of rules (b) of the algorithms on the heart disease dataset for different support thresholds

Figure 7:a shows the execution times using a Dell Precision T7500 machine with an Intel Xeon 3GHz CPU and 16GB of RAM. All algorithms are implemented in matlab. The "without pruning" chart corresponds to the execution of the Apriori algorithm that relies only on the support of the patterns to prune the search space. The "with pruning" chart corresponds to the execution of the algorithm that applies the additional MPR pruning technique in section 2.5.

We can see that the execution time of Apriori exponentially blows up for low support values. On the other hand, the MPR pruning controls the complexity and the execution time increases very slowly for low support values. For example, when the absolute support threshold is 15, which corresponds to min_sup = 5% on this dataset, applying the MPR pruning makes the algorithm about 6 times faster.

Figure 7:b shows the number of rules generated by the different methods. To improve the visibility, we did not include *closed*, *prior_binom* and *corr_chi* in this graph. We can see that the output of MPR does not change much when min_sup is very low. For example, by changing min_sup from 10% (absolute support = 30) to 5% (absolute support=15), the number of MPRs increases from 79 to 83 rules. In comparison, the same change causes the number of all association rules to increase from 5,475 to about 20,000 rules! Clearly, this large number of rules is overwhelming the user.

To summarize, our framework relieves the user from the burden of deciding the optimal min_sup by allowing him to conservatively set the support very low without drastically affecting the performance or the results of the algorithm.

4 Related Research

Several research attempts have been made to reduce the large number of association rules in order to make the results more suitable for knowledge discovery. Constrained associations rules methods [22] allow the user to define his own constraints and retrieve the rules that match these constraints. An opposite approach [23] mines the rules that are most different from the user's expectations. Maximal frequent itemsets [19] is a lossy compression of the frequent itemsets and cannot be used to generate rules. The profile based approach [27] is another lossy compression method.

The work in [6] aimed to reduce the number of class association rules by defining the confidence improvement constraint. This constraint was adopted by [18,17]. As we showed in the analysis and experiments, this approach can still generate many redundant rules. [21] defined the concept of direction setting (DS) rules in order to make browsing the class association rules easier for the user. However, their objective is different from ours because non-DS rules can indeed be significant MPRs. [5] extends the problem of association rule mining to the problem of mining contrasting sets. [26] defines a measure to rank the patterns by predicting the support of a pattern from the support of its subpatterns and measuring the deviation between the actual support and the prediction.

5 Conclusion

In this paper, we have developed a new framework for mining association rules based on the minimal predictive rules (MPR) concept. We showed that our method can produce a small set of predictive rules. Most importantly, each rule in the result is important because it concisely describes a distinct pattern that cannot be explained by any other rule in the set.

Motivated by our results, we plan to investigate the benefits of MPR for classification. In particular, we plan on incorporating the MPRs as additional features with the SVM classifier and comparing it against the state-of-the-art classifiers [13,10]. In addition, we plan to apply our method for anomaly detection in categorical data [12].

Acknowledgment. This research work was supported by grants 1R21LM009102-01A1, 1R01LM010019-01A1, and 1R01GM088224-01 from the NIH. Its content is solely the responsibility of the authors and does not necessarily represent the official views of the NIH.

References

1. Agrawal, R., Imielinski, T., Swami, A.: Mining association rules between sets of items in large databases. In: Proceedings of SIGMOD, pp. 207–216 (1993)
2. Asuncion, A., Newman, D.J.: UCI machine learning repository (2007)
3. Bastide, Y., Pasquier, N., Taouil, R., Stumme, G., Lakhal, L.: Mining minimal non-redundant association rules using frequent closed itemsets. In: Palamidessi, C., Moniz Pereira, L., Lloyd, J.W., Dahl, V., Furbach, U., Kerber, M., Lau, K.-K., Sagiv, Y., Stuckey, P.J. (eds.) CL 2000. LNCS (LNAI), vol. 1861, p. 972. Springer, Heidelberg (2000)
4. Batal, I., Sacchi, L., Bellazzi, R., Hauskrecht, M.: Multivariate time series classification with temporal abstractions. In: FLAIRS (2009)
5. Bay, S., Pazzani, M.: Detecting group differences: Mining contrast sets. Data Mining and Knowledge Discovery 5(3), 213–246 (2001)
6. Bayardo, R.J., Agrawal, R., Gunopulos, D.: Constraint-based rule mining in large, dense databases. In: Proceedings of ICDE, pp. 188–197 (1999)
7. Benjamini, Y., Hochberg, Y.: Controlling the false discovery rate: A practical and powerful approach to multiple testing. Journal of the Royal Statistical Society. Series B (Methodological) 57(1), 289–300 (1995)

8. Brin, S., Motwani, R., Silverstein, C.: Beyond market baskets: Generalizing association rules to correlations. In: Proceedings of SIGMOD (1997)
9. Castelo, R., Feelders, A.J., Siebes, A.: Mambo: Discovering association rules based on conditional independencies. In: Hoffmann, F., Adams, N., Fisher, D., Guimarães, G., Hand, D.J. (eds.) IDA 2001. LNCS, vol. 2189, p. 289. Springer, Heidelberg (2001)
10. Cheng, H., Yan, X., Han, J., Hsu, C.: Discriminative frequent pattern analysis for effective classification. In: Proceedings of ICDE (2007)
11. Cohen, E., Datar, M., Fujiwara, S., Gionis, A., Indyk, P., Motwani, R., Ullman, J.D., Yang, C.: Finding interesting associations without support pruning. In: Proceedings of ICDE (2000)
12. Das, K., Schneider, J., Neill, D.: Anomaly pattern detection in categorical datasets. In: Proceedings of SIGKDD (2008)
13. Fan, W., Zhang, K., Cheng, H., Gao, J., Yan, X., Han, J., Yu, P., Verscheure, O.: Direct mining of discriminative and essential frequent patterns via model-based search tree. In: Proceedings of SIGKDD (2008)
14. Fayyad, U., Irani, K.: Multi-interval discretization of continuous-valued attributes for classification learning. In: Proceedings of IJCAI (1993)
15. Geng, L., Hamilton, H.: Interestingness measures for data mining: A survey. ACM Comput. Surv. 38(3) (2006)
16. Kuramochi, M., Karypis, G.: Frequent subgraph discovery. In: Proceedings of ICDM (2001)
17. Li, J., Shen, H., Topor, R.: Mining optimal class association rule set. In: Cheung, D., Williams, G.J., Li, Q. (eds.) PAKDD 2001. LNCS (LNAI), vol. 2035, p. 364. Springer, Heidelberg (2001)
18. Li, W., Han, J., Pei, J.: CMAR: Accurate and efficient classification based on multiple class-association rules. In: Proceedings of ICDM (2001)
19. Lin, D., Kedem, Z.: Pincer-search: A new algorithm for discovering the maximum frequent set. In: Proceedings of EDBT, pp. 105–119 (1997)
20. Liu, B., Hsu, W., Ma, Y.: Integrating classification and association rule mining. In: Knowledge Discovery and Data Mining, pp. 80–86 (1998)
21. Liu, B., Hsu, W., Ma, Y.: Pruning and summarizing the discovered associations. In: Proceedings of SIGKDD (1999)
22. Ng, R., Lakshmanan, L., Han, J., Pang, A.: Exploratory mining and pruning optimizations of constrained associations rules. In: Proceedings of SIGMOD (1998)
23. Padmanabhan, B., Tuzhilin, A.: A belief-driven method for discovering unexpected patterns. In: Proceedings of SIGKDD (1998)
24. Piatetsky-Shapiro, G.: AAAI 1991 Workshop on Knowledge Discovery in Databases (1991)
25. Shaffer, J.P.: Multiple hypothesis testing: A review. Annual Review of Psychology (1995)
26. Tatti, N.: Maximum entropy based significance of itemsets. Knowledge Information System 17(1), 57–77 (2008)
27. Yan, X., Cheng, H., Han, J., Xin, D.: Summarizing itemset patterns: a profile-based approach. In: Proceedings of SIGKDD (2005)
28. Zaki, M.J.: Spade: an efficient algorithm for mining frequent sequences. Machine Learning Journal, 31–60 (2001)

Euclidean Distances, Soft and Spectral Clustering on Weighted Graphs

François Bavaud

University of Lausanne, Department of Geography
Department of Computer Science and Mathematical Methods
Bâtiment Anthropole, CH-1015 Lausanne, Switzerland

Abstract. We define a class of Euclidean distances on weighted graphs, enabling to perform thermodynamic soft graph clustering. The class can be constructed form the "raw coordinates" encountered in spectral clustering, and can be extended by means of higher-dimensional embeddings (Schoenberg transformations). Geographical flow data, properly conditioned, illustrate the procedure as well as visualization aspects.

Keywords: average commute time distance, metastability, migratory flow, multidimensional scaling, Schoenberg transformations, shortest-path distance, spectral clustering, thermodynamic clustering, quasi-symmetry.

1 Introduction

In a nutshell (see e.g. Shi and Malik (2000); Ng, Jordan and Weiss (2002); von Luxburg (2007) for a review), spectral graph clustering consists in

A) constructing a features-based similarity or affinity matrix between n objects
B) performing the spectral decomposition of the normalized affinity matrix, and representing the objects by the corresponding eigenvectors or *raw coordinates*
C) applying a clustering algorithm on the raw coordinates.

The present contribution focuses on (C) thermodynamic clustering (Rose et al. 1990; Bavaud 2009), an aggregation-invariant soft K-means clustering based upon *Euclidean distances between objects*. The latter constitute *distances on weighted graphs*, and are constructed from the raw coordinates (B), whose form happens to be justified from presumably new considerations on equivalence between vertices (Section 3.3). Geographical *flow data* illustrate the theory (Section 4). Once properly symmetrized, endowed with a sensible diagonal and normalized, flows define an *exchange matrix* (Section 2), that is an affinity matrix (A) which might be positive definite or not.

A particular emphasis is devoted to the definition of Euclidean distances on weighted graphs and their properties (Section 3). For instance, diffusive and chi-square distances are *focused*, that is zero between *equivalent* vertices. Commute-time and absorption distances are not focused, but their values between equivalent vertices possess an universal character. All these distances,

J.L. Balcázar et al. (Eds.): ECML PKDD 2010, Part I, LNAI 6321, pp. 103–118, 2010.

whose relationships to the shortest-path distance on weighted graphs is partly elucidated, differ in the way eigenvalues are used to scale the raw coordinates. Allowing further *Schoenberg transformations* (Definition 3) of the distances still extends the class of admissible distances on graphs, by means of a high-dimensional embedding familiar in the Machine Learning community.

2 Preliminaries and Notations

Consider n objects, together with an *exchange* matrix $E = (e_{ij})$, that is a $n \times n$ non-negative, symmetric matrix, whose components add up to unity (Berger and Snell 1957). E can be obtained by normalizing an affinity of similarity matrix, and defines the normalized adjacency matrix of a weighted undirected graph (containing loops in general), where e_{ij} is the weight of edge (ij) and $f_i = \sum_{j=1}^n e_{ij}$ is the *relative degree* or *weight* of vertex i, assumed strictly positive.

2.1 Eigenstructure

$P = (p_{ij})$ with $p_{ij} = e_{ij}/f_i$ is the transition matrix of a reversible Markov chain, with stationary distribution f. The t-step exchange matrix is $E^{(t)} = \Pi P^t$, where Π is the diagonal matrix containing the weights f. In particular, assuming the chain to be regular (see e.g. Kijima 1997)

$$E^{(0)} = \Pi \qquad E^{(2)} = E\Pi^{-1}E \ . \qquad E^{(\infty)} = ff'$$

P is similar to the symmetric, *normalized exchange matrix* $\Pi^{-\frac{1}{2}}E\Pi^{-\frac{1}{2}}$ (see e.g. Chung 1997), and share the same eigenvalues $1 = \lambda_0 \geq \lambda_1 \geq \lambda_2 \geq \ldots \lambda_{n-1} \geq -1$. It is well-known that the second eigenvalue λ_1 attains its maximum value 1 iff the graph contains disconnected components, and $\lambda_{n-1} = -1$ iff the graph is bipartite. We note $U'\Lambda U$ the spectral decomposition of the normalized exchange matrix, where Λ is diagonal and contains the eigenvalues, and $U = (u_{i\alpha})$ is orthonormal and contains the normalized eigenvectors. In particular, $u_0 = \sqrt{f}$ is the eigenvector corresponding to the trivial eigenvalue $\lambda_0 = 1$. Also, the spectral decomposition of higher-order exchange matrices reads $\Pi^{-\frac{1}{2}}E^{(t)}\Pi^{-\frac{1}{2}} = U\Lambda^t U'$.

2.2 Hard and Soft Partitioning

A *soft partition* of the n objects into m groups is specified by a $n \times m$ *membership matrix* $Z = (z_{ig})$, whose components (obeying $z_{ig} \geq 0$ and $\sum_{g=1}^m z_{ig} = 1$) quantify the membership degree of object i in group g. The relative *volume* of group g is $\rho_g = \sum_i f_i z_{ig}$. The components $\theta_{gh} = \sum_i f_i z_{ig} z_{ih}$ of the $m \times m$ matrix $\Theta = Z'\Pi Z$ measure the *overlap* between groups g and h. In particular, $\theta_{gg}/\rho_g \leq 1$ measures the *hardness* of group g. The components $a_{gh} = \sum_{ij} e_{ij} z_{ig} z_{jh}$ of the $m \times m$ matrix $A = Z'EZ$ measure the *association* between groups g and h.

A group g can also be specified by the objects it contains, namely by the *distribution* π^g with components $\pi_i^g = f_i z_{ig}/\rho_g$, obeying $\sum_i \pi_i^g = 1$ by construction. The object-group *mutual information*

$$I(O,Z) = H(O) + H(Z) - H(O,Z) = -\sum_i f_i \ln f_i - \sum_g \rho_g \ln \rho_g + \sum_{ig} f_i z_{ig} \ln(f_i z_{ig})$$

measures the object-group dependence or cohesiveness (Cover and Thomas 1991).

A partition is *hard* if each object belongs to an unique group, that is if the memberships are of the form $z_{ig} = I(i \in g)$, or equivalently if $z_{ig}^2 = z_{ig}$ for all i, g, or equivalently if $\theta_{gg} = \rho_g$ for all g, or still equivalently if the *overall softness* $H(Z|O) = H(Z) - I(O, Z)$ takes on its minimum value of zero.

Also, $H(O) \leq \ln n$, with equality iff $f_i = 1/n$, that is if the graph is regular.

2.3 Spectral versus Soft Membership Relaxation

In their presentation of the Ncut-driven spectral clustering, Yu and Shi (2003) (see also Nock et al. 2009) determine the hard $n \times m$ membership Z maximizing

$$\epsilon[Z] = \sum_{g=1}^{m} \frac{a_{gg}}{\rho_g} = \sum_{g} \frac{a_{gg}}{\theta_{gg}} = \mathrm{tr}(X'EX) \qquad \text{where} \quad X[Z] = Z\,\Theta^{-\frac{1}{2}}[Z]$$

under the constraint $X'\Pi X = I$. Relaxing the hardness and non-negativity conditions, they show the solution to be $\epsilon[Z_0] = 1 + \sum_{\alpha=1}^{m-1} \lambda_\alpha$, attained with an optimal "membership" of the form $Z_0 = X_0 R \Theta^{\frac{1}{2}}$ where R is any orthonormal $m \times m$ matrix and $X_0 = (\mathbf{1}, x_1, \ldots, x_\alpha, \ldots, x_{m-1})$ is the $n \times m$ matrix formed by the unit vector followed by of the first *raw coordinates* (Sec. 3.3). The above spectral relaxation of the memberships, involving the eigenstructure of the normalized exchange matrix, completely differs from the soft membership relaxation which will be used in Section 3.2, preserving positivity and normalization of Z.

3 Euclidean Distances on Weighted Graphs

3.1 Squared Euclidean Distances

Consider a collection of n objects together with an associated pairwise distance. A successful clustering consists in partitioning the objects into m groups, such that the average distances between objects belonging to the same (different) group are small (large). The most tractable pairwise distance is, by all means, the *squared Euclidean distance* $D_{ij} = \sum_{c=1}^{q}(x_{ic} - x_{jc})^2$, where x_{ic} is the coordinate of object i in dimension c. Its virtues follow from *Huygens principles*

$$\sum_j p_j D_{ij} = D_{ip} + \Delta_p \qquad \Delta_p = \sum_j p_j D_{jp} = \frac{1}{2} \sum_{ij} p_i p_j D_{ij} \tag{1}$$

where p_i represents a (possibly non positive) *signed distribution*, i.e. obeying $\sum_i p_i = 1$, D_{ip} is the squared Euclidean distance between i and the centroid of coordinates $\bar{x}_{pc} = \sum_i p_i x_{ic}$, and Δ_p the average pairwise distance or *inertia*. Equations (1) are easily checked using the coordinates, although the latter do *not* explicitly appear in the formulas. To that extent, squared Euclidean distances enable a feature-free formalism, a property shared with the kernels of Machine Learning, and to the "kernel trick" of Machine Learning amounts an equivalent "distance trick" (Schölkopf 2000; Williams 2002), as expressed by the well-known

Classical Multidimensional Scaling (MDS) procedure. Theorem 1 below presents a weighted version (Bavaud 2006), generalizing the uniform MDS procedure (see e.g. Mardia et al. 1979). Historically, MDS has been developed from the independent contributions of Schoenberg (1938b) and Young and Householder (1938). The algorithm has been popularized by Torgeson (1958) in Data Analysis.

Theorem 1 (weighted classical MDS). *The dissimilarity square matrix D between n objects with weights p is a squared Euclidean distance iff the* scalar product matrix $B = -\frac{1}{2}HDH'$ *is (weakly) positive definite (p.d.), where H is the $n \times n$ centering matrix with components $h_{ij} = \delta_{ij} - p_j$. By construction, $B_{ij} = -\frac{1}{2}(D_{ij} - D_{ip} - D_{jp})$ and $D_{ij} = B_{ii} + B_{jj} - 2B_{ij}$. The object coordinates can be reconstructed as $x_{i\beta} = \mu_\beta^{\frac{1}{2}} p_i^{-\frac{1}{2}} v_{i\beta}$ for $\beta = 1, 2, \ldots$, where the μ_β are the decreasing eigenvalues and the $v_{i\beta}$ are the eigenvectors occurring in the spectral decomposition $K = VMV'$ of the* weighted scalar product *or* kernel *K with components $K_{ij} = \sqrt{p_i p_j} B_{ij}$. This reconstruction provides the optimal low-dimensional reconstruction of the inertia associated to p*

$$\Delta = \frac{1}{2} \sum_{ij} p_i p_j D_{ij} = tr(K) = \sum_{\beta \geq 1} \mu_\beta \ .$$

Also, the Euclidean (or not) character of D is independent of the choice of p.

3.2 Thermodynamic Clustering

Consider the overall objects weight f, defining a centroid denoted by 0, together with m soft groups defined by their distributions π^g for $g = 1, \ldots, m$, with associated centroids denoted by g. By (1), the overall inertia decomposes as

$$\Delta = \sum_i f_i D_{i0} = \sum_{ig} f_i z_{ig} D_{i0} = \sum_g \rho_g \sum_i \pi_i^g D_{i0} = \sum_g \rho_g [D_{g0} + \Delta_g] = \Delta_B + \Delta_W$$

where $\Delta_B[Z] = \sum_g \rho_g D_{g0}$ is the between-groups inertia, and $\Delta_W[Z] = \sum_g \rho_g \Delta_g$ the within-groups inertia. The optimal clustering is then provided by the $n \times m$ membership matrix Z minimizing $\Delta_W[Z]$, or equivalently maximizing $\Delta_B[Z]$. The former functional can be shown to be *concave* in Z (Bavaud 2009), implying the minimum to be attained for *hard* clusterings.

Hard clustering is notoriously computationally intractable and some kind of regularization is required. Many authors (see e.g. Huang and Ng (1999) or Filippone et al. (2008)) advocate the use of the *c-means clustering*, involving a power transform of the memberships. Despite its efficiency and popularity, the c-means algorithm actually suffers from a serious formal defect, questioning its very logical foundations: its objective function is indeed *not aggregation-invariant*, that is generally changes when two groups g and h supposed equivalent in the sense $\pi^g = \pi^h$ are merged into a single group $[g \cup h]$ with membership $z_{i[h \cup g]} = z_{ih} + z_{jh}$ (Bavaud 2009).

An alternative, aggregation-invariant regularization is provided by the *thermodynamic clustering*, minimizing over Z the *free energy* $F[Z] = \Delta_W[Z] + TI[Z]$, where $I[Z] \equiv I(O,Z)$ is the objects-groups mutual information and $T > 0$ the *temperature* (Rose et al. 1990; Rose 1998; Bavaud 2009). The resulting membership is determined iteratively through

$$z_{ig} = \frac{\rho_g \exp(-D_{ig}/T)}{\sum_{h=1}^{m} \rho_h \exp(-D_{ih}/T)} \tag{2}$$

and converges towards a local minimum of the free energy. Equation (2) amounts to fitting Gaussian clusters in the framework of *model-based clustering*.

3.3 Three Nested Classes of Squared Euclidean Distances

Equation (2) solves the K-way soft graph clustering problem, given of course the availability of a sound class of squared Euclidean distances on weighted graphs. Definitions 2 and 3 below seem to solve the latter issue.

Consider a graph possessing two distinct but equivalent vertices in the sense their relative exchange is identical with the other vertices (including themselves). Those vertices somehow stand as duplicates of the same object, and one could as a first attempt require their distance to be zero.

Definition 1 (Equivalent vertices; focused distances). *Two distinct vertices i and j are* equivalent, *noted $i \sim j$, if $e_{ik}/f_i = e_{jk}/f_j$ for all k. A distance is* focused *if $D_{ij} = 0$ for $i \sim j$.*

Proposition 1. *$i \sim j$ iff $x_{i\alpha} = x_{j\alpha}$ for all $\alpha \geq 1$ such that $\lambda_\alpha \neq 0$, where $x_{i\alpha} = u_{i\alpha}/\sqrt{f_i}$ is the* raw coordinate *of vertex i in dimension α.*

The proof directly follows from the substitution $e_{ik} \rightarrow f_i e_{jk}/f_j$ in the identity $\sum_k f_i^{-\frac{1}{2}} e_{ik} f_k^{-\frac{1}{2}} u_{k\alpha} = \lambda_\alpha u_{i\alpha}$. Note that the condition trivially holds for the trivial eigenvector $\alpha = 0$, in view of $f_i^{-\frac{1}{2}} u_{i0} \equiv 1$ for all i. It also holds trivially for the "completely connected" weighted graph $e_{ij}^{(\infty)} = f_i f_j$, where all vertices are equivalent, and all eigenvalues are zero, except the trivial one.

Hence, any expression of the form $D_{ij} = \sum_{\alpha \geq 1} g_\alpha (f_i^{-\frac{1}{2}} u_{i\alpha} - f_j^{-\frac{1}{2}} u_{j\alpha})^2$ with $g_\alpha \geq 0$ constitutes an admissible squared Euclidean distance, obeying $D_{ij} = 0$ for $i \sim j$, *provided* $g_\alpha = 0$ *if* $\lambda_\alpha = 0$. The quantities g_α are non-negative, but otherwise arbitrary; however, it is natural to require the latter to depend upon the sole parameters at disposal, namely the eigenvalues, that is to set $g_\alpha = g(\lambda_\alpha)$.

Definition 2 (Focused and Natural Distances on Weighted Graphs). *Let E be the exchange matrix associated to a weighted graph, and define $E^s := \Pi^{-\frac{1}{2}}(E - E^{(\infty)})\Pi^{-\frac{1}{2}}$, the* standardized *exchange matrix. The class of* focused squared Euclidean distances on weighted graphs *is*

$$D_{ij} = B_{ii} + B_{jj} - 2B_{ij}, \quad \text{where} \quad B = \Pi^{-\frac{1}{2}} K \Pi^{-\frac{1}{2}} \quad \text{and} \quad K = g(E^s)$$

where $g(\lambda)$ is any non-negative sufficiently regular real function with $g(0) = 0$. Dropping the requirement $g(0) = 0$ defines the more general class of natural *squared Euclidean distances on weighted graphs.*

If $g(1)$ is finite, K can also be defined as $K = g(\Pi^{-\frac{1}{2}} E \Pi^{-\frac{1}{2}}) = U g(\Lambda) U'$.

First, note the standardized exchange matrix to result from a "centering" (eliminating the trivial eigendimension) followed by a "normalization":

$$e^s_{ij} = \frac{e_{ij} - f_i f_j}{\sqrt{f_i f_j}} = \sum_{\alpha \geq 1} \lambda_\alpha u_{i\alpha} u_{j\alpha} \ . \tag{3}$$

Secondly, B is the matrix of scalar products appearing in Theorem 1. The resulting optimal reconstruction coordinates are $\sqrt{g(\lambda_\alpha)}\, x_{i\alpha}$, where the quantities $x_{i\alpha} = f_i^{-\frac{1}{2}} u_{i\alpha}$ are the *raw coordinates* of vertex i in dimension $\alpha = 1, 2, \ldots$ appearing in Proposition 1 - which yields a general rationale for their widespread use in clustering and low-dimensional visualization. Thirdly, the matrix $g(E^s)$ can be defined, for $g(\lambda)$ regular enough, as the power expansion in $(E^s)^t$ with coefficients given by the power expansion of $g(\lambda)$ in λ^t, for $t = 0, 1, 2, \ldots$. Finally, the two variants of B appearing in Definition 2 are identical up to a matrix $g(1)\mathbf{1}_n\mathbf{1}'_n$, leaving D unchanged.

If $g(1) = \infty$, the distance between vertices belonging to distinct irreducible components becomes infinite: recall the graph to be disconnected iff $\lambda_1 = 1$. Such distances will be referred to as *irreducible.*

Natural distances are in general not focused. The distances between equivalent vertices are however *universal*, that is independent of the details of the graph or of the associated distance (Proposition 2). To demonstrate this property, consider first an equivalence class $J := \{k \mid k \sim j\}$ containing at least two equivalent vertices. Aggregating the vertices in J results in a new $\tilde{n} \times \tilde{n}$ exchange matrix $\tilde{E}$ with $\tilde{n} = (n - |J| - 1)$, with components $\tilde{e}_{JJ} = \sum_{ij \in J} e_{ij}$, $\tilde{e}_{Jk} = \tilde{e}_{kJ} = \sum_{j \in J} e_{jk}$ for $k \notin J$ and $\tilde{f}_J = \sum_{j \in J} f_j$, the other components remaining unchanged.

Proposition 2. *Let D be a natural distance and consider a graph possessing an equivalence class J of size $|J| \geq 2$. Consider two distinct elements $i \sim j$ of J and let $k \notin J$. Then*

$$D_{ij} = g(0)\,(\frac{1}{f_i} + \frac{1}{f_j}) \qquad D_{jJ} = g(0)\,(\frac{1}{f_i} - \frac{1}{\tilde{f}_J}) \qquad \Delta_J = g(0)\,\frac{|J| - 1}{\tilde{f}_J} \ .$$

Moreover, the Pythagorean relation $D_{kj} = D_{kJ} + D_{jJ}$ holds.

Proof: consider the eigenvalues $\tilde{\lambda}_\beta$ and eigenvectors $\tilde{u}_\beta$, associated to the aggregated graph $\tilde{E}$, for $\beta = 0, \ldots, \tilde{n}$. One can check that, due to the collinearity generated by the $|J|$ equivalent vertices,

- $\tilde{n}$ among the original eigenvalues λ_α coincide with the set of aggregated eigenvalues $\tilde{\lambda}_\beta$ (non null in general), with corresponding eigenvectors $u_{j\beta} = f_j^{\frac{1}{2}} \tilde{f}_J^{-\frac{1}{2}} \tilde{u}_{J\beta}$ for $j \in J$ and $u_{k\beta} = \tilde{u}_{k\beta}$ for $k \notin J$
- $|J - 1|$ among the original eigenvalues λ_α are zero. Their corresponding eigenvectors are of the form $u_{j\gamma} = h_{j\gamma}$ for $j \in J$ and $u_{k\gamma} = 0$ for $k \notin J$, where the h_γ constitute the $|J| - 1$ columns of an orthogonal $|J| \times |J|$ matrix, the remaining column being $(f_j^{\frac{1}{2}} \tilde{f}_J^{-\frac{1}{2}})_{j \in J}$.

Identities in Proposition 2 follow by substitution. For instance,

$$D_{ij} = \sum_{\beta=1}^{\tilde{n}} g(\lambda_\beta)(\frac{u_{i\beta}}{\sqrt{f_i}} - \frac{u_{j\beta}}{\sqrt{f_j}})^2 + g(0) \sum_{\gamma=1}^{|J|-1} (\frac{h_{i\gamma}}{\sqrt{f_i}} - \frac{h_{j\gamma}}{\sqrt{f_j}})^2 = g(0)\,(\frac{1}{f_i} + \frac{1}{f_j}) \ .$$

General at it is, the class of squared Euclidean distances on weighted graphs of Definition 2 can still be extended: a wonderful result of Schoenberg (1938a), still apparently little known in the Statistical and Machine Learning community (see however the references in Kondor and Lafferty (2002); Hein et al. (2005)) asserts that the componentwise correspondence $\tilde{D}_{ij} = \phi(D_{ij})$ transforms any squared Euclidean distance D into another squared Euclidean distance $\tilde{D}$, provided that

i) $\phi(D)$ is positive with $\phi(0) = 0$
ii) odd derivatives $\phi'(D)$, $\phi'''(D)$,... are positive
iii) even derivatives $\phi''(D)$, $\phi''''(D)$,... are negative.

For example, $\phi(D) = D^a$ (for $0 < a \leq 1$) and $\phi(D) = 1 - \exp(-bD)$ (for $b > 0$) are instances of such *Schoenberg transformations* (Bavaud 2010).

Definition 3 (Extended Distances on Weighted Graphs). *The class of* extended squared Euclidean distances on weighted graphs *is*

$$\tilde{D}_{ij} = \phi(D_{ij})$$

where $\phi(D)$ is a Schoenberg transformation *(as specified above), and D_{ij} is a natural squared Euclidean distance associated to the weighted graph E, in the sense of Definition 2.*

3.4 Examples of Distances on Weighted Graphs

The chi-square distance. The choice $g(\lambda) = \lambda^2$ entails, together with (3)

$$\Delta = \mathrm{tr}(K) = \mathrm{tr}((E^s)^2) = \sum_{ij} \frac{(e_{ij} - f_i f_j)^2}{f_i f_j} = \chi^2$$

which is the familiar chi-square measure of the overall rows-columns dependency in a (square) contingency table, with distance $D^\chi_{ij} = \sum_k f_k^{-1}(f_i^{-1}e_{ik} - f_j^{-1}e_{jk})^2$, well-known in the *Correspondence Analysis* community (Lafon and Lee 2006; Greenacre 2007 and references therein). Note that $D^\chi_{ij} = 0$ for $i \sim j$, as it must.

The diffusive distance. The choice $g(\lambda) = \lambda$ is legitimate, provided the exchange matrix is purely *diffusive*, that is p.d. Such are typically the graphs resulting from inter-regional migrations (Sec. 4) or social mobility tables (Bavaud 2008). As most people do not change place or status during the observation time, the exchange matrix is strongly dominated by its diagonal, and hence p.d.

Positive definiteness also occurs for graphs defined from the affinity matrix $\exp(-\beta D_{ij})$ (Gaussian kernel), as in Belkin and Niyogi (2003), among many others. Indeed, distances derived from the Gaussian kernel provide a prototypical example of Schoenberg transformation (see Definition 3). By contrast, the affinity $I(D_{ij} \leq \epsilon^2)$ used by Tenenbaum et al. (2000) is not p.d.

The corresponding distance, together with the inertia, plainly read

$$D^{dif}_{ij} = \frac{e_{ii}}{f_i^2} + \frac{e_{jj}}{f_j^2} - 2\frac{e_{ij}}{f_i f_j} \qquad \Delta^{dif} = \sum_i \frac{e_{ii}}{f_i} - 1 \ .$$

The "frozen" distance. The choice $g(\lambda) \equiv 1$ produces, for any graph, a result identical to the application of any function $g(\lambda)$ (with $g(1) = 1$) to the purely diagonal "frozen" graph $E^{(0)} = \Pi$, namely (compare with Proposition 2):

$$D^{fro}_{ij} = \frac{1}{f_i} + \frac{1}{f_j} \quad \text{for } i \neq j \qquad D^{fro}_{i0} = \frac{1}{f_i} - 1 \qquad \Delta^{fro} = n - 1 \ .$$

This "star-like" distance (Critchley and Fichet 1994) is embeddable in a tree.

The average commute time distance. The choice $g(\lambda) = (1-\lambda)^{-1}$ corresponds to the *average commute time distance*; see Fouss et al. (2007) for a review and recent results. The amazing fact that the latter constitutes a squared Euclidean distance has only be recently explicitly recognized as such, although the key ingredients were at disposal ever since the seventies.

Let us sketch a derivation of this result: on one hand, consider a random walk on the graph with probability transition matrix $P = \Pi^{-1}E$, and let T_j denotes the first time the chain hits state j. The average time to go from i to j is $m_{ij} = E_i(T_j)$, with $m_{ii} = 0$, where $E_i(.)$ denotes the expectation for a random walk started in i. Considering the state following i yields for $i \neq j$ the relation $m_{ij} = 1 + \sum_k p_{ik} m_{kj}$, with solution (Kemeny and Snell (1976); Aldous and Fill, draft chapters) $m_{ij} = (y_{jj} - y_{ij})/f_j$, where $Y = \Pi^{-1}\sum_{t\geq 0}(E^{(t)} - E^{(\infty)}) = (E^{(0)} - E + E^{(\infty)})^{-1}\Pi$ is the so-called *fundamental matrix* of the Markov chain. On the other hand, Definition 2 yields $K = (I - E^s)^{-1} = \Pi^{\frac{1}{2}}(E^{(0)} - E + E^{(\infty)})^{-1}\Pi^{\frac{1}{2}} = \Pi^{\frac{1}{2}} Y \Pi^{-\frac{1}{2}}$, and thus $B = Y\Pi^{-1} = \Pi^{-1}Y$. Hence

$$D^{com}_{ij} = B_{ii} + B_{jj} - 2B_{ij} = \frac{y_{ii}}{f_i} + \frac{y_{jj}}{f_j} - \frac{y_{ij}}{f_j} - \frac{y_{ij}}{f_j} = m_{ij} + m_{ji}$$

which is the average time to go from i to j and back to i, as announced.

Consider, for future use, the *Dirichlet form* $\mathcal{E}(y) = \frac{1}{2}\sum_{ij} e_{ij}(y_i - y_j)^2$, and denote by y^0 the solution of the "electrical" problem $\min_{y \in C_{ij}} \mathcal{E}(y)$, where C_{ij} denotes the set of vectors y such that $y_i = 1$ and $y_j = 0$. Then $y^0_k = P_k(T_i < T_j)$, where $P_k(.)$ denotes the probability for a random walk started at k. Then $D^{com}_{ij} = 1/\mathcal{E}(y^0)$ (Aldous and Fill, chapter 3).

The shortest-path distance. Let Γ_{ij} be the set of paths with extremities i and j, where a path $\gamma \in \Gamma_{ij}$ consists of a succession of consecutive unrepeated edges denoted by $\alpha = (k, l) \in \gamma$, whose weights e_α represent *conductances*.

Their inverses are *resistances*, whose sum is to be minimized by the *shortest path* $\gamma^0 \in \Gamma_{ij}$ (not necessarily unique) on the weighted graph E. This setup generalizes the unweighted graphs framework, and defines the *shortest path distance*

$$D^{sp}_{ij} = \min_{\gamma \in \Gamma_{ij}} \sum_{\alpha \in \gamma} \frac{1}{e_\alpha} \ .$$

We believe the following result to be new - although its proof simply combines a classical result published in the fifties (Beurling and Deny 1958) with the above "electrical" characterization of the average commute time distance.

Proposition 3. $D^{sp}_{ij} \geq D^{com}_{ij}$ *with equality for all* i, j *iff* E *is a weighted tree.*

Proof: let $\gamma^0 \in \Gamma_{ij}$ be the shortest-path between i and j. Consider a vector y and define $dy_\alpha = y_l - y_k$ for an edge $\alpha = (k, l)$. Then

$$|y_i - y_j| \overset{(a)}{\leq} \sum_{\alpha \in \gamma^0} |dy_\alpha| = \sum_{\alpha \in \gamma^0} \sqrt{e_\alpha} \frac{|dy_\alpha|}{\sqrt{e_\alpha}} \overset{(b)}{\leq} (\sum_{\alpha \in \gamma^0} e_\alpha (dy_\alpha)^2)^{\frac{1}{2}} (\sum_{\alpha \in \gamma^0} \frac{1}{e_\alpha})^{\frac{1}{2}} \overset{(c)}{\leq} \sqrt{\mathcal{E}(y)} \sqrt{D^{sp}_{ij}}$$

Hence $D^{sp}_{ij} \geq (y_i - y_j)^2 / \mathcal{E}(y)$ for all y, in particular for y^0 defined above, showing $D^{sp}_{ij} \geq D^{com}_{ij}$. Equality holds iff (a) y^0 is monotonously decreasing along the path γ^0, (b) for all $\alpha \in \gamma^0$, $dy^0_\alpha = c/e_\alpha$ for some constant c, and (c) $dy^0_\alpha e_\alpha = 0$ for all $\alpha \notin \gamma^0$. (b), expressing Ohm's law $U = RI$ in the electrical analogy, holds for y^0, and (a) and (c) hold for a *tree*, that is a graph possessing no closed path.

The shortest-path distance is unfocused and irreducible. Seeking to determine the corresponding function $g(\lambda)$ involved in Definition 2, and/or the Schoenberg transformation $\phi(D)$ involved in Definition 3, is however hopeless:

Proposition 4. D^{sp} *is* not *a squared Euclidean distance.*

Proof: a counter-example is provided (Deza and Laurent (1997) p. 83) by the complete bipartite graph $K_{2,3}$ of Figure 1:

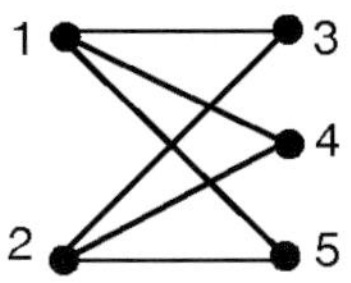

$$E = \frac{1}{12} \begin{pmatrix} 0 & 0 & 1 & 1 & 1 \\ 0 & 0 & 1 & 1 & 1 \\ 1 & 1 & 0 & 0 & 0 \\ 1 & 1 & 0 & 0 & 0 \\ 1 & 1 & 0 & 0 & 0 \end{pmatrix} \qquad D^{sp} = 12 \begin{pmatrix} 0 & 2 & 1 & 1 & 1 \\ 2 & 0 & 1 & 1 & 1 \\ 1 & 1 & 0 & 2 & 2 \\ 1 & 1 & 2 & 0 & 2 \\ 1 & 1 & 2 & 2 & 0 \end{pmatrix}$$

Fig. 1. Bipartite graph $K_{2,3}$, associated exchange matrix and shortest-path distance

The eigenvalues occurring in Theorem 1 are $\mu_1 = 3$, $\mu_2 = 2.32$, $\mu_3 = 2$, $\mu_4 = 0$ and $\mu_5 = \mathbf{-0.49}$, thus ruling out the possible squared Euclidean nature of D^{sp}.

The absorption distance. The choice $g(\lambda) = (1-\rho)/(1-\rho\lambda)$ where $0 < \rho < 1$ yields the *absorption distance*: consider a modified random walk, where, at each discrete step, a particle at i either undergoes with probability ρ a transition

$i \to j$ (with probability p_{ij}) or is forever absorbed with probability $1-\rho$ into some additional "cemetery" state. The quantities $v_{ij}(\rho)$ = "average number of visits from i to j before absorption" obtain as the components of the matrix (see e.g. Kemeny and Snell (1976) or Kijima (1997))

$$V(\rho) = (I - \rho P)^{-1} = (\Pi - \rho E)^{-1}\Pi \text{ with } f_i v_{ij} = f_j v_{ji} \text{ and } \sum_i f_i v_{ij} = \frac{f_j}{1-\rho} \ .$$

Hence $K = g(\Pi^{-\frac{1}{2}} E \Pi^{-\frac{1}{2}}) = (1-\rho)\Pi^{\frac{1}{2}} V \Pi^{-\frac{1}{2}}$ and $B_{ij} = (1-\rho)v_{ij}/f_j$, measuring the ratio of the average number of visits from i to j over its expected value over the initial state i. Finally,

$$D^{abs}_{ij}(\rho) = \frac{v_{ii}(\rho)}{f_i} + \frac{v_{jj}(\rho)}{f_j} - 2\frac{v_{ij}(\rho)}{f_j} \ .$$

By construction, $\lim_{\rho\to 0} D^{abs}(\rho) = D^{fro}$ and $\lim_{\rho\to 1}(1-\rho)^{-1}D^{abs}(\rho) = D^{com}$. Also, $\lim_{\rho\to 1} D^{abs}(\rho) \equiv 0$ for a connected graph.

The "sif" distance. The choice $g(\lambda) = \lambda^2/(1-\lambda)$ is the simplest one insuring an *irreducible and focused* squared Euclidean distance. Identity $\lambda^2/(1-\lambda) = 1/(1-\lambda) - \lambda - 1$ readily yields (wether D^{dif} is Euclidean or not)

$$D^{sif}_{ij} = D^{com}_{ij} - D^{dif}_{ij} - D^{fro}_{ij} \ .$$

4 Numerical Experiments

4.1 Inter-cantonal Migration Data

The first data set consists of the numbers $N = (n_{ij})$ of people inhabiting the Swiss canton i in 1980 and the canton j in 1985 ($i, j = 1, \ldots, n = 26$), with a total count of $6'039'313$ inhabitants, 93% of which are distributed over the diagonal. N can be made brutally symmetric as $\frac{1}{2}(n_{ij} + n_{ji})$ or $\sqrt{n_{ij}n_{ji}}$, or, more gently, by fitting a *quasi-symmetric model* (Bavaud 2002), as done here. Normalizing the maximum likelihood estimate yields the exchange matrix E. Raw coordinates $x_{i\alpha} = u_{i\alpha}/\sqrt{f_i}$ are depicted in Figure 2. By construction, they do not depend of the form of the function $g(\lambda)$ involved in Definition 2, but they do depend on the form of the Schoenberg transformation $\tilde{D} = \phi(D)$ involved in Definition 3, where they obtain as solutions of the weighted MDS algorithm (Theorem 1) on $\tilde{D}$, with unchanged weights f (Figure 3 (a) and (b)).

Iterating (2) from an initial $n \times m$ membership matrix Z_{init} (with $m \le n$) at fixed T yields a membership $Z_0(T)$, which is by construction a local minimizer of the free energy $F[Z, T]$. The number $M(Z_0) \le m$ of independent columns of Z_0 measures the number of *effective groups*: equivalent groups, that is groups whose columns are proportional, could and should be aggregated, thus resulting in M

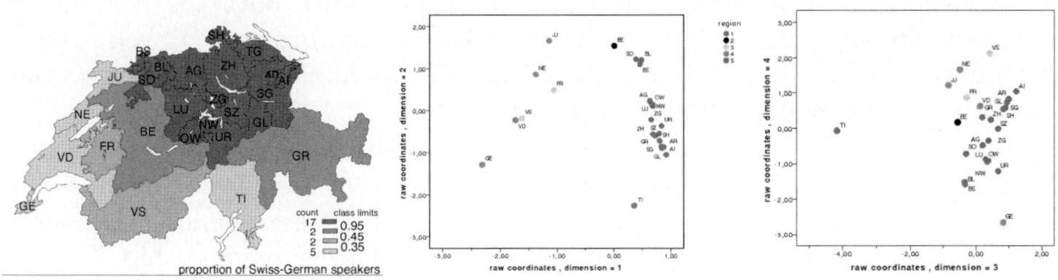

Fig. 2. Proportion of Swiss-German speakers in the 26 Swiss cantons (left), and raw coordinates $x_{i\alpha}$ associated to the inter-cantonal migrations, in dimensions $\alpha = 1, 2$ (center) and $\alpha = 3, 4$ (right). Colours code the linguistic regions, namely: 1 = German, 2 = mainly German, 3 = mainly French, 4 = French and 5 = Italian. The central factorial map reconstructs fairly precisely the geographical map, and emphasizes the linguistic German-French barrier, known as "Röstigraben". The linguistic isolation of the sole Italian-speaking canton, intensified by the Alpine barrier, is patent.

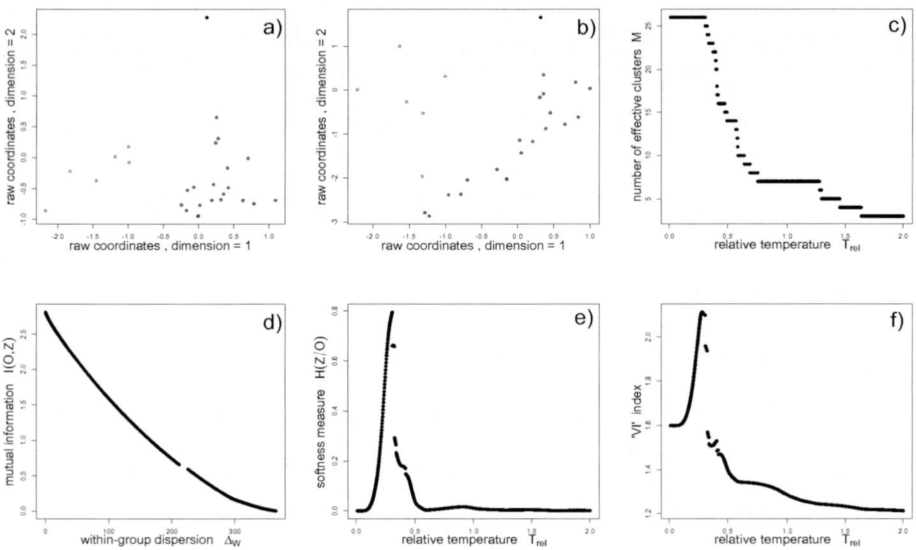

Fig. 3. Raw coordinates extracted from weighted MDS after applying Schoenberg transformations $\tilde{D} = \phi(D^{com})$ with $\phi(D) = D^{0.7}$ (a), and $\phi(D) = 1 - \exp(-bD)$ with $b = 1/(4\Delta^{com})$ (b). Decrease of the number of effective groups with the temperature (c); beside the main component, two microscopic groups of size $\rho_2 = 6 \cdot 10^{-4}$ and $\rho_3 = 2 \cdot 10^{-45}$ survive at $T_{\text{rel}} = 2$. (d) is the so-called *rate-distortion function* of Information Theory; its discontinuity at $T_{\text{crit}} = 0.406$ betrays a *phase transition* between a cold regime with numerous clusters and a hot regime with few clusters (Rose et al. 1990; Bavaud 2009). Behaviour of the *overall softness* $H(Z|O)$ (e) (Section 2.2) and of the clusters-regions *variation of information* (f) (see text).

distinct groups, without changing the free energy, since both the intra-group dispersion and the mutual information are aggregation-invariant (Bavaud 2009). In practice, groups g and h are judged as equivalent if their relative overlap (Section 2.2) obeys $\theta_{gh}/\sqrt{\theta_{gg}\theta_{hh}} \geq 1 - 10^{-10}$.

Define the relative temperature as $T_{\text{rel}} = T/\Delta$. One expects $M = 1$ for $T_{\text{rel}} \gg 1$, and $M = n$ for $T_{\text{rel}} \ll 1$, provided of course that the initial membership matrix contains at least n columns. We operate a *soft hierarchical descendant clustering* scheme, consisting in starting with the identity membership $Z_{\text{init}} = I$ for some $T_{\text{rel}} \ll 1$, iterating (2) until convergence, and then aggregating the equivalent columns in $Z_0(T)$ into M effective groups. The temperature is then slightly increased, and, choosing the resulting optimum $Z_0(T)$ as the new initial membership, (2) is iterated again, and so forth until the emergence of a single effective group ($M = 1$) in the high temperature phase $T_{\text{rel}} \geq 1$.

Numerical experiments (Figure 3) actually conform to the above expectations, yet with an amazing propensity for tiny groups $\rho_g \ll 1$ to survive at high temperature, that is before to be aggregated in the main component. This metastable behaviour is related to the locally optimal nature of the algorithm; presumably unwanted in practical applications, it can be eliminated by forcing group coalescence if, for instance, $H(Z)$ or $F[Z] - \Delta$ become small enough.

The softness measure of the clustering $H(Z|O)$ is expected to be zero in both temperature limits, since both the identity matrix and the single-group membership matrix are hard. We have attempted to measure the quality of the clustering Z with respect to the regional classification R of Figure 2 by the "variation of information" index $H(Z) + H(R) - 2I(Z, R)$ proposed by Meila (2005). Further investigations, beyond the scope of this paper, are obviously still to be conducted in this direction.

The stability of the effective number of clusters around $T_{\text{rel}} = 1$ might encourage the choice of the solution with $M = 7$ clusters. Rather disappointingly, the latter turns out (at $T_{\text{rel}} = 0.8$, things becoming even worse at higher temperature) to consist of one giant main component of $\rho_1 > 0.97$, together with 6 other practically single-object groups (UR, OW, NW, GL, AI, JU), totalizing less than three percent of the total mass (see also Section 5).

4.2 Commuters Data

The second data set counts the number of commuters $N = n_{ij}$ between the $n = 892$ French speaking Swiss communes, living in commune i and working in commune j in 2000. A total of 733′037 people are involved, 49% of which are distributed over the diagonal. As before, the exchange matrix E is obtained after fitting a quasi-symmetric model to N. The first two dimensions $\alpha = 1, 2$ of the raw coordinates $x_{i\alpha} = u_{i\alpha}/\sqrt{f_i}$ are depicted in Figure 4 a). The objects cloud consists of all the communes (up, left) except a single one (down, right), namely "Roche d'Or" (JU), containing 15 active inhabitants, 13 of which work in Roche d'Or. Both the very high value of the proportion of stayers e_{ii}/f_i and the low value of the weight f_i make Roche d'Or (together with other communes, to a lesser extent) quasi-disconnected from the rest of the system, hence producing,

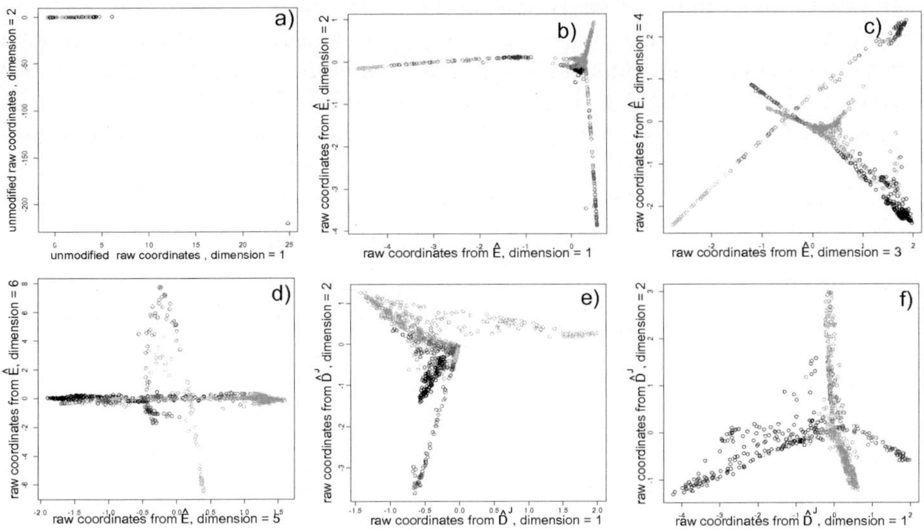

Fig. 4. Raw coordinates associated to the unmodified exchange matrix E are unable to approximate the geographical map (a), in contrast to (b), (c) and (d), based upon the diagonal-free exchange matrix $\hat{E}$. Colours code the cantons, namely BE=brown, FR=black, GE=orange, JU=violet, NE=blue, VD=green, VS=red. In particular, the central position of VD (compare with Figure 2) is confirmed. (e) and (f) represent the low-dimensional coordinates obtained by MDS from $\hat{D}^{jump}$ (4).

in accordance to the theory, eigenvalues as high as $\lambda_1 = .989$, $\lambda_2 = .986$, ... , $\lambda_{30} > .900$...

Theoretically flawless as is might be, this behavior stands as a complete geographical failure. As a matter of fact, commuters (and migration)-based graphs are *young*, that is E is much closer to its short-time limit $E^{(0)}$ than to its equilibrium value $E^{(\infty)}$. Consequently, diagonal components are huge and equivalent vertices in the sense of Definition 1 cannot exist: for $k = i \neq j$, the proportion of stayers e_{ii}/f_i is large, while e_{ij}/f_j is not.

Attempting to consider the *Laplacian* $E - E^{(0)}$ instead of E does not improve the situation: both matrices indeed generate the same eigenstructure, keeping the order of eigenvalues unchanged. A brutal, albeit more effective strategy consists in plainly destroying the diagonal exchanges, that is by replacing E by the *diagonal-free exchange matrix* $\hat{E}$, with components and associated weights

$$\hat{e}_{ij} = \frac{e_{ij} - \delta_{ij} e_{ii}}{1 - \sum_k e_{kk}} \qquad \hat{f}_i = \frac{f_i - e_{ii}}{1 - \sum_k e_{kk}} \ .$$

Defining $\hat{E}$ as the new exchange matrix yields (Sections 2 and 3) new weights $\hat{f}$, eigenvectors $\hat{U}$, eigenvalues $\hat{\Lambda}$ (with $\hat{\lambda}_n = 0$), raw coordinates $\hat{X}$ and distances $\hat{D}$, as illustrated in Figure 4 b), c) and d).

However, an example of equivalent nodes in the sense of Definition 1 is still unlikely to be found, since $0 = \hat e_{ii}/\hat f_i \neq \hat e_{ij}/\hat f_j$ in general. A weaker concept of equivalence consists in comparing $i \neq j$ by means of their transition probabilities towards the *other* vertices $k \neq i, j$, that is by means of the Markov chain conditioned to the event that the next state is *different*. Such Markov transitions approximate the so-called *jump* process, if existing (see e.g. Kijima (1997) or Bavaud (2008)). Their associated exchange matrix is precisely given by $\hat E$.

Definition 4 (Weakly equivalent vertices; weakly focused distances). *Two distinct vertices i and j are* weakly equivalent, *noted $i \overset{w}{\sim} j$, if $\hat e_{ik}/\hat f_i = \hat e_{jk}/\hat f_j$ for all $k \neq i, j$,. A distance is* weakly focused *if $D_{ij} = 0$ whenever $i \overset{w}{\sim} j$.*

By construction, the following "jump" distance is squared Euclidean and weakly focused:

$$\hat D^{jump}_{ij} = \sum_{k\,|\,k\neq i,j} \hat f_k\left(\frac{\hat e_{ik}}{\hat f_i \hat f_k} - \frac{\hat e_{jk}}{\hat f_j \hat f_k}\right)^2 = \sum_k \frac{1}{\hat f_k}\left(\frac{\hat e_{ik}}{\hat f_i} - \frac{\hat e_{jk}}{\hat f_j}\right)^2 - \frac{\hat e_{ij}^2}{\hat f_i \hat f_j}\left(\frac{1}{\hat f_i}+\frac{1}{\hat f_j}\right) . \quad (4)$$

The restriction $k \neq i, j$ in (4) complicates the expression of D^{jump} in terms of the eigenstructure $(\hat U, \hat \Lambda)$, and the existence of raw coordinates $\hat x_{i\alpha}$, adapted to the diagonal-free case, and justified by an analog of Proposition 1, remains open. In any case, jump distances (4) are well defined, and yield low-dimensional coordinates of the 892 communes by weighted MDS (Theorem 1) with weights $\hat f$, as illustrated in Figure 4 e) and f).

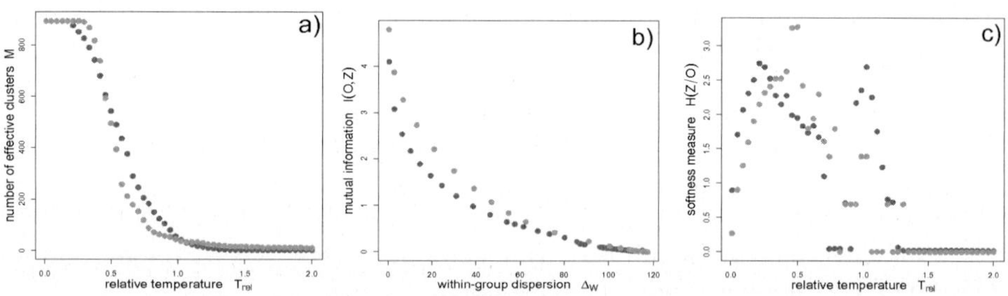

Fig. 5. Comparison between the clustering obtained from $\hat D^{sif}$ (in red) and $\hat D^{jump}$ (in green): evolution of the number of effective clusters with the temperature (a), rate-distortion function (b) and overall softness measure (c). In (b), $\hat\Delta^{jump}$ has been multiplied by a factor five to fit to the scale.

5 Conclusion

Our first numerical results confirm the theoretical coherence and the tractability of the clustering procedure presented in this paper. Yet, further investigations are certainly required: in particular, the precise role that the diagonal components of the exchange matrix should play into the construction of distances on graphs deserves to be thoroughly elucidated. Also, the presence of fairly small clusters in the clustering solutions of Section 4, from which the normalized cut algorithm `Ncut` was supposed to prevent, should be fully understood. Our present guess is that small clusters are inherent to the spatial nature of the data under consideration: elongated and connected clouds as those of Figure 4 *cannot* miraculously split into well-distinct groups, irrespectively of the details of the clustering algorithm (classical chaining problem). This being said, squared Euclidean are closed under addition and convex mixtures. Hence, an elementary yet principled remedy could simply consist in adding *spatial squared Euclidean distances* to the *flow-induced distances* investigated in the present contribution.

References

Aldous, D., Fill, J.: Reversible Markov Chains and Random Walks on Graphs. Draft chapters, `http://www.stat.berkeley.edu/users/aldous/RWG/book.html`

Bavaud, F.: The quasi-symmetric side of gravity modelling. Environment and Planning A 34, 61–79 (2002)

Bavaud, F.: Spectral clustering and multidimensional scaling: a unified view. In: Batagelj, V., Bock, H.-H., Ferligoj, A., Ziberna, A. (eds.) Data science and classification, pp. 131–139. Springer, Heidelberg (2006)

Bavaud, F.: The Endogenous analysis of flows, with applications to migrations, social mobility and opinion shifts. Journal of Mathematical Sociology 32, 239–266 (2008)

Bavaud, F.: Aggregation invariance in general clustering approaches. Advances in Data Analysis and Classification 3, 205–225 (2009)

Bavaud, F.: On the Schoenberg Transformations in Data Analysis: Theory and Illustrations (submitted, 2010)

Belkin, M., Niyogi, P.: Laplacian Eigenmaps for Dimensionality Reduction and Data Representation. Neural Computation 15, 1373–1396 (2003)

Berger, J., Snell, J.L.: On the concept of equal exchange. Behavioral Science 2, 111–118 (1957)

Beurling, A., Deny, J.: Espaces de Dirichlet. I. Le cas élémentaire. Acta Mathematica 99, 203–224 (1958)

Chung, F.R.K.: Spectral graph theory. In: CBMS Regional Conference Series in Mathematics, vol. 92. American Mathematical Society, Washington (1997)

Cover, T.M., Thomas, J.A.: Elements of Information Theory. Wiley, Chichester (1991)

Critchley, F., Fichet, B.: The partial order by inclusion of the principal classes of dissimilarity on a finite set, and some of their basic properties. In: van Cutsem, B. (ed.) Classification and dissimilarity analysis. Lecture Notes in Statistics, pp. 5–65. Springer, Heidelberg (1994)

Deza, M., Laurent, M.: Geometry of cuts and metrics. Springer, Heidelberg (1997)

Dunn, G., Everitt, B.: An Introduction to Mathematical Taxonomy. Cambridge University Press, Cambridge (1982)

Filippone, M., Camastra, F., Masulli, F., Rovetta, S.: A survey of kernel and spectral methods for clustering. Pattern Recognition 41, 176–190 (2008)

Fouss, F., Pirotte, A., Renders, J.-M., Saerens, M.: Random-walk computation of similarities between nodes of a graph with application to collaborative recommendation. IEEE Transactions on Knowledge and Data Engineering 19, 355–369 (2007)

Greenacre, M.J.: Correspondence analysis in practice. Chapman and Hall, Boca Raton (2007)
Hein, M., Bousquet, O., Schölkopf, B.: Maximal margin classification for metric spaces. Journal of Computer and System Sciences 71, 333–359 (2005)
Huang, Z., Ng, M.K.: A fuzzy k-modes algorithm for clustering categorical data. IEEE Transactions on Fuzzy Systems 7, 446–452 (1999)
Kemeny, J.G., Snell, J.L.: Finite Markov Chains. Springer, Heidelberg (1976)
Kijima, M.: Markov processes for stochastic modeling. Chapman and Hall, Boca Raton (1997)
Kondor, R.I., Lafferty, J.D.: Diffusion kernels on graphs and other discrete input spaces. In: Sammut, C., Hoffmann, A.G. (eds.) Proceedings of the Nineteenth International Conference on Machine Learning, pp. 315–322 (2002)
Lafon, S., Lee, A.B.: Diffusion Maps and Coarse-Graining: A Unified Framework for Dimensionality Reduction, Graph Partitioning, and Data Set Parameterization. IEEE Trans. on Pattern Analysis and Machine Intelligence 28, 1393–1403 (2006)
von Luxburg, U.: A tutorial on spectral clustering. Statistics and Computing 17, 395–416 (2007)
Mardia, K.V., Kent, J.T., Bibby, J.M.: Multivariate analysis. Academic Press, London (1979)
Meila, M.: Comparing clusterings: an axiomatic view. In: ACM International Conference Proceeding Series, vol. 119, pp. 577–584 (2005)
Ng, A., Jordan, M., Weiss, Y.: On spectral clustering: analysis and an algorithm. In: Advances in Neural Information Processing Systems, vol. 14, pp. 849–856. MIT Press, Cambridge (2002)
Nock, R., Vaillant, P., Henry, C., Nielsen, F.: Soft memberships for spectral clustering, with application to permeable language distinction. Pattern Recognition 42, 43–53 (2009)
Rose, K., Gurewitz, E., Fox, G.C.: Statistical mechanics and phase transitions in clustering. Phys. Rev. Lett. 65, 945–948 (1990)
Rose, K.: Deterministic Annealing for clustering, compression, classification, regression, and related optimization problems. Proceedings of the IEEE 86, 2210–2239 (1998)
Schoenberg, I.J.: Metric Spaces and Completely Monotone Functions. The Annals of Mathematics 39, 811–841 (1938a)
Schoenberg, I.J.: Metric Spaces and Positive Definite Functions. Transactions of the American Mathematical Society 44, 522–536 (1938b)
Schölkopf, B.: The Kernel Trick for Distances. In: Advances in Neural Information Processing Systems, vol. 13, pp. 301–307 (2000)
Shi, J., Malik, J.: Normalized cuts and image segmentation. IEEE Transactions on Pattern Analysis and Machine Intelligence 22, 888–905 (2000)
Tenenbaum, J.B., de Silva, V., Langford, J.C.: A Global Geometric Framework for Nonlinear Dimensionality Reduction. Science 22, 2319–2323 (2000)
Torgerson, W.S.: Theory and Methods of Scaling. Wiley, Chichester (1958)
Williams, C.K.I.: On a Connection between Kernel PCA and Metric Multidimensional Scaling. Machine Learning 46, 11–19 (2002)
Young, G., Householder, A.S.: Discussion of a set of points in terms of their mutual distances. Psychometrika 3, 19–22 (1938)
Yu, S., Shi, J.: Multiclass Spectral Clustering. In: Proceedings of the Ninth IEEE International Conference on Computer Vision, pp. 313–319 (2003)

Adaptive Parallel/Serial Sampling Mechanisms for Particle Filtering in Dynamic Bayesian Networks

Eva Besada-Portas[1], Sergey M. Plis[2,3], Jesus M. de la Cruz[1], and Terran Lane[2]

[1] Universidad Complutense de Madrid, 28040 Madrid, Spain
[2] University of New Mexico, Albuquerque NM 87131, USA
[3] The Mind Research Network, Albuquerque NM 87106, USA

Abstract. Monitoring the variables of real world dynamical systems is a difficult task due to their inherent complexity and uncertainty. Particle Filters (PF) perform that task, yielding probability distribution over the unobserved variables. However, they suffer from the curse of dimensionality problem: the necessary number of particles grows exponentially with the dimensionality of the hidden state space. The problem is aggravated when the initial distribution of the variables is not well known, as happens in global localization problems. In this paper we present two new adaptive sampling mechanisms for PFs for systems whose variable dependencies can be factored into a Dynamic Bayesian Network. The novel PFs, developed over the proposed sampling mechanisms, exploit the strengths of other existing PFs. Their adaptive mechanisms 1) modify or establish probabilistic links among the subspaces of hidden variables that are independently explored to build particles consistent with the current measurements and past history, and 2) tune the performance of the new PFs toward the behaviors of several existing PFs. We demonstrate their performance on some complex dynamical system estimation problems, showing that our methods successfully localize and track hidden states, and outperform some of the existing PFs.

1 Introduction

Estimating the hidden state of a multivariate dynamical system remains a large challenge. While the Dynamic Bayesian Network (DBN) formalism provides an excellent way to *represent* such systems, performing *inference* in these models is difficult, and approximations are usually necessary. Among the most popular approximate inference methods for DBNs are *Particle Filters* (PFs): point-mass approximations of the hidden state distribution, which are calculated with sequential Monte Carlo simulation techniques, usually combining importance sampling and weighted resampling steps [1].

In spite of their popularity, PFs can be difficult to apply to new problems. One of the core challenges arises from the familiar curse of dimensionality: in even modest dimensionality spaces, there are high chances that many or most particles will fall into near-zero probability regions of the state space, leading to serious particle depletion and quickly driving the PF off track. This problem is evident, for example, in multi-object tracking tasks [2]. The difficulty is exacerbated when the distribution used to initialize the values of the particles is unlikely to generate particles that have high-probability.

The dimensionality/depletion difficulty can be reduced with a careful choice of the initialization, importance sampling and weighted resampling distributions, but doing so

J.L. Balcázar et al. (Eds.): ECML PKDD 2010, Part I, LNAI 6321, pp. 119–134, 2010.

is not always straightforward. For example, in global localization problems it is difficult to define a tight initialization proposal distribution. Efficient sampling and resampling proposals developed around specific properties of the dynamics of the problem require extensive domain knowledge and/or generate PFs for specific types of problems [3,4].

The general approaches that have recently appeared in [5,6] overcome some of the inherent difficulties of the standard PF for DBNs [7] using two complementary strategies. The PF of [5] follows the ancestral ordering of the variables to *serialize sampling and resampling steps*. Whereas, the PFs in [6] *parallelize the sampling step* to sample different subspaces of hidden variables independently. The serialization in [5] leads to elimination of those particles whose already sampled hidden variables are not probabilistically consistent with some of the measurements. This process avoids sampling their remaining hidden variables. However, it can drive the PF off track when a premature specialization of the survived particles makes it impossible to sample values for the remaining hidden variables consistent with the measurements. The parallelization presented by the authors in [6] lets their PFs create particles that are probabilistically consistent with the measurements associated with each subspace of hidden variables. However, it can continuously reset the PFs when the parallel sampled measurement likely values of their hidden variables are inconsistent with the past history.

The two sampling mechanisms presented in this paper combine the benefits of the previous strategies [5,6,7] by automatically adapting the serialization/parallelization level of the proposal distributions used to sample the values of the hidden variables. They are developed within the serial importance sampling methodology and so the developed PFs that incorporate those mechanisms have a firm probabilistic foundation.

The characteristics of the new mechanisms, which represent two new points in the spectrum of already existing sampling strategies, are the following. First, they allow an automatic adaptation of the PF behavior towards the PFs in [5,6,7]. Second, they help to overcome problems of poor particle initialization, improving state localization and convergence when the initial particle sample is far from the target distribution. And third, they need fewer particles than some of the existing methods.

Finally, this paper also analyzes in depth the characteristics of the PFs presented in [5,6,7] and states the old and new techniques under a unified formalism.

2 Particle Filter Fundamentals

Our core contribution is the development of two new proposal distributions that support the definition of PFs for DBNs that adapt their behaviors towards the serial PF of [5], the parallel PF of [6], and the standard PF of [7]. In this section we introduce a uniform notation and formalism in which to state the previous and new PFs.

2.1 Definitions and Notation

In this paper, a capital letter U represents a random variable, a boldface capital letter $\boldsymbol{U}$ – a set of random variables, a lowercase letter u – the specific value of the corresponding random variable U, and a lowercase bold letter $\boldsymbol{u}$ – an assignment of the values to the variables of its set $\boldsymbol{U}$. We also define $\mathcal{P}(\boldsymbol{U})$ as a partition of $\boldsymbol{U}$.

A Bayesian Network (BN) is an annotated directed graph that encodes the probability distribution of a set of random variables $\boldsymbol{V}$. DBNs model the distribution of a sequence of sets of variables. A set of variables belonging to the kth time slice is represented as $\boldsymbol{V}_k$ and the total set of variables up to the current time slice t is $\boldsymbol{V}_{0:t}$. To distinguish hidden and observed variables, we use X, $\boldsymbol{X}_t$ and $\boldsymbol{X}_{0:t}$ to denote the former, and Y, $\boldsymbol{Y}_t$ and $\boldsymbol{Y}_{0:t}$ for the latter. The graph is completely defined by the sets of parents of all its variables: $\text{Pa}_k(V)$ represents the subset of the parents of variable V that belong to time slice k and $\text{pa}_k(V)$ their assignment to particular values. Similarly, we also define the set of children of a variable and their assignments by $\text{Ch}_k(V)$ and $\text{ch}_k(V)$.

Probability distributions will be represented as $\text{p}(\cdot)$, $\text{q}(\cdot)$ and $\text{r}(\cdot)$: the first related with the probabilities of the problem and the others with the proposal distributions used to sample the values of the particles from. The operation $a \sim \text{q}(\cdot)$ represents sampling a according to $\text{q}(\cdot)$. The operation $E_{\text{p}(X|\cdot)}[X]$ represents the expected value of X with respect to $\text{p}(X|\cdot)$. And $\delta(\cdot)$ and δ_i^j are respectively the Dirac and Kronecher delta functions.

A PF approximates the probability $\text{p}(\boldsymbol{X}|\boldsymbol{y})$ of a set of hidden variables $\boldsymbol{X}$ given the values of the measurements $\boldsymbol{Y}$ by the point mass distribution $\sum_{i=1}^{N} w(\boldsymbol{x}^{(i)})\delta(\boldsymbol{X} - \boldsymbol{x}^{(i)})$, where $\boldsymbol{x}^{(i)}$ are the values of the variables in $\boldsymbol{X}$ in the i-th particle, $w(\boldsymbol{x}^{(i)})$ their weights, and N the number of particles. Additionally, $x^{(i)}$, $\boldsymbol{x}_{0:t}^{(i)}$ and $\text{pa}_k^{(i)}(V)$ represent the assignments of X, $\boldsymbol{X}_{0:t}$ and $\text{Pa}_k(V)$ to the values they have in the i-th particle.

Finally, in this paper we only consider DBNs whose variables have parents belonging to the current or previous time slice $(\forall V \in \boldsymbol{V}_t \wedge \forall k \notin \{t, t-1\}\ \text{Pa}_k(V) = \emptyset)$. This restriction upon the structure of the DBN factors the joint probability of the set of hidden and observation variables up to time slice t as Eq. (1) shows.

$$
\begin{aligned}
\text{p}(\boldsymbol{X}_{0:t}, \boldsymbol{Y}_{0:t}) &= \text{p}(\boldsymbol{X}_t, \boldsymbol{Y}_t | \boldsymbol{X}_{0:t-1}, \boldsymbol{Y}_{0:t-1})\, \text{p}(\boldsymbol{X}_{0:t-1}, \boldsymbol{Y}_{0:t-1}) \\
\text{p}(\boldsymbol{X}_t, \boldsymbol{Y}_t | \boldsymbol{X}_{0:t-1}, \boldsymbol{Y}_{0:t-1}) &= \prod_{X \in \boldsymbol{X}_t} \text{p}(X | \text{Pa}_{t-1}(X), \text{Pa}_t(X)) \prod_{Y \in \boldsymbol{Y}_t} \text{p}(Y | \text{Pa}_{t-1}(Y), \text{Pa}_t(Y))
\end{aligned}
\quad (1)
$$

2.2 Importance Sampling (IS) and Weighted Resampling (WR)

The values of the particles $\boldsymbol{x}^{(i)}$ and their weights $w(\boldsymbol{x}^{(i)})$ used to approximate the probability distribution $\text{p}(\boldsymbol{X}|\boldsymbol{y})$ as $\sum_{i=1}^{N} w(\boldsymbol{x}^{(i)})\delta(\boldsymbol{X} - \boldsymbol{x}^{(i)})$ are obtained by means of a sequential combination of importance sampling and weighted resampling steps [8]. The way of proceeding for each step is detailed in the following subsections, in order to state the serial, parallel, standard and adaptive PFs under a unified perspective. In short, IS is used to create the particles while WR is carried out to increment the number of particles in the regions of high interest and reduce them in the others.

Importance Sampling is used to (1) create new particles by means of a proposal distribution $\text{q}(\boldsymbol{X}|\boldsymbol{y})$ that generates their values $(\boldsymbol{x}^{(i)} \sim \text{q}(\boldsymbol{X}|\boldsymbol{y}))$ and (2) calculate their weights as $w(\boldsymbol{x}^{(i)}) \propto \text{p}(\boldsymbol{x}^{(i)}|\boldsymbol{y}) / \text{q}(\boldsymbol{x}^{(i)}|\boldsymbol{y})$.

The IS operation can be carried out sequentially exploiting the independence assumptions imposed by the structure of the DBN and therefore the recursive factorization of the joint presented in Eq. (1). The following two IS steps, whose development is detailed in [6,8,9], emerge from different choices of posterior distributions $\text{p}(\boldsymbol{X}|\boldsymbol{y})$.

- When $p(\boldsymbol{X}_{0:t}|\boldsymbol{y}_{1:t})$ is the posterior distribution to be approximated by the point mass representation $\sum_{i=1}^{N} w(\boldsymbol{x}_{0:t}^{(i)})\delta(\boldsymbol{X}_{0:t}-\boldsymbol{x}_{0:t}^{(i)})$ the IS step can obtain the values $\boldsymbol{x}_t^{(i)}$ (of the hidden variables $\boldsymbol{X}_t$ of the current time slice t) associated with each particle $\boldsymbol{x}_{0:t}^{(i)}$ by means of the proposal $q(\boldsymbol{x}_t^{(i)}|\boldsymbol{x}_{0:t-1},\boldsymbol{y}_{1:t})$ and calculate their weights $w(\boldsymbol{x}_{0:t}^{(i)})$ with Eq. (2). As this IS step is responsible for estimating the probability of the trajectory $\boldsymbol{X}_{0:t}$ of the hidden variables, we will call it hereafter *"Trajectory" IS* or TIS.
- When $p(\boldsymbol{X}_t|\boldsymbol{y}_{1:t})$ is the posterior distribution to be approximated by the point mass representation $\sum_{i=1}^{N} w(\boldsymbol{x}_t^{(i)})\delta(\boldsymbol{X}_t-\boldsymbol{x}_t^{(i)})$ the IS step can obtain the values $\boldsymbol{x}_t^{(i)}$ (of the hidden variables $\boldsymbol{X}_t$ of the current time slice t) of the particle $\boldsymbol{x}_t^{(i)}$ by means of the proposal $q(\boldsymbol{x}_t^{(i)}|\boldsymbol{y}_{1:t})$ and calculate their weights $w(\boldsymbol{x}_t^{(i)})$ with Eq. (3). As this IS step is responsible for estimating the probability of the hidden state $\boldsymbol{X}_t$ at the current time slice t, we will call it hereafter *"Instantaneous"* IS or IIS.

$$w(\boldsymbol{x}_{0:t}^{(i)}) \propto \frac{\prod_{Y\in\boldsymbol{Y}_t} p\left(y|\mathrm{pa}_{t-1}^{(i)}(Y),\mathrm{pa}_t^{(i)}(Y)\right)\prod_{X\in\boldsymbol{X}_t} p\left(x^{(i)}|\mathrm{pa}_{t-1}^{(i)}(X),\mathrm{pa}_t^{(i)}(X)\right)}{q\left(\boldsymbol{x}_t^{(i)}|\boldsymbol{x}_{0:t-1},\boldsymbol{y}_{1:t}\right)} w(\boldsymbol{x}_{0:t-1}^{(i)}) \tag{2}$$

$$w(\boldsymbol{x}_t^{(i)}) \propto \frac{\sum_{j=1}^{N}\left(w(\boldsymbol{x}_{t-1}^{(j)})\prod_{Y\in\boldsymbol{Y}_t} p\left(y|\mathrm{pa}_{t-1}^{(j)}(Y),\mathrm{pa}_t^{(i)}(Y)\right)\prod_{X\in\boldsymbol{X}_t} p\left(x^{(i)}|\mathrm{pa}_{t-1}^{(j)}(X),\mathrm{pa}_t^{(i)}(X)\right)\right)}{q\left(\boldsymbol{x}_t^{(i)}|\boldsymbol{y}_{1:t}\right)} \tag{3}$$

Note that both IS use a proposal $q(\boldsymbol{x}_t^{(i)}|\cdot)$ in each time iteration t to sequentially generate only the values $\boldsymbol{x}_t^{(i)}$ associated with the hidden variables $\boldsymbol{X}_t$ that belong to the time slice t. They also obtain the weights of the whole particle ($w(\boldsymbol{x}_{0:t}^{(i)})$ in TIS and $w(\boldsymbol{x}_t^{(i)})$ in IIS) based on the values of the weights of the particle up to the previous time slice ($w(\boldsymbol{x}_{0:t-1}^{(i)})$ in TIS and $w(\boldsymbol{x}_{t-1}^{(j)})$ in IIS), and the values of the hidden and observed variables at t and $t-1$ ($\boldsymbol{x}_t^{(i)}$, y, $\mathrm{pa}_t^{(i)}(V)$ and $\mathrm{pa}_{t-1}^{(k)}(V)$). Finally, in both numerators also appears the joint distribution (introduced in Eq. (1)) of the current time variables given their parents.

The main computational difference, which affects the number of operations of the selected IS step, appears in the numerator of the expressions used to calculate the weights: the IIS Eq. (3) has a summation that doesn't appear in the TIS Eq. (2). The total computational cost also depends on the proposal distribution $q(\boldsymbol{x}_t^{(i)}|\cdot)$ used to sample the values of $\boldsymbol{x}_t^{(i)}$, and so, IIS is not necessary computationally more costly than TIS. Moreover, selecting the appropriated proposals both IS steps can be equivalent. Besides, Eq. (3) can be simplified taking out of the summation those variables that don't have hidden parents belonging to the previous time slice ($V\in\boldsymbol{V}_t$ s.t. $\mathrm{Pa}_{t-1}(V)\cap\boldsymbol{X}_{t-1}=\emptyset$).

It is also important to mention that although the two types of IS estimate the probability of a different set of hidden variables ($\boldsymbol{X}_{0:t}$ in TIS and $\boldsymbol{X}_t$ in IIS), both can be used to estimate the values of the current state variables $\boldsymbol{X}_t$, because the trajectory $\boldsymbol{X}_{0:t}$ also includes them. However, the search space of TIS is significantly larger, and so the number of necessary particles in IIS should be smaller. Another interesting fact to consider is that TIS can be used to estimate any $\boldsymbol{X}_k$ with $k \leq t$ from the approximated distribution $\mathrm{p}(\boldsymbol{X}_{0:t}|\boldsymbol{y}_{1:t})$ and so the estimated values of $\boldsymbol{X}_k$ can take into account the measurements up to t. In IIS the values of $\boldsymbol{X}_k$ have to be estimated from $\mathrm{p}(\boldsymbol{X}_k|\boldsymbol{y}_{1:k})$ and so they can only take into account the measurements up to k.

Weighted Resampling is used to approximate a $\mathrm{p}(\boldsymbol{X}|\boldsymbol{y})$ that is already approximated by an existing point-mass representation $\sum_{i=1}^{N} w(\boldsymbol{x}^{(i)})\delta(\boldsymbol{X}-\boldsymbol{x}^{(i)})$ by another point-mass representation $\sum_{i=1}^{N} w(\boldsymbol{x}'^{(i)})\delta(\boldsymbol{X}-\boldsymbol{x}'^{(i)})$. The WR operation works by redistributing the set of existing weighted particles $(\boldsymbol{x}^{(i)}, w(\boldsymbol{x}^{(i)}))$ using a resampling strategy (such as multinomial, systematic or residual sampling) that for each $\boldsymbol{x}'^{(i)}$ picks a $\boldsymbol{x}^{(j)}$ according to the $\mathrm{r}(\boldsymbol{x}^{(j)})$ values obtained with a selected $\mathrm{r}(\boldsymbol{X})$ function. The weights $w(\boldsymbol{x}'^{(i)})$ of the new set of particles $\boldsymbol{x}'^{(i)}$ are calculated as $w(\boldsymbol{x}^{(j)})/\mathrm{r}(\boldsymbol{x}^{(j)})$.

The usual election of $\mathrm{r}(\boldsymbol{X})$ in TIS-based PFs is $\mathrm{r}(\boldsymbol{x}_{0:t}^{(j)}) = w(\boldsymbol{x}_{0:t}^{(j)})$ because this WR step applied after TIS eliminates the particles with negligible $w(\boldsymbol{x}_{0:t}^{(j)})$ and populates the hidden space $\boldsymbol{X}_{0:t}$ with equally weighted particles distributed according to $w(\boldsymbol{x}_{0:t}^{(j)})$. Thus, it focuses the PF to explore those regions that look more promising so far. The same approach can be used in IIS-based PFs making $\mathrm{r}(\boldsymbol{x}_t^{(j)}) = w(\boldsymbol{x}_t^{(j)})$.

Finally, it is worth mentioning that a combined sequential implementation of IS with the usual WR steps can delete those particles that were so far negligible for the WR step according to $w(\boldsymbol{x}_{0:t}^{(j)})$ in TIS and $w(\boldsymbol{x}_t^{(j)})$ in IIS but whose information could be important taking into account the future measurements ($\boldsymbol{y}_f$ with $f > t$). For that reason some PFs avoid the WR step or perform it every now and then [10].

3 General PFs for DBN

In this section we present the three general PFs whose behavior we want to dynamically approximate with the new adaptive parallel/serial sampling proposals. We also analyze their behavior to justify later the benefits we expect from the new adaptive mechanisms.

3.1 Standard PF (KLPF)

The PF presented in [7] (that we will call KLPF after their authors Koller and Lenner) approximates the posterior $\mathrm{p}(\boldsymbol{X}_{0:t}|\boldsymbol{y}_{1:t})$ using for each time step a TIS step followed by a WR step. The TIS proposal $\mathrm{q}(\boldsymbol{x}_t^{(i)}|\cdot)$ is the product of the transition priors presented in Eq. (4) and so the values of each X in $\boldsymbol{X}_t$ can be sampled from $\mathrm{p}(X|\mathrm{pa}_{t-1}^{(i)}(X), \mathrm{pa}_t^{(i)}(X))$ using the ancestral ordering of the hidden variables within the time slice t. The WR $\mathrm{r}(\boldsymbol{X}_{0:t})$ function is the usual choice: $\mathrm{r}(\boldsymbol{x}_{0:t}^{(j)}) = w(\boldsymbol{x}_{0:t}^{(j)})$. So, according to Eq. (2), the selected proposal and WR step, $w(\boldsymbol{x}_{0:t}^{(i)}) = \prod_{Y \in \boldsymbol{Y}_t} \mathrm{p}(y|\mathrm{pa}_{t-1}^{(i)}(Y), \mathrm{pa}_t^{(i)}(Y))$.

$$\mathrm{q}(\boldsymbol{x}_t^{(i)}|\cdot) = \prod_{X \in \boldsymbol{X}_t} \mathrm{p}(X|\mathrm{pa}_{t-1}^{(i)}(X), \mathrm{pa}_t^{(i)}(X)) \tag{4}$$

The choices of $\mathrm{q}(\cdot)$ and $\mathrm{r}(\cdot)$ make the TIS generation of the values $\boldsymbol{x}_t^{(i)}$ probabilistically consistent with the values of $\boldsymbol{x}_{t-1}^{(i)}$ and the WR selection of the survival particles probabilistically consistent with the current measurements. Once KLPF falls off track the chances to recover are small because the values of the future $\boldsymbol{x}_f^{(i)}$ $(f > t)$ will depend on the current off-track $\boldsymbol{x}_t^{(i)}$. That is, KLPF is good for building particles $\boldsymbol{x}_{0:t}^{(i)}$ probabilistically consistent firstly with the past hidden history $\boldsymbol{x}_{0:t-1}^{(i)}$ and secondary with the new measurements $\boldsymbol{y}_t$.

3.2 Serial PF (SPF)

The PF presented in [5] approximates $\mathrm{p}(\boldsymbol{X}_{0:t}|\boldsymbol{y}_{1:t})$ dividing the operations of each time step t in a serial of alternating *partial* TIS and WR steps. The alternation is imposed by the ancestral ordering of the variables $V \in \boldsymbol{V}_t$: when V is hidden $v^{(i)} \sim \mathrm{p}(V|\mathrm{pa}_{t-1}^{(i)}(V), \mathrm{pa}_t^{(i)}(V))$, when V is observed the particles built up to the moment are weighted resampled according to $\mathrm{p}(v|\mathrm{pa}_{t-1}^{(i)}(V), \mathrm{pa}_t^{(i)}(V))$.

Overall SPF uses the same $\mathrm{q}(\boldsymbol{x}_t^{(i)}|\cdot)$ as KLPF with WR steps based on the likelihood of each $Y \in \boldsymbol{Y}_t$ between the sampling of the $X \in \boldsymbol{X}_t$ according to the transition priors. Each *partial* TIS makes the sampled $\boldsymbol{x}^{(i)}$ probabilistically consistent with the already sampled hidden variables. Each *partial* WR selects the particles which are probabilistically consistent with the given y. Thus, SPF is good for building particles $\boldsymbol{x}_{0:t}^{(i)}$ alternatively probabilistically consistent with the already sampled hidden history and used measurements. SPF will also fall offtrack easily when the likelihood of each measurement and the product of the likelihoods take significantly different values.

3.3 Parallel PFs (PPF)

The PFs described in [6] approximate either $\mathrm{p}(\boldsymbol{X}_{0:t}|\boldsymbol{y}_{1:t})$ or $\mathrm{p}(\boldsymbol{X}_t|\boldsymbol{y}_{1:t})$ using for each time step respectively a TIS or IIS step, followed optionally by a WR step. The TIS or IIS proposal $\mathrm{q}(\boldsymbol{x}_t^{(i)}|\cdot,)$ is presented in Eq. (5) and (6). It groups the $X \in \boldsymbol{X}_t$ in subsets $\boldsymbol{X}$ belonging to a defined[1] disjoint partition $\mathcal{P}(\boldsymbol{X}_t)$. And it independently samples the values of the hidden variables within each subset $\boldsymbol{X}$ from a mixture model of the product of transition priors where each mixture component is weighted with the product of the expected likelihoods[2] of the measurements Y associated with the $X \in \boldsymbol{X}$. When applied, the WR $\mathrm{r}(\cdot)$ function is the usual choice: $\mathrm{r}(\boldsymbol{x}_{0:t}^{(j)}) = w(\boldsymbol{x}_{0:t}^{(j)})$ or $\mathrm{r}(\boldsymbol{x}_t^{(j)}) = w(\boldsymbol{x}_t^{(j)})$.

$$\mathrm{q}\left(\boldsymbol{x}_t^{(i)}|\cdot\right) = \prod_{\boldsymbol{X} \in \mathcal{P}(\boldsymbol{X}_t)} \left(\sum_{j=1}^{N} \left(\alpha_j^{\boldsymbol{X}} \prod_{X \in \boldsymbol{X}} \mathrm{p}\left(x|\mathrm{pa}_{t-1}^{(j)}(X), \mathrm{pa}_t^{(i)}(X)\right) \right) \right) \tag{5}$$

[1] The partition is constructed based on the DBN structure with the rules presented in [6].

[2] While the value of any hidden $X \in \mathrm{Pa}_t(V)$ in $\mathrm{pa}_t^{(j)}(V)$ is the value of X in the j-th particle, the value associated with the same $X \in \mathrm{Pa}_t(V)$ in $\mathrm{pa}_t^{*(j)}(V)$ is $E_{p(X|\mathrm{pa}_{t-1}^{(j)}(X), \mathrm{pa}_t^{*(j)}(X))}[X]$. The expectations are needed because the value of X won't be sampled until the component of the mixture is selected according to $\alpha_j^{\boldsymbol{X}}$.

$$\alpha_j^{\boldsymbol{X}} \propto \prod_{Y \in \mathrm{Ch}_t(X) \wedge X \in \boldsymbol{X}} \mathrm{p}\left(y | \mathrm{pa}_{t-1}^{(j)}(Y), \mathrm{pa}_t^{*(j)}(Y)\right) \tag{6}$$

The choice of $\mathrm{q}(\cdot)$ makes the IS generation of the values $\boldsymbol{x}_t^{(i)}$ probabilistically consistent with the expected likelihood of the measurements associated with each of the disjoint subsets of hidden variables $\boldsymbol{X}$ defined by $\mathcal{P}(\boldsymbol{X}_t)$. However, the independence imposed by the factorization of the proposal can lead to creation of particles that are not consistent with the previous time step particles ($\boldsymbol{x}_{0:t-1}^{(i)}$ or $\boldsymbol{x}_{t-1}^{(i)}$). When no particle is consistent, the numerators of Eq. (2) or Eq. (3) will make all the weights zero and the PPF will be automatically reset. This behavior is positive when the PPF is losing track but unnecessary when that is not the case. So the PPFs are good for building particles ($\boldsymbol{x}_{0:t}^{(i)}$ or $\boldsymbol{x}_t^{(i)}$) probabilistically consistent firstly with different subsets of current measurements and secondly with the previous time step particles ($\boldsymbol{x}_{0:t-1}^{(i)}$ or $\boldsymbol{x}_{t-1}^{(i)}$).

The choice of $\mathrm{r}(\cdot)$ influences the elimination of particles with negligible weights by WR. However, this choice also decrements the sampling possibilities of the independent mixture proposals. So disabling the WR step increases the PPFs sampling possibilities facilitating the PPFs reset when the measurements subsets contradict the hidden history.

4 Adaptive Sampling Mechanisms for DBN

Optimal IS steps are those that minimize the variance of the weights [10]. In the design of PFs for DBNs this requirement implies that the sampling proposal $\mathrm{q}(\boldsymbol{x}_t^{(i)}|\cdot)$ samples the new values of $\boldsymbol{X}_t$ simultaneously taking into account the values of $\boldsymbol{x}_{t-1}^{(i)}$ and $\boldsymbol{y}_t$. However, this is rarely possible and so non-optimal proposals are usually selected.

The sampling proposals used in KLPF, SPF and PPFs are not optimal. Nevertheless, these proposals exhibit several complementary positive probabilistic behaviors. KLPF and SPF build particles probabilistically consistent firstly with $\boldsymbol{x}_{t-1}^{(i)}$ and secondly with the current measurements $\boldsymbol{y}_t$, while the PPFs build particles probabilistically consistent firstly with $\boldsymbol{y}_t$ and secondly with $\boldsymbol{x}_{t-1}^{(i)}$. Besides, KLPF builds particles consistent with all the current measurements $\boldsymbol{y}_t$, SPF builds particles incrementally consistent with each measurement $y \in \boldsymbol{y}_t$, and PPFs build them probabilistically consistent with subsets of measurements and of hidden variables.

We believe that a PF that simultaneously takes advantage of several of those behaviors should be closer to the optimal PF. Based on that intuition we have developed two novel sampling mechanisms that adapt the proposal towards the behavior of some of the general PFs while the PF is running. In the following section we present the adapting proposals, the mechanisms that control the adaptation and the PFs based on these two.

4.1 Adaptive Proposals

In this section we present two proposals that let the PFs based on them adapt the behavior towards two different types of PFs. Both are based on the adaptation of the parallel proposal[3] of Eq. (5) towards the behavior obtained in KLFP or SPF by the simultaneous use of TIS based on the proposal of Eq. (4) and WR based on the values of the weights.

[3] We adapt the parallel proposal toward the other behaviors due to its higher complexity.

Grouped Parallel Proposal: The parallel proposal of Eq. (5) divides the $\boldsymbol{Y}_t$ used to pick the values $\boldsymbol{x}_{t-1}^{(i)}$ to sample from the disjoint subsets $\boldsymbol{X}$ according to a given $\mathcal{P}(\boldsymbol{X}_t)$. The $\mathcal{P}(\boldsymbol{X}_t)$ definition is not unique, but it must at least fulfill a set of restrictions imposed by the DBN structure [6]. The basic disjoint subsets of $\boldsymbol{X} \in \boldsymbol{X}_t$ and $\boldsymbol{Y} \in \boldsymbol{Y}_t$ imposed by the DBN structure and defined by the partitions, named hereafter as $\mathcal{P}^{DBN}(\boldsymbol{X}_t)$ and $\mathcal{P}^{DBN}(\boldsymbol{Y}_t)$, can be grouped to define new pairs of partitions $(\mathcal{P}'(\boldsymbol{X}_t), \mathcal{P}'(\boldsymbol{Y}_t))$.

As different pairs of partitions modify the behavior of the parallel proposal presented in Eq. (5), the *grouping* of the existing subsets to define new partitions and the use of the different partition pairs $(\mathcal{P}(\boldsymbol{X}_t), \mathcal{P}(\boldsymbol{Y}_t))$ in each time step of the PPFs makes the PPFs achieve a dynamic sampling behavior.

This simple idea is exploited by our first adaptive proposal. Originally we use the proposal of Eq. (5) with the pair $(\mathcal{P}^{DBN}(\boldsymbol{X}_t), \mathcal{P}^{DBN}(\boldsymbol{Y}_t))$, and in a posterior time step we join two of the subsets of the previous partition pair to define a new pair $(\mathcal{P}'(\boldsymbol{X}_t), \mathcal{P}'(\boldsymbol{Y}_t))$ to be used in Eq. (5). The grouping adaptation can continue until all the sets are joined and the final possible pair $(\mathcal{P}^F(\boldsymbol{X}_t) = \boldsymbol{X}_t, \mathcal{P}^F(\boldsymbol{Y}_t) = \boldsymbol{Y}_t)$ is reached. Or stopped to restart the process from the pair $(\mathcal{P}^{DBN}(\boldsymbol{X}_t), \mathcal{P}^{DBN}(\boldsymbol{Y}_t))$ when the mechanism that controls the adaptation considers it beneficial.

This adaptive proposal makes a PPF based on $(\mathcal{P}^{DBN}(\boldsymbol{X}_t), \mathcal{P}^{DBN}(\boldsymbol{Y}_t))$ adapt its behavior toward the KLPF. This is due to the fact that by each grouping step we make a subset of hidden and observation variables bigger and sample the hidden variables within that subspace according to its associated measurements. In the limit it samples all the current hidden values from the same previous particle and picks the selected ones according to the product of likelihoods[4] of all the measurements.

Linked Serial Proposal: Another way of adapting the behavior of the parallel proposal of Eq. (5) consists of imposing an ordering $\{\boldsymbol{O}_1, \boldsymbol{O}_2, ..., \boldsymbol{O}_{|\mathcal{P}(\boldsymbol{X}_t)|}\}$ on the subsets of hidden variables defined by the partition $(\boldsymbol{O}_k = \boldsymbol{X} \in \mathcal{P}(\boldsymbol{X}_t))$, and sample the hidden variables within each subset following that order and a mixture model of the product of the prior transitions whose weights take into account the product of the expected likelihoods and which particles were sampled in the previous subsets. This way of proceeding, that *links* the chances of picking the elements of a particle with the already selected ones, is defined by Eq. (7) and (8), where j_z represent the particle selected for the same particle in the previous subsets, $\alpha_j^{\boldsymbol{O}_k}$ is the product of expected likelihoods defined in (6), and $\gamma \in [0,1]$ the proposal parameter used to modify the *linking level*.

$$\mathrm{q}\left(\boldsymbol{x}_t^{(i)}|\cdot\right) = \prod_{k=1:|\mathcal{P}(\boldsymbol{X}_t)|} \left(\sum_{j=1}^{N} \left(\beta_j^{\boldsymbol{O}_k} \prod_{X \in \boldsymbol{O}_k} \mathrm{p}\left(x|\mathrm{pa}_{t-1}^{(j)}(X), \mathrm{pa}_t^{(i)}(X)\right) \right) \right) \tag{7}$$

$$\beta_j^{\boldsymbol{O}_k} = \gamma \alpha_j^{\boldsymbol{O}_k} + \frac{(1-\gamma)}{k} \sum_{z=1:k-1} \delta_j^{j_z} \tag{8}$$

This adaptive proposal makes a PF based on it adapt its behavior from PPF to SPF and vice versa. This is due to the fact that when $\gamma = 1$ the proposal behaves as the original parallel proposal, and as the values of γ is decreased the particles used to sample

[4] There is a small additional difference: this new proposal based on Eq. (5) uses the "expected" likelihoods $\mathrm{p}(y|\mathrm{pa}_{t-1}^{(j)}(Y), \mathrm{pa}_t^{*(j)}(Y))$ while KLPF uses the actual ones $\mathrm{p}(y|\mathrm{pa}_{t-1}^{(j)}(Y), \mathrm{pa}_t^{(j)}(Y))$.

the hidden variables within each $\boldsymbol{O}_k$ has a higher dependency on which particles were already used to sample the previously sampled subset of hidden variables. As the subsets of observed variables used to calculate the $\alpha_j^{\boldsymbol{X}}$ don't change, this adaptive proposal samples the values within each subset taking into account the different subsets of observations and so it makes the PFs based on it behave more closely to SPF than to KLPF. In the limit, when $\gamma = 0$, the only set of measurements considered to select the particles is that associated with the first subspace of the ordering, ignoring the effect of the rest.

4.2 Adaptive Control Mechanism

The PFs based on the two adapting proposals presented in the previous section need a mechanism that modifies their behavior by means of the grouping process used in the first case or the linking parameter γ in the second. The current mechanisms consist of:

1. Initializing the new proposals to work according to the original parallel one. That is, the grouped parallel proposal uses Eq. (5) with $(\mathcal{P}^{DBN}(\boldsymbol{X}_t), \mathcal{P}^{DBN}(\boldsymbol{Y}_t))$ and the linked serial proposal uses Eq. (7) with $(\mathcal{P}^{DBN}(\boldsymbol{X}_t), \mathcal{P}^{DBN}(\boldsymbol{Y}_t))$ and $\gamma = 1$.
2. In each time step of the PF, if there is a $w(\cdot) \neq 0$, the proposal adapts its behavior away from the original parallel one. In the grouped parallel proposal two of the current independent subsets are picked randomly and joined. In the linked serial proposal the ordering of the subsets is randomly determined and γ is decremented a selected amount $\Delta\gamma$ up to a selected threshold $\gamma_{\min}$.
3. In each time step of the PF, if all $w(\cdot) = 0$, the proposals are reset to its original value (step 1).

The insight behind the selected mechanism is the following. On one hand, starting with the original parallel proposal and returning to it when the PF have just lost track ($w(\cdot) = 0$) increments the chances to start from a better initial distribution or recover based on the measurements. On the other hand, moving away from the original parallel proposal when the PF is keeping track lets it increase their capacity of being probabilistically consistent with the past history. The random selection of the sets to group or the ordering of the existing sets is used to avoid prioritizing some possibilities over others.

4.3 The Complete Adaptive Particle Filters

The first step, before carrying out the filtering steps of the novel PFs, consists of 1) defining the disjoint partitions $(\mathcal{P}^{DBN}(\boldsymbol{X}_t), \mathcal{P}^{DBN}(\boldsymbol{Y}_t))$, 2) creating the original particles $x_0^{(i)}$ using an initialization proposal, and 3) setting up the original values of the selected proposal (either Eq. (5) and (6) with $\mathcal{P}^{DBN}(\boldsymbol{X}_t)$, or Eq. (7) and (8) with a random ordering $\boldsymbol{O}_k = \boldsymbol{X} \in \mathcal{P}^{DBN}(\boldsymbol{X}_t)$ and $\gamma = 1$). Next, and in each time step t:

1. The PF samples the values of the hidden variables $\boldsymbol{x}_t^{(i)}$ using one of the new proposals as its currently setup. In short, we calculate either in parallel $\alpha_j^{\boldsymbol{X}}$ or sequentially $\beta_j^{\boldsymbol{O}_r}$ and use those values to select a component of the mixtures and sample the hidden values associated with that component from the product of transition priors.
2. The PF calculates the $w(\cdot)$ of each particle using either Eq. (2) or (3) depending on what posterior we are approximating and Eq. (5) or (7) depending on what proposal

we are using. The numerator can be zero when the values of the hidden variables in the current time slice are non probabilistically consistent with the ones of the previous time slice and/or the measurements. The denominator can't, because we have generated it with the proposal. So, the final $w(\cdot) \geq 0$.

3. If any $w(\cdot) > 0$ then adapt the selected proposal away from the original parallel one. In the grouped parallel proposal two of the current independent subsets are picked randomly and joined. In the linked serial proposal the next ordering of the subsets is randomly determined and γ is decremented by a selected amount $\Delta\gamma$ up to a selected threshold $\gamma_{\min}$. Go to step 1.
4. If all $w(\cdot) = 0$, the proposals are reset to its original value ($\mathcal{P}^{DBN}(\boldsymbol{X}_t)$ in the grouped parallel and $\gamma = 1$ in the linked serial). Go to step 1.

The computational cost of the new adaptive PFs depends on the expressions used for calculating the numerator (TIS or IIS) and denominator (group parallel or linked serial proposal) of (2) or (3). The new TIS PFs need less computation than their IIS counterparts. The corresponding PPFs need more/less computations than the ones with the grouping/linked proposals. KLPF and SPF are less computational demanding.

5 Related Work

Our novel PFs, called GPF and LPF after their *"Grouping"* and *'Linked"* sampling proposals, represent new points in the spectrum of Monte Carlo sequential importance sampling strategies that adapt their behavior from the parallel PF in [6] to either the standard PF in [7] or the serial PF in [5], as Fig. 1 represents.

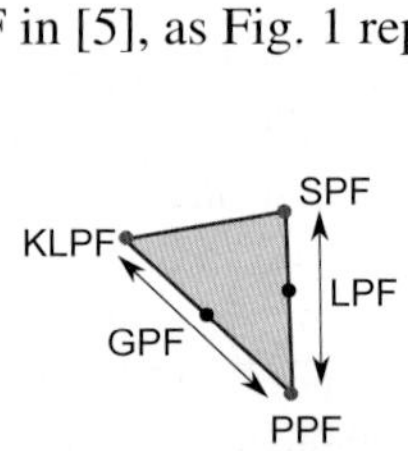

Fig. 1. Relationships among the behaviors of the different PFs

They are not the first adaptive PFs as there exist others that dynamically change the number of particles [11,12,13], change the transition model $p(X|\text{pa}_{t-1}^{(i)}(X), \text{pa}_t^{(i)}(X))$ [13,14], enable and disable the WR step [10,13], or use a WR step with an $r(\cdot)$ function whose values are adapted/smoothed with the previous time step weights [15,16]. Our approach is different: we make use of adaptive sampling proposals controlled by an adaptive mechanism to modify the novel PF behaviors towards several general PFs.

Finally, the idea of using a proposal that samples form the past history and current measurements simultaneously has also been used in the mixture Monte Carlo PF [17], which can't be generally used because sampling new values from the current measurements is not always possible.

6 Experimental Results

The following results compare the new PFs (GPF and LPF with $\Delta\gamma = 0.05$ and $\gamma_{min} = 0.1$) with KLPF, SPF and PPFs using two sets of experiments. The first set, performed with simulated data compare the performance of the PFs using the Root Mean Square Error (RMSE) between the mean value of the particle estimates for each PF and the true value of the hidden variables at the last step of the simulation. In the second, carried out with real world data, we use the PFs to score the structure of a DBN given the measurements. In both cases, the PFs with parallel sampling possibilities (PPF, GPF, LPF) outperform the rest (KLPF and SPF) on average. Besides, the novel GPF and LPF and the old PPF can change their relative performance based on the task.

6.1 Simulated Experiments

This section compares all the PFs, working with a limited number of particles, in two complex real problems modeled by different DBNs. We have selected them because the structures of their DBNs, with multiple hidden variables that can be sampled in parallel, let them benefit significantly from PFs with parallel sampling possibilities. In both cases, we want to localize a set of G mobile objects given the information provided by the existing sensors. To be able to distinguish the variables related with the l-th object, in this section some variable names have a subindex (V_l).

In the first problem, each mobile object (M_l) is repelled from another (M_k) by a common unknown force (F), and the position of each mobile is observed by a different sensor (S_l). The PFs have to estimate the probability of the hidden variables (M_l, F) given the measurements (S_l) for the DBN whose structure[5] is defined by $\text{Pa}_{t-1}(M_l) = \{M_l, M_{rem(l,G)+1}, F\}$, $\text{Pa}_{t-1}(F) = \{F\}$ and $\text{Pa}_t(S_l) = \{M_l\}$. The structure of this DBN lets us divide the hidden variables in $G+1$ groups, each with a hidden variable, and assign to the $\alpha_j^{\boldsymbol{X}}$ of each group the estimated likelihoods of the measurement associated with each hidden variable ($\alpha_j^{\{F\}} \propto 1$ and $\alpha_j^{\{M_l\}} \propto \text{p}(s_l|\text{pa}_t^{*(j)}(S_l))$). The difficulty of this problem originates from the high coupling of the M_l variables imposed by $\text{Pa}_{t-1}(M_l)$.

The second problem is an abstraction of a sea rescue problem where an Unmanned Air Vehicle (UAV) has to track several objects that are in the water and whose initial positions are not completely known. The objects (M_l) move with the direction of the sea current and wind (E) and the sensors, which are onboard the UAV (U), are only able to detect (D_l) the mobiles that are within a circular area of the UAV and provide their position (S_l) when detected. The PFs have to estimate the probability of the hidden variables (M_l, E) given the observed ones (U, D_l, S_l) for the DBN whose structure is defined by $\text{Pa}_{t-1}(M_l) = \{M_l, E\}$, $\text{Pa}_{t-1}(E) = \{E\}$, $\text{Pa}_t(D_l) = \{M_l, U\}$ and $\text{Pa}_t(S_l) = \{M_l, D_l\}$. Again, the structure of this DBN let us divide the hidden variables in $G+1$ groups, each with a hidden variable, and assign to the $\alpha_j^{\boldsymbol{X}}$ of each group the product of the estimated likelihoods of the measurements associated with each hidden variable ($\alpha_j^{\{E\}} \propto 1$ and $\alpha_j^{\{M_l\}} \propto \text{p}(d_l|\text{pa}_t^{*(j)}(D_l))\text{p}(s_l|\text{pa}_t^{*(j)}(S_l))$). This problem is less coupled than the previous, but it is also difficult: it has multiple measurements per hidden variable and when the object is outside the circular area its position is not observed.

[5] In the following, $rem(a,b)$ represents the remainder of dividing a by b.

For each problem, we obtain the observations and the real values of the hidden variables from a simulation based on its probability model. Then we run each PF with a given number of particles and 50 different sets of initial particles (the same for each PF to compare their behavior under the same initial conditions). And finally, we measure at the last step of the simulation the performance of each PF using the Root Mean Square Error (RMSE) between the mean value of the particle estimates and the true value.

The complete experiment setup consists of running for each problem the 8 PFs with 3 different number of particles, 3 different initialization region proposals and the same 50 sets of initial particles. The 8 PFs configurations are: KLPF, "Instantaneous" GPF, "Trajectory" GPF, "Instantaneous" LPF, "Trajectory" LPF, "Instantaneous" PPF, "Trajectory" PPF, and SPF. The tested number of particles N are 20, 50, 100. The 3 initialization proposals are uniform rectangular regions of different sizes around the actual initial value: the second quadruples the area of the first, and the third quadruples the area of the second. Each of the 50 sets of initial particles is created for each problem for a combination of initialization proposal and number of particles.

Figure 2 shows the results as ratios of the RMSE obtained by each PF (x-axis) and the RMSE obtained by the "Instantaneous" PPF for each problem (Fig. 2a and Fig. 2b), number of particles (figure row), initialization region (figure column) and initial particle set (y-axis). Mid-grey (orange) ratios represent the equality. They are shown as reference in the sixth column of each figure as the ratio of RMSE("Instantaneous" PPF)/RMSE("Instantaneous" PPF)=1. Darker gray (red) shows that RMSE of the "Instantaneous" GPF is better and lighter (yellow) that it is worse. Ratios bigger that 10 or smaller than 1/10 are represented in black and white, and the difference between ratios$>$1 and ratios$<$1 is easily observed by not using the gray levels (colors) around the mid-gray (orange) one.

As the ratios in Fig. 2 show the improvement with respect to the "Instantaneous" PPF, we can't quantify how well each of the PFs is actually doing. This information appears in Fig. 3 that divides the results of each PF in each configuration in "convergent" (white) and "divergent" (black) based on their RMSE values and a fixed threshold.

Fig. 2 shows that the selected PF usually outperforms, many times with ratios$>$10, KLPF (first dark column) and SPF (last dark column) in both problems and all configurations. The same happens when any of the PFs with parallel sampling possibilities is selected: KLPF and SPF rarely outperform GPF, LPF, and PPF. The worst behavior of KLPF and SPF with respect to the others is also illustrated in Fig. 3, where the results of KLPF and SPF are also usually classified as divergent.

The relations among the PFs with parallel sampling possibilities change with the problem and number of particles: For problem 1, Fig. 2a shows that the PFs compete: the LPFs (columns 4 and 5) are usually worst, and the GPF (columns 2 and 3) behavior improves as we increment the number of particles and decrement the area of the initialization proposal. This behavior is not followed by 20 particles and the third initialization proposal configuration. However, many of the solutions are divergent in the corresponding graphic in Fig. 3a and for that reason we don't consider the ratios that meaningful in that case. For problem 2, Fig. 2b shows that the adaptive PFs are usually better than the original PPF. Moreover, the behavior of GPF is maintained through all the configurations, while the behavior of the LPF improves with the number of particles.

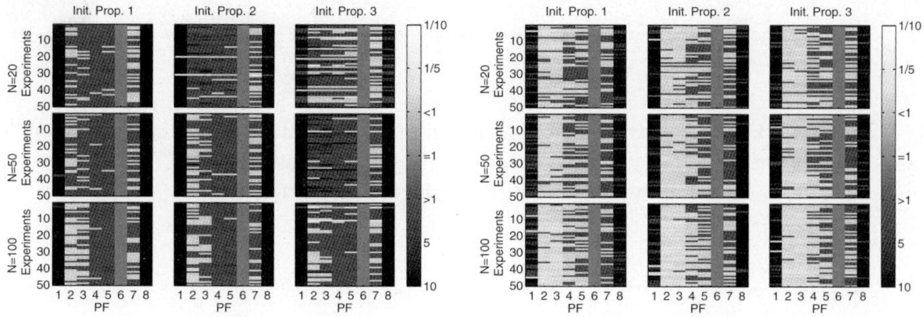

(a) Problem 1 (b) Problem 2

Fig. 2. Ratios of the RMSE of each PF and the RMSE of the "Instantaneous" PPF. In the x-axis, 1 is KLPF, 2 -"Instantaneous" GPF, 3 - "Trajectory" GPF, 4 - "Instantaneous" LPF, 5 -"Trajectory" LPF, 6 - "Instantaneous" PPF, 7 - "Trajectory" PPF, and 8 - SPF

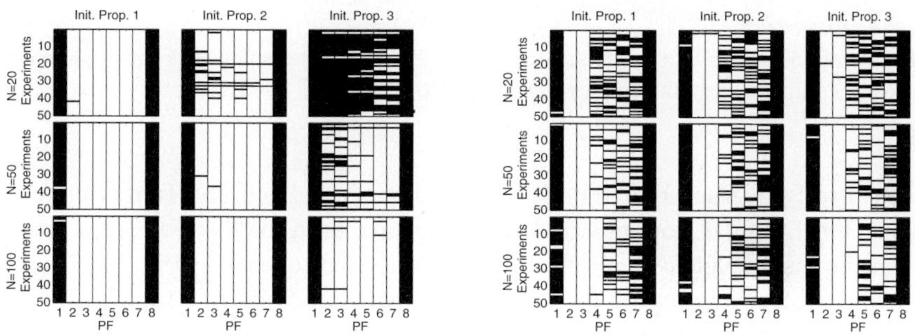

(a) Problem 1 (b) Problem 2

Fig. 3. "Convergent" (white) and "divergent" (black) experiments for all the PFs

Finally, Fig. 3 shows that GPF usually converges (and does it more often than the others except for problem 1 with small number of particles and wider initialization proposal), and that LPF and SPF have similar convergent numbers of experiments.

So, for the selected problems GPF is usually the best (except in problem 1 with just a few particles and extremely wide initialization regions), followed in problem 1 by PPF and in problem 2 by sometimes PPF and others LPF. Besides, the number of convergent runs in a configuration for the adaptive PFs is rarely worse than for the fixed PPF, and so, although the latter can obtain punctually better values, overall the adaptive PFs produce a convergent run at least as often as the PPF.

6.2 Real Data

For the real data experiment we have used a functional Magnetic Resonance Imaging (fMRI) dataset [18]. FMRI signal represents the Blood Oxygenation Level Dependent (BOLD) response measured in a small rectangular region of the brain (voxel). BOLD

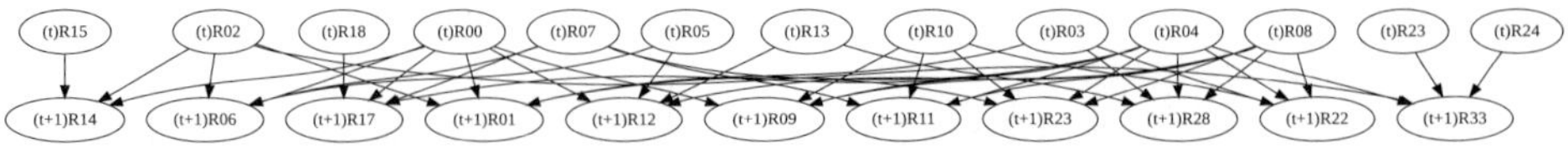

Fig. 4. The network for a subset of regions of interest (ROIs) from the Talairach anatomical atlas database. ROI names are replaced by short labels since they are not significant for our work. Note that each ROI is observed through its indirect measurement by fMRI.

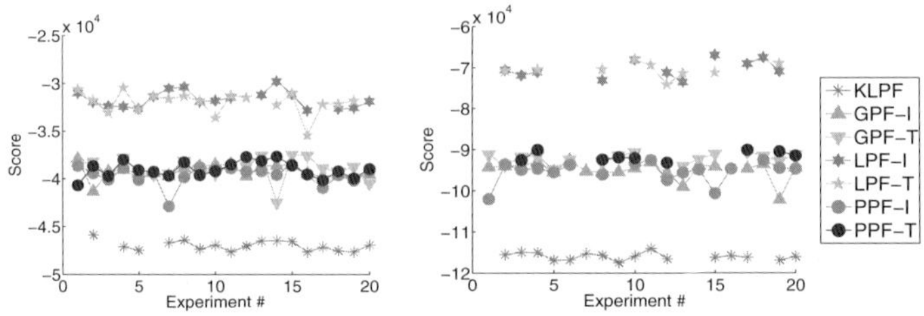

(a) 20 particles (b) 50 particles

Fig. 5. Cross validation log likelihood score plots for the PFs. Not displayed points mean that the kernel density estimator did not produce a valid result.

response is itself governed by the underlying hidden neural activity. To model generation of the BOLD signal from neural activity we have used the hemodynamic forward model based on a coupled system of ordinary differential equations [19].

The number of voxel measurements collected per time point in a typical fMRI experiment is too large to model directly. Thus voxel time courses were averaged on a per Region of Interest (ROI) basis. ROIs were selected according to the widely used Talairach database [20].

In order to obtain the underlying structure of the DBN, we have used the approach that treats fMRI data as fully observed and quantizes it into categorical representation [18]. Among the discovered DBN families we have used several most significant ones according to the cross validation procedure (t-test with p-value of 0.05). This resulted in a DBN of 42 hidden variables per slice. Figure 4 shows a small portion of the hidden structure of the DBN we use. The observation nodes are not shown.

All PFs were run on this dataset for 50 time points (100 seconds with 2 seconds fMRI sampling rate). Since the ground truth for neural activity of ROIs is unknown in this dataset, for evaluation we have used cross validation log likelihood based on kernel density estimators with Gaussian kernels and automatically chosen variance value using a cross validation procedure [21].

The results[6] for 20 runs using 20 and 50 particles are shown in Figure 5. LPFs (stars) show higher score (better), followed by GPFs (triangles) and PPFs (circles). KLPF (asterisks) has considerably lower mean score. Note how the score decreases as the number of particles grows. This is an expected behavior with Parzen window estimators, that is

[6] SPF is missing because the kernel density estimator didn't produce valid results.

due to the smoother and generally wider estimates in the cases of more data that lead to lower probability values.

In part these results support the observations obtained on simulated data: the PFs with parallel sampling possibilities (GPFs, LPFs, and PPFs) have improved performance over the other tested PFs. However, as Figure 5 shows, the adaptive LPFs outperform GPFs and PPFs. This means that the benefit of the use of adaptive PFs depends on the problem. Besides, in this case, the PPF performance can be considered a lower bound of GPFs and LPFs, because the results of the adaptive PFs are not usually worse than the results of the fixed PPFs.

7 Conclusions

We present a set of adaptive PFs to estimate the trajectory or current state of the hidden variables of a DBN. They use two adaptive proposals to sample in parallel/serial the values of changing/static-linked subsets of hidden variables to build a whole particle whose weight is updated according to the proposal and the estimation problem.

Our PFs explore different subspaces of the state space while performing the sampling step, and the whole space as a block while updating the weights. In the grouping parallel proposal the variables in the subsets change dynamically while in the linked serial proposal the variables are maintained and the subsets randomly linked. Each proposals is used to develop two different PFs that adapt their behavior from the parallel PF in [6] to the standard PF in [7] or the serial PF in [5].

The tests used to compare the new and the old PFs show that, with a reduced number of particles when the particles are initialized in spread regions, all the PFs with parallel sampling possibilities (the novel PFs and the old parallel PF) usually outperform the standard and serial PF. Besides, the new PFs usually demonstrate a behavior similar or better than the parallel PF. The different behavior of the adaptive PFs in the different problems implies that different adaptive PFs are good for different DBNs. Determining the best method for each DBN type requires further study which is left for future work.

Finally, it is worth mentioning that the results of the adaptive PFs also depend on their adaptation mechanism. The one that we are testing so far moves more slowly from the parallel PF to the others than the other way around (it resets completely). Besides, we decide the direction of the movement only taking into account if there is at least one particle whose weight is not zero. Although it is enough for letting the novel PFs outperform the already existing PFs, we are planning to explore the use of better-balanced and softer decision (for instance, based in the number of effective particles [10]) mechanisms in our future work.

Acknowledgments

This work was supported by the Spanish Grant DPI2009-14552-C02-01. Further, Dr. Besada-Portas was supported by the Spanish post-doctoral Grant EX-2007-0915, Dr. Plis by NIH 1 R01 EB 006841, 1 R01 EB 005846, and Dr. Lane by NSF Grant IIS-0705681. The authors also thank the Aula Sun-UCM for providing access to their computational resources for doing parts of the experiments.

References

1. Doucet, A., Freitas, N., Gordon, N. (eds.): Sequential Monte Carlo methods in practice. Springer, Heidelberg (2001)
2. MacCormick, J., Isard, M.: Partitioned sampling, articulated objects, and interface-quality hand tracking. In: Vernon, D. (ed.) ECCV 2000. LNCS, vol. 1843, pp. 3–19. Springer, Heidelberg (2000)
3. Doucet, A., de Freitas, N., Murphy, K., Russell, S.: Rao-blackwellised particle filtering for dynamic bayesian networks. In: 16th Conf. on Uncertainty in Artificial Intelligence (2000)
4. Vaswani, N.: Particle filters for infinite (or large) dimensional state spaces-part 2. In: IEEE ICASSP (2006)
5. Rose, C., Saboune, J., Charpillet, F.: Reducing particle filtering complexity for 3D motion capture using dynamic bayesian networks. In: Proc. of AAAI 2008 (2008)
6. Besada-Portas, E., Pliz, S.M., de la Cruz, J.M., Lane, T.: Parallel subspace sampling for particle filtering in dynamic bayesian networks. In: Proc. of ECML 2009 (2009)
7. Koller, D., Lerner, U.: Sampling in factored dynamic systems. In: Sequential Monte Carlo in Practice, pp. 445–464 (2001)
8. Maccormick, J., Blake, A.: A probabilistic exclusion principle for tracking multiple objects. Intenational Journal of Computer Vision 39(1), 57–71 (2000)
9. Klaas, M., de Freitas, N., Doucet, A.: Toward practical n^2 Monte Carlo: the Marginal Particle Filter. In: Proceedings of UAI 2005 (2005)
10. Doucet, A., Godsill, S., Andrieu, C.: On sequential monte carlo sampling methods for bayesian filtering. Statistics and Computing 10, 197–208 (2000)
11. Fox, D.: Kld-sampling: Adaptive particle filters. In: NIPS 2001 (2001)
12. Soto, A.: Self adaptive particle filter. In: International Joint Conference on Artificial Intelligence (2005)
13. Bolie, M., Hong, S., Djurie, P.M.: Performance and complexity analysis of adaptive particle filtering for tracking applications. In: 36th Conf. on Signals, Sys. and Comput. (2002)
14. Duan, Z., Cai, Z., Yu, J.: Robust position tracking for mobile robots with adaptive evolutionary particle filter. In: Proc. of ICNC 2007 (2007)
15. Nistico, W., Hebbel, M.: Temporal smoothing particle filter for vision based autonomous mobilde robot localization. In: Proc. of ICINCO 2008 (2008)
16. Roefer, T., Jungel, M.: Vision-based fast and reactive monte-carlo localization. In: Proc of ICRA 2003 (2003)
17. Thrun, S., Fox, D., Burgard, W., Dellaert, F.: Robust monte carlo localization for mobile robots. Artificial Intelligence 128 (2001)
18. Burge, J., Lane, T., Link, H., Qiu, S., Clark, V.P.P.: Discrete dynamic bayesian network analysis of fMRI data. Human Brain Mapping (November 2007)
19. Friston, K.J., Harrison, L., Penny, W.: Dynamic Causal Modelling. NeuroImage 19(4) (2003)
20. Lancaster, J.L., Woldorff, M.G., Parsons, L.M., Liotti, M., Freitas, C.S., Rainey, L., Kochunov, P.V., Nickerson, D., Mikiten, S.A., Fox, P.T.: Automated Talairach atlas labels for functional brain mapping. Human Brain Mapping 10(3), 120–131 (2000)
21. Hofmann, R., Tresp, V.: Discovering structure in continuous variables using bayesian networks. In: Advances in Neural Information Processing Systems, vol. 8 (1996)

Leveraging Bagging for Evolving Data Streams

Albert Bifet, Geoff Holmes, and Bernhard Pfahringer

University of Waikato, Hamilton, New Zealand
{abifet,geoff,bernhard}@cs.waikato.ac.nz

Abstract. Bagging, boosting and Random Forests are classical ensemble methods used to improve the performance of single classifiers. They obtain superior performance by increasing the accuracy and diversity of the single classifiers. Attempts have been made to reproduce these methods in the more challenging context of evolving data streams. In this paper, we propose a new variant of bagging, called *leveraging bagging*. This method combines the simplicity of bagging with adding more randomization to the input, and output of the classifiers. We test our method by performing an evaluation study on synthetic and real-world datasets comprising up to ten million examples.

1 Introduction

Data Stream real time analytics are needed to manage data generated at an increasing rate from sensor applications, measurements in network monitoring and traffic management, log records or click-streams in web exploring, manufacturing processes, call detail records, email, blogging, twitter posts and others. In fact, all data generated can be considered as streaming data or as a snapshot of streaming data, since it is obtained from an interval of time.

In the data stream model, data arrive at high speed, and algorithms that process them must do so under very strict constraints of space and time. Consequently, data streams pose several challenges for data mining algorithm design. First, algorithms must make use of limited resources (time and memory). Second, they must deal with data whose nature or distribution changes over time.

Bagging and Boosting are ensemble methods used to improve the accuracy of classifier methods. Non-streaming bagging [7] builds a set of M base models, training each model with a bootstrap sample of size N created by drawing random samples with replacement from the original training set. Each base model's training set contains each of the original training example K times where $P(K = k)$ follows a binomial distribution. This binomial distribution for large values of N tends to a Poisson(1) distribution, where Poisson(1)= $\exp(-1)/k!$. Using this fact, Oza and Russell [25,24] proposed *Online Bagging*, an online method that instead of sampling with replacement, gives each example a weight according to Poisson(1).

Boosting algorithms combine multiple base models to obtain a small generalization error. Non-streaming boosting builds a set of models sequentially, with

J.L. Balcázar et al. (Eds.): ECML PKDD 2010, Part I, LNAI 6321, pp. 135–150, 2010.

the construction of each new model depending on the performance of the previously constructed models. The intuitive idea of boosting is to give more weight to misclassified examples, and reducing the weight of the correctly classified ones.

From studies appearing in the literature [25,24,6], Online Bagging seems to perform better than online boosting methods. Why bagging outperforms boosting in the data stream setting is still an open question. Adding more random weight to all instances seems to improve accuracy more than adding weight to misclassified instances. In this paper we focus on randomization as a powerful tool to increase accuracy and diversity when constructing an ensemble of classifiers. There are three ways of using randomization:

- Manipulating the input data
- Manipulating the classifier algorithms
- Manipulating the output targets

In this paper we focus on randomizing the input data and the output prediction of online bagging. The paper is structured as follows: related work is presented in Section 2. Leveraging bagging is discussed in Section 3. An experimental evaluation is conducted in Section 4. Finally, conclusions and suggested items for future work are presented in Section 5.

2 Related Work

Breiman [7] introduced bagging classification using the notion of an *order-correct* learner. An order-correct learner ϕ at the input x is a predictor that if input x results in a class more often than any other class, then ϕ will also predict this class at x more often than any other class. An order-correct learner is not necessarily an accurate predictor but its aggregated predictor is optimal. If a predictor is good because it is order-correct for most inputs x then aggregation can transform it into a nearly optimal predictor. The vital element to gain accuracy is the instability of the prediction method. A learner is unstable if a small change in the input data leads to large changes in the output.

Friedman [16] explained that bagging works by reducing variance without changing the bias. There are several definitions of bias and variance for classification, but the common idea is that bias measures average error over many different training sets, and variance measures the additional error due to the variation in the model produced by using different training sets.

Domingos [13] claimed that Breiman's line of reasoning is limited, since we may never know *a priori* whether a learner is order-correct for a given example or not, and what regions of the instance space will be order-correct or not. He explained bagging's success showing that bagging works by effectively changing a single-model learner to another single-model learner, with a different implicit prior distribution over models, one that is less biased in favor of simple models.

Some work in the literature shows that bagging asymptotically performs some smoothing on the estimate. Friedman and Hall [15] used an asymptotic truncated Taylor series of the estimate to show that in the limit of infinite samples, bagging reduces the variance of non-linear components.

Bühlmann and Yu [10], analyzed bagging using also asymptotic limiting distributions, and they proposed *subagging* as a less expensive alternative to bagging. Subagging uses subsampling as an alternative aggregation scheme. They claimed that subagging is as accurate as bagging but uses less computation.

Grandvalet[19] explained the performance of bagging by the goodness and badness of highly influential examples, in situations where the usual variance reduction argument is questionable. He presented an experiment showing that bagging increases the variance of decision trees, and claimed that bagging does not simply reduce variance in its averaging process.

In [6] two new state-of-the-art bagging methods were presented: ASHT Bagging using trees of different sizes, and `ADWIN` Bagging using a change detector to decide when to discard underperforming ensemble members.

Breiman [8] proposed Random Forests as a method to use randomization on the input and on the internal construction of the decision trees. Random Forests are ensembles of trees with the following characteristics: the input training set is obtained by sampling with replacement, the nodes of the tree only may use a fixed number of random attributes to split, and the trees are grown without pruning. Abdulsalam et al. [1] presented a streaming version of random forests and Saffari et al. [26] presented an online version.

3 Leveraging Bagging

In this section, we present a new online leveraging bagging algorithm, improving Online Bagging of Oza and Russell. The pseudo-code of Online Bagging of Oza and Russell is listed in Algorithm 1.

We leverage the performance of bagging, with two randomization improvements: increasing resampling and using output detection codes.

Resampling with replacement is done in Online Bagging using Poisson(1). There are other sampling mechanisms:

- Lee and Clyde [23] uses the Gamma distribution (Gamma(1,1)) to obtain a Bayesian version of Bagging. Note that Gamma(1,1) is equal to Exp(1).
- Bulhman and Yu [10] proposes subagging, using resampling without replacement.

Algorithm 1. Oza and Russell's *Online Bagging* for M models

```
Initialize base models h_m for all m ∈ {1, 2, ..., M}
for all training examples do
    for m = 1, 2, ..., M do
        Set w = Poisson(1)
        Update h_m with the current example with weight w
anytime output:
return hypothesis: h_fin(x) = arg max_{y∈Y} Σ_{t=1}^{T} I(h_t(x) = y)
```

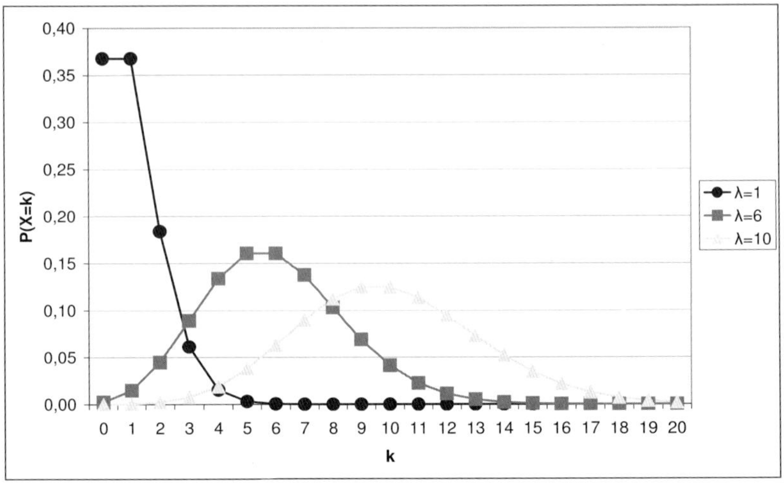

Fig. 1. Poisson Distribution

Our proposal is to increase the weights of this resampling using a larger value λ to compute the value of the Poisson distribution. The Poisson distribution is used to model the number of events occurring within a given time interval.

Figure 1 shows the probability function mass of the distribution of Poisson for several values of λ. The mean and variance of a Poisson distribution is λ. For $\lambda = 1$ we see that 37% of the values are zero, 37% are one, and 26% are values greater than one. Using a weight of Poisson(1) we are taking out 37% of the examples, and repeating 26% of the examples, in a similar way to non streaming bagging. For $\lambda = 6$ we see that 0.25% of the values are zero, 45% are lower than six, 16% are six, and 39% are values greater than six. Using a value of $\lambda > 1$ for Poisson(λ) we are increasing the diversity of the weights and modifying the input space of the classifiers inside the ensemble. However, the optimal value of λ may be different for each dataset.

Our second improvement is to add randomization at the output of the ensemble using output codes. Dietterich and Bakiri [12] introduced a method based on error-correcting output codes, which handles multiclass problems using only a binary classifier. The classes assigned to each example are modified to create a new binary classification of the data induced by a mapping from the set of classes to {0,1}. A variation of this method by Schapire [27] presented a form of boosting using output codes.

We assign to each class a binary string of length n and then build an ensemble of n binary classifiers. Each of the classifiers learns one bit for each position in this binary string. When a new instance arrives, we assign x to the class whose binary code is closest. We can view an error-correcting code as a form of voting in which a number of incorrect votes can be corrected.

We use random output codes instead of deterministic codes. In standard ensemble methods, all classifiers try to predict the same function. However, using output codes each classifier will predict a different function. This may reduce the effects of correlations between the classifiers, and increase diversity of the ensemble.

We implement random output codes in the following way: we choose for each classifier m and class c a binary value $\mu_m(c)$ in a uniform, independent, and random way. We ensure that exactly half of the classes are mapped to 0. The output of the classifier for an example is the class which has more votes of its binary mapping classes. Table 1 shows an example for an ensemble of 6 classifiers in a classification task of 3 classes.

Table 1. Example matrix of random output codes for 3 classes and 6 classifiers

	Class 1	Class 2	Class 3
Classifier 1	0	0	1
Classifier 2	0	1	1
Classifier 3	1	0	0
Classifier 4	1	1	0
Classifier 5	1	0	1
Classifier 6	0	1	0

We use the strategy of [6] to deal with concept drift. `ADWIN` [3] is a change detector and estimator that solves in a well-specified way the problem of tracking the average of a stream of bits or real-valued numbers. `ADWIN` keeps a variable-length window of recently seen items, with the property that the window has the maximal length statistically consistent with the hypothesis "there has been no change in the average value inside the window".

`ADWIN` is parameter- and assumption-free in the sense that it automatically detects and adapts to the current rate of change. Its only parameter is a confidence bound δ, indicating how confident we want to be in the algorithm's output, inherent to all algorithms dealing with random processes.

Also important for our purposes, `ADWIN` does not maintain the window explicitly, but compresses it using a variant of the exponential histogram technique.

Algorithm 2. *Leveraging Bagging* for M models

Initialize base models h_m for all $m \in \{1, 2, ..., M\}$
Compute coloring $\mu_m(y)$
for all training examples (x, y) **do**
    for $m = 1, 2, ..., M$ **do**
        Set $w = Poisson(\lambda)$
        Update h_m with the current example with weight w and class $\mu_m(y)$
if `ADWIN` detects change in error of one of the classifiers **then**
    Replace classifier with higher error with a new one
anytime output:
return hypothesis: $h_{fin}(x) = \arg\max_{y \in Y} \sum_{t=1}^{T} I(h_t(x) = \mu_t(y))$

This means that it keeps a window of length W using only $O(\log W)$ memory and $O(\log W)$ processing time per item.

Algorithm 2 shows the pseudo-code of our Leveraging Bagging. First we build a matrix with the values of μ for each classifier and class. For each new instance that arrives, we give it a random weight of $Poisson(k)$. We train the classifier with this weight, and when a change is detected, the worst classifier of the ensemble of classifiers is removed and a new classifier is added to the ensemble. To predict the class of an example, we compute for each class c the sum of the votes for $\mu(c)$ of all the ensemble classifiers, and we output as a prediction the class with the most votes.

4 Comparative Experimental Evaluation

Massive **O**nline **A**nalysis (MOA) [4] is a software environment for implementing algorithms and running experiments for online learning from data streams. All algorithms evaluated in this paper were implemented in the Java programming language by extending the MOA software.

We use the experimental framework for concept drift presented in [6]. Considering data streams as data generated from pure distributions, we can model a concept drift event as a weighted combination of two pure distributions that characterizes the target concepts before and after the drift. This framework defines the probability that a new instance of the stream belongs to the new concept after the drift based on the sigmoid function.

Definition 1. *Given two data streams a, b, we define $c = a \oplus^W_{t_0} b$ as the data stream built by joining the two data streams a and b, where t_0 is the point of change, W is the length of change,* $\Pr[c(t) = b(t)] = 1/(1 + e^{-4(t-t_0)/W})$ *and* $\Pr[c(t) = a(t)] = 1 - \Pr[c(t) = b(t)]$.

In order to create a data stream with multiple concept changes, we can build new data streams joining different concept drifts, i. e. $(((a \oplus^{W_0}_{t_0} b) \oplus^{W_1}_{t_1} c) \oplus^{W_2}_{t_2} d) \ldots$.

4.1 Datasets for Concept Drift

Synthetic data has several advantages – it is easier to reproduce and there is little cost in terms of storage and transmission. For this paper we use the data generators most commonly found in the literature.

SEA Concepts Generator. This artificial dataset contains abrupt concept drift, first introduced in [28]. It is generated using three attributes, where only the two first attributes are relevant. All the attributes have values between 0 and 10. The points of the dataset are divided into 4 blocks with different concepts. In each block, the classification is done using $f_1 + f_2 \leq \theta$, where f_1 and f_2 represent the first two attributes and θ is a threshold value. The most frequent values are 9, 8, 7 and 9.5 for the data blocks. In our framework, SEA concepts are defined as follows:

$$(((SEA_9 \oplus^W_{t_0} SEA_8) \oplus^W_{2t_0} SEA_7) \oplus^W_{3t_0} SEA_{9.5})$$

Rotating Hyperplane. This data was used as a testbed for CVFDT versus VFDT in [22]. Examples for which $\sum_{i=1}^{d} w_i x_i \geq w_0$ are labeled positive, and examples for which $\sum_{i=1}^{d} w_i x_i < w_0$ are labeled negative. Hyperplanes are useful for simulating time-changing concepts, because we can change the orientation and position of the hyperplane in a smooth manner by changing the relative size of the weights.

Random RBF Generator. This generator was devised to offer an alternate complex concept type that is not straightforward to approximate with a decision tree model. The RBF (Radial Basis Function) generator works as follows: A fixed number of random centroids are generated. Each center has a random position, a single standard deviation, class label and weight. New examples are generated by selecting a center at random, taking weights into consideration so that centers with higher weight are more likely to be chosen. A random direction is chosen to offset the attribute values from the central point. Drift is introduced by moving the centroids with constant speed.

LED Generator. This data source originates from the CART book [9]. An implementation in C was donated to the UCI [2] machine learning repository by David Aha. The goal is to predict the digit displayed on a seven-segment LED display, where each attribute has a 10% chance of being inverted. The particular configuration of the generator used for the experiments (led) produces 24 binary attributes, 17 of which are irrelevant.

4.2 Real-World Data

The UCI machine learning repository [2] contains some real-world benchmark data for evaluating machine learning techniques. We consider three of the largest: Forest Covertype, Poker-Hand, and Electricity.

Forest Covertype. Contains the forest cover type for 30 x 30 meter cells obtained from US Forest Service (USFS) Region 2 Resource Information System (RIS) data. It contains 581, 012 instances and 54 attributes, and it has been used in several papers on data stream classification [18,25].

Poker-Hand. Consists of 1, 000, 000 instances and 11 attributes. Each record of the Poker-Hand dataset is an example of a hand consisting of five playing cards drawn from a standard deck of 52. Each card is described using two attributes (suit and rank), for a total of 10 predictive attributes. There is one class attribute that describes the "Poker Hand".

Electricity is another widely used dataset described by M. Harries [20] and analysed by Gama [17]. This data was collected from the Australian New South Wales Electricity Market. In this market, prices are not fixed and are affected by demand and supply of the market. They are set every five minutes. The ELEC dataset contains 45, 312 instances. The class label identifies the change of the price relative to a moving average of the last 24 hours.

Table 2. Comparison of Hoeffding Tree, Online bagging and ADWIN bagging. Accuracy is measured as the final percentage of examples correctly classified over the 1 or 10 million test/train interleaved evaluation. Time is measured in seconds, and memory in MB. The best individual accuracies are indicated in boldface.

	Hoeffding Tree			Online Bagging			ADWIN Bagging		
	Time	Acc.	Mem.	Time	Acc.	Mem.	Time	Acc.	Mem.
RBF(0,0)	0.97	88.10 ± 0.34	141.37	27.35	**91.59** ± 0.11	2656.28	27.16	91.58 ± 0.11	3311.22
RBF(50,0.001)	0.97	30.93 ± 0.03	178.30	25.48	32.89 ± 0.04	2894.04	0.25	**41.64** ± 0.04	5481.12
RBF(10,0.001)	0.97	80.23 ± 0.15	137.30	26.90	83.39 ± 0.10	2759.74	26.07	**83.41** ± 0.08	3579.72
RBF(50,0.0001)	0.98	35.25 ± 0.09	166.22	27.85	44.48 ± 0.07	3245.18	0.73	**60.54** ± 0.07	5519.06
RBF(10,0.0001)	0.97	80.95 ± 0.14	132.80	27.86	84.59 ± 0.12	2682.15	26.83	**84.78** ± 0.11	3481.96
LED(50000)	1.94	68.50 ± 0.29	22.99	50.93	69.00 ± 0.16	544.15	5.10	**73.08** ± 0.08	541.42
SEA(50)	0.49	86.48 ± 0.06	5.32	12.13	86.83 ± 0.06	86.66	10.23	**87.59** ± 0.29	107.13
SEA(50000)	0.49	86.45 ± 0.07	5.55	12.12	86.79 ± 0.06	91.43	6.32	**88.32** ± 0.14	99.49
HYP(10,0.001)	1.45	80.70 ± 1.44	85.62	57.85	83.05 ± 1.49	2017.98	28.08	**90.74** ± 0.21	2822.05
HYP(10,0.0001)	1.26	84.12 ± 0.75	85.43	48.91	85.88 ± 0.80	1913.85	36.05	**91.23** ± 0.12	3145.88
CovType	2.65	80.70	27.46	59.83	83.93	345.62	4.80	**84.91**	486.00
Electricity	0.09	79.20	0.98	3.12	82.66	5.88	1.16	**84.51**	7.13
Poker	0.59	77.10	11.62	13.85	**82.29**	171.13	0.39	70.68	202.99
CovPokElec	5.69	77.65	62.63	138.09	**82.67**	1247.47	7.74	76.40	1367.09
	74.03 Acc. 0.01 RAM-Hours 2.86 avg. rank			77.15 Acc. 2.98 RAM-Hours 1.79 avg. rank			**79.24** Acc. 1.48 RAM-Hours 1.36 avg. rank		

Nemenyi significance: Online Bagging≻ Hoeffding Tree; ADWIN Bagging ≻ Hoeffding Tree;

Table 3. Comparison of ADWIN Half subagging, ADWIN subagging, and ADWIN bagging using all instances. Accuracy is measured as the final percentage of examples correctly classified over the 1 or 10 million test/train interleaved evaluation. Time is measured in seconds, and memory in MB. The best individual accuracies are indicated in boldface.

	ADWIN Half subagging			ADWIN Subagging			ADWIN Bagging WT		
	Time	Acc.	Mem.	Time	Acc.	Mem.	Time	Acc.	Mem.
RBF(0,0)	23.99	91.22 ± 0.09	3986.69	24.07	**91.43** ± 0.11	3896.28	31.50	91.36 ± 0.16	3198.21
RBF(50,0.001)	0.16	43.81 ± 0.03	4887.80	0.16	**43.93** ± 0.02	5244.13	0.45	43.66 ± 0.03	6627.15
RBF(10,0.001)	23.28	83.14 ± 0.09	4133.80	23.26	**83.29** ± 0.08	4132.29	30.11	83.04 ± 0.16	3353.40
RBF(50,0.0001)	0.37	56.61 ± 0.05	4927.20	0.41	57.07 ± 0.06	5296.05	2.55	**72.31** ± 0.08	6682.20
RBF(10,0.0001)	23.14	85.07 ± 0.16	4090.61	23.18	**85.12** ± 0.12	4091.19	32.88	84.57 ± 0.11	3272.87
LED(50000)	1.50	73.05 ± 0.07	478.39	1.83	**73.11** ± 0.08	515.13	12.28	73.06 ± 0.07	616.17
SEA(50)	4.83	87.43 ± 0.27	82.08	6.13	87.51 ± 0.28	98.79	19.95	**87.88** ± 0.36	153.28
SEA(50000)	2.91	87.98 ± 0.17	75.05	3.60	88.17 ± 0.18	92.18	12.46	**88.55** ± 0.24	136.75
HYP(10,0.001)	17.81	90.27 ± 0.17	2510.85	20.50	90.35 ± 0.17	2739.65	50.60	**90.59** ± 0.20	3786.17
HYP(10,0.0001)	22.51	90.65 ± 0.09	2605.25	23.49	90.73 ± 0.11	2886.55	59.75	**91.10** ± 0.12	4075.18
CovType	1.82	82.03	507.03	2.18	82.55	525.32	17.74	**88.47**	505.22
Electricity	0.57	82.45	6.91	0.62	83.38	7.42	1.99	**86.67**	8.82
Poker	0.38	69.24	215.07	0.38	69.99	230.15	1.61	**77.35**	213.41
CovPokElec	3.27	74.16	1647.06	3.37	74.94	1411.78	22.30	**82.18**	1527.23
	78.36 Acc. 1.04 RAM-Hours 2.79 avg. rank			78.68 Acc. 1.13 RAM-Hours 1.64 avg. rank			**81.49** Acc. 2.74 RAM-Hours 1.57 avg. rank		

Nemenyi significance: ADWIN Subagging≻ADWIN Half subagging; ADWIN Bagging WT≻ADWIN Half subagging;

We use normalized versions of these datasets, so that the numerical values are between 0 and 1. With the Poker-Hand dataset, the cards are not ordered, i.e. a hand can be represented by any permutation, which makes it very hard for propositional learners, especially for linear ones. We use a modified version, where cards are sorted by rank and suit, and have removed duplicates. This dataset loses about 171,799 examples, and comes down to 829,201 examples.

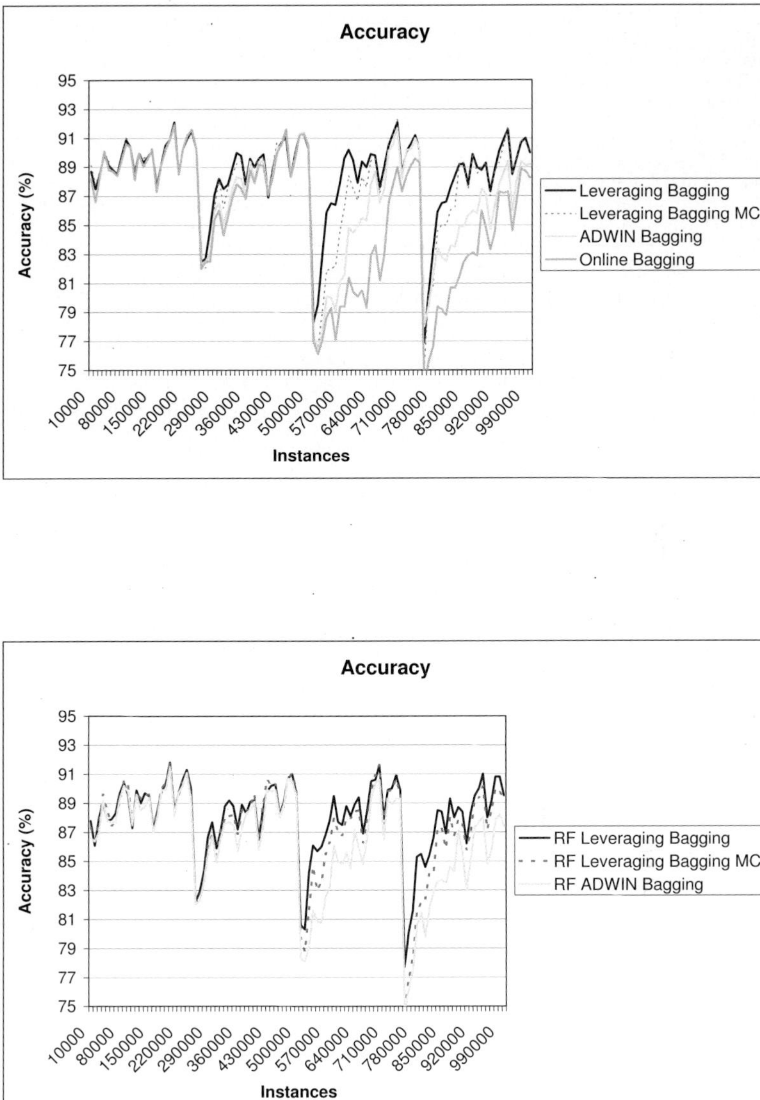

Fig. 2. Accuracy on the SEA data with three concept drifts

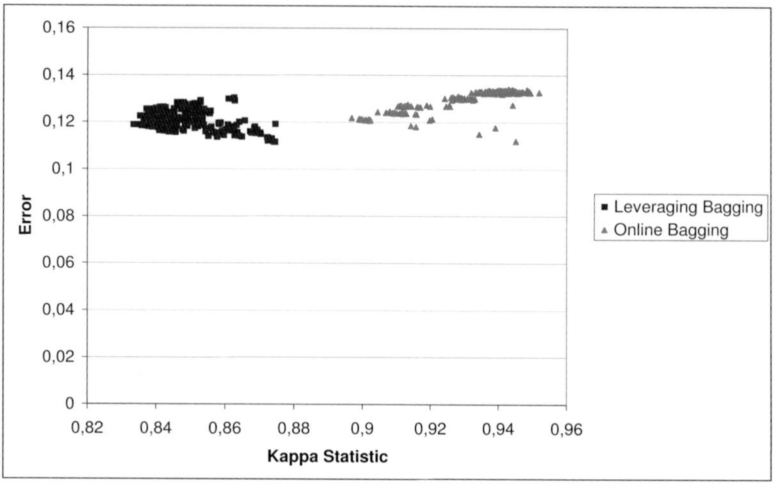

Fig. 3. Kappa-Error diagrams for Leveraging Bagging and Online bagging (bottom) on on the SEA data with three concept drifts, plotting 576 pairs of classifiers

These datasets are small compared to synthetic datasets we consider. Another important fact is that we do not know when drift occurs or indeed if there is any drift. We may simulate concept drift, joining the three datasets, merging attributes, and supposing that each dataset corresponds to a different concept

$$\text{CovPokElec} = (\text{CoverType} \oplus^{5{,}000}_{581{,}012} \text{Poker}) \oplus^{5{,}000}_{1{,}000{,}000} \text{ELEC}$$

As all examples need to have the same number of attributes, we simply concatenate all the attributes, and set the number of classes to the maximum number of classes of all the datasets.

4.3 Results

We ran three experimental evaluations to test our new leveraging method using 25 classifiers. In the first, in order to understand why online bagging works, we compare the original Online Bagging, with the `ADWIN` Bagging presented in [6], and with two different resampling stategies

- half subagging
- without replacement or subagging
- without taking out all instances (WT)

The second experiment measures accuracy improvement of leveraging bagging. And finally, the third experiment compares our method against online Random Forests.

We use the datasets explained in the previous sections for evaluation. The experiments were performed on 2.66 GHz Core 2 Duo E6750 machines with 4

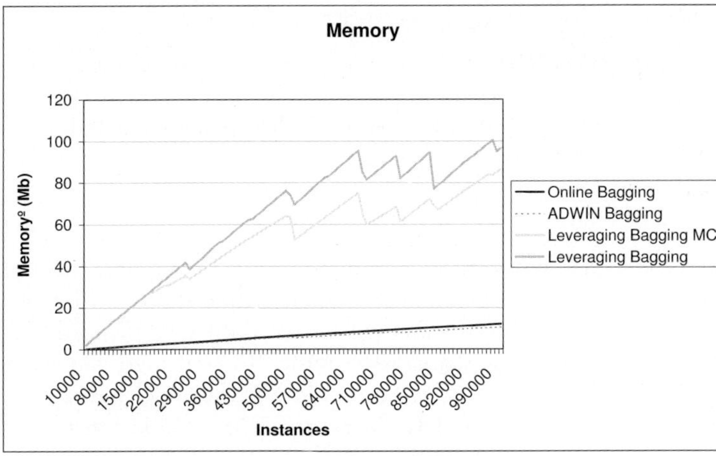

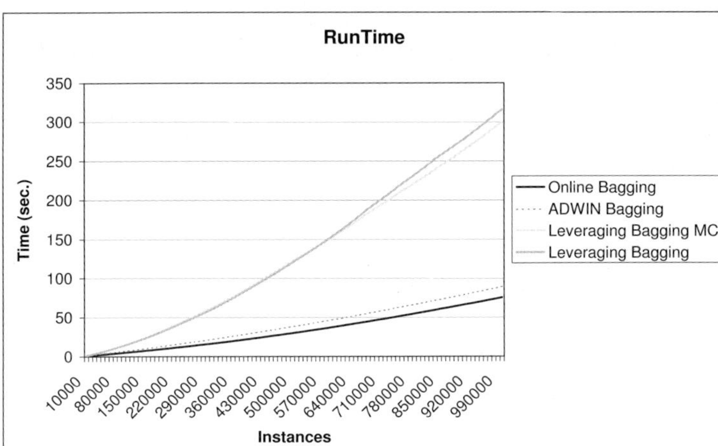

Fig. 4. Runtime and memory on the SEA data with three concept drifts

GB of memory. The evaluation methodology used was Interleaved Test-Then-Train on 10 runs and 25 classifiers: every example was used for testing the model before using it to train. This interleaved test followed by train procedure was carried out on 10 million examples from the hyperplane and RandomRBF datasets, and one million examples from the SEA dataset. The parameters of these streams are the following:

- RBF(x,v): RandomRBF data stream of 5 classes with x centroids moving at speed v.
- HYP(x,v): Hyperplane data stream of 5 classes with x attributes changing at speed v.
- SEA(v): SEA dataset, with length of change v.
- LED(v): LED dataset, with length of change v.

The Nemenyi test [11] is used for computing significance: it is an appropriate test for comparing multiple algorithms over multiple datasets, being based on the average ranks of the algorithms across all datasets. We use a p-value of 0.05. Under the Nemenyi test, $\{x,y\} \succ \{z\}$ indicates that algorithms x and y are statistically significantly more likely to be more favourable than z.

In [5] we introduced the use of RAM-Hours as an evaluation measure of the resources used by streaming algorithms. Every GB of RAM deployed for 1 hour equals one RAM-Hour.

Tables 2, 3 and 4 report the final accuracy, and speed of the classification models induced on the synthetic data and the real datasets: Forest CoverType, Poker Hand, Electricity and CovPokElec. Accuracy is measured as the final percentage of examples correctly classified over the test/train interleaved evaluation. Time is measured in seconds, and memory in MB. All experiments are performed using leveraging bagging with Poisson(6), since empirically this was determined to be the best value. We implemented Online Bayesian Bagging, but we don't report the results as they are similar to those using Poisson(1).

We use as a base learner for our experiments the *Hoeffding Naive Bayes Tree* (**hnbt**). Hoeffding trees [14] are state-of-the-art in classification for data streams and they perform prediction by choosing the majority class at each leaf. Their predictive accuracy can be increased by adding naive Bayes models at the leaves of the trees. A Hoeffding Naive Bayes Tree [21] works by performing a naive Bayes prediction per training instance, and comparing its prediction with the majority class. Counts are stored to measure how many times the naive Bayes prediction gets the true class correct as compared to the majority class. When performing a prediction on a test instance, the leaf will only return a naive Bayes prediction if it has been more accurate overall than the majority class, otherwise it resorts to a majority class prediction.

Table 2 reports the accuracy, speed and memory of a Hoeffding Naive Bayes Tree(**hnbt**), compared with online bagging of Hoeffding Naive Bayes Tree and `ADWIN` bagging of Hoeffding Naive Bayes Trees. We observe that online bagging improves the accuracy of a single Hoeffding Naive Bayes Tree from 74.03% to 77.15%. `ADWIN` bagging improves this result getting 79.24%. In terms of memory and speed, the single Hoeffding Naive Bayes Tree is much faster and needs less memory.

Table 4. Comparison of Leveraging Bagging without using Random Output Codes, Leveraging Bagging using Random Output Codes, and Leveraging Bagging giving more weight to missclassified examples. The best individual accuracies are indicated in boldface.

	Leveraging Bagging			Leveraging Bagging MC			Leveraging Bagging ME		
	Time	Acc.	Mem.	Time	Acc.	Mem.	Time	Acc.	Mem.
RBF(0,0)	56.51	91.05 ± 0.06	2862.03	133.63	**91.07** ± 0.03	3980.25	23.22	90.11 ± 0.38	5492.83
RBF(50,0.001)	3.31	**58.21** ± 0.05	7333.89	181.04	56.64 ± 0.07	6511.17	0.17	44.67 ± 0.02	7674.30
RBF(10,0.001)	61.77	82.45 ± 0.07	2997.72	212.74	82.62 ± 0.06	6518.94	22.89	**83.77** ± 0.15	5153.45
RBF(50,0.0001)	21.55	**80.48** ± 0.02	7421.81	73.84	78.42 ± 0.04	5155.42	0.38	58.58 ± 0.07	7375.62
RBF(10,0.0001)	64.86	83.94 ± 0.07	2899.91	249.36	86.00 ± 0.17	7847.99	22.30	**86.35** ± 0.24	5180.18
LED(50000)	27.64	73.10 ± 0.09	714.17	121.15	71.67 ± 0.16	491.14	1.56	**73.11** ± 0.08	647.63
SEA(50)	92.88	**88.65** ± 0.15	354.51	96.59	88.50 ± 0.18	362.56	1.27	87.64 ± 0.16	58.92
SEA(50000)	75.12	**88.69** ± 0.11	324.51	80.78	88.51 ± 0.10	336.27	1.02	87.31 ± 0.13	59.10
HYP(10,0.001)	409.89	88.66 ± 0.38	11307.98	189.75	88.01 ± 0.43	5537.68	3.89	**91.02** ± 0.07	2047.55
HYP(10,0.0001)	405.58	89.36 ± 0.13	11838.65	207.66	88.63 ± 0.27	5873.24	4.28	**91.19** ± 0.05	2014.43
CovType	49.45	91.29	559.57	43.88	**92.53**	368.29	1.62	90.96	479.49
Electricity	6.23	**88.41**	11.11	6.85	87.98	11.56	0.16	**88.41**	5.88
Poker	32.42	98.03	194.48	47.25	**98.76**	194.62	0.27	75.87	208.08
CovPokElec	167.47	95.23	1610.18	185.44	**95.83**	1204.34	2.25	81.80	1360.60
		85.54 Acc. 20.17 RAM-Hours 1.79 avg. rank			85.37 Acc. 22.04 RAM-Hours 2.00 avg. rank			80.77 Acc. 0.87 RAM-Hours 2.14 avg. rank	

We ran experiments to test three different bagging strategies: subagging (resampling without replacement), half subagging (resampling without replacement half of the instances), and bagging without taking out any instance. We implement this third strategy using $1+Poisson(1)$ instead of Poisson. We tested the three strategies on `ADWIN` bagging. Table 3 shows, for these bagging strategies, their accuracy, speed and memory. We observe that using subagging we get faster methods but less accurate. If we use all instances, it seems that we improve accuracy but not speed or the memory used.

The learning curves and model growth curves for the Sea dataset are plotted in Figures 2 and 4. We observe that for this data stream the new leveraging bagging methods need more time and memory than the other methods, but they are more accurate.

We use the Kappa statistic κ [6] to show how using Leveraging Bagging with $\lambda = 6$, we increase the diversity of the ensemble. When two classifiers agree on every example then $\kappa = 1$, and when their predictions coincide purely by chance, then $\kappa = 0$.

The Kappa-Error diagram is a scatterplot where each point corresponds to a pair of classifiers. The x coordinate of the pair is the κ value for the two classifiers. The y coordinate is the average of the error rates of the two classifiers.

Figure 3 shows the Kappa-Error diagram for the SEA dataset with three concept drifts and 25 classifiers. We observe that for this dataset the Kappa statistic for Leveraging Bagging is lower than for Online Bagging, showing the higher diversity of the output of the classifiers of the Leveraging Bagging method.

Table 4 reports the accuracy, speed and memory of the new Levering Bagging methods using **hnbt**. We compare Levering Bagging with Levering Bagging MC without using Random Ouput Codes, and Levering Bagging ME giving more weight to missclassified examples. In this last method, if an instance is

Table 5. Comparison of methods using Random Forests: Leveraging Bagging without using Random Output Codes, Online Bagging, and ADWIN bagging. Accuracy is measured as the final percentage of examples correctly classified over the 1 or 10 million test/train interleaved evaluation. Time is measured in seconds, and memory in MB. The best individual accuracies are indicated in boldface.

	RF Leveraging Bagging			RF Online Bagging			RF ADWIN Bagging		
	Time	Acc.	Mem.	Time	Acc.	Mem.	Time	Acc.	Mem.
RBF(0,0)	45.75	**90.70** ± 0.05	2193.41	24.99	90.33 ± 0.11	1624.37	24.59	90.31 ± 0.10	1939.25
RBF(50,0.001)	3.54	**55.56** ± 0.06	1835.02	24.32	31.31 ± 0.03	1453.76	0.11	34.18 ± 0.02	1177.31
RBF(10,0.001)	49.21	**82.13** ± 0.06	2152.77	24.84	81.84 ± 0.09	1691.17	23.93	81.81 ± 0.08	2020.98
RBF(50,0.0001)	23.92	**77.77** ± 0.03	2478.19	24.64	39.72 ± 0.08	1783.48	0.39	48.53 ± 0.11	1497.54
RBF(10,0.0001)	51.93	**83.45** ± 0.07	2181.24	24.92	82.74 ± 0.05	1702.03	23.95	82.73 ± 0.05	2065.70
LED(50000)	20.90	**67.83** ± 0.74	286.48	11.46	60.22 ± 0.71	148.12	3.02	66.12 ± 0.66	166.40
SEA(50)	189.36	**87.86** ± 0.13	760.45	63.31	86.86 ± 0.06	176.44	60.39	86.98 ± 0.06	190.24
SEA(50000)	186.40	**87.74** ± 0.09	728.12	63.30	86.78 ± 0.06	168.49	59.96	86.88 ± 0.06	185.48
HYP(10,0.001)	143.84	**86.09** ± 0.36	5059.25	27.86	80.45 ± 1.47	1484.37	25.39	83.18 ± 0.68	1562.69
HYP(10,0.0001)	132.58	**86.73** ± 0.37	4826.40	27.73	83.43 ± 0.89	1460.90	26.30	84.08 ± 0.57	1607.33
CovType	6.98	**87.81**	162.88	16.05	74.71	130.50	1.07	76.47	140.29
Electricity	2.55	**86.85**	5.52	1.36	80.08	2.94	0.60	82.44	4.32
Poker	7.38	**75.72**	92.94	22.69	74.07	88.94	0.44	65.96	81.23
CovPokElec	11.12	**73.43**	448.82	32.28	68.22	383.78	2.11	69.73	429.11
	80.69 Acc. 5.51 RAM-Hours 1.00 avg. rank			72.91 Acc. 1.30 RAM-Hours 2.71 avg. rank			74.24 Acc. 0.89 RAM-Hours 2.29 avg. rank		

Nemenyi significance: RF Leveraging Bagging$\succ$RF Online Bagging; RF Leveraging Bagging$\succ$RF ADWIN Bagging;

misclassified it is accepted with a weight of one. If not, it is accepted with probability $e_T/(1-e_T)$, where the error estimate e_T is computed as a smoothed version of the proportion of misclassified examples using the estimation of ADWIN that is monitoring the error. We observe that the accuracy of the two Leveraging bagging methods are similar and that they are 6% more accurate than the ADWIN bagging. When leveraging bagging is used to give more weight to missclassified examples, it does not seem to increase accuracy. However it improves the need for RAM-Hours, so the Leveraging Bagging ME is a very good classifier when resources are scarce.

Finally, we compare our methods with Random Forests. We build Random Forests using Hoeffding Naive Bayes Trees in the following way: let n be the number of attributes, we select for each node, $\sqrt{(n)}$ attributes randomly, and we only keep statistics of these attributes. The splits of the node are made using the best of these attributes, and the predictions at the leaves are made using only the statistics of these attributes.

We compare using the three bagging methods: online bagging, ADWIN bagging, and leveraging bagging using 25 classifiers. The results are shown in Table 5. We see that RandomForests are, for many datasets, twice as fast and use half of the memory. However their accuracy is 5% below that of using standard Hoeffding Naive Bayes trees. To obtain the same accuracy as the Hoeffding Naive Bayes trees, we only need to increase the number of classifiers of the ensemble. We can observe that using our new leveraging bagging we increase the accuracy of Random Forests.

5 Conclusions

We have presented a new leveraging bagging method, that uses randomization on the weights of the instances of the input stream, to improve the accuracy of the ensemble classifier. Using random output codes, we may use a binary classifier without losing accuracy. We tested subagging, half subagging, and bagging without replacement, and these methods performed faster but they are marginally less accurate. Finally, we have compared our method with Random Forests with improved results.

References

1. Abdulsalam, H., Skillicorn, D.B., Martin, P.: Streaming random forests. In: IDEAS 2007: Proceedings of the 11th International Database Engineering and Applications Symposium, Washington, DC, USA, pp. 225–232. IEEE Computer Society, Los Alamitos (2007)
2. Asuncion, A., Newman, D.: UCI machine learning repository (2007)
3. Bifet, A., Gavaldà, R.: Learning from time-changing data with adaptive windowing. In: Jonker, W., Petković, M. (eds.) SDM 2007. LNCS, vol. 4721. Springer, Heidelberg (2007)
4. Bifet, A., Holmes, G., Kirkby, R., Pfahringer, B.: MOA: Massive Online Analysis. Journal of Machine Learning Research, JMLR (2010), `http://moa.cs.waikato.ac.nz/`
5. Bifet, A., Holmes, G., Pfahringer, B., Frank, E.: Fast perceptron decision tree learning from evolving data streams. In: Zaki, M.J., Yu, J.X., Ravindran, B., Pudi, V. (eds.) Advances in Knowledge Discovery and Data Mining. LNCS, vol. 6119, pp. 299–310. Springer, Heidelberg (2010)
6. Bifet, A., Holmes, G., Pfahringer, B., Kirkby, R., Gavaldà, R.: New ensemble methods for evolving data streams. In: KDD, pp. 139–148 (2009)
7. Breiman, L.: Bagging predictors. Machine Learning 24(2), 123–140 (1996)
8. Breiman, L.: Random forests. Machine Learning 45(1), 5–32 (2001)
9. Breiman, L., Friedman, J.H., Olshen, R.A., Stone, C.J.: Classification and Regression Trees. Wadsworth, Belmont (1984)
10. Bühlmann, P., Yu, B.: Analyzing bagging. Annals of Statistics (2003)
11. Demšar, J.: Statistical comparisons of classifiers over multiple data sets. The Journal of Machine Learning Research 7, 1–30 (2006)
12. Dietterich, T.G., Bakiri, G.: Solving multiclass learning problems via error-correcting output codes. J. Artif. Intell. Res. (JAIR) 2, 263–286 (1995)
13. Domingos, P.: Why does bagging work? A bayesian account and its implications. In: KDD, pp. 155–158 (1997)
14. Domingos, P., Hulten, G.: Mining high-speed data streams. In: KDD, pp. 71–80 (2000)
15. Friedman, J., Hall, P.: On bagging and nonlinear estimation. Technical report, Stanford University (1999)
16. Friedman, J.H.: On bias, variance, 0/1—loss, and the curse-of-dimensionality. Data Min. Knowl. Discov. 1(1), 55–77 (1997)
17. Gama, J., Medas, P., Castillo, G., Rodrigues, P.P.: Learning with drift detection. In: Bazzan, A.L.C., Labidi, S. (eds.) SBIA 2004. LNCS (LNAI), vol. 3171, pp. 286–295. Springer, Heidelberg (2004)

18. Gama, J., Rocha, R., Medas, P.: Accurate decision trees for mining high-speed data streams. In: KDD, pp. 523–528 (2003)
19. Grandvalet, Y.: Bagging equalizes influence. Machine Learning 55(3), 251–270 (2004)
20. Harries, M.: Splice-2 comparative evaluation: Electricity pricing. Technical report, The University of South Wales (1999)
21. Holmes, G., Kirkby, R., Pfahringer, B.: Stress-testing Hoeffding trees. In: Jorge, A.M., Torgo, L., Brazdil, P.B., Camacho, R., Gama, J. (eds.) PKDD 2005. LNCS (LNAI), vol. 3721, pp. 495–502. Springer, Heidelberg (2005)
22. Hulten, G., Spencer, L., Domingos, P.: Mining time-changing data streams. In: KDD, pp. 97–106 (2001)
23. Lee, H.K.H., Clyde, M.A.: Lossless online bayesian bagging. J. Mach. Learn. Res. 5, 143–151 (2004)
24. Oza, N., Russell, S.: Online bagging and boosting. In: Artificial Intelligence and Statistics 2001, pp. 105–112. Morgan Kaufmann, San Francisco (2001)
25. Oza, N.C., Russell, S.J.: Experimental comparisons of online and batch versions of bagging and boosting. In: KDD, pp. 359–364 (2001)
26. Saffari, A., Leistner, C., Santner, J., Godec, M., Bischof, H.: On-line random forests. In: 3rd IEEE - ICCV Workshop on On-line Learning for Computer Vision (2009)
27. Schapire, R.E.: Using output codes to boost multiclass learning problems. In: ICML 1997: Proceedings of the Fourteenth International Conference on Machine Learning, pp. 313–321. Morgan Kaufmann Publishers Inc., San Francisco (1997)
28. Street, W.N., Kim, Y.: A streaming ensemble algorithm (SEA) for large-scale classification. In: KDD, pp. 377–382 (2001)

ITCH: Information-Theoretic Cluster Hierarchies

Christian Böhm[1], Frank Fiedler[1], Annahita Oswald[1], Claudia Plant[2], Bianca Wackersreuther[1], and Peter Wackersreuther[1]

[1] University of Munich, Munich, Germany
{boehm,fiedler,oswald,wackersb,wackersr}@dbs.ifi.lmu.de
[2] Florida State University, Tallahassee, FL, USA
cplant@fsu.edu

Abstract. Hierarchical clustering methods are widely used in various scientific domains such as molecular biology, medicine, economy, etc. Despite the maturity of the research field of hierarchical clustering, we have identified the following four goals which are not yet fully satisfied by previous methods: First, to guide the hierarchical clustering algorithm to identify only meaningful and valid clusters. Second, to represent each cluster in the hierarchy by an intuitive description with e.g. a probability density function. Third, to consistently handle outliers. And finally, to avoid difficult parameter settings. With ITCH, we propose a novel clustering method that is built on a hierarchical variant of the information-theoretic principle of Minimum Description Length (MDL), referred to as hMDL. Interpreting the hierarchical cluster structure as a statistical model of the data set, it can be used for effective data compression by Huffman coding. Thus, the achievable compression rate induces a natural objective function for clustering, which automatically satisfies all four above mentioned goals.

1 Introduction

Since dendrograms and similar hierarchical representations provide extremely useful insights into the structure of a data set, hierarchical clustering has become very popular in various scientific disciplines, such as molecular biology, medicine, or economy. However, well-known hierarchical clustering algorithms often either fail to detect the true clusters that are present in a data set, or they identify invalid clusters, which are not existing in the data set. These problems are particularly dominant in the presence of noise and outliers and result in the questions "How can we decide if a given representation is really natural, valid, and therefore meaningful?" and "How can we enforce a hierarchical clustering algorithm to identify only the meaningful cluster structure?"

Information Theory for Clustering. We give the answer to these questions by relating the hierarchical clustering problem to that of information theory and data compression. Imagine you want to transfer the data set via an extremely expensive and small-banded communication channel. Then you can interpret the cluster hierarchy as a statistical model of the data set, which defines more or less likely areas of the data space. The knowledge of these probabilities can be used for an efficient compression of the data set: Following the idea of (optimal) Huffman coding, we assign few bits to points in areas of high probability and more bits to areas of low probability. The interesting observation is the following: the compression becomes the more effective, the better our

J.L. Balcázar et al. (Eds.): ECML PKDD 2010, Part I, LNAI 6321, pp. 151–167, 2010.

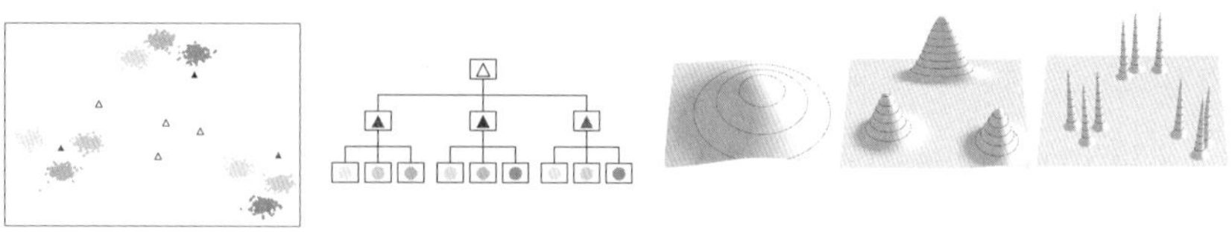

(a) Data set: Δ display outliers.
(b) Hierarchical representation.
(c) Cluster contents on each level modeled by PDFs.

Fig. 1. Contributions of ITCH

statistical model, the hierarchical cluster structure, fits to the data. This so-called *Minimum Description Length* (MDL) principle has recently received increasing attention in the context of partitioning (i.e. non-hierarchical) clustering methods. Note that it can not only be used to assess and compare the quality of the clusters found by different algorithms and/or varying parameter settings. Rather, we use this concept as an objective function to implement clustering algorithms directly using simple but efficient optimization heuristics.

In this paper, we extend the idea of MDL to the hierarchical case and develop hMDL for *hierarchical* cluster structures. Whereas previous MDL approaches can only evaluate the result of partitioning clustering methods, our new hMDL criterion is able to assess a complete cluster hierarchy. Moreover, hMDL can be used in combination with an optimization heuristic to exactly determine that cluster hierarchy, which optimizes the data compression according to the MDL criterion.

Challenges and Contributions. With an assessment condition for cluster hierarchies, we develop a complete hierarchical clustering approach on top of the idea of hMDL. This proposed algorithm ITCH (**I**nformation-**T**heoretic **C**luster **H**ierarchies) yields numerous advantages, out of which we demonstrate the following four:

1. All single clusters as well as their hierarchical arrangement are guaranteed to be **meaningful**. Nodes only are placed in the cluster hierarchy if they improve the data compression. This is achieved by optimizing the hMDL criterion. Moreover, a maximal consistency with partitioning clustering methods is obtained.
2. Each cluster is represented by an **intuitive description** of its content in form of a Gaussian probability density function (PDF). Figure 1(c) presents an example of a 3-stage hierarchy. The output of conventional methods is often just the (hierarchical) cluster structure and the assignment of points.
3. ITCH is **outlier-robust**. Outliers are assigned to the root of the cluster hierarchy or to an appropriate inner node, depending on the degree of outlierness. For example, in Figures 1(a) and 1(b) the outlier w.r.t. the three red clusters at the bottom level is assigned to the parent cluster in the hierarchy, marked by a red triangle.
4. ITCH is **fully automatic** as no difficult parameter settings are necessary.

To the best of our knowledge, ITCH is the only clustering algorithm that meets *all* of the above issues by now. The remainder of this paper is organized as follows: Section 2 gives a brief survey of related work. Section 3 presents a derivation of our hMDL criterion and introduces the ITCH algorithm. Section 4 documents the advantages of ITCH on synthetic and real data sets. Section 5 summarizes the paper.

2 Related Work

Hierarchical Clustering. One of the most widespread approaches to hierarchical clustering is the Single Link algorithm [12] and its variants [14]. The resulting hierarchy obtained by the merging order is visualized as a tree, which is called dendrogram. Cuts through the dendrogram at various levels obtain partitioning clusterings. However, for complex data sets it is hard to define appropriate splitting levels, which correspond to meaningful clusterings. Furthermore, outliers may cause the well-known Single Link effect. Also, for large data sets, the fine scale visualization is not appropriate. The algorithm OPTICS [1] avoids the Single Link effect by requiring a minimum object density for clustering, i.e. $MinPts$ number of objects are within a hyper-sphere with radius ϵ. Additionally, it provides a more suitable visualization, the so-called reachability plot. However, the problem that only certain cuts represent useful clusterings still remains unsolved.

Model-based Clustering. For many applications and further data mining steps, it is essential to have a model of the data. Hence, clustering with PDFs, which goes back to the popular EM algorithm [8], is a widespread alternative to hierarchical clustering. After a suitable initialization, EM iteratively optimizes a mixture model of K Gaussian distributions until no further significant improvement of the log-likelihood of the data can be achieved. Usually a very fast convergence is observed. However, the algorithm may get stuck in a local maximum of the log-likelihood. Moreover, the quality of the clustering result strongly depends on an appropriate choice of K, which is a non-trivial task in most applications. And even with a suitable choice of K the algorithm is very sensitive w.r.t. noise and outliers.

Model-based Hierarchical and Semi-supervised Clustering. [22] proposes a hierarchical extension of EM to speed up query processing in an object recognition application. In [6] a hierarchical variant of EM is applied for image segmentation. Goldberger and Roweis [9] focus on reducing the number of clusters in a mixture model. The consistency with the initial clustering is assured by the constraint that objects belonging to the same initial cluster must end up after the reduction in the same new cluster as well. Each of these approaches needs a suitable parameter setting for the number of hierarchy levels. Clustering respecting some kind of hierarchy can also be regarded as *semi-supervised clustering*, i.e. clustering with side information. In most of some recently published papers [13,4,3], this information is introduced by constraints on the objects and typically consists of strong expert knowledge. In contrast, ITCH does not consider any external information. The data itself is our single source of knowledge.

Information Theory in the Field of Clustering. Only a few papers on compression based clustering, that avoid difficult parameter settings have been published so far. X-Means [16], G-Means [11] and RIC [5] focus on avoiding the choice of K in partitioning clustering by trying to balance data likelihood and model complexity. This sensitive trade-off can be rated by model selection criteria, among them the Akaike Information

Criterion (AIC), the Bayesian Information Criterion (BIC) and Minimum Description Length (MDL) [10]. X-Means provides a parameter-free detection of spherical Gaussian clusters by a top-down splitting algorithm, which integrates K-Means clustering and BIC. G-Means extends this idea to detect non-spherical Gaussian clusters. The model selection criterion of RIC is based on MDL, which allows to define a coding scheme for outliers and to identify non-Gaussian clusters.

There is a family of further closely related ideas, such as Model-based Clustering [2], the work of Still and Bialek [20] and the so-called Information Bottleneck Method [21], introduced by Tishby *et al.* This technique aims at providing a quantitative notation of *meaningful* or *relevant* information. The authors formalize this perception by finding the best tradeoff between accuracy and complexity when clustering a random variable X, given a joint probability distribution between X and an observed relevant variable Y. It is used for clustering terms and documents [19]. However, all parameter-free algorithms mentioned so far, do not provide any cluster hierarchy. One recent paper [7] presents a new method for clustering based on compression. In the first step, this method determines a parameter-free, universal, similarity distance, computed from the lengths of compressed data files. Afterwards a hierarchical clustering method is applied. In contrast to ITCH, this work was not designed to handle outliers in an appropriate way. Furthermore, no description of the content of a cluster is available.

3 Information-Theoretic Hierarchical Clustering

The clustering problem is highly related to that of data compression: The detected cluster structure can be interpreted as a PDF $f_\Theta(x)$ where $\Theta = \{\theta_1, \theta_2, ...\}$ is a set of parameters, and the PDF can be used for an efficient compression of the data set n. It is well-known that the compression by *Huffman coding* is optimal if the data distribution really corresponds to $f_\Theta(x)$. Huffman coding represents every point x by a number of bits which is equal to the negative binary logarithm of the PDF:

$$C_{\text{data}}(x) = -\log_2(f_\Theta(x)).$$

The better the point set corresponds to $f_\Theta(x)$, the smaller the coding costs $C_{\text{data}}(x)$ are. Hence, $C_{\text{data}}(x)$ can be used as the objective function in an optimization algorithm. However, in data compression, Θ serves as a *code book* which is required to decode the compressed data set again. Therefore, we need to complement the compressed data set with a coding of this code book, the parameter set Θ. When, for instance, a Gaussian Mixture Model (GMM) is applied, Θ corresponds to the weights, the mean vectors and the variances of the single Gaussian functions in the GMM. Considering Θ in the coding costs is also important for the clustering problem, because neglecting it leads to overfitting. For partitioning (non-hierarchical) clustering structures, several approaches have been proposed for the coding of Θ [16,17,18] (cf. Section 2). These approaches differ from each other because there is no unambiguous and natural choice for a distribution function, which can be used for the Huffman coding of Θ, and different assumptions lead to different objective functions. In case of the hierarchical cluster structure in ITCH, a very natural distribution function for Θ exists: With the only exception of the root node, every node in the hierarchy has a parent node. This parent node is associated with a PDF which can naturally be used as a code book for the mean vector (and indirectly also for the variances) of the child node. The coding costs of the root node, however, are not important, because every valid hierarchy has exactly one root node with a constant number of parameters, and therefore, the coding costs of the root node are always constant.

3.1 Hierarchical Cluster Structure

In this Section, we formally introduce the notion of a hierarchical cluster structure (HCS). A HCS contains clusters $\{A, B, ...\}$ each of which is represented by a Gaussian distribution function. These clusters are arranged in a tree:

Definition 1 (Hierarchical Cluster Structure). *(1) A HCS is a tree* $\mathcal{T} = (\mathcal{N}, \mathcal{E})$ *consisting of a set of nodes* $\mathcal{N} = \{A, B, ...\}$ *and a set of directed edges* $\mathcal{E} = \{e_1, e_2, ...\}$ *where* A *is a parent of* B *(*B *is a child of* A*) iff* $(A, B) \in \mathcal{E}$*. Every node* $C \in \mathcal{N}$ *is associated with a weight* W_C *and a Gaussian PDF defined by the parameters* μ_C *and* Σ_C *such that the sum of the weights equals one:*

$$\sum_{C \in \mathcal{N}} W_C = 1.$$

(2) If a path from A *to* B *exists in* $\mathcal{T}$ *(or* $A = B$*) we call* A *an ancestor of* B *(*B *a descendant of* A*) and write* $B \sqsubseteq A$*.*
(3) The level l_C *of a node* C *is the height of the descendant subtree. If* C *is a leaf, then* C *has level* $l_C = 0$*. The root has the highest level (length of the longest path to a leaf).*

The PDF which is associated with a cluster C is a multivariate Gaussian in a d-dimensional data space which is defined by the parameters μ_C and Σ_C (where $\mu_C = (\mu_{C,1}, ..., \mu_{C,d})^{\mathbf{T}}$ is a vector from a d-dimensional space, called the location parameter, and Σ_C is a $d \times d$ covariance matrix) by the following formula:

$$N(\mu_C, \Sigma_C, x) = \frac{1}{\sqrt{(2\pi)^d \cdot |\Sigma_C|}} \cdot \mathbf{e}^{-\frac{1}{2}(x-\mu_C)^{\mathbf{T}} \cdot \Sigma_C^{-1} \cdot (x-\mu_C)}.$$

For simplicity we restrict $\Sigma_C = \text{diag}(\sigma_{C,1}^2, ..., \sigma_{C,d}^2)$ to be diagonal such that the multivariate Gaussian can also be expressed by the following product:

$$N(\mu_C, \Sigma_C, x) = \prod_{1 \le i \le d} N(\mu_{C,i}, \sigma_{C,i}^2, x_i)$$

$$= \prod_{1 \le i \le d} \frac{1}{\sqrt{2\pi\sigma_{C,i}^2}} \cdot \mathbf{e}^{-\frac{(x_i - \mu_i)^2}{2\sigma_{C,i}^2}}.$$

Since we require the sum of all weights in a HCS to be 1, a HCS always defines a function whose integral is ≤ 1. Therefore, the HCS can be interpreted as a complex, multimodal, and multivariate PDF, defined by the mixture of the Gaussians of the HCS $\mathcal{T} = (\mathcal{N}, \mathcal{E})$:

$$f_{\mathcal{T}}(x) = \max_{C \in \mathcal{N}} \{W_C \, N(\mu_C, \Sigma_C, x)\} \text{ with} \int_{\mathbb{R}^d} f_{\mathcal{T}}(x) \mathbf{d}x \le 1.$$

If the Gaussians of the HCS do not overlap much, then the integral becomes close to 1.

The operations, described in Section 3.3, assign each point $x \in DB$ to a cluster of the HCS $\mathcal{T} = (\mathcal{N}, \mathcal{E})$. We distinguish between the *direct* and the *indirect* association. A point is directly associated with that cluster $C \in \mathcal{N}$ the probability density of which is maximal at the position of x, and we write $C = Cl(x)$ and also $x \in C$, with:

$$Cl(x) = \arg\max_{C \in \mathcal{N}} \{W_C \cdot N(\mu_C, \Sigma_C, x)\} .$$

As we have already stated in the introduction, one of the main motivations of our hierarchical, information-theoretic clustering method ITCH is to represent a sequence of clustering structures which range from a very coarse (unimodal) view to the data distribution to a very detailed (multi-modal) one, and that all these views are meaningful and represent an individual complex PDF. The ability to cut a HCS at a given level L is obtained by the following definition:

Definition 2 (Hierarchical Cut). *A HCS $\mathcal{T}' = (\mathcal{N}', \mathcal{E}')$ is a hierarchical cut of a HCS $\mathcal{T} = (\mathcal{N}, \mathcal{E})$ at level L (in symbols: $\mathcal{T}' = HC_L(\mathcal{T})$), if the following properties hold:*
(1) $\mathcal{N}' = \{A \in \mathcal{N} | l_A \geq L\}$,
(2) $\mathcal{E}' = \{(A, B) \in \mathcal{E} | l_A > l_B \geq L\}$,
(3) For each $A \in \mathcal{N}'$ the following properties hold:

$$W'_A = \begin{cases} W_A & \text{if } l_A > L \\ \sum_{B \in \mathcal{N}, B \sqsubseteq A} W_B & \text{otherwise,} \end{cases}$$

where W_C and W'_C is the weight of node C in $\mathcal{T}$ and $\mathcal{T}'$, respectively.
(4) Analogously, for the direct association of points to clusters the following property holds: Let x be associated with Cluster B in $\mathcal{T}$, i.e. $Cl(x) = B$. Then in $\mathcal{T}'$, x is associated with:

$$Cl'(x) = \begin{cases} B & \text{if } l_B \geq L \\ A | B \sqsubseteq A \wedge l_A = L & \text{otherwise.} \end{cases}$$

Here, the weights of the pruned nodes are automatically added to the leaf nodes of the new HCS, which used to be the ancestors of the pruned nodes. Therefore, the sum of all weights is maintained (and still equals 1), and the obtained tree is again a HCS according to Definition 1. The same holds for the point-to-cluster assignments.

3.2 Generalization of the MDL Principle

Now we explain how the MDL principle can be generalized for hierarchical clustering and develop the new objective function hMDL. Following the traditional MDL principle, we compress the data points according to their negative log likelihood corresponding to the PDF which is given by the HCS. In addition, we penalize the model complexity by adding the code length of the HCS parameters to the negative log likelihood of the data. The better the PDFs of child nodes fit into the PDFs of the parent, the less the coding costs will be. Therefore, the overall coding costs corresponds to the natural, intuitive notion of a good hierarchical representation of data by distribution functions. The discrete assignment of points to clusters allows us to determine the

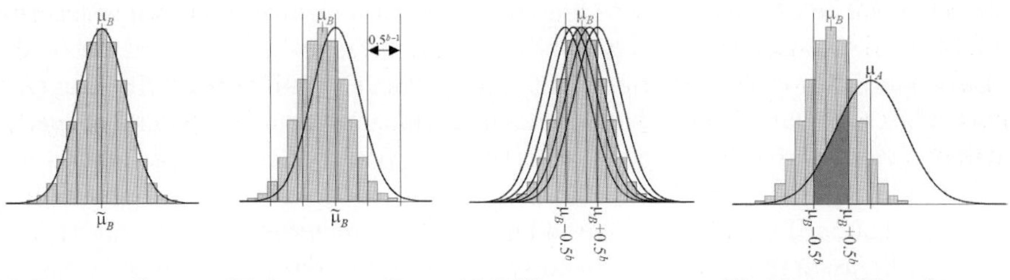

(a) Exact coding of μ_B. (b) Inexact coding of μ_B. (c) Different Gaussians w.r.t. different grid positions. (d) All possible values for recovered μ_B.

Fig. 2. Optimization of the grid resolution for the hMDL criterion

coding costs of the points clusterwise and dimensionwise, as explained in the following: The coding costs of the points associated with the clusters $C \in \mathcal{N}$ of the HCS $\mathcal{T} = (\mathcal{N}, \mathcal{E})$ corresponds to:

$$C_{\text{data}} = -\log_2 \prod_{x \in DB} \max_{C \in \mathcal{N}} \left\{ W_C \prod_{1 \le j \le d} N(\mu_{C,j}, \sigma^2_{C,j}, x_j) \right\}.$$

Since every point x is associated with that cluster C in the HCS which has maximum probability density, we can rearrange the terms of the above formula and combine the costs of all points that are in the same cluster:

$$= -\sum_{x \in DB} \log_2 \left(W_{Cl(x)} \prod_{1 \le j \le d} N(\mu_{Cl(x),j}, \sigma^2_{Cl(x),j}, x_j) \right)$$

$$= -\left(\left(\sum_{C \in \mathcal{N}} nW_C \log_2 W_C \right) + \left(\sum_{x \in DB;\, 1 \le j \le d} \sum \log_2 N(\mu_{Cl(x),j}, \sigma^2_{Cl(x),j}, x_j) \right) \right)$$

$$= -\sum_{C \in \mathcal{N}} \left(nW_C \log_2 W_C + \sum_{x \in C;\, 1 \le j \le d} \sum \log_2 N(\mu_{C,j}, \sigma^2_{C,j}, x_j) \right).$$

The ability to model the coding costs of each cluster separately allows us now, to focus on a single cluster, and even on a single dimension of a single cluster. A common interpretation of the term $-nW_C \log_2 W_C$, which actually comes from the weight a single Gaussian contributes to the GMM, is a Huffman coding of the cluster ID. We assume that every point carries the information which cluster it belongs to, and a cluster with many points gets a shortly coded cluster ID. These costs are referred to the *ID cost* of a cluster C. Lets consider two clusters, A and B, where $B \sqsubseteq A$. We now want to derive the coding scheme for the cluster B and its associated points. Several points are associated with B, where the overall weight of assignment sums up to W_B. When coding the parameters of the associated PDF of B, i.e. μ_B, and σ_B, we have to consider two aspects: (1) The *precision* both parameters should be coded to minimize the overall description length depends on W_B, as well as on σ_B. For instance, if only few points are associated with cluster B and/or the variance σ_B is very large, then it is not necessary to know the position of μ_B very precisely and vice versa. (2) The knowledge of the PDF of cluster A can be exploit for the coding of μ_B, because for likely positions (according to the PDF of A) we can assign fewer bits. Basically, model selection criteria, such as the Bayesian Information Criterion (BIC) or the Aikake Information Criterion (AIC) already address the first aspect, but not the hierarchical aspect. To make this paper self contained, we consider both aspects. In contrast to BIC, which uses the natural logarithm, we use the binary logarithm to represent the code length in *bits*. For simplicity, we assume that our PDF is univariate and the only parameter is its mean value μ_B. We neglect σ_B by assuming e.g. some fixed value for all clusters. We drop these assumptions at the end of this section. When the true PDF of cluster B is coded inexactly by some parameter $\widetilde{\mu}_B$, the coding costs for each point x (which truly belongs to the distribution $N(\mu_B, \sigma_B^2, x)$) in B is increased compared to the *exact* coding of μ_B, which would result in c_{ex} bits per point:

$$c_{\text{ex}} = \int_{-\infty}^{+\infty} -\log_2(N(\mu_B, \sigma_B^2, x)) \cdot N(\mu_B, \sigma_B^2, x) \; \mathbf{d}x = \log_2(\sigma_B \sqrt{2\pi \cdot \mathbf{e}}).$$

If $\widetilde{\mu}_B$ instead of μ_B is applied for compression, we obtain:

$$c(\widetilde{\mu}_B, \mu_B) = \int_{-\infty}^{+\infty} -\log_2(N(\widetilde{\mu}_B, \sigma_B^2, x)) \cdot N(\mu_B, \sigma_B^2, x)\, \mathbf{d}x.$$

The difference is visualized in Figure 2(a) and 2(b) respectively: In 2(a) $\widetilde{\mu}_B$ of the coding PDF, depicted by the Gaussian function, fits exactly to μ_B of the data distribution, represented by the histogram. This causes minimum code lengths for the compressed points but also a considerable effort for the coding of μ_B. In Figure 2(b) μ_B is coded by some regular quantization grid. Thereby, the costs for the cluster points slightly increase, but the costs for the location parameter decreases. The difference between $\widetilde{\mu}_B$ and μ_B depends on the bit resolution and on the position of the quantization grid. One example is depicted in Figure 2(b) by five vertical lines, the Gaussian curve is centered by the vertical line closest to μ_B. We derive lower and upper limits of $\widetilde{\mu}_B \in [\mu_B - 1/2^b ... \mu_B + 1/2^b]$ from the number of bits b, spent for coding $\widetilde{\mu}_B$. The real difference between μ_B and $\widetilde{\mu}_B$ depends again on the grid position. Not to prefer clusters that are incidentally aligned with the grid cells, we average over *all* possible positions of the discretization grid. Figure 2(c) presents five different examples of the infinitely many Gaussians that could be recovered w.r.t. different grid positions. Note that all positions inside the given interval have equal probability. Hence, the average coding costs for every possible position of $\widetilde{\mu}_B$ can be expressed by the following integral:

$$\begin{aligned} c_{\text{appx}}(b) &= 2^{b-1} \int_{\mu_B - 1/2^b}^{\mu_B + 1/2^b} c(\widetilde{\mu}_B, \mu_B)\, \mathbf{d}\widetilde{\mu}_B \\ &= \frac{1}{2}\log_2(\pi \cdot \mathbf{e} \cdot \sigma_B^2) + \frac{1}{2} + \frac{\log_2 \mathbf{e}}{6\sigma_B^2} \cdot 4^{-b}. \end{aligned}$$

Coding all $n \cdot W_B$ coordinates of the cluster points as well as the parameter μ_B (neglecting the ID cost) requires then the following number of bits:

$$C_{\text{appx}}(B) = c_{\text{appx}}(b) \cdot n \cdot W_B + b.$$

The optimal number b_{opt} of bits is determined by setting the derivation of the above term to zero.

$$\frac{\mathbf{d}}{\mathbf{d}b} C_{\text{appx}}(B) = 0 \Longrightarrow b_{\text{opt}} = \frac{1}{2}\log_2(\frac{n \cdot W_B}{3 \cdot \sigma_B^2}).$$

The unique solution to this equation corresponds to a *minimum*, as can easily be seen by the second derivative.

Utilization of the Hierarchical Relationship. We do not want to code the (inexact) position of μ_B without the prior knowledge of the PDF associated with cluster A. Without this knowledge, we would have to select a suitable range of values and code μ_B at the determined precision b assuming e.g. a uniform distribution inside this range. In contrast, μ_B is a value taken from the distribution function of cluster A. Hence, the number of bits used for coding of μ_B corresponds to the overall density around the imprecise interval defined by μ_B, i.e.

$$c_{\text{hMDL}}(\mu_B) = -\log_2 \int_{\mu_B - 1/2^b}^{\mu_B + 1/2^b} N(\mu_A, \sigma_A^2, x)\, \mathbf{d}x.$$

Figure 2(d) visualizes the complete interval of all possible values for the recovered mean value (marked in red) and illustrates the PDF of the cluster$A$, which is the predecessor of cluster B. $\widetilde{\mu}_B$ can be coded by determining the whole area under the PDF of A where $\widetilde{\mu}_B$ could be. The area actually corresponds to a probability value. The negative logarithm of this probability represents the required code length for μ_B. The costs for coding all points of cluster B and μ_B then corresponds to

$$c_{\text{appx}}(b) \cdot n \cdot W_B + c_{\text{hMDL}}(\mu_B).$$

Note, that it is also possible to optimize b directly by setting the derivative of this formula to zero. However, this is impossible in an analytic way, and the difference to the optimum which is obtained by minimizing $C_{\text{appx}}(B)$ is negligible. In addition, if the parent A of B is not the root of the HCS, μ_B causes some own ID cost. In this case, μ_B is a sample from the complex distribution function of the hierarchical cut (cf. Definition 2), which prunes the complete level of B and all levels below. Hence, the weight of these levels is added to the new leaf nodes (after cutting), and the ID costs of μ_B correspond to:

$$-\log_2\Big(\sum_{X \sqsubseteq A} W_X\Big).$$

A similar analysis can be done for the second parameter of the distribution function, σ_B. Since it is not straightforward to select a suitable distribution function for the Huffman coding of variances, one can apply a simple trick: Instead of coding σ_B, we code $y_B = \mu_B \pm v \cdot \sigma_B$, where v is a constant close to zero. Then, y_B is also a sample from the distribution function $N(\mu_A, \sigma_A^2, x)$ and can be coded similar to μ_B. Therefore, $c_{\text{hMDL}}(\sigma_B) = c_{\text{hMDL}}(\mu_B)$, and we write $c_{\text{hMDL}}(\text{param})$ for the coding costs per parameter instead. In general, if the PDF, which is associated with a cluster has r parameters, then the optimal number of bits can be obtained by the formula:

$$b_{\text{opt}} = \frac{1}{2}\log_2(\frac{n \cdot W_B}{3 \cdot r \cdot \sigma_B^2}).$$

And the overall coding costs are:

$$C_{\text{hMDL}}(B) = c_{\text{appx}}(b) \cdot n \cdot W_B + r \cdot c_{\text{hMDL}}(\text{param})$$

Until now, only the trade-off between coding costs of points and the parameters of the assigned cluster are taken into account. If we go above the lowest level of the HCS, we have to trade between coding costs of parameters at a lower level and coding costs of the parameters at the next higher level. This can be done in a similar way as before: Let b_B be the precision, which has already been determined for the representation of μ_B and σ_B, the parameters for cluster B, which is a subcluster of A. However, this is the minimum coding costs assuming that μ_A and σ_A have been stored at maximum precision, and that μ_B and σ_B are also given. Now, we assume that μ_B is an arbitrary point selected from the distribution function $N(\mu_A, \sigma_A^2, x)$ and determine an expectation for the cost:

$$\int_{-\infty}^{+\infty} -\log_2 \int_{\mu_B - 1/2^{b_B}}^{\mu_B + 1/2^{b_B}} N(\mu_A, \sigma_A^2, x)\mathbf{d}x \; N(\mu_A, \sigma_A^2, \mu_B)\mathbf{d}\mu_B.$$

Finally, we assume that μ_A is also coded inexactly by its own grid with resolution b_A. Then the expected costs are:

$$2^{b_A - 1} \int_{\mu_A - 1/2^{b_A}}^{\mu_A + 1/2^{b_A}} \int_{-\infty}^{+\infty} \left(-\log_2 \int_{\mu_B - 1/2^{b_B}}^{\mu_B + 1/2^{b_B}} N(y, \sigma_A^2, x) \, \mathbf{d}x \right) \cdot \\ \cdot N(\mu_A, \sigma_A^2, \mu_B) \, \mathbf{d}\mu_B \, \mathbf{d}y.$$

Since it is analytically impossible to determine the optimal value of b_A, we can easily get an approximation of the optimum by simply treating μ_B and σ_B like the points which are directly associated with the cluster A. The only difference is the following. While the above integral considers that the PDF varies inside the interval $[\mu_B - 1/2^{b_B}, \mu_B + 1/2^{b_B}]$ and determines the average costs in this interval, treating the parameters as points only considers the PDF value at one fixed position. This difference is negligible provided that $\sigma_B < \sigma_A$, which makes sense as child clusters should usually be much smaller (in terms of σ) than their parent cluster.

Coding Costs for a Cluster. Summarizing, the coding costs for a cluster can be obtained as follows: (1) Determine the optimal resolution parameter for each dimension according to the formula:

$$b_{\text{opt}} = \frac{1}{2} \log_2 \left(\frac{n \cdot W_B + r \cdot \#\text{ChildNodes}(B)}{3 \cdot r \cdot \sigma_B^2} \right).$$

(2) Determine the coding costs for the data points and the parameters according to:

$$C_{\text{hMDL}}(B) = c_{\text{appx}}(b) \cdot n \cdot W_B + r \cdot c_{\text{hMDL}}(\text{param})$$

(3) Add the costs obtained in step (2) to the ID costs of the points $(-nW_B \log_2(W_B))$ and of the parameters $(-\log_2(\sum_{X \sqsubseteq A} W_X))$. Whereas the costs determined in (2) are individual in each dimension the costs in (3) occur only once per stored point or parameter set of a cluster.

Coding Costs for the HCS. The coding costs for all clusters sum up to the overall coding costs of the hierarchy where we define constant ID costs for the parameters of the root:

$$hMDL = \sum_{C \in \mathcal{N}} \left(C_{\text{hMDL}}(C) - nW_C \log_2(W_C) - \log_2(\sum_{X \sqsubseteq \text{parent of } C} W_X) \right).$$

3.3 Obtaining and Optimizing the HCS

We optimize our objective function in an EM-like clustering algorithm ITCH. Reassignment of objects and re-estimation of the parameters of the HCS are done interchangeably until convergence. Starting from a suitable initialization, ITCH periodically modifies the HCS.

Initialization of the HCS. Clustering algorithms that follow the EM-scheme have to be suitable initialized before starting with the actual iterations of E- and M-step. An established method is to initialize with the result of a K-Means clustering. This is typically repeated several times with different seeds and the result with best mean squared overall deviation from the cluster centers is taken. Following this idea, ITCH uses a initialization hierarchy determined by a bisecting K-Means algorithm taking the hMDL value of the HCS as a stopping criterion for partitioning. First, a root node that contains all points is created. Then this root node is partitioned into two subclusters by applying K-Means with $K = 2$. This is done recursively until the hMDL of the binary HCS does not improve anymore within three steps. This ensures not to get stuck in a local minimum. Finally, after the best hierarchy is selected, μ_C and Σ_C are determined for each node C according to Section 3.2, and equal weights are assigned to the nodes, to ensure that clusters compeed likewise for the data points.

E-step and M-step. Whenever an object is associated directly to a cluster C then it is also indirectly associated with every ancestor of C. Nevertheless, points can also be directly associated not only to leaf nodes but also to inner nodes of the HCS. For instance, if a point P_i is an outlier w.r.t. any of the clusters at the bottom level of the HCS, then P_i has to be associated with an inner node or even the root. As established in Section 3.1, the clusters at all levels of the HCS compete for the data points. A point x is directly associated with that Cluster $C \in \mathcal{N}$ the probability density function of which is maximal:

$$Cl(x) = \arg\max_{C \in \mathcal{N}} \{W_C \cdot N(\mu_C, \Sigma_C, x)\}.$$

In the E-step of our hierarchical clustering algorithm, the direct association $Cl(x)$ for every object x is updated. Whereas, in the E-step only the direct association is used in the M-step which updates the location and scale parameters of all clusters we use both the direct and indirect association. The motivation is the following: The distribution function of every node in the HCS should always represent the whole data set in this branch of the tree, and the root node should even represent the complete data set. Therefore, for the location and scale parameters, all directly and indirectly associated objects are considered, as in the following formulas:

$$\mu_C = \frac{\sum_{B \in \mathcal{N}, B \sqsubseteq C} \left(\sum_{x \in B} x\right)}{\sum_{B \in \mathcal{N}, B \sqsubseteq C} |B|}, \sigma^2_{C,j} = \frac{\sum_{B \in \mathcal{N}, B \sqsubseteq C} \left(\sum_{x \in B} (x_j - \mu_{C,j})^2\right)}{\sum_{B \in \mathcal{N}, B \sqsubseteq C} |B|}$$

$$\Sigma_C = \text{diag}(\sigma^2_{C,1}, ..., \sigma^2_{C,d}).$$

In contrast, the weight W_C of each cluster should reflect the strenght of the individual Gaussian in the overall mixture of the HCS and sum up to 1 in order to define a valid

PDF with an integral over the complete data space of 1. Therefore, we use the direct associations for calculating the cluster weight with $W_C = |C|$.

Rearrangement of the HCS
The binary HCS that results from the initialization, does not limit our generality. ITCH is flexible enough to convert the initial HCS into a general one. Given a binary hierarchy, which is deeper than any n-ary hierarchy with $n > 2$, ITCH aims in flattening the HCS as far as the rearrangement improves our hMDL criterion. Therefore we trade off the two operations *delete* or *collapse* a node to eliminate clusters that do not pay off any more. Figure 3 visualizes the operations for an extract of a given HCS. By deleting a cluster C, the child nodes of C become child nodes of the parent of C (Figure 3(b)). By collapsing C, all of its child nodes are merged into a new cluster C' (including C), and therefore all of their child nodes become child nodes of C' (Figure 3(c)). Afterwards all points are redistributed, and E- and M-step are performed alternately until convergence. ITCH rearranges the HCS in an iterative way. In each iteration we tentatively delete/collapse each node in the HCS and perform E- and M-steps. Then first, the node and the operation that improves the hMDL criterion best is selected and second, the corresponding local neighborhood (parent, child and sibling nodes) is processed. These two steps are performed alternately until convergence.

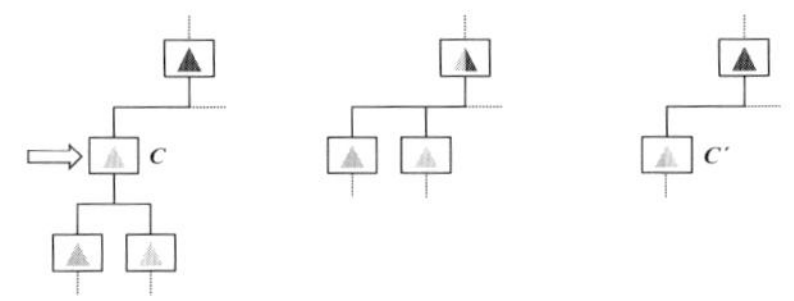

(a) Initial. (b) Deletion. (c) Collapse.

Fig. 3. Restructure operations of ITCH

4 Experimental Evaluation

Since ITCH is a hybrid approach combining the benefits of hierarchical *and* model-based clustering, we compare to algorithms of both classes to demonstrate the effectiveness of ITCH. We selected Single Link (SL) which probably is the most common approach to hierarchical clustering. As especially on noisy data, SL suffers from the so-called Single Link effect, we additionally compare to OPTICS, a more outlier-robust hierarchical clustering algorithm. Unless otherwise mentioned, OPTICS is parameterized with $\epsilon = 10,000$ and $MinPts = 10$. For an extensive description of parameterization strategies, we refer to [1]. Furthermore, we compare to RIC, an outlier-robust and parameter-free state-of-the-art algorithm to model-based clustering. In all plots, we mark cluster points by circles and outliers by triangles respectively. To relieve evaluation w.r.t. outliers, we added a color bar below the dendrograms of SL and the reachability plots of OPTICS, where colors refer to the class labels in the original data.

4.1 Synthetic Data

Experiments on DS_1 demonstrate the superiority of ITCH on hierarchical data sets. DS_1 comprises about 3,500 2-dimensional points that form a hierarchy of 12 clusters with outliers at different levels of the hierarchy. Seven Gaussian clusters are located at

(a) Synthetic data set DS_1.

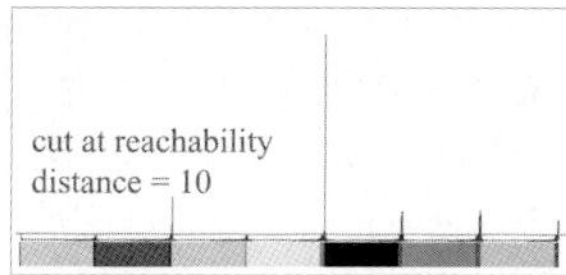

(b) Result of OPTICS.

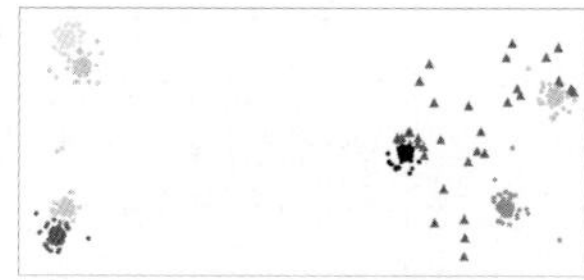

(c) Result of RIC.

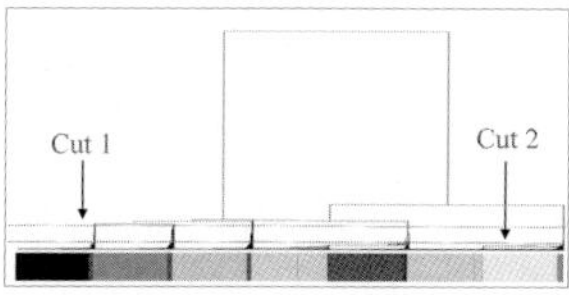

(d) Result of SL.

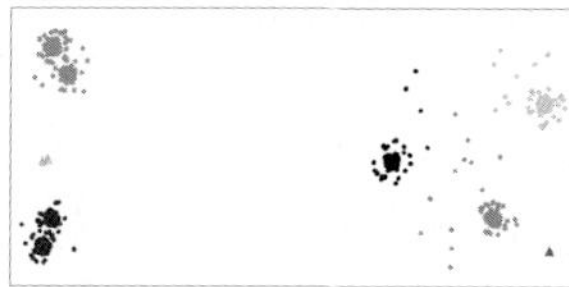

(e) Cut 1.

(f) Cut 2.

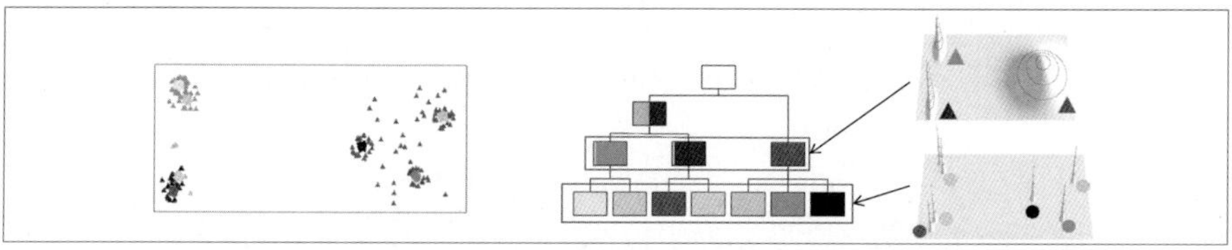

(g) Result of ITCH including the PDFs for each cluster at level 0 and 1.

Fig. 4. Experimental evaluation on synthetic data set $\mathbf{DS_1}$

the bottom level (Figure 4(a)). Experiments on DS_2 indicate the limitations of existing approaches to form meaningful clusters in extremely noisy non-hierarchical data. DS_2 is composed of two Gaussian clusters with 1,650 points each, overlapping in the marginal area without any global outliers. The quantitative evaluation of the results is always performed w.r.t. the "true" hierarchies present in these data sets.

Experimental Evaluation on $\mathbf{DS_1}$. As Clusters can be recognized as valleys in the reachability plot, OPTICS yields a satisfactory result (Precision: 94.8% Recall: 95.4% w.r.t. reachability distance < 10). But without our added color bar it would be impossible to spot the outliers since high distance peaks can also be caused by the usual jumps (Figure 4(b)). At a first glance, the SL-hierarchy (Figure 4(d)) reflects the true hierarchy quite well. However, a closer look at the data partitioning w.r.t. different cuts does not lead to meaningful clusters. Figure 4(e) illustrates the data that refers to a cut resulting in seven clusters. SL identifies only five clusters and three outliers (Precision: 70.0% Recall: 85.9%). The four clusters on the left side are wrongly combined into two clusters. Even at a much deeper split (Figure 4(f)) this effect remains for the orange cluster. Actually, the cluster quality is getting worse (Precision: 8.5% Recall: 9.0%) as the multiple outliers w.r.t. the three subclusters on the right side cause the well-known SL effect. Even though, each outlier is assigned to a single cluster the points marked by a red circle are not identified as outliers. Altogether, it is extremely hard to find the right parameter to cut through the dendrogram which gives a meaningful cluster representation. In order to apply RIC to the hierarchical data set, we preprocessed DS_1 with SL and applied RIC as postprocessing step in each level of the hierarchy. Figure 4(c) demonstrates the result when applying RIC to Cut 1 of the SL dendrogram (Precision:

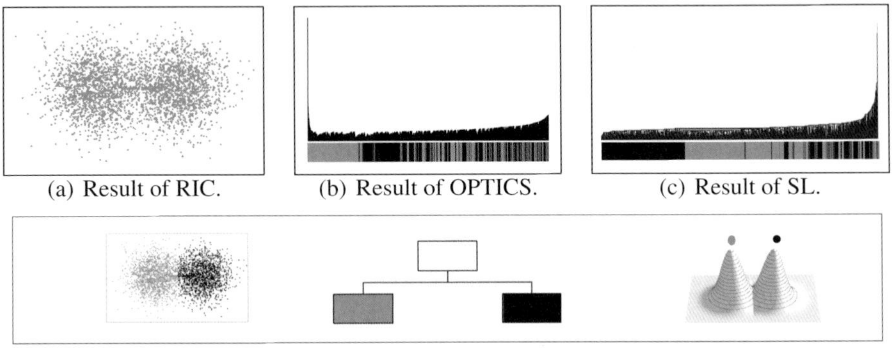

(a) Result of RIC. (b) Result of OPTICS. (c) Result of SL.

(d) Result of ITCH including the PDFs for both clusters at level 0.

Fig. 5. Experimental evaluation on synthetic data set $\mathbf{DS_2}$

94,9% Recall: 92,2%). It is obvious that even RIC fails to successfully filter out all outliers. More precisely, RIC assigns points (marked by a dark blue and orange triangle in the original data) that obviously are outliers w.r.t. two clusters on the left upper and lower side misleadingly to clusters. Also a majority of the red outliers are incorrectly identified as cluster points. ITCH is the best method to detect the true cluster hierarchy including outliers fully automatically (Precision: 93.8% Recall: 97.5%), and ITCH provides meaningful models on the data for each level of the hierarchy (Figure 4(g)).

Experimental Evaluation on $\mathbf{DS_2}$. Figure 5(a) demonstrates that RIC merges the two Gaussian clusters into only one cluster (Precision: 50.0% Recall: 100.0%). Also with OPTICS, it is impossible to detect the true structure of DS_2. The color bar in Figure 5(b) indicates that OPTICS assigns the points in an almost arbitrary order. Even when increasing the parameter for the minimum object density per cluster to a large value, OPTICS fails in detecting two clusters. SL miscarries due to the massive SL effect (Figure 5(c)). Here, OPTICS is not suitable to cure that problem. Moreover, the hierarchies generated by OPTICS and SL are overly complex but do not capture any cluster structure. Hence, it is not possible to evaluate these results in a quantitative fashion. Only ITCH discovers a meaningful result without requiring any input parameters (Precision: 99.2% Recall: 99.7%). All clusters that do not pay off w.r.t. our hMDL are pruned and hence, only two Gaussian clusters remain in the resulting flat hierarchy which are described by an intuitive description in form of a PDF (Figure 5(d)).

4.2 Real World Data

Finally, we show the practical application of ITCH on real data sets available at UCI[1].

Glass Data. The *Glass Identification* data set comprises nine numerical attributes representing different glass properties. 214 instances are labelled according to seven different types of glass that form a hierarchy as presented in Figure 6(a). ITCH perfectly separates *window glass* from *non window glass*. Additionally, *tableware* and *containers* are

[1] `http://archive.ics.uci.edu/ml/`

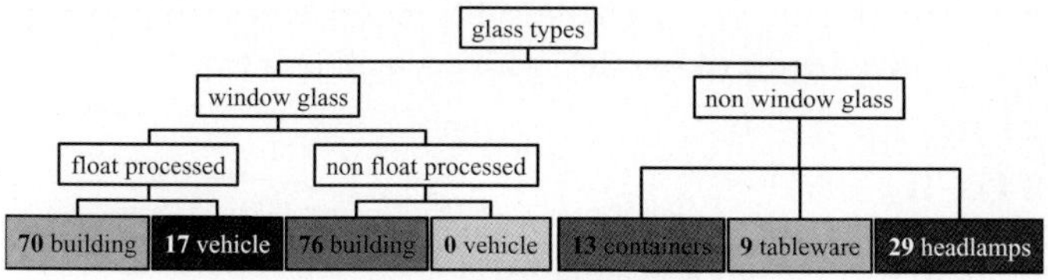

(a) Hierarchy in the original data set.

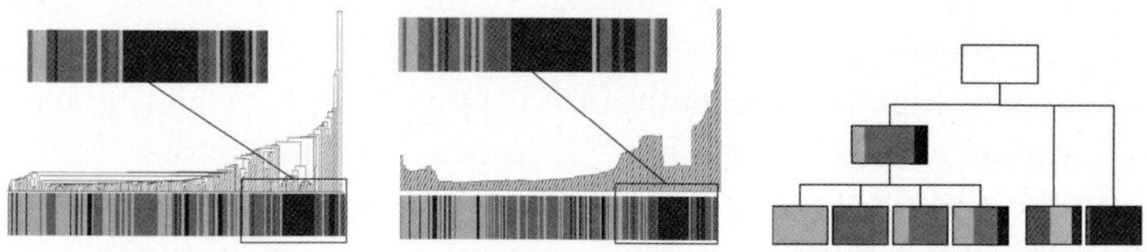

(b) Result of SL. (c) Result of OPTICS. (d) Result of ITCH.

Fig. 6. Hierarchical clustering of 9-dimensional glass data (214 instances)

almost perfectly separated from *headlamps*. The four subclusters of *window glass* are very similar. Hence, ITCH arranges them at the same level. Some outliers are directly assigned to *window glass*. In contrast to ITCH, neither SL nor OPTICS separates *window glass* from *non window glass* perfectly (Figures 6(b) and 6(c)). *Containers* and *tableware* do not form discrete clusters but are constituted as outliers instead. In the dendrogram only the *headlamps* can be identified, whereas in the reachability plot two clusters are visible. Nevertheless, both approaches do not reflect the original hierarchy successfully. As it is not clear where to define an adequate cut through the dendrogram we applied RIC at the bottom level. This results in only two clusters without any separation between *window glass* or *non window glass*.

Cancer Data. The high-dimensional *Breast Cancer Wisconsin* data set contains 569 instances each describing 30 different characteristics of the cell nuclei, where each instance is either labelled by *benign* (blue) or *malignant* (red). OPTICS and SL both fail to detect a clear cluster structure in this data set (Figures 7(a) and 7(b)). Hence, we applied RIC on top of a K-Means clustering with K=15. As stated by the authors we chose K large enough compared to the number of classes. However, RIC also fails and results in three mixed clusters. In contrast, despite the high dimensionality of the data, ITCH almost perfectly separates the *benign* from the *malignant* objects which are then split

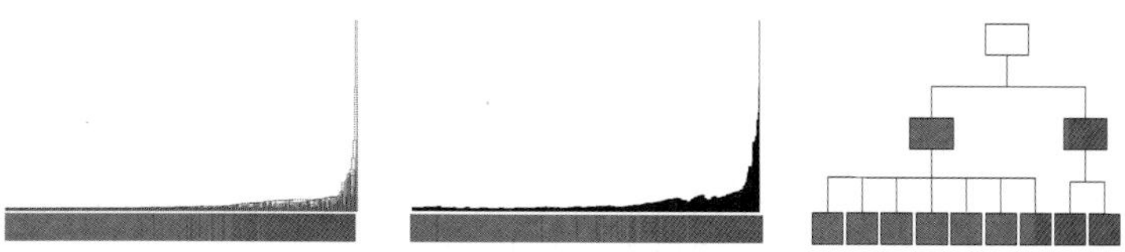

(a) Result of SL. (b) Result of OPTICS. (c) Result of ITCH.

Fig. 7. Hierarchical clustering of 30-dimensional breast cancer data (569 instances) into different subclusters (Figure 7(c)). This result is consistent with previous findings as the two classes exhibit a degree of overlap with each other [15].

4.3 Stability of ITCH

Since we do not want to rely on single results we additionally tested the stability of ITCH over 20 runs for each data set. Figure 8 shows the variance of the hMDL value in percent depending on the mean value. The result of ITCH is highly stable within DS1, DS2 having only a variance of 0.03% and 0.12%, respectively. Also in the real world data sets the result of ITCH shows only little variance.

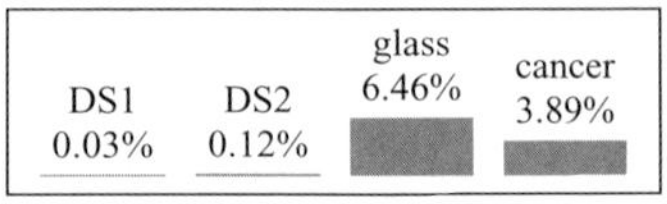

Fig. 8. Stability of the ITCH result over 20 runs

5 Conclusions

We have introduced a new hierarchical clustering method to arrange only natural, valid, and meaningful clusters in a hierarchical structure – ITCH. ITCH is based on an objective function for clustering that was guided by the information-theoretic idea of data compression. We have shown that without difficult parameter settings ITCH finds the *real* cluster hierarchy effectively, and that it provides accurate and intuitive interpretable information in a wide variety of domains, even in the presence of outliers.

Acknowledgements. We thank Johannes Huber for assisting us with the evaluation.

References

1. Ankerst, M., Breunig, M.M., Kriegel, H.P., Sander, J.: OPTICS: Ordering Points To Identify the Clustering Structure. In: SIGMOD, pp. 49–60 (1999)
2. Banfield, J.D., Raftery, A.E.: Model-Based Gaussian and Non-Gaussian Clustering. Biometrics 49(3), 803–821 (1993)
3. Basu, S., Bilenko, M., Mooney, R.J.: A Probabilistic Framework for Semi-supervised Clustering. In: KDD, pp. 59–68 (2004)
4. Bilenko, M., Basu, S., Mooney, R.J.: Integrating Constraints and Metric Learning in Semi-supervised Clustering. In: ICML (2004)
5. Böhm, C., Faloutsos, C., Pan, J.Y., Plant, C.: Robust Information-theoretic Clustering. In: KDD. pp. 65–75 (2006)
6. Chardin, A., Pérez, P.: Unsupervised Image Classification with a Hierarchical EM Algorithm. In: ICCV, pp. 969–974 (1999)
7. Cilibrasi, R., Vitányi, P.M.B.: Clustering by Compression. IEEE Transactions on Information Theory 51(4), 1523–1545 (2005)
8. Dempster, A.P., Laird, N.M., Rubin, D.B.: Maximum Likelihood from Incomplete Data via the EM Algorithm. J. Roy. Stat. Soc. 39, 1–31 (1977)
9. Goldberger, J., Roweis, S.T.: Hierarchical Clustering of a Mixture Model. In: NIPS (2004)
10. Grünwald, P.: A Tutorial Introduction to the Minimum Description Length Principle. CoRR math.ST/0406077 (2004)
11. Hamerly, G., Elkan, C.: Learning the K in K-means. In: NIPS (2003)

12. Jain, A.K., Dubes, R.C.: Algorithms for Clustering Data. Prentice-Hall, Englewood Cliffs (1988)
13. Lu, Z., Leen, T.K.: Semi-supervised Learning with Penalized Probabilistic Clustering. In: NIPS (2004)
14. Murtagh, F.: A Survey of Recent Advances in Hierarchical Clustering Algorithms. Comput. J. 26(4), 354–359 (1983)
15. Pantazi, S., Kagolovsky, Y., Moehr, J.R.: Cluster analysis of wisconsin breast cancer dataset using self-organizing maps (2002)
16. Pelleg, D., Moore, A.W.: X-means: Extending K-means with Efficient Estimation of the Number of Clusters. In: ICML, pp. 727–734 (2000)
17. Rissanen, J.: Stochastic Complexity in Statistical Inquiry Theory. World Scientific Publishing Co., Inc., River Edge (1989)
18. Rissanen, J.: Information and Complexity in Statistical Modeling. Springer Publishing Company, Incorporated (2007)
19. Slonim, N., Tishby, N.: Document Clustering using Word Clusters via the Information Bottleneck Method. In: SIGIR, pp. 208–215 (2000)
20. Still, S., Bialek, W.: How Many Clusters? An Information-Theoretic Perspective. Neural Computation 16(12), 2483–2506 (2004)
21. Tishby, N., Pereira, F.C., Bialek, W.: The information bottleneck method. CoRR physics/0004057 (2000)
22. Vasconcelos, N., Lippman, A.: Learning Mixture Hierarchies. In: NIPS. pp. 606–612 (1998)

Coniunge et Impera: Multiple-Graph Mining for Query-Log Analysis

Ilaria Bordino[1,⋆], Debora Donato[2], and Ricardo Baeza-Yates[3]

[1] Sapienza Università di Roma, Rome, Italy
bordino@dis.uniroma1.it
[2] Yahoo! Labs, California, US
debora@yahoo-inc.com
[3] Yahoo! Research, Spain
rbaeza@acm.org

Abstract. Query logs of search engines record a huge amount of data about the actions of the users who search for information on the Web. Hence, they contain a wealth of valuable knowledge about the users' interests and preferences, as well as the implicit feedback that Web searchers provide when they click on the results obtained for their queries.

In this paper we propose a general and completely unsupervised methodology for query-log analysis, which consists of aggregating multiple graph representations of a query log, tailored to capturing different semantic information. The combination is carried out by applying simple but efficient graph-mining techniques. We show that our approach achieves very good performance for two different applications, which are classifying query transitions and recognizing spam queries.

1 Introduction

The Web-search experience is nowadays part of the life of a continuously growing number of people. According to a recent study released by ComScore[1], the global Web-Search Market has increased by 41% in 2009, reaching more than 113 billion searches. Unique users and penetration[2] have experienced an impressive growth in the last decade. The most up-to-date estimations report the Internet is currently used by more than one fourth of the whole world population. Moreover, the highest growth rates have been observed within those segments of the world market where penetration is still very low.

The results of these studies provide clear evidence of the fact that, despite the massive user adoption and the maturity of the services offered by the Web industry, further efforts are still needed to conquer those fractions of the market where the Web experience is newer and penetration less exhaustive.

Commercial search engines attempt to improve the appeal and the usability of their services by offering tools like query recommendation, user profiling and

⋆ Part of this work was done while visiting Yahoo! Research Labs, Barcelona.

[1] http://www.comscore.com/

[2] http://www.internetworldstats.com/stats.htm

J.L. Balcázar et al. (Eds.): ECML PKDD 2010, Part I, LNAI 6321, pp. 168–183, 2010.

spam detection. They invest considerable amounts of resources in the development of instruments for gathering novel knowledge about the users' interests.

In this scenario, query-log analysis is broadly applied to obtain useful insights about the way users refine their queries, and the search strategies they apply to satisfy their information needs. Recent studies [7,17,20] have shown that the wealth of information stored in search logs can be successfully used to build a very accurate characterization of query reformulation types, which is a key step towards improving the service provided by search engines and developing innovative web-search paradigms. However, the most commonly adopted approaches share a major limitation, which is to be found in the usage of supervised learning, be it as a sole learning mechanism, or combined with classifiers and exact look-up on dictionaries. These editorial resources are very costly, and thus difficult to obtain for specific languages or cultures.

In this paper, we study the problem of developing effective approaches for query-log analysis. We aim to develop methods that are simple and at the same time able to deal with huge data volumes. The approach we propose is completely unsupervised and based on the map/reduce paradigm.

We use graph structures to build compact and navigable representations of the information extracted from query logs. The idea of inferring graphs from query logs has been extensively explored by recent research [5,6,7,12,14,24]. The above works focus on the analysis of a single graph, which tipically captures only some particular aspects of the interactions between users and search engines. None of these approaches is able to exploit exhaustively the huge amount of hidden information available in the log.

In this work, we propose to analyze query-log data through the joint mining of multiple graphs, as the combination of them is the ultimate wisdom-of-crowds approach. More specifically, our main contributions are listed below.

- We present *Coniunge et Impera*, a general framework for query-log analysis, designed to provide an effective support for the execution of many tasks that are relevant from a search-engine point of view. The fundamental building blocks of our framework consist of (*i*) a collection of graph projections extracted from a query log according to different semantic criteria, and (*ii*) a set of operations to be used for mining and maintenance purposes.
- Concerning the choice of the graph representations, we build three graphs that relate queries according to various types of information: common words, common clicked results, session information. The definitions we use are introduced by Baeza-Yates [2].
- The toolbox built for graph analysis includes set operations and more complex graph operations, like extraction of subgraphs, connected components and articulation points. The choice of the operations was driven by the applications we had in mind, which are extensively described in the paper. Although interesting results can be obtained by applying operations as simple as set operations, we will show that more complex graph algorithms are needed to exploit all the wealth of information enclosed in the graphs.

- The operations are also used for the maintenance of the framework: for example, the union of two graphs can be computed to merge the data extracted from consecutive snapshots of a query log. The extraction of subgraphs can be applied to restrict the analysis to a subset of queries satisfying some particular conditions.
- To demonstrate the practical applicability and usefulness of our method, we analyze a large query log provided by a commercial search engine and we show how to customize our approach for two different applications, namely, classifying query transitions and recognizing spam queries.

We remark the fact that our work aims at providing a general instrument that can be useful for many different problems. For this reason, we impose no restrictions on the building blocks of our system, and we intentionally leave room for extensions. For example, more complex graph definitions could be introduced to analyze the semantics behind queries at a finer granularity.

It is also worth to observe that the approach we follow is completely unsupervised: it does not require the usage of any knowledge bases or other expensive editorial resources. This characteristic makes the method very general and applicable to different data, without any kind of linguistic and culture-specific issues.

The remainder of the paper is organized as follows. Section 2 discusses previous work. In section 3, we present the definitions used to extract different graphs from raw logs. Section 4 introduces the set of instruments that we included in our framework for mining and maintenance purposes. Various examples of the practical usefulness of the chosen operations are provided. Next, we show how to customize the *Coniunge et Impera* methodology for classifying query transitions (see Section 5) and for detecting spam queries (Section 6). Finally, Section 7 offers our concluding remarks.

2 Related Work

Inferring graphs from query logs. The idea of extracting a graph structure from query-logs has been introduced in the last decade. The attention on the possibilities offered by using graphs to mine different semantic information implicitly contained in query-logs was formalized by the work of Baeza-Yates [2]. Here five different types of such graphs, relying on different criteria to connect queries by edges (e.g. common words, common clicked URLs), are studied.

Click graphs [6,12,24] have been extensively used for various purposes. A click graph is a query-document bipartite graph, where a query is connected to the documents that were clicked in the associated result list.

Inspired by the works on click graphs, Baeza-Yates and Tiberi [5] propose a novel way to represent queries in a vector space based on a graph derived from the query-click bipartite graph. This graph is then used to find similar queries.

Starting from the click graph, Castillo et al. [10] define two alternative graphs, the *view graph* and the *anti-click graph*. In the view graph the edge set of the click graph is replaced with the one containing edges that relate a query to the documents whose URL has been viewed in the answer list returned to the user, but not necessarily clicked. The view graph is a generalization of the click graph since each click is also a view. The anti-click graph intends to capture the negative feedback that users implicitly give to a top-ranked document when they ignore it by clicking on documents ranked below. This graph contains one edge between a query and any not-clicked documents ranked higher than one clicked by the user.

Boldi et al. [7] introduce the *query-flow graph*, which models user behavioral patterns and query dependencies. In this graph, a directed edge from one query to another means that the two queries are likely to be part of the same search mission. Any path over the query-flow graph may be seen as a searching behavior, whose likelihood is given by the strength of the edges along the path.

Similarity or distance between queries. Projecting query-logs on graphs is not the only way to infer semantic relations from implicit user feedback. The most explored approach consists of defining a similarity function between queries.

Raghavan and Sever [21] propose to measure the similarity between two queries through the mining of the differences in the ordering of the documents retrieved. Due to a matter of scalability, this technique is not of practical use when one has to deal with the whole Web.

Wen et al. [24] cluster similar queries to support recommendations for queries frequently submitted to search engines. They use several notions of query distance, based either on the keywords composing the query text or on the set of common clicked URLs.

Fonseca et al. [15] use association rules to discover related queries. Their approach views the log as a set of transactions, thus failing in discovering the most interesting related queries, which are the ones submitted by different users. It also encounters problems in determining successive query sessions that belong to the same search process.

Baeza-Yates et al. [3,4] introduce a term-weight vector model to represent queries. In this model, a weight is assigned to each term occurring in the content of the Web pages clicked after a query, depending on the number of occurrences of the query within the log and on the number of clicks of the documents which the term appears in.

3 Preliminaries

3.1 Query Graphs

We analyze query-log data by building different *query graphs* to capture the various semantic aspects of the relations between queries. We use the term *query graph* to refer to a graph $G = (V, E, w_V, w_E)$ extracted from a snapshot of a search-engine log, such that the set of nodes V comprises all the distinct queries appearing in the log, whereas the edges in the set E denote the existence of a particular semantic relation between queries, depending on a specific criterion considered in the graph definition. The weighting functions $W_V : V \to \mathcal{N}$ and $W_E : E \to \mathcal{R}$ are respectively used to associate nodes and edges with a weight whose meaning changes depending on the information used to generate the graph.

Table 1. Graph definitions: Summary

Graph feature	Word graph	Session graph	Click graph
Edge	Common words	Consecutive appearance in a session	Common clicked results
Node weight	# Occurrences	# Sessions	# Occurrences
Edge weight	# Common words in the text of q_i and q_j	# Sessions in which q_i and q_j appear sequentially	Cosine similarity of the vectors of URLs clicked for q_i and q_j

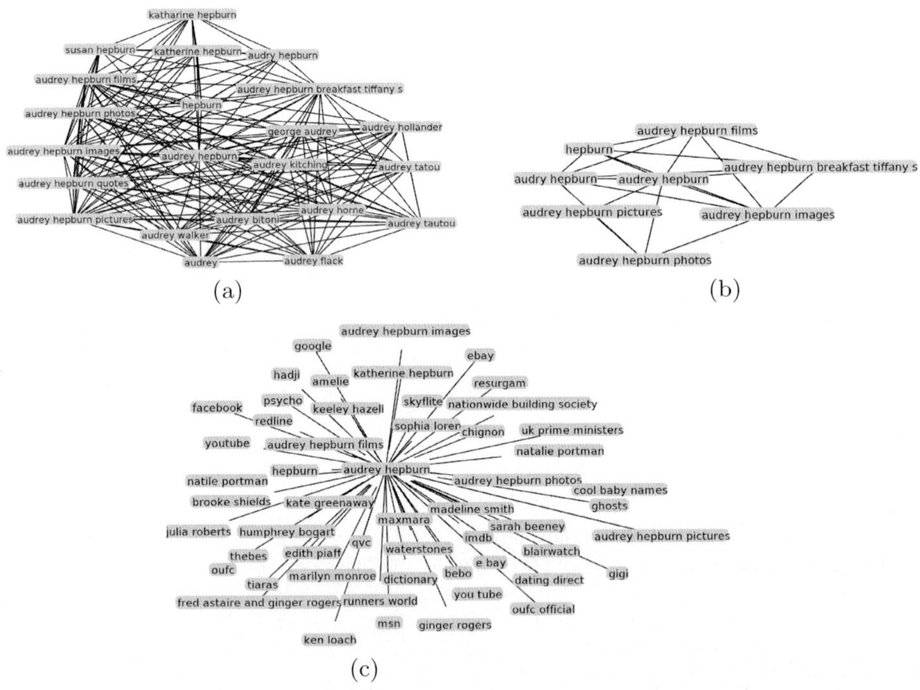

Fig. 1. Sample graphs for *audrey hepburn*: Word (a), Click (b) and Session (c) graph

In this paper, we generate three query graphs using the definitions of *Word Graph*, *Session Graph* and *Click Graph* (called *Url Cover Graph* in the original paper) formalized by Baeza-Yates [2]. We summarize the definitions in Table 1 and give an example of each graph in Figure 1.

Dataset. We studied a sample log of early 2008 from the Yahoo! UK search engine. The raw data consisted of a sequence of annotated sessions, containing more than 20 million distinct queries. Following [2] and [7], we removed the queries containing non-English characters. Motivated by preliminary assessments, we

Table 2. Basic statistics

Graph	# Nodes	# Edges	# Isolated nodes	# CC	Size of largest CC
Word graph	801 876	565 095 123	121 423	135 942	649 298
Click graph	801 876	21 971 751	161 344	195 379	549 217
Session graph	801 876	6, 654 614	25 657	26 199	775 096

filtered out the queries with less than 5 occurrences in the log, and we retained a common subset of queries to avoid the dishomogeneties caused by the different definitions. Table 2 reports statistics about the three graphs we built.

4 A Software Framework for Query-Log Graphs

We now describe the algorithms included in our framework. We carefully studied the operations that were necessary for the purposes of (*i*) data mining, and (*ii*) manteinance of the framework. We selected a collection of initially minimal and crucial mining tools that are listed below. The presentation is intentionally informal and descriptive. The list should by no means be considered exhaustive. Different operations might be needed for different applications.

A recent work [16] proposes a formal algebra that manipulates graphs as basic units of information, maintaining the same basic characteristics and the same expressive power as the relational algebra. We don't follow a similar approach because we aim at handling complex graph problems, which cannot be efficiently represented and studied by means of relational algebra.

4.1 Operations

We use both *binary* operations and *unary* operations.

Binary operations. The binary operations are the fundamental set operations that can be applied on graphs: union, intersection, difference.

- **Union.** Given two query graphs G and H, their union is represented by a graph $F = G \cup H$ such that $V(F) = V(G) \cup V(H)$ and $E(F) = E(G) \cup E(H)$.
  Example. The union of two graphs is a critical operation for the maintenance of our framework: it is necessary to merge data extracted from different snapshots of the log. Updates are needed from time to time to regenerate the graphs, given that the information extracted from query logs suffers a slow aging effect, as demonstrated in [18].
- **Difference.** The set difference of two graphs G and H is a graph $F = G \setminus H$ such that $V(F) = V(G) \setminus V(H)$ and $E(F) = E(G) \setminus E(H)$.
  Example: Similar queries. Identifying queries that express closely related information needs is crucial task for generating query recommendations [4]. In the *Coniunge et Impera* framework, a very simple solution for the problem of detecting queries that are semantically, but not sintactically similar consists of computing the set difference between the Click graph and the Word

graph: the edges included in the result connect queries that have common clicked results, and thus are likely to be related to the same topic despite the fact that their texts have no common words. The edge weights of the click graph can be used to filter out the least significant relations.

- **Intersection.** The intersection of two graphs G and H is a graph $F = G \cap H$ such that $V(F) = V(G) \cap V(H)$ and $E(F) = E(G) \cap E(H)$.
  Example: Finding logical sessions. Identifying changes in the *search mission*, that is, in the information need [7,17] expressed by a user within a given session, is a critical issue for modeling user behavior and predicting user satisfaction. To detect different search missions within a user session, we compute the set difference between the Session graph and the intersection of the Click graph and the Word graph.

Note. The set operations can be applied on graphs that have weights both on the nodes and on the edges. We impose no restrictions on how to combine the weights, as the best choice depends on the particular task that one has in mind.

Unary operations. We consider the following unary operations:

- **Node filter.** Given a query graph G and a mathematical condition c, this operation returns a query graph H corresponding to the induced subgraph of G whose vertex set is formed by all the vertices in $V(G)$ associated with weights satisfying c.
- **Edge filter.** Given a query graph G and a mathematical condition c, this operation returns the (non-induced) subgraph of G whose vertex set includes all the vertices in $V(G)$, while the set of edges is formed by the edges in $E(G)$ associated with a weight that satisfies the specified condition.
  Examples. We used the filter operations in the preliminary phases of our study, when we extensively analyzed the structural properties of the graphs extracted from the log. We tested various thresholds on the weights of nodes and edges, to study the differences between the resulting denser or sparser versions of the graphs.
  The filter operations allow to select portions of the graphs that are relevant for a particular application. For example, if we aim to detect *similar queries*, we can apply an edge filter on the Click graph, selecting the edges with a large weight. In the Click graph, a large-weight edge indicates a significant degree of similarity in the results clicked for the two queries. We applied this idea in our methods for detecting error corrections (see Section 5).
  An opposite filter is needed to identify *spam queries*, i.e., queries that attract a high number of spam pages in their result set. The heuristic we developed for this application, which is presented in Section 6, exploits the fact that a low-weight edge in the Click graph is indication of a relation of very poor quality between the queries connected by such edge. The clicked results shared by these queries are usually spam or multi-topical URLs. The queries containing these URLs in their set of answers are included into a base set of spam candidates.

- **Connected components.** This operation takes a query graph G and returns a set of graphs $S = \{G_1, G_2, \ldots, G_k\}$, such that, for each $i = 1, 2, \ldots, k$, G_i is isomorphic to a maximal connected subgraph of G.
  Example: Clustering queries. The computation of the connected components in the Session graph and/or in the Click graph is required for the task of clustering queries, which is useful for many purposes, such as re-ranking, query recommendation, and finding logical sessions.
- **Biconnected components.** This operation takes a query graph G and returns a set of graphs $S = \{G_1, G_2, \ldots, G_k\}$ such that, for each $i = 1, 2, \ldots, k$, G is isomorphic to a maximal biconnected subgraph of G.
- **Articulation points.** Given a query graph G, this operation returns a set of query instances $S = \{q_1, q_2, \ldots, q_k\}$ such that $S \subseteq V(G)$ and, for $i = 1, 2, \ldots, k$ q_i is an articulation point in G. An articulation point may belong to one or more biconnected component.
  Example: Polysemic queries. Polysemic queries [19,23] are related by a syntactic point of view (for instance, they share one or more common words in their text representations), but they cover unrelated semantic aspects. Articulation points in the Click graph and/or in the Word graph are natural candidates for word polysemy.

4.2 Graph Extraction and Representation

The raw data we analyzed consisted of a set of annotated sessions, containing 20 million distinct queries.

For efficiency and scalability, we used Hadoop's MapReduce[3] to extract the query graphs. MapReduce [13] is a popular programming model for processing large-scale data. This model allows users to write map/reduce components that are scheduled to distributed resources for processing, with a transparent handling of problems like parallelization, network communication, and fault tolerance.

Due to lack of space, we do not describe the three algorithms we implemented to generate the graphs. Our procedures cascade a sequence of 3-5 MapReduce jobs, which are used to gather the specific data that is needed to interconnect queries and to create the adjacency list of each query.

We ran the MapReduce routines on a server with 3G RAM. We needed 2 hours to generate the Session graph and 4 hours to create the Word graph and the Click graph. The time required for the Session graph was lower because the particular format of the input data made the extraction of the information required for this graph immediate. Altogether, the time used for generating the three graphs is less than half a day for our sample log. We also observe that our toolbox allows an incremental construction of the query graphs.

The operations included in our framework are complex and procedural by nature; they require global computations that involve all the nodes in a graph. This consideration oriented us towards the choice of compressed graph representations to store and manipulate our data.

[3] `http://hadoop.apache.org/common/docs/current/mapred_tutorial.html`

Compressed graph representations have become very attractive, as they allow to overcome the limitations encountered by many modern applications, whose storage requirements exceed the capacity of the faster memories.

Web search uses graphs as natural models for the Web structure. Several algorithms that are used by search engines to rank pages, discover communities and so on are run on these Web graphs.

For graph representation, we use Webgraph [9], a framework that obtains state-of-the-art results in terms of compression through the combination of several mechanisms, such as node reordering, differential encoding, compact interval representations, and references to similar adjacency lists.

Starting from a raw input of $\sim$ 30GB, we obtained three representations of the following sizes: $500M$ for the Word graph, $500M$ for the Click graph, and $20M$ for the Session graph.

Exploting the fact that our graphs are compact and can be efficiently kept in main memory for navigation purposes, we developed main-memory implementations of our algorithms. Due to lack of space, we omit details about the specific realizations, which are all based on standard algorithms. All the operations require time linear in the size of the input graphs. In our experiments, we reported running times in the order of minutes for all the algorithms implemented.

We also explored the possibility of developing MapReduce realizations for our algorithms. We decomposed each operation into a series of MapReduce processes, adopting an approach similar to [11]. The obtained implementations are still inefficient, mostly because many operations are based on graph traversal, which requires a large number of iterations because a mapper can only read a random record for each map operation. J.Ekanayake [1] has recently developed *i-MapReduce*, a streaming-based framework that supports iterative MapReduce computations. We plan to explore the usage of this instrument in future work.

5 Classifying Query Transitions

We now show how to apply our methodology to the task of classifying query transitions. By *transition* we mean a pair of queries that a user submitted consecutively to a search engine. The analysis of these query sequences is extremely important for the purpose of understanding how the users reformulate their queries to gather information of better quality.

Retrieving information from the Web is a process that requires a continual interaction between the user and the search engine. Only in half of the cases an information need is satisfied with just a single query [22]. In the other cases, the user submits a refined query, because she is not satisfied with the documents obtained in the first attempt. This might happen for various reasons, for example because the former query contained errors, or it was either too general or too specific for the purpose the user had in mind. A transition between queries that are part of the same search mission is usually referred to as a *query reformulation*.

Characterizing query reformulation patterns is a task of critical importance to improve the relevance of web-search results, or to refine tools for query recommendation. Following the taxonomy introduced and used in [7,8], we focus on the task of recognizing the following types of query transitions:

- **Error correction:** the user is trying a different spelling or capitalization of a query. Example: *audry hepburn, audrey hepburn.*
- **Generalization:** the second query is more general than the first one. Example: *audrey hepburn quotes, audrey hepburn.*
- **Specialization:** the first query is more specific than the second one. Example: *audrey hepburn films, audrey hepburn breakfast tiffany s.*
- **Parallel move:** the user is modifying her query to search for something related but not equivalent. Example: *audrey hepburn, sophia loren.*
- **Different mission:** the user is trying to satisfy a completely different information need. Example: *audrey hepburn, runners world.*

The last two types of query transitions, *Parallel move* and *Different mission*, can be considered as belonging to a broader category, which comprises the transitions that represent a change in the information need expressed by the user. In the work of Jones et al. [17], these two transition types are both labeled as *Different chain.* We adopt the same approach, aggregating Parallel move and Different mission into a more general category, which we call *Different goal.* In the remainder of this section we describe how to extract and aggregate information from the Word graph, the Click graph and the Session graph to classify query transitions.

5.1 Unsupervised Approach to Query Transition Classification

Our method for classifying query transitions combines three independent heuristics, which are tailored to labelling non-overlapping sets of transitions. This is an extremely advantageous characteristic, which implies that the three building blocks of our approach can be either applied in parallel, for example on a grid, or they can be executed consecutively in a sequential setting, from the most to the least aggressive approach, so that the amount of data to be analyzed is drastically reduced at each step.

Algorithm 1: Identifying different goals. Our first algorithm aims at recognizing transitions between two queries that express a different search goal. The algorithm computes the set difference between the Session graph and the union of the Click graph and the Word graph. This corresponds to filtering the edges of the Session graph to retain only the transitions connecting two queries that have no common clicked results and no common words in their text representations.

Algorithm 2: Error corrections with no common words. An *Error correction* is a query reformulation that the user submits to fix a typo, usually trying a different spelling or capitalization of a query. Our second algorithm detects error corrections involving queries with no words in common.
Step 1: Edge filter on the Click Graph. We start from the Click graph and we filter out the edges whose weight is less than 0.3. In this way, we only retain

the top 10% strongest semantic relations, which is useful because in the case of an error correction the search goal remains the same.
Step 2: Intersection with the Session graph. We intersect the subset of the Click graph that was computed in the previous step with the Session graph: This is needed to identify query transitions.
Step 3: Set difference with the Word graph. We compute the set difference between the subgraph obtained at the previous step and the Word graph: this is required to remove transitions with common words. (These transitions are considered by our third algorithm).
Step 4: Filter by Click labels. We refine the partial result obtained in the previous step by selecting the edges whose type label in the Click graph is equal to 2, meaning that the set of results clicked for the first query is strictly contained in the set of answers associated with the second query. Upon detecting errors in the text of the query submitted by a user, search engines suggest the proper spelling, presenting the top-ranked answers of the correct formulation in the first positions of the result list. This practice makes the above condition on click labels very likely to be satisfied by error-correction transitions.
Step 4: Filter by edit distance. Finally, we use Levehnstein distance to isolate error corrections from other cases. We retain only the transitions for which the edit distance between the two queries is no greater than 0.2. The threshold was chosen after experimenting with different values (0.2, 0.3, 0.4): As expected, the lower the threshold, the more we are effective in isolating error corrections from other reformulations.

Algorithm 3: Transitions involving queries with common words. Our last algorithm analyzes the transitions involving queries with common words in their text representations.
Step 1: Intersection between the Session graph and the Word graph. This is needed to isolate the set of transitions we want to focus on. Many types of reformulations are characterized by overlapping query texts. A number of tests are applied to distinguish the various cases.
Step 2. The text of the first query is a subset of the text of the second query: then we label the transition as Specialization.
Step 3. The text of the first query is a superset of the text of the second query: then we label the transition as Generalization.
Step 4. We use edit distance to isolate error corrections.

5.2 Evaluation

We used an annotated query-flow graph [8] extracted from the same log snapshot as the ground truth for evaluating the quality of results. This graph originally contained $\sim$ 21M nodes and $\sim$ 43M edges. Restricting to the nodes included in our query graphs, we extracted a subgraph that contains $\sim$ 3.5M edges.

The model introduced by Boldi et al. [8] and used to annotate the edges of the query-flow graph is able to produce an automatic classification of query

Table 3. Classification of query transitions: quality of results

Transition type	TP	TN	FP	FN	Precision	Recall	Accuracy
Different Goal	2 594 560	318 125	516 501	4 970	0.834	0.998	0.848
Generalization	58 627	3 366 295	2 660	6 574	0.956	0.899	0.997
Specialization	179 774	2 827 724	19	426 639	0.999	0.296	0.878
Error Correction	60 330	3 249 459	21 685	102 682	0.736	0.370	0.964

reformulation types with an accuracy as high as 92%. This method represents the state of the art for classifying query transitions: However, we remark the fact that this is a supervised approach, based on machine learning from a human-labeled log sample.

To evaluate the quality of our method, we compared the classification generated by our algorithms with the labels associated to the same transitions in the query-flow graph. We computed True Positives, True Negatives, False Positives, False Negatives, Precision, Recall, and Accuracy. Results are shown in Table 3.

We report the global evaluation of our methodology in Table 3. We remark two main results. The first one is that every part of our methodology relies on merging the information coming from at least two different query graphs. The second one is that the approach is totally unsupervised, and uses operations which are linear in the size of the graphs involved. Moreover, since the subsets of transitions labeled by each heuristic do not overlap, the number of edges to consider decreases at each step.

Altogether, the labels obtained with our method agreed with those assigned to the same transitions in the query-flow graph in 84% of the cases. More specifically, we obtained very good results in the cases of generalization and different goal, for which we measured an accuracy respectively of 99 and 85%. On the other end, we noticed a very low recall in the case of specialization and error correction. Since the query-flow graph was labeled using a model, we wondered whether this poor accuracy could be induced by errors in the query-flow graph rather than in our methodology.

To investigate this hypothesis, we manually evaluated a random sample of the transitions for which our method and the one based on the query-flow graph did not agree. We selected 1K transitions labeled as specialization in the query-flow graph, and $1K$ transitions labeled as error correction. We considered the transitions divided into buckets according to their frequency of occurrence in the log. Three human assessors were asked to evaluate the transitions, assigning them one of the reformulation types considered in this study.

We assumed the manual assessment to be the golden truth for the sample, and we measured how well the two automatic classification methods (ours, and the one using the query-flow graph) agreed with the golden truth. The evaluation gave the following results: the labels produced by our method agreed with the ones assigned by human assessors in 80.4% of cases, while the query-flow-graph labels resulted equal to the ground truth in 12.8% of cases.

To evaluate the statistical significance of this result, we associated each automatic method with a vector that has a cell for each sampled transition, containing a 1 if the label assigned by the method agreed with the ground truth, and a 0 otherwise. We them performed a Wilcoxon signed-rank test, comparing the positive differences between the judgements obtained for the two methods. The test determined a statistical significance at $p \ll 1\%$ for the difference between our methodology and the query-flow graph.

Hence, we believe that the actual precision of our approach is higher than the one obtained with this experiment, and that the disagreement with the query-flow graph is due to errors in the model used by the latter method to train the classifier.

6 A Heuristic for Detecting Spam Queries

Spam queries [10] are queries that generate a high number of spam pages in the top positions of their lists of answers. Identifying these queries can uncover meaningful information for the task of designing more robust spam-detection algorithms. We now focus on detecting queries that have collected many spam results. Spam pages typically include many unrelated keywords or links, advertising and machine-generated content. These characteristics make spam pages likely to be selected as answers for very different queries, which may express unrelated or poorly related information needs.

We use the Click graph to identify candidate spam pages. Given that spam pages often cover a large number of unrelated topics, we follow the approach suggested in [5] to identify multitopical URLs. In the Click graph, low-weight edges indicate a poor-quality relation between the involved queries. In our experiment, we consider all the edges whose weight is < 0.005; in this way, we isolate the least significant 25% edges in the graph.

We count each edge as a spamicity vote for all the URLs in the intersection of the result sets associated with the two queries. We consider the top 200 URLs that obtain the highest number of votes. All the queries that have one of these URLs in their result set form a clique in the Click graph. We retain the groups of queries that are also mutually connected in the other graphs.

As a result, we obtain a list of 9 140 queries. We filter out the queries that have more than $1K$ occurrences to get rid of navigational queries. We then sort the queries by decreasing degree in the Click graph and we select the top 10% of the queries with highest degree: the intuition is that queries that form small dense subgraphs in the three graphs, while maintaining many connections to other nodes in the graph, are good spam candidates. In the end, we extract a set of 928 queries.

6.1 Experimental Evaluation

The above method led us to isolate 928 queries. We tested the effectiveness of our strategy by conducting an experimental evaluation aimed to assess the

Table 4. (a): Top categories obtained from Yahoo! Directory (b): Examples of spam queries extracted from the graphs

(a)

Level 1	Level 2
Regional	Shopping
Business and Economy	US States
Entertainment	Countries
Arts	Business
Society and Culture	Music
Recreation	Movies
Computers and Internet	TV Shows
News and Media	Humanities
Government	Actors

(b)

Query	Spam	1st Category	Cat.
filma shqiptar	5	Video	1
pro wresting	1	Entertainment	11
shapely figures	5	Entertaniment	9
video eyewear	3	Business	11
ana visi	2	Entertainment	1
make money	6	Entertainment	10
spanked cutie	5	Recreation	10
suger babes	4	Adult	1
pin up girls	4	Art/Shopping	11

spamicity of the selected queries. For each query, we manually evaluated its ability of attracting spam results by resubmitting it to the Yahoo! search engine and counting the number of spam pages obtained within the top ten results. We marked as spam every query containing at least one spam result in the top answers.

Given that no agreement on a univocal definition of *spam* page has been reached so far, we enforced the strength of our judgements following the labeling guidelines used in the construction of the WEBSPAM-UK2007[4] collection, which is a Web spam dataset built through the collaborative effort of a team of volunteers to advance research on Web spam detection.

We matched the top results obtained for each query selected by our algorithm against the Yahoo! Directory. Table 4(a) shows the most frequent categories associated with the results returned for the queries. Only the top levels are considered: Level 1 is the top category, Level 2 the second highest category. The main categories obtained include Shopping, Business, Movies, Sex and Adult galleries.

The results obtained in the assessment phase are really encouraging: two thirds of the evaluated queries collected at least one spam result, and thus were marked as spam queries. A few examples are shown in Table 4(b), which reports, for each query, the number of spam results, the main category assigned by Yahoo! Directory and the number of different categories associated.

The inspection of the categories obtained for our sample set makes us notice that most of the queries analyzed, even those that are not classified as spam, are associated with (top) results related to adult, celebrities, video or shopping. We believe this is a confirmation of the effectiveness of our strategy for selecting queries that are characterized by a significant degree of spamicity, or, if not spam, by the covering a broad set of topics.

The peculiar nature of the queries discovered by our heuristic suggests an immediate application of the tool to developing parental filters.

[4] `http://www.yr-bcn.es/webspam/datasets/uk2007/guidelines/`

6.2 Conclusions

In this work we have introduced the idea of combining different graph representations of query logs coming from different relevance signals, to be able to tackle different problems related to queries in a completely unsupervised manner. This is very important in cases where we do not have enough labeled data such as in non-popular languages.

The combination of the graph is done with very simple operations over graphs. In spite of the simplicity of the approach we show that we can obtain very good results for two different applications as detecting the type of a query transition or query spam. The former can be used to adapt the ranking of new queries, whereas the results obtained in the latter case could be used directly in the design of parental filters.

Further work includes scalability issues of the methods proposed, testing our implementation in an iterative map/reduce scenario as well as apply these ideas to other applications. In fact, the same ideas could be applied to the identification of polysemic queries, the recognition of logical sessions (missions) or for query recommendations. We believe that the approach introduced in this paper will be a valid support for the analysis of the huge amount of data stored in the logs of search engines. Our methodology can provide efficient methods for combining different aspects of the same multi-faceted scenario and acquiring novel knowledge that might not be obtained through other methods, in particular if we have the restriction of having to use unsupervised methods.

References

1. Architecture and Performance of Runtime Environments for Data Intensive Scalable Computing, Portland, OR, 11/09 (2009)
2. Baeza-Yates, R.: Graphs from search engine queries. In: van Leeuwen, J., Italiano, G.F., van der Hoek, W., Meinel, C., Sack, H., Plášil, F. (eds.) SOFSEM 2007. LNCS, vol. 4362, pp. 1–8. Springer, Heidelberg (2007)
3. Baeza-Yates, R., Hurtado, C., Mendoza, M.: Query Clustering for Boosting Web Page Ranking. In: Favela, J., Menasalvas, E., Chávez, E. (eds.) AWIC 2004. LNCS (LNAI), vol. 3034, pp. 164–175. Springer, Heidelberg (2004)
4. Baeza-Yates, R., Hurtado, C., Mendoza, M.: Query Recommendation Using Query Logs in Search Engines. In: Lindner, W., Mesiti, M., Türker, C., Tzitzikas, Y., Vakali, A.I. (eds.) EDBT 2004. LNCS, vol. 3268, pp. 588–596. Springer, Heidelberg (2004)
5. Baeza-Yates, R., Tiberi, A.: Extracting semantic relations from query logs. In: KDD 2007, pp. 76–85. ACM, New York (2007)
6. Beeferman, D., Berger, A.: Agglomerative clustering of a search engine query log. In: KDD 2000, pp. 407–416. ACM Press, New York (2000)
7. Boldi, P., Bonchi, F., Castillo, C., Donato, D., Gionis, A., Vigna, S.: The query-flow graph: model and applications. In: CIKM 2008, pp. 10+ (October 2008)
8. Boldi, P., Bonchi, F., Castillo, C., Vigna, S.: From 'dango' to 'japanese cakes': Query reformulation models and patterns. In: WI 2009, IEEE, Los Alamitos (September 2009)

9. Boldi, P., Vigna, S.: The webgraph framework: Compression techniques. In: WWW 2004. ACM Press, New York (2004)
10. Castillo, C., Corsi, C., Donato, D., Ferragina, P., Gionis, A.: Query-log mining for detecting spam. In: AIRWeb 2008 (2008)
11. Cohen, J.: Graph twiddling in a mapreduce world. Computing in Science and Engg. 11(4), 29–41 (2009)
12. Craswell, N., Zoeter, O., Taylor, M., Ramsey, B.: An experimental comparison of click position-bias models. In: WSDM 2008, pp. 87–94. ACM Press, New York (2008)
13. Dean, J., Ghemawat, S.: Mapreduce: simplified data processing on large clusters. In: OSDI 2004: Proceedings of the 6th conference on Symposium on Opearting Systems Design & Implementation, p. 10. USENIX Association, Berkeley (2004)
14. Diemert, E., Vandelle, G.: Unsupervised query categorization using automatically-built concept graphs. In: WWW 2009, pp. 461–470. ACM, New York (2009)
15. Fonseca, B.M., Golgher, P.B., de Moura, E.S., Ziviani, N.: Using association rules to discover search engines related queries. In: LA-WEB 2003. IEEE Computer Society, Washington (2003)
16. He, H., Singh, A.K.: Graphs-at-a-time: query language and access methods for graph databases. In: SIGMOD 2008, pp. 405–418. ACM, New York (2008)
17. Jones, R., Klinkner, K.L.: Beyond the session timeout: automatic hierarchical segmentation of search topics in query logs. In: CIKM 2008 (2008)
18. Nardini, F.M., Perego, R., Silvestri, F., Castillo, C., Donato, D., Baraglia, R.: Aging effects on query flow graph for query suggestion. In: CIKM 2009 (2009)
19. Qiu, G., Liu, K., Bu, J., Chen, C., Kang, Z.: Quantify query ambiguity using odp metadata. In: SIGIR 2007, pp. 697–698. ACM, New York (2007)
20. Radlinski, F., Joachims, T.: Query chains: learning to rank from implicit feedback. In: KDD (2005)
21. Raghavan, V.V., Sever, H.: On the reuse of past optimal queries. In: SIGIR 2008, pp. 344–350. ACM Press, New York (1995)
22. Rieh, S.Y., Xie, H.: Analysis of multiple query reformulations on the web: the interactive information retrieval context. Inf. Process. Manage. 42(3), 751–768 (2006)
23. Song, R., Luo, Z., Wen, J.-R., Yu, Y., Hon, H.-W.: Identifying ambiguous queries in web search. In: WWW 2007, pp. 1169–1170. ACM Press, New York (2007)
24. Wen, J.-R., Nie, J.-Y., Zhang, H.-J.: Clustering user queries of a search engine. In: WWW 2001, pp. 162–168. ACM, New York (2001)

Process Mining Meets Abstract Interpretation

J. Carmona and J. Cortadella

Universitat Politècnica de Catalunya, Spain

Abstract. The discovery of process models out of system traces is a problem that has received significant attention in the last years. In this work, a theory for the derivation of a Petri net from a set of traces is presented. The method is based on the theory of abstract interpretation, which has been applied successfully in other areas. The principal application of this theory is Process Mining, an area that tries to incorporate the use of formal models both in the design and use of information systems.

1 Introduction

Traces are everywhere: from information systems that store their continuous executions, to any type of health care applications that record each patient's history. The transformation of a set of traces into a mathematical model that can be used for a formal reasoning is therefore of great value.

This paper proposes methods to build a process model representing the causal relations between the events in the trace, i.e., whether the event a occurs before b and after c or d. The goal is to construct a graph modeling all these orderings in a concise form. Among many of the graph formalisms that exist nowadays, we have selected Petri nets (PN) [14] for representing a set of traces. The reasons for this selection are: sound mathematical model, clear semantics, succinctness, ability of representing concurrent and conflict behavior among others.

The problem of deriving a PN out of a set of traces (called *log*) is one of the main areas of Process Mining [19]. More concretely, the goal is to obtain a PN whose behavior contains all the traces in the log, but maybe more. Within this area, several algorithms have been proposed to accomplish this task [4,20,5], most of them based on the theory of regions [10]. Informally, the theory of regions tries to map structures in the state-based or language-based representation of a system into *places* of the derived PN. However, given the well-known *state explosion problem*, algorithms that are defined at the level of the states will suffer when dealing with large systems exhibiting a high degree of concurrency.

Abstract interpretation [8] is a generic approach for the static analysis of complex systems. The underlying notion in abstract interpretation is that of *upper approximation*: to provide an abstraction of a complex behavior with less details. A property about a system such as an invariant is in some way an abstraction: it represents all the states of the system that satisfy the property.

Intuitively, abstract interpretation defines a procedure to compute an upper approximation for a given behavior of a system. This definition guarantees (a) the termination of the procedure and (b) that the result is conservative. An

J.L. Balcázar et al. (Eds.): ECML PKDD 2010, Part I, LNAI 6321, pp. 184–199, 2010.

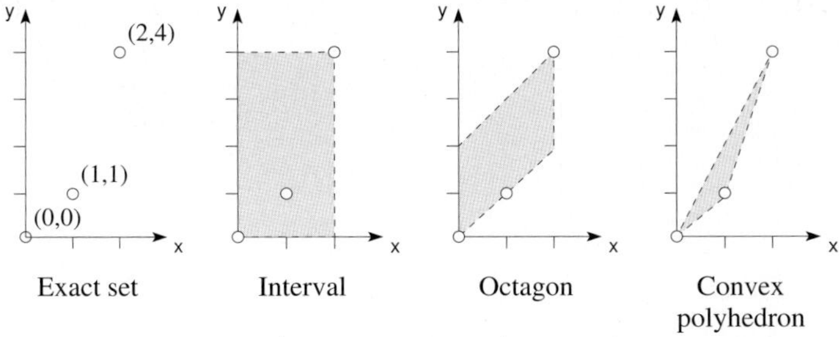

Interval	**Octagon**	**Convex polyhedron**
$0 \leq x \leq 2$	$0 \leq x \leq 2$	$y - x \geq 0$
$0 \leq y \leq 4$	$0 \leq y - x \leq 2$	$x - 2y \geq 0$
		$3x - y \leq 2$

Fig. 1. Approximating a set of values (left) with several abstract domains

important decision is the choice of the kind of upper approximation to be used, which is called the *abstract domain*. For a given problem, there are typically several abstract domains available. Each abstract domain provides a different trade-off between precision (proximity to the exact result) and efficiency.

There are many problems where abstract interpretation can be applied, several of them oriented towards the compile-time detection of run-time errors in software. For example, some analysis based on abstract interpretation can discover numeric invariants among the variables of a program. Also, it has been applied to extract invariants from a PN [6]. Several abstract domains can be used to describe the invariants: intervals [7], octagons [13], convex polyhedra [9], among others. These abstract domains provide different ways to approximate sets of values of numeric variables. For example, Figure 1 shows how these abstract domains can represent the set of values of a pair of variables x and y.

In this work we present an approach for deriving a PN from a log, based on the theory of abstract interpretation. The contributions are: 1) a theory for deriving PNs out of a set of traces, 2) a technique to allow for the partitioning of the set of events into groups. The relations inside the groups and between groups can be detected and the corresponding causalities computed, 3) a sampling strategy that can be applied to detect the relations on a small set of instances instead of the whole set, and 4) a prototype tool implementing all the theory of the paper.

1.1 An Introductory Example

Let us provide a simple example to illustrate the theory of this paper. The example is taken from [17] and considers the process of handling customer orders. The starting point in Process mining is a set of traces representing the log of a system. In our example, the log contains seven traces with the following activities: *r=register*, *s=ship*, *sb=send_bill*, *p=payment*, *ac=accounting*, *ap=approved*,

1	r,s,sb,p,ac,ap,c
2	r,sb,em,p,ac,ap,c
3	r,sb,p,em,ac,rj,rs,c
4	r,em,sb,p,ac,ap,c
5	r,sb,s,p,ac,rj,rs,c
6	r,sb,p,s,ac,ap,c
7	r,sb,p,em,ac,ap,c

(a)

$p \geq ac$

$r \geq em + s$

$ap + rs \geq c$

$\vdots$

(b)

$r \geq em + s$

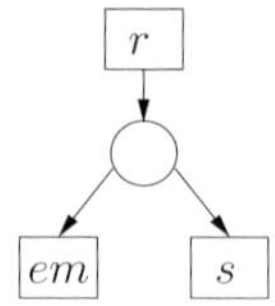

(c)

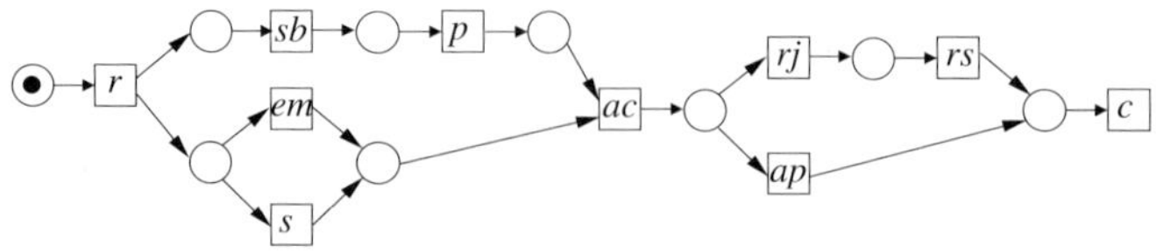

(d)

Fig. 2. Derivation of PNs using abstract interpretation: (a) log, (b) some invariants obtained, (c) from invariants to PN arcs, (d) mined Petri net

c=close, *em=express_mail*, *rj=rejected*, and *rs=resolve*. Part of these traces is shown in Figure 2(a), whilst Figure 2(b) shows some invariants that have been extracted from these traces using the theory of abstract interpretation. These inequalities can be obtained under the domain of convex polyhedra (see Figure 1), and relate the number of occurrences between events, e.g., $r \geq em + s$ indicates that the number of occurrences of r is always greater or equal than the sum of occurrences of em and s. Each invariant can be converted into a set of arcs in a PN, as it is shown in Figure 2(c). The final PN that covers all the traces in the log is presented in Figure 2(d) (see Section 2.1 for the formal semantics of a PN). It accepts the language defined by the expression[1]: $r; (sb; p)||(em|s); ac; (rj; rs)|ap; c$, where $||$, $|$ and ; denote interleaving, union and concatenation operators.

1.2 Related Work

Besides the work related to the theory of regions cited above [4, 20, 5], there are other approaches for process mining. In [19], an algorithm (called α-algorithm) to derive a restricted class of Petri nets was presented. The α-algorithm has been extended in [22] to enable a wider class of nets. Other techniques like [21] derive models that are easily transformed to a Petri net.

2 Preliminaries

Some mathematical notation is provided for the understanding of the paper. Given a set T, we denote $\mathcal{P}(T)$ as the powerset over T, i.e. the set of possible

[1] For the reader not familiar with Petri nets: a transition (box) in a PN is enabled if every input place (circle) holds a token (black dot). If enabled, the transition can fire, removing tokens from its input places and adding tokens to its output places.

subsets of elements of T. A sequence $\sigma \in T^*$ is a called *trace*. Given a trace $\sigma = t_1, t_2, \ldots, t_n$, and a natural number $0 \leq k \leq n$, the trace $t_1, t_2, \ldots, t_k$ is called the *prefix* of length k in σ. Given a set of traces L, we denote $Pref(L)$ the set of all prefixes for traces in L. Finally, given a trace σ, $\#(\sigma, e)$ computes the number of times that event e occurs in σ.

2.1 Logs and Petri Nets

Definition 1 (Log). *A log over a set of activities T is a set $L \in \mathcal{P}(T^*)$.*

Definition 2 (Petri net [14]). *A Petri net is a tuple (P, T, F, M_0) where P and T represent finite sets of places and transitions, respectively, and $F : (P \times T) \cup (T \times P) \rightarrow \mathbb{N}$ is the weighted flow relation. The initial marking $M_0 \in \mathbb{N}^{|P|}$ defines the initial state of the system.*

The sets of input and output transitions of place p in PN N are denoted by $^\bullet p$ and $p^\bullet$, respectively. A transition $t \in T$ is *enabled* in a marking M if $\forall p \in P : M[p] \geq F(p, t)$. Firing an enabled transition t in a marking M leads to the marking M' defined by $M'[p] = M[p] - F(p, t) + F(t, p)$, for $p \in P$, and is denoted by $M \xrightarrow{t} M'$. The set of all markings reachable from the initial marking m_0 is called its *Reachability Set*. The *Reachability Graph* of PN ($\mathsf{RG}(\mathsf{PN})$) is an automaton in which the set of states is the Reachability Set, the arcs are labeled with the transitions of the net and an arc (m_1, t, m_2) exists if and only if $m_1 \xrightarrow{t} m_2$. We use $L(\mathsf{PN})$ as a shortcut for $L(\mathsf{RG}(\mathsf{PN}))$, i.e. the language of the reachability graph of the net. Finally, a place p in a PN is *redundant* if its removal does not changes $L(\mathsf{PN})$. Figure 2(d) contains an example of a PN such that $\sigma = r, s, sb, p, ap, c \in L(\mathsf{PN})$.

2.2 Convex Polyhedra

As suggested in Section 1.1, the convex polyhedra domain provides the necessary inequalities for the purposes of this paper. It can be described as the set of solutions of a set of *linear inequality constraints* with rational ($\mathbb{Q}$) coefficients. Let P be a polyhedron over $\mathbb{Q}^n$, then it can be represented as the solution to the system of m inequalities $P = \{X | AX \leq B\}$ where $A \in \mathbb{Q}^{m \times n}$ and $B \in \mathbb{Q}^m$.

The domain of convex polyhedra provides the operations required in abstract interpretation. In this paper, we will mainly use the following two operations:

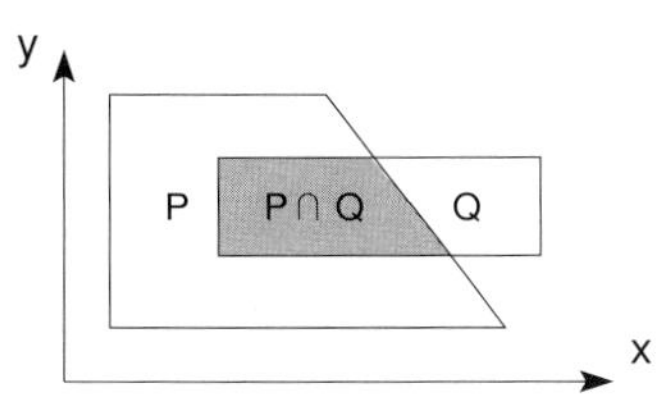

Meet ($\cap$)**:** Given convex polyhedra P and Q, computes $R = P \cap Q$. Notice that this operation is exact, e.g., the intersection of two convex polyhedra is always a convex polyhedra, implying that R does not contain any point outside $P \cap Q$.

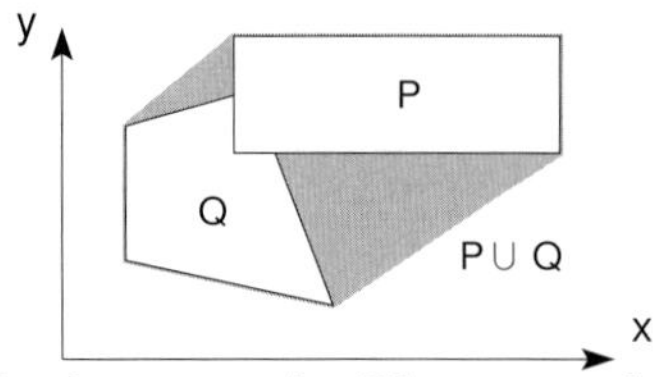

Join ($\cup$)**:** Given convex polyhedra P and Q, computes $R = P \cup Q$. Unfortunately the union of convex polyhedra is not necessarily a convex polyhedron. Therefore, the union of two convex polyhedra is approximated by the *convex hull*, the smallest convex polyhedron that includes both operands. The example on the left shows in gray the zone added by computing the convex hull of P and Q.

3 From Logs to Petri Nets via Extraction of Invariants

This section will set the basis for the approach presented in this paper. The underlying idea can be stated informally: for each trace of the log and each prefix of the trace, a vector describing the number of firings of each event for the prefix is computed[2]. All these vectors are then inserted as n-dimensional points in the theory of convex polyhedra, where n is the number of events considered. Finally, a polyhedron is computed such that contains all these points, and its set of constraints represents invariants for the system.

3.1 Derivation of Invariants from Logs

We introduce the main element to link traces from a log L and convex polyhedra:

Definition 3 (Parikh vector). *Given a trace* $\sigma \in \{t_1, t_2, \ldots, t_n\}^*$*, the* Parikh vector *of* σ *is defined as* $\widehat{\sigma} = (\#(\sigma, t_1), \#(\sigma, t_2), \ldots, \#(\sigma, t_n))$.

Any component of a Parikh vector can be seen as a constraint for the n-dimensional point that it defines. Hence, a Parikh vector $\widehat{\sigma} = (\#(\sigma, t_1), \#(\sigma, t_2), \ldots, \#(\sigma, t_n))$ can be seen as the following polyhedron:

$$P_{\widehat{\sigma}} = (x_1 = \#(\sigma, t_1)) \ \cap \ (x_2 = \#(\sigma, t_2)) \ \cap \ \ldots \ \cap \ (x_n = \#(\sigma, t_n))$$

where each variable x_i denotes the number of occurrences of t_i in σ, i.e., $x_i = \#(\sigma, t_i)$[3]. For each prefix σ of a trace in L, a polyhedron $P_{\widehat{\sigma}}$ can be obtained. Given all possible prefixes $\sigma_1, \sigma_2, \ldots, \sigma_m$ of traces in L, the polyhedra $P_{\widehat{\sigma_1}}, P_{\widehat{\sigma_2}}, \ldots, P_{\widehat{\sigma_k}}$ can be found[4]. Finally, the polyhedron

$$P = \bigcup_{i \in \{1 \ldots m\}} P_{\widehat{\sigma_i}}$$

can be seen as the *convex-hull* of the points represented by the polyhedra $P_{\widehat{\sigma_1}}, P_{\widehat{\sigma_2}}, \ldots, P_{\widehat{\sigma_m}}$, thus representing completely the behavior of the log. As Section 2.2 explains, a polyhedron can be described as the set of solutions of a

---

[2] We use the terms *event* and *transition* as synonyms in this paper.

[3] Hence a point $\widehat{\sigma}$ is represented as the polyhedron $P_{\widehat{\sigma}}$ that defines it.

[4] Here k is in practice significantly smaller than $\sum_{\sigma \in L} |\sigma|$ since many prefixes of different traces in L share the same Parikh vector.

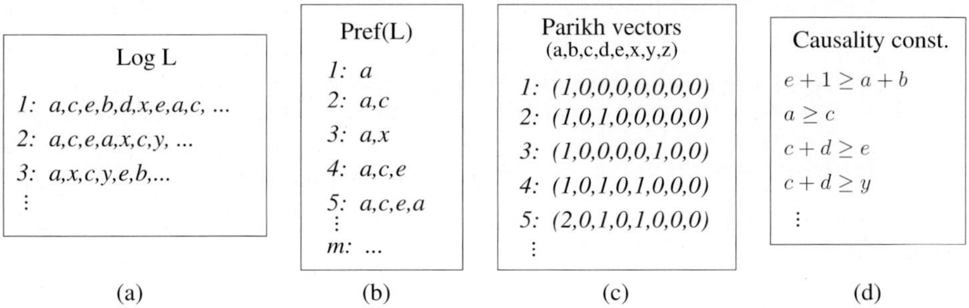

Fig. 3. From traces to invariants: (a) Initial log, (b) corresponding m prefixes of the log, (c) Parikh vectors associated to the prefixes, and (d) derived causality constraints

conjunction of linear inequality constraints. These constraints can be obtained from P in state-of-the-art libraries for convex polyhedra [11]. Hence from P one can obtain the set of m constraints representing it:

$$\begin{aligned} a_{11} \cdot x_1 + a_{12} \cdot x_2 + \ldots + a_{1n} \cdot x_n &\leq b_1 \\ a_{21} \cdot x_1 + a_{22} \cdot x_2 + \ldots + a_{2n} \cdot x_n &\leq b_2 \\ \vdots \quad &\leq \vdots \\ a_{m1} \cdot x_1 + a_{m2} \cdot x_2 + \ldots + a_{mn} \cdot x_n &\leq b_m \end{aligned}$$

each one of these constraints models invariants that the system (i.e., the log) satisfy.

Example 1. Figure 3(a) shows part of a log containing several traces on the events a, b, c, d, e, x, y and z[5]. Once the prefixes of the traces are found (Figure 3(b)), corresponding Parikh vectors are converted into polyhedra. A unique polyhedron is derived by performing a join operation on the polyhedra, and the related invariants are extracted, some of them shown in Figure 3(d).

As it was said in Section 2.2, the join operator to obtain the convex-hull may introduce extra points not belonging to any prefix in L, which may in turn invalidate some of the invariants that hold only for the points in L. How severe is this limitation and how can be alleviated is a topic for future investigation. In practice, however, the effects of these spurious points where not observed in the experiments shown in Section 6.

3.2 From Invariants to Petri Nets

If we split the coefficients into positive and negative coefficients, constraint i can be represented in the following way:

$$\sum_{a_{ij}>0} a_{ij} \cdot x_j + \sum_{a_{ij}<0} a_{ij} \cdot x_j \leq b_i$$

[5] This log contains 100 traces of length 50 each. The reader can inspect the log by following the reference provided in [2].

that can be transformed into:

$$\sum_{a_{ij}>0} a_{ij} \cdot x_j - b_i \leq \sum_{a_{ij}<0} -a_{ij} \cdot x_j$$

A constraint i is a *causality constraint* if the following conditions hold:

- There is at least one positive coefficient, and
- $b_i \leq 0$

Hence causality constraints can be described as:

$$\sum_{a_{ij}>0} a_{ij} \cdot x_j + c_i \leq \sum_{a_{ij}<0} -a_{ij} \cdot x_j \tag{1}$$

with $c_i = -b_i \geq 0$. The intuition behind causality constraints is that they represent real causalities observed in the log which can be explicit in the derived PN. Hence if we assume indices $n_1, \ldots n_k$ range over the indices of variables with negative coefficients and $p_1, \ldots p_l$ range over the variables with positive coefficients, (1) can be modeled in a PN as:

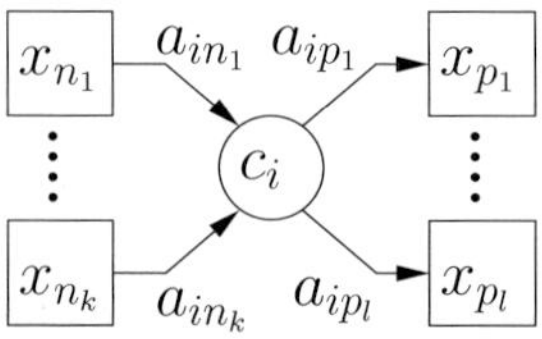

where c_i inside the place denotes c_i tokens, and a_{ij} in an arc represents the weighted flow relation F for the arc (see Def. 2).

Example 2. Following the example in the previous section (shown in Figure 3), causality constraints can be selected and the corresponding places and arcs introduced, deriving the Petri net shown in Figure 4. For instance the place labeled p is obtained from the constraint $c + d \geq y$.

Finally, a necessary property in the area of Process Mining that relates the set of traces possible in the PN and the ones in the log can be established:

Theorem 1. *Let* PN $= (P, T, F, M_0)$ *and* L *be a Petri net and a log, respectively, such that* $L(\mathsf{PN}) \supseteq L$*, and the i-th causal constraint from* L *as described in (1). Then the* PN$'$ $= (P', T, F', M_0')$ *defined as*

$$P' = P \cup \{p\}$$
$$F' = F \cup \{t_j \xrightarrow{a_{ij}} p \mid a_{ij} < 0\} \cup \{p \xrightarrow{a_{ij}} t_j \mid a_{ij} > 0\}$$
$$M_0'[q] = \begin{cases} M_0[q] \text{ if } q \neq p \\ c_i \quad \text{otherwise} \end{cases}$$

where $p \notin P$*, satisfies* $L(\mathsf{PN}) \supseteq L(\mathsf{PN}') \supseteq L$.

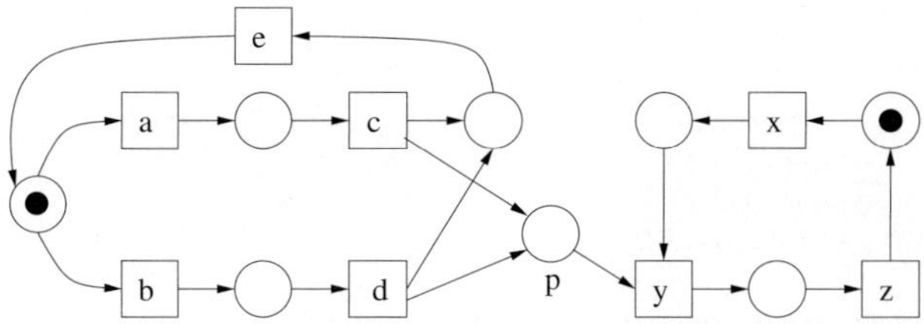

Fig. 4. Petri net derived from the causality constraints shown in Figure 3(d)

Proof. The inclusion $L(\mathsf{PN}) \supseteq L(\mathsf{PN}')$ is well-known in Petri net theory from the fact that $P \subset P'$, $F \subset F'$ and $M_0 \leq M_0'$. The inclusion $L(\mathsf{PN}') \supseteq L$ can be shown by induction on the length of traces in L, and we sketch here the proof. First, if a trace $\sigma = \sigma' t \in L$ satisfies $\sigma' \in L(\mathsf{PN}')$ but $\sigma \notin L(\mathsf{PN}')$, then $t \in p^\bullet$ because transitions not in the postset of the new place inserted p will also be enabled by firing σ' in PN'. Second, the induction can now be used to prove that p will have enough tokens to also enable t, hence contradicting the hypothesis $\sigma \notin L(\mathsf{PN}')$. For $|\sigma| = 1$ it trivially holds. Assume it is true for $|\sigma| \leq n-1$, let us consider $|\sigma| = n$, with $\sigma = \sigma' x t$. If $x \notin {}^\bullet p$ or $t \notin p^\bullet$, applying the induction hypothesis on σ' the statement on p holds. If $x \in {}^\bullet p$ and $t \in p^\bullet$, the induction hypothesis guarantees that after σ', either some other place $q \neq p$ is disabling t or t is enabled. Hence, by firing x the enabling state of t cannot change, contradicting the disabling of t after σ' in PN'. □

The addition of places and arcs corresponding to causality constraints is applied starting from the net $\mathsf{PN}_{init} \stackrel{def}{=} (\emptyset, T, \emptyset, \emptyset)$, which accepts the language T^*. In summary, the flow for Process Mining will follow the steps Log $\stackrel{abstract\ interpretation}{\longrightarrow}$ Convex Polyhedra $\stackrel{causality\ constraints}{\longrightarrow}$ PN. The next corollary follows from Theorem 1 and $L \subseteq L(\mathsf{PN}_{init})$:

Corollary 1. *Let* PN *be the net obtained after adding to* PN_{init} *all the places and arcs corresponding to causality constraints in the polyhedron* P *derived from* L*. Then* $L(\mathsf{PN}) \supseteq L$*.*

3.3 Derivation of Unbounded Places

Perhaps one of the main theoretical results of this work has been already presented in the example of the previous section. Informally, the derivation of places and arcs from causality constraints may produce *unbounded* places in the Petri net, i.e. places where no bound is possible on their number of tokens. For instance, the place p in Figure 4 may have k tokens when k firings of the sequence ac occur and no firing of y occurs, for any natural number k.

4 Process Mining of Large Logs

The approach presented in the previous section cannot be applied for logs extracted from industrial/real-life applications, where either the number of events

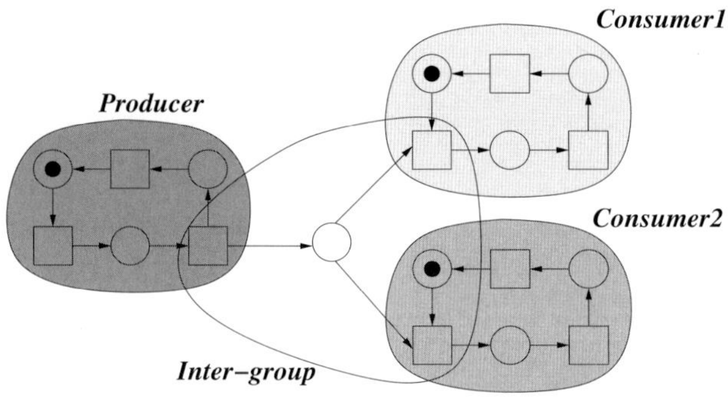

Fig. 5. Detection of groups of related events: *Producer*, *Consumer1* and *Consumer2* are tightly related, whereas the transitions within the *Inter-group* area are loosely related

or the number of Parikh vectors in the traces or both might be too large for growing polyhedra straightaway. For these situations, a divide-and-conquer strategy is required. A possible strategy is presented in this section: instead of a blind search for causality constraints on the whole set of events T, groups of events that are tightly related are identified, and causality constraints are divided into *intra-group* and *inter-group*. For instance, on a log representing a producer and a pair of consumers, intra-group relations might provide the causalities within the three gray zones depicted in Figure 5, whereas inter-group relations might derive the causalities within the corresponding area shown in the figure.

4.1 Identification of Groups of Tightly Coupled Events

For determining the partition of T into groups, several techniques can be applied. In this paper, two different techniques are used:

- *Principal Component Analysis (PCA) [12]* is an exploratory data analysis technique that, given a data set of possibly correlated variables, tries to select a subset of variables that is uncorrelated (called principal components) and which accounts for as much of the variability in the data as possible.
- *Firing causalities* is an ad hoc technique to extract causalities between two events from the Parikh vectors considered in the previous section.

In the remainder of this section we explain them in detail:

Principal Component Analysis can be applied to select the partition on $T = \{t_1, \ldots, t_n\}$. The steps are i) the set of Parikh vectors $\widehat{\sigma_1}, \ldots, \widehat{\sigma_m}$ is transformed to the set $\widehat{\sigma'_1}, \ldots, \widehat{\sigma'_m}$ so that $\widehat{\sigma'_i} = (\#(\sigma_i, t_1)/\overline{t_1}, \ldots, \#(\sigma_i, t_n)/\overline{t_n})$, where $\overline{t_i}$ is the mean for number of occurrences of t_i in the set of Parikh vectors of L, ii) compute the *correlation matrix* $A \in [-1 \ldots + 1]^{n \times n}$ using the data set found at i) [12]. This matrix measures the amount of correlation between variables t_i and t_j: when $|A(i,j)| \simeq 1$ then both variables are highly correlated. Finally, iii) the number of groups is decided by finding the *eigenvalues* and *eigenvectors* of A: the eigenvalues are sorted according to their value (the highest eigenvalue explains the highest correlation and so on), and only the most important (those that explain the important amount of correlation) are taken[6]. For each selected eigenvalue λ_i, we can select the *leader* of the group for λ_i by looking at the corresponding eigenvector $\alpha_1 \cdot x_1 + \ldots + \alpha_n \cdot x_n$: the leader will be the transition t_i for which absolute value of the coefficient α_i is maximal [12]. A transition t_j such that $|A(i,j)| \backsimeq 1$ will be incorporated to the group led by t_i. Transitions not assigned to any group can be considered as independent events that may be left out of the analysis. This way, a natural noise filtering is accomplished by using this method.

Firing causalities between two events t_i and t_j can be extracted by considering the maximal distance (in number of firings) between both events in any possible Parikh vector. Formally, we build the matrix $M \in \mathbb{Z}^{n \times n}$ such that $M(i,j) = \max\{\#(\sigma_k, t_i) - \#(\sigma_k, t_j) \mid 1 \leq k \leq m\}$. There is a causality between t_i and t_j if $M(i,j) > 0$ and $M(j,i) \leq 0$.

4.2 Intra-group Causality Constraints

The information obtained from the two previous techniques can be combined to form the groups. Intuitively, events t_i and t_j will belong to the same group if

- t_i leads a group and has a high correlation with t_j or vice versa, or
- there is a firing causality relating t_i and t_j

Once a group is identified, the Parikh vectors can be projected into the events of the group and the technique presented in Section 3 can be applied for the projected Parikh vectors.

Example 3. Following with the running example used in the previous section (see the resulting PN in Figure 4), we will show how the same Petri net can be obtained by the hierarchical approach presented in this section. Using the firing causalities, we will find the pairwise causalities $a \rightarrow c$, $b \rightarrow d$, $x \rightarrow y$ and $y \rightarrow z$. With PCA, more complex relations will be detected: e related with a and b, and also e related with c and d. Hence, two groups are selected: $g_1 = \{a, b, c, d, e\}$ and $g_2 = \{x, y, z\}$. Projecting the Parikh vectors into each group of events will give the causality constraints only relating the events in the group, e.g., for group g_1 the constraints $a \leq c$, $b \leq d$, $e + 1 \leq a + b$ and $c + d \leq e$ will be obtained. These constraints correspond to the subnet to the left of place p in Figure 4. The right subnet corresponds to group g_2.

4.3 Inter-group Causality Constraints

The causalities between different groups might be detected by applying a hierarchical approach: for each group $g_i = \{t_1^i, \ldots, t_{|g_i|}^i\}$, a new variable h_i is created

[6] Threshold values are used both for deciding whether $|A(i,j)| \backsimeq 1$ and for selecting the correlation ratio explained by the selected eigenvalues.

such that it represents the sum of firings of the transitions in the group for each Parikh vector. By using the sum of the firings, relations between group's firings might be revealed. Afterwards, the same strategy of Section 3 can be applied to detect causalities between these new variables introduced.

Formally, given the groups $g_1, \ldots, g_k$ and a set of Parikh vectors $\widehat{\sigma_1}, \ldots, \widehat{\sigma_m}$, a new set of hierarchical Parikh vectors $\widehat{\sigma_1^h}, \ldots, \widehat{\sigma_m^h}$ is created such that

$$\widehat{\sigma_i^h} = (\sum_{t \in g_1} \#(\sigma_i, t), \ldots, \sum_{t \in g_k} \#(\sigma_i, t))$$

and now convex polyhedra can be created representing the hierarchical Parikh vectors:

$$P_{\widehat{\sigma_i^h}} = (h_1 = \sum_{t \in g_1} \#(\sigma_i, t)) \cap \ldots \cap (h_k = \sum_{t \in g_k} \#(\sigma_i, t))$$

And in the same way as Section 3.1, a set of invariants can be extracted from the union of the m polyhedra build as explained above.

$$\begin{aligned} a_{11} \cdot h_1 + a_{12} \cdot h_2 + \ldots + a_{1k} \cdot h_k &\leq b_1 \\ a_{21} \cdot h_1 + a_{22} \cdot h_2 + \ldots + a_{2k} \cdot h_k &\leq b_2 \\ \vdots \quad &\leq \vdots \\ a_{m1} \cdot h_1 + a_{m2} \cdot h_2 + \ldots + a_{mk} \cdot h_k &\leq b_m \end{aligned}$$

These invariants provide relations between groups of variables. Intuitively, invariants where the constant b_i is small denote relevant causalities between groups, whilst invariants with a large constant represent loose causalities possibly originated by the length of the traces in the log. Hence, only these invariants with small constant are used[7].

When a set of groups are identified to be related, the same technique of Section 4.2 applied for a group can be now applied for the set of groups: the Parikh vectors are projected into the variables that belong to any of the groups related, and causality constraints that relate these variables can be extracted.

The general algorithm is presented as Algorithm 1. The functions used in the algorithm are next defined:

- InvariantMining is the invariant derivation technique from Section 3.1.
- ComputeGroups is the group derivation technique explained in Section 4.1.
- SelectLowConstant is a function that given a set of invariants, chooses those ones having a small constant.
- NonZeroCoefs is a function that given an invariant, return these variables that have non-zero coefficients, i.e., the variables that define the invariant.

[7] Several threshold criteria can be applied to limit the number of invariants to consider. For instance, one can greedily take invariants as far as the constant lies within the order of the previous one.

Algorithm 1. GroupMining

Input: Parikh vectors $\widehat{\sigma_1}, \dots, \widehat{\sigma_m}$,
Output: Invariant set I containing inter and intra-group causality constraints

begin
  $I = \emptyset$
  $\{g_1, \dots, g_k\} = \text{ComputeGroups}(\widehat{\sigma_1}, \dots, \widehat{\sigma_m})$
  foreach *group* g_i **do**
    $I = I \cup \text{InvariantMining}(\widehat{\sigma_1}|_{g_i}, \dots, \widehat{\sigma_m}|_{g_i})$
  end
  $H = \text{SelectLowConstant}(\text{InvariantMining}(\widehat{\sigma_1^h}, \dots, \widehat{\sigma_m^h}))$
  foreach *invariant* $i \in H$ **do**
    $\{g_1, \dots, g_l\} = \text{NonZeroCoefs}(i)$
    $I = I \cup \text{InvariantMining}(\widehat{\sigma_1}|_{g_1,\dots,g_l}, \dots, \widehat{\sigma_m}|_{g_1,\dots,g_l})$
  end
end

Example 4. Let us show the relation between the two only groups g_1 and g_2 found in Example 3. By creating two sum variables h_1 and h_2 as explained in Section 4.3 and building the polyhedron that corresponds to the union of the polyhedra representing the projection of the Parikh vectors into these variables, the constraint $h_2 \leq h_1$ is detected, meaning that the number of firings in the group g_2 is always less or equal than the number of firings of group g_1. By projecting now the Parikh vectors into these groups and extracting the causality constraints that relate both groups of variables, the constraint $y \leq c + d$ will be extracted, which corresponds to the place p shown in Figure 4. Notice that although for this toy example we ended up by building polyhedra for the whole set of events, in general this will not be the case for real systems. For instance, we experimented with several systems like the one used in our running example, working in parallel. The approach presented in this paper was able to find the intra and inter-group relations for each individual system, thus avoiding to project into the whole set of events. In section 6 we provide such experiments.

5 Sampling

Orthogonal to the approach presented in the previous section, this section introduces a technique to avoid dealing with a large number of polyhedra and use instead a limited amount that might be enough for extracting the important relations between the events. For instance, if the log contains ten thousand traces of length a hundred, then in the worst case the techniques presented in the previous sections will be dealing with a million of polyhedra that must be joined, a scenario that often can not be completed successfully with existing libraries for abstract interpretation.

The general algorithm for applying sampling is shown as Algorithm 2. In order to avoid operations with a large number of polyhedra, one can randomly select

Algorithm 2. Sampling

Input: Parikh vectors $\widehat{\sigma_1}, \ldots, \widehat{\sigma_m}$, number of samplings p, sampling size s
Output: Invariant set I

1 **begin**
2     $I = \emptyset$
3   **for** $i \leftarrow 1$ **to** p **do**
4         $P =$ empty domain
5   **for** $j \leftarrow 1$ **to** s **do**
6             $r =$Random$(1 \ldots m)$
7   compute $P_{\widehat{\sigma_r}}$
8             $P = P \cup P_{\widehat{\sigma_r}}$
9   **end**
10         $I_1 =$ Invariants(P)
11   **foreach** *invariant* $i \in I_1$ **do**
12   **if** i *satisfies* $\widehat{\sigma_1}, \ldots, \widehat{\sigma_m}$ **then** $I = I \cup \{i\}$
13   **end**
14   **end**
15 **end**

with uniform probability a small set (s) of Parikh vectors that will be converted to polyhedra and joined (lines 5-9). Once the join operation for the s vectors has been done, the set of invariants *that denote properties for the Parikh vectors considered* must be verified on each one of the Parikh vectors not considered in the join, and only those invariants that are true for all the Parikh vectors will be accepted (lines 10-13). This sampling technique can be applied more than once, i.e., one can apply p samplings in order to find the relations on a set of events (external loop starting at line 3).

Sampling and the strategy presented in the previous section can be applied jointly. This will be accomplished by simply substituting the calls to Invariant-Mining in Algorithm 1 by calls to the function Sampling with a user-defined sampling size and number of samplings. In the experiments, this joint use of these strategies has enabled dealing with large specifications.

6 Experiments

The theory has been implemented in the prototype tool `aim`, which is written in `C/C++` and uses the `Apron` library for Convex Polyhedra manipulation [11]. For the PCA method which requires computation of eigenvalues and eigenvectors, the `ALGLIB` library [1] was used. Some conclusions can be drawn from applying the tool on some well-known benchmarks within the Process Mining domain.

The benchmarks applied are synthetic logs publicly available within the website [3]. These logs have been used by other algorithms and therefore will be considered in this paper to perform a comparison with two other tools for the same purpose. The tools are: `genet`, which implements algorithms based on the theory of regions and supports the mining of k-bounded PNs [5], and the

Table 1. PN derivation from logs

Log	Log Information			genet		ILPMiner		aim	
	$\|T\|$	#traces	#Parikh	P/F	Time	P/F	Time	P/F	Time
a12f0n00_1	12	200	17	11/25	0.1	11/25	1	11/27	0
a12f0n00_5	12	1800	17	11/25	0.1	11/25	0.7	12/30	0
a22f0n00_1	22	100	750	19/49	0.3	19/49	3	19/48	20
a22f0n00_5	22	900	3290	19/49	0.3	19/49	23	16/38	2
a32f0n00_1	32	100	1377	32/75	718	31/73	25	34/84	33
a32f0n00_5	32	900	5543	31/73	1	31/73	112	31/68	6
a42f0n00_1	42	100	1211	memout		44/109	154	41/88	16
a42f0n00_5	42	900	4326	timeout		44/101	1557	49/118	77

`ILPMiner` [20] (within ProM), that uses the language version of the theory of regions for the same purpose. For using `genet`, an automaton representing all the traces is the input of the tool. Several algorithms exists to transform the log into an automaton [18]. For both tools we used the default parameters.

The comparison is shown in Table 1. For each log, we report the number of events ($|T|$), the number of traces and the number of Parikh vectors obtained after removing repetitions. The number of places discovered (P) and the number of arcs (F) is then provided, together with the CPU time (measured in a desktop computer) in seconds. For testing each tool, we limited the amount of memory and time that could be used to 1Gb and 10000 seconds respectively.

For the experiments, we run the tool applying 5 samplings with sampling size a number between 50 and 100, depending on the log. This light sampling application allowed to derive PNs sometimes within two orders of magnitude less CPU time than other methods. Notice that `genet` has both memory (`memout`) and time (`timeout`) problems with the last two logs. On the other hand, `aim` invests considerably more time in deriving a PN for a22f0n00_1, which may be due to the particular structure of the polyhedra built on that log.

A second point to consider is the quality of the information obtained. The PNs derived with `aim` most of the time have the same arcs and places of the other tools. Sometimes extra causalities might be obtained like in a12f0n00_1 or a42f0n00_5. These denote redundant causalities (unnecessary places in the model) that can be removed by a final application of well-known PN methods for redundant places removal [16]. More elaborated quality measures, like the one presented in [15], are restricted to a particular class of Petri nets and hence cannot be used in our general setting.

Table 2 reports experiments with two logs that represent the activity of a system of producers and consumers where components are synchronized through unbounded places (see Figure 5). For ProdCons_1, the PN derived by `aim` is the one shown in Figure 4. The traces for ProdCons_3 contain the interleaving of three independent instances of PNs like the one in Figure 4. Both `genet` and the `Parikh Miner` have problems in dealing with these logs: `genet` cannot derive the unbounded place in ProdCons_1 and received a `timeout` for ProdCons_3,

Table 2. PN derivation from two logs obtained from a Producers/Consumers system

	Log Information			genet		ILPMiner		aim	
Log	$\|T\|$	#traces	#Parikh	P/F	Time	P/F	Time	P/F	Time
ProdCons_1	8	50	3756	7/16	14	0/0	5	8/19	7
ProdCons_3	24	50	4910	timeout		0/0	182	24/57	36

whereas the ILPMiner did not obtain any relation between the activities of the log[8]. In contrast, aim was able to discover the exact PN in both logs.

7 Conclusions and Future Work

A novel theory for deriving a PN from a set of traces has been presented. The results obtained are promising when compared with some of the approaches in the literature for the same task. The current work is mainly focused in obtaining a mature implementation of the first prototype. Also, other strategies to complement the ones described in this paper will be investigated. Finally, the derivation of other graph formalisms will be explored.

Acknowledgements

This work has been supported by the project FORMALISM (TIN2007-66523), and a grant by Intel Corporation.

References

1. ALGLIB library, http://www.alglib.net
2. Example log, http://www.lsi.upc.edu/jcarmona/prodcons1000.tr
3. Process mining, http://www.processmining.org
4. Bergenthum, R., Desel, J., Lorenz, R., Mauser, S.: Process mining based on regions of languages. In: Alonso, G., Dadam, P., Rosemann, M. (eds.) BPM 2007. LNCS, vol. 4714, pp. 375–383. Springer, Heidelberg (2007)
5. Carmona, J., Cortadella, J., Kishinevsky, M.: New region-based algorithms for deriving bounded Petri nets. IEEE Transactions on Computers 59(3), 371–384 (2009)
6. Clarisó, R., Rodríguez-Carbonell, E., Cortadella, J.: Derivation of non-structural invariants of petri nets using abstract interpretation. In: Ciardo, G., Darondeau, P. (eds.) ICATPN 2005. LNCS, vol. 3536, pp. 188–207. Springer, Heidelberg (2005)
7. Cousot, P., Cousot, R.: Static determination of dynamic properties of programs. In: Proc. of the 2nd Int. Symposium on Programming, Dunod, Paris, France, pp. 106–130 (1976)

[8] By changing the default parameters of the ILPMiner, 5 places and 11 arcs are derived for ProdCons_1, but for ProdCons_3 the net is degraded (49 places, 239 arcs).

8. Cousot, P., Cousot, R.: Abstract interpretation: a unified lattice model for static analysis of programs by construction or approximation of fixpoints. In: Proc. ACM SIGPLAN-SIGACT Symp. on Principles of Programming Languages, pp. 238–252. ACM Press, New York (1977)
9. Cousot, P., Halbwachs, N.: Automatic discovery of linear restraints among variables of a program. In: Proc. ACM SIGPLAN-SIGACT Symp. on Principles of Programming Languages, pp. 84–97. ACM Press, New York (1978)
10. Ehrenfeucht, A., Rozenberg, G.: Partial (Set) 2-Structures. Part I, II. Acta Informatica 27, 315–368 (1990)
11. Jeannet, B., Miné, A.: Apron: A library of numerical abstract domains for static analysis. In: Bouajjani, A., Maler, O. (eds.) CAV 2009. LNCS, vol. 5643, pp. 661–667. Springer, Heidelberg (2009)
12. Jolliffe, I.T.: Principal Component Analysis. Springer, Heidelberg (2002)
13. Miné, A.: The octagon abstract domain. In: Analysis, Slicing and Tranformation (in Working Conference on Reverse Engineering), pp. 310–319. IEEE, IEEE CS Press (October 2001)
14. Murata, T.: Petri Nets: Properties, analysis and applications. Proceedings of the IEEE, 541–580 (April 1989)
15. Rozinat, A., van der Aalst, W.M.P.: Conformance checking of processes based on monitoring real behavior. Inf. Syst. 33(1), 64–95 (2008)
16. Silva, M., Teruel, E., Colom, J.M.: Linear algebraic and linear programming techniques for the analysis of place or transition net systems. In: Reisig, W., Rozenberg, G. (eds.) APN 1998. LNCS, vol. 1491, pp. 309–373. Springer, Heidelberg (1998)
17. van der Aalst, W.M.P., Günther, C.W.: Finding structure in unstructured processes: The case for process mining. In: Basten, T., Juhás, G., Shukla, S.K. (eds.) ACSD, pp. 3–12. IEEE Computer Society, Los Alamitos (2007)
18. van der Aalst, W.M.P., Rubin, V., Verbeek, H.M.W.E., van Dongen, B.F., Kindler, E., Günther, C.W.: Process mining: a two-step approach to balance between underfitting and overfitting. In: Software and Systems Modeling (2009)
19. van der Aalst, W.M.P., Weijters, T., Maruster, L.: Workflow mining: Discovering process models from event logs. IEEE Trans. Knowl. Data Eng. 16(9), 1128–1142 (2004)
20. van der Werf, J.M.E.M., van Dongen, B.F., Hurkens, C.A.J., Serebrenik, A.: Process discovery using integer linear programming. In: van Hee, K.M., Valk, R. (eds.) PETRI NETS 2008. LNCS, vol. 5062, pp. 368–387. Springer, Heidelberg (2008)
21. Weijters, A., van der Aalst, W., de Medeiros, A.A.: Process mining with the heuristics miner-algorithm. Technical Report WP 166, BETA Working Paper Series, Eindhoven University of Technology (2006)
22. Wen, L., van der Aalst, W.M.P., Wang, J., Sun, J.: Mining process models with non-free-choice constructs. Data Min. Knowl. Discov. 15(2), 145–180 (2007)

Smarter Sampling in Model-Based Bayesian Reinforcement Learning

Pablo Samuel Castro and Doina Precup

School of Computer Science
McGill University
{pcastr,dprecup}@cs.mcgill.ca

Abstract. Bayesian reinforcement learning (RL) is aimed at making more efficient use of data samples, but typically uses significantly more computation. For discrete Markov Decision Processes, a typical approach to Bayesian RL is to sample a set of models from an underlying distribution, and compute value functions for each, e.g. using dynamic programming. This makes the computation cost per sampled model very high. Furthermore, the number of model samples to take at each step has mainly been chosen in an ad-hoc fashion. We propose a principled method for determining the number of models to sample, based on the parameters of the posterior distribution over models. Our sampling method is local, in that we may choose a different number of samples for each state-action pair. We establish bounds on the error in the value function between a random model sample and the mean model from the posterior distribution. We compare our algorithm against state-of-the-art methods and demonstrate that our method provides a better trade-off between performance and running time.

1 Introduction

Reinforcement learning (RL) attempts to find an optimal way of behaving in an unknown environment, often assumed to be a Markov Decision Process (MDP). By interacting with the environment, an RL agent learns more about its dynamics and rewards; this information can be used to shape the agent's behavior or policy. Traditional RL methods are typically based on estimating the expected value of the agent's long-term return (also called a value function). However, in some applications one would also like to have an estimate of the agent's uncertainty, in addition to the expectation of the returns. Bayesian RL methods attempt to provide such estimates by maintaining either a distribution over models, e.g.(Asmuth et al., 2009; Dearden et al., 1999; Poupart et al., 2006; Price & Boutilier, 2003; Strens, 2000), or a distribution over value functions, e.g. (Engel et al., 2003). The first category of methods is often called model-based Bayesian RL and will be the focus of our paper. The Bayesian approach allows incorporating prior knowledge about the MDP (if available), in the form of a prior. Bayesian approaches have also been proposed as a method that can potentially use small amounts of data; hence, they could be useful in applications where the data is difficult to obtain. Several Bayesian RL method emphasize the use of the distribution as a tool for formulating good exploration policies, e.g. (Duff, 2003; Asmuth et al., 2009; Wang et al.,

J.L. Balcázar et al. (Eds.): ECML PKDD 2010, Part I, LNAI 6321, pp. 200–214, 2010.

2005). However, recent work (Kolter & Ng, 2009) has cast doubt on the effectiveness of Bayesian exploration. However, even if exploration is not the main emphasis, having uncertainty estimates in the performance of a policy is still quite useful.

Most model-based Bayesian RL methods work by maintaining a distribution over model parameters, which is initialized with a prior and updated based on data obtained from the real environment. The algorithms usually sample a batch of models from this distribution and the value function is computed for each of these sampled models. This can make the computation cost for each iteration very high, especially for large systems. The number of sampled models is often chosen in an ad-hoc fashion, based on computation time constraints. Models are re-sampled periodically, either at fixed intervals, or when their likelihood drops below a certain threshold. Recent work has attempted to make this choice in a more principled way. In (Asmuth et al., 2009), the authors propose an algorithm (BOSS) which aims to ensure enough exploration; they prove that if one samples $\Theta(\frac{1}{\delta}\ln\frac{1}{\delta})$ models, then with probability at least $1-\delta$ one of these models will yield a value function that is optimistic compared to the true underlying model. They also provide theoretically a value for the number of data samples that have to be gathered before new models should be sampled. However, the computation of both the number of models and the number of samples is difficult and for their empirical results, the authors resort to choosing these values manually.

In this paper, we propose a new method for determining the number of model samples to take, in the context of discrete MDPs, based on the parameters of the posterior distribution over models. Furthermore, we argue that one should sample locally: the number of transition models to sample for each state-action pair should depend on the statistical properties of the posterior distribution for that particular state-action pair, rather than being fixed over the entire MDP. We overcome the difficulty of combining these different numbers of transition samples by constructing a *merged* MDP as in (Asmuth et al., 2009). We use an optimistic approach to resolve the exploration-exploitation dilemma. We provide a method for dynamically determining when to re-sample and re-compute the solution to the merged MDP. This decision is based on the difference between the mean of the posterior last used for sampling and the mean of the current posterior distribution. Our method exploits the statistical properties of the posterior distribution and the proposed score can be computed very quickly.

We motivate our approach by first providing theoretical bounds for the difference in value function between a random model sampled from a Dirichlet distribution and the mean of the distribution. The tightness of the bounds depends directly on the variance of the posterior distribution; hence, the variance directly affects the decision of when and how much to sample in our proposed algorithms.

The paper is structured as follows. Sec. 2 presents necessary background and notation. In Sec. 3 we present bounds on the difference between the value function of the mean of a distribution over MDPs, compared to a randomly sampled model. Sec. 4 presents two algorithms for determining the number of models and the frequency of sampling them, based on these bounds. Sec. 5 contains an empirical comparison of the proposed algorithms against other Bayesian RL methods. Sec. 6 contains a discussion and avenues for future work.

2 Background

A finite Markov Decision Process (MDP) $M = \langle S, A, P, R \rangle$ consists of a finite set of states S; a finite set of actions A; a transition function $P : S \times A \rightarrow Dist(S)$, where $Dist(X)$ is the set of distributions over X; and a reward function $R : S \times A \times S \rightarrow \mathbb{R}$ (see (Puterman, 1994) for more details). A *policy* is a function $\pi : S \rightarrow Dist(A)$. The value of a state $s \in S$ given a policy π is defined as $V^{\pi}(s) = \mathbb{E}^{\pi}\left(\sum_{i=0}^{\infty} \gamma^i r_i\right)$, where $0 \leq \gamma < 1$ is a discount factor and r_i a random variable representing the reward received at time step i when starting from state s and choosing actions according to policy π. The values for all states can be computed by solving the following system of Bellman linear equations:

$$V^{\pi}(s) = \sum_{s' \in S} P(s, \pi(s))(s') \left[R(s, \pi(s), s') + \gamma V^{\pi}(s')\right]$$

The optimal value function is defined as: $V^*(s) = \max_{\pi} V^{\pi}(s), \forall s \in S$ and obeys the following system of non-linear equations:

$$V^*(s) = \max_{a \in A} \sum_{s' \in S} P(s, a)(s') \left[R(s, a, s') + \gamma V^*(s')\right]$$

If the model of the MDP (i.e. R and P) is known, several well-known methods can be used to solve this system, e.g. dynamic programming and linear programming.

In reinforcement learning (RL) (Sutton & Barto, 1998) the model of the MDP is usually assumed to be unknown. Traditional model-based RL methods rely on maintaining a model estimate which is updated based on the data obtained by the agent in its interaction with the environment. In Bayesian RL, instead of maintaining just one model, one maintains a distribution over models. In this paper we assume that the reward model is known; this is often the case in RL, as rewards are provided by the designer of the system to define the task at hand. The transition model P, on the other hand, is unknown.

For any $s, s' \in S$ and $a \in A$ let $\theta(s, a, s')$ be a random variable representing the probability of a particular model for the transition probability from s to s' under a. Given a prior distribution $Pr(\theta)$ (representing initial knowledge about θ) and an observed transition $s \rightarrow^a s'$, one can compute the posterior distribution $Pr(\theta | s, a, s')$ using Bayes' Theorem. Usually, the prior and posterior distributions are chosen to be in the same family of distributions. For discrete MDPs, the preferred choice is to use a Dirichlet distribution as a prior/posterior, which is a parameterized distribution conjugate to the multinomial distribution. This is a natural choice since the parameters of a Dirichlet distribution function as "counts" (i.e. the parameter for $\theta(s, a, s')$ can be interpreted as the number of times the transition $s \rightarrow^a s'$ has been observed). A typical approach is to sample at each step a number of models from the distribution over models, compute the optimal value function for each sample model, and then use these values to choose the next action.

The Dirichlet distribution is parameterized by a set of real numbers $\alpha = (\alpha_1, \cdots, \alpha_N)$, where N is the order of the Dirichlet distribution. Let $\alpha_0 = \sum_{i=1}^{N} \alpha_i$. The Dirichlet is the conjugate prior of the Multinomial distribution (a multivariate generalization of the binomial distribution) with parameters $p_1 = \alpha_1 / \alpha_0$, $\cdots$, $p_N = \alpha_N / \alpha_0$. In a finite MDP, a Dirichlet distribution $Dir(\alpha^{(s,a)})$ of order $N = |S|$ can be maintained for each state-action pair (s,a), where $\alpha_{s'}^{(s,a)}$ is the parameter for transition $s \rightarrow^{a} s'$. We refer to the *mean (or mean model)* of a Dirichlet distribution $Dir(\alpha^{(s,a)})$ as the multinomial distribution $\mu_{(s,a)} \in Dist(S)$ where $\mu_{(s,a)(s_i)} = \alpha_i^{(s,a)}/\alpha_0^{(s,a)}$. The marginals of this multinomial have variance given by $\sigma^2_{(s,a)(s_i)} = \frac{\alpha_{(s,a)(s_i)}(\alpha_0^{(s,a)} - \alpha_{(s,a)(s_i)})}{\left(\alpha_0^{(s,a)}\right)^2(\alpha_0^{(s,a)}+1)}$. Throughout the learning process we will maintain a set of Dirichlet distributions parameterized by state-action pairs: $\{Dir(\alpha_{(s,a)})\}_{s\in S, a\in A}$.

3 Bounding the Value Function

As the agent gathers experience by interacting with the environment, the expected distance between a random sample and the mean model should go down. As this expected distance goes down, the number of required samples should also go down. Furthermore, once enough experience has been obtained, the change in model estimates should change less and less, implying that the need to resample should decrease. These are the main ideas of the algorithms presented in the next section, but we begin by providing a probabilistic bound on the difference between a random sample and the mean model, formalizing the intuition just mentioned.

Let $\alpha^{(s,a)}$ be the parameters for $s \in S$ and $a \in A$ with $\alpha_0^{(s,a)} = \sum_{s'\in S}\alpha_{s'}^{(s,a)}$; $\mu_{(s,a)(s')}$ and $\sigma^2_{(s,a)(s')}$ are the expected value and variance, respectively, for any $s' \in S$.

A well-known way to measure the difference between two probability distributions over the same domain is the total variation distance. Given two state distributions $P, Q \in Dist(S)$, their total variation distance (TV) is defined as:

$$TV(P,Q) = \frac{1}{2}\sum_{s\in S}|P(s) - Q(s)|$$

Clearly $0 \leq TV(P,Q) \leq 1$ for any P, Q. We can now easily obtain the following two inequalities (stated without proof), where V_1 and V_2 are the value functions for two MDPs that differ only in their transition probabilities (P_1 and P_2, respectively):

Lemma 1. *For any policy* π,

$$|V_1^{\pi}(s) - V_2^{\pi}(s)| \leq \frac{2\gamma R_{max}}{1-\gamma}\max_{s\in S}\max_{a\in A} TV(P_1(s,a), P_2(s,a))$$

For the optimal value function:

$$|V_1^{*}(s) - V_2^{*}(s)| \leq \frac{2\gamma R_{max}}{1-\gamma}\max_{s\in S}\max_{a\in A}\max_{b\in A} TV(P_1(s,a), P_2(s,b))$$

Both these inequalities can be bounded by the trivial upper bound $\frac{2\gamma R_{max}}{1-\gamma}$, which is very loose. We would like to be able to tighten the bounds of Lemma 1 by using information from the Dirichlet distribution. In particular, given a sampled transition model $X_{(s,a)}$

from $Dir(\alpha^{(s,a)})$ and a real number $m > 1$, we would like to know with what probability $2TV(\mu_{(s,a)}, X_{(s,a)}) < 1/m$. One approach is to examine, for each $s' \in S$, with what probability $|X_{(s,a)(s')} - \mu_{(s,a)(s')}| < 1/(Nm)$, where $N = |S|$.

A simple application of Chebyshev's inequality yields the following result.

$$Pr\left(|X_{(s,a)(s')} - \mu_{(s,a)(s')}| \geq \frac{1}{mN}\right) \leq \sigma^2_{(s,a)(s')}(mN)^2$$

Now we can obtain a lower bound for the total variation being less than $1/2m$ as follows:

$$\begin{aligned} Pr\left(2TV(X_{(s,a)}, \mu_{(s,a)}) < \frac{1}{m}\right) &\geq \prod_{s' \in S} Pr\left(|X_{(s,a)(s')} - \mu_{(s,a)(s')}| < \frac{1}{mN}\right) \\ &\geq \prod_{s' \in S}\left(1 - \sigma^2_{(s,a)(s')}(mN)^2\right) \geq 1 - \sum_{s' \in S}\left(\sigma^2_{(s,a)(s')}(mN)^2\right) \\ &= 1 - (mN)^2 \sum_{s' \in S} \sigma^2_{(s,a)(s')} \end{aligned} \quad (1)$$

where the second to last line follows by Weierstrass' product inequality. Let us examine the sum of the variances:

$$\begin{aligned} \sum_{s' \in S} \sigma^2_{(s,a)(s')} &= \sum_{s' \in S} \frac{\alpha_{s'}^{(s,a)}\left(\alpha_0^{(s,a)} - \alpha_{s'}^{(s,a)}\right)}{\left(\alpha_0^{(s,a)}\right)^2\left(\alpha_0^{(s,a)} + 1\right)} = \frac{\sum_{s' \in S} \alpha_{s'}^{(s,a)}\left(\alpha_0^{(s,a)} - \alpha_{s'}^{(s,a)}\right)}{\left(\alpha_0^{(s,a)}\right)^2\left(\alpha_0^{(s,a)} + 1\right)} \\ &= \frac{\alpha_0^{(s,a)} \sum_{s' \in S} \alpha_{s'}^{(s,a)}}{\left(\alpha_0^{(s,a)}\right)^2\left(\alpha_0^{(s,a)} + 1\right)} - \frac{\sum_{s' \in S}\left(\frac{\alpha_{s'}^{(s,a)}}{\alpha_0^{(s,a)}}\right)^2}{\alpha_0^{(s,a)} + 1} = \frac{1}{\alpha_0^{(s,a)} + 1}\left(1 - \sum_{s' \in S} \mu^2_{(s,a)(s')}\right) \end{aligned}$$

Therefore we have:

$$Pr\left(2TV(X_{(s,a)}, \mu_{(s,a)}) < \frac{1}{m}\right) \geq 1 - \frac{(mN)^2}{\alpha_0^{(s,a)} + 1}\left(1 - \sum_{i=1}^{N} \mu^2_{(s,a)(s')}\right) \quad (2)$$

Note that if $\alpha_0^{(s,a)} + 1 \gg (mN)^2$ we get better bounds on the total variation distance. In other words, in order to have tighter bounds on the Total Variation distance between a sampled model and the mean model, we need to have chosen action a from state s enough times in the learning process.

4 Algorithms for Smart Sampling

In this section we use the result above to construct two algorithms that determine the number of sampled models, as well as to decide when to sample new models.

The previous section indicates that the number of sampled models should depend directly on the variance of the underlying model distribution. Given a multinomial sample $X = (X_1, \cdots, X_N) \sim Dir(\alpha^{(s,a)})$, the marginals $X_{s'}$ are Binomial distributions with expected value $\mu_{(s,a)(s')}$ and variance $\sigma^2_{(s,a)(s')}$. Given a set of K model samples $\{X^i\}_{i=1}^K$, for each of the marginals the mean model can be computed as follows:

$$\hat{\mu}^K_{(s,a)(s')} = \frac{1}{K}\sum_{i=1}^{K} X^i_{s'}$$

Additionally, for each of the marginals the variance can be computed as follows:

$$\begin{aligned}
\left(\hat{\sigma}^K_{(s,a)(s')}\right)^2 &= Var\left(\frac{1}{K}\sum_{i=1}^{K} X^i_{s'}\right) = \frac{1}{K^2} Var\left(\sum_{i=1}^{K} X^i_{s'}\right) \\
&= \frac{1}{K^2}\sum_{i=1}^{K}\left(\sigma^2_{(s,a)(s')}\right) \qquad \text{Since the } X^i\text{s are drawn independently} \\
&= \frac{\sigma^2_{(s,a)(s')}}{K} \qquad \text{Since the } X^i\text{s are identically distributed}
\end{aligned} \tag{3}$$

By the strong law of large numbers, $\lim_{K\to\infty} \hat{\mu}^K_{(s,a)(s')} = \mu_{(s,a)(s')}$, but we would like to determine how many model samples would be "good enough". To achieve this we try to bring the variance for each of the marginals below some constant ε: $\left(\hat{\sigma}^K_{(s,a)(s')}\right)^2 \leq \varepsilon$. To obtain this, for each state-action pair (s,a) we must then set the number of sampled models as follows:

$$K_{(s,a)} = \max_{s' \in S} \left\lceil \frac{\sigma^2_{(s,a)(s')}}{\varepsilon} \right\rceil$$

Although the mean model is readily obtainable from the known posterior parameters, a lower variance is a good indication that we have a good enough approximation of the posterior distribution.

Based on this bound, each state-action pair may have a different number of required samples. In previous methods, if K was the desired number of samples, then K model samples were taken from each state-action pair, and these were combined in the obvious way to form K sampled models. Our situation is slightly more complicated and we cannot form a set of distinct models. We follow the approach of (Asmuth et al., 2009) and construct a merged MDP $m^\#$. The state space in $m^\#$ is the same as the original MDP, but we expand the actions. For each state-action pair (s,a), let $\{T_{(s,a)}(i)\}_{1\leq i\leq K_{(s,a)}}$ be the set of sampled models from (s,a). Each state s in MDP $m^\#$ has $\sum_{a\in A} K_{(s,a)}$ actions. Action $a_{i,j}$, for $1 \leq i \leq K_{(s,j)}$ and $1 \leq j \leq A$, corresponds to the sampled transition $T_{(s,j)}(i)$. As in (Asmuth et al., 2009), we assume the rewards are known in advance for simplicity and for sake of comparison, but the uncertainty about them is encoded in the transitions.

In (Asmuth et al., 2009), if in $m^\#$ the optimal action choice from state s is $a_{i,j}$, then the agent chooses action a_j in the real MDP. This is an optimistic approach that is the basis of their exploration strategy. With this exploration policy the authors can determine a formal choice of K to obtain a near-optimal behavior with high probability. We follow this optimistic exploratory strategy in our approach.

As in most previous work, we do not sample models at every iteration. Instead, we fix a number B beforehand and resample every time a state-action pair has been tried B times during learning. In (Asmuth et al., 2009) the authors determine what the value of B needs to be in order for the model distribution to be "close" to the true model. Although proven theoretically, in practice they set this value manually. We follow this approach for our initial algorithm: SmartSampler.

Algorithm 1. SmartSampler(ε, B)

Choose a starting state s_1
For all $(s,a) \in S \times A$, $qCounts(s,a) \leftarrow 0$
$reSample \leftarrow TRUE$
for all timesteps $t = 1,2,3,\cdots$ **do**
    if $reSample$ **then**
        For all $(s,a) \in S \times A$, sample $K_{(s,a)}$ transitions from the posterior
        Combine all the samples into the merged MDP $m^{\#}$
        Solve $m^{\#}$ and extract $\pi^{\#}$
        $reSample \leftarrow FALSE$
    end if
    Extract a_t from $\pi^{\#}(s_t)$
    Perform action a_t and observe reward r_t and next state s_{t+1}
    $qCounts(s_t,a_t) \leftarrow qCounts(s_t,a_t)+1$
    Update posterior based on (s_t,a_t,r_t,s_{t+1})
    if $qCounts(s_t,a_t) = B$ **then**
        $reSample \leftarrow TRUE$
    end if
end for

The SmartSampler algorithm chooses how many model samples to take for each state-action pair, based on the parameters of the posterior distribution. Establishing *when* we should resample and re-compute a solution to the resulting MDP $m^{\#}$ is still unresolved. We propose determining when to sample by examining the change in the distribution of the mean models of the posteriors. We use the standard score as inspiration for determining this change. Specifically, we wish to determine how many standard deviations the mean model of the current posterior is from the mean model of the posterior used the last time model sampling was performed. For any $(s,a) \in S \times A$, let $P_t(s,a)$ be the mean transition distribution for (s,a) of the posterior at time t. Let timestep t be the last time we sampled a model from our posterior to obtain $m^{\#}$. Given a maximum threshold parameter δ, we will resample at time step t' only when

$$\sum_{s' \in S} \frac{|P_t(s,a)(s') - P_{t'}(s,a)(s')|}{\sigma_{(s,a)(s')}} > \delta$$

Note that the standard deviation $\sigma_{(s,a)(s')}$ used is the one computed from the posterior at timestep t (*i.e.* when we last sampled models). This measure of change takes into account both changes in the transition distribution as well as changes in the variance. In this way, if we start with a confident prior that is already quite close to the true

Algorithm 2. SmarterSampler(ε, δ)

Choose a starting state s_1
$reSample \leftarrow TRUE$
$lastSamp \leftarrow 1$
for all timesteps $t = 1, 2, 3, \cdots$ **do**
  if $reSample$ **then**
    For all $(s,a) \in S \times A$, sample $K_{(s,a)}$ transitions from $\alpha^{(s,a)}$
    Combine all the samples into the merged MDP $m^{\#}$
    Solve $m^{\#}$ and extract $\pi^{\#}$
    $lastSamp \leftarrow t$
  end if
  Extract a_t from $\pi^{\#}(s_t)$
  Perform action a_t and observe reward r_t and next state s_{t+1}
  Update posterior based on (s_t, a_t, r_t, s_{t+1})
  if $\sum_{s' \in S} \frac{|P_t(s_t,a_t)(s') - P_{lastSamp}(s_t,a_t)(s')|}{\sigma_{(s,a)(s')}} > \delta$ **then**
    $reSample \leftarrow TRUE$
  end if
end for

model, we will resample very infrequently. Furthermore, as the algorithm gathers more information, the number of times we need to resample decreases. Algorithm SmarterSampler summarizes this approach.

5 Experimental Results

We compare our algorithm only against the BOSS algorithm of (Asmuth et al., 2009), because it is similar in spirit and the efficacy of BOSS against other algorithms was already established in (Asmuth et al., 2009). Additionally, we compare against a Naive Bayesian algorithm that uses the same parameters as BOSS. For this approach, K indicates the number of global model samples to take (as in BOSS); however, the sampled models are not combined into a merged MDP $m^{\#}$; instead, each one is solved independently and the maximum value for each state-action pair is used. The B parameter indicates how many iterations should pass before resampling; thus, every B iterations the naive algorithm will sample K models. All algorithms were implemented in Matlab; for fairness and consistency, the algorithms use the same code where possible. For all domains, we started with a uniform prior; that is, for every pair of states $s, s' \in S$ and

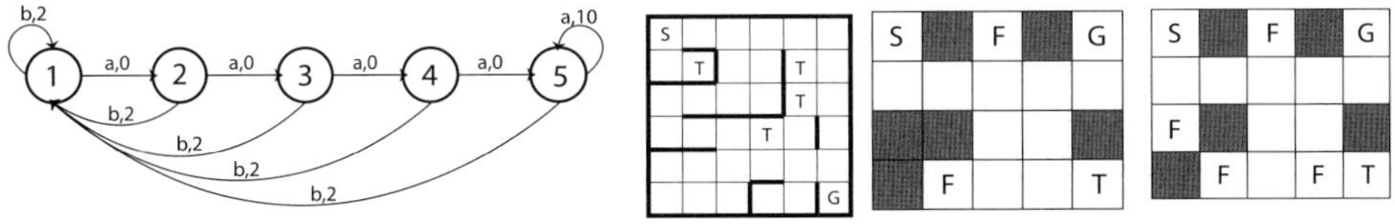

Fig. 1. Problem domains, from left to right: Chain, 6x6 maze, Dearden maze, larger maze

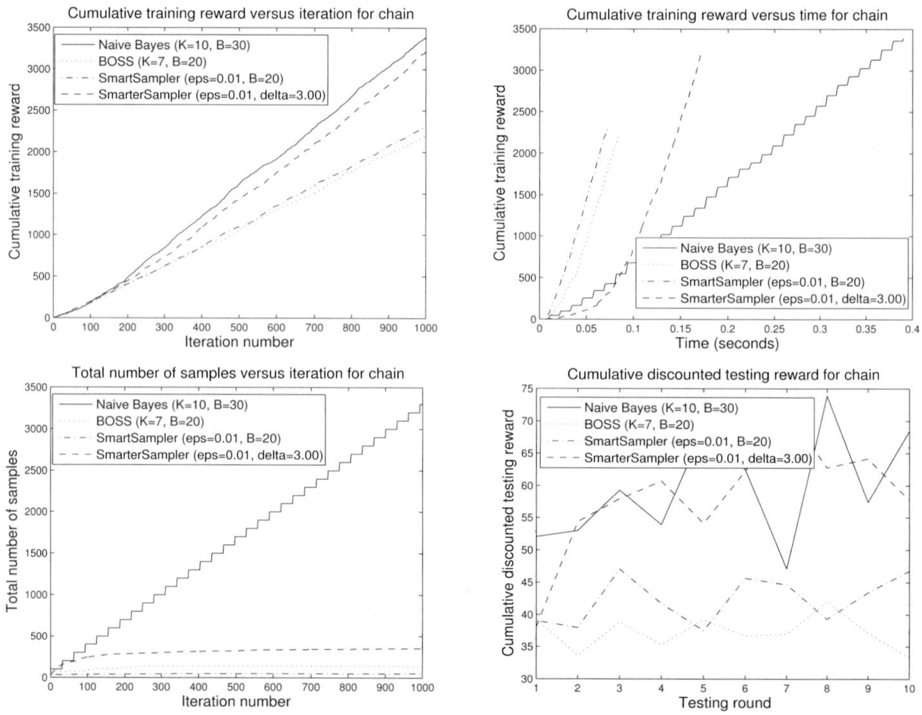

Fig. 2. Cumulative reward versus iteration (top left), cumulative reward versus time (top right), number of samples per iteration (bottom left), sum of discounted rewards during testing (bottom right). Chain problem.

action a, the Dirichlet parameter was set to $\alpha_{s'}^{(s,a)} = 1/|S|$. All results are averaged over 10 runs.

We ran experiments on a number of domains shown in Figure 1. The chain domain from (Strens, 2000) has 5 states and 2 actions; the action has the desired effect with 0.8 probability and has the opposite effect with 0.2 probability; in the figure the transitions are labeled with the action name and the reward. The 6x6 maze domain from (Russell & Norvig, 1994) has 36 states and 4 actions (up, down, left, right); the agent starts at the cell marked 'S' and aims to reach the goal 'G', trying to avoid the trap states 'T'; the action moves the agent in the desired direction with probability 0.8, and the rest of the time the agent moves in a direction perpendicular to the desired direction, where moving into a wall will maintain the agent in the same state; each move has a cost of 0.001, and the agent receives a terminal reward of -1 or $+1$ for entering a trap or the goal state, respectively, whereupon the agent is placed back in the start state. The first maze domain is from (Dearden et al., 1999) and has 56 states; the agent moves in the desired direction with probability 0.9, and the rest of the time the agent moves in a direction perpendicular to the desired direction, where moving into a wall will maintain the agent in the same state; the agent receives a penalty of -10 for entering a trap state 'T' and upon entering the goal state, receives a terminal reward of 1 for every flag 'F'

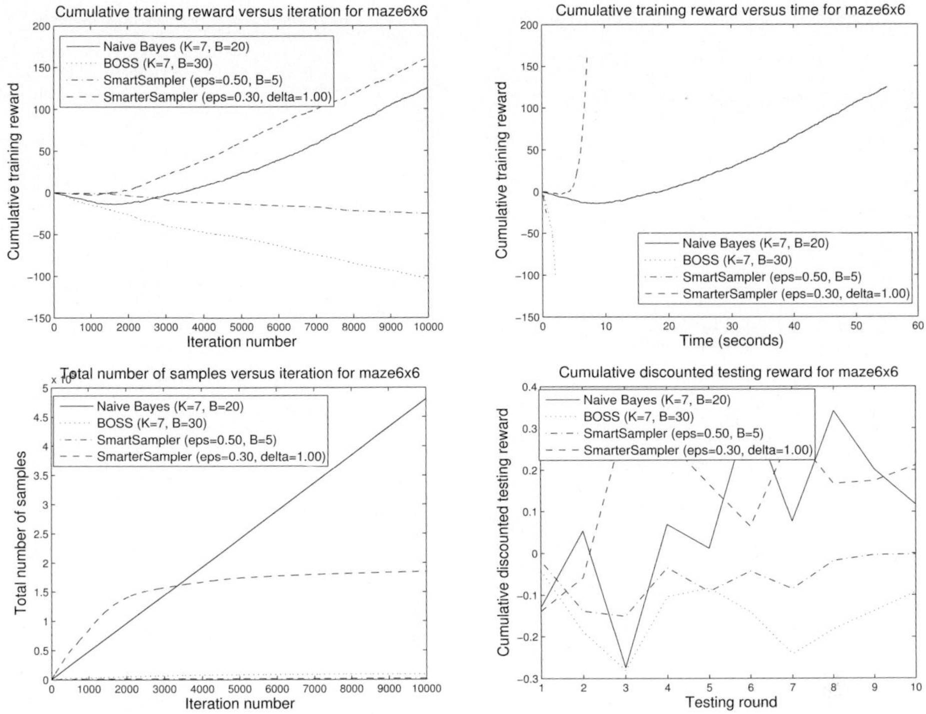

Fig. 3. Cumulative reward versus iteration (top left), cumulative reward versus time (top right), number of samples per iteration (bottom left), sum of discounted rewards during testing (bottom right). 6x6 maze problem.

the agent has caught. The larger maze domain is similar, but with extra flags, which increases the state size to 240. We plot the cumulative reward received during learning versus iteration and the cumulative reward received throughout learning versus time (top 2 graphs in each figure). In previous algorithms, the number of transition models sampled for each state-action pair was the same. In our methods they vary, so we also plot the total number of transition models sampled versus iteration. Finally, we tested the policy at 10 evenly spaced intervals for 1000 iterations throughout the learning process and plot the sum of discounted rewards for each test episode. The results for the chain problem are illustrated in Figure 2; for the 6x6 maze in Figure 3; for the maze problem in Figure 4; and for the larger maze in Figure5.

We ran each of the algorithms with varying parameters and picked the ones that gave the best balance between training reward accumulation, testing reward accumulation and running time. We can see that both the SmartSampler and SmarterSampler have a clear advantage over BOSS in terms of return and time; although the Naive Bayes algorithm seems to generate fairly good returns, it is extremely slow compared to the other algorithms. In the larger domains, the advantage of using SmarterSampler becomes more evident. In both the 6x6 maze and the maze problem of (Dearden et al., 1999) we

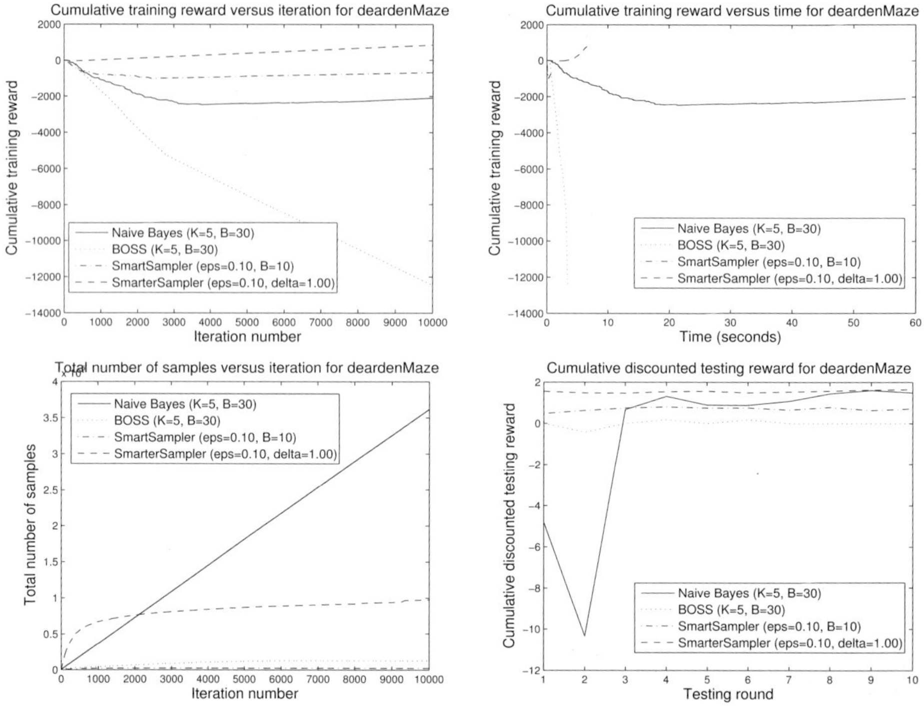

Fig. 4. Cumulative reward versus iteration (top left), cumulative reward versus time (top right), number of samples per iteration (bottom left), sum of discounted rewards during testing (bottom right). Maze from (Dearden et al., 1999).

can see that SmarterSampler is performing the best exploration in the least amount of time; both BOSS and SmartSampler appear to get stuck on a "safe" policy with little returns. SmarterSampler is once again able to obtain higher returns both during training and during testing, with a reasonable running time.

We also plot the effect of varying the parameters of the SmarterSampler algorithm in Figures 6 and 7. As expected, there is a tradeoff between rewards obtained (both during training and during testing) and the running time of the algorithm. We can also see that the algorithm is not too sensitive to small changes in the parameters, which makes it a robust choice for model-based Bayesian exploration.

6 Discussion and Future Work

We presented an approach to Bayesian RL based on a bound on the difference between the value function of a model sampled from a Dirichlet distribution and the mean of that distribution. There has already been work on bounding the difference in value functions between two systems with equal state spaces but different dynamics. In (Müller, 1997), a generalized theory for bounding the difference in value function (up to a finite horizon) is presented, using various probability metrics (such as the Kantorovich metric) along with structural properties of the value function. The result is theoretically pleasing, but it is somewhat involved and does not provide an algorithm for computing the bounds. The paper by (Ferns et al., 2004) introduces bisimulation metrics for MDPs. These metrics are based on the Kantorovich probability metric, and in fact, can be viewed as a slightly modified instance of the bounds of (Müller, 1997). Other related papers performed sensitivity analysis of the effect of varying dynamics on a stochastic system, e.g. (Hinderer, 2005) and (Smith & McCardle, 2002). We relied on total variation in this paper because it is easy to compute (a big asset in the context of Bayesian RL) and still gives powerful results in the particular case of the Dirichlet distribution.

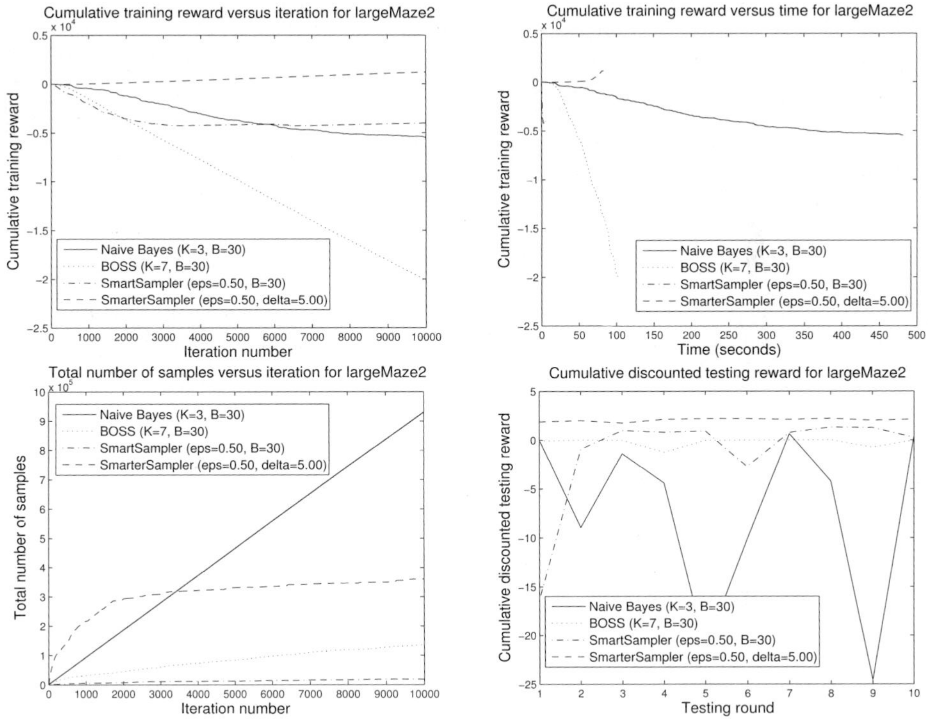

Fig. 5. Cumulative reward versus iteration (top left), cumulative reward versus time (top right), number of samples per iteration (bottom left), sum of discounted rewards during testing (bottom right). Larger maze.

Although in (Asmuth et al., 2009) the authors demonstrated how to choose these values in order to guarantee near-optimal exploration with high probability, the values are very difficult to compute, and the authors resorted to choosing them manually in practice. We presented a method for choosing these values dynamically by using statistics

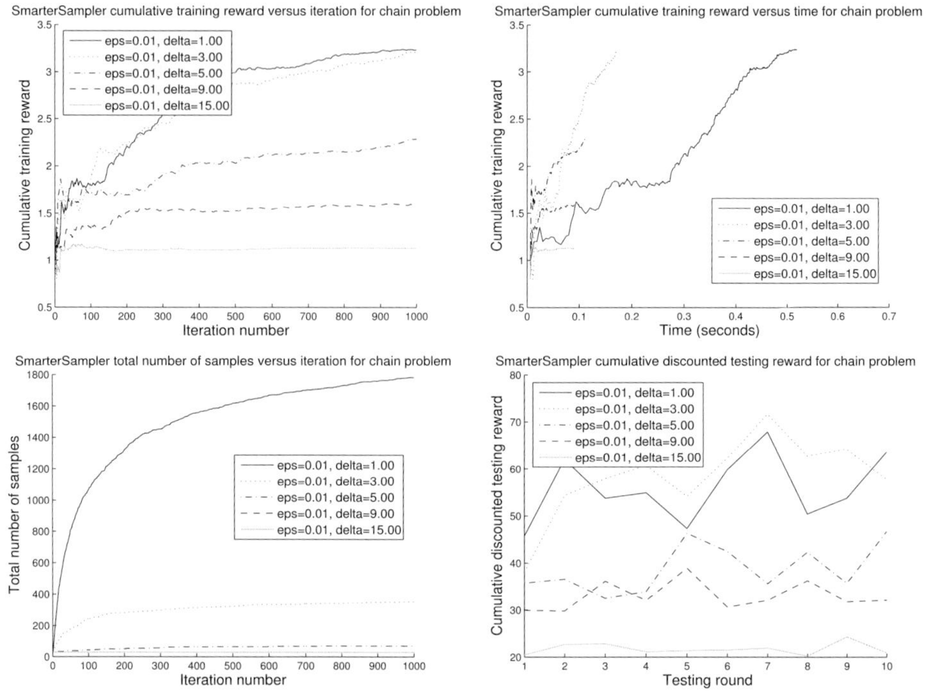

Fig. 6. SmarterSampler. Cumulative reward versus iteration (top left), cumulative reward versus time (top right), number of samples per iteration (bottom left), sum of discounted rewards during testing (bottom right). Chain problem.

of the posterior distribution. again, computing these values is cheap and our experimental results demonstrate that they produce superior performance, both in terms of returns and running time. Although the UCRL algorithm from (Auer, 2000) is intended for the undiscounted reward case, our algorithms also have strong connections to it and it merits further investigation and an empirical comparison.

We motivated our algorithms by first providing theoretical bounds on the difference in value function between the mean model and a random sample from the posterior distribution over models, when this posterior is a Dirichlet distribution. Although we used the Dirichlet for our experimental results, our algorithm can hold for any other type of distribution, as long as we can compute the means and variances of the marginals of the mean model.

For simplicity, we assumed that the reward function was known beforehand, but our method can be adapted easily to the case when the rewards are not known. If we assume there is a finite set of possible rewards, we can also maintain a distribution over reward models and use similar techniques for determining when and how much to sample. This situation would be of particular interest when the rewards are stochastic.

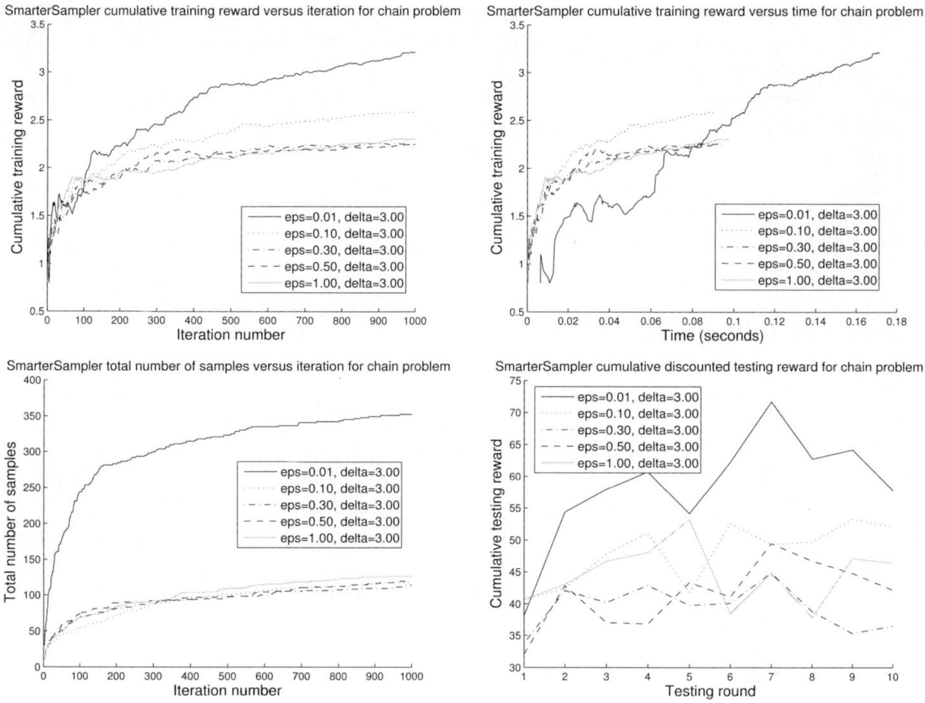

Fig. 7. SmarterSampler. Cumulative reward versus iteration (top left), cumulative reward versus time (top right), number of samples per iteration (bottom left), sum of discounted rewards during testing (bottom right). Chain problem.

Acknowledgements

This work was funded by NSERC and ONR. We would also like to thank the reviewers for their helpful comments.

References

Asmuth, J., Li, L., Littman, M.L., Nouri, A., Wingate, D.: A Bayesian Sampling Approach to Exploration in Reinforcement Learning. In: Proceedings of The 25th Conference on Uncertainty in Artificial Intelligence, UAI 2009 (2009)

Auer, P.: Using upper confidence bounds for online learning. In: Proceedings of the 41st Annual Symposium on Foundations of Computer Science (2000)

Dearden, R., Friedman, N., Andre, D.: Model based Bayesian exploration. In: Proceedings of the 15th Conference on Uncertainty in Artificial Intelligence, UAI 1999 (1999)

Duff, M.: Design for an optimal probe. In: Proceedings of the Twentieth International Conference on Machine Learning, pp. 131–138 (2003)

Engel, Y., Mannor, S., Meir, R.: Bayes Meets Bellman: The Gaussian Process Approach to Temporal Difference Learning. In: Proceedings of the Twentieth International Conference on Machine Learning, pp. 154–161 (2003)

Ferns, N., Panangaden, P., Precup, D.: Metrics for finite Markov decision processes. In: Proceedings of the 20th Annual Conference on Uncertainty in Artificial Intelligence, pp. 162–169 (2004)

Hinderer, K.: Lipschitz Continuity of Value Functions in Markovian Decision Processes. Mathematical Methods of Operations Research 62, 3–22 (2005)

Kolter, J.Z., Ng, A.Y.: Near-Bayesian Exploration in Polynomial Time. In: Proceedings of the Twenty-Sixth International Conference on Machine Learning (2009)

Müller, A.: How does the value function of a Markov Decision Process depend on the transition probabilities? Mathematics of Operations Research 22, 872–885 (1997)

Poupart, P., Vlassis, N., Hoey, J., Regan, K.: An analytic solution to discrete Bayesian reinforcement learning. In: Proceedings of the 23rd International Conference on Machine Learning, pp. 697–704 (2006)

Price, B., Boutilier, C.: A Bayesian approach to imitation in reinforcement learning. In: Proceedings of the 18th International Joint Conference on Artificial Intelligence, IJCAI 2003 (2003)

Puterman, M.L.: Markov Decision Processes. John Wiley & Sons, New York (1994)

Russell, S., Norvig, P.: Artificial Intelligence: A Modern Approach. Prentice-Hall, Englewood Cliffs (1994)

Smith, J.E., McCardle, K.F.: Structural Properties of Stochastic Dynamic Programs. Operations Research 50, 796–809 (2002)

Strens, M.: A Bayesian Framework for Reinforcement Learning. In: Proceedings of the 17th International Conference on Machine Learning, ICML 2000 (2000)

Sutton, R.S., Barto, A.G.: Reinforcement Learning: An Introduction. MIT Press, Cambridge (1998)

Wang, T., Lizotte, D., Bowling, M., Schuurmans, D.: Bayesian sparse sampling for on-line reward optimization. In: Procedings of the 22nd International Conference on Machine Learning, ICML 2005 (2005)

Predicting Partial Orders: Ranking with Abstention

Weiwei Cheng[1], Michaël Rademaker[2], Bernard De Baets[2], and Eyke Hüllermeier[1]

[1] Department of Mathematics and Computer Science
University of Marburg, Germany
{cheng,eyke}@mathematik.uni-marburg.de

[2] Department of Applied Mathematics, Biometrics and Process Control
Ghent University, Belgium
{michael.rademaker,bernard.debaets}@ugent.be

Abstract. The prediction of structured outputs in general and rankings in particular has attracted considerable attention in machine learning in recent years, and different types of ranking problems have already been studied. In this paper, we propose a generalization or, say, relaxation of the standard setting, allowing a model to make predictions in the form of partial instead of total orders. We interpret such kind of prediction as a ranking with partial abstention: If the model is not sufficiently certain regarding the relative order of two alternatives and, therefore, cannot reliably decide whether the former should precede the latter or the other way around, it may abstain from this decision and instead declare these alternatives as being incomparable. We propose a general approach to ranking with partial abstention as well as evaluation metrics for measuring the correctness and completeness of predictions. For two types of ranking problems, we show experimentally that this approach is able to achieve a reasonable trade-off between these two criteria.

1 Introduction

The problem of "learning to rank" has recently attracted considerable attention in machine learning, and different types of ranking problems have been studied, both theoretically and empirically. Roughly speaking, the goal of methods developed in this field is to learn a "ranker" that outputs predictions in the form of a ranking of a set of alternatives. Thus, learning to rank can be seen as a specific type of structured output prediction [1].

A ranking is commonly understood as a strict total order, i.e., an irreflexive, asymmetric, and transitive relation. In this paper, we propose a generalization of the standard setting, allowing a model to make predictions in the form of *partial* instead of *total* orders. We interpret such kind of prediction as a ranking with partial abstention: If the ranker is not sufficiently certain regarding the relative order of two alternatives and, therefore, cannot reliably decide whether the former should precede the latter or the other way around, it may abstain from this decision and instead declare these alternatives as being incomparable.

J.L. Balcázar et al. (Eds.): ECML PKDD 2010, Part I, LNAI 6321, pp. 215–230, 2010.

The notion of abstention is actually well-known for conventional classification, and the corresponding extension is usually referred to as *classification with a reject option* [2,3,4]: The classifier is allowed to abstain from a prediction for a query instance in case it is not sure enough. An abstention of this kind is an obvious means to avoid unreliable predictions. Needless to say, the same idea does also make sense in the context of ranking. In fact, one may even argue that a reject option becomes even more interesting here: While a conventional classifier has only two choices, namely to predict a class or to abstain, a ranker can abstain *to a certain degree*: The order relation predicted by the ranker can be more or less complete or, stated differently, more or less partial, ranging from a total order (conventional ranking) to the empty relation in which all alternatives are incomparable. Later on, we will express the degree of abstention of a ranker more precisely in terms of a degree of completeness of the partial order it predicts.

The main contribution of this paper is a general approach to ranking with partial abstention, which is applicable to different types of ranking problems. In a nutshell, our approach consists of two main steps. First, a preference relation is derived that specifies, for each pair of alternatives $\boldsymbol{a}$ and $\boldsymbol{b}$, a degree of preference for $\boldsymbol{a}$ over $\boldsymbol{b}$ and, vice versa, a degree of preference for $\boldsymbol{b}$ over $\boldsymbol{a}$. The idea is that, the more similar these two degrees are, the more uncertain the learner is. Then, in a second step, a partial order maximally compatible with this preference relation, in a sense to be specified later on, is derived as a prediction. In order to realize the first step, we make use of ensemble learning techniques, although other possibilities are conceivable.

The remainder of the paper is organized as follows. In the next section, we briefly review some important ranking problems. Our approach to ranking with partial abstention is then detailed in Section 3. In Section 4, we address the question of how to evaluate predictions in the form of partial orders and propose suitable performance metrics for measuring the correctness and completeness of such predictions. Section 5 is devoted to experimental studies. For two types of ranking problems, we show that our approach is indeed able to achieve a reasonable trade-off between these two criteria. The paper ends with a couple of concluding remarks in Section 6.

2 Ranking Problems

Following [5], we distinguish three types of ranking problems that have been studied extensively in the machine learning literature, namely label ranking [6,7,8], instance ranking [9], and object ranking [10], to be described in more detail in the following.

2.1 Label Ranking

Like in the conventional setting of supervised learning (classification), we assume to be given an instance space $\mathbf{X}$ and a finite set of labels $\mathbf{Y} = \{y_1, y_2, \ldots, y_k\}$.

In label ranking, the goal is to learn a "label ranker" in the form of an $\mathbf{X} \rightarrow S_{\mathbf{Y}}$ mapping, where the output space $S_{\mathbf{Y}}$ is given by the set of all total orders (permutations) of the set of labels $\mathbf{Y}$ (the notation is leaned on the common notation S_k for the symmetric group of order k). Thus, label ranking can be seen as a generalization of conventional classification, where a complete ranking

$$y_{\pi_{\boldsymbol{x}}^{-1}(1)} \succ_{\boldsymbol{x}} y_{\pi_{\boldsymbol{x}}^{-1}(2)} \succ_{\boldsymbol{x}} \ldots \succ_{\boldsymbol{x}} y_{\pi_{\boldsymbol{x}}^{-1}(k)}$$

is associated with an instance $\boldsymbol{x}$ instead of only a single class label. Here, $\pi_{\boldsymbol{x}}$ is a permutation of $\{1, 2, \ldots, k\}$ such that $\pi_{\boldsymbol{x}}(i)$ is the position of label y_i in the ranking associated with $\boldsymbol{x}$.

The training data $\mathcal{T}$ used to induce a label ranker typically consists of a set of pairwise preferences of the form $y_i \succ_{\boldsymbol{x}} y_j$, suggesting that, for instance $\boldsymbol{x}$, y_i is preferred to y_j. In other words, a single "observation" consists of an instance $\boldsymbol{x}$ together with an ordered pair of labels (y_i, y_j).

To measure the predictive performance of a label ranker, a loss function on rankings is needed. In principle, any distance or correlation measure on rankings (permutations) can be used for that purpose. An important example is Kendall's tau, which counts the number of pairs of labels that are incorrectly ordered and normalizes this number to the interval $[-1, +1]$: For two permutations π and σ, let c be the number of correctly ordered pairs $(i, j) \in \{1, \ldots, k\}^2$, i.e., the pairs (i, j) with $i < j$ and $(\pi(i) - \pi(j))(\sigma(i) - \sigma(j)) > 0$. Likewise, let d be the number of incorrectly ordered pairs, i.e., the pairs (i, j) with $(\pi(i) - \pi(j))(\sigma(i) - \sigma(j)) < 0$. Kendall's tau, expressing a degree of correlation between π and σ, is then given by

$$\tau = \frac{c - d}{k(k-1)/2} \tag{1}$$

This coefficient assumes the extreme value 1 if $\sigma = \pi$ and the value -1 if σ is the reversal of π.

2.2 Instance Ranking

This setting proceeds from the setting of *ordinal classification*, where an instance $\boldsymbol{x} \in \mathbf{X}$ belongs to one among a finite set of classes $\mathbf{Y} = \{y_1, y_2, \ldots, y_k\}$ and, moreover, the classes have a natural order: $y_1 < y_2 < \ldots < y_k$. Training data consists of a set $\mathcal{T}$ of labeled instances. As an example, consider the assignment of submitted papers to categories *reject*, *weak reject*, *weak accept*, and *accept*.

In contrast to conventional classification, the goal is not to learn a classifier but a ranking function $f(\cdot)$. Given a subset $X \subset \mathbf{X}$ of instances as an input, the function produces a ranking, i.e., a (strict) total order $\succ$, of these instances as an output (typically by assigning a score to each instance and then sorting by scores).

For the case $k = 2$, this problem is well-known as the *bipartite ranking* problem. The case $k > 2$ has recently been termed *multipartite ranking* [9]. As an example, consider the task of a reviewer who has to rank the papers according to their quality, possibly though not necessarily with the goal of partitioning this ranking into the above four categories.

Thus, the goal of *instance ranking* is to produce a ranking $\succ$ in which instances from higher classes precede those from lower classes. Different types of accuracy measures have been proposed for predictions of this kind. Typically, they count the number of ranking errors, that is, the number of pairs $(\boldsymbol{x}, \boldsymbol{x}') \in X \times X$ such that $\boldsymbol{x}$ is ranked higher than $\boldsymbol{x}'$ even though the former belongs to a lower class than the latter. In the two-class case, this amounts to the well-known AUC, the area under the ROC-curve [11]:

$$\mathrm{AUC}(\succ, X) = \frac{1}{|P| \cdot |N|} \sum_{\boldsymbol{x} \in P} \sum_{\boldsymbol{x}' \in N} \begin{cases} 1 & \text{if } \boldsymbol{x} \succ \boldsymbol{x}' \\ 0 & \text{if } \boldsymbol{x}' \succ \boldsymbol{x} \end{cases} , \tag{2}$$

where $P \subset X$ is the set of positive and $N \subset X$ the set of negative examples in X.[1] Its generalization to multiple (ordered) classes is known as the concordance index or C-index in statistics [12].

2.3 Object Ranking

In the setting of object ranking, there is no supervision in the sense that no output or class label is associated with an object. The goal in object ranking is to learn a ranking function $f(\cdot)$ which, given a subset Z of an underlying referential set $\mathbf{Z}$ of objects as an input, produces a ranking of these objects as an output. Again, this is typically done by assigning a score to each instance and then sorting by scores.

Objects $\boldsymbol{z} \in \mathbf{Z}$ are commonly though not necessarily described in terms of an attribute-value representation. As training information, an object ranker has access to exemplary rankings or pairwise preferences of the form $\boldsymbol{z} \succ \boldsymbol{z}'$ suggesting that $\boldsymbol{z}$ should be ranked higher than $\boldsymbol{z}'$. This scenario is also known as "learning to order things" [10].

The performance of an object ranker can again be measured in terms of a distance function or correlation measure on rankings. In contrast to the setting of label ranking, however, the number of items to be ordered in the context of object ranking is typically much larger. Therefore, one often prefers measures that put more emphasis on the top of a ranking while paying less attention to the bottom [13]. In Web search, for example, people normally look at the top-10 results while ignoring the rest. Besides, the target is often not a "true" ranking but instead a single object or a subset of relevant objects, for example a set of documents relevant to a query. Evaluation measures especially tailored toward these types of requirements have been proposed in information retrieval. Typical examples include precision and recall as well as normalized discounted cumulative gain (NDCG) [14].

[1] Note that we assume $\succ$ to be a strict order. If ties are allowed, then these are typically counted by 1/2.

3 Ranking with Partial Abstention

As explained above, the set of alternatives, say, $\mathbf{A}$, to be ordered by a ranker depends on the type of ranking problem. In label ranking, $\mathbf{A}$ is a fixed set of labels $\mathbf{Y}$, whereas in instance ranking, it is a subset X of the instance space $\mathbf{X}$. A ranking on $\mathbf{A}$ is a strict, total, asymmetric, and transitive relation $\succ$, specifying for all pairs $\boldsymbol{a}, \boldsymbol{b} \in \mathbf{A}$ whether $\boldsymbol{a}$ precedes $\boldsymbol{b}$, denoted $\boldsymbol{a} \succ \boldsymbol{b}$, or $\boldsymbol{b}$ precedes $\boldsymbol{a}$. The key property of transitivity can be seen as a principle of consistency: If $\boldsymbol{a}$ is preferred to $\boldsymbol{b}$ and $\boldsymbol{b}$ is preferred to $\boldsymbol{c}$, then $\boldsymbol{a}$ must be preferred to $\boldsymbol{c}$.

A partial order $\sqsupset$ on $\mathbf{A}$ is a generalization that sticks to this consistency principle but is not necessarily total. If, for two alternatives $\boldsymbol{a}$ and $\boldsymbol{b}$, neither $\boldsymbol{a} \sqsupset \boldsymbol{b}$ nor $\boldsymbol{b} \sqsupset \boldsymbol{a}$, then these alternatives are considered as incomparable, written $\boldsymbol{a} \perp \boldsymbol{b}$. Note that, in the following, we still assume strictness of $\sqsupset$, even of this is not always mentioned explicitly.

3.1 Partial Orders in Learning to Rank

As mentioned before, our idea is to make use of the concept of a partial order in a machine learning context, namely to generalize the problem of learning to rank. More specifically, the idea is that, for each pair of alternatives $\boldsymbol{a}$ and $\boldsymbol{b}$, the ranker can decide whether to make a prediction about the order relation between these labels, namely to hypothesize that $\boldsymbol{a}$ precedes $\boldsymbol{b}$ or that $\boldsymbol{b}$ precedes $\boldsymbol{a}$, or to abstain from this prediction. We call a ranker having this possibility of abstention a ranker with partial reject option. Note, however, that for different pairs of alternatives, the reject decisions cannot be made independently of each other. Instead, the pairwise predictions should of course be consistent in the sense of being transitive and acyclic. In other words, a ranker with a (partial) reject option is expected to make a prediction in the form of a (strict) partial order $\sqsupset$ on the set of alternatives. This partial order is considered as an incomplete estimation of an underlying (ground-truth) order relation $\succ$: For alternatives $\boldsymbol{a}, \boldsymbol{b} \in \mathbf{A}$, $\boldsymbol{a} \sqsupset \boldsymbol{b}$ corresponds to the prediction that $\boldsymbol{a} \succ \boldsymbol{b}$ (and not $\boldsymbol{b} \succ \boldsymbol{a}$) holds, whereas $\boldsymbol{a} \perp \boldsymbol{b}$ indicates an abstention on this pair of alternatives.

In this section, we propose a method that enables a ranker to make predictions of such kind. Roughly speaking, our approach consists of two main steps, to be detailed in the forthcoming subsections:

- The first step is the prediction of a preference relation P that specifies, for each pair of alternatives $\boldsymbol{a}$ and $\boldsymbol{b}$, a degree of uncertainty regarding their relative comparison.
- In the second step, a (strict) partial order maximally compatible with this preference relation is derived.

3.2 Prediction of a Binary Preference Relation

Let P be an $\mathbf{A} \times \mathbf{A} \rightarrow [0, 1]$ mapping, so that $P(\boldsymbol{a}, \boldsymbol{b})$ is a measure of support for the order (preference) relation $\boldsymbol{a} \succ \boldsymbol{b}$. We assume P to be reciprocal, i.e.,

$$P(\boldsymbol{b}, \boldsymbol{a}) = 1 - P(\boldsymbol{a}, \boldsymbol{b})$$

for all $\boldsymbol{a}, \boldsymbol{b} \in \mathbf{A}$. A relation of that kind can be produced in different ways. For example, some ranking methods explicitly train models that compare alternatives in a pairwise way, e.g., by training a single classifier for each pair of alternatives [15]. If these models are able to make probabilistic predictions, these can be used directly as preference degrees $P(\boldsymbol{a}, \boldsymbol{b})$.

However, since probability estimation is known to be a difficult problem, we like to emphasize that our method for predicting strict partial orders does only assume an *ordinal* structure of the relation P. In fact, as will be seen below, the partial order induced by P is invariant toward monotone transformations of P. In other words, only the order relation of preference degrees is important, not the degrees themselves: If $P(\boldsymbol{a}, \boldsymbol{b}) > P(\boldsymbol{a}', \boldsymbol{b}')$, then $\boldsymbol{a} \succ \boldsymbol{b}$ is considered as more certain than $\boldsymbol{a}' \succ \boldsymbol{b}'$.

Here, we propose a generic approach that allows one to turn every ranker into a partial ranker. To this end, we resort to the idea of ensembling. Let L be a learning algorithm that, given a set of training data, induces a model M that in turn makes predictions in the form of rankings (total orders) $\succ$ of a set of alternatives $\mathbf{A}$. Now, instead of training a single model, our idea is to train k such models $M_1, \ldots, M_k$ by resampling from the original data set, i.e., by creating k bootstrap samples and giving them as input to L. Consequently, by querying all these models, k rankings $\succ_1, \ldots, \succ_k$ will be produced instead of a single prediction.

For each pair of alternatives $\boldsymbol{a}$ and $\boldsymbol{b}$, we then define the degree of preference $P(\boldsymbol{a}, \boldsymbol{b})$ in terms of the fraction of rankings in which $\boldsymbol{a}$ precedes $\boldsymbol{b}$:

$$P(\boldsymbol{a}, \boldsymbol{b}) = \frac{1}{k} \left| \{ i \,|\, \boldsymbol{a} \succ_i \boldsymbol{b} \} \right| \tag{3}$$

Thus, $P(\boldsymbol{a}, \boldsymbol{b}) = 1$ suggests a consensus among the ensemble members, since all of them agree that $\boldsymbol{a}$ should precede $\boldsymbol{b}$. On the other hand, $P(\boldsymbol{a}, \boldsymbol{b}) \approx 1/2$ indicates a highly uncertain situation.

3.3 Prediction of a Strict Partial Order Relation

On the basis of the preference relation P, we seek to induce a (partial) order relation $\sqsupset$ on $\mathbf{A}$, that we shall subsequently also denote by $\mathcal{R}$. Thus, $\mathcal{R}$ is an $\mathbf{A} \times \mathbf{A} \rightarrow \{0, 1\}$ mapping or, equivalently, a subset of $\mathbf{A} \times \mathbf{A}$, where $\mathcal{R}(\boldsymbol{a}, \boldsymbol{b}) = 1$, also written as $(\boldsymbol{a}, \boldsymbol{b}) \in \mathcal{R}$ or $\boldsymbol{a} \mathcal{R} \boldsymbol{b}$, indicates that $\boldsymbol{a} \sqsupset \boldsymbol{b}$.

The simplest idea is to let $\boldsymbol{a} \mathcal{R} \boldsymbol{b}$ iff $P(\boldsymbol{a}, \boldsymbol{b}) = 1$. The relation $\mathcal{R}$ thus defined is indeed a (strict) partial order, but since a perfect consensus ($P(\boldsymbol{a}, \boldsymbol{b}) \in \{0, 1\}$) is a strong requirement, most alternatives will be declared incomparable. Seeking a prediction that is as informative as possible, it is therefore natural to reduce the required degree of consensus. We therefore proceed from an "α-cut" of the relation P, defined as

$$\mathcal{R}_\alpha = \{ (\boldsymbol{a}, \boldsymbol{b}) \mid P(\boldsymbol{a}, \boldsymbol{b}) \geq \alpha \} \tag{4}$$

for $0 < \alpha \leq 1$. A cut of that kind provides a reasonable point of departure, as it comprises the most certain preference statements while ignoring those comparisons $(\boldsymbol{a}, \boldsymbol{b})$ with $P(\boldsymbol{a}, \boldsymbol{b}) < \alpha$. However, it is not necessarily transitive and may even contain cycles. For example, suppose $\boldsymbol{a} \succ_1 \boldsymbol{b} \succ_1 \boldsymbol{c}$, $\boldsymbol{b} \succ_2 \boldsymbol{c} \succ_2 \boldsymbol{a}$ and $\boldsymbol{c} \succ_3 \boldsymbol{a} \succ_3 \boldsymbol{b}$. Clearly, $P(\boldsymbol{a},\boldsymbol{b}) = P(\boldsymbol{b},\boldsymbol{c}) = P(\boldsymbol{c},\boldsymbol{a}) = 2/3$, rendering $\mathcal{R}_{2/3}$ a cyclical relation. While transitivity is easily enforced by computing the transitive closure of $\mathcal{R}_\alpha$, absence of cycles is not as easily obtained. Intuitively, it seems natural that for larger α, cycles become less probable. However, as the example shows, even for $\alpha > 1/2$, cycles can still occur. Furthermore, the larger α, the less informative the corresponding $\mathcal{R}_\alpha$.

Consequently, we propose to look for a minimal α (denote it as α^*) such that the transitive closure of $\mathcal{R}_\alpha$ (denote it as $\overline{\mathcal{R}}_\alpha$) is a strict partial order relation [16]. This $\overline{\mathcal{R}}_{\alpha^*}$ will be the predicted strict partial order relation $\mathcal{R}$, and we call α^* the consensus threshold. By minimizing this threshold, we maximize $\mathcal{R}_\alpha$ as well as its transitive closure $\overline{\mathcal{R}}_\alpha$, and thereby also the information extracted from the ensemble on the basis of which P was computed. In the remainder of this section, we deal with the problem of computing α^* in an efficient way.

3.4 Determination of an Optimal Threshold

Suppose that P can assume only a finite number of values. In our case, according to (3), this set is given by $D = \{0, 1/k, 2/k, \ldots, 1\}$, and its cardinality by $k+1$, where k is the ensemble size. Obviously, the domain of α can then be restricted to D. The simplest approach, therefore, it to test each value in D, i.e., to check for each value whether $\mathcal{R}_\alpha$ is acyclic, and hence $\overline{\mathcal{R}}_\alpha$ a partial order. Of course, instead of trying all values successively, it makes sense to exploit a monotonicity property: If $\mathcal{R}_\alpha$ is not acyclic, then $\mathcal{R}_\beta$ cannot be acyclic either, unless $\beta > \alpha$. Consequently, α^* can be found in at most $\log_2(k+1)$ steps using bisection. More specifically, by noting that α^* is lower-bounded by

$$\alpha_l = \frac{1}{k} + \max_{\boldsymbol{a},\boldsymbol{b}} \min\left(P(\boldsymbol{a},\boldsymbol{b}), P(\boldsymbol{b},\boldsymbol{a})\right) \tag{5}$$

and trivially upper-bounded by $\alpha_u = 1$, one can repeatedly update the bounds as follows, until $\alpha_u - \alpha_l < 1/k$:

(i) set α to the middle point between α_l and α_u
(ii) compute $\mathcal{R}_\alpha$
(iii) compute $\overline{\mathcal{R}}_\alpha$ (e.g., using the Floyd-Warshall's algorithm [17])
(iv) if $\overline{\mathcal{R}}_\alpha$ is a partial order, set α_u to α
(v) else set α_l to α

This procedure stops with $\alpha^* = \alpha_l$. The complexity of this procedure is not worse than the transitive closure operation, i.e., it is at most $\mathcal{O}(|\mathbf{A}|^3)$.

As shown in [16], the same result can be computed with another algorithm that is conceptually simpler (though equally costly in terms of complexity, at least theoretically). This algorithm operates on an $|\mathbf{A}| \times |\mathbf{A}|$ matrix $\mathbf{R}$ initialized with the entries $P(\boldsymbol{a},\boldsymbol{b})$ (recall that $\mathbf{A}$ is the set of alternatives). It repeatedly performs a transitive closure operation at all the levels of D simultaneously:

$$\mathbf{R}(\boldsymbol{a},\boldsymbol{b}) \leftarrow \max\left(\mathbf{R}(\boldsymbol{a},\boldsymbol{b}), \max_{\mathbf{c}\in\mathbf{A}}(\min(\mathbf{R}(\boldsymbol{a},\boldsymbol{c}), \mathbf{R}(\boldsymbol{c},\boldsymbol{b})))\right) \tag{6}$$

for all $\boldsymbol{a}, \boldsymbol{b} \in \mathbf{A}$, until no further changes occur. These transitive closure operations can be seen as a correction of inconsistences in P ($\boldsymbol{a}$ is to some degree preferred to $\boldsymbol{b}$, which in turn is to some degree preferred to $\boldsymbol{c}$, but $\boldsymbol{a}$ is not sufficiently preferred to $\boldsymbol{c}$). Since these inconsistencies do not occur very often, the number of update operations needed to stabilize $\mathbf{R}$ is normally quite small; in practice, we found that we rarely need more then one or two iterations.

Algorithm 1

Require: training data $\mathcal{T}$, test data $\mathcal{D}$, ensemble size k, base learner L
Ensure: a matrix $\mathbf{R}$ encoding partial order information for alternatives in $\mathcal{D}$ ($\mathbf{R}(i,j) = 1$ means $d_i \succ d_j$, where $d_i, d_j \in \mathcal{D}$)

initialize $\mathbf{R}$ as zero matrix
generate k bootstrap samples from $\mathcal{T}$
constitute the ensemble with k rankers trained using L
get k rankings of alternatives in $\mathcal{D}$
for each of k rankings **do**
    for every pair of alternatives $d_i, d_j \in \mathcal{D}$ **do**
        if $d_i \succ d_j$ **then**
            set $\mathbf{R}(i,j) := \mathbf{R}(i,j) + 1/k$
        end if
    end for
end for
repeat
    for every entry in $\mathbf{R}$ **do**
        $\mathbf{R}(i,j) := \max\left(\mathbf{R}(i,j), \max_{k\in\mathcal{D}}(\min(\mathbf{R}(i,k), \mathbf{R}(k,j)))\right)$
    end for
until No entry in $\mathbf{R}$ is changed.
for every entry in $\mathbf{R}$ **do**
    $\alpha := \max_{i,j} \min(\mathbf{R}(i,j), \mathbf{R}(j,i))$
end for
for every entry in $\mathbf{R}$ **do**
    if $\mathbf{R}(i,j) > \alpha$ **then**
        $\mathbf{R}(i,j) := 1$
    end if
end for

By construction, thresholding the final relation $\mathbf{R}$ at a level α will yield the transitive closure of relation $\mathcal{R}_\alpha$ in (4). Therefore, α^* can be taken as

$$\alpha^* = \frac{1}{k} + \max\left(\mathbf{R}(\boldsymbol{a},\boldsymbol{b}) \,|\, \mathbf{R}(\boldsymbol{a},\boldsymbol{b}) \leq \mathbf{R}(\boldsymbol{b},\boldsymbol{a})\right) , \tag{7}$$

which is obviously the smallest α that avoids cycles. The whole procedure is summarized in Algorithm 1.

Finally, we note that, as postulated above, α^* in (7) yields a maximal partial order as a prediction. In principle, of course, any larger value can be used as well, producing a less complete relation and, therefore, a more "cautious" prediction. We shall come back to this issue in Section 4.

3.5 Illustrating Example

We illustrate our approach by means of a small two-dimensional toy example for the case of bipartite ranking. Suppose that the conditional class distributions of the positive and the negative class are two overlapping Gaussians. A training data set may then look like the one depicted in Fig. 1 (left), with positive examples as black and negative examples as white dots. Given a new set of query instances X to be ranked, one may expect that a learner will be uncertain for those instances lying close to the overlap region, and may hence prefer to abstain from comparing them.

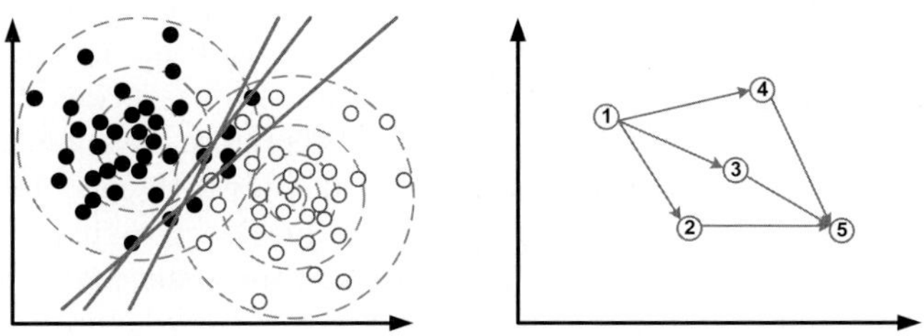

Fig. 1. Left: training data and ensemble models; right: partial order predicted for a set of five query instances

More specifically, suppose that a linear model is used to train a ranker. Roughly speaking, this means fitting a separating line and sorting instances according to their distance from the decision boundary. Fig. 1 (left) shows several such models that may result from different bootstrap samples. Now, consider the five query instances shown in the right picture of Fig. 1. Whereas all these models will rank instance 1 ahead of 2, 3 and 4, and these in turn ahead of 5, instances 2, 3 and 4 will be put in various orders. Applying our approach as outlined above, with a proper choice of the threshold α, may then yield the strict partial order indicated by the arrows in the right picture of Fig. 1. A prediction of that kind agrees with our expectation: Instance 1 is ranked first and instance 5 last; instance 2, 3 and 4 are put in the middle, but the learner abstains from comparing them in a mutual way.

4 Evaluation Measures

If a model is allowed to abstain from making predictions, it is expected to reduce its error rate. In fact, it can trivially do so, namely by rejecting all predictions, in which case it avoids any mistake. Clearly, this is not a desirable solution. Indeed, in the setting of prediction with reject option, there is always a trade-off between two criteria: *correctness* on the one side and *completeness* on the other side. An ideal learner is correct in the sense of making few mistakes, but also complete in the sense of abstaining but rarely. The two criteria are normally conflicting: increasing completeness typically comes along with reducing correctness and vice versa.

4.1 Correctness

As a measure of correctness, we propose a quantity that is also known as the *gamma rank correlation* [18] in statistics, although it is not applied to partial orders. Instead, it is used as a measure of correlation between rankings (with ties). As will be seen, however, it can also be used in a more general way.

Let $\sqsupset_*$ be the ground-truth relation on the set of alternatives $\mathbf{A}$. If this relation is a total order, like in label ranking, then $\boldsymbol{a} \sqsupset_* \boldsymbol{b}$ if $\boldsymbol{a}$ precedes $\boldsymbol{b}$ and $\boldsymbol{b} \sqsupset_* \boldsymbol{a}$ if $\boldsymbol{b}$ precedes $\boldsymbol{a}$; exactly one of these two cases is true, i.e., we never have $\boldsymbol{a} \perp_* \boldsymbol{b}$. Interestingly, in the case of instance ranking, it is not entirely clear whether the ground-truth is a total or a partial order. The goal of most learning algorithms for AUC maximization is to sort instances $\boldsymbol{x}$ according to their probability of belonging to the positive class, $\mathbf{P}(y = 1 \,|\, \boldsymbol{x})$.[2] Seen from this point of view, the underlying ground-truth is assumed to be a complete order. On the other hand, this complete order is never known and, therefore, can never be used as a reference for evaluating a prediction. Instead, only the class information is provided, and given a concrete test sample, evaluation measures like AUC do not care about the relative order of instances from the same class. In that sense, the ground-truth is treated like a partial order: $\boldsymbol{a} \sqsupset_* \boldsymbol{b}$ whenever $\boldsymbol{a}$ is positive and $\boldsymbol{b}$ negative (or, in the multi-class case, if the class of $\boldsymbol{a}$ is higher than the class of $\boldsymbol{b}$), while $\boldsymbol{a} \perp_* \boldsymbol{b}$ when $\boldsymbol{a}$ and $\boldsymbol{b}$ belong to the same class.

Now, let $\sqsupset$ be a predicted (strict) partial order, i.e., a prediction of $\sqsupset_*$. We call a pair of alternatives $\boldsymbol{a}$ and $\boldsymbol{b}$ *concordant* if they ought to be compared, because $\neg(\boldsymbol{a} \perp_* \boldsymbol{b})$, and are indeed compared in the correct way, that is,

$$(\boldsymbol{a} \sqsupset_* \boldsymbol{b} \wedge \boldsymbol{a} \sqsupset \boldsymbol{b}) \vee (\boldsymbol{b} \sqsupset_* \boldsymbol{a} \wedge \boldsymbol{b} \sqsupset \boldsymbol{a}) \; .$$

Likewise, we call $\boldsymbol{a}$ and $\boldsymbol{b}$ *discordant* if they ought to be compared, but the comparison is incorrect, that is,

$$(\boldsymbol{a} \sqsupset_* \boldsymbol{b} \wedge \boldsymbol{b} \sqsupset \boldsymbol{a}) \vee (\boldsymbol{b} \sqsupset_* \boldsymbol{a} \wedge \boldsymbol{a} \sqsupset \boldsymbol{b}) \; .$$

Note that, if $\boldsymbol{a} \perp_* \boldsymbol{b}$ (there is no need to compare $\boldsymbol{a}$ and $\boldsymbol{b}$) or $\boldsymbol{a} \perp \boldsymbol{b}$ (abstention on $\boldsymbol{a}$ and $\boldsymbol{b}$), then the two alternatives are neither concordant nor discordant.

Given these notions of concordance and discordance, we can define

$$\mathrm{CR}(\sqsupset, \sqsupset_*) = \frac{|C| - |D|}{|C| + |D|} \; , \tag{8}$$

where C and D denote, respectively, the set of concordant and discordant pairs of alternatives. Obviously, $\mathrm{CR}(\sqsupset, \sqsupset_*) = 1$ for $\sqsupset_* = \sqsupset$ and $\mathrm{CR}(\sqsupset, \sqsupset_*) = -1$ if $\sqsupset$ is the inversion of $\sqsupset_*$.

It is also interesting to mention that (8) is indeed a proper generalization of commonly used measures for the complete (non-partial) case, in the sense of reducing to these measures if $\sqsupset$ is a total order. In particular, it is easy to see that (8) reduces to Kendall's tau (1) in the case of label ranking (where $\sqsupset_*$ is a total order, too), and to the AUC measure (2) in the case of instance ranking (where $\sqsupset_*$ is a partial order).

---

[2] Indeed, this prediction maximizes the *expected* AUC on a test set.

4.2 Completeness

To measure the degree of completeness of a prediction, a straightforward idea is to punish the abstention from comparisons that should actually be made (while ignoring or, say, tolerating comparisons that are made despite not being necessary). This leads to the following measure of completeness:

$$\mathrm{CP}(\sqsupset) = \frac{|C| + |D|}{|\sqsupset_*|} \tag{9}$$

5 Experimental Results

As mentioned before, our method described in Section 3 can be applied to different ranking problems in a generic way. In this section, we present experimental results for two of the ranking problems outlined in Section 2, namely instance ranking and label ranking.

5.1 Instance Ranking

To test our method in the instance ranking scenario, we have selected a set of 16 binary classification data sets from the UCI repository and the Statlog collection[3]. We have used logistic regression as a base learner and produced ensembles of size 10.

Table 1. Results for instance ranking: mean values and standard deviations for correctness and completeness

data set	#attr.	#inst.	correctness with abstention	correctness w/o abstention	completeness
breast	9	286	0.330±0.150	0.318±0.141	0.578±0.074
breast-w	9	699	0.988±0.014	0.987±0.015	0.982±0.015
horse colic	22	368	0.734±0.135	0.697±0.142	0.790±0.044
credit rating	15	690	0.858±0.062	0.827±0.065	0.888±0.038
credit german	20	1000	0.610±0.088	0.568±0.084	0.741±0.060
pima diabetes	8	768	0.684±0.084	0.666±0.086	0.819±0.047
heart statlog	13	270	0.811±0.102	0.797±0.101	0.890±0.060
hepatitis	19	155	0.709±0.292	0.697±0.271	0.797±0.084
ionosphere	34	351	0.771±0.174	0.722±0.190	0.814±0.098
kr-vs-kp	36	3196	0.992±0.006	0.980±0.007	0.991±0.006
labor	16	57	0.990±0.049	0.985±0.060	0.989±0.052
mushroom	22	8124	1.000±0.000	1.000±0.000	0.808±0.017
thyroid disease	29	3772	0.890±0.071	0.883±0.070	0.928±0.040
sonar	60	206	0.684±0.224	0.575±0.271	0.575±0.056
tic-tac-toe	9	958	0.253±0.127	0.221±0.120	0.908±0.013
vote	16	435	0.981±0.032	0.976±0.036	0.913±0.035

[3] `www.ics.uci.edu/~mlearn/MLRepository.html`, `lib.stat.cmu.edu/`

The values reported in Table 1 are averages over five repetitions of a 10-fold cross-validation. Comparing the correctness of predictions in terms of (8), it can be seen that our approach of partial abstention generally leads to improved performance. In fact, it is never worse and yields better results most of the time, sometimes with significant margins. Moreover, this gain in performance comes with an acceptable loss in terms of completeness. Indeed, the degrees of completeness are quite high throughout, often significantly above 90%.

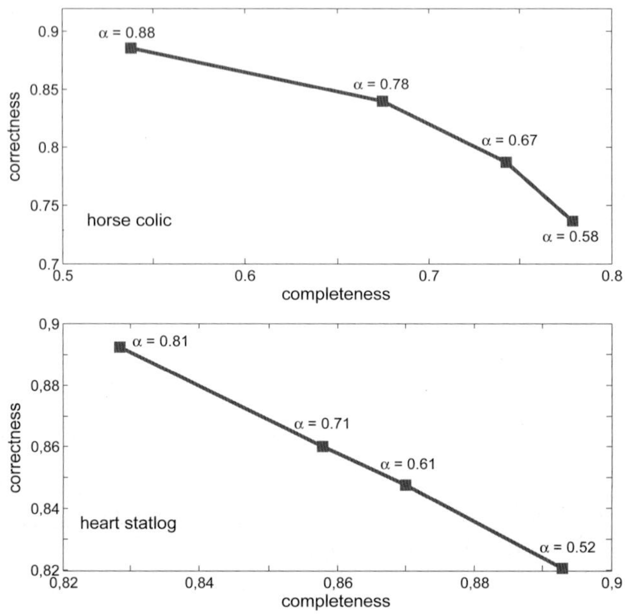

Fig. 2. Instance ranking with partial abstention: Trade-off between correctness and completeness for selected data sets

We conducted a second experiment with the aim to investigate the trade-off between correctness and completeness. As was mentioned earlier, and to some extent already confirmed by our first experiment, we expect a compromise between both criteria insofar as it should be possible to increase correctness at the cost of completeness. To verify this conjecture, we varied the threshold α in (4) in the range $[\alpha^*, 1]$. Compared to the use of α^*, larger thresholds will make the predictions increasingly incomplete; at the same time, however, they should also become more correct. Indeed, the results we obtained are well in agreement with these expectations. Fig. 2 shows typical examples of the trade-off between correctness and completeness for two data sets.

Finally, it is interesting to look at the maximal chains of a predicted partial order. A maximal chain of a partially ordered set X is a maximal subset $C \subset X$ that is totally ordered, like the set of elements depicted as black nodes in the following partial order:

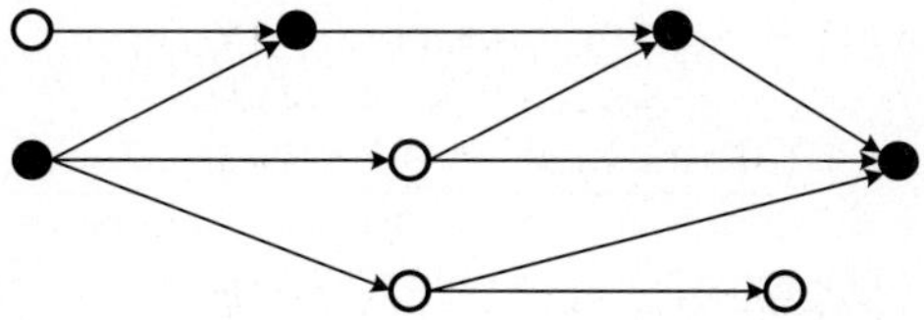

The remaining elements $X \setminus C$ can then be considered as those that cannot be inserted into the order in a reliable way, and are hence ignored. Since each maximal chain is a total order, it can be visualized in terms of an ROC curve. A typical example for the sonar data set is shown in Fig. 3. As can be seen, the curves for the maximal chains tend to dominate the original ROC curve for this data, suggesting that the ranking of elements in the individual chains is indeed more reliable than the ranking of the complete data.

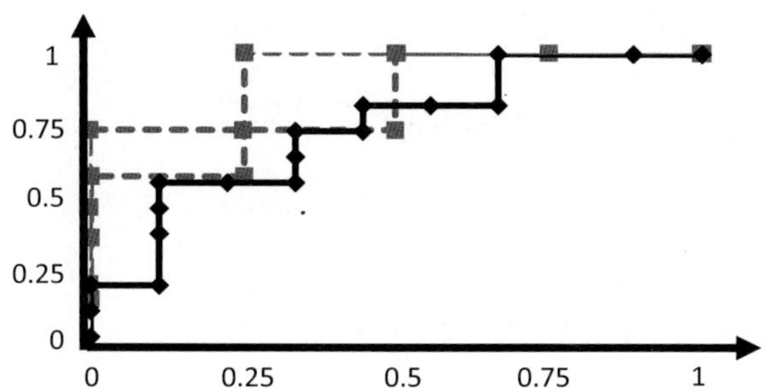

Fig. 3. ROC curves for the maximal chains of a predicted partial order (dashed lines) and the complete data (solid line) for the sonar data

5.2 Label Ranking

In view of a lack of benchmark data for label ranking, we resorted to multi-class data sets from the UCI repository and turned them into label ranking data by following the procedure proposed in [15]: A naive Bayes classifier is first trained on the complete data set. Then, for each example, all the labels present in the data set are ordered with respect to the predicted class probabilities (in the case of ties, labels with lower index are ranked first).[4]

The setting of this experiment is similar to the one we did for instance ranking. We performed five repetitions of a 10-fold cross-validation and used an ensemble size of 10. As a label ranking method, we used the *ranking by pairwise comparison* approach [15], again with logistic regression as a base learner.

The results, summarized in Table 2, convey the same message as before: Correctness can be improved at the cost of completeness, and compared to the case of instance ranking, the loss in completeness is even much smaller here; see also Fig. 4, in which we show the same kind of trade-off curves as for the case of instance ranking.

[4] The data sets are available at `www.uni-marburg.de/fb12/kebi/research`

Table 2. Results for label ranking: mean values and standard deviations for correctness and completeness

data set	#attr.	#classes	#inst.	correctness with abstention	correctness w/o abstention	completeness
iris	4	3	150	0.910±0.062	0.885±0.068	0.991±0.063
wine	13	3	178	0.940±0.051	0.921±0.053	0.988±0.067
glass	9	6	214	0.892±0.039	0.882±0.042	0.990±0.030
vowel	10	11	528	0.657±0.019	0.647±0.019	0.988±0.016
vehicle	18	4	846	0.858±0.026	0.854±0.025	0.992±0.039
authorship	70	4	841	0.941±0.016	0.910±0.015	0.989±0.043
pendigits	16	10	10992	0.933±0.002	0.932±0.002	0.999±0.005
segment	18	7	2310	0.938±0.006	0.934±0.006	0.998±0.011

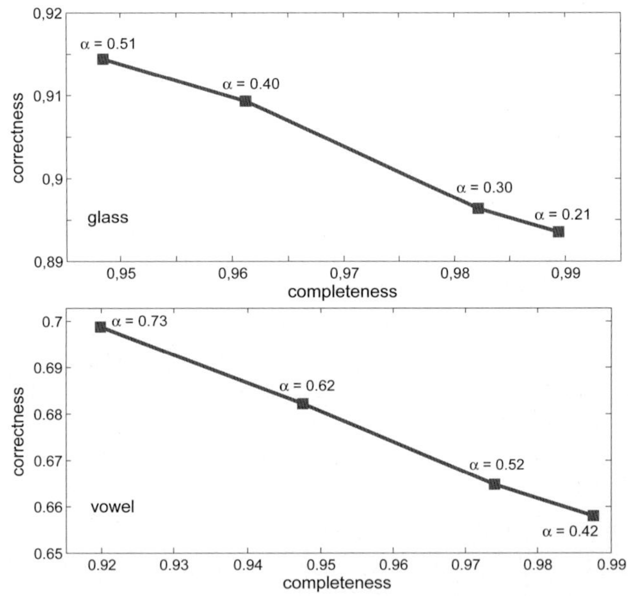

Fig. 4. Label ranking with partial abstention: Trade-off between correctness and completeness for selected data sets

6 Conclusions and Future Work

In this paper, we have addressed the problem of "reliable" prediction in the context of learning to rank. In this regard, we have made the following main contributions:

- Based on the idea of allowing a learner to abstain from an uncertain comparison of alternatives, together with the requirement that predictions are consistent, we have proposed a relaxation of the conventional setting in which predictions are given in terms of partial instead of total orders.

- We have proposed a generic approach to predicting partial orders or, according to our interpretation, ranking with partial abstention, which is applicable to different types of ranking problems.
- We have introduced reasonable measures for evaluating the performance of a ranker with (partial) reject option, namely measures of correctness and completeness. These measures are proper generalizations of conventional and commonly used measures for total orders.
- Empirically, we have shown that our method is indeed able to trade off accuracy against completeness: The correctness of a prediction can be increased at the cost of reducing the number of alternatives that are compared.

The extension from predicting total to predicting partial orders as proposed in this paper opens the door for a multitude of further studies. Here, we mention just one example of an interesting direction for future work, which concerns the type of target order $\sqsupset_*$ to be predicted. In this paper, we have essentially assumed that the target is a complete order, and a prediction in terms of a partial order $\sqsupset$ an incomplete estimation thereof, even though it was already mentioned that, in the case of instance ranking (AUC maximization), the target may also be considered as a partial order. However, even in that case, our evaluation measure does not penalize the prediction of an order relation between two instances $\boldsymbol{a}$ and $\boldsymbol{b}$ from the same class. In other words, we do not penalize the case where $\boldsymbol{a} \sqsupset \boldsymbol{b}$ even though $\boldsymbol{a} \perp_* \boldsymbol{b}$. Now, if $\sqsupset_*$ is a true partial order, it clearly makes sense to request, not only the correct prediction of order relations $\boldsymbol{a} \sqsupset_* \boldsymbol{b}$ between alternatives, but also of incomparability relations $\boldsymbol{a} \perp_* \boldsymbol{b}$. Although the difference may look subtle at first sight, the changes will go beyond the evaluation of predictions and instead call for different learning algorithms. In particular, in this latter scenario, $\boldsymbol{a} \perp \boldsymbol{b}$ will be interpreted as a prediction that $\boldsymbol{a}$ and $\boldsymbol{b}$ are incomparable ($\boldsymbol{a} \perp_* \boldsymbol{b}$), and not as a rejection of the decision whether $\boldsymbol{a} \sqsupset_* \boldsymbol{b}$ or $\boldsymbol{b} \sqsupset_* \boldsymbol{a}$. Nevertheless, the two settings are of course related, and we plan to elaborate on their connection in future work.

References

1. Bakir, G., Hofmann, T., Schölkopf, B., Smola, A., Taskar, B., Vishwanathan, S. (eds.): Predicting structured data. MIT Press, Cambridge (2007)
2. Chow, C.: On optimum recognition error and reject tradeoff. IEEE Transactions on Information Theory 16(1), 41–46 (1970)
3. Herbei, R., Wegkamp, M.H.: Classification with reject option. Canadian Journal of Statistics 34(4), 709–721 (2006)
4. Bartlett, P.L., Wegkamp, M.H.: Classification with a reject option using a hinge loss. Journal of Machine Learning Research 9, 1823–1840 (2008)
5. Fürnkranz, J., Hüllermeier, E. (eds.): Preference Learning. Springer, Heidelberg (2010)
6. Har-Peled, S., Roth, D., Zimak, D.: Constraint classification for multiclass classification and ranking. In: Becker, S., Thrun, S., Obermayer, K. (eds.) Advances in Neural Information Processing Systems 15 (NIPS 2002), pp. 785–792 (2003)

7. Fürnkranz, J., Hüllermeier, E.: Pairwise preference learning and ranking. In: Lavrač, N., Gamberger, D., Todorovski, L., Blockeel, H. (eds.) ECML 2003. LNCS (LNAI), vol. 2837, pp. 145–156. Springer, Heidelberg (2003)
8. Dekel, O., Manning, C., Singer, Y.: Log-linear models for label ranking. In: Advances in Neural Information Processing Systems (2003)
9. Fürnkranz, J., Hüllermeier, E., Vanderlooy, S.: Binary decomposition methods for multipartite ranking. In: Proceedings ECML/PKDD–2009, European Conference on Machine Learning and Knowledge Discovery in Databases, Bled, Slovenia (2009)
10. Cohen, W., Schapire, R., Singer, Y.: Learning to order things. Journal of Artificial Intelligence Research 10, 243–270 (1999)
11. Fawcett, T.: An introduction to ROC analysis. Pattern Recognition Letters 27(8), 861–874 (2006)
12. Gnen, M., Heller, G.: Concordance probability and discriminatory power in proportional hazards regression. Biometrika 92(4), 965–970 (2005)
13. Fagin, R., Kumar, R., Sivakumar, D.: Comparing top k lists. SIAM Journal of Discrete Mathematics 17(1), 134–160 (2003)
14. Manning, C., Raghavan, P., Schütze, H.: Introduction to Information Retrieval. Cambridge University Press, Cambridge (2008)
15. Hüllermeier, E., Fürnkranz, J., Cheng, W., Brinker, K.: Label ranking by learning pairwise preferences. Artificial Intelligence 172, 1897–1917 (2008)
16. Rademaker, M., De Baets, B.: A threshold for majority in the context of aggregating partial order relations. In: Proc. WCCI 2010, World Congress on Computational Intelligence, Barcelona, Spain (2010)
17. Floyd, R.: Algorithm 97: Shortest path. Communications of the ACM 5 (1962)
18. Goodman, L., Kruskal, W.: Measures of Association for Cross Classifications. Springer, New York (1979)

Predictive Distribution Matching SVM for Multi-domain Learning

Chun-Wei Seah[1], Ivor W. Tsang[1], Yew-Soon Ong[1], and Kee-Khoon Lee[2]

[1] School of Computer Engineering, Nanyang Technological University, Singapore
{Seah0116,IvorTsang,asYSOng}@ntu.edu.sg
[2] Institute of High Performance Computing, Singapore
leekk@ihpc.a-star.edu.sg

Abstract. Domain adaptation (DA) using labeled data from related source domains comes in handy when the labeled patterns of a target domain are scarce. Nevertheless, it is worth noting that when the predictive distribution $P(y|\mathbf{x})$ of the domains differs, which establishes *Negative Transfer* [19], DA approaches generally fail to perform well. Taking this cue, the Predictive Distribution Matching SVM (PDM-SVM) is proposed to learn a robust classifier in the target domain (referred to as the target classifier) by leveraging the labeled data from only the relevant regions of multiple sources. In particular, a k-nearest neighbor graph is iteratively constructed to identify the regions of relevant source labeled data where the predictive distribution maximally aligns with that of the target data. Predictive distribution matching regularization is then introduced to leverage these relevant source labeled data for training the target classifier. In addition, progressive transduction is adopted to infer the label of target unlabeled data for estimating the predictive distribution of the target domain. Finally, extensive experiments are conducted to illustrate the impact of Negative Transfer on several existing state-of-the-art DA methods, and demonstrate the improved performance efficacy of our proposed PDM-SVM on the commonly used multi-domain Sentiment and Reuters datasets.

Keywords: Domain Adaptation, Negative Transfer, Predictive Distribution Matching, Progressive Transduction.

1 Introduction

Sentiment classification is an important task [3] for the marketer to predict sentiment polarity (e.g. positive or negative) of user reviews collected for different products. For instance, there are different categories of products from Amazon: books, DVDs, electronics and kitchen appliances. Users' comments of these products are usually described by some common words. Traditional machine learning algorithms can be used to train a sentiment classifier from manually labeled feedbacks for each of these reviews. When a category of products does not have much labeled reviews (referred to as target domain), Domain Adaptation (DA) methods come into hand as these methods can leverage labeled reviews from some

J.L. Balcázar et al. (Eds.): ECML PKDD 2010, Part I, LNAI 6321, pp. 231–247, 2010.

related products referred to as source domains. Besides sentiment classification, DA methods have also been applied in many real applications ranging from Natural Language Processing [13,4,9], text categorization [17], visual concept detection [11,10], WiFi localization [17] and remote sensing [5].

One of the major challenges of leveraging source domains to train a target classifier lies on the dissimilarity of predictive distribution among different domains which will be illustrated by an example in Section 3.2. However, many works on DA are assuming that the predictive distribution between target and source domains is the same [12,8,16,20]. However, Bruzzone and Marconcini [5] explained that when there are limited labeled data, these labeled data do not represent the general population especially for imbalance problem, and introduce a bias in estimating the predictive distribution (e.g. by Naïve Bayes Classifier). In most cases, the target domain has very few labeled data and source domains might have different class distribution from the target one. Thus, this might easily lead to dissimilarity of the predictive distribution between the domains.

In addition, the true class distribution of the target domain is unknown as the labeled data are limited, therefore re-sampling strategies (e.g. SMOTE [6]) for adjusting the source domain to have the same class distribution with the target domain might not be directly applicable in this setting.

Since the predictive distribution of the source domains might differ from the target domain, the classifier directly trained with all labeled data from multiple source domains might not classify well on unlabeled data in the target domain. Direct transferring of knowledge from the source domains to the target domain may also lead to adverse effects, which is referred to as *negative transfer* [19]. In this work, we propose a novel DA method, namely Predictive Distribution Matching Support Vector Machine (PDM-SVM), to address the challenges arisen from the difference in predictive distribution among multiple domains.

The main contributions in this paper are as follows: 1) A k-nearest neighbor graph is iteratively constructed to identify relevant source labeled data that have high similarity in predictive distribution of the target data. Then we exploit this dependency to define the so-called predictive distribution matching regularization that leverages only relevant source labeled patterns to train the target classifier. 2) We demonstrate how to infer the pseudo-labels of target unlabeled patterns by the use of progressive transduction which eventually learns the predictive distribution of the target domain. 3) We illustrate how the negative transfer affects SVMs trained with source domains, Semi-Supervised Learning (SSL) and DA methods when the source and target domains have dissimilar class distribution in Section 3.2. We show that our PDM-SVM approach can handle this problem and significantly outperform those methods in the comprehensive experiments on Sentiment and Reuters datasets.

2 Related Works

Initial work of leveraging labeled patterns from a source domain for the target domain was proposed to minimize the weighted empirical risk for both the source and target domains [21], which does not consider the distribution difference between the two domains. To address this, several instance weighting methods [13] had been proposed, but these methods usually require considerable amount of target labeled data in order to robustly re-weighting the training instances.

However, target labeled patterns are scarce. Instead of using many target labels, several methods [4,9,17,18] had been proposed to extract some useful features to be augmented in the original feature space. For example, a heuristic method was proposed in [4] to identify some *pivot features* representing common feature structure between different domains to learn an embedded space. Then this space is augmented to the original input feature vector for domain adaptation. Another example is Feature Augmentation (FA) [9], which augments features belonging to the same domain by twice that of the original features so that data within the same domains would be treated as more similar than data in different domains.

Recent DA works [8,16,20,10] are taking up the challenge of learning from multiple source domains. Crammer *et al.* [8] assumed that the distribution of multiple sources is the same, and the change of output labels is a result of varying noise. Luo *et al.* [16] maximized the consensus of predictions from multiple sources. In [20], the authors proposed a Multiple Convex Combination of SVM (M-SVM) trained from multiple source domains and a target domain. However, some source domains may not be useful for knowledge transfer. In [10], a domain-dependent regularizer was proposed to enforce that the prediction of the target classifier on target unlabeled data is close to the prediction of source classifiers from similar sources only. Recently, Domain Adaptation SVM (DASVM) [5] was proposed to tackle the mismatch of the predictive distribution between the domains by removing all source labeled patterns progressively; meanwhile, using Progressive Transductive SVM(PTSVM) [7] to infer the label of target unlabeled patterns by using all remaining source and target labeled data. However, when there are many overlapping sources and target data, and the label of some source labeled data are not consistent with the label of the target data, all these methods might not perform well.

DA is also similar to several learning paradigms such as multi-task learning and multi-view learning. The major difference between DA and multi-task learning is that DA learns a classifier for a specific task in the target domain whereas multi-task simultaneously learns for multiple tasks. For multi-view learning, a classifier is trained for each source domain and labels the unlabeled data when all these source classifiers agree on the predicted output of the unlabeled data to a certain degree, thus assuming the source domains have the same distribution.

3 Predictive Distribution Matching SVM

3.1 Preliminaries and Problem Statements

Throughout the rest of this paper, whenever superscript ^{s} and ^{t} appear in the contents, they represent source domain and target domain respectively. A summary of all important symbols can be found in Table 1.

Table 1. Symbol Definition

Symbol	Definition
m	Total number of domains, the first (m-1) domains represent source domains and the last domain, mth domain, is the target domain
$\mathbf{x}$	Feature vector of a data
y	Class Label for the data x or pseudo-label which is a class label that can be learned for a particular $\mathbf{x}$ that is described in Section 3.5,
n_r	Number of labeled data in rth domain. For the target domain, it is the combination of labeled and pseudo-labeled data
n	$\sum_{r=1}^{m} n_r$
D_L^s	$\cup_{r=1}^{m-1}\{\mathbf{x}_i^r, y_i^r\}^{n_r}$, all labeled data in all source domains
D_L^t	$\{\mathbf{x}_i, y_i\}^{n_m}$, all labeled data in target domain
$\mathcal{D}_L$	$D_L^s \cup D_L^t$
$\mathbf{x}_i^u$	Feature vector of ith unlabeled data in target domain
D_U	All unlabeled data in target domain
$P(\mathbf{x})$	Marginal distribution
$P(y\|\mathbf{x})$	Predictive distribution

Recall that labeled patterns of one domain can be drawn from the joint distribution $P(\mathbf{x}, y) = P(y|\mathbf{x})P(\mathbf{x})$. We also let $P(\mathbf{x})$ be the marginal distribution of the input sets $\{\mathbf{x}_i\}^{n_r}$ in the rth domains. DA methods usually assume that $P^t(\mathbf{x})$ of the target domain and $P^s(\mathbf{x})$ of the source domain are different. Then the task of DA is to predict the labels y_i^t's corresponding to the inputs $\mathbf{x}_i^t$'s in the target domain. Notice that DA is different from Semi-Supervised Learning (SSL). SSL methods employ both labeled and unlabeled data to achieve better prediction performance, in which the labeled and unlabeled data are usually assumed to be drawn from the same domain. It is also worth noting that the common assumption in many DA methods [12,8,16,20] is that $P^s(\mathbf{x}) \neq P^t(\mathbf{x})$, but the source and target domains share the same predictive distribution, *i.e.*, $P^s(y|\mathbf{x}) = P^t(y|\mathbf{x})$, where $P^s(y|\mathbf{x})$ and $P^t(y|\mathbf{x})$ are the predictive distribution of the source and target domains, respectively. This is also referred to as covariate shift [2]. Hence in this work, we attempt to solve domain adaptation in the setting where the predictive distribution is not to be preserved, i.e. $P^s(y|\mathbf{x}) \neq P^t(y|\mathbf{x})$. This can be materializing by diverse class distribution and limited samples in each domain, and by class label inconsistency among different domains.

3.2 An Illustrating Example

Before we introduce our proposed method, in this subsection, we first study how the dissimilarity of the class distribution between the target and source domains affects Domain Adaptation (DA) and Semi-Supervised Learning (SSL) methods. Suppose that there are very few labeled data but a lot of unlabeled data in the target domain, we vary Positive Class Ratio (PCR) of the source domains. Here, PCR defines the percentage of positive class data in the source domains. For example, PCR = 0.1 implies that 10% of the data are positive. Note that when PCR is skewed towards either extremes, the class distribution of the source domains becomes very imbalanced. As mentioned in Section 1, when the data is imbalanced, the limited labeled data might introduce a bias in estimating the predictive distribution. Thus, the difference in the predictive distribution between the target and source domains might occur. To study this, we train two SVM models: SVM_T trained with only labeled data $\mathcal{D}_L^t$ from target domain; SVM_S trained with only labeled data $\mathcal{D}_L^s$ from all source domains. We include two SSL methods: Transduction SVM (TSVM) [15], trained with the labeled data from all the source and target domains, $\mathcal{D}_L$, and unlabeled data in target domain, $\mathcal{D}_U$; LapSVM [1], the training set is the same as TSVM. We also compare with a DA algorithm: SVM_ST [21], is a SVM trained with $\mathcal{D}_L$.

Here, we will demonstrate the trends of different SVM-based algorithms according to different PCR settings on Sentiment dataset. The task is to classify whether the reviews are positive or negative. The target domain is the reviews of *Electronics* while the source domains are the reviews of *Book*, *DVDs* and *Kitchen appliances*. In the present setup, the test set is designed to contain equal amount of positive and negative data. The source domains however possess different PCR values. The details of other experimental setup will be described in Section 4.

Testing accuracy of different methods against varying PCR in the source domains are reported in Figure 1. Firstly, it is worth to observe that by using both the source and target data, the TSVM, LapSVM and SVM_ST can perform better than SVM_S and SVM_T which use only source and target data respectively. Secondly, one can observe that most methods perform optimally for PCR of 0.5 (i.e. the source and target domains have the same ratio of positive and negative data). However, the performances of all methods dropped sharply when the PCR is skewed toward either extremes (i.e. the source domains are very imbalanced, or the source and target domains have different class distribution), which implies $P^s(y|\mathbf{x})$ and $P^t(y|\mathbf{x})$ would most likely be dissimilar [5]. In addition, leveraging source labeled data from other domains lead to adverse effects on domain adaptation, which is regarded as negative transfer [19]. It is clear that those values below the line of SVM_T can be indicated as negative transfer, as the accuracy of a classifier borrowing labeled data from other source domains performs worse than just using the available target labeled data. The possible reason is that the source and target domains have different predictive distribution, when the source and target domains are combined together as a training set, which represents another predictive distribution and does not reflect the true population of its own domain. Therefore, all classifiers trained with this training set might have poorer generalization performance.

Interestingly, it can be observed that most of the values reported for the SSL methods are above the SVM_T value. A possible reason might be these two methods use target unlabeled data as the regularization. TSVM enforces unlabeled data to have same class ratio as labeled dataset. Therefore when PCR is 0.5, TSVM will classify the unlabeled data into half of them as positive and another half as negative which is the true class distribution for the unlabeled data. Hence this might cause it to perform the best when PCR setting is 0.5. In other PCR settings, TSVM classifies the unlabeled data into the same class ratio as the PCR setting, and suffers poorer classification performance.

Note, LapSVM assumes that if two patterns are close together in high density region, these patterns should have similar predictive outputs. If the manifold assumption holds strongly, LapSVM should perform well in all PCR settings.

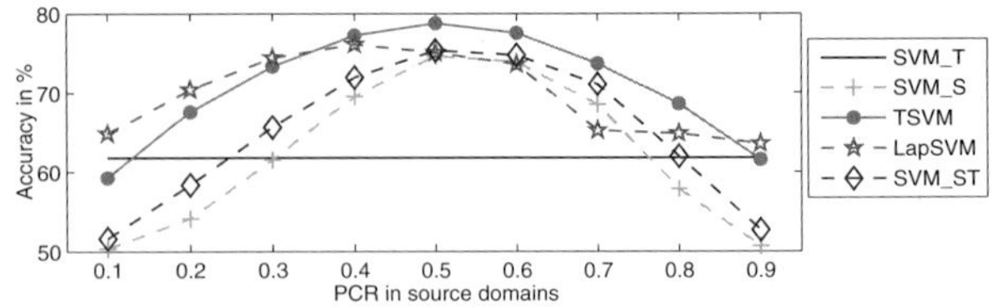

Fig. 1. Testing accuracies on Sentiment Data with varying PCR in source domains and *Electronics* as target domain

However, from our experiences, this is not the case on Sentiment dataset because two reviews with similar comments can have different meanings. For example, "I really like this" and "I dare you would really like this". For the former sentence, it is a positive feedback whereas the latter sentence is a totally negative comment. Thus, LapSVM achieves lower accuracy than just using supervised labeled information (i.e. SVM_ST) in some PCR settings. We also observe that LapSVM performs better than TSVM at both extreme ends of PCR settings, it is possibly because manifold regularization [1] on imbalanced data might be more robust than *cluster assumption* in TSVM [15].

3.3 Predictive Distribution Matching across Multiple Domains

From all the observations in Section 3.2, we are motivated to introduce a new DA method by using target unlabeled data and explicitly considering the predictive distribution of both the source and target data for multi-domain learning. Here, we define a regularizer such that two similar patterns $\mathbf{x}_i^r$ and $\mathbf{x}_j^c$ from the rth and cth domains respectively would produce similar predictive outputs for a *positive transfer* (i.e. two patterns have a high relevance measured by W_{ij}^{rc}):

$$\Omega(f) = \frac{1}{n^2} \sum_{r,c=1}^{m} \sum_{i=1}^{n_r} \sum_{j=1}^{n_c} (f(\mathbf{x}_i^r) - f(\mathbf{x}_j^c))^2 W_{ij}^{rc}, \tag{1}$$

where n_r and n_c are the number of patterns in the rth and cth domains respectively. Here, $\mathbf{x}_i^r$ is the ith data in the rth domain and $\mathbf{x}_j^c$ is the jth data in the cth domain. The similarity W_{ij}^{rc} to measure a *positive transfer* of two patterns $\mathbf{x}_i^r$ and $\mathbf{x}_j^c$ is defined as follows:

$$W_{ij}^{rc} = \sum_{z=1}^{v} P^r(y_z|\mathbf{x}_i^r) P^c(y_z|\mathbf{x}_j^c) I[y_i = y_j] D[r \neq c] S(\mathbf{x}_i^r, \mathbf{x}_j^c), \tag{2}$$

where v is the number of classes, $P^r(y|\mathbf{x}_i^r)$ is the predictive distribution of the rth domain on pattern $\mathbf{x}_i^r$ which can be estimated by means of Naïve Bayes Classifier on labeled data, while $I(\cdot)$ and $D(\cdot)$ are indicator functions. Here, $S(\mathbf{x}_i^r, \mathbf{x}_j^c)$ measures the similarity between patterns $\mathbf{x}_i^r$ and $\mathbf{x}_j^c$, and is defined as the weight of an edge in a graph constructed by k nearest neighbors.

In this work, we also define the *predictive distribution matching score* for two nearby patterns $\mathbf{x}_i^r$ and $\mathbf{x}_j^c$ as $\sum_{z=1}^{v} P^r(y_z|\mathbf{x}_i^r)P^c(y_z|\mathbf{x}_j^c)$ for measuring the similarity of the predictive distribution of two patterns, where $\sum_{z=1}^{v} P^r(y_z|\mathbf{x}_i^r)$ and

$\sum_{z=1}^{v} P^c(y_z|\mathbf{x}_j^c)$ are both equal to 1. Moreover, those patterns not in the same class will be disconnected, as the indicator function $I[y_i = y_j]$ returns a logic of 1 if both labels are the same, otherwise it returns 0. Intuitively, this indication function can be viewed as a pairwise constraint that two patterns in the same class can be linked together whereas two patterns belonging to different class should not be linked [22]. As we do not assume manifold assumption in each domain, we use an indicator function $D[r \neq c]$ to allow only data in different domains to be connected. Note that if the target domain follows manifold assumption, the manifold regularizer can be easily added into our formulation. But in this paper, we do not assume manifold property in any dataset and hence our method can apply in general cases.

From the definition of W_{ij}^{rc} in (2), two similar patterns from different domains having a high response of the predictive distribution matching score and the same class label would share similar predictive outputs. Therefore, we can identify relevant source labeled data from the data having high similar predictive distribution for domain adaptation.

3.4 Proposed Formulation

In our regularization framework, the decision function is learned by minimizing the following regularized risk functional: $f^* = \arg\min_f \gamma_A \|f\|^2 + \frac{1}{n}\sum_{i=1}^{n} \ell(y_i, f(x_i))$, where $\ell(y_i, f(x_i))$ denotes the loss function and γ_A controls the smoothness of the solutions. In this paper, we employ hinge loss function of SVM as $\ell(\cdot)$. Together with our proposed data-dependent regularizer in (1), the regulated risk functional for domain adaptation is then formulated as:

$$\min_f \gamma_I \Omega(f) + \gamma_A \|f\|^2 + \frac{1}{n}\sum_{i=1}^{n} \ell(y_i, f(x_i)), \tag{3}$$

where γ_I regulates the decision function $f(\mathbf{x})$ according to our proposed regularizer (1) for multiple source domain adaptation. We refer our proposed method to as Predictive Distribution Matching SVM (PDM-SVM). Note that our regularizer can be easily added into other standard regularization frameworks. We use SVM as our formulation since we are investigating and comparing with other SVM-based methods.

By defining a Laplacian matrix $L = D - W$ where W is a $n \times n$ matrix with entries defined in (2) and D is a $n \times n$ diagonal matrix with diagonal entry $D_{ii} = \sum_{j=1}^{n} W_{ij}$, the resultant optimization problem (3) can be formulated as LapSVM formulation [1]. Thus allowing us to take advantage of existing LapSVM algorithm to solve (3) by using (2) as the Laplacian matrix L in the algorithm. By duality, the minimization problem (3) is equivalent to the dual problem:

$$\max_{\alpha \in \Re^n} \sum_{i=1}^{n} \alpha_i - \frac{1}{2}\alpha' Y K \left(2\gamma_A I + \frac{2\gamma_I}{n^2} L K\right)^{-1} Y\alpha, \tag{4}$$

where $Y = diag(y_1, \ldots, y_n) \in \Re^{(n\times n)}$, K is the $n \times n$ kernel matrix and I is an identity matrix . The decision function is defined as follows:

$$f(x) = \sum_{i=1}^{n} \beta_i K(x, x_i), \tag{5}$$

where $\beta = (2\gamma_A I + 2\frac{\gamma_I}{n^2} LK)^{-1} Y\alpha$. For more details of the derivations, interested reader may refer to [1].

3.5 Progressive Transduction on $\mathcal{D}_U$

One of major challenges of domain adaptation is that the prediction distribution $P^t(y|\mathbf{x})$ cannot be well-estimated with limited labeled data in the target domain. Therefore, many existing DA algorithms [12,17] assume that the target domain and the sources domains share the same prediction distribution, i.e., $P^s(y|\mathbf{x}) = P^t(y|\mathbf{x})$. However, as illustrated in Figure 1, when the predictive distribution varies across domains, DA and SSL methods may have impaired performance. In this subsection, we propose to use progressive transduction method for acquiring the additional labeled data to estimate $P^t(y|\mathbf{x})$.

Progressive transduction is to progressively label certain number of unlabeled data with pseudo-label which are the most confident predicted outputs in current iteration. These learned pseudo-labeled data are then used to estimate the predictive distribution $P^t(y|\mathbf{x})$. After that, we apply the learned $P^t(y|\mathbf{x})$ to our proposed regularizer (1) for multiple source domains adaptation and train a new classifier with the newly added pseudo-labeled data using (4). The progressive transduction step is then repeated until it reaches the stopping criterion.

In jth iteration, a classifier is trained using the available labeled and pseudo-labeled data using (4). Then the classifier predicts the unlabeled data using decision function (5). Let us group these unlabeled data into their predicted classes and assign them with labels accordingly before sorting the positive set in decreasing order and negative set in increasing order as follows:

$$T_+^j = \{(x_i^u, +)|x_i^u \in D_u^j, f^j(x_i^u) \geq f^j(x_{i+1}^u) \geq 0\}, \tag{6}$$

$$T_-^j = \{(x_i^u, -)|x_i^u \in D_u^j, f^j(x_i^u) \leq f^j(x_{i+1}^u) < 0\}, \tag{7}$$

where $D_u^j = D_u \backslash B^{j-1}$ and B^j are all the pseudo-labeled data from the start of initialization to the current jth iteration, which is defined as follows:

$$B^j = B^{j-1} \cup B_+^j \cup B_-^j, \tag{8}$$

where B^{j-1} is all the pseudo-labeled data from the start of initialization to the $(j-1)$th iteration and the current jth iteration's pseudo-labeled data are from B_+^j and B_-^j which are defined as follows:

$$B_+^j = \{(x_i^u, y_i^u) \in T_+^j | 1 \leq i \leq P_+^j\}, \tag{9}$$

$$B_-^j = \{(x_i^u, y_i^u) \in T_-^j | 1 \leq i \leq P_-^j\}, \tag{10}$$

where $P_+^j = min(p, |T_+^j|)$, $P_-^j = min(p, |T_-^j|)$, p is a hyper-parameter. Hence, the pseudo-labeled set B_+^j contains data with the highest p predictive values in T_+^j and B_-^j contains data with the lowest p predictive values in T_-^j. Therefore, the pseudo-labeled data learned from the current iteration are the most confident predicted labels as they are the furthest away from the decision boundary. Then these pseudo-labeled patterns are being incorporated as part of training set in (4). As these pseudo-labeled data and the target labeled domain are from

Algorithm 1(PDM-SVM)
1. Initialize $B^0 = \emptyset, F^t = \emptyset$, M=m% of unlabeled data
2. While $\|B^{j-1}\| < M$
3. Build B^j using (8)
4. Train the classifier F^t in (4) using D_L and B^j
5. if $\|D_L\| + \|B^j\| >= \Theta$
6. Train a classifier F^t in (4) using D_L^t and B^j

the same domain, their predictive distribution $P^t(y|\mathbf{x})$ is re-estimated to compute W_{ij}^{rc} in (2) for the target domain in each iteration. When m% of the entire unlabeled data are incorporated as part of the training set, the whole progressive transduction process terminates. After that, if the size of the combination of target labeled dataset and pseudo-labeled dataset is larger than a certain threshold Θ, then SVM is trained using only the target labeled data and the pseudo-labeled patterns so the final classifier would consist only target data to represent the true distribution for the target data. The detailed algorithm of PDM-SVM is presented in Algorithm 1. Finally, the final trained classifier is used to classify the test data using decision function (5).

3.6 Demonstration of PDM-SVM on a Synthetic Dataset

Besides the dissimilarity of class distribution among multi-domains, here we also consider the class label inconsistence from multi-domains, where each domain has its own modality. We generate a synthetic dataset (Figure 2) to depict how PDM-SVM uses the predictive distribution similarities between two patterns from different domains to construct a graph. In this dataset, there are three labeled source domains and a target domain that contain only unlabeled data.

Figure 2(a) depicts the target domain. Figure 2(b) shows the first two source domains having their positive and negative data overlapping with the target positive and negative data respectively and the third source domain having its negative data overlapping with the target negative data, but its positive data are near to the target negative data. Intuitively, those target data close to the positive data of the third source will be classified as positive, which is undesirable. However, this is the case for SVM_ST and its decision boundary is depicted in Figure 2(c) by a thinner curve line. As the entire dataset is formed from several domains, and each domain has its own modality, the dataset could become a multi-modality problem. Hence, traditional DA methods which cannot handle multi-modal datasets would fail as the classifier trained with all labeled data has incorrect decision boundary. Whereas, PDM-SVM can classify all the target data correctly, and its decision boundary is shown by the thicker curve line. This is because PDM-SVM can iteratively construct a graph to identify relevant source data as shown in Figure 2(d).

Figure 2 (d) shows that PDM-SVM can construct a graph using the predictive distribution similarity between two patterns from different domains. The lower rectangular box depicts the connections of one target pseudo-labeled learned by PDM-SVM with several source data points which demonstrates that PDM-SVM

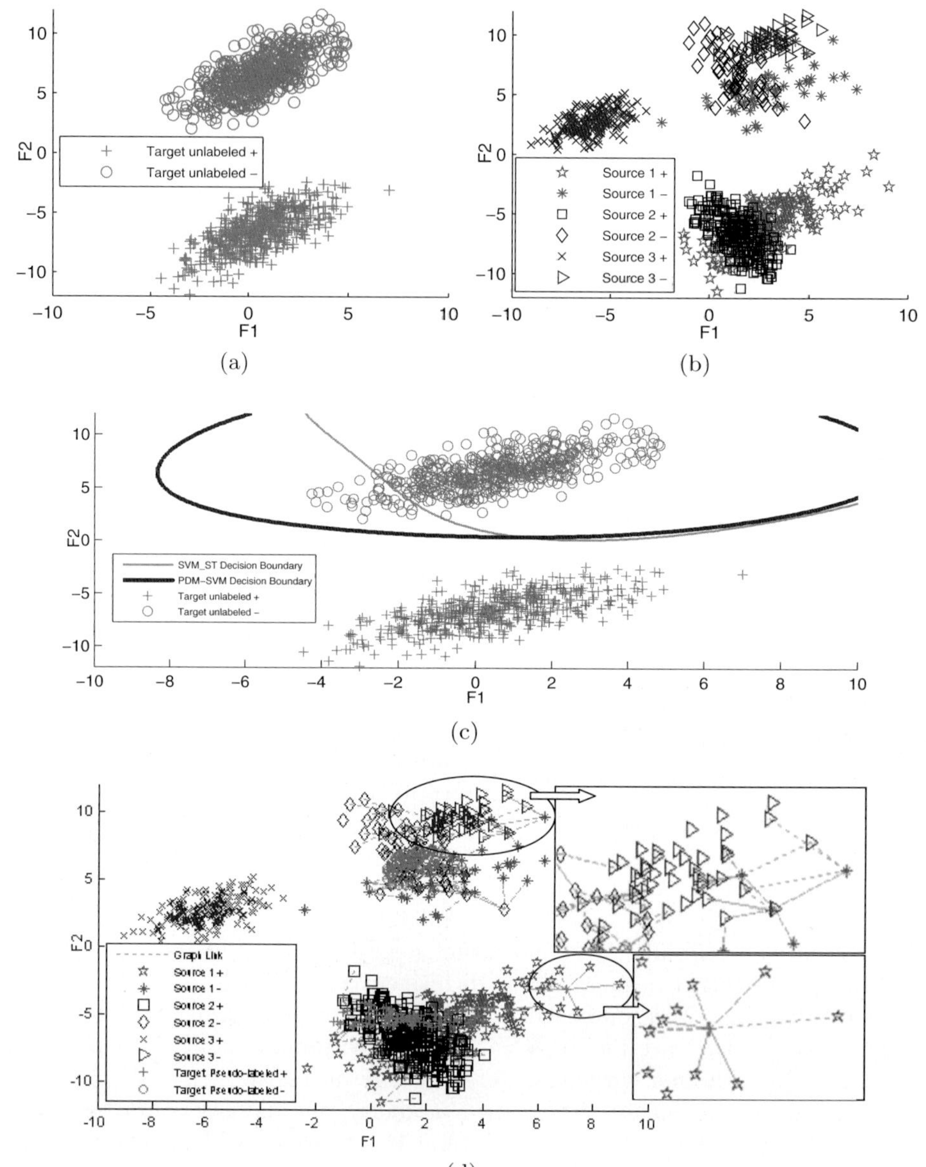

Fig. 2. PDM-SVM demonstration using a Synthetic dataset. The data are in 2-Dimension represented by two features, F1 and F2 respectively.
(a) Target domain with 500 positive and 500 negative instances. (b)Three source domains related to the target domain except the positive data of the third source domain near to the target negative data. Each domain consists of 150 positive and 50 negative instances (c) The decision boundary of SVM_ST and PDM-SVM. The space below the decision boundary is classified as positive, whereas the other side is classified as negative. (d)The graph's connections between labeled and pseudo-labeled data which are learned by PDM-SVM in the final iteration.

can identify pseudo-labeled data from the constructed graph. The upper rectangular box depicts negative data mainly in the third domain where data closer to other data in different domains are connected and data that are further apart from the data in other domains are not connected. When certain regions having many nodes with high predictive distribution values are connected together, these regions can be used to reflect the predictive output of the target regions. Hence, the pseudo-labeled data can be learned from these regions and eventually PDM-SVM can learn a classifier using these pseudo-labeled data.

4 Experiments

In this Section, we investigate several existing state-of-the-art SVM-based methods, DA methods and the proposed PDM-SVM under a multi-domain setting of differing predictive distribution and scare target labels. SSL methods are also considered in the present study to see how they perform when they use target unlabeled data as part of their training. Note that DA methods (e.g. [13]) that require considerable number of target labels to function and cater only to single source domain are omitted. Here, apart from investigating the methods considered in Section 3.2, we also include additional DA learning algorithms such as: M-SVM, [20], is a linear combination of SVMs trained with $\mathcal{D}_L^s$'s and $\mathcal{D}_L^t$; FA, [9], is trained with $\mathcal{D}_L$; DASVM, [5], is trained with $\mathcal{D}_L$ and $\mathcal{D}_U$.

4.1 Experimental Setup

The parameters of the DA methods are configured by means of k-fold cross-source domains validation as suggested in [14] (an extension of k-fold cross validation for domain adaptation) and are tabulated in Table 2. For methods using only labeled data, i.e. SVM_S, SVM_B, SVM_ST, M-SVM and FA, each partition represents a source domain in k-fold cross-source domains validation. For methods using both labeled and unlabeled data, i.e. TSVM, LapSVM, DASVM and PDM-SVM, the kth source domain is used as the labeled data in the kth fold evaluation, while the rest are used as unlabeled data and for validation.

Table 2. Parameter Settings

Classifiers	Parameter Settings
SVM_S, TSVM, SVM_ST, M-SVM, FA & DASVM	C is chosen by cross-validation.
LapSVM & PDM-SVM	γ_A and γ_I are chosen by cross-validation. Using 6 nearest neighbors and normalized Laplacian matrix to construct the graph. The weight for Laplacian matrix is based on cosine distances, as commonly used in text classification.
SVM_T & SVM_B	C is fixed as 1 since the labeled data are limited. For example, in Reuters dataset setting, the target domain will only consist two labeled data.
Other parameters	p in DASVM and PDM-SVM is fixed as 5. β is fixed as 3.10^{-2} for DASVM which is the same value used in [5]. m and Θ for PDM-SVM are set as 20% and 50, respectively.

4.2 Datasets

In the present study, we consider two datasets namely Sentiment and Reuters-21578. Sentiment is a popular multi-domain benchmark dataset, defined in [3]. It is typically used in the context of DA and consisted of even positive and negative class distribution, hence it is used here to synthesize diverse PCR settings for investigating the robustness of SSL and DA methods. Reuters dataset, on the other hand, allows us to study the efficacy of SSL and DA methods in the presence of uneven class distribution in each domain.

In the experimental study, we further pre-process the datasets by extracting only the single-terms, removing all stopwords, normalizing each feature and performing stemming. Finally, each feature of a review is represented by its respective *tf-idf* value, and linear kernel is used in the experiments.

Multi-Domain Sentiment Dataset. It is generated from *Amazon.com* comprising four categories of product reviews: *Book*, *DVDs*, *Electronics* and *Kitchen appliances*. Each category of product review is considered as a domain and comprises of 1000 positive and 1000 negative reviews. For each task, we used one dataset as target domain while the rest as related source domains. For a target domain, we randomly selected 10 positive and 10 negative instances as labeled data and keeping the rest as unlabeled data. In regards to each source domain, we randomly selected 200 to form the labeled data. To study the mismatch in predictive distribution between the source and target domains, 9 different PCR settings are generated for investigations. The 9 PCR settings are chosen from 0.1 to 0.9 at an incremental of 0.1 step size. For example, in a setting of PCR of 0.1, out of the 200 data selected for each source domain, 20 positive data are selected while the rest make up the negative data. To study the performance of an ideal SVM_ST with prior knowledge on class distribution, we consider here additional SVM_B classifier. For each source domain, SVM_B re-samples the data to have the same PCR as the target domain. Let ρ and η be the number of positive and negative samples in each source domain respectively. Since our target unlabeled data has equal number of positive and negative samples, then for each PCR setting, the classifier re-samples both positive and negative samples as $min(\rho, \eta)$ in each source domain. Thus all domains have the same class distribution.

Multi-Domain Reuters Dataset. 3 out of 4 main categories of the dataset namely *People*, *Organizations* and *Exchanges* are considered in the present study, thus resulting 3 tasks being experimented: *People* versus *Organizations*, *People* versus *Exchanges* and *Organizations* versus *Exchanges*. *Places* category is not used due to the vast instances belonging to this category that overwhelms all other categories, thus making the study fruitless. Further, in each main category, the subcategory with largest dataset is used as target domain while the remaining 4 largest subcategories as related source domains. Then in each task, the xth largest subcategory of a main category is labeled as positive while the xth largest subcategory from another main category is labeled as negative. All data in the source domains are used as labeled data and for the target domain, one positive and one negative data are randomly selected to form the labeled data while the rest are used as unlabeled data. Note that this dataset has uneven positive and negative samples in each subcategory, hence the testing distribution is imbalanced and the predictive distribution of the source domains is quite diverse with respect to one another. Furthermore, since it is not always feasible to re-sample the source domains to match the SVM_B setting, it is not considered in the study of this dataset.

4.3 Results and Discussions

We first study the performance of various classifiers with varying PCR in the source domains on Sentiment dataset. For the sake of conciseness, Figure 3 depicts the testing accuracies on Sentiment data for 9 different PCR settings in the source domains, with Electronics and Kitchen Appliances as the target domain. Note that PCR of the target unlabeled dataset is confirmed at approximately 0.5, hence when the PCR of the source domains is also in the region of 0.5, the predictive distribution of the source domains is most likely to be similar to the target unlabeled dataset. The rest of the PCR settings on the other hand would likely result in mismatch of predictive distribution between the source and target domains. Each of the four domains in the Sentiment dataset will take turns to be used as the target domain. Their detailed results for PCR at 0.3, 0.5 and 0.7 are then reported in Table 3.

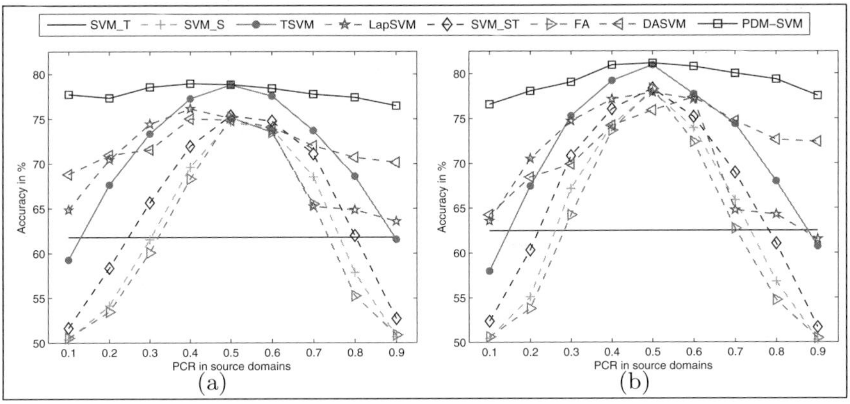

Fig. 3. Testing accuracies on Sentiment dataset for varying PCR in source domains. (a) Target domain is Electronics (b) Target domain is Kitchen Appliances.

As shown in Figure 3, at both extreme ends of the PCR settings, for the same labeled and unlabeled data, LapSVM and TSVM are found to underperform DASVM and PDM-SVM. This is expected since LapSVM and TSVM do not consider the predictive distribution mismatch between different domains. Furthermore, TSVM is shown to underperform SVM_T since $P(y)$ of unlabeled differs significantly from that of labeled data at PCR = 0.1 and 0.9. Apart from the extreme ends of PCR settings, it can also be observed that DA methods including SVM_ST and FA underperform SVM_T on some of the imbalance PCR settings, indicating the presence of negative transfer. This is because both SVM_ST and FA require predictive distribution of the source and target domains to be similar, but the predictive distribution of imbalance PCR settings is quite diverse. From Table 3, when the predictive distribution of the source domains is similar to target domain (i.e. PCR≈0.5), SVM_S, SVM_ST and FA are shown to outperform SVM_T on all datasets. This implies that additional source labeled data can be useful for improving testing accuracy when the predictive distribution between source and target domain matches.

In all PCR settings, SVM_B is reported with better accuracies than many DA methods including SVM_ST, FA, M-SVM and DASVM. This implies re-sampling of the source domains to match the target predictive distribution is important for transfer learning to work well. In contrast, PDM-SVM can be observed to outperform all other classifiers, implying the predictive distribution matching of source and target domains in the PDM-SVM is deemed to be effective. In particular, even under extreme conditions of PCR settings in the source domain, PDM-SVM reported up to 28% accuracy improvements over the other classifiers.

As shown in Figure 3, each classifier displayed similar performance trends on the subgraph where most of the classifiers (excluding LapSVM, DASVM and PDM-SVM) showed sharp declining accuracies when the PCR is skewed toward either extremes. It can be observed that SVM_ST, FA and SVM_S gave the best accuracy of around 75% at PCR=0.5 and the worst accuracy in the

Table 3. Testing accuracies on Sentiment data set for PCR at 0.3, 0.5 and 0.7 in the source domains. The values below the accuracy results are the standard deviation.

Target	PCR	SVM_T	SVM_S	SVM_B	TSVM	LapSVM	SVM_ST	M-SVM	FA	DASVM	PDM-SVM
Book	0.3	58.51	59.91	69.11	68.65	68.89	62.4	57.8	56.27	66.25	**74.47**
		±2.31	±1.25	±1.34	±0.71	±1.24	±1.37	±3.12	±1.48	±4.61	±2.06
	0.5	58.51	70.88	71.77	72.8	71.14	71.77	69.08	71.48	65.35	**74.23**
		±2.31	±1.57	±1.24	±1.19	±1.28	±1.24	±1.55	±1.73	±2.93	±1.17
	0.7	58.51	58.98	69.11	69.39	61.15	61.05	58.96	55.9	62.4	**72.71**
		±2.31	±1.18	±1.34	±0.93	±2.57	±1.27	±2.23	±1.11	±1.11	±1.58
DVDs	0.3	60.1	60.73	72.1	70.05	70.74	63.7	59.74	56.74	67.86	**75.64**
		±2.40	±2.16	±1.24	±0.94	±1.21	±2.24	±3.02	±1.84	±3.61	±1.32
	0.5	60.1	73	73.24	74.3	73.06	73.24	68.75	73.18	71.58	**75.94**
		±2.40	±1.07	±0.93	±1.34	±1.18	±0.93	±2.05	±0.90	±3.51	±1.46
	0.7	60.1	61.04	72.1	71.38	63.03	63.45	60.9	57.53	67.01	**75.19**
		±2.40	±1.86	±1.24	±0.88	±2.55	±2.01	±1.76	±1.44	±6.94	±3.32
Electronic	0.3	61.78	61.54	74.37	73.35	74.42	65.64	61.7	60.07	71.55	**78.55**
		±2.56	±2.34	±1.10	±1.10	±0.93	±1.88	±1.69	±1.95	±2.48	±0.98
	0.5	61.78	74.67	75.36	78.79	75.08	75.36	70.03	75.17	74.88	**78.84**
		±2.56	±1.85	±1.67	±1.14	±1.48	±1.67	±2.59	±1.63	±1.65	±1.05
	0.7	61.78	68.52	74.37	73.7	65.22	71.11	63.67	65.47	72.01	**77.76**
		±2.56	±1.70	±1.10	±1.17	±1.29	±1.64	±2.26	±2.68	±3.11	±1.03
Kitchen	0.3	62.47	67.16	75.8	75.29	74.72	70.84	63.49	64.23	69.91	**79.06**
		±2.62	±1.83	±1.38	±0.75	±1.32	±1.75	±2.59	±1.91	±2.01	±1.12
	0.5	62.47	77.96	78.34	80.94	77.88	78.34	74.85	78.29	75.91	**81.15**
		±2.62	±1.01	±0.96	±1.28	±0.86	±0.96	±0.98	±1.16	±2.06	±1.34
	0.7	62.47	65.88	75.8	74.39	64.75	68.97	62.26	62.7	74.72	**80.01**
		±2.62	±1.96	±1.38	±0.91	±2.37	±1.50	±1.37	±2.27	±3.81	±1.68

region of 50% at PCR=0.1 and 0.9. Note that this marks a large difference in accuracies of up to 25%. Other methods also displayed significant variance in accuracy under the diverse PCR settings. As opposed to existing approaches suffering in performances due to the effect of negative transfer, on the other hand, PDM-SVM can effectively identify useful knowledge from multi-source domains by means of prediction distribution matching, thus achieving robust prediction performances in the target domain on the Sentiment data. In particular, PDM-SVM can still give readily stable results, and the accuracies are within 5% across all the PCR settings. This demonstrates the robustness of PDM-SVM under different PCR settings by benefiting from the positive transfer.

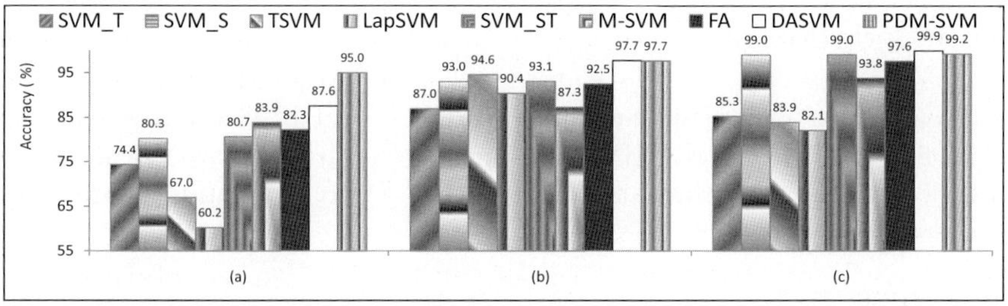

Fig. 4. Testing accuracies on Reuters data sets. (a)*People* versus *Organizations* (b)*People* versus *Exchanges* (c)*Organizations* versus *Exchanges*

Last but not least, we further experiment the classifiers on Reuters dataset with uneven class distribution in each domain. The results are reported in Figure 4. It can be observed that both PDM-SVM and DASVM had outperformed all other classifiers considered, see Figure 4(b,c). PDM-SVM on the other hand is competitive to DASVM. Interestingly, Figure 4(a) also indicated that PDM-SVM attained significant improvement in accuracy over DASVM. In all the experiments, SVM_S is shown to be competitive to some DA methods: SVM_ST, M-SVM and FA. It appears that most of the labels in the source domains are consistent with the target domain. This may be the reason why DASVM had performed well on the Reuters dataset while most DA methods outperforming SSL methods and SVM_T, whereas LapSVM and TSVM outperformed the other counterparts on Sentiment dataset. SSL methods on the other hand had performed much worse than the others on Reuters dataset. This is likely due to the manifold assumption and cluster assumption failing to hold on Reuters dataset. Overall, PDM-SVM is able to perform robustly and outperform all classifiers considered on both datasets, due to success of the predictive distribution matching regularizer in the identification of relevant data from source domains.

5 Conclusion

In this paper, we have presented a formalization of predictive distribution matching for addressing the effects of differing predictive distributions between related domains. We address this problem by leveraging multiple domains to identify high predictive density regions, in which the class label represents the target class label in the same regions. Furthermore, we also present how to estimate the predictive distribution $P^t(y|\mathbf{x})$ of the target domain by using progressive transduction. On the other hand, empirical results obtained showed that while most DA methods suffer from the effect of negative transfer when the problem domains have mismatched predictive distributions, the proposed PDM-SVM reported robust prediction accuracy for diverse levels of PCR results on the dataset considered. In addition, PDM-SVM is shown capable of generating a substantial improvement over existing methods in most cases.

Acknowledgement

This research was supported by Singapore NTU AcRF Tier-1 Research Grant (RG15/08) and NTU and IHPC joint project entitled "Large Scale Domain Adaptation Machines: Information Integration, Revolution and Transfer".

References

1. Belkin, M., Niyogi, P., Sindhwani, V.: Manifold regularization: A geometric framework for learning from labeled and unlabeled examples. JMLR 12, 2399–2434 (2006)
2. Bickel, S., Brückner, M., Scheffer, T.: Discriminative learning under covariate shift. JMLR 10, 2137–2155 (2009)
3. Blitzer, J., Dredze, M., Pereira, F.: Biographies, bollywood, boom-boxes and blenders: Domain adaptation for sentiment classification. In: ACL (2007)
4. Blitzer, J., McDonald, R., Pereira, F.: Domain adaptation with structural correspondence learning. In: EMNLP (2006)
5. Bruzzone, L., Marconcini, M.: Domain adaptation problems: A dasvm classification technique and a circular validation strategy. IEEE Trans. on PAMI 32(5), 770–787 (2010)
6. Chawla, N.V., Bowyer, K.W., Hall, L.O., Kegelmeyer, W.P.: SMOTE: Synthetic minority over-sampling technique. JAIR 16, 321–357 (2002)
7. Chen, Y., Wang, G., Dong, S.: Learning with progressive transductive support vector machine. In: ICDM (2002)
8. Crammer, K., Kearns, M., Wortman, J.: Learning from multiple sources. JMLR 9, 1757–1774 (2008)
9. Daumé III., H.: Frustratingly easy domain adaptation. In: ACL (2007)
10. Duan, L., Tsang, I.W., Xu, D., Chua, T.S.: Domain adaptation from multiple sources via auxiliary classifiers. In: ICML (2009)
11. Duan, L., Tsang, I.W., Xu, D., Maybank, S.J.: Domain transfer svm for video concept detection. In: CVPR (2009)
12. Huang, J., Smola, A., Gretton, A., Borgwardt, K.M., Schölkopf, B.: Correcting sample selection bias by unlabeled data. In: NIPS (2006)
13. Jiang, J., Zhai, C.: Instance weighting for domain adaptation in NLP. In: ACL (2007)

14. Jiang, J., Zhai, C.: A two-stage approach to domain adaptation for statistical classifiers. In: CIKM (2007)
15. Joachims, T.: Transductive inference for text classification using support vector machines. In: ICML (1999)
16. Luo, P., Zhuang, F., Xiong, H., Xiong, Y., He, Q.: Transfer learning from multiple source domains via consensus regularization. In: CIKM (2008)
17. Pan, S.J., Tsang, I.W., Kwok, J.T., Yang, Q.: Domain adaptation via transfer component analysis. In: IJCAI (2009)
18. Pan, S.J., Ni, X., Sun, J.T., Yang, Q., Chen, Z.: Cross-domain sentiment classification via spectral feature alignment. In: WWW (2010)
19. Rosenstein, M.T., Marx, Z., Kaelbling, L.P.: To transfer or not to transfer. In: NIPS 2005 Workshop on Inductive Transfer: 10 Years Later (2005)
20. Schweikert, G., Widmer, C., Schölkopf, B., Rätsch, G.: An empirical analysis of domain adaptation algorithm for genomic sequence analysis. In: NIPS (2009)
21. Wu, P., Dietterich, T.G.: Improving SVM accuracy by training on auxiliary data sources. In: ICML (2004)
22. Zhu, X.: Semi-supervised learning literature survey. Technical report, Computer Sciences Technique Report 1530, University of Wisconsin-Madison (2009)

Kantorovich Distances between Rankings with Applications to Rank Aggregation

Stéphan Clémençon[1] and Jérémie Jakubowicz[1]

LTCI, Telecom Paristech (TSI) - UMR Institut Telecom/CNRS 5141
stephan.clemencon@telecom-paristech.fr,
jeremie.jakubowicz@telecom-paristech.fr

Abstract. The goal of this paper is threefold. It first describes a novel way of measuring disagreement between rankings of a finite set $\mathcal{X}$ of $n \geq 1$ elements, that can be viewed as a (mass transportation) *Kantorovich metric*, once the collection rankings of $\mathcal{X}$ is embedded in the set $\mathcal{K}_n$ of $n \times n$ *doubly-stochastic* matrices. It also shows that such an embedding makes it possible to define a natural notion of *median*, that can be interpreted in a probabilistic fashion. In addition, from a computational perspective, the convexification induced by this approach makes median computation more tractable, in contrast to the standard metric-based method that generally yields NP-hard optimization problems. As an illustration, this novel methodology is applied to the issue of ranking aggregation, and is shown to compete with state of the art techniques.

1 Introduction

Formulated more than two centuries ago in the context of emerging social sciences and voting theories [Fis73], the problem of aggregating binary relations, (pre-) orders in particular, has recently received much attention in the machine-learning literature, see [HFCB08], [FKM+03] or [MPPB07] for instance. Various modern applications sparked off the revival of interest in this issue, ranging from e-commerce to information retrieval through spam-fighting and database middleware. Indeed, in a wide variety of information systems now, input or output data take the form of an ordered list of items: search-engines, recommending systems, *etc.* Numerous tasks such as the design of meta-search engines, collaborative filtering, or combining results from multiple databases have motivated the development of new results in this domain, dedicated to three topics essentially: the extension of the notion of consensus among rankings [FKM+06], the design of efficient algorithmic procedures for computing such median rankings, see [MM09] or [BFB09], and the building of probabilistic models on sets of rankings [LL03].

The present paper addresses all these aspects of the consensus problem, from an original angle. Its primary purpose is to show how the problem of measuring disagreement between rankings can be cast in terms of discrete *mass transportation problems*, by embedding the set of permutations in a convenient convex set of matrices. We prove that the continuum of metrics thus defined includes some

J.L. Balcázar et al. (Eds.): ECML PKDD 2010, Part I, LNAI 6321, pp. 248–263, 2010.

classical permutation metrics, such as the Hamming distance, the Spearman ρ distance or the Spearman footrule distance. From the perspective of rank aggregation, a novel (probabilistic) notion of median is next defined and related computational issues are tackled, taking advantage of the convexification step.

The paper is organized as follows. Notations are set out in Section 2, where most concepts involved in the subsequent analysis are introduced and an example motivating the present approach is also discussed. A novel way of measuring agreement between rankings is then proposed in Section 3, together with a definition of a probabilistic version of the notion of median ranking in Section 4. Results describing the computational complexity of the aggregation method proposed are stated in Section 5, while an illustrative application is presented in Section 6 Technical details are deferred to the Appendix.

2 Preliminary Background

It is the purpose of this section to introduce the main concepts and definitions that shall be used throughout the paper.

2.1 First Definitions and Notation

We start off by recalling some definitions and setting out the notations needed in the subsequent analysis. Here and throughout, $\mathbb{I}\{\mathcal{E}\}$ denotes the indicator function of any event $\mathcal{E}$.

Rankings and matrix spaces. Let $n \geq 1$. We denote by $\mathfrak{S}_n$ the symmetric group of order n, *i.e.* the group of permutations of $\{1, \ldots, n\}$, and by $\mathcal{M}_n(\mathbb{R})$ the space of $n \times n$ matrices with real entries. Any permutation $\sigma \in \mathfrak{S}_n$ can be classically represented by the matrix

$$M^\sigma = (\mathbb{I}\{\sigma(i) = j\})_{1 \leq i,\, j \leq n}\,,$$

in $\mathcal{M}_n(\mathbb{R})$, whose entry $M^\sigma_{i,j}$ indicates whether rank j is assigned to the object indexed by i or not. The elements of the set $\Sigma_n = \{M_\sigma :\ \sigma \in \mathfrak{S}_n\}$ are called *permutation matrices.*

Medians. Given a collection $\Pi = \{\sigma_1, \ldots, \sigma_K\} \subset \mathfrak{S}_n$ of permutations (one commonly uses the term *profile* in social choice theory), the issue of summarizing the orders defined by Π's elements, by a "consensual" (pre-) order, is called the *aggregation problem.* The so-termed *metric approach* is the most popular method for defining such a consensus. It assumes that a certain distance δ on the set $\mathfrak{S}_n$ is given. One calls a *median ranking* for the profile Π with respect to a subset $\mathcal{R} \subset \mathfrak{S}_n$ any ranking $\sigma^* \in \mathcal{R}$ such that:

$$\sum_{k=1}^{K} d(\sigma^*, \sigma_k) = \min_{\sigma \in \mathcal{R}} \sum_{k=1}^{K} d(\sigma, \sigma_k). \tag{1}$$

The study of metrics on rankings has a long history, for instance one may refer to Chapter 11 in [DD09] for an excellent account of distances on permutations. The following distances, originally introduced in the context of nonparametric hypothesis testing, are among the most widely used.

- **The Kendall τ distance.** Counting the number of "discording pairs", it is given by: $\forall(\sigma_1, \sigma_2) \in \mathfrak{S}_n^2$,

$$d_\tau(\sigma_1, \sigma_2) = \sum_{1 \leq i < j \leq n} \mathbb{I}\{(\sigma_1(i) - \sigma_2(i)) \cdot (\sigma_1(j) - \sigma_2(j)) < 0\}.$$

- **The Spearman ρ distance.** It corresponds to the l_2-metric: $\forall(\sigma_1, \sigma_2) \in \mathfrak{S}_n^2$,

$$d_2(\sigma_1, \sigma_2) = \left(\sum_{i=1}^{n} (\sigma_1(i) - \sigma_2(i))^2\right)^{1/2}.$$

- **The Spearman footrule distance.** This is actually the l_1-distance between rank vectors: $\forall(\sigma_1, \sigma_2) \in \mathfrak{S}_n^2$,

$$d_1(\sigma_1, \sigma_2) = \sum_{i=1}^{n} |\sigma_1(i) - \sigma_2(i)|.$$

- **The Hamming distance.** This is the l_0-distance between rank vectors: $\forall(\sigma_1, \sigma_2) \in \mathfrak{S}_n^2$,

$$d_0(\sigma_1, \sigma_2) = \sum_{i=1}^{n} \mathbb{I}\{\sigma_1(i) \neq \sigma_2(i)\}.$$

Many other distances could be considered, such as the Cayley/Kemeny distance [Kem59], or so-termed *word metrics* more generally [How00]. The major barrier to practical implementation of this approach lies in the fact that it generally leads to NP-hard problems, see [Hud08] or [Wak98]. Notice in addition that uniqueness of the median is not guaranteed in general. One may easily check for instance that, considering the Kendall τ distance, any permutation $\sigma \in \mathfrak{S}_n$ is a median with respect to the set $\mathfrak{S}_n$ (see also the example given below).

Remark 1. (The ordinal approach) Metric-based techniques are by no means the sole approach to rank aggregation. The so-termed "ordinal approach" includes a wide variety of techniques for combining rankings or, more generally, binary relations. They return to the famous "Arrow's voting paradox" and consist of a series of duels (i.e. pairwise comparisons) as in Condorcet's methods or successive tournaments as in the celebrated proportional voting Hare system. Special attentions has recently been paid to such techniques in the context of preference learning ("Ranking by Pairwise Comparison" methods); see [HFCB08] for instance.

2.2 A Simple Example

The following example shows that, beyond the computational difficulties above mentioned, the metric-based approach may have important drawbacks. Let us regard the problem of aggregating/summarizing the permutations described by the rank vectors (1, 2, 3) and (3, 2, 1) in $\mathfrak{S}_3$ for instance. Considering Kendall τ medians with respect to $\mathfrak{S}_3$ is clearly not informative, any permutation except those two permutations being a median. Looking at the hexagon in Fig. 1, providing a natural representation of $\mathfrak{S}_3$ (adjacent vertices are at Kendall τ distance one from each other), one inevitably longs to define the median in the middle of the line segment connecting the opposite vertices. In other terms, the major drawback of the aforementioned metric-based approach does not lie in the metric considered itself, but rather in the fact that the search for a "barycenter" is restricted to the "curve-shaped" set $\mathfrak{S}_n$. The view developed subsequently provides a rigorous meaning to a definition of a median in the interior of the hexagon. We are going to incorporate some uncertainty/fuzziness to the notion of median ranks by enlarging/convexifying the original ensemble $\mathfrak{S}_n$, and next define well adapted metrics on the larger space thus obtained.

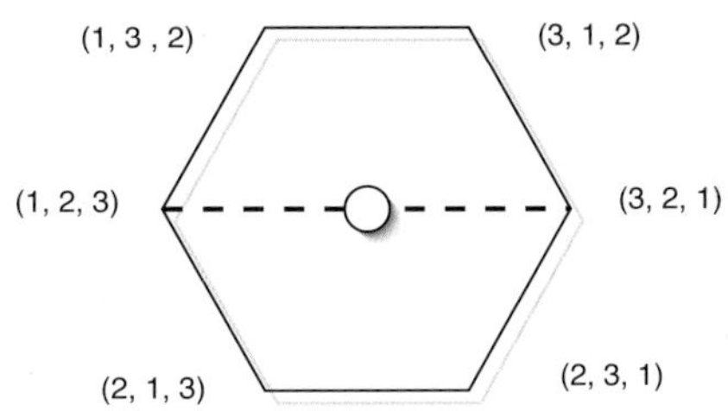

Fig. 1. Representation of the symmetric group $\mathfrak{S}_3$ as a regular hexagon

2.3 Convexification/Randomization

For clarity, we first recall the following definition.

Definition 1. (Double stochasticity) *A matrix* $A = (a_{i,j}) \in \mathcal{M}_n(\mathbb{R})$ *with nonnegative entries is said to be doubly stochastic if and only if*

$$\forall i \in \{1, \ldots, n\}, \ \sum_{j=1} a_{i,j} = \sum_{j=1} a_{j,i} = 1.$$

The set $\mathcal{K}_n$ *of such doubly stochastic matrices is a convex subset of* $\mathcal{M}_n(\mathbb{R})$.

Permutation matrices are special cases of doubly stochastic matrices. For clarity, let us recall the following celebrated result (see, for instance, [HJ85, p.539]).

Theorem 1 (Birkhoff–Von Neumann). *The set* $\mathcal{K}_n$ *is the convex hull of the set of permutation matrices* Σ_n*:*

$$\mathcal{K}_n = \mathrm{conv}\left(\Sigma_n\right).$$

In addition, Σ_n *corresponds to the set of* $\mathcal{K}_n$*'s extremal points.*

Identifying $\mathfrak{S}_n$ with Σ_n, its embedding in $\mathcal{K}_n$ is a natural way of "convexifying" the rank aggregation problem. From the ranking perspective, the entry $a_{i,j}$ of a doubly stochastic matrix A can be interpreted as a marginal probability that rank j be assigned to the object No. i. Two standard ways of randomly generating a ranking from such a matrix are reviewed in subsections 4.1 and 4.2.

3 Kantorovich Distances

We now introduce a general framework for measuring dissimilarity between rankings, following in the footsteps of the so-termed *mass transportation* approach to defining metrics between probability measures [Rac91].

3.1 Definitions and Properties

We suppose now that we are given a certain (nonnegative) *cost function*, that is to say a mapping $c : \{1, \ldots, n\}^2 \times \{1, \ldots, n\}^2 \to \mathbb{R}_+ \cup \{+\infty\}$, $c((i,j),(k,l))$ representing the cost of moving one mass unit from (i,j) to (k,l). The technical conditions listed below shall be required in the subsequent analysis.

(*i*) (Diagonal terms) For all (i,j) in $\{1, \ldots, n\}^2$,

$$c\left((i,j),(i,j)\right) = 0.$$

(*ii*) (Symmetry) For all (i,j), (k,l) in $\{1, \ldots, n\}^2$,

$$c\left((i,j),(k,l)\right) = c\left((k,l),(i,j)\right).$$

(*iii*) (Triangular inequality) The cost function c on $\{1, \ldots, n\}^2$ fulfills the condition: for all (i,j), (k,l) and (s,t) in $\{1, \ldots, n\}^2$,

$$c((i,j),(k,l)) \leq c((i,j),(s,t)) + c((s,t),(k,l)).$$

(*iv*) (Non diagonal terms) For all $(i,j) \neq (k,l)$ in $\{1, \ldots, n\}^2$,

$$c\left((i,j),(k,l)\right) > 0.$$

Remark 2. Condition (*iii*) guarantees that the cost function c satisfies the (stronger in appearance) *reduction property*, meaning that

$$c((i,j),(k,l)) = \inf_{h \geq 1} c^h((i,j),(k,l)),$$

where, denoting by $\mathcal{P}_m((i,j),(k,l))$ the set of all paths of length m, $\{(u_m, v_m) : m = 0, \ldots, \}$, connecting (i,j) to (k,l), *i.e.* such that $(u_0, v_0) = (i,j)$ and $(u_{h+1}, v_{h+1}) = (k,l)$, we set: $\forall h \geq 1$,

$$c^h((i,j),(k,l)) = \inf \left\{ \sum_{m=1}^{h+1} c((u_{m-1}, v_{m-1})(u_m, v_m)) : (u,v) \in \mathcal{P}_m((i,j),(k,l)) \right\}.$$

Roughly, reduction amounts to state that no mass movement should be cheaper whenever performed in several steps rather than in one step.

Equipped with the notion of (symmetric and reduced) cost function, we may now define the concept of Kantorovich pseudo-metric on $\mathcal{K}_n$.

Proposition 1. (MASS TRANSPORTATION DISTANCE) *Let c be a cost function on $\{1, \ldots, n\}^2$ fulfilling conditions (i) – (iii), $A = (a_{i,j})$ and $A' = (a'_{i,j})$ two elements of $\mathcal{K}_n^2$. If one defines the Kantorovich optimal transportation cost related to cost function c and real value $p \geq 1$ by:*

$$d_{c,p}(A, A') = \min_{\Phi \in \mathcal{M}(A,A')} \mu_{c,p}^{1/p}(\Phi), \tag{2}$$

with

$$\mu_{c,p}(\Phi) = \sum_{\substack{(i,j)\in\{1,\ldots,n\}^2 \\ (k,l)\in\{1,\ldots,n\}^2}} c\left((i,j),(k,l)\right)^p \Phi\left((k,l),(i,j)\right),$$

and where the set $\mathcal{M}(A, A')$ denotes the collection of mappings ("transportation plans") $\Phi : \{1, \ldots, n\}^2 \times \{1, \ldots, n\}^2 \to [0, 1]$ such that: $\forall (i,j) \in \{1, \ldots, n\}^2$,

$$\sum_{(k,l)\in\{1,\ldots,n\}^2} \Phi((i,j),(k,l)) = a_{i,j} \text{ and } \sum_{(k,l)\in\{1,\ldots,n\}^2} \Phi((k,l),(i,j)) = a'_{i,j}. \tag{3}$$

Then $d_{c,p}$ is a pseudo-metric on $\mathcal{K}_n$: it satisfies the separability, symmetry and triangular inequality properties, but might fail to be always finite.

Obviously, $d_{c,p}$ being a pseudo-metric on $\mathcal{K}_n$, it is a pseudo-metric on $\mathfrak{S}_n$ (identified as Σ_n) as well, we set in this case $d_{c,p}(M^{\sigma_1}, M^{\sigma_2}) = d_{c,p}(\sigma_1, \sigma_2)$, with a slight abuse of notation. Before showing several important examples of such pseudo-metrics, a few remarks are in order.

Remark 3. (NORMALIZATION) For the sake of simplicity, this definition above uses a slightly different convention than in the classical mass transportation setting (see, for instance, [RR98, Vil09]). Indeed A and A' do not define probability measures on $\{1, \ldots, n\}^2$, their mass with respect to the counting measure being equal to n (it would simply suffice to divide the latter by n for leading back to the usual setup).

Remark 4. (MONGE VS. KANTOROVICH) When the search for transportation plans with minimum cost is restricted to plans Φ taking their values in $\{0, 1\}$ (one does not try to divide the mass described by the entries of the initial matrix to transport it, assigning new locations to the original entries being sufficient in this situation), the problem is said of *Monge*'s type. We point out that, even when both the initial and final distributions of mass are described by permutation matrices M^{σ_1} and M^{σ_2}, OTP's for the Kantorovich problem are not Monge transportation plans in general. Indeed, consider for instance the simple case where $n = 2$ and the cost is constant, equal to some fixed scalar $\gamma > 0$, except on the diagonal $\{(i,j) = (k,l)\}$ where it is 0 (as required by condition (*i*)). It is easy to see that the optimal transportation cost between $\iota = (1\ 2)$ and $\tau = (2\ 1)$ is $2^{1/p}\gamma$ (identifying Σ_n with $\mathfrak{S}_n$). Additionally, observe that every transportation plan achieves this cost and only two of them are of Monge type.

Remark 5. (ON UNIQUENESS) It should be pointed up that optimal transportation plans are not necessarily unique (refer for instance to the example mentioned in the preceding remark).

Remark 6. (RANKING STABILITY) In the same way as Wasserstein-Kantorovich probability metrics turned to be quite adapted to the study of stability of stochastic models such as queuing systems (see [Rac91]), the optimal properties of distances $d_{c,p}(.,.)$ make them very useful for investigating the stability of ranking algorithms/models in a proper way. By a ranking algorithm, we mean here a mapping $\sigma : \mathcal{D} \mapsto \sigma_{\mathcal{D}}$ that assigns a permutation $\sigma_{\mathcal{D}} \in \mathfrak{S}_n$ to any training data sample $\mathcal{D}_N$ of size $N \geq 1$, allowing for ranking n objects, indexed by $i = 1, \ldots, n$. The nature of the sample may vary depending on the application considered (collaborative filtering or nonparametric scoring for instance), it can be made of preferences, rankings, binary data, *etc.* (see [CV09] and the references therein for instance). The definition below does not require to specify the nature of the training data however. Given a cost function c on $\{1, \ldots, n\}^2$, we define the instability measure as:

$$\mathbf{Instab}_N(\sigma) = \mathbb{E}_{\mathcal{D}_N, \mathcal{D}'_N}\left[d_{c,p}(\sigma_{\mathcal{D}_N}, \sigma_{\mathcal{D}'_N})\right],$$

where $\mathcal{D}'_N$ denotes an independent copy of the sample and $\mathbb{E}_{\mathcal{D}_N, \mathcal{D}'_N}[.]$ denotes the expectation taken with respect to $(\mathcal{D}, \mathcal{D}')$. We mention incidentally that such an instability measure can be estimated through a standard resampling scheme. By drawing data with replacement among the original sample, one may get $B \geq 1$ bootstrap replicates $\mathcal{D}^{*(1)}, \ldots, \mathcal{D}^{*(B)}$ of the sample $\mathcal{D}_N$. A bootstrap estimate of $\mathbf{Instab}_N(\sigma)$ is then given by:

$$\widehat{\mathbf{Instab}}_N(\sigma) = \frac{2}{B(B-1)} \sum_{1 \leq b < b' \leq B} d_{c,p}(\sigma_{\mathcal{D}^{*(b)}}, \sigma_{\mathcal{D}^{*(b')}}).$$

3.2 Examples

As proof of relevance of the approach embraced in this paper, we now show that some widely used metrics for measuring disagreement between rankings on $\{1, \ldots, n\}$, can be viewed as restrictions to Σ_n of a Kantorovich distance (for an adequate choice of the cost function). A few important examples are listed below.

1. **Hamming distance.** It corresponds to the cost function

$$c_{\mathcal{H}}((i,j),(k,l)) = \begin{cases} 0 & \text{if } i = k, j = l \\ 1 & \text{if } i = k, j \neq l \\ +\infty & \text{otherwise} \end{cases},$$

with $p = 1$ in the sense that: $\forall (\sigma_1, \sigma_2) \in \mathfrak{S}_n^2$, $\delta_{\mathcal{H}}(\sigma_1, \sigma_2) = d_{c_{\mathcal{H}},1}(\sigma_1, \sigma_2)$.

2. **Spearman footrule distance.** It corresponds to the cost function

$$c((i,j),(k,l)) = \begin{cases} |j-l| & \text{if } i=k \\ +\infty & \text{otherwise} \end{cases},$$

with $p = 1$.
3. **Spearman ρ metric.** It corresponds to the same cost function as above, except that $p = 2$ here.

The assertions above are easy to prove, $\mathcal{M}(M^{\sigma_1}, M^{\sigma_2})$ containing, in each case, a single element only.

Beyond the fact they can be seen as extensions of numerous permutation distances, the major advantage of the collection of Kantorovich pseudo-metrics lies in the considerable flexibility it provides for measuring disagreement between rankings. By choosing the cost properly, one may attach much more importance to the top ranks than to the others for instance, which makes sense in various rank aggregation tasks.

However, we suspect that not every classical distance on permutations can be recovered as a Kantorovich distance. Let us first introduce the following notions.

Definition 2. *A function f defined on $\mathcal{K}_n^2$ is said 'right-invariant' (respectively, 'left-invariant'), when: $\forall \sigma \in \mathfrak{S}_n$, $\forall A \in \mathcal{K}_n$,*

$$f(A\cdot\sigma, A'\cdot\sigma) = f(A, A') \quad (\textit{respectively,} \quad f(\sigma\cdot A, \sigma\cdot A') = f(A, A'))$$

where $A\cdot\sigma = (A_{i,\sigma(j)})$ (respectively, $\sigma\cdot A = (A_{\sigma(i),j})$). The function f is said bi-invariant when it is right-invariant and left-invariant both at the same time.

Equipped with these definitions, we state the following result, relating invariance properties of a cost function to those of the related pseudo-metric. Owing to space limitations, the proof is omitted and left to the reader.

Proposition 2. *Let c be a cost function fulfilling conditions $(i)-(iv)$ and n denote a large enough integer. The pseudo-metric d_c is bi-invariant if and only if the function c is bi-invariant when $\mathfrak{S}_n$ acts on $\{1,\ldots,n\}^2$ on the right by $(i,j)\cdot\sigma = (i,\sigma(j))$ (respectively, on the left by $\sigma\cdot(i,j) = (\sigma(i),j)$).*

Corollary 1. *There exists a nonnegative integer n_0, such that for all $n \geq n_0$, the Cayley distance between two permutations in $\mathfrak{S}_n$ (defined as the minimum number of transpositions to be composed with one of them to turn it into the other) is not the restriction to Σ_n of any Kantorovich distance on $\mathcal{K}_n$.*

4 From Medians in $\mathcal{K}_n$ to Median Rankings

Now the concept of Kantorovich metric between rankings has been introduced, our main goal is to use it in order to define and compute *medians*, summarizing a profile of rankings, in $\mathcal{K}_n$ first, and in $\mathfrak{S}_n$ next.

Definition 3. *Let $n \geq 1$ and let $A_1, \ldots, A_N$ be $N \geq 1$ elements of $\mathcal{K}_n$. Any matrix $A^* \in \mathcal{K}_n$ such that*

$$\sum_{m=1}^{N} d_{c,p}(A^*, A_m) = \inf_{A \in \mathcal{K}_n} \sum_{m=1}^{N} d_{c,p}(A, A_m) \tag{4}$$

is called a median matrix for the profile $\{A_1, \ldots, A_N\}$.

Remark 7. (On existence) We point out that medians, in the sense of Definition 3, always exist. Indeed, for any $(A_1, \ldots, A_N) \in \mathcal{K}_n^N$, $N \geq 1$, the mapping $A \in \mathcal{M}_n(\mathbb{R}) \mapsto \sum_{m=1}^{N} d_c(A, A_m)$ is continuous, the infimum over the compact set $\mathcal{K}_n$ being thus achieved. In contrast, regarding uniqueness, we underline that in general several medians may exist.

By means of this definition, given a profile $(A_m)_{1 \leq m \leq N}$ in $\mathfrak{S}_n$, we end up with a summary median matrix A^* in $\mathcal{K}_n$, which, in general, does not lie in Σ_n. This is the convexification step. When trying to summarize the statistical properties of the profile, all the useful information is encoded in this 'central matrix'. However, when willing to perform certain specific tasks related to rank aggregation, it is desirable to recover a ranking, not a matrix in $\mathcal{K}_n$. We review below two popular approaches for building a 'median ranking' based on a median matrix.

4.1 The Mallows Model

Let $A^* = (a^*_{i,j}) \in \mathcal{K}_n$ be fixed. A flexible approach is to generate a ranking σ, of which permutation matrix M^σ is 'close to A^*' (in the sense of a Kantorovich pseudo-metric $d_{c,p}$), consists in drawing at random an element from $\mathfrak{S}_n$ so that the smaller the distance $d_{c,p}(A^*, M^\sigma)$, the larger the probability of occurence. This is exactly the purpose of the celebrated Mallows model [Mal57, Dia89], that consists, in our context, to consider the distribution given by: $\forall \sigma \in \mathfrak{S}_n$,

$$\mathbb{P}\{\sigma\} = \frac{1}{Z} \exp\left(-\theta d_{c,p}(M^\sigma, A^*)\right),$$

where Z is a normalization constant and θ is a positive parameter. When $1/\theta$ is small compared to the distance $d_{c,p}(A^*, \Sigma_n)$ only the nearest profiles are given a chance to be drawn, when it is large more distant profile are likely to appear. The main drawback of this model lies in its huge computational complexity, in $O(n!)$ namely.

When $1/\theta$ tends to 0, the Mallows model degenerates towards the uniform distribution on the set $\{\sigma \in \mathfrak{S}_n : d_{c,p}(A^*, M^\sigma) = d_{c,p}(A^*, \Sigma_n)\}$, where we set $d_{c,p}(A^*, \Sigma_n) = \min_{M \in \Sigma_n} d_{c,p}(A^*, M)$ by definition.

4.2 Variants of the Luce Model

Probabilistic approach. A more lightweight approach consists in reinterpreting the median matrix entries as scores and relying then on an adaptation of the *Luce model*, see [Luc59, Pla75]. More precisely, we start off with choosing randomly the object i_1 ranked first, by drawing according to the distribution on $\{1, \ldots, n\}$ defined by the column $(a^*_{i,1})_{1\le i\le n}$ of A^*, then we generate the object ranked second among the remaining ones $\{1, \ldots, n\} \setminus \{i_1\}$ according to the distribution $(a^*_{i,1} + a^*_{i,2})/(2 - a^*_{i_1,1})$ for $i \neq i_1$, and so on and so forth. Formally, the distribution of the ranking σ drawn from this model can be written as,

$$\mathbb{P}\left\{\sigma^{-1}(1,\ldots,n) = (i_1, \ldots, i_n)\right\} = \prod_{k=1}^{n} f_k(i_k;\, i_{k-1}, \ldots, i_1),$$

for all permutation $(i_1, \ldots, i_n)$ of $(1, \ldots, n)$, where

$$f_k(i_k;\, i_{k-1}, \ldots, i_1) = \frac{\sum_{j=1}^{k} a^*_{i_k,j}}{\sum_{i\notin\{i_1,\ldots,i_{k-1}\}} \sum_{j=1}^{k} a^*_{i,j}}.$$

Of course, in a dual fashion, we could draw at random the rank assigned to the object 1, according to the distribution $(a^*_{1,j})_{1\le j\le n}$, *etc.*

Greedy approach. A greedy version of the approach described above can also be considered. Precisely, it consists in using the model

$$\mathbb{P}\left\{\sigma^{-1}(1, \ldots, n) = (i_1, \ldots, i_n)\right\} = \prod_{k=1}^{n} g_k(i_k;\, i_{k-1}, \ldots, i_1)$$

where

$$g_k(i_k;\, i_{k-1}, \ldots, i_1) = \frac{\mathbb{I}\left\{\sum_{j=1}^{k} a^*_{i_k,j} = \max_{i\notin\{i_1,\ldots,i_{k-1}\}} \sum_{j=1}^{k} a^*_{i,j}\right\}}{\sum_{i'\notin\{i_1,\ldots,i_{k-1}\}} \mathbb{I}\left\{\sum_{j=1}^{k} a^*_{i',j} = \max_{i\notin\{i_1,\ldots,i_{k-1}\}} \sum_{j=1}^{k} a^*_{i,j}\right\}}$$

The computational cost of both procedures is linear in the size of A^* (*i.e.* $O(n^2)$), insofar as the sums $s_{i,k} = \sum_{j=1}^{k} a_{i,j}$ are precomputed.

5 Computational Aspects

It is the purpose of this section to investigate the computional complexity of the key ingredients of the methodology proposed precedingly: distances and medians.

5.1 Computing Kantorovich Distances in $\mathcal{K}_n$

To compute $d_{c,p}$ with no other assumptions on the cost c than (i), (ii), (iii) and (iv), we make use of linear programming via an interior-point method [BGLS03], which solves the minimization problem (2) in weak polynomial time. If c only takes integer values, then the Edmonds-Karp algorithm is known to solve the problem in $O(n^2(m + n^2 \log n))$ time, m denoting the number of entries $((i,j),(k,l))$ where the cost is finite (there are at most n^4 such entries), see [KV00]. However, for some specific choices of c, the computational cost may reduce to $O(n)$, as it is the case for the Hamming and Spearman distances restricted to Σ_n, as already seen in Section 3.

5.2 Computing Median Matrices

In absence of any further assumptions on c (except conditions $(i)-(iv)$) and p, we resort to general non-linear programming method for solving the minimization problem (4), with no guarantee of convergence to an optimal solution. For specific cost functions and configurations, one can compute the median matrix efficiently, in polynomial time, as claimed in the next proposition.

Proposition 3. *Let $N \in \mathbb{N}^*$ and σ_m are permutations of $\mathfrak{S}_n$ for $m \in 1 \ldots n$. Assume also that $\forall i \in 1 \ldots n$, $m \neq m'$ implies $\sigma_m(i) \neq \sigma_{m'}(i)$. Then, equipped with the cost $c_{\mathcal{H}}$, $\bar{M} = 1/n \sum_{m=1}^{N} M^{\sigma_m}$ is a median matrix of $(M^{\sigma_m})_{m \in 1 \ldots N}$ for $d_{c_{\mathcal{H}}}$. It can be computed with complexity $O(n^2 N \wedge nN \log(nN))$.*

6 Applications to Rank Aggregation

The advantages of the approach to the 'consensus issue' proposed in the preceding sections are now illustrated on numerical experiments related to the so-termed 'rank aggregation task' for meta-search engines in the context of Information Retrieval (IR) applications.

6.1 Datasets

Our experimental study is based on the LETOR database ([LXQ+07]). It is a public data repository, created for evaluating ranking algorithms. There are two 'rank aggregation' datasets in LETOR: `MQ2007-agg` and `MQ2008-agg`. Both datasets are of the same format, but differ in size and in number of ranks to be aggregated. A query q is submitted to N search engines; each engine outputs a list of pairs (document, score). Each line of the database is of the form:

```
Relevance   QueryId   ScoreEngine1 ...   ScoreEngineN   DocId
```

where *Relevance* is an integer between 0 (irrelevant) and 2 (highly relevant). With our previous notations, n changes at each query, we denote it by n_q, whereas N is constant within the dataset (21 for `MQ2007-agg`, 25 for `MQ2008-agg`). Table 1 contains some basic statistics about the data: the number of queries, engines, the average number of documents by query and the max number of documents by query. The goal is to find, for each query in the dataset, the best possible ranking, ranking accuracy being assessed by the means of the Normalized Discounted Cumulative Gain (NDCG) measure ([JK02]). Recall that the DCG, up to rank r, is defined by

$$\mathrm{DCG_r}(\sigma) = \sum_{\mathrm{i}=1}^{\mathrm{r}} \frac{2^{\mathrm{rel}\left(\sigma^{-1}(\mathrm{i})\right)} - 1}{\log_2(1+\mathrm{i})},$$

where rel denotes the relevance of document i for a given query, and σ the sought aggregated ranking. Now $\mathrm{NDCG_r}$ is the same quantity normalized so that we have $\mathrm{NDCG_r}(\sigma) = 1$ for the best ranking.

LETOR provides a benchmark with the `BordaCount` method ([AM01]).

Table 1. Datasets statistics

dataset	MQ2007-agg	MQ2008-agg
#queries	1692	784
#engines	21	25
#avgdocs	41.1	19.4
#maxdocs	147	121

6.2 Implementation Details

The (huge) size of the datasets considered lead us to rule out general cost functions and the use of the Mallows model. Instead, we chose the Hamming cost $c_{\mathcal{H}}$ and used Proposition 3 for median computations (when the hypothesis of disjoint supports of Proposition 3 is not satisfied, the computed 'median' solely consists of an approximation of the optimum). We then tested and compared the models of section 4.2 (except the Mallows model, too demanding in regards to the dataset size) to extract a ranking from the computed median matrix. For the degenerate Mallows model, we used the Hungarian method (Kuhn-Munkres algorithm) [Mun57]. It has complexity $O(n_q^3)$, where n_q denotes the number of documents associated to query q.

6.3 Results

LETOR datasets are organized into training, validation and test set. However, since, like `BordaCount` our method is unsupervised, we used the whole dataset as a test set without restriction.

The tables above show that, for both datasets, rank aggregation based on the Hamming-Kantorovich distance in $\mathcal{K}_n$ lead to competitive results, compared to the `BordaCount` procedure.

Table 2. Results comparisons on the `MQ2007-agg` dataset

NDCG	@1	@2	@3	@4	@5	@6	@7	@8	@9	@10
BordaCount	0.1902	0.2014	0.2081	0.2128	0.2188	0.2247	0.2312	0.2377	0.2444	0.2507
LUCE Greedy	0.1980	0.2058	0.2137	0.2229	0.2301	0.2379	0.2441	0.2505	0.2575	0.2648
LUCE Random	0.2275	0.2328	0.2406	0.2450	0.2515	0.2578	0.2623	0.2683	0.2745	0.2814
Mallows-∞	0.1920	0.2044	0.2100	0.2170	0.2226	0.2283	0.2346	0.2419	0.2471	0.2535

Table 3. Results comparisons on the MQ2008-agg dataset

NDCG	@1	@2	@3	@4	@5	@6	@7	@8	@9	@10
BordaCount	0.2368	0.2806	0.3080	0.3432	0.3713	0.3888	0.3992	0.3724	0.1643	0.1694
LUCE Greedy	0.2026	0.2563	0.3058	0.3426	0.3752	0.3936	0.4030	0.3749	0.1567	0.1644
LUCE Random	0.2188	0.2726	0.3005	0.3279	0.3498	0.3691	0.3833	0.3579	0.1456	0.1508
Mallows-∞	0.1937	0.2374	0.2787	0.3176	0.3487	0.3703	0.3841	0.3587	0.1459	0.1541

7 Conclusion

In this paper, we have provided a novel family of distances between rankings of a finite number of elements, which can be viewed as mass transportation distances, by the means of an embedding of the set of permutation matrices Σ_n in the set $\mathcal{K}_n$ of doubly-stochastic matrices. This convexification step is also shown to be a key ingredient for defining a new and flexible concept of median, reflecting a consensus among a finite number of rankings. Although the freedom in the choice of the cost function may lead to optimize a variety of tasks in the ranking context such as stability evaluation or ranking prediction, a simple application of this approach based on the Hamming cost yielded promising results, competing with those produced by the `BordaCount` method on LETOR benchmark datasets. Truth be told, this choice has been mainly motivated by computational convenience. Algorithmic issues concerning distance/median computation and properties of the median (Pareto efficiency, *etc.*), depending on the conditions fulfilled by the underlying cost, will be the subject of further research.

Appendix - Technical Proofs

Proof of Proposition 1

Observe first that, for any $(A, A') \in \mathcal{K}_n^2$, the quantity $d_c(A, A')$ is well-defined as a minimum, since the functional μ_c is linear (hence continuous) on $\{\Phi : \{1, \dots, n\}^4 \to \mathbb{R}\}$ and thus continuous on $\mathcal{M}(A, A')$ as well, which space is compact. Therefore a transportation plan Φ^* achieves the minimum (2). It is called an *optimal transportation plan* (OTP). The symmetry of d_c immediately results from the symmetry of the cost function c. The separability of d_c is an easy consequence of hypothesis (i) and (iv) for c. Let us finally prove the triangular inequality for d_c. Assume that A, A' and A'' are three given matrices in $\mathcal{K}_n$. Let us denote by Φ_1 an OTP from A to A'', and by Φ_2 an OTP from A'' to A'. From the *gluing lemma* [Vil09, p.23] there exists a map Φ_{132} from $\{1, \dots, n\}^3$ to $[0, 1]$ such that $\forall (i, j, k, l) \in \{1, \dots, n\}^4$,

$$\sum_{(k,l)\in\{1,\dots,n\}^2} \Phi_{132}\left((i,j),(r,s),(k,l)\right) = \Phi_1\left((i,j),(r,s)\right)$$

and

$$\sum_{(i,j)\in\{1,\ldots,n\}^2} \Phi_{132}\left((i,j),(r,s),(k,l)\right) = \Phi_1\left((r,s),(k,l)\right)$$

Let $\Phi\left((i,j),(k,l)\right) = \sum_{(r,s)\in\{1,\ldots,n\}^2} \Phi_{132}\left((i,j),(r,s),(k,l)\right)$, then, triangular inequality of c and Minkowski inequality implies:

$$\begin{aligned} d_{c,p}(A, A') &\leq \mu_{c,p}^{1/p}(\Phi) \\ &= \left(\sum_{\substack{(i,j)\in\{1,\ldots,n\}^2 \\ (k,l)\in\{1,\ldots,n\}^2 \\ (r,s)\in\{1,\ldots,n\}^2}} c\left((i,j),(k,l)\right)^p \Phi_{132}\left((k,l),(r,s),(i,j)\right) \right)^{1/p} \\ &\leq d_c(A, A'') + d_c(A'', A') \end{aligned}$$

Proof of Corollary 1

The Caley distance is bi-invariant. By virtue of Proposition 2, the cost c itself is bi-invariant. Now, under the action

$$(\sigma,\sigma') \in \mathfrak{S}_n \times \mathfrak{S}_n \mapsto \left(\left((i,j),(k,l)\right) \in \{1,\ldots,n\}^4 \mapsto \left((\sigma(i),\sigma'(j)),(\sigma(k),\sigma'(l))\right)\right),$$

the set $\{1,\ldots,n\}^4$ has 4 distinct orbits. The first orbit is the diagonal $D_n = \{\left((i,j),(i,j)\right) : (i,j) \in \{1,\ldots,n\}^2\}$, the 3 other orbits are $H_n = \{\left((i,j),(i,l)\right) : j \neq l\}$, $V_n = \{\left((i,j),(k,j)\right)\}$ and $O_n = \{\left((i,j),(k,l)\right) : i \neq k, j \neq l\}$. On D_n, the cost c is necessarily zero due to condition (i). From $\mathfrak{S}_n^2$ invariance of c, we know that c takes a constant value h over H_n, (over V_n and over O_n respectively). Triangular inequality implies that $o \leq h + v$, but also that $h \leq 2o$ and $v \leq 2o$. We now distinguish two cases. Either $o < +\infty$ or else $o = +\infty$. If $o = +\infty$ then either v or h is infinite, in which case we have already seen that it corresponds to the Hamming distance on permutations (which is obviously bi-invariant too and different from the Cayley distance), since we rule out the case where $o = v = h = +\infty$. If $o < +\infty$, then h and v are also finite. Since both h and v are finite, invariance also leads to $h = v$ using $(M^\sigma)^T = (M^\sigma)^{-1}$. Considering the distance between a transposition matrix and the identity, we deduce that $h = v = 1/2$. Now, considering the matrix corresponding to a length 3-cycle we deduce that $3o \leq 2$ (otherwise the cost of transportation from the identity to the cycle using only horizontal or vertical movements is larger than n whereas Cayley distance is equal to $n - 1$). But now there is a contradiction with the length n cycle whose transportation cost to the identity is less than $2n/3$ instead of $n - 1$ for the Cayley distance.

Proof of Proposition 3

It easy to see that the transportation distance between two vectors v and w in $\mathbb{R}^n$ induced by the cost $c(i,j) = \mathbb{I}\{i \neq j]\}$ is $\sum_{i=1}^n |w_i - v_i|/2$. Consider a fixed row index i in $1 \dots n$. By definition of $c_{\mathcal{H}}$, the transportation plan should not mix rows (otherwise the transportation cost would be infinite). The cost induced by row i between the matrix M^{σ} and M, where σ denotes any permutation of $\mathfrak{S}_n$ and M any matrix in $\mathcal{K}_n$, is then: $1 - M(i, \sigma(i))$. Hence, we have:

$$\sum_m d_{c_{\mathcal{H}}}(M, M^{\sigma_m}) \geq \sum_i \sum_m (1 - M(i, \sigma_m(i))) .$$

Now, since M is doubly stochastic, we may write

$$\sum_i \sum_m (1 - M(i, \sigma_m(i))) = nN - \sum_i \sum_m M(i, \sigma_m(i)) \geq Nn - \sum_m \sum_i \bar{M}(i, \sigma_m(i))$$

. Since all the $\sigma_m(i)$'s are distinct (i being fixed), we have

$$\sum_m d_{c_{\mathcal{H}}}(M, M^{\sigma_m}) \geq \sum_m d_{c_{\mathcal{H}}}(\bar{M}, M^{\sigma_m}).$$

If $\log(nN) \leq n$, one may store all the matrices M^{σ_m} using a dictionary structure, where each lookup costs at most $\log(nN)$. Otherwise, one can simply sum up the matrices M^{σ_m}.

References

[AM01] Aslam, J.A., Montague, M.: Models for metasearch. In: SIGIR 2001: Proceedings of the 24th Annual International ACM SIGIR Conference on Research and Development in Information Retrieval. ACM Press, New York (2001)

[BFB09] Bansal, M.S., Fernandez-Baca, D.: Computing distances between partial rankings. Information Processing Letters 109, 238–241 (2009)

[BGLS03] Bonnans, J.F., Gilbert, J.C., Lemaréchal, C., Sagastizábal, C.A.: Numerical optimization. Universitext. Springer, Berlin (2003)

[CV09] Clémençon, S., Vayatis, N.: Tree-based ranking methods. IEEE Transactions on Information Theory 55(9), 4316–4336 (2009)

[DD09] Deza, M.M., Deza, E.: Encyclopedia of Distances. Springer, Heidelberg (2009)

[Dia89] Diaconis, P.: A generalization of spectral analysis with application to ranked data. The Annals of Statistics 17(3), 949–979 (1989)

[Fis73] Fishburn, P.: The Theory of Social Choice. University Press, Princeton (1973)

[FKM+03] Fagin, R., Kumar, R., Mahdian, M., Sivakumar, D., Vee, E.: Comparing and aggregating rankings with ties. In: Proceedings of the 12th WWW conference, pp. 366–375 (2003)

[FKM+06] Fagin, R., Kumar, R., Mahdian, M., Sivakumar, D., Vee, E.: Comparing partial rankings. SIAM J. Discrete Mathematics 20(3), 628–648 (2006)

[HFCB08] Hüllermeier, E., Fürnkranz, J., Cheng, W., Brinker, K.: Label ranking by learning pairwise preferences. Artificial Intelligence 172, 1897–1917 (2008)

[HJ85] Horn, R.A., Johnson, C.R.: Matrix Analysis. Cambridge University Press, Cambridge (1985)

[How00] Howie, J.: Hyperbolic groups. In: Metaftsis, V. (ed.) Groups and Applications, Ekdoseis Ziti, Thessaloniki, pp. 137–160 (2000)

[Hud08] Hudry, O.: NP-hardness results for the aggregation of linear orders into median orders. Ann. Oper. Res. 163, 63–88 (2008)

[JK02] Järvelin, K., Kekäläinen, J.: Cumulated gain-based evaluation of IR techniques. ACM Transactions on Information Systems (TOIS) 20(4), 446 (2002)

[Kem59] Kemeny, J.G.: Mathematics without numbers. Daedalus 88, 571–591 (1959)

[KV00] Korte, B., Vygen, J.: Combinatorial optimization. Algorithms and Combinatorics, vol. 21. Springer, Berlin (2000)

[LL03] Lebanon, G., Lafferty, J.: Conditional models on the ranking poset. In: Proceedings of NIPS 2003 (2003)

[Luc59] Luce, R.D.: Individual Choice Behavior. Wiley, Chichester (1959)

[LXQ$^+$07] Liu, T.Y., Xu, J., Qin, T., Xiong, W., Li, H.: Letor: Benchmark dataset for research on learning to rank for information retrieval. In: Proceedings of SIGIR 2007 Workshop on Learning to Rank for Information Retrieval, pp. 3–10 (2007)

[Mal57] Mallows, C.L.: Non-null ranking models. i. Biometrika 44(1-2), 114–130 (1957)

[MM09] Mandhani, B., Meila, M.: Tractable search for learning exponential models of rankings. In: Proceedings of AISTATS. JMLR:W&CP 5, vol. 5 (2009)

[MPPB07] Meila, M., Phadnis, K., Patterson, A., Bilmes, J.: Consensus ranking under the exponential model. In: Conference on Artificial Intelligence (UAI), pp. 729–734 (2007)

[Mun57] Munkres, J.: Algorithms for the assignment and transportation problems. Journal of the Society for Industrial and Applied Mathematics 5(1), 32–38 (1957)

[Pla75] Plackett, R.L.: The analysis of permutations. Applied Statistics 2(24), 193–202 (1975)

[Rac91] Rachev, S.T.: Probability Metrics and the Stability of Stochastic Models. Wiley, Chichester (1991)

[RR98] Rachev, S.T., Rüschendorf, L.: Mass Transportation Problems. Theory, vol. I. Springer, Heidelberg (1998)

[Vil09] Villani, C.: Optimal transport. Grundlehren der Mathematischen Wissenschaften, vol. 338. Springer, Heidelberg (2009)

[Wak98] Wakabayashi, Y.: The complexity of computing medians of relations. Resenhas 3(3), 323–349 (1998)

Characteristic Kernels on Structured Domains Excel in Robotics and Human Action Recognition

Somayeh Danafar[1], Arthur Gretton[2], and Jürgen Schmidhuber[1]

[1] Istituto Dalle Molle di Studi sull'Intelligenza Artificiale (IDSIA)
Galleria 2, 6928 Manno-Lugano, Switzerland
[2] Carnegie Mellon University, Pittsburgh, USA
and MPI for Biological Cybernetics, Tübingen, Germany
{somayeh,juergen}@idsia.ch, arthur.gretton@gmail.com

Abstract. Embedding probability distributions into a sufficiently rich (characteristic) reproducing kernel Hilbert space enables us to take higher order statistics into account. Characterization also retains effective statistical relation between inputs and outputs in regression and classification. Recent works established conditions for characteristic kernels on groups and semigroups. Here we study characteristic kernels on periodic domains, rotation matrices, and histograms. Such structured domains are relevant for homogeneity testing, forward kinematics, forward dynamics, inverse dynamics, etc. Our kernel-based methods with tailored characteristic kernels outperform previous methods on robotics problems and also on a widely used benchmark for recognition of human actions in videos.

Keywords: Characteristic kernels, Locally compact Abelian groups, Rotation matrices, Semigroups, Recognition of human actions in videos.

1 Introduction

Kernel methods solve difficult non-parametric problems by embedding data points in higher-dimensional reproducing kernel Hilbert spaces (RKHS). This property makes kernel methods useful and strong tools to be used in different tasks. They were successfully applied to a wide range of learning tasks such as regression and classification [16]. Recent studies focused on mapping random variables into RKHS to collect linear statistics in RKHS which in turn were used to derive their meaning in the original space [8], [19], [20]. When the embedding is injective, the RKHS is said to be *characteristic* [5]. Such mappings allow for testing whether two distributions coincide [8],[9], or for finding the most predictive subspace in regression [6]. The most predictive (effective) subspace in regression is obtained by isolating the features that capture the statistical relationship between inputs and targets.

Characteristic kernels are defined on non-compact and complex domains. Sriperumbudur et al. [20] showed that a continuous shift-invariant $\mathbb{R}$-valued positive definite kernel on $\mathbb{R}^n$ is characteristic if and only if the support of its

J.L. Balcázar et al. (Eds.): ECML PKDD 2010, Part I, LNAI 6321, pp. 264–279, 2010.

Fourier transform is the entire $\mathbb{R}^n$. Fukumizu et al. [7] extended Fourier analysis to groups and semigroups and obtained necessary conditions for defining characteristic kernels on spaces other than $\mathbb{R}^n$.

The main contribution of this paper is empirical evaluation of characteristic kernels. We investigate characteristic kernels on structured domains (groups/semigroups) for various kernel-based methods: Maximum Mean Discrepancy (MMD) [8], [9] as a non-parametric hypothesis test, Support Vector Regression with ϵ-insensitive loss function (ϵ-SVR) [18], Gaussian Process Regression (GPR) [14] as a non-parametric regression method, and Support Vector Machines (SVM) [16] to classify human actions in videos. We provide experimental evidence that these kernel-based methods with appropriate kernels lead to significant performance gains.

Section 2 briefly reviews kernel-based methods. Section 3 introduces novel characteristic kernels on periodic data, the orthogonal group SO(3), and histograms. Section 4 experimentally confirms their theoretical advantages: we obtain state-of-the-art results in homogeneity testing, forward kinematics, forward dynamics, inverse dynamics, and recognition of human actions in videos.

2 Kernel-Based Learning Methods

In this section we briefly review some of kernel-based methods that we use to investigate our characteristic kernels.

2.1 A Non-parametric Statistical Test

One basic statistic on Euclidean space is the *mean*. By embedding the distributions into RKHS, the corresponding factor is the *mean element*. The distance between mapped mean elements is known as Maximum Mean Discrepancy (MMD) [8], [9]. The definition of MMD is given in the following theorem:

Theorem 1. *Let $(\mathcal{X}, \mathcal{B})$ be a metric space, and let P, Q be two Borel probability measures defined on $\mathcal{X}$. Then $P = Q$ if and only if* $\mathrm{MMD}(P, Q) = 0$, where

$$\begin{aligned} \mathrm{MMD}(P,Q) &:= \| \mu_P - \mu_Q \|_{\mathcal{H}} \\ &= \| E_P[k(x,.)] - E_Q[k(y,.)] \|_{\mathcal{H}} \\ &= (E_{x,x'' \sim P}[k(x,x'')] + E_{y,y'' \sim Q}[k(y,y'')] - 2E_{x\sim P, y\sim Q}[k(x,y)])^{\frac{1}{2}} \end{aligned} \quad (1)$$

One application of MMD is homogeneity testing, which tests whether the samples were drawn from different distributions. We compare MMD to another two-sample test suited for periodic distributions, namely, the Uniform Scores Test (UST) [4]. UST is not a kernel-based method.

Uniform Scores Test (UST). UST [4] is a two-sample test which tests whether distributions of circular data coincide. In UST each distribution is represented by a radius. The null hypothesis is rejected if the summation of radii is too large. Here we define UST more precisely.

Suppose we have n_i samples where $i = 1, 2, .., r$. We treat sample $n_1 = \{\theta_1, ..., \theta_n\}$ as linear data, re-arrange them in ascending order, and assign rank r_i to each θ_i. The *circular rank* of θ_i is then defined as $\gamma_i = 2\pi r_i/n$, for $i = 1, ..., n$. We denote γ_i as the *uniform score* corresponding to θ_i. We take all $N = n_1 + ... + n_r$ data values as a single sample and calculate their circular ranks. Let γ_{ij} denote the circular rank of θ_{ij} among all the data. For each sample $n_i, i = 1, ..., r$, we calculate

$$C_i = \sum_{j=1}^{n_i} cos\gamma_{ij}, S_i = \sum_{j=1}^{n_i} sin\gamma_{ij} \tag{2}$$

and hence the test statistics

$$W_r = 2\sum_{i=1}^{r} (C_i^2 + S_i^2)/n_i \tag{3}$$

If W_r is too large, we reject the null hypothesis that the distributions are identical.

2.2 Non-parametric Regression Methods for Model Learning

The task of regression is to learn the input/target mapping, to predict target values for query inputs.

Support Vector Regression with ϵ-Insensitive Loss function (ϵ-SVR) The goal of ϵ-SVR regression is to find a mapping function $f(x)$ which for each training input x deviates from its target by at most ϵ, and simultaneously is as flat as possible. According to [19], $f(x)$ is

$$\sum_{i=1}^{l} (\alpha_i - \alpha_i^*) K(x_i, x) + b. \tag{4}$$

where $K(x_i, x) = \phi(x_i)^T \phi(x)$, and i ranges over the training points. The solution of a quadratic optimization problem determines the quantities α_i^*, α_i, and b.

Gaussian Processes Regression (GPR). Gaussian Process Regression (GPR) [14] uses a linear model to find a latent function $f(x)$. Uncertainty is modeled probabilistically by:

$$f \sim N(0, \Phi\Sigma\Phi^T) \sim N(0, K) \tag{5}$$

where matrix Φ describes transformation columns $\phi(x)$ for all cases in the training set, Σ is the covariance matrix of the weights, and K is a positive semidefinite matrix with elements $K_{i,j} = k(x_i, x_j)$ for some covariance function $k(., .)$.

2.3 Classification: Support Vector Machines

Consider the problem of separating the training set into two classes. If we assume that the two classes can be separated by a hyperplane $w.x + b = 0$ in some space $\mathcal{H}$, and that we have no prior knowledge about the data distribution, then the optimal hyperplane maximizes the margin [16]. Optimal values for w and b can be found by solving a constrained minimization problem, using Lagrange multipliers $\alpha_i (i = 1, .., l)$. The classifier is defined as:

$$f(x) = \operatorname{sgn}\left(\sum_{i=1}^{l} \alpha_i y_i K(x_i, x) + b\right) \tag{6}$$

where K is the kernel mapping data points to RKHS $\mathcal{H}$, and α_i and b are found using an SVC learning algorithm. Those x_i with nonzero α_i are called the *support vectors.*

3 Characteristic Kernels on Structured Domains

Characteristic kernels were defined on non-compact domain like entire $\mathbb{R}^n$. Sriperumbudur et al. [20] showed that if and only if the support of Fourier transform of a shift invariant positive definite kernel is the entire $\mathbb{R}^n$, this kernel is characteristic. A question that naturally arises is whether characteristic kernels can be defined on spaces besides $\mathbb{R}^n$. Several such domains constitute topological groups/semigroups. Fukumizu et al. [7] based on extensions of Fourier analysis to groups and semigroups established necessary and sufficient conditions of introducing characteristic kernels. Our main contribution in this paper is to study these characteristic kernels defined by their algebraic structure and assess them in relevant applications. For the sake of this purpose, thanks to the established conditions and theorems by Fukumizu et al. [7] we define our proper characteristic kernels. We investigate characteristic kernels on Locally Compact Abelian (LCA) groups (periodic data), Compact Groups (rotation matrices), and Abelian Semigroups (histogram-based data). In this section we clarify our characteristic kernels, thereafter relevant experiments and evaluations will be discussed in section 4.

3.1 Shift Invariant Characteristic Kernels on LCA Groups

Periodic domains are examples of Locally Compact Abelian groups which we consider in this study. To define our proper characteristic kernels on periodic domains, we use Theorems 7 and 8 of [7] which describe necessary and sufficient conditions for kernels on LCA groups to be characteristic, as well as Corollary 9 of [7] on the multiplication of shift-invariant characteristic kernels, which is again a characteristic kernel. Our novel characteristic kernels on periodic domains are:

1. $k_1(x, y) = \prod_{i=1}^{l} (\pi - (x_i - y_i)_{mod\ 2\pi})^2$,
2. $k_2(x, y) = \prod_{i=1}^{l} (\cosh(\pi - (x_i - y_i)_{mod 2\pi})$,

3. $k_3(x,y) = \prod_{i=1}^{l}(-\log(1-2\alpha\cos(x_i-y_i)+\alpha^2)$,
4. $k_4(x,y) = \prod_{i=1}^{l}(1-\alpha^2)/(1-2\alpha\cos(x_i-y_i)+\alpha^2)$,

where l denotes the input dimension. Periodic domains are relevant for two-sample testing, and in regression tasks like forward kinematics, forward dynamics, and inverse dynamics. In the case of forward dynamics besides periodic data we have torques which do not belong to periodic domain. We work with the following justified characteristic kernels in that case:

1. $k_5(x,y) = \prod_{i=1}^{m}(\pi - (x_i-y_i)_{mod\ 2\pi})^2 \cdot \text{Gaussian}(x_{m,..,l}, y_{m,..,l})$,
2. $k_6(x,y) = \prod_{i=1}^{m}(\cosh(\pi-(x_i-y_i)_{mod 2\pi}) \cdot \text{Gaussian}(x_{m,..,l}, y_{m,..,l})$,
3. $k_7(x,y) = \prod_{i=1}^{m}(-\log(1-2\alpha\cos(x_i-y_i)+\alpha^2) \cdot \text{Gaussian}(x_{m,..,l}, y_{m,..,l})$,
4. $k_8(x,y) = \prod_{i=1}^{m}(1-\alpha^2)/(1-2\alpha\cos(x_i-y_i)+\alpha^2) \cdot \text{Gaussian}(x_{m,..,l}, y_{m,..,l})$.

3.2 Characteristic Kernels on Compact Groups

Famous examples of non-Abelian topological groups are the ones consisting of matrices, such as the orthogonal group SO(3). According to Theorems 11 and 12 of [7], we define proper kernels on rotation matrices $\{A, B\} \in \mathbb{R}^3$. Let $\cos\theta = \frac{1}{2}Tr[B^{-1}A]$, and $0 \leq \theta \leq \pi$, we formulate the characteristic kernels as follows:

$$k_1(A,B) = \frac{1}{\sin\theta}\sum_{n=0}^{\infty}\frac{\sin((2n+1)\theta)}{(2n+1)^3} = \frac{\pi\theta(\pi-\theta)}{8\sin\theta}. \tag{7}$$

$$k_2(A,B) = \sum_{n=0}^{\infty}\frac{\alpha^{2n+1}\sin((2n+1)\theta)}{(2n+1)\sin\theta} = \frac{1}{2\sin\theta}\arctan\left(\frac{2\alpha\sin\theta}{1-\alpha^2}\right). \tag{8}$$

3.3 Characteristic Kernels on Abelian Semigroups

Now consider histograms as an example of Abelian semigroups such as $(\mathbb{R}_+^n, +)$. Theorems 13 and 14 of [7] obtain necessary and sufficient conditions for tailored kernels for histogram-based information. Let $a = (a_i)_{i=1}^n$ and $b = (b_i)_{i=1}^n$, $(a_i \geq 0, b_i \geq 0)$ be non-negative measures on n points. We use the following characteristic kernel:

$$k(a,b) = e^{-\beta\sum_{i=1}^{n}\sqrt{a_i+b_i}}. \tag{9}$$

where $\beta \geq 0$ and $\mathcal{X} \in \mathbb{R}$. Another tailored kernel for histogram-based data which is not a characteristic kernel is *Generalized Histogram Intersection (GHI)* kernel. In [1] GHI was introduced as a positive-definite kernel:

$$K_{\text{GHI}}(a,b) = \sum_{i=1}^{m}\min\{|\ a_i^\beta\ |, |\ b_i^\beta\ |\}, \quad (a,b) \in \mathcal{X}\times\mathcal{X} \tag{10}$$

We compare the results of these two kernels in human action classification task in section 4.4.

4 Experiments and Evaluations

Now we confirm theoretical advantages of characteristic kernels on various practical applications.

4.1 MMD for Two-Sample Testing

One application of MMD is for two-sample tests, which involve testing the null hypothesis $H_0 : P = Q$ versus $H_1 : P \neq Q$. Two-sample tests require a measure of distance between probabilities and a notion of whether this distance is statistically significant. Our MMD test determines the test threshold by the bootstrap procedure [8]. In this study we consider this application of MMD to compare two artificially generated distributions of periodic nature. Suppose we obtain the first sample from a uniform distribution P. The other sample is drawn from a perturbed uniform distribution $Q : 1+\sin(\omega x)$. For higher perturbation frequencies ω (where $1/\omega$ is smaller), it becomes harder to discriminate Q from P— see Figure 1.

Figure 2 shows the acceptance percentage of null hypothesis with MMD during 1000 runs with a user-defined significance level 0.05. The quality of MMD as a

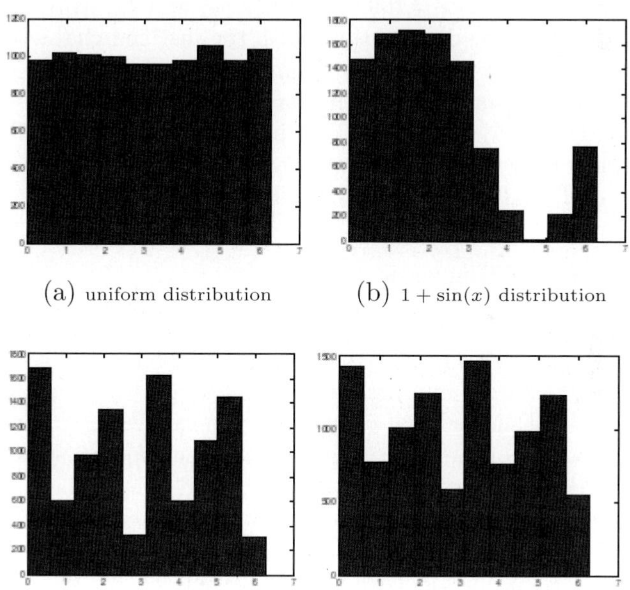

(a) uniform distribution (b) $1 + \sin(x)$ distribution

(c) $1 + \sin(3x)$ (d) $1 + \sin(6x)$

Fig. 1. (a)represents an example of circular data $[0, 2\pi)$ with uniform distribution, (b), (c), and (d) are periodic data with distribution $1 + \sin(\omega x)$ and ω equal to 1, 3, and 6 respectively. Higher perturbation frequencies make the perturbed distribution much closer to the uniform distribution, and the discrimination more difficult.

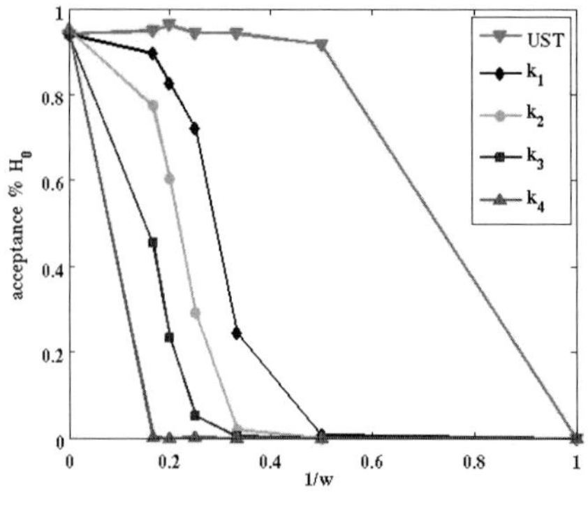

Fig. 2. Acceptance percentage of $H_0 : P = Q$ for MMD and UST, with user-defined significance level of 0.05 during 1000 runs, and $\frac{1}{\omega} = 0, \frac{1}{6}, \frac{1}{5}, \frac{1}{4}, \frac{1}{3}, \frac{1}{2}$, and 1. P is a uniform distribution of circular data, and Q is $1 + \sin(\omega x)$.

statistic depends on the richness of RKHS $\mathcal{H}$ which is defined by a measurable kernel. Characteristic kernels [5], [6] yield an RKHS for which probabilities have unique images. Here we use characteristic kernels k_1, k_2, k_3, and k_4 in MMD with $l = 1$ and hyper-parameter $\alpha = 0.9$ for kernels k_3 and k_4. MMD discriminated subtle changes between the distributions with the justified characteristic kernels on periodic domain. This can be seen from different acceptance percentage of H_0 in Figure 2. MMD has the best performance with k_4 which needs tuning a hyper-parameter α. We compared the result of MMD with UST. We observe that UST can not deal with subtle nonlinear changes in distributions. It gives true results when P and Q are either completely similar or dissimilar.

4.2 Applications of Regression

Given a training set of data points $\mathcal{D} = \{(x_1, y_1), ...(x_l, y_l)\}$ where the $x_i \in \mathbb{R}^n$ are inputs and the $y_i \in \mathbb{R}^1$ are the corresponding targets, the task of regression is to learn the input/target mapping, to predict target values for query inputs. Fukimuzu et al. [5], [6] showed that characterization allows us to derive a contrast function for estimation of the effective subspace. The effective subspace can help to retain the statistical relationship between x and y by isolating the features that capture this relation. We evaluated characteristic kernels k_1, k_2, k_3, and k_4 in forward kinematics and inverse dynamics for datasets with periodic nature. For forward dynamics problem, characteristic kernels k_5, k_6, k_7, and k_8 are used.

Forward kinematics. Kinematics studies motion ignoring the forces which cause it. The forward kinematics of a revolute robot arm are described by the function $T = f(\theta, \phi)$, where θ is the vector of joint angles, ϕ are the parameters describing the kinematics of the arm, and T is the 4×4 homogeneous transformation matrix [2].

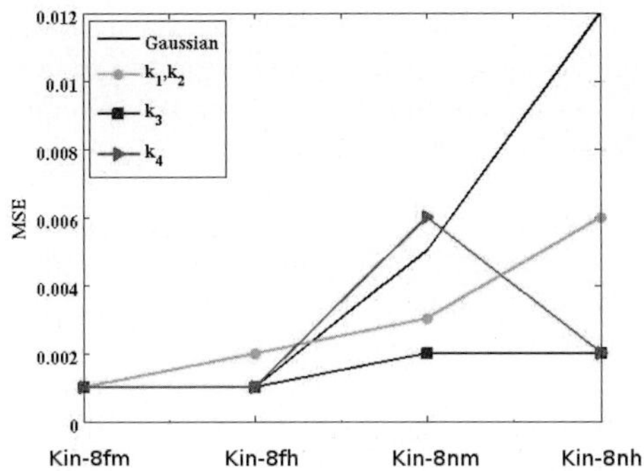

Fig. 3. The result of ϵ-SVR for forward kinematics task on Kin-8fm, Kin-8fh, Kin-8nm, and Kin-8nh datasets. The results of ϵ-SVR is based on characteristic kernels k_1, k_2, k_3, k_4 (section 3.1), and Gaussian kernel with $\epsilon = 0.01$, $\alpha = 0.9$, for k_3, and k_4, and $\sigma = 10$ for Gaussian kernel.

We use the 8 input *Kin* dataset (`http://www.cs.utoronto.ca/~delve/data/kin/desc.html`). It is generated from a realistic simulation of the forward kinematics of an 8 link all-revolute robot arm. The task is to predict the distance of the end-effector from a target, given the angular position of the 8 joints, the link twist angles, link lengths and link offset distances. Combinations of the following attributes are considered in datasets:

1. output : highly nonlinear (n) vs. fairly linear (f)
2. predicted value : medium noise (m) vs. high noise(h)

We use a training set of size 1024, 4096 test instances and the validation set of size 3072. The hyper-parameters α and σ in kernels k_3, k_4, and Guassian kernel respectively were tuned during the 5-fold leave-one-out cross validation procedure. Support vector regression with ϵ insensitive loss function ($\epsilon - SVR$) is used as our non-parametric regression method. A run consisted of model selection, training and testing, and the confidence interval over Mean Squared Errors (MSE) results are obtained over 10 runs. In this task the input dimension is 8. l is set to 8 in the formula of our characteristic kernels of section 3.1. In Figure 3, the results of $\epsilon - SVR$ with $\epsilon = 0.01$ on four datasets of 8-input Kin (Kin-8fm, Kin-8fh, Kin-8nm, and Kin-8nh) are depicted. Figure 3 demonstrates that tailored characteristic kernels on the LCA group work better than Gaussian kernel which is just characteristic. We compared our best results on the above datasets to the results given by GPR [14], K-Nearest Neighbor (K-NN), Linear Regression (LR), Multi-Layer Perceptrons (MLP) with single hidden layer and early stopping [14], and mixtures of experts trained by Bayesian methods (HME)[22]. The results reported in Table 1. Results of 22 methods (by Ghahramani) on the same datasets are available at `http://www.cs.toronto.edu/~delve/data/kin/desc.html`. The reported results show that GPR obtained better results

Table 1. Best results of ϵ-SVR with characteristic kernel k_4 in comparison with reported results on the Kin family of datasets. Our results are obtained with $\epsilon = 0.01$ in ϵ-SVR and the kernel k_4 with hyper-parameter $\alpha = 0.9$. Rounded standard deviation of MSEs are also reported.

Method	Kin-8fm	Kin-8fh	Kin-8nm	Kin-8nh
ϵ-SVR	0.001 ± 0.0001	0.001± 0.0001	0.002± 0.0001	0.002± 0.0001
GPR	0.25 ± 0.0001	0.02 ± 0.01	0.43 ± 0.1	0.1 ± 0.2
HME	0.26 ± 0.0001	0.03 ± 0.01	0.48 ± 0.3	0.28± 0.2
KNN	0.29 ± 0.0001	0.08 ± 0.01	0.65 ± 0.1	0.45 ± 0.2
LR	0.28 ± 0.0001	0.06 ± 0.01	0.65 ± 0.1	0.45 ± 0.2
MLP	0.26 ± 0.0001	0.03 ± 0.02	0.42 ± 0.01	0.1 ± 0.2

than LR, as it captures the nonlinear relationship between data points by a Gaussian kernel and the affect of noise with probabilistic nature of the method. This draws the attention to our datasets which are generated by fairly linear and nonlinear movements of robot arm in combination with noise. Moreover the results of GPR in comparison with HME as another Bayesian based method is better which shows the superiority of kernel-based methods. The nonlinearities captured by MLP and GPR produced comparable results with better performance for GPR. Our results showed that $\epsilon - SVR$ and a tailored characteristic kernel on periodic data outperforms the other methods. This highlights the fact that in kernel-based methods selection of an appropriate kernel according to the nature of available data leads to significant performance gains. Our results with tailored characteristic kernels for periodic data confirm this fact.

Forward dynamics. To simulate robot control systems, forward dynamics computes joint accelerations and actuator torques, given position and velocity state [2]. We used the 8 input *Pumadyn* dataset at `http://www.cs.utoronto.ca/~delve/data/pumadyn/desc.html`. It was synthetically generated from a realistic simulation of the dynamics of a Puma560 robot arm. The task is to predict the angular acceleration of the robot arm links, given angular positions, velocities, torques. The combination of fairly linear and nonlinear movements of robot arm with unpredictability captured by medium or high amount of noise generate 4 datasets (Pdyn-8fm, Pdyn-8fh, Pdyn-8nm, and Pdyn-8nh). We used characteristic kernels k_5, k_6, k_7, and k_8 in this task. All the settings are like in the forward kinematics case. Figure 4 shows the justified characteristic kernels have better performance than Gaussian kernel. We compared our best results to those obtained by GPR [14], K-Nearest Neighbor (K-NN), Linear Regression (LR), MLP with early stopping and single hidden layer [14], mixtures of experts trained by Bayesian methods (HME) [22] in Table 2. Results of 25 methods (by Ghahramani) are available at `http://www.cs.toronto.edu/~delve/data/pumadyn/desc.html`. Like the reported results in the forward kinematics case,

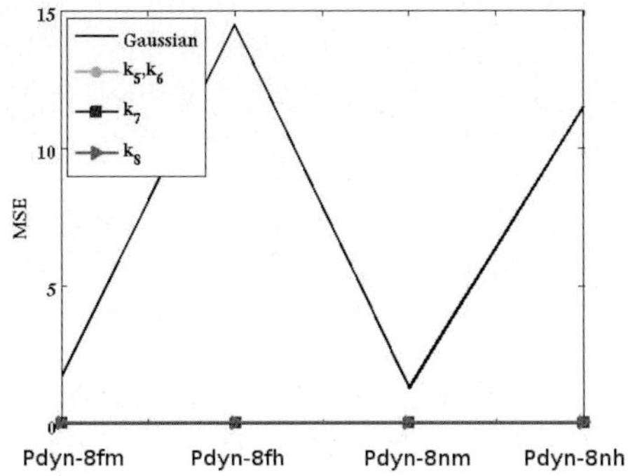

Fig. 4. The result of ϵ-SVR for forward dynamics task on Pdyn-8fm, Pdyn-8fh, Pdyn-8nm, and Pdyn-8nh. The results of ϵ-SVR is based on characteristic kernels k_5, k_6, k_7, and k_8 (section 3.1) with $\epsilon = 0.01$, $\alpha = 0.9$ for k_7, and k_8 respectively, and $\sigma = 10$ for Gaussian kernel.

Table 2. Best results of our ϵ-SVR with characteristic kernel k_8, as well as earlier reported results on the Pumadyn family of datasets. The results are obtained with $\epsilon = 0.01$ in ϵ-SVR and the kernel k_8 with hyper-parameter $\alpha = 0.9$. Rounded standard deviation of MSEs are also reported.

Method	P-8fm	P-8fh	P-8nm	P-8nh
ϵ-SVR	0.01±0.0001	0.01 ±0.0001	0.01 ±0.0001	0.01 ±0.0001
GPR	0.39 ± 0.001	0.05 ± 0.1	0.32± 0.01	0.03 ± 0.2
HME	0.41 ± 0.001	0.06 ± 0.1	0.37 ± 0.5	0.04 ± 0.3
KNN	0.41 ± 0.001	0.15 ± 0.1	0.52 ± 0.01	0.3 ± 0.1
LR	0.48 ± 0.001	0.08 ± 0.1	0.55 ± 0.01	0.48 ± 0.1
MLP	0.4 ± 0.001	0.06 ± 0.2	0.35 ± 0.01	0.033± 0.1

the results of kernel based method GPR are better than those of linear Regression, and is better than HME method which is a Bayesian method. The results of GPR and MLP are comparable although the performance of GPR is better. The best outcome is for our ϵ-SVR method with justified characteristic kernels on datasets. ϵ-SVR captures the nonlinearity, and the relation of observations with tailored characteristic kernels.

Inverse Dynamics. Finding sufficiently accurate dynamic models of rigid body equations in automatic robot control is difficult due to unmodeled nonlinearities, complex friction and actuator dynamics. Imprecise prediction of joint torques leads to poor control performance and may even damage the system. Learning more precise inverse dynamics models from measured data by regression is an interesting alternative. Here we compare $\epsilon - SVR$ and GPR as regression methods for computing inverse dynamics, which could be used for automatic robot control (e.g., [13]).

The inverse dynamic model [2] is given in the rigid-body formulation $u = M(q)\ddot{q} + F(q, \dot{q})$, where $q, \dot{q}, \ddot{q}$ are joint angles, angular velocities and angular accelerations of the robot. $M(q)$ denotes the inertia matrix and $F(q, \dot{q})$ the internal forces. Let us define the inverse dynamic model by $u = f(q, \dot{q}, \ddot{q})$; the regression task is to learn f.

We use the 7-DOF *SARCOS* anthropomorphic robot arm data `http://www.gaussianprocess.org/gpml/data`. Each observation in the data set consists of 21 input features (7 joint positions, 7 joint velocities, and 7 joint accelerations) and the corresponding 7 joint torques for the 7-DOF. There are two disjoint sets, one for training and one for testing. We use only 1100 examples of the training set for training, but the entire test set for testing. Results are shown in terms of normalized Mean Squared Errors (nMSEs) defined as MSE divided by target variance. Results of $\epsilon - SVR$ and GPR are shown in Figure 5. $\epsilon - SVR$ and GPR with tailored characteristic kernels work better than with the Gaussian kernel. Their results are comparable with slightly better performance in ϵ-SVR. The larger errors for the 5th and 6th DOF show that nonlinearities (e.g. hydrolic cables, complex friction) can not be approximated well using just the rigid body functions. This is an example of the difficulty of using an analytical model for control in practice.

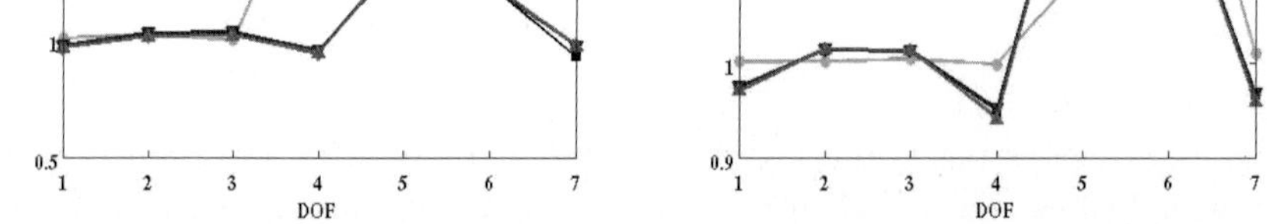

(a) Results of $\epsilon - SVR$ with tailored characteristic kernels on periodic domains and Gaussian kernel with $\epsilon = 0.01$ and $\alpha = 0.5$ for k_3, and k_4, and $\sigma = 21$ for Gaussian kernel.

(b) Results of GPR with the same tailored characteristic kernels are used in $\epsilon - SVR$ and Gaussian kernel.

Fig. 5. The results of ϵ-SVR and GPR with characteristic kernels k_1, k_2, k_3, k_4, and Gaussian kernel on SARCOS dataset

Yeung et al. [21] investigated different training sample sizes for GPR. They achieved the same result as reported in the current paper with GPR and Gaussian kernel over training set of size 1100 with the mean of nMSE = 1.06 and the standard deviation of nMSE = 0.12. They further improved their results by multi-task learning and reported the mean of nMSE =0.35 and the standard deviation of nMSE= 0.49 for multi-task GPR. From our improvement for both ϵ-SVR and GPR with tailored characteristic kernels in comparison with Gaussian kernel (Figure 5) we expect to see a performance boost in multi-task learning, but this is a topic of future work.

4.3 Rotation Matrices in Forward Kinematics

As mentioned before the task in forward kinematics is to find $T = f(\theta, \phi)$, where T a the 4×4 homogeneous rotation matrix (an example of SO(3) group). We considered the solution of the regression task with ϵ-SVR and the tailored characteristic kernels on rotation matrices of formula 7, 8, and Gaussian kernel on Kin-8nh dataset. We obtained the following results:

1. $k_1(A, B) \Rightarrow \text{MSE} = 0.009$

2. $k_2(A, B)$ with $\alpha = 0.9 \Rightarrow \text{MSE} = 0.006$

3. Gaussian kernel with $\sigma = 0.05 \Rightarrow \text{MSE} = 0.005$

Unexpectedly, Gaussian kernel worked better than justified kernels on the SO(3) group.

4.4 Abelian Semigroups: Classification of Human Actions

One example of Abelian semigroups are histograms. As many authors in computer vision area are working with kernel-based methods and histograms (for example, see the recent VOC2006 object classification challenge), it is worth studying kernel classes suitable for histogram-based information. We use the action descriptors introduced by Danafar and Gheissari [3], which are histograms of optical flow and capture both local and global information about actions. These feature vectors are described in Figure 6. We use the challenging human action video database of KTH [17]. It contains 6 types of human actions: walking, jogging, running, boxing, hand waving, and hand clapping, performed by 25 people in four different scenarios: outdoors (s1), outdoors with scale variations (s2), outdoors with different clothes (s3), and indoors (s4). Some samples from this dataset are shown in Figure 7.

Our action recognition approach is based on SVM. The database is divided into three parts: training, testing and validation. 8 subjects were used for training, 9 for test and 8 for validation. The validation data is first used to tune the hyper-parameter β of GHI kernel and our defined characteristic kernel with a 5-fold leave-one-out cross validation procedure. Danafar and Gheissari [3] recognized actions with SVMs and the GHI Kernel.

The crucial condition an SVM kernel should satisfy is to be positive definite, meaning that the SVM problem is convex, and hence the solution of its objective function is unique. Characteristic kernels have positive definite property and have been shown to be more discriminative; because characterization can capture

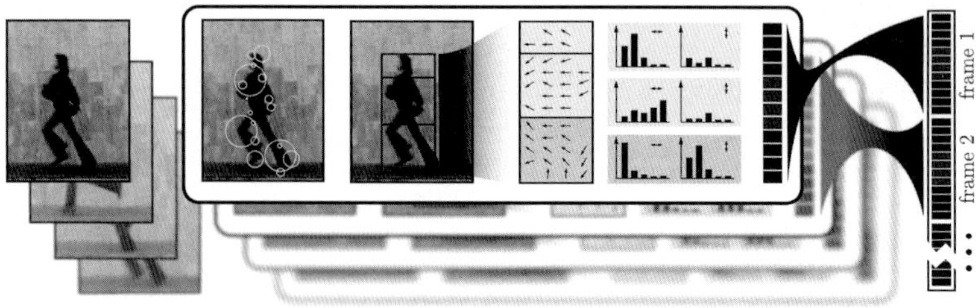

Fig. 6. The features used in our supervised classifier are described in [3]; a single feature vector (right) is computed for each sequence by concatenating data coming from each frame of the sequence (left). In each frame, Harris interest points are used to recover a tight bounding box, which is vertically partitioned in three regions. The topmost 1/5 of the bounding box approximately contains the head and the neck. The middle 2/5 contains the torso and hands, and the bottom 2/5 of the bounding box contains the legs. Such segmentation is obviously approximated, and the resulting features would still be usable in cases where the assumptions are not met. Flow data in each region is summarized in separate histograms for the horizontal and vertical directions.

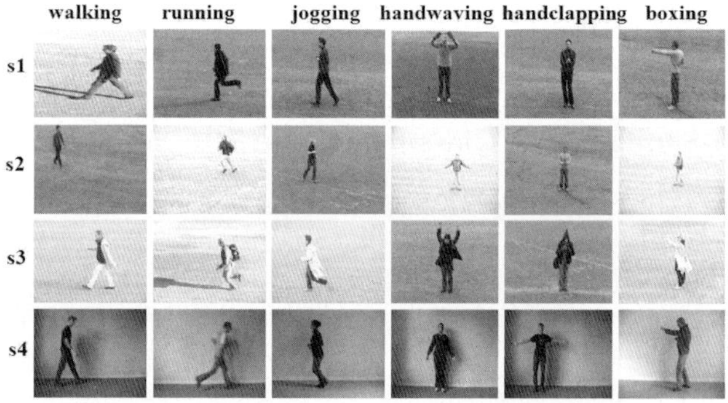

Fig. 7. Example images from video sequences in KTH dataset

effective statistical and discriminative relationship between response variables from an explanatory variables [5], [6]. Our reported accuracy of 93.1% obtained with characteristic kernels is a very significant improvement with respect to the accuracy of 83.5% reported in [3], obtained using Histogram Intersection Kernel in the same setting. We also compared our characteristic kernel for histogram-based data to the Gaussian kernel, which is also characteristic but is not tailored to histogram-based data. In our experiments, the accuracy of Gaussian kernel is 33.8% which is much lower than our result of 93.1%. Confusion matrices in the three cases are reported in Figure 8. Therefore we conclude that our experimental results are due to our kernel being both characteristic and suitable for histogram-based data; removing any of the two properties results in a significant performance loss.

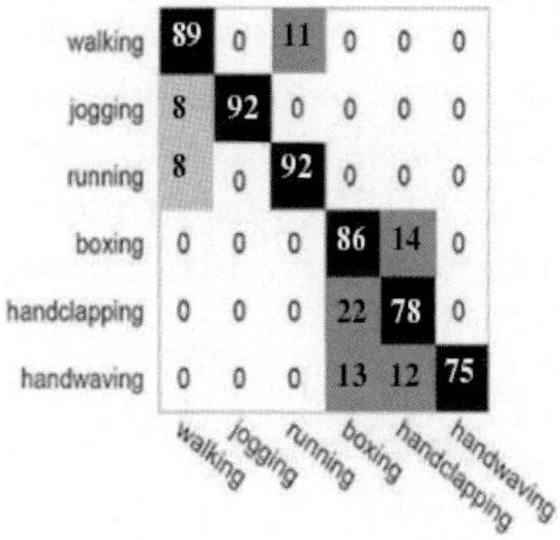

(a) Results of GHI kernel as a positive definite kernel with $\beta = 1$ and overall accuracy rate of 85.3%.

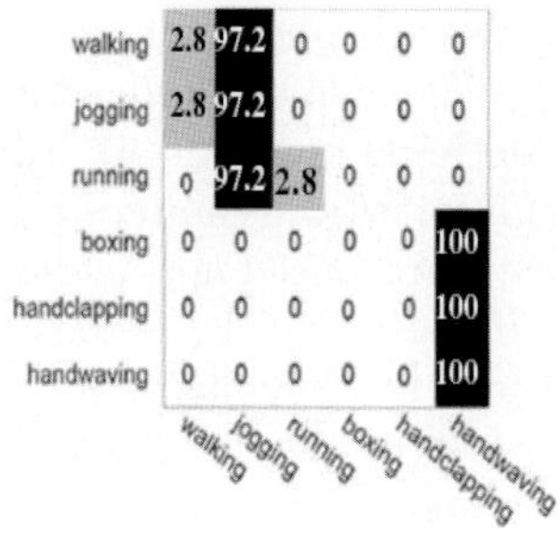

(b) Results of Gaussian kernel as a characteristic kernel with $\sigma = 21$ and overall accuracy rate of 33.8%.

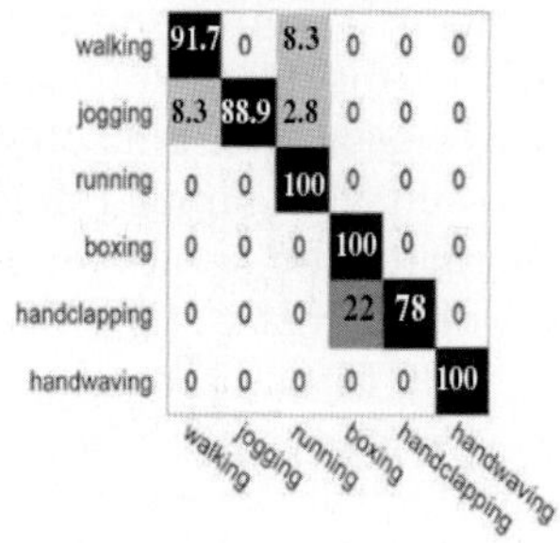

(c)Results of the tailored characteristic kernel on histogram-based data with $\beta = 0.001$ and overall accuracy rate of 93.1%.

Fig. 8. Confusion matrices obtained on the KTH dataset with descriptors [3], using SVM and the indicated kernels. Figure(a) shows the recognition rates of histogram Intersection kernel which is a positive definite but not a characteristic kernel. Figure (b) denotes the result of a characteristic kernel (Gaussian) which is not tailored to histogram-based information. Figure (c) is the result of characteristic kernel which is tailored to histograms.

Table 3. Recognition results of various methods on the KTH dataset. The recognition rate reported by Jhuang et al. (2007) is obtained on video sequences from scenarios 1 and 4 only. Other reported rates are on all scenarios.

Method	Recognition rate %
SVM by charac. Kernel	93.1
Lin et al. [11]	93.4
Schindler and Van Gool [15]	92.7
Jhuang et al. [10]	91.7
Danafar & Gheissari [3]	85.3
Niebles et al. [12]	83.3
Schüldt et al. [17]	71.7

In Table 3 recognition results of various methods on the KTH dataset are compared. Our overall rate exceeds previously reported results and is comparable to 93.4% reported rate in [11], demonstrating superiority of our method. In [15] and [11], the authors benefited from stronger feature vectors as combination of shape and motion and reported high accuracy rates rather than motion feature which is used here. This concludes that the achievement of higher recognition rate with stronger histogram-based feature vector is promising.

5 Conclusion

We studied empirically characteristic kernels on structured domains, yielding powerful kernel-based methods for structured data. Characteristic kernels for periodic domains and SO(3) were applied to homogeneity testing, forward kinematics, forward dynamics, and inverse dynamics for robotics. Our methods outperformed other published methods on the 8-input Kin forward kinematics data set, and the 8-input Pumadyn forward dynamics data set. We also used tailored characteristic kernels on histogram-based action descriptors to recognize human actions in video. Our results on the KTH database of human actions are comparable to or better than those of previous state-of-the-art methods. Ongoing work aims at improving inverse dynamics results through multi-task kernel-based learning with our tailored characteristic kernels.

References

1. Boughorbel, S., Tarel, J.P., Boujemaa, N.: Generalized Histogram Intersection Kernel for image recognition. In: IEEE International Conference on Image Processing, vol. 3, pp. 161–164 (2005)
2. Craig, J.I.: Introduction to Robotics, mechanics and control, 3rd edn. Prentice-Hall, Englewood Cliffs (2004)
3. Danafar, S., Gheissari, N.: Action recognition for surveillance application using optic flow and SVM. In: Yagi, Y., Kang, S.B., Kweon, I.S., Zha, H. (eds.) ACCV 2007, Part II. LNCS, vol. 4844, pp. 457–466. Springer, Heidelberg (2007)
4. Fisher, N.I.: Statistical analysis of circular data. Cambridge University Press, Cambridge (1993)
5. Fukumizu, K., Gretton, A., Sun, X., Schölkopf, B.: Kernel Measures of Conditional Dependence. In: Platt, J.C., Koller, D., Singer, Y., Roweis, S. (eds.) Advances in Neural Information Processing Systems 20: Proceedings of the 2007 Conference, pp. 489–496. MIT Press, Cambridge (2008)
6. Fukumizu, K., Bach, F.R., Jordan, M.I.: Dimensionality reduction for supervised learning with reproducing kernel Hilbert Spaces. JMLR 5, 73–99 (2004)
7. Fukumizu, K., Sriperumbudur, B.K., Gretton, A., Schölkopf, B.: Characteristic kernels on groups and semigroups. In: Koller, D., Schuurmans, D., Bengio, Y., Bottou, L. (eds.) Proceedings of the 2008 Conference Advances in Neural Information Processing Systems 21, Curran, Red Hook, NY, USA, pp. 473–480 (June 2009)
8. Gretton, A., Borgwadt, B.K., Rasch, M., Schölkopf, B., Smola, A.: In: Schölkopf, B., Platt, J., Hofmann, T. (eds.) Proceedings of the 2006 Conference Advances in Neural Information Processing Systems 19, pp. 513–520. MIT Press, Cambridge (2007)
9. Gretton, A., Fukumizu, K., Teo, C.H., Song, L., Schölkopf, B., Smola, A.: A Kernel Statistical Test of Independence. In: Platt, J.C., Koller, D., Singer, Y., Roweis, S. (eds.) Proceedings of the 2007 Conference Advances in Neural Information Processing Systems, vol. 20, pp. 585–592. MIT Press, Cambridge (2008)
10. Jhuang, H., Serre, T., Wolf, L., Poggio, T.: A biologically inspired system for action recognition. In: Internation Conference on Computer Vision, pp. 1–8 (2007)
11. Lin, Z., Jiang, Z., Davis, L.S.: Recognizing Actions by Shape-Motion Prototype Trees. In: IEEE International Conference on Computer Vision (ICCV 2009), Kyoto, Japan (2009)
12. Niebles, J.C., Wang, H., Fei-Fei, L.: Unsupervised learning of human action categories using spatial-temporal words. IJCV 79(3), 299–318 (2008)

13. Nguyen-Tuong, D., Peters, J., Seeger, M., Schölkopf, B.: Learning Inverse Dynamics: a Comparison. In: Verleysen, M. (ed.) Advances in Computational Intelligence and Learning: Proceedings of the European Symposium on Artificial Neural Networks (ESANN 2008), d-side, Evere, Belgium, pp. 13–18 (2008)
14. Rasmussen, C.E., Williams, K.A.: Gaussian processes for machine learning. MIT-Press, Cambridge (2006)
15. Schindler, K., Van Gool, L.: Action snippets: How many frames does human action recognition require? In: International conference on Computer Vison and Pattern Recognition (CVPR), pp.1–8 (2008)
16. Schölkopf, B., Smola, A.: Learning with kernels. MIT Press, Cambridge (2002)
17. Schüldt, C., Laptev, I., Caputo, B.: Recognizing human actions: a local SVM approach. In: Proceedings of the 17th International Conference on In Pattern Recognition, ICPR 2004, vol. 3, pp. 32–36 (2004)
18. Smola, A.J., Schölkopf, B.: A tutorial on Support Vector Regression.: NeuroCOLT Technical Report TR-98-030 (1998)
19. Smola, A.J., Gretton, A., Song, L., Schölkopf, B.: A Hilbert Space Embedding for Distributions. In: Hutter, M., Servedio, R.A., Takimoto, E. (eds.) ALT 2007. LNCS (LNAI), vol. 4754, pp. 13–31. Springer, Heidelberg (2007)
20. Sriperumbudur, B.K., Gretton, A., Fukumizu, K., Lanckeriet, G.R.G., Schölkopf, B.: Injective Hilbert space embeddings of probability measures. In: Servedio, R.A., Zhang, T. (eds.) Proceedings of the 21st Annual Conference on Learning Theory (COLT 2008), pp. 111–122. OmnioPress, Madison (2008)
21. Yeung, D.Y., Zhang, Y.: Learning inverse dynamics by Gaussian Process Regression under the multi-task learning framework. In: Sukatme, G.S. (ed.) The Path to Autonomous Robots, pp. 131–142. Springer, Heidelberg (2009)
22. Waterhouse, S.: PhD Thesis on Mixtures of Experts Available (1998)

Regret Analysis for Performance Metrics in Multi-Label Classification: The Case of Hamming and Subset Zero-One Loss

Krzysztof Dembczyński[1,3], Willem Waegeman[2], Weiwei Cheng[1], and Eyke Hüllermeier[1]

[1] Department of Mathematics and Computer Science, Marburg University, Hans-Meerwein-Str., 35039 Marburg, Germany
{cheng,dembczynski,eyke}@informatik.uni-marburg.de
[2] Department of Applied Mathematics, Biometrics and Process Control, Ghent University, Coupure links 653, B-9000 Ghent, Belgium
willem.waegeman@ugent.be
[3] Institute of Computing Science, Poznań University of Technology, Piotrowo 2, 60-965 Poznań, Poland

Abstract. In multi-label classification (MLC), each instance is associated with a subset of labels instead of a single class, as in conventional classification, and this generalization enables the definition of a multitude of loss functions. Indeed, a large number of losses has already been proposed and is commonly applied as performance metrics in experimental studies. However, even though these loss functions are of a quite different nature, a concrete connection between the type of multi-label classifier used and the loss to be minimized is rarely established, implicitly giving the misleading impression that the same method can be optimal for different loss functions. In this paper, we elaborate on risk minimization and the connection between loss functions in MLC, both theoretically and empirically. In particular, we compare two important loss functions, namely the Hamming loss and the subset 0/1 loss. We perform a regret analysis, showing how poor a classifier intended to minimize the subset 0/1 loss can become in terms of Hamming loss and vice versa. The theoretical results are corroborated by experimental studies, and their implications for MLC methods are discussed in a broader context.

1 Introduction

The setting of *multi-label classification* (MLC) which, in contrast to conventional (single-label) classification, allows an instance to belong to several classes simultaneously, has received increasing attention in machine learning in recent years [1,2,3,4,5,6]. In particular, several approaches aiming at the exploitation of dependencies between class labels have been proposed. Even though the goal itself is clearly worthwhile, and empirically, many approaches have indeed been shown to improve predictive performance, a thorough theoretical analysis of the MLC setting is still missing.

J.L. Balcázar et al. (Eds.): ECML PKDD 2010, Part I, LNAI 6321, pp. 280–295, 2010.

Indeed, the notion of "label dependence" is often used in a purely intuitive manner. In this paper, we will argue that a careful distinction should be made between two different but related forms of statistical dependence in MLC, namely conditional and unconditional dependence. Moreover, we will establish a close connection between conditional label dependence and loss minimization. In MLC, a multitude of loss functions can be considered, and indeed, a large number of losses has already been proposed and is commonly applied as performance metrics in experimental studies. However, even though these loss functions are of a quite different nature, a concrete connection between the type of multi-label classifier used and the loss to be minimized is rarely established, implicitly giving the misleading impression that the same method can be optimal for different loss functions.

More specifically, this paper extends our previous work [7], in which we analyzed the connection between conditional label dependence and risk minimization for three loss functions commonly used in MLC problems: Hamming, rank and subset 0/1 loss. According to our results, the first two losses can in principle be minimized without taking conditional label dependence into account, which is not the case for the subset 0/1 loss.

In this paper, we further elaborate on the relationship between the Hamming and subset 0/1 loss. Our main theoretical result states that, even though we can establish mutual bounds for these loss functions, the bounds are not very tight. On the contrary, we can show that the minimization of subset 0/1 loss may come along with a very high regret in terms of Hamming loss and vice versa. As will be discussed in more detail later on, these results have important implications and suggest that previous experimental studies have often been interpreted in an incorrect way.

Let us also remark that the analysis performed in this paper is simplified by assuming an unconstrained hypothesis space. This allows for an analysis with respect to the joint conditional distribution alone. Regarding related work, we mention that generalization bounds have already been considered for problems with structured outputs. Some of these results apply directly to MLC as a special case [8,9]. Moreover, it is worth mentioning that a similar problems can be found in information theory, namely bitwise and codeword decoding [10]. One can easily notice that the bitwise and codeword decoding correspond to Hamming loss and subset 0/1 loss minimization, respectively.

The structure of the paper is the following. Section 2 introduces the MLC problem in a formal way. Section 3 contains the main theoretical results concerning the bound and regret analysis. In Section 4, we present some experimental results confirming our theoretical claims. The last section concludes the paper.

2 Multi-Label Classification

In this section, we describe the MLC problem in more detail and formalize it within a probabilistic setting. Along the way, we introduce the notation used throughout the paper.

2.1 Problem Statement

Let $\mathcal{X}$ denote an instance space, and let $\mathcal{L} = \{\lambda_1, \lambda_2, \ldots, \lambda_m\}$ be a finite set of class labels. We assume that an instance $\boldsymbol{x} \in \mathcal{X}$ is (non-deterministically) associated with a subset of labels $L \in 2^{\mathcal{L}}$; this subset is often called the set of relevant labels, while the complement $\mathcal{L} \setminus L$ is considered as irrelevant for $\boldsymbol{x}$. We identify a set L of relevant labels with a binary vector $\boldsymbol{y} = (y_1, y_2, \ldots, y_m)$, in which $y_i = 1 \Leftrightarrow \lambda_i \in L$. By $\mathcal{Y} = \{0,1\}^m$ we denote the set of possible labelings.

We assume observations to be generated independently and randomly according to a probability distribution $\mathbf{p}(\mathbf{X}, \mathbf{Y})$ on $\mathcal{X} \times \mathcal{Y}$, i.e., an observation $\boldsymbol{y} = (y_1, \ldots, y_m)$ is the realization of a corresponding random vector $\mathbf{Y} = (Y_1, Y_2, \ldots, Y_m)$. We denote by $\mathbf{p}_{\boldsymbol{x}}(\mathbf{Y}) = \mathbf{p}(\mathbf{Y} \,|\, \boldsymbol{x})$ the conditional distribution of $\mathbf{Y}$ given $\mathbf{X} = \boldsymbol{x}$, and by $\mathbf{p}_{\boldsymbol{x}}^{(i)}(Y_i) = \mathbf{p}^{(i)}(Y_i \,|\, \boldsymbol{x})$ the corresponding marginal distribution of Y_i:

$$\mathbf{p}_{\boldsymbol{x}}^{(i)}(b) = \sum_{\boldsymbol{y} \in \mathcal{Y} : y_i = b} \mathbf{p}_{\boldsymbol{x}}(\boldsymbol{y})$$

A multi-label classifier $\mathbf{h}$ is an $\mathcal{X} \to \mathcal{Y}$ mapping that assigns a (predicted) label subset to each instance $\boldsymbol{x} \in \mathcal{X}$. Thus, the output of a classifier $\mathbf{h}$ is a vector

$$\mathbf{h}(\boldsymbol{x}) = (h_1(\boldsymbol{x}), h_2(\boldsymbol{x}), \ldots, h_m(\boldsymbol{x})).$$

The problem of MLC can be stated as follows: Given training data in the form of a finite set of observations $(\boldsymbol{x}, \boldsymbol{y}) \in \mathcal{X} \times \mathcal{Y}$, drawn independently from $\mathbf{p}(\mathbf{X}, \mathbf{Y})$, the goal is to learn a classifier $\mathbf{h} : \mathcal{X} \to \mathcal{Y}$ that generalizes well beyond these observations in the sense of minimizing the risk with respect to a specific loss function.

2.2 Label Dependence

As already announced in the introduction, we propose to distinguish two types of dependence. We call the labels $\mathbf{Y}$ *unconditionally independent* if and only if

$$\mathbf{p}(\mathbf{Y}) = \prod_{i=1}^{m} \mathbf{p}^{(i)}(Y_i). \tag{1}$$

On the other hand, the labels are *conditionally independent* if the joint posterior distribution is the product of the marginals:

$$\mathbf{p}_{\boldsymbol{x}}(\mathbf{Y}) = \prod_{i=1}^{m} \mathbf{p}_{\boldsymbol{x}}^{(i)}(Y_i)$$

Obviously, both types of dependence are related to each other, since

$$\mathbf{p}(\mathbf{Y}) = \int_{\mathcal{X}} \mathbf{p}_{\boldsymbol{x}}(\mathbf{Y}) d\mathbf{P}(\boldsymbol{x}) .$$

Nevertheless, unconditional dependence does *not* imply *nor* is implied by conditional dependence.

It has been widely established in statistics that exploiting unconditional label dependence can improve the generalization performance, because unconditional label dependence mainly originates from a similar structure of the different models [11]. The same arguments have played a key role in the development of related areas like multi-task learning and transfer learning, where task i and task j as well their models are assumed to be related [12]. As we will show the conditional dependence is rather connected with loss functions and their minimizers.

Let us remind that the joint distribution of a random vector $\mathbf{Y} = (Y_1, \ldots, Y_m)$ can be expressed by the product rule of probability:

$$\mathbf{p}(\mathbf{Y}) = \mathbf{p}(Y_1) \prod_{i=2}^{m} \mathbf{p}(Y_i | Y_1, \ldots, Y_{i-1})$$

If $Y_1, \ldots, Y_m$ are independent, then the product rule simplifies to (1).

2.3 Loss Functions

The performance in MLC is perhaps most frequently reported in terms of the Hamming loss, which is defined as the fraction of labels whose relevance is incorrectly predicted:[1]

$$L_H(\boldsymbol{y}, \mathbf{h}(\boldsymbol{x})) = \frac{1}{m} \sum_{i=1}^{m} [\![y_i \neq h_i(\boldsymbol{x})]\!] . \qquad (2)$$

Another natural loss function in the MLC setting is generalization of the well-known 0/1 loss from the conventional to the multi-label setting:

$$L_s(\boldsymbol{y}, \mathbf{h}(\boldsymbol{x})) = [\![\boldsymbol{y} \neq \mathbf{h}(\boldsymbol{x})]\!] . \qquad (3)$$

This loss function is referred to as *subset 0/1* loss. Admittedly, it may appear overly stringent, especially in the case of many labels. Moreover, since making a mistake on a single label is punished as hardly as a mistake on all labels, it does not discriminate well between "almost correct" and completely wrong predictions. However, mainly because of the fact that it is so extreme, it is especially relevant for our discussion about label dependence. Besides, as will be seen in more detail later on, it is a strong complement to the Hamming loss.

3 Analysis of Hamming and Subset 0/1 Loss

In this section, we analyze the Hamming and the subset 0/1 loss. The analysis is performed by assuming an unconstrained hypothesis space. This allows us to simplify the analysis by considering the conditional distribution for a given $\boldsymbol{x}$. First, we recall the risk minimizers of the two loss functions, already presented in [7], and then show that, despite being different in general, they may coincide under specific conditions. Further, we derive mutual bounds for the two loss functions. Finally, we will show how poorly a classifier intended to minimize the subset 0/1 loss can perform in terms of Hamming loss and vice versa.

[1] For a predicate P, the expression $[\![P]\!]$ evaluates to 1 if P is true and to 0 if P is false.

3.1 Risk Minimization

The risk of a classifier $\mathbf{h}$ is defined as the expected loss over the joint distribution $\mathbf{p}(\mathbf{X}, \mathbf{Y})$:

$$R_L(\mathbf{h}) = \mathbb{E}_{\mathbf{XY}} L(\mathbf{Y}, \mathbf{h}(\mathbf{X})), \tag{4}$$

where $L(\cdot)$ is a loss function on multi-label predictions. A risk-minimizing model $\mathbf{h}^*$ is given by

$$\mathbf{h}^* = \arg\min_{\mathbf{h}} \mathbb{E}_{\mathbf{XY}} L(\mathbf{Y}, \mathbf{h}(\mathbf{X})) = \arg\min_{\mathbf{h}} \mathbb{E}_{\mathbf{X}} [\mathbb{E}_{\mathbf{Y}|\mathbf{X}} L(\mathbf{Y}, \mathbf{h}(\mathbf{X}))] \tag{5}$$

and determined in a pointwise way by the *Bayes optimal decisions*

$$\mathbf{h}^*(\boldsymbol{x}) = \arg\min_{\boldsymbol{y}} \mathbb{E}_{\mathbf{Y}|\mathbf{X}} L(\mathbf{Y}, \boldsymbol{y}). \tag{6}$$

For the Hamming loss (2), it is easy to see that the risk minimizer (6) is obtained by

$$\boldsymbol{h}^*_H(\boldsymbol{x}) = (h_{H_1}(\boldsymbol{x}), \ldots, h_{H_m}(\boldsymbol{x})),$$

where

$$h_{H_i}(\boldsymbol{x}) = \arg\max_{b \in \{0,1\}} \mathbf{p}^{(i)}_{\boldsymbol{x}}(b) \qquad (i = 1, \ldots, m). \tag{7}$$

The Bayes prediction for (3) is also straight-forward. As for any other 0/1 loss, it simply consists of predicting the mode of the distribution:

$$\boldsymbol{h}^*_s(\boldsymbol{x}) = \arg\max_{\boldsymbol{y} \in \mathcal{Y}} \mathbf{p}_{\boldsymbol{x}}(\boldsymbol{y}) \tag{8}$$

As one of the most important consequences of the above results we note that, according to (7), a risk-minimizing prediction for the Hamming loss can be obtained from the marginal distributions $\mathbf{p}^{(i)}_{\boldsymbol{x}}(Y_i)$ $(i = 1, \ldots, m)$ alone. In other words, it is not necessary to know the joint label distribution $\mathbf{p}_{\boldsymbol{x}}(\mathbf{Y})$ on $\mathcal{Y}$. For finding the minimizer (8), on the other hand, the joint distribution must obviously be known. These results suggest that taking conditional label dependence into account is less important for Hamming loss than for subset 0/1 loss.

Despite the differences noted above, we can show that the two risk minimizers coincide under specific conditions. More specifically, we can show the following proposition.

Proposition 1. *The Hamming loss and subset 0/1 have the same risk minimizer, i.e., $\boldsymbol{h}^*_H(\boldsymbol{x}) = \boldsymbol{h}^*_s(\boldsymbol{x})$, if one of the following conditions holds:*

(1) Labels $Y_1, \ldots, Y_m$ are conditionally independent, i.e., $\mathbf{p}_{\boldsymbol{x}}(\mathbf{Y}) = \prod_{i=1}^m \mathbf{p}_{\boldsymbol{x}}(Y_i)$.
(2) The probability of the mode of the joint probability is greater or equal than 0.5, i.e., $\mathbf{p}_{\boldsymbol{x}}(\boldsymbol{h}^*_s(\boldsymbol{x})) \geq 0.5$.

Proof. (1) Since the joint probability of any combination of $\boldsymbol{y}$ is given by the product of marginal probabilities, the highest value of this product is given by the highest values of the marginal probabilities. Thus, the joint mode is composed of the marginal modes.
(2) If $\mathbf{p}_{\boldsymbol{x}}(\boldsymbol{h}^*_s(\boldsymbol{x})) \geq 0.5$, then $\mathbf{p}_{\boldsymbol{x}}(h^*_{s_i}(\boldsymbol{x})) \geq 0.5$, $i = 1, \ldots, m$, and from this it follows that $h^*_{s_i}(\boldsymbol{x}) = h^*_{H_i}(\boldsymbol{x})$. □

As a simple corollary of this proposition, we have the following.

Corollary 1. *In the separable case (i.e., the joint conditional distribution is deterministic,* $\mathbf{p}_{\boldsymbol{x}}(\mathbf{Y}) = [\![\mathbf{Y} = \boldsymbol{y}]\!]$*, where* $\boldsymbol{y}$ *is a binary vector of size* m*), the risk minimizers of the Hamming loss and subset 0/1 coincide.*

Proof. If $\mathbf{p}_{\boldsymbol{x}}(\mathbf{Y}) = [\![\mathbf{Y} = \boldsymbol{y}]\!]$, then $\mathbf{p}_{\boldsymbol{x}}(\mathbf{Y}) = \prod_{i=1}^{m} \mathbf{p}_{\boldsymbol{x}}(Y_i)$. In this case, we also have $\mathbf{p}_{\boldsymbol{x}}(\boldsymbol{h}_s^*(\boldsymbol{x})) \geq 0.5$. Thus, the result follows from both (1) and (2) in Proposition 1. □

3.2 Bound Analysis

So far, we have looked at the minimizers of Hamming and subset 0/1 loss and we have seen that these minimizers may coincide under special conditions. In general, however, they are different and, therefore, will call for different classifiers. Another natural question one may ask is the following: If we fix a classifier and we know, say, its subset 0/1 loss, can we say anything about its Hamming loss? This question is answered by the following proposition.

Proposition 2. *For all distributions of* $\mathbf{Y}$ *given* $\boldsymbol{x}$*, and for all models* $\boldsymbol{h}$*, the expectation of the subset 0/1 loss can be bounded in terms of the expectation of the Hamming loss as follows:*

$$\frac{1}{m}\mathbb{E}_{\mathbf{Y}}[L_s(\mathbf{Y}, \boldsymbol{h}(\boldsymbol{x}))] \leq \mathbb{E}_{\mathbf{Y}}[L_H(\mathbf{Y}, \boldsymbol{h}(\boldsymbol{x}))] \leq \mathbb{E}_{\mathbf{Y}}[L_s(\mathbf{Y}, \boldsymbol{h}(\boldsymbol{x}))] \tag{9}$$

Proof. For a fixed $\boldsymbol{x} \in \mathcal{X}$, we can express the expected loss as follows:

$$\mathbb{E}_{\mathbf{Y}}[L(\mathbf{Y}, \boldsymbol{h}(\boldsymbol{x}))] = \sum_{\boldsymbol{y} \in \mathcal{Y}} \mathbf{p}(\boldsymbol{y}) L(\boldsymbol{y}, \boldsymbol{h}(\boldsymbol{x}))$$

Suppose we can express an MLC loss in terms of an aggregation $G : \{0,1\}^m \to [0,1]$ of the standard zero-one losses $L_{0/1}$ on individual labels (as used in conventional classification):

$$L(\boldsymbol{y}, \boldsymbol{h}(\boldsymbol{x})) = G(L_{0/1}(y_1, h_1(\boldsymbol{x})), \ldots, L_{0/1}(y_m, h_m(\boldsymbol{x}))) \,. \tag{10}$$

Indeed, the subset 0/1 loss and the Hamming loss can be written, respectively, as

$$G_{\max}(\boldsymbol{a}) = G_{\max}(a_1, \ldots, a_m) = \max\{a_1, \ldots, a_m\}$$
$$G_{\mathrm{mean}}(\boldsymbol{a}) = G_{\mathrm{mean}}(a_1, \ldots, a_m) = \frac{1}{m}(a_1 + \ldots + a_m).$$

This immediately leads to the above lower and upper bound for the Hamming loss. The proposition then immediately follows from the fact that $\frac{1}{m}G_{\max}(\boldsymbol{a}) \leq G_{\mathrm{mean}}(\boldsymbol{a}) \leq G_{\max}(\boldsymbol{a})$ for all $\boldsymbol{a} \in [0,1]^m$. □

Interestingly, it turns out that the location of the Hamming loss between the bounds in (9) is in direct correspondence with the conditional dependence between the labels, and when the dependence structure of the conditional distribution of $\boldsymbol{Y}$ is given, the difference between Hamming loss and subset 0/1 loss can be determined in a more precise way. Roughly speaking, the less (more) dependent the labels are, the more the Hamming loss moves toward the lower (upper) bound. Without going into detail, we note that more precise estimations of the difference between subset 0/1 loss and Hamming loss can be derived with the help of copulas [13].

Nevertheless, we need to emphasize that a complete analysis of the relationship between bound (9) and conditional label dependence has to take additional factors into account. One of these factors is the hypothesis space one considers; the lower bound will become more tight when restricting to certain hypothesis spaces. Another important factor is the interplay between conditional and unconditional label dependence. For example, in case of full positive dependence between all labels, estimating the Hamming loss might still be more simple than estimating the subset 0/1 loss, implying that the upper bound will not behave as an equality. The analysis of this section mainly provides insights on the behavior of the different loss functions for the underlying distribution. Yet, we realize that the picture might look different in terms of estimated performance after training on a finite set of examples.

3.3 Regret Analysis

Some of the previous results may suggest that, for learning a risk minimizing classifier, either loss functions can be used as a proxy of the other one. For example, the bounds in (9) may suggest that a low subset 0/1 loss will also imply a low Hamming loss. On the other hand, one may argue that the bounds themselves are rather weak, and reckon that the concrete difference in terms of Hamming and subset 0/1 loss may become quite high. In this section, we present a regret analysis showing that minimization of Hamming loss does not guarantee good performance in terms of subset 0/1 loss and vice versa.

The regret of a classifier $\boldsymbol{h}$ with respect to a loss function L_z is defined as follows:

$$r_{L_z}(\boldsymbol{h}) = R_{L_z}(\boldsymbol{h}) - R_{L_z}(\boldsymbol{h}^*_z), \tag{11}$$

where R is the risk given by (4), and $\boldsymbol{h}^*_z$ is the Bayes-optimal classifier with respect to the loss function L_z.

In the following, we consider the regret with respect to the Hamming loss, given by

$$r_H(\boldsymbol{h}) = \mathbb{E}_{\mathbf{XY}} L_H(\mathbf{Y}, \boldsymbol{h}(\mathbf{X})) - \mathbb{E}_{\mathbf{XY}} L_H(\mathbf{Y}, \boldsymbol{h}^*_H(\mathbf{X})),$$

and the subset 0/1 loss, given by

$$r_H(\boldsymbol{h}) = \mathbb{E}_{\mathbf{XY}} L_s(\mathbf{Y}, \boldsymbol{h}(\mathbf{X})) - \mathbb{E}_{\mathbf{XY}} L_s(\mathbf{Y}, \boldsymbol{h}^*_s(\mathbf{X})).$$

Since both loss functions are decomposable with respect to individual instances, we analyze the expectation over $\mathbf{Y}$ for a given $\boldsymbol{x}$. The first result concerns the highest value of the regret in terms of the subset 0/1 loss for $\boldsymbol{h}_H^*(\mathbf{X})$, the optimal strategy for the Hamming loss.

Proposition 3. *The following upper bound holds:*

$$\mathbb{E}_{\mathbf{Y}} L_s(\mathbf{Y}, \boldsymbol{h}_H^*(\boldsymbol{x})) - \mathbb{E}_{\mathbf{Y}} L_s(\mathbf{Y}, \boldsymbol{h}_s^*(\boldsymbol{x})) < 0.5 .$$

Moreover, this bound is tight, i.e.,

$$\sup_{\mathbf{P}} \left(\mathbb{E}_{\mathbf{Y}} L_s(\mathbf{Y}, \boldsymbol{h}_H^*(\boldsymbol{x})) - \mathbb{E}_{\mathbf{Y}} L_s(\mathbf{Y}, \boldsymbol{h}_s^*(\boldsymbol{x}))\right) = 0.5,$$

where the supremum is taken over all probability distributions on $\mathcal{Y}$.

Proof. Since the risk of any classifier $\boldsymbol{h}$ is within the range $[0, 1]$, the maximal value of the regret is 1. However, according to the second part of Proposition 1, both risk minimizers coincide if $\mathbb{E}_{\mathbf{Y}} L_s(\mathbf{Y}, \boldsymbol{h}_s^*(\boldsymbol{x})) \leq 0.5$. Consequently, the regret must be (strictly) smaller than 0.5. To prove the tightness of the bound, we show that, for any $\delta \in (0, \frac{1}{6})$, there is a probability distribution $\mathbf{p}$ that yields the regret $0.5 - \delta$. Define $\mathbf{p}$ as follows:

$$\mathbf{p}(\boldsymbol{y}) = \begin{cases} \frac{1}{2} - \delta, & \text{if } \boldsymbol{y} = (a_1, \ldots, a_{k-1}, \bar{a}_{k+1}, \ldots, \bar{a}_m) \\ \frac{1}{2} - \delta, & \text{if } \boldsymbol{y} = (\bar{a}_1, \ldots, \bar{a}_{k-1}, a_{k+1}, \ldots, a_m) \\ 2\delta, & \text{if } \boldsymbol{y} = (a_1, \ldots, a_{k-1}, a_{k+1}, \ldots, a_m) \end{cases}$$

where $a_i \in \{0, 1\}$ and $\bar{a}_i = 1 - a_i$. Such a distribution can be constructed for all $m > 1$. Obviously,

$$\begin{aligned} \boldsymbol{h}_s^*(\boldsymbol{x}) &= (a_1, \ldots, a_{k-1}, \bar{a}_{k+1}, \ldots, \bar{a}_m) \quad \text{or} \\ \boldsymbol{h}_s^*(\boldsymbol{x}) &= (\bar{a}_1, \ldots, \bar{a}_{k-1}, a_{k+1}, \ldots, a_m) \end{aligned}$$

and

$$\boldsymbol{h}_H^*(\boldsymbol{x}) = (a_1, \ldots, a_{k-1}, a_{k+1}, \ldots, a_m)$$

Finally, we thus obtain

$$\mathbb{E}_{\mathbf{Y}} L_s(\mathbf{Y}, \boldsymbol{h}_H^*(\boldsymbol{x})) = 1 - 2\delta$$

and

$$\mathbb{E}_{\mathbf{Y}} L_s(\mathbf{Y}, \boldsymbol{h}_s^*(\boldsymbol{x})) = 0.5 - \delta,$$

which immediately proves the proposition. □

The second result concerns the highest value of the regret in terms of the Hamming loss for $\boldsymbol{h}_s^*(\mathbf{X})$, the optimal strategy for the subset 0/1 loss.

Proposition 4. *The following upper bound holds for $m > 3$:*

$$\mathbb{E}_{\mathbf{Y}} L_H(\mathbf{Y}, \boldsymbol{h}_s^*(\boldsymbol{x})) - \mathbb{E}_{\mathbf{Y}} L_H(\mathbf{Y}, \boldsymbol{h}_H^*(\boldsymbol{x})) < \frac{m-2}{m+2}.$$

Moreover, this bound is tight, i.e.

$$\sup_{\mathbf{P}} (\mathbb{E}_{\mathbf{Y}} L_H(\mathbf{Y}, \boldsymbol{h}_s^*(\boldsymbol{x})) - \mathbb{E}_{\mathbf{Y}} L_H(\mathbf{Y}, \boldsymbol{h}_H^*(\boldsymbol{x}))) = \frac{m-2}{m+2},$$

where the supremum is taken over all probability distributions on $\mathcal{Y}$.

Proof. Because of space limitations, we only show how to construct the distribution for which the regret is close to the given bound.

Let $a_i \in \{0,1\}$ and $\bar{a}_i = 1 - a_i$. If $\boldsymbol{a}_m = (a_1, a_2, \ldots, a_m)$ is a $\{0,1\}$-vector of length m, then $\bar{\boldsymbol{a}}_m$ denotes the vector $(\bar{a}_1, \bar{a}_2, \ldots, \bar{a}_m)$. Furthermore, let $d_H(\boldsymbol{a}, \boldsymbol{b})$ denote the Hamming distance, given by

$$d_H(\boldsymbol{a}, \boldsymbol{b}) = \sum_{i=1}^{m} |a_i - b_i|$$

for all $\boldsymbol{a}, \boldsymbol{b} \in \{0,1\}^m$. Now, consider a joint probability distribution defined as follows:

$$\mathbf{p}(\boldsymbol{y}) = \begin{cases} \frac{1}{m+2} + \delta & \text{if } \boldsymbol{y} = \boldsymbol{a}_m \\ \frac{1}{m+2} - \frac{\delta}{m+1} & \text{if } d_H(\boldsymbol{y}, \bar{\boldsymbol{a}}_m) \leq 1 \\ 0 & \text{otherwise} \end{cases},$$

where $\delta > 0$. Hence, we obtain:

$$\mathbb{E}_{\mathbf{Y}} L_H(\mathbf{Y}, \boldsymbol{h}_s^*(\boldsymbol{x})) = \frac{1}{m+2} - \frac{\delta}{m+1} + m(\frac{1}{m+2} - \frac{\delta}{m+1})\frac{m-1}{m},$$

$$\mathbb{E}_{\mathbf{Y}} L_H(\mathbf{Y}, \boldsymbol{h}_H^*(\boldsymbol{x})) = \frac{1}{m+2} + \delta + m(\frac{1}{m+2} - \frac{\delta}{m+1})\frac{1}{m}.$$

The difference is then given by

$$\mathbb{E}_{\mathbf{Y}} L_H(\mathbf{Y}, \boldsymbol{h}_s^*(\boldsymbol{x})) - \mathbb{E}_{\mathbf{Y}} L_H(\mathbf{Y}, \boldsymbol{h}_H^*(\boldsymbol{x})) = \frac{m-2}{m+2} - \delta(\frac{m-1}{m+1} + 1).$$

Since this holds for any $\delta > 0$, the regret is close to the bound. □

As we can see, the regret is quite high in both cases, suggesting that a single classifier will not be able to perform equally well in terms of both loss functions. Instead, a classifier specifically tailored for the Hamming (subset 0/1) loss will indeed perform much better for this loss than a classifier trained to minimize the subset 0/1 (Hamming) loss.

3.4 Summary and Implications of Theoretical Results

Our theoretical results so far can be summarized as follows:

- The risk minimizers of Hamming and subset 0/1 loss have a different structure: In the latter case, the minimizer is the mode of a joint distribution on the label space, whereas in the former, it is a combination of the modes of (one-dimensional) marginal distributions.
- Under specific conditions, these two types of loss minimizers are provably equivalent, though in general, they will produce different predictions.
- The Hamming loss is upper-bounded by the subset 0/1 loss, which in turn is bounded by the Hamming loss multiplied by the number of labels m.

- Minimization of the subset 0/1 loss may cause a high regret for the Hamming loss and vice versa.

These results have a number of implications, not only from a theoretical but also from a methodological and empirical point of view:

- The idea to exploit label dependencies, one of the main research topics in MLC, should be reconsidered in a careful way, distinguishing the two types of dependence mentioned above. The conditional dependence, for example, is arguably more related to non-decomposable (with respect to labels) losses like subset 0/1 loss than to decomposable ones like Hamming. This distinction is largely ignored in papers on that topic.
- A careful distinction between loss functions seems to be even more important for MLC than for standard classification, and one cannot expect the same MLC method to be optimal for different types of losses. Surprisingly, new methods are often proposed without explicitly saying what loss they intend to minimize. Instead, they are typically shown to perform well across a wide spectrum of different loss functions, casting some doubts on the reliability of such studies.

4 Experimental Studies

To corroborate our theoretical results by means of empirical evidence, this section presents a number of experimental studies, using both synthetic and benchmark data. As MLC methods, two meta-techniques will be employed, namely the Binary Relevance (BR) and the Label Power-set (LP) classifier. These methods are commonly used as baselines in experimental studies and are of a quite complementary nature [4]. Besides, we will also propose a simple modification of LP that allows for adapting this approach to any loss function. We present results on three artificial data sets pointing to some important pitfalls often encountered in experimental studies of MLC. Finally, we present some results on benchmark data sets and discuss them in the light of these pitfalls.

In the experimental study, we used the WEKA [14] and Mulan [6] packages.

4.1 Binary Relevance and Label Power-Set Classifier

BR is arguably the simplest approach to MLC. It trains a separate binary classifier $h_i(\cdot)$ for each label λ_i. Learning is performed independently for each label, ignoring all other labels. At prediction time, a query instance $\boldsymbol{x}$ is submitted to all binary classifiers, and their outputs are combined into an MLC prediction.

Obviously, BR is tailored for Hamming loss minimization or, more generally, every loss whose risk minimizer can be expressed solely in terms of marginal distributions; as shown in [7], this also includes the rank loss. However, BR does not take label dependence into account, neither conditional nor unconditional, and this is what it is most often criticized for. Indeed, as suggested by our theoretical results, BR will in general not be able to yield risk minimizing predictions for losses like subset 0/1. Moreover, one may suspect that, even though it can minimize Hamming loss theoretically, exploiting label dependencies may still be beneficial practically.

LP reduces the MLC problem to multi-class classification, considering each label subset $L \in \mathcal{L}$ as a distinct meta-class. The number of these meta-classes may become as large as $|\mathcal{L}| = 2^m$, although it is often reduced considerably by ignoring label combinations that never occur in the training data. Nevertheless, the large number of classes produced by this reduction is generally seen as the most important drawback of LP.

Since prediction of the most probable meta-class is equivalent to prediction of the mode of the joint label distribution, LP is tailored for the subset 0/1 loss. Interestingly, however, it can easily be extended to any other loss function, given that the underlying multi-class classifier $\boldsymbol{f}(\cdot)$ does not only provide a class prediction but a reasonable estimate of the probability of all meta-classes (label combinations), i.e., $f(\boldsymbol{x}) \approx \mathbf{p}_{\boldsymbol{x}}(\mathbf{Y})$. Given a loss function $L(\cdot)$ to be minimized, an optimal prediction can then be derived in an explicit way:

$$\mathbf{h}^*(\boldsymbol{x}) = \arg\min_{\boldsymbol{y}} \mathbb{E}_{\mathbf{Y}|\mathbf{X}} L(\mathbf{Y}, \boldsymbol{y})$$

In particular, LP can be improved for the Hamming loss, simply by computing the marginal distributions and combining the marginal modes into a single MLC prediction. In this regard, we note that computing margins is not harder than searching the mode of $\mathbf{p}_{\boldsymbol{x}}(\mathbf{Y})$. We refer to this modification of LP as LP+.

Practically, we improve LP+ by regularizing the (joint) probability estimation. To this end, we make use of shrinking. In Bayesian inference, it is well-known that the estimated parameters are shrunk toward the prior distribution. We mimic such kind of shrinkage by means of a regularized probability estimate:

$$\tilde{p}_{\boldsymbol{x}}(\boldsymbol{y}) = \alpha \hat{p}_{\boldsymbol{x}}(\boldsymbol{y}) + (1 - \alpha)\hat{p}(\boldsymbol{y}),$$

where $\hat{p}_{\boldsymbol{x}}(\boldsymbol{y})$ is given by LP, $\hat{p}(\boldsymbol{y})$ is a prior estimated from the training data, and α is the shrinkage parameter. This parameter can be determined empirically so as to maximize performance on the test set: For given α, the accuracy of the classifier is estimated on a validation set, and an optimal α is found through line-search. In the following, we use $\alpha = 0.95$.

4.2 Artificial Data

We consider three artificial data sets, each one reflecting a typical situation for MLC. In each case, we generated 30 training and testing folds, each containing 1000 instances.

The first data set represents the case of conditional independence. Data are drawn uniformly from the square $\boldsymbol{x} \in [-0.5, 0.5]^2$. The label distribution is given by the product of the marginal distributions defined by $\mathbf{p}_{\boldsymbol{x}}(y_i) = 1/(1 + \exp(-f_i(\boldsymbol{x}))$, where the f_i are linear functions: $f_1(\boldsymbol{x}) = x_1 + x_2$, $f_2(\boldsymbol{x}) = -x_1 + x_2$, $f_3(\boldsymbol{x}) = x_1 - x_2$. The cardinality of labels (the average number of relevant labels for an instance) is 1.503.

In the second data set, the labels are dependent. Data are drawn from the univariate uniform distribution $\boldsymbol{x} \in [-0.5, 0.5]$. The joint distribution is obtained by applying the product rule of probability:

$$\mathbf{p}_{\boldsymbol{x}}(\mathbf{Y}) = \mathbf{p}_{\boldsymbol{x}}(Y_1) \prod_{i=2}^{3} \mathbf{p}_{\boldsymbol{x}}(Y_i | Y_1, \ldots, Y_{i-1}),$$

where the probabilities are modeled by linear functions in a similar way as before: $f_1(x) = x$, $f_2(y_1, x) = -x - 2y_1 + 1$, $f_3(y_2, y_1, x) = x + 12y_1 - 2y_2 - 11$. The cardinality of labels for this data set is 1.314.

The results of the experiment are reported for both data sets in Table 1. All approaches are used with linear support vector machine as base learner. In the case of LP+, we used the approach of [15] to turn SVM scores into probabilities, thus obtaining an estimation of the joint distribution and its marginals. Since the true data generating process is known, we also report the loss of the Bayes-optimal classifier. In the case of the independent data, we observe that both approaches perform equally well in terms of both loss functions. For the dependent data, however, we see that BR and LP are tailored toward different loss functions. As expected, the former performs well in terms of Hamming loss, whereas the latter is superior in terms of subset 0/1 loss. As expected, LP+ is able to adapt to both loss functions. Overall, the results are in complete agreement with our theoretical findings.

Table 1. Results on two artificial data sets: conditionally independent (left) and conditionally dependent (right). Standard errors are given in parentheses.

	Conditional independence		Conditional dependence	
classifier	Hamming loss	subset 0/1 loss	Hamming loss	subset 0/1 loss
BR	0.4208(.0014)	0.8088(.0020)	0.3900(.0015)	0.7374(.0021)
LP	0.4212(.0011)	0.8101(.0025)	0.4227(.0019)	0.6102(.0033)
LP+	0.4181(.0013)	0.8093(.0021)	0.3961(.0033)	0.6135(.0034)
B-O	0.4162	0.8016	0.3897	0.6029

In the literature, LP is often shown to outperform BR even in terms of Hamming loss. Given our results so far, this is somewhat surprising and calls for an explanation. We argue that results of that kind should be considered with caution, mainly because a meta learning technique (such as BR and LP) must always be considered in conjunction with the underlying base learner. In fact, differences in performance should not only be attributed to the meta but also to the base learner. In particular, since BR uses binary and LP multi-class classification, they are typically applied with different base learners, and hence are not directly comparable.

We illustrate this by means of an example. For simplicity, suppose that data is generated without noise (whence the risk of the Bayes optimal classifier for both

Hamming and subset 0/1 loss is 0), and consider a problem with two-dimensional instances $\boldsymbol{x} = (x_1, x_2) \in \mathcal{X} = [-1, 1]^2$ and two labels: $y_1 = [\![x_1 < 0]\!]$ and $y_2 = [\![x_1 > 0]\!] \oslash [\![x_2 > 0]\!]$, where $\oslash$ is the exclusive (logical) disjunction. Obviously, using a linear base learner, BR is not able to solve this problem properly, whereas LP, using a multi-class extension of the linear support vector machine (based on a one-vs-one decomposition) yields almost perfect predictions. However, this multi-class extension is no longer a truly linear classifier. Instead, several linear classifiers are wrapped in a decomposition and an aggregation procedure, yielding a more complex classifier that can produce non-linear decision boundaries. And indeed, giving BR access to a more complex base learner, like a rule ensemble [16], it is able to solve the problem equally well; see results and the scatter plot of data in Fig. 1.

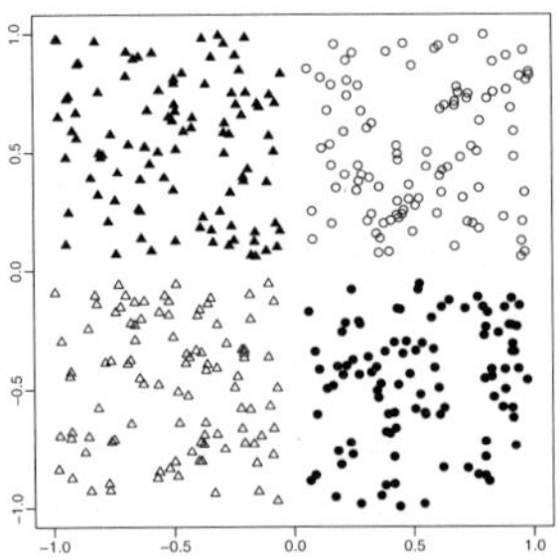

classifier	Hamming loss	subset 0/1 loss
BR Linear SVM	0.2399(.0097)	0.4751(.0196)
BR MLRules	0.0011(.0002)	0.0020(.0003)
LP Linear SVM	0.0143(.0020)	0.0195(.0011)
B-O	0	0

Fig. 1. Plot of the data set composed of two labels: the first label is obtained by a linear model, while the second label represents the exclusive disjunction. The table contains results of three classifiers on this data set.

4.3 Benchmark Data

The second part of the experiments was performed on a collection of 8 MLC data sets.[2] In the case of the *Reuters* data, we used the preprocessed version as in [5]. A summary of the data sets and their properties are given in Table 2.

Table 2. Data sets used in the experiment

data set	# inst.	# attr.	# labels	card.	data set	# inst.	# attr.	# labels	card.
image	2000	135	5	1.236	yeast	2417	103	14	4.237
scene	2407	294	6	1.074	genbase	662	1186	27	1.252
emotions	593	72	6	1.868	slashdot	3782	1079	22	1.181
reuters	7119	243	7	1.241	medical	978	1449	45	1.245

[2] Data sets are taken from `http://mlkd.csd.auth.gr/multilabel.html` and `http://www.cs.waikato.ac.nz/~jmr30/#datasets`

Table 3. Results for Hamming loss. Ranks of classifiers are given in parentheses

	BR SVM	BR MLRules	LP+ pSVM	LP pSVM	LP SVM
image	0.1980 (4)	0.1928(2)	0.1888(1)	0.2021(5)	0.1954(3)
scene	0.1071 (5)	0.0871(1)	0.0919(3)	0.0950(4)	0.0891(2)
emotions	0.2049 (1)	0.2080(2)	0.2091(3)	0.2232(5)	0.2119(4)
reuters	0.0663 (5)	0.0479(1)	0.0565(2)	0.0596(3)	0.0628(4)
yeast	0.2016 (1)	0.2086(3)	0.2156(4)	0.2523(5)	0.2075(2)
genbase	0.0008 (1)	0.0015(5)	0.0011(3)	0.0012(4)	0.0010(2)
slashdot	0.0480 (2)	0.0402(1)	0.0534(4)	0.0631(5)	0.0481(3)
medical	0.0102 (1)	0.0106(2)	0.0132(4)	0.0135(5)	0.0115(3)
Avg. Rank	2.7	2.1	2.75	4.25	3.2

Table 4. Results for subset 0/1 loss. Ranks of classifiers are given in parentheses

	BR SVM	BR MLRules	LP+ pSVM	LP pSVM	LP SVM
image	0.7670 (5)	0.6705(4)	0.5595(2)	0.5600(3)	0.5315(1)
scene	0.4757 (5)	0.4221(4)	0.3299(2)	0.3303(3)	0.3008(1)
emotions	0.7538 (4)	0.7622(5)	0.7353(2)	0.7386(3)	0.6846(1)
reuters	0.3735 (5)	0.2684(4)	0.2391(1)	0.2406(2)	0.2676(3)
yeast	0.8552 (4)	0.8643(5)	0.8155(2)	0.8159(3)	0.7460(1)
genbase	0.0211 (1.5)	0.0332(5)	0.0257(3.5)	0.0257(3.5)	0.0211(1.5)
slashdot	0.6560 (2)	0.6721(3)	0.6819(4)	0.6835(5)	0.5460(1)
medical	0.3405 (2)	0.3497(3)	0.3630(4.5)	0.3630(4.5)	0.3119(1)
Avg. Rank	3.56	4.13	2.63	3.38	1.31

We used BR and LP with linear support vector machines as base learner, and additionally BR with MLRules and LP+ based on the probabilistic SVM. Results of a 3-fold cross-validation are given for Hamming loss in Table 3 and for subset 0/1 loss in Table 4. Overall, the results are again in agreement with our expectations. In particular, LP achieves better results for the subset 0/1 loss, while BR is on average superior in terms of Hamming loss.

Let us have a closer look at the results for the *scene* data set. As reported in [17], LP outperforms BR on this data set in terms of Hamming loss; both methods were used with linear SVM as base learner. Although our results here give the same picture, note that BR with MLRules outperforms both approaches. As pointed out above, comparing LP and BR with the same base learner is questionable and may lead to unwarranted conclusions.

Let us underline that LP+ outperforms LP using probabilistic SVM in terms of the Hamming loss on all datasets. This confirms our theoretical claims and justifies the modification of LP. Also shrinking used in LP+ improves the results for the subset 0/1 loss. However, let us notice that probabilistic SVM performs worse than classical SVM in the case of classification. We can observe that, in terms of the subset 0/1 loss, the latter is much better than the former. Moreover, the latter receives good results for Hamming loss minimization on some data sets. We suppose that, for these data sets, one of the conditions that imply equivalence of the risk minimizers will hold (cf. Proposition 1).

5 Conclusions

In this paper, we have addressed a number of issues related to loss minimization in multi-label classification. In our opinion, this topic has not received enough attention so far, despite the increasing interest in MLC in general. However, as we have argued in this paper, empirical studies of MLC methods are often meaningless or even misleading without a careful interpretation, which in turn requires a thorough understanding of underlying theoretical conceptions.

In particular, by looking at the current literature, we noticed that papers proposing new methods for MLC, and for exploiting label dependencies, rarely distinguish between the type of loss function to be minimized. Instead, a new method is often shown to be better than existing ones "on average", evaluating on a number of different loss functions. Our technical results in this paper, already summarized in Section 3.4 and therefore not repeated here, indicate that studies of that kind might be less illuminative than they could be. First, we have shown that the type of loss function has a strong influence on whether or not, and perhaps to what extent, an exploitation of conditional label dependencies can be expected to yield a true benefit. Consequently, some loss functions will be more suitable than others for showing the benefit of label dependencies. Second, using the example of Hamming and subset 0/1 loss, we have shown that loss functions in MLC cover a broad spectrum, and that minimizing different losses will normally require different estimators. Consequently, one cannot expect an MLC method to perform equally well for various losses of different type.

Our focus on Hamming and subset 0/1 loss can be justified by their complementarity, and by noting that these losses can be considered representative of decomposable and non-decomposable loss functions, respectively. Besides, they are among the most well-known and frequently used performance measures in MLC. Nevertheless, looking at other loss functions is of course worthwhile and holds the promise to gain further insight into the nature of MLC. Expanding our studies in this direction is therefore on our agenda for future work.

References

1. Boutell, M., Luo, J., Shen, X., Brown, C.: Learning multi-label scene classification. Pattern Recognition 37(9), 1757–1771 (2004)
2. Ghamrawi, N., McCallum, A.: Collective multi-label classification. In: CIKM 2005, pp. 195–200 (2005)
3. Amit, Y., Dekel, O., Singer, Y.: A boosting algorithm for label covering in multilabel problems. In: JMLR W&P, vol. 2, pp. 27–34 (2007)
4. Tsoumakas, G., Katakis, I.: Multi label classification: An overview. Int. J. Data Warehousing and Mining 3(3), 1–13 (2007)

5. Cheng, W., Hüllermeier, E.: Combining instance-based learning and logistic regression for multilabel classification. Machine Learning 76(2-3), 211–225 (2009)
6. Tsoumakas, G., Katakis, I., Vlahavas, I.: Mining multi-label data. In: Maimon, O., Rokach, L. (eds.) Data Mining and Knowledge Discovery Handbook. Springer, Heidelberg (2010)
7. Dembczyński, K., Cheng, W., Hüllermeier, E.: Bayes optimal multilabel classification via probabilistic classifier chains. In: ICML 2010 (2010)
8. Taskar, B., Guestrin, C., Koller, D.: Max-margin markov networks. In: NIPS 16. MIT Press, Cambridge (2004)
9. McAllester, D.: Generalization bounds and consistency for structured labeling. In: Predicting Structured Data. MIT Press, Cambridge (2007)
10. MacKay, D.J.C.: Information Theory, Inference, and Learning Algorithms. Cambridge University Press, Cambridge (2003)
11. Breiman, L., Friedman, J.: Predicting multivariate responses in multiple linear regression. J. R. Stat. Soc. Ser. B 69, 3–54 (1997)
12. Caruana, R.: Multitask learning: A knowledge-based source of inductive bias. Machine Learning 28, 41–75 (1997)
13. Nelsen, R.: An Introduction to Copulas, 2nd edn. Springer, Heidelberg (2006)
14. Witten, I.H., Frank, E.: Data Mining: Practical machine learning tools and techniques, 2nd edn. Morgan Kaufmann, San Francisco (2005)
15. Platt, J.C.: Probabilistic outputs for support vector machines and comparisons to regularized likelihood methods. In: Advances in Large Margin Classifiers, pp. 61–74. MIT Press, Cambridge (1999)
16. Dembczyński, K., Kotłowski, W., Słowiński, R.: Maximum likelihood rule ensembles. In: ICML 2008, pp. 224–231 (2008)
17. Tsoumakas, G., Vlahavas, I.: Random k-labelsets: An ensemble method for multilabel classification. In: Kok, J.N., Koronacki, J., Lopez de Mantaras, R., Matwin, S., Mladenič, D., Skowron, A. (eds.) ECML 2007. LNCS (LNAI), vol. 4701, pp. 406–417. Springer, Heidelberg (2007)

Clustering Vessel Trajectories with Alignment Kernels under Trajectory Compression

Gerben de Vries and Maarten van Someren

Informatics Institute, University of Amsterdam, Sciencepark 107, 1098 XG Amsterdam, The Netherlands
{G.K.D.deVries,M.W.vanSomeren}@uva.nl

Abstract. In this paper we apply a selection of alignment measures, such as dynamic time warping and edit distance, to the problem of clustering vessel trajectories. Vessel trajectories are an example of moving object trajectories, which have recently become an important research topic. The alignment measures are defined as kernels and are used in the kernel k-means clustering algorithm. We investigate the performance of these alignment kernels in combination with a trajectory compression method. Experiments on a gold standard dataset indicate that compression has a positive effect on clustering performance for a number of alignment measures. Also, soft-max kernels, based on summing all alignments, perform worse than classic kernels, based on taking the score of the best alignment.

Keywords: alignment kernels, trajectory compression, trajectory clustering.

1 Introduction

Largely due to the ubiquity of GPS receivers, moving object trajectories have become an important research topic. To gain insight into the behavior of moving objects, such as vessels, it can be useful to cluster their trajectories into groups of similar movement patterns. Essentially, moving object trajectories are a kind of multivariate time-series. Furthermore, they have the property that they are usually different in temporal length, distance traveled and the number of data points. Alignment methods, such as dynamic time warping or edit distance, are designed to handle these kinds of variations. Thus, such methods can make a suitable similarity measure for clustering moving object trajectories.

We cannot use these similarities directly in the standard, and very popular, k-means clustering method, because this method requires an explicitly defined feature space. Kernel k-means [17] is a more recent k-means based clustering algorithms that deals with similarities directly, provided they are defined as kernels. Defining these alignments as kernels also has the advantage that they can potentially be used in other kernel based machine learning settings.

Computing alignment kernels can be expensive, because of the number of elements in a trajectory. However, vessel trajectories have the property that

J.L. Balcázar et al. (Eds.): ECML PKDD 2010, Part I, LNAI 6321, pp. 296–311, 2010.

they are very regular. Without losing much information they can be compressed very well with trajectory compression techniques such as [10]. Computing an alignment between two compressed trajectories can be a lot cheaper, because of the reduction in data points. But, the quality of the alignment does not necessarily have to be the same. Compression loses information and may have a bad effect on the alignment.

In this paper we define different alignment kernels for moving object trajectories. The kernels are based on a number of well-known alignment measures. We consider two kernel versions of these measures, a classic and a soft-max one. Not all of these kernels are proper kernels, since they are not positive semi-definite. However, for simplicity, we refer to all of them as kernels. Furthermore, all kernels can be used with the kernel k-means clustering algorithm that we use in our experiment. The goal of this experiment is to investigate the performance of the different defined kernels on the task of clustering vessel trajectories. Kernels are computed on uncompressed and compressed trajectories to discover the influence of trajectory compression on this task. A number of the alignment kernels show better performance when the trajectories are compressed first.

The rest of the paper is organized as follows. In Sect. 2 we discuss related work in the fields of trajectory clustering and kernel methods for time-series. Section 3 contains the technical definitions: three kinds of alignments, two kinds of kernel definitions based on these alignments, kernel k-means clustering and trajectory compression. Our experiment and the results of clustering a hand-labeled set of vessel trajectories is described in Sect. 4. We end with some conclusions and directions for future work.

2 Related Work

We have not seen work on clustering moving object trajectories using an alignment kernel based approach. A number of studies address clustering time series data, see [14] for an overview. More specifically, there are a number of papers researching the problem of clustering trajectories, mostly coming from the Moving Object Database community [8,13,15,16,19]. None of these studies consider the influence of compression on the clustering task. Also, these papers are not comparative studies between similarity measures. The authors of [19] take an alignment based approach, as we do in this paper, but use a density based clustering algorithm. Also, [15] takes a density based approach but computes the distance between trajectories based on the area between them. In [13], compression based on the minimum description length principle is applied to trajectories first and then density based clustering is used to discover sub-trajectories. Trajectories are first converted into a grid based representation and then clustered using fuzzy c-means in [16]. A route similarity function very similar to dynamic time warping is described in [1]. Very different is the somewhat older [8], in which the authors use a mixture of regression models based approach. Most of the above papers use a density based clustering algorithm, which assumes that clusters are not (densely) connected. However, we do not know whether this is actually the case when using alignment based similarities, therefore, we prefer kernel k-means.

There is also a body of work on using alignment kernels for time-series, for instance [5,9,11,12]. This work is mostly in the context of classification of handwriting data using support vector machines. Both [11] and [12] use an alignment kernel based on Dynamic Time Warping (DTW). However, [12] uses the soft-max version defined in [5], whereas [11] uses a more classic version of DTW. In the following we will consider both types.

3 Trajectory Alignment Kernels

In this section we first define the notion of a moving object trajectory. Then we look at alignments and define different kernels for them. We also briefly review the kernel k-means clustering algorithm. Finally, we give the trajectory compression algorithm that we have used.

3.1 Trajectories

A moving object trajectory is defined in Definition 1 below. Note that trajectories include time.

Definition 1. *A moving object trajectory in 2-dimensional space is represented by a sequence of vectors:* $T = \langle x_1, y_1, t_1 \rangle, \ldots, \langle x_n, y_n, t_n \rangle$. *Where* x_i *and* y_i *represent the position of the object at time* t_i. *The length of a trajectory, i.e. the number of vectors, is denoted as:* $|T|$. *Furthermore, let* $T_i = \langle x_i, y_i, t_i \rangle$.

The sample rate of trajectories is not fixed, thus the difference between consecutive values t_i, t_{i+1} is not the same. Also, there are more dimensions to trajectories that can be derived from the x, y, t information that we do not consider, such as speed and direction. In some tasks and applications these attributes might be more relevant than the absolute position x, y and time t. In principle these dimensions could just be added to the $\langle x, y, t \rangle$ vector, but we have not considered this for the clustering task that we will define in Sect. 4. In the following we refer to a vector $\langle x_i, y_i, t_i \rangle$ as trajectory element or point.

3.2 Alignments

We define three types of alignments between two trajectories. Based on these alignment we define kernels that express similarity between trajectories in Sect. 3.3. We start with the simplest alignment, which can be considered to be the baseline alignment. The goal of an alignment is to bring the elements of two sequences, e.g. trajectories, in a correspondence, typically with the aim to maximize (or minimize) a scoring function defined on this correspondence.

Assume in the following that there is a function $\mathbf{sub}(S_i, T_j)$ that gives a score to substituting the trajectory element S_i with T_j. This score represents the similarity between these two elements. To enhance readability, we will not make use of superscripts to indicate the specific type of alignment when this cannot lead to confusion.

We define different alignments for two trajectories below, following the notation and definitions from [18].

Definition 2. *An alignment* π*, possibly with gaps, of* $p \geq 0$ *positions between two trajectories* S *and* T *is a pair of* p*-tuples:*

$$\pi = ((\pi_1(1), \ldots, \pi_1(p)), (\pi_2(1), \ldots, \pi_2(p))) \in \mathbb{N}^{2p} \ .$$

The idea behind this definition is that the alignment encodes p elements in each trajectory that are aligned to each other. That is, the $\pi_1(i)$th element in S is aligned to the $\pi_2(i)$th element in T. Multiple elements of S can be aligned to one element in T and vice versa. Furthermore, not all elements have to be aligned to an element in the other trajectory, in this case an element is a *gap*.

Shortest Sequence Alignment. One of the simplest alignments defined on two sequences with discrete elements is the Longest Common SubSequence (LCSS) measure. It is defined as the length of the longest sequence of elements existing in both sequences, whereby gaps are allowed. Definition 3 differs from other definitions [9,19] of LCSS for sequences with continuous elements, i.e. elements with continuous values, because we will not use a threshold in our substitution function. On first glance the resulting measure does not have much in common with LCSS, thus to avoid confusion we redub it: Shortest Sequence Alignment (SSA).

Definition 3. *An SSA alignment* π^{SSA} *is an alignment according to Definition 2, with the additional constraints that:*

$$p = \min(|S|, |T|)$$

and

$$1 \leq \pi_1(1) < \pi_1(2) < \ldots < \pi_1(p) \leq |S| \ ,$$
$$1 \leq \pi_2(1) < \pi_2(2) < \ldots < \pi_2(p) \leq |T| \ .$$

Intuitively this means that all elements of the shortest sequence are aligned with different unique elements in the other sequence, hence Shortest Sequence Alignment.

Dynamic Time Warping. A very popular alignment method more specifically designed for time-series is Dynamic Time Warping (DTW). In Definition 4 we follow [5].

Definition 4. *A DTW alignment* π^{DTW} *is an alignment according to Definition 2, but has the additional constraints that there are unitary increments and no simultaneous repetitions, thus* $\forall(1 \leq i \leq p-1)$*,*

$$\pi_1(i+1) \leq \pi_1(i) + 1, \ \ \pi_2(j+1) \leq \pi_2(j) + 1$$

and

$$(\pi_1(i+1) - \pi_1(i)) + (\pi_2(i+1) - \pi_2(i)) \geq 1 \ .$$

Furthermore,

$$1 = \pi_1(1) \leq \pi_1(2) \leq \ldots \leq \pi_1(p) = |S| \ ,$$
$$1 = \pi_2(1) \leq \pi_2(2) \leq \ldots \leq \pi_2(p) = |T| \ .$$

This means that all elements in both trajectories are aligned, which might require repeating elements from a trajectory, but in the alignment we cannot simultaneously repeat an element in both trajectories. Furthermore, the start and end of trajectories are aligned by default.

Edit Distance. Edit distances are a popular method for comparing similarity of strings. This similarity is computed in terms of the number of substitutions, deletions and insertions that are required to transform one string into another string. We define (Definition 5 and 6) a version for continuous elements similar to how [4] defines it for time-series. However we chose to consider fixed gap penalties, i.e. deletion and insertion costs, because this seems more natural in the trajectory case.[1]

Definition 5. *An edit distance alignment* π^{ED} *is an alignment according to Definition 2, but also has the constraints:*

$$1 \leq \pi_1(1) < \pi_1(2) < \ldots < \pi_1(p) \leq |S| \ ,$$
$$1 \leq \pi_2(1) < \pi_2(2) < \ldots < \pi_2(p) \leq |T| \ .$$

This means that not all elements have to be aligned and there is no repetition of elements.

We define a score function for the three types of alignments below in Definition 6. Note that the function is the same for DTW and SSA alignments.

Definition 6. *The score for an alignment* π *of type* ϕ *of two trajectories* S *and* T *is equal to:*

$$s^{\phi}(\pi) = \left(\sum_{i=1}^{|\pi|} \mathbf{sub}(S_{\pi_1(i)}, T_{\pi_2(i)}) \right) + g(|S| - |\pi|) + g(|T| - |\pi|) \ ,$$

where $g = 0$ *if* $\phi = \mathrm{SSA, DTW}$ *and* $g < 0$ *if* $\phi = \mathrm{ED}$.

[1] In [4] the gap penalties are essentially the value of the gapped element minus a fixed constant. For trajectories this would mean that something on the edge of the map would have a higher penalty than in the middle, which we think is counter-intuitive.

Considering the three types of alignments that are defined above, we note that the most important difference is in how they treat gaps. In the SSA case gaps get no penalty, in the DTW case gaps are treated by repeating a trajectory element, and thus get a score according to the substitution function **sub**, and in the edit distance case gaps have a fixed penalty g.

3.3 Alignment Kernels

Below we will give two different ways to create a kernel to express similarity between trajectories, using the alignments defined above. The first type of kernel is based on taking the score of the alignment that maximizes the score function. This differs from the second kernel which takes the soft-max of the scores of all alignments.

First, we define the substitution function to be the negative of the L^2 norm, i.e. the regular Euclidean distance, in Definition 7. Other functions are possible here, but we have not experimented with this. Note that x and y are of the same dimension, but t is not directly comparable to x and y, hence in the experiments we apply a weight to t to make it comparable. In practice, this weight will depend on the domain and goal of one's application.

Definition 7.

$$\mathbf{sub}(\langle x_i, y_i, t_i \rangle, \langle x_j, y_j, t_j \rangle) = -\|\langle x_i - x_j, y_i - y_j, t_i - t_j \rangle\| \ .$$

Above, in Sect. 3.2, we have defined a number of alignments between two trajectories and their score functions. However, this does not give a similarity between two trajectories yet, because there are a lot of possible alignments, and corresponding scores, for one type of alignment.

One option for similarity between two trajectories is to take the score of the alignment that maximizes the respective score function s. We will call this similarity $\mathit{Sim}_{\max}$ and define it below in Definition 8. This similarity corresponds to the typical way that DTW and edit distance are defined.

Definition 8. *Given two trajectories S and T, let $\Pi^{\phi}_{S,T}$ be the set of all possible alignments between these trajectories under a certain alignment measure ϕ, then*

$$\mathit{Sim}^{\phi}_{\max}(S,T) = \max_{\pi \in \Pi^{\phi}_{S,T}} s^{\phi}(\pi) \ .$$

However, it has been argued [5,18] that taking the soft-max[2] of the scores of all possible alignments leads to better kernels, because all scores are taken into account. We will call this alignment score $\mathit{Sim}_{\mathrm{sum}}$, because the sum over all possible alignments is taken. It is given in Definition 9 below.

Definition 9. *Given two trajectories S and T, let $\Pi^{\phi}_{S,T}$ be the set of all possible alignments between these trajectories under a certain alignment measure ϕ, then*

$$\mathit{Sim}^{\phi}_{\mathrm{sum}}(S,T) = \sum_{\pi \in \Pi^{\phi}_{S,T}} \exp(s^{\phi}(\pi)) \ .$$

---

[2] Given a set of positive scalars $Z = z_1, \ldots, z_n$, the soft-max of Z is $\log \sum e^{z_i}$.

These similarities are not kernels yet. We define a kernel based on $Sim_{\max}$ in Definition 10.

Definition 10. *For all trajectories T^i and T^j in a set of trajectories $\mathcal{T}$, we compute the $K^{\phi}_{\max}$-kernel for an alignment measure ϕ as:*

$$K^{\phi}_{\max}(i,j) = Sim^{\phi}_{\max}(T^i, T^j) \ ,$$

furthermore we normalize and make a kernel out of $K^{\phi}_{\max}$ by:

$$K^{\phi}_{\max} = 1 - \frac{K^{\phi}_{\max}}{\min(K^{\phi}_{\max})} \ .$$

The kernel based on Sim_{sum} is given in Definition 11.

Definition 11. *For all trajectories T^i and T^j in a set of trajectories $\mathcal{T}$, we compute, and normalize, the K^{ϕ}_{sum}-kernel for an alignment measure ϕ as:*

$$K^{\phi}_{\text{sum}}(i,j) = \frac{Sim^{\phi}_{\text{sum}}(T^i, T^j)}{\sqrt{Sim^{\phi}_{\text{sum}}(T^i, T^i) Sim^{\phi}_{\text{sum}}(T^j, T^j)}} \ .$$

The $K_{\max}$ kernels are not proper kernels in the sense that they are not Positive Semi-Definite (PSD). However it has been observed that non-PSD kernels can still work well in practice (for DTW for instance in [11]). We notice this in our experiment in Sect. 4 as well. The soft-max version of the DTW alignment kernel was proven to be PSD in [5] given a proper substitution function (e.g. the one we have defined). Furthermore, we conjecture that similar proofs are possible for the other two soft-max kernels we define, because they are very similar to either DTW or Smith-Waterman, for which there is a proof in [18].

The authors of [5,18] also notice that the soft-max type of kernel often suffers from the diagonal dominance issue and they remedy this by taking the logarithm of this kernel. We have experimented with taking the logarithm, but this did not improve results.

The classic alignment similarities, $Sim_{\max}$, are efficiently computable via a dynamic programming approach. By replacing the max-sum algebra with a sum-product algebra we can do the same for the Sim_{sum} versions. This is shown for DTW in [5]. We can see that this works when we, for instance, work out $Sim^{\text{DTW}}_{\text{sum}}$ as in (1).

$$\begin{aligned} Sim^{\text{DTW}}_{\text{sum}}(S,T) &= \sum_{\pi \in \Pi_{S,T}} \exp\Big(\sum_{i=1}^{|\pi|} \mathbf{sub}(S_{\pi_1(i)}, T_{\pi_2(i)})\Big) \\ &= \sum_{\pi \in \Pi_{S,T}} \prod_{i=1}^{|\pi|} \exp(\mathbf{sub}(S_{\pi_1(i)}, T_{\pi_2(i)})) \ . \end{aligned} \tag{1}$$

To sum up, we have now defined a number of kernel functions. Two for the SSA measure: $K^{\text{SSA}}_{\max}$ and $K^{\text{SSA}}_{\text{sum}}$. Two for the DTW measure: $K^{\text{DTW}}_{\max}$ and $K^{\text{DTW}}_{\text{sum}}$. And a family of kernels for the edit distance measure, since these are parameterized by the value of the gapping penalty g: $K^{\text{ED}}_{\max}$ and $K^{\text{ED}}_{\text{sum}}$.

3.4 Kernel K-Means

Because the focus of the paper is on alignments, we use the relatively simple kernel k-means algorithm [17]. More advanced kernel based algorithms exist, such as weighted kernel k-means [6] and support vector clustering [2].

The kernel version of k-means works in exactly the same way as regular k-means. This means that we start out with a random initialization of k clusters and at each iteration try to minimize the distance between the cluster centers and objects in the clusters until we arrive at a stable clustering (Definition 12).

Definition 12. *A clustering C of a set of objects $\mathcal{O}$ into k partitions is:*

$$C = \{c_1, \ldots, c_k\} ,$$

such that for all $o \in \mathcal{O}$, o is exactly in one c_i, where $c_i \subseteq \mathcal{O}$.

The difference lies in how the distance from an object to a cluster center is calculated. This is done using the assumption that a kernel represents a dot-product in a higher dimensional feature space. Without explicitly knowing this feature space, the distance from an object to a cluster center can be calculated with these dot-products, as in Definition 13.

Definition 13. *Let K be a kernel computed for a set of objects $\mathcal{O}$ and C a clustering of $\mathcal{O}$. The distance from an object $o_i \in \mathcal{O}$ to the center of a cluster $c \in C$ is computed by:*

$$K(i,i) - \frac{2}{|c|} \sum_{o_j \in c} K(i,j) + \frac{1}{|c|^2} \sum_{o_j, o_k \in c} K(j,k) .$$

Because different random initializations can lead to different stable partitions C, the kernel k-means clustering is run a number of times. The partitioning with the lowest intra cluster spread is kept as the final clustering.

3.5 Trajectory Compression

For trajectory compression we use the Piecewise Linear Segmentation (PLS) method proposed in [10], which is an advanced version of a classic line simplification algorithm [7]. The algorithm, given in Algorithm 1, compresses a trajectory T by finding the point $\langle x_i, y_i, t_i \rangle$ that has the largest error $\mathbf{E}_\mu$, as defined in Definition 14. If this error is larger than a given threshold ϵ then we keep this point and recursively apply the same procedure on the two sub-trajectories created by splitting T at that point. Thus the result of applying trajectory compression to a trajectory T is that the compressed trajectory T^C is a subset of the points that are in T, always including the start and end points. This subset contains the most salient points of the trajectory. Thus, contrary to smoothing or resampling, existing points are not altered and no new points are created.

In Definition 14, the error $\mathbf{E}_\mu$ is defined as the Euclidean distance between a point $\langle x_i, y_i, t_i \rangle$ and the closest point on the line $\langle x_1, y_1, t_1 \rangle, \langle x_n, y_n, t_n \rangle$, where the influence of the temporal dimension is weighed using the parameter μ.

Algorithm 1. $\mathbf{pls}(T, \epsilon)$

We use *end* to indicate the index of the last element of a trajectory.
$d_{max} = 0$, $i_{max} = 0$
for $i = 2$ **to** $end - 1$ **do**
  $d = \mathbf{E}_\mu(T_i, T_1, T_{end})$
  if $d > d_{max}$ **then**
    $i_{max} = i$, $d_{max} = d$
  end
end
if $d_{max} \geq \epsilon$ **then**
  $A = \mathbf{pls}(T_1, \ldots, T_{i_{max}}, \epsilon)$, $B = \mathbf{pls}(T_{i_{max}}, \ldots, T_{end}, \epsilon)$
  $T^C = A, B_2, \ldots, B_{end}$
else
  $T^C = T_1, T_{end}$
end
return T^C

Definition 14

$$\mathbf{E}_\mu(\langle x_i, y_i, t_i\rangle, \langle x_1, y_1, t_1\rangle, \langle x_n, y_n, t_n\rangle) = \|\langle x_i - x'_i, y_i - y'_i, \mu(t_i - t'_i)\rangle\| \ ,$$

where $\langle x'_i, y'_i, t'_i\rangle$ *is the closest point on the line-segment* $\langle x_1, y_1, t_1\rangle, \langle x_n, y_n, t_n\rangle$ *in terms of the Euclidean distance.*

4 Evaluation

In this section we will present a number of experiments to test the performance of the different combinations of alignment measures and kernels on the task of clustering a set of vessel trajectories. We especially investigate the effect of applying a compression algorithm to these trajectories as a preprocessing step.

4.1 Dataset

Our experimental dataset consist of 716 vessel trajectories originally gathered using the Automatic Identification System (AIS). A domain expert has partitioned these trajectories into 8 different clusters, creating a gold standard $G = g_1, \ldots, g_8$. The clusters are very different in size, ranging from 8 to 348. The average length of a trajectory is a sequence of 300 $\langle x, y, t\rangle$ vectors. Trajectories are delimited either by the vessel leaving the area of observation or the vessel stopping. For all trajectories t_1 is set to 0. We illustrate the clustering in Fig. 1. The general direction of movement for each of the 8 clusters is indicated with an arrow. Note that we only plotted the x and y dimensions and not t. The position data of the vessels was originally collected in a latitude, longitude format but has been converted, using a suitable map projection, to allow for normal Euclidean geometry.

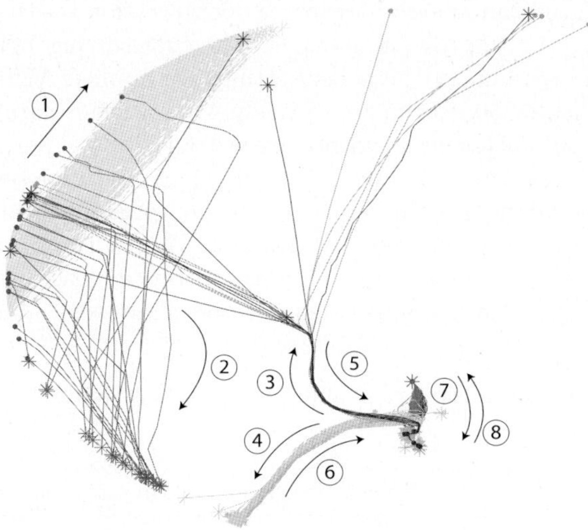

Fig. 1. An overview of the vessel trajectory dataset with the 8 different clusters of the gold standard indicated with shades of gray and an arrow giving the general direction. A solid dot indicates the start of a trajectory and an asterix indicates the end.

4.2 Experimental Setup

As test datasets we constructed 4 random subsets from the 716 vessels set. From each cluster in the gold standard we randomly extracted up to 20 trajectories. Some clusters contained fewer elements in which case we took them all. This resulted in 4 datasets of 138 trajectories. We did this to allow for reasonable computation times for the experiments.

Because it is unclear how to weigh the time dimension against the space dimensions, we have defined a number of settings for this. In the first setting we treated the trajectories as if they were sampled with a fixed sample rate. This means that some positions had to be linearly interpolated between two positions from the original data. In this setting a trajectory is a sequence of $\langle x, y\rangle$ points and time is implicitly represented by the fixed time interval between these points. We introduce this condition, because DTW is commonly applied to fixed sample rate time-series. Furthermore, by decreasing the sample rate we get a baseline form of compression to compare to Algorithm 1. In the other settings we tested 4 weights w for the time dimension. So that a trajectory essentially becomes: $T = \langle x_1, y_1, wt_1\rangle, \ldots, \langle x_n, y_n, wt_n\rangle$. As weights we took $w = 0, \frac{1}{3}, \frac{2}{3}, 1$. With $w = 0$ we ignore the time dimension and with the other weights we increasingly weigh the time dimension more heavily. The $w = 1$ setting means that the average difference between two points in the space dimension is roughly equal to the average difference in the time dimension.[3]

[3] To enhance readability we do not give the actual values for the weights, because they depend on the units that the dimensions are in.

To investigate the influence of trajectory compression on the clustering performance we used 4 compression settings. The first setting being $\epsilon = 0$, thus, we apply no compression. In the other settings we applied Algorithm 1 to each trajectory with $\epsilon = 25, 50, 100$ m, respectively. We set the μ parameter to a value that we determined earlier and did not vary it for these experiments, however, it is a potential parameter to tune. In the fixed sample rate setting we set the sample rate to be (very) roughly equal to the compression rate achieved under the different compression settings, see Table 1. We set the sample rate for the no compression case to the average temporal distance between two consecutive trajectory samples in the uncompressed dataset, again see Table 1.

Table 1. Average compression rate and corresponding sample rate for different ϵ settings

	No compression	$\epsilon = 25$ m	$\epsilon = 50$ m	$\epsilon = 100$ m
Compression rate (%)	0	96	97	98
sample rate (Hz)	0.1	0.01	0.002	0.001

For each of the different combinations of settings, i.e. for each combination of subset, time setting and compression setting, we computed a number of kernels. Two for the SSA measure: K^{SSA}_{max} and K^{SSA}_{sum}. Two for the DTW measure: K^{DTW}_{max} and K^{DTW}_{sum}. And 10 for the Edit measure; five times K^{ED}_{max} and five times K^{ED}_{sum}, with $g = -0.01, -0.025, -0.05, -0.075, -0.1$.

We used each of these kernels as input for the kernel k-means clustering algorithm. The kernel k-means algorithm is run 100 times with random initializations and we keep the clustering with minimal intra cluster spread. This process is repeated 10 times. We set $k = 8$, i.e. the same amount of clusters as in the gold standard. Thus for each kernel we get 10 clusterings that we evaluate against the gold standard G according to the scoring function in Definition 15, taken from [14]. The intuition behind this definition is that we take the best F1-score[4] for each cluster g_i in the gold standard G and average over these scores.

Definition 15. *The score of a clustering C of size k with respect to a gold standard G of size k is:*

$$\mathbf{score}(C, G) = \frac{1}{k} \sum_{i=1}^{k} \max_{1 \leq j \leq k} \frac{2|g_i \cap c_j|}{|g_i| + |c_j|} \; .$$

4.3 Results

In the results below the mean of a kernel is computed over the scores for the 4 subsets with 10 repeats each, thus for each kernel $N = 40$. Per kernel we statistically compare the scores for the high compression/low sample rates settings

[4] The F1-score is the harmonic mean between precision and recall.

to the scores for the no compression/high sample rate setting using a two-tailed student t-test with $p < 0.05$.

We do not give all the mean scores because of space constraints. Also, discussion of the presented results is postponed to Sect. 5. First we will look at the performance of the different kernels under trajectory compression. Therefore we look at the time setting for which the highest scores are achieved. In Table 2 we give the results for the 14 kernels under the time setting $w = 0$. Scores in **bold** indicate a significant difference between that clustering score, according to the above defined test, and the clustering score for the no compression case, with the bold score being higher. Scores in *italic* indicate a significant difference between that score and the score for the no compression case, with the score in italic being lower. For completeness we give the mean score of 40 random clusterings, which is 0.24.

The best performing kernels for our clustering task are the $K_{\text{max}}^{\text{ED}}$ kernels. With the right g setting they can achieve a perfect clustering, e.g. a score of 1.0.[5] These kernels perform better on the compressed trajectories than on the uncompressed trajectories. However, the $K_{\text{sum}}^{\text{ED}}$ kernels perform worse on the compressed trajectories. The $K_{\text{max}}^{\text{DTW}}$ kernel also performs better on the compressed data. However, it does not perform as much better on the compressed data as the $K_{\text{max}}^{\text{ED}}$ kernels. Both shortest sequence alignment kernels perform a lot worse than the other kernels. The $K_{\text{sum}}^{\text{ED}}$ and $K_{\text{sum}}^{\text{DTW}}$ kernels perform almost identical in the uncompressed case.

Table 2. Value of **score** for 14 different kernels under time setting $w = 0$

Kernel	No compression	$\epsilon = 25$ m	$\epsilon = 50$ m	$\epsilon = 100$ m
max, SSA	0.52	*0.34*	*0.35*	*0.42*
max, DTW	0.87	**0.90**	**0.91**	**0.91**
max, ED, $g = -0.01$	0.91	0.92	0.93	**0.94**
max, ED, $g = -0.025$	0.88	**1.0**	**1.0**	**1.0**
max, ED, $g = -0.05$	0.88	**1.0**	**1.0**	**1.0**
max, ED, $g = -0.075$	0.88	**0.93**	**0.97**	**0.99**
max, ED, $g = -0.1$	0.87	**0.90**	**0.89**	**0.96**
sum, SSA	0.52	*0.34*	*0.35*	*0.33*
sum, DTW	0.87	*0.82*	*0.85*	0.87
sum, ED, $g = -0.01$	0.87	*0.62*	*0.63*	*0.65*
sum, ED, $g = -0.025$	0.87	*0.64*	*0.64*	*0.63*
sum, ED, $g = -0.05$	0.87	*0.63*	*0.63*	*0.66*
sum, ED, $g = -0.075$	0.87	*0.64*	*0.63*	*0.66*
sum, ED, $g = -0.1$	0.87	*0.64*	*0.63*	*0.67*

Next we will look at the effect of the different time settings. In Table 3 we give a selection of kernel results for the fixed sample rate time setting. We have included

[5] This is the rounded average of all 40 runs, not all runs get a score of 1.0.

the two edit distance kernels with the best performance and also added the best performing kernel[6] ($K^{\mathrm{ED}}_{\max,g=-0.05}$) for the $w = 0$ setting. Bold and italic have a similar meaning as in the previous table. Furthermore, $^+$ indicates a significant difference with the corresponding value from Table 2 if this difference is positive. This is similar for $^-$ but the difference is negative. We can see that there are better performing kernels in the $w = 0$ setting. Table 3 also shows that low frequency fixed sample rates (≤ 0.01 Hz) do not degrade the performance of a number of kernels and sometimes even improve it.

Table 3. Value of **score** for a selection of 7 different kernels under the fixed sample rate setting

Kernel	0.1 Hz	0.01 Hz	0.002 Hz	0.001 Hz
max, SSA	0.47^-	0.46^+	0.47^+	0.45^+
max, DTW	0.87	0.87^-	0.87^-	0.87^-
max, ED, $g = -0.01$	0.95^+	0.94	**0.97^+**	0.94
max, ED, $g = -0.05$	0.87^-	**0.87^-**	**0.87^-**	0.87^-
sum, SSA	0.48^-	*0.39^-*	*0.29^-*	*0.26^-*
sum, DTW	0.86^-	*0.71^-*	*0.70^-*	*0.75^-*
sum, ED, $g = -0.01$	0.86^-	*0.62*	*0.61^-*	*0.59^-*

In Table 4 we give the results for a selection of kernels for the setting $w = \frac{1}{3}$. We have taken the same edit distance kernels as in Table 3, which also includes the best performing one for this setting. Again, $^+$ and $^-$ indicate statistical significance compared to the same kernel for the $w = 0$ setting. We see that there are better performing kernels in the $w = 0$ setting.

Table 4. Value of **score** for a selection of 7 different kernels under time setting $w = \frac{1}{3}$

Kernel	No compression	$\epsilon = 25$ m	$\epsilon = 50$ m	$\epsilon = 100$ m
max, SSA	0.52	*0.36^+*	*0.42^+*	0.52^+
max, DTW	0.87^-	**0.89^-**	**0.89^-**	**0.90**
max, ED, $g = -0.01$	0.89^-	*0.82^-*	*0.80^-*	*0.80^-*
max, ED, $g = -0.05$	0.88	**0.95^-**	**0.98^-**	**0.99^-**
sum, SSA	0.51	*0.33*	*0.35*	*0.33*
sum, DTW	0.83^-	*0.82*	0.84	**0.86**
sum, ED, $g = -0.01$	0.83^-	*0.63*	*0.64*	*0.65*

The performance for the time settings $w = \frac{2}{3}, 1$ is worse than the results for $w = \frac{1}{3}$. Thus, we did not consider it necessary to include them. We furthermore remark that the difference in mean score between the best performing kernel in

[6] Based on the non-rounded scores.

the $w = 0$ setting ($K^{\text{ED}}_{\max,g=-0.05}, \epsilon = 50\,\text{m}, \textbf{score} = 1.0$) and the best performing kernel in the fixed sample rate setting ($K^{\text{ED}}_{\max,g=-0.01}, 10\,\text{Hz}, \textbf{score} = 0.95$) is significant.

We give some results on computation time in Table 5. These computation times are for running our MatLab implementation of the DTW alignment kernels[7] on an AMD Phenom II X4 955 (3.2 Ghz) cpu system with 4 GB of ram. The computation time for a kernel on the full dataset would be around a factor 40 larger. This does not make it intractable, but did not allow us enough flexibility in performing experiments.

Table 5. Running time for computing one DTW kernel. In the compression cases, the computation time needed for compression is included.

Kernel type	No compression	$\epsilon = 25$m	$\epsilon = 50$m	$\epsilon = 100$m
max	330s	2.4s	1.9s	1.7s
sum	356s	2.6s	2.1s	1.8s

4.4 Performance on the Full Dataset

We have also computed a set of kernels on the full dataset. Because of long computation time we did not do this for the no compression/high sample rate setting. We used the same kernel k-means settings as before, which results in 10 clusterings per kernel. In general we see the same performance differences as above, but the absolute numbers are lower. The highest performance, 0.86, is achieved by $K^{\text{ED}}_{\max}$ with $g = -0.075$ under the $w = \frac{2}{3}$ and $\epsilon = 50\,\text{m}$ settings. However, this score does not differ significantly from the best score, 0.85, in the $w = 0$ setting, which is for $K^{\text{ED}}_{\max}$ with $g = -0.025$ and $\epsilon = 25\,\text{m}$.

5 Conclusions and Future Work

The main result is that the $K^{\text{ED}}_{\max}$ and $K^{\text{DTW}}_{\max}$ kernels perform better on the compressed trajectories, even though we see from Tables 1 and 5 that the dataset is reduced by 96% or more, and kernel computation time is at least 100 times faster. The trajectory compression algorithm works by recursively retaining the most salient elements of a trajectory, reducing noise. Removing so many points from the trajectories can have a negative influence on the alignments. However, in our dataset, similar trajectories have similar salient elements, vessels stop and turn in the same places due to geographical and legislative constraints. This means that similar compressed trajectories can be aligned well. In such a setting applying trajectory compression as a preprocessing step can reduce computational costs significantly and, at the same time, improve performance.

[7] The results are comparable for other kernels.

That the $K_{\max}^{\mathrm{ED}}$ kernels perform better than the $K_{\max}^{\mathrm{DTW}}$ kernels on the compressed trajectories can be explained by how gaps are treated. In the edit distance case only a relatively small gap penalty is added for unaligned points. However, in the DTW case, all points have to be aligned. Because of compression, consecutive points in a trajectory can be very far apart, and repeating an element can lead to a high penalty. For the $K_{\mathrm{sum}}^{\mathrm{ED}}$ kernels the relatively small gap penalty actually seems to work against them, since bad alignments still have a relatively high score: the worst score is a summing of just gap penalties. These high scores are a relatively big, but meaningless, part of the sum. This does not happen in the $K_{\mathrm{sum}}^{\mathrm{DTW}}$ case, bad alignments do not get such a relatively high score. So, we see that the $K_{\mathrm{sum}}^{\mathrm{DTW}}$ kernels perform almost as well as $K_{\max}^{\mathrm{DTW}}$ versions. The almost identical performance of the $K_{\mathrm{sum}}^{\mathrm{ED}}$ and $K_{\mathrm{sum}}^{\mathrm{DTW}}$ kernels in the uncompressed case is because the amount of points in a trajectory is so high that there are very little gaps compared to aligned elements, and thus the influence of how the DTW and edit kernels treat gaps is very small.

The main reason for shortest sequence alignment kernels to perform so much worse is that a trajectory of a small number of elements can still be very similar to a trajectory with a lot more elements, because we do not penalize gaps. In most cases, such trajectories are actually not very similar.

Inspection of the dataset shows that the best performance being in the $w = 0$ setting is not so strange. Within one cluster most trajectories actually have similar speeds, and time is thus not so important, the sequence of the $\langle x, y \rangle$ points is enough. We have not investigated what happens if we do not set each start time of the trajectory to 0, but, for instance, temporally align them differently, this is something for future work.

In future work we would also like to look at other types of movement data to test the influence of trajectory compression. Especially data with less structure than vessel trajectories and a larger role of the temporal component. Furthermore, it would be interesting to look at other types of clustering methods with these alignments, for instance density based ones. Finally, we would like to look at similarity measures from computational geometry, as for instance in [3,15].

Acknowledgments

This work has been carried out as a part of the Poseidon project in collaboration with Thales Nederland, under the responsibilities of the Embedded Systems Institute (ESI). This project is partially supported by the Dutch Ministry of Economic Affairs under the BSIK program.

References

1. Andrienko, G., Andrienko, N., Wrobel, S.: Visual analytics tools for analysis of movement data. SIGKDD Explor. Newsl. 9(2), 38–46 (2007)
2. Ben-Hur, A., Horn, D., Siegelmann, H.T., Vapnik, V.: Support vector clustering. Journal of Machine Learning Research 2, 125–137 (2001)

3. Buchin, K., Buchin, M., van Kreveld, M.J., Luo, J.: Finding long and similar parts of trajectories. In: Agrawal, D., Aref, W.G., Lu, C.T., Mokbel, M.F., Scheuermann, P., Shahabi, C., Wolfson, O. (eds.) GIS, pp. 296–305. ACM, New York (2009)
4. Chen, L., Ng, R.: On the marriage of lp-norms and edit distance. In: VLDB 2004: Proceedings of the Thirtieth international conference on Very large data bases, pp. 792–803. VLDB Endowment (2004)
5. Cuturi, M., Vert, J.P., Birkenes, O., Matsui, T.: A kernel for time series based on global alignments. In: IEEE International Conference on Acoustics, Speech and Signal Processing, ICASSP 2007, vol. 2, pp. II–413–II–416 (2007)
6. Dhillon, I.S., Guan, Y., Kulis, B.: Weighted graph cuts without eigenvectors – a multilevel approach. IEEE Transactions on Pattern Analysis and Machine Intelligence 29, 1944–1957 (2007)
7. Douglas, D.H., Peucker, T.K.: Algorithms for the reduction of the number of points required to represent a digitized line or its caricature. Cartographica 10(2), 112–122 (1973)
8. Gaffney, S., Smyth, P.: Trajectory clustering with mixtures of regression models. In: Knowledge Discovery and Data Mining, pp. 63–72 (1999)
9. Gruber, C., Gruber, T., Sick, B.: Online signature verification with new time series kernels for support vector machines. In: Zhang, D., Jain, A.K. (eds.) ICB 2005. LNCS, vol. 3832, pp. 500–508. Springer, Heidelberg (2005)
10. Gudmundsson, J., Katajainen, J., Merrick, D., Ong, C., Wolle, T.: Compressing spatio-temporal trajectories. Computational Geometry 42(9), 825–841 (2009)
11. Gudmundsson, S., Runarsson, T.P., Sigurdsson, S.: Support vector machines and dynamic time warping for time series. In: IJCNN, pp. 2772–2776. IEEE, Los Alamitos (2008)
12. Joder, C., Essid, S., Richard, G.: Alignment kernels for audio classification with application to music instrument recognition. In: European Signal Processing Conference (EUSIPCO), Lausanne, Suisse (2008)
13. Lee, J.G., Han, J., Whang, K.Y.: Trajectory clustering: a partition-and-group framework. In: SIGMOD 2007: Proceedings of the 2007 ACM SIGMOD international conference on Management of data, pp. 593–604. ACM Press, New York (2007)
14. Liao, T.W.: Clustering of time series data – a survey. Pattern Recognition 38(11), 1857–1874 (2005)
15. Nanni, M., Pedreschi, D.: Time-focused clustering of trajectories of moving objects. J. Intell. Inf. Syst. 27(3), 267–289 (2006)
16. Pelekis, N., Kopanakis, I., Kotsifakos, E., Frentzos, E., Theodoridis, Y.: Clustering trajectories of moving objects in an uncertain world. In: ICDM 2009: Proceedings of the 2009 Ninth IEEE International Conference on Data Mining. pp. 417–427. IEEE Computer Society, Washington (2009)
17. Shawe-Taylor, J., Cristianini, N.: Kernel Methods for Pattern Analysis. Cambridge University Press, New York (2004)
18. Vert, J.P., Saigo, H., Akutsu, T.: Local alignment kernels for biological sequences. In: Scholkopf, B., Tsuda, K., Vert, J.P. (eds.) Kernel Methods in Computational Biology, pp. 131–154. The MIT Press, Cambridge (2004)
19. Vlachos, M., Gunopoulos, D., Kollios, G.: Discovering similar multidimensional trajectories. In: ICDE 2002: Proceedings of the 18th International Conference on Data Engineering, Washington, DC, USA, p. 673. IEEE Computer Society Press, Los Alamitos (2002)

Adaptive Bases for Reinforcement Learning

Dotan Di Castro and Shie Mannor

Department of Electrical Engineering
Technion - Israel Institute of Technology
{dot,shie,tx,ee}@technion.ac.il

Abstract. We consider the problem of reinforcement learning using function approximation, where the approximating basis can change dynamically while interacting with the environment. A motivation for such an approach is maximizing the value function fitness to the problem faced. Three errors are considered: approximation square error, Bellman residual, and projected Bellman residual. Algorithms under the actor-critic framework are presented, and shown to converge. The advantage of such an adaptive basis is demonstrated in simulations.

1 Introduction

Reinforcement Learning (RL) [4] is an approach for solving Markov Decision Processes (MDPs), when interacting with an unknown environment. One of the main obstacles in applying RL methods is how to cope with a large state space. In general, the underlying methods are based on dynamic programming, and include adaptive schemes that mimic either value iteration, such as Q-learning, or policy iteration, such as Actor-Critic (AC) methods. While the former attempts to directly learn the optimal value function, the latter are based on quickly learning the value of the currently used policy, followed by a slower policy improvement step. In this paper we focus on AC methods.

There are two major problems when solving MDPs with a large state space. The first is the storage problem, i.e., it is impractical to store the value function and the optimal action explicitly for each state. The second is generalization: some notion of similarity between states is needed since most states are not visited or visited only a few times. Thus, these issues are addressed by the Function Approximation (FA) approach [4] that involves approximating the value function by functional approximators with a smaller number of parameters in comparison to the original number of states. The success of this approach rests mainly on selecting appropriate features and on a proper choice of the approximation architecture. In a linear approximation architecture, the value of a state is determined by linear combination of the low dimensional feature vector. In the RL context, linear architectures enjoy convergence results and performance guarantees (e.g., [4]).

The approximation quality depends on the choice of the basis functions. In this paper we consider the possibility of tuning the basis functions on-line, under the AC framework. As mentioned before, an agent interacting with the environment

J.L. Balcázar et al. (Eds.): ECML PKDD 2010, Part I, LNAI 6321, pp. 312–327, 2010.

is composed of two sub-systems. The first is a critic, that estimates the value function for the states encountered. This sub-system acts on a fast time scale. The second is an actor that based on the critic output, and mainly the *temporal-difference* (TD) signal, improves the agent's policy using gradient methods. The actor operates on a second time scale, slower than the time-scale of the critic. Bhatnagar et al. [5] proved that such an algorithm with an appropriate relation between the time scales, converges[1].

The main contributions of this work are the following. First, we suggest an AC algorithm, based on [5], where a third time scale is added that is slower than both the critic and the actor, minimizing some error criteria while adapting the critic's basis functions to better fit the problem. Convergence of such an architecture is guaranteed, and simulations show that a dramatic improvement can be achieved using basis adaptation. Second, in a more general view, we suggest a framework that converts algorithms with a linear FA to adaptive basis algorithms, where the original algorithm and its convergence proof are used in the adaptive base algorithm.

There are several works done in the area of adaptive bases. We mention here several noticeable works which are similar in spirit to our work. The first work is of Menache et al. [12]. Two algorithms were suggested for adaptive bases by the authors: one algorithm is based on gradient methods for *least-squares TD* (LSTD) of Bradtke and Barto [2], and the other algorithm is based on the cross entropy method. Both algorithms were demonstrated in simulations to achieve better performance than their fixed basis counterparts but no convergence guarantees were supplied. Yu and Bertsekas [18] suggested several algorithms for two main problem classes: policy evaluation and optimal stopping. The former is closer to our work than the latter so we focus on this class. Three target functions were considered in that work: mean TD error, Bellman error, and projected Bellman error. The main difference between [18] and our work (besides the policy improvement) is the following. The algorithmic variants suggested in [18] are in the flavor of the LSTD and LSPE algorithms [3], while in our work the algorithms are TD based, thus, in our work no matrix inversion is involved. Another related class of algorithms with adaptive bases are those concerning of *direct policy improvement* (or *actor only* algorithms) [8,9]. In these works, rather than adapting the basis functions of the value function, the basis functions of the policy function are adapted, yet not in multiple time scales fashion as in our work.

The paper is organized as follows. In Section 2 we define some preliminaries and outline the framework. In Section 3 we introduce the algorithms suggested

[1] Using multiple time scales may pose a convergence drawback at first sight. Two approaches may be applied in order to overcome this problem. First, a recent work of Mokkadem and Pelletier [13] have demonstrated that combining the algorithm iterates with an averaging method leads to convergence rate in distribution that is the same as the optimal rate. Second, in multiple time scales the rate between the time steps of the slower and faster time scales should converge to 0. Thus, time scales which are close, operate on the fast time scale, and satisfy the condition above, are easy to find for any practical needs.

for adaptive bases. In Section 4 we show the convergence of the algorithms suggested, while in Section 5 we demonstrate the algorithms in simulations. In Section 6 we discuss the results.

2 Preliminaries

In this section, we introduce the framework, review actor-critic algorithms, overview multiple time scales stochastic approximation (MTS-SA), and state a related theorem which will be used later in proving the main results.

2.1 The Framework

We consider an agent interacting with an unknown environment that is modeled by a *Markov Decision Process* (MDP) in discrete time with a finite state set X and an action set U where $N \triangleq |X|$. Each selected action $u \in U$ of the agent determines a stochastic transition matrix $P_u = [P_u(y|x)]_{x,y \in X}$, where y is the state followed the state x.

For each state $x \in X$ the agent receives a corresponding reward $g(x)$ that depends only on the current state[2]. The agent maintains a parameterized *policy function* which is a probabilistic function, denoted by $\mu_\theta(u|x)$, mapping an observation $x \in X$ into a probability distribution over the controls U. The parameter $\theta \in \mathrm{IR}^{K_\theta}$ is a tunable parameter where $\mu_\theta(u|x)$ is a differentiable function w.r.t. θ. We note that for different values of θ, different probability distributions over $\mathcal{U}$ may be associated for each $x \in \mathcal{X}$. We denote by $x_0, u_0, g_0, x_1, u_1, g_1, \ldots$ a state-action-reward trajectory where the subindex specifies time.

Under each policy induced by $\mu_\theta(u|x)$, the environment and the agent induce together a Markovian transition function, denoted by $P_\theta(y|x)$, satisfying $P_\theta(y|x) = \sum_u \mu_\theta(u|x) P_u(y|x)$. The Markovian transition function $P_\theta(y|x)$ induces a stationary distribution over the state space X, denoted by $D(\theta)$. This distribution induces a natural norm, denoted by $\|\cdot\|_{D(\theta)}$, which is a weighted norm and is defined by $\|x\|^2_{D(\theta)} \triangleq x^\top D(\theta) x$. Note that when the parameter θ changes, the norm changes as well. We denote by $\mathrm{E}_\theta[\cdot]$ the expectation operator w.r.t. the measures $P_\theta(y|x)$ and $D(\theta)$. There are several performance criteria investigated in the RL literature that differ mainly on their time horizon and the treatment of future rewards [4]. In this work we focus on *average reward* criteria defined by $\eta_\theta = \mathrm{E}_\theta[g(x)]$. The agent's goal is to find the parameter θ that maximizes η_θ. Similarly, define the *(differential) value function* as

$$J(x) \triangleq \mathrm{E}_\theta \left[\sum_{n=0}^{\tau} (g(x_n) - \eta_\theta) \middle| x_0 = x \right], \tag{1}$$

where $\tau \triangleq \min\{k > 0 | x_k = x^*\}$ and x^* is some recurrent state for all policies, we assume to exist. Define the *Bellman operator* as $TJ(x) = r - \eta + \mathrm{E}_\theta[J(y)|x]$.

[2] Generalizing the results presented here to state-action rewards is straight forward.

Thus, based on (1) it is easy to show the following connection between the average reward to the value function under a given policy [3], i.e.,

$$J(x) = g(x) - \eta + \mathrm{E}_\theta[J(y)|x] \triangleq TJ(x), \tag{2}$$

For later use, we denote by TJ and J the column representations of $J(x)$ and $TJ(x)$, respectively.

We define the Temporal Difference (TD) [4,15] signal of the state x followed by the state y as $d(x,y) = g(x) - \eta + J(y) - J(x)$, where for a specific time n we abbreviate $d(x_n, x_{n+1})$ as d_n. Based on (2) we can see that

$$\mathrm{E}_\theta[d(x,y)|x] = 0, \quad \text{and} \quad \mathrm{E}_\theta[d(x,y)] = 0. \tag{3}$$

Based on this property, a wide family of algorithms known as TD algorithms exist [4], where common to all these algorithms is solving (3) iteratively.

Notational comment: from now on, we omit the dependency on θ whenever it is clear from the context.

2.2 Actor-Critic Algorithms

A well known class of RL approaches is the so called actor-critic (AC) algorithms, where the agent is divided into two components, an actor and a critic. The critic functions as a state value estimator using the so called *TD-learning* algorithm, whereas the actor attempts to select actions based on the TD signal estimated by the critic. These two components solve their own optimization problems separately interacting with each other.

The critic typically uses a function approximator which approximates the value function in a subspace of a reduced dimension $\mathbb{R}^{K_r}$. Define the basis matrix

$$\Phi \triangleq [\phi_k(x_n)]_{1\le n\le N, 1\le k\le K_r} \in \mathbb{R}^{N\times K_r}, \tag{4}$$

where its columns span the subspace $\mathbb{R}^{K_r}$. Thus, the approximation of the value function is $\tilde{J}(x,r) \triangleq \phi(x)^\top r$, where r is the solution of the following quadratic program $r = \arg\min_{r'\in\mathbb{R}^{K_r}} \|\Phi r' - J\|_D^2$. This solution yields the linear projection operator,

$$\Pi = \Phi\left(\Phi^\top D_\theta \Phi\right)^{-1} \Phi^\top D_\theta \tag{5}$$

that satisfies

$$\tilde{J}(r) = \Pi J. \tag{6}$$

where $\tilde{J}(r)$ is the vector representation of $\tilde{J}(x,r)$. Abusing notations, we define the (state dependent) projection operator on $J(x)$ as $\tilde{J}(x) = \Pi J(x)$.

As mentioned above, the actor receives the TD signal from the critic, where based on this signal, the actor tries to select the optimal action. As described in Section 2.1, the actor maintains a policy function $\mu_\theta(u|x)$. In the following, we state a theorem that serves as the foundation for the policy gradient algorithms described later. The theorem relates the gradient w.r.t. θ of the average reward, $\nabla_\theta \eta_\theta$, to the TD signal, $d(x,y)$. Define the *likelihood ratio derivative* as $\psi_\theta(x,u) \triangleq \nabla_\theta \mu_\theta(u|x)/\mu_\theta(u|x)$. We omit the dependency of ψ on x, u, and θ through the paper. The following assumption states that ψ is bounded.

Assumption 1. *For all $x \in X$, $u \in U$, and $\theta \in \mathbb{R}^{K_\theta}$, there exists a positive constant, B_ψ, such that $\|\psi\|_2 \leq B_\psi$, $\|\nabla_\theta \psi\|_2 \leq B_\psi$.*

Based on this, we present the following lemma that relates the gradient of η to the TD signal [5].

Lemma 2. *The gradient of the average reward (w.r.t. to θ) can be expressed by* $\nabla_\theta \eta = \mathrm{E}[\psi_\theta(x,u) d(x,y)]$.

2.3 Multiple Time Scales Stochastic Approximation

Stochastic approximation (SA), and in particular the ODE approach [10], is a widely used method for investigating the asymptotic behavior of stochastic iterates. For example, consider the following stochastic iterate

$$\varphi_{n+1} = \varphi_n + \alpha_n G(\varphi_n, \zeta_{n+1}), \tag{7}$$

where $\{\zeta_{n+1}\}$ is some random process and $\{\alpha_n\}$ are step sizes that form a positive series satisfying conditions to be defined later. The key idea of the technique is the following. Suppose that the iterate (7) can be decomposed into a mean function, denoted by $F(\cdot)$, and a noise term (usually a martingale difference noise), denoted by M_{n+1},

$$\varphi_{n+1} = \varphi_n + \alpha_n \left(F(\varphi_n) + M_{n+1} \right), \tag{8}$$

and suppose that the effect of the noise weakens due to repeated averaging. Consider the following ordinary differential equation (ODE)

$$\dot{\varphi}_t = F(\varphi_t), \tag{9}$$

where the dot above a variable stands for a time derivative. Then, a typical result of the ODE method in the SA theory suggests that the asymptotic limit of (8) and (9) are identical.

The theory of SA considers an iterate, which may be in some finite dimensional Euclidean space. Sometimes, we need to deal with several multidimensional iterates, dependent one on the other, and where each iterate operates on different timescale. Surprisingly, this type of SA, called *multiple time scale SA* (MTS-SA), is sometimes easier to analyze, with respect to the same iterates operate on single timescale. The first analysis of two time-scales SA algorithms was given by Borkar in [6] and later expanded to MTS by Leslie and Collins in [11]. In the following we describe the problem of MTS-SA, state the related ODEs, and finally state the conditions under which MTS-SA iterates converge. We follow the definitions of [11].

Consider L dependent SA iterates as the following

$$\varphi_{n+1}^{(i)} = \varphi_n^{(i)} + \alpha_n^{(i)} \left(F^{(i)} \left(\varphi_n^{(1)}, \ldots, \varphi_n^{(N)} \right) + M_{n+1}^{(i)} \right), \quad 1 \leq i \leq L, \tag{10}$$

where $\varphi_n^{(i)} \in \mathbb{R}^{d_i}$, and $F^{(i)} : \mathbb{R}^{\otimes_{j=1}^{L} d_j} \to \mathbb{R}^{d_i}$. The following assumption contains a standard requirement for MTS-SA step size.

Assumption 3. *(MTS-SA step size assumptions)*

1. *For* $1 \leq n \leq L$, *we have* $\sum_{n=0}^{\infty} \alpha_n^{(i)} = \infty, \quad \sum_{n=0}^{\infty} \left(\alpha_n^{(i)}\right)^2 < \infty$,
2. *For* $1 \leq n \leq L-1$, *we have* $\lim_{n\to\infty} \alpha_n^{(i)}/\alpha_n^{(i+1)} = 0$.

We interpret the second requirement in the following way: the higher the index i of an iterate, it operates on higher time scale. This is because that there exists some n_0 such that for all $n > n_0$ the step size of the i-th iterate is larger uniformly then the step size of the iterates $1 \leq j \leq i-1$. Thus, the i-th iterate advances more than any of the iterates $1 \leq j \leq i-1$, or in other words, it operates on faster time scale. The following assumption aggregates the main requirement for the MTS-SA iterates.

Assumption 4. *(MTS-SA iterate assumptions)*

1. $F^{(i)}(\cdot)$ *are globally Lipschitz continuous,*
2. *For* $1 \leq i \leq L$, *we have* $\sup_n \left\|\varphi_n^{(i)}\right\| < \infty$.
3. *For* $1 \leq i \leq L$, $\sum_{k=0}^{n} a_k^{(i)} M_{k+1}^{(i)}$ *converges a.s.*
4. *(The ODEs requirements)*
   (a) *Denote* $\varphi^{(i\to j)} \triangleq (\varphi^{(i)}, \ldots, \varphi^{(j)})$. *Define the L-th ODE system to be*

$$\begin{cases} \dot{\varphi}_t^{(1\to L-1)} = 0, \\ \quad\dot{\varphi}_t^{(L)} \quad = F^{(L)}(\varphi_t^{(1)}, \ldots, \varphi_t^{(L)}), \end{cases} \tag{11}$$

   and suppose the initial condition $\left.\varphi_t^{(1\to L-1)}\right|_{t=0} = \varphi_0$. *Then, there exists a Lipschitz continuous function* $\xi^{(L)}(\varphi_0)$ *such that the ODE system* (11) *converges to the point* $(\varphi_0, \xi^{(L)}(\varphi_0))$.

   (b) *Define the i-th ODE system,* $i = L-1, \ldots, 1$, *to be* .

$$\begin{cases} \dot{\varphi}_t^{(1\to i-1)} = 0, \\ \quad\dot{\varphi}_t^{(i)} \quad = F^{(i)}(\varphi^{(1)}, \ldots, \varphi^{(i-1)}, \varphi^{(i)}, \xi^{(i+1)}(\varphi_0, \varphi^{(i)})), \end{cases} \tag{12}$$

   where $\xi^{(i+1)}(\cdot,\cdot)$ *is determined by the* $(i+1)$*-th ODE system, and suppose the initial condition* $\left.\varphi_t^{(1\to i-1)}\right|_{t=0} = \varphi_0$. *Then, there exists a Lipschitz continuous function* $\xi^{(i)}(\varphi_0)$ *such that the ODE system* (12) *converges to the point* $(\varphi_0, \xi^{(i)})$.

The requirements 1 and 2 are common conditions for SA iterates to converge. Requirement 3 ensures that the noise term asymptotically vanishes. Requirement 4 is defined recursively where requirement (a) is the initial requirement related to the L-th ODE, and requirement (b) describes the i-th ODE system that is recursively based on the $(i+1)$-th ODE system, going from $i = L-1$ to $i = 1$. Requirement 4 ensures that for each time scale i there exists a Lipschitz convergent function, where the slower time scales $1, \ldots, i-1$ are static and where for the faster time scales $i+1, \ldots, L$ there exists a function $\xi^{(j+1\to L)}(\cdot)$ (which is the solution of the $i+1$ ODE system). Based on these requirements, we cite the following theorem due to Leslie and Collins [11].

Theorem 5. *Consider the iterates* (10) *and suppose Assumption 3 and 4 hold. Then, the asymptotic behavior of the iterates* (10) *converge to the invariant set of the dynamic system*

$$\dot{\varphi}_t^{(1)} = F^{(1)}\left(\varphi_t^{(1)}, \xi^{(2)}\left(\varphi_t^{(1)}\right)\right), \tag{13}$$

where $\xi^{(2)}(\cdot)$ *is determined by requirement 4 of Assumption 4.*

3 Main Results

In this section we present the main theoretical results of the work. We start by introducing adaptive bases and show the algorithms that are derived by choosing different approximating schemes. We note that below we adopt different types of criteria in order to adapt the basis functions. A discussion on the different approximation schemes and their relation to a linear FA can be found in Schoknecht [14].

3.1 Adaptive Bases

The motivation for adaptive bases is the following. Consider a domain expert that chooses a basis for the critic in order to approximate the value function. The basis which the domain expert chooses with no prior knowledge might not be suitable for the problem at hand, thus, it may lead the agent to poor performance. Therefore, the domain expert might prefer to choose a parameterized basis that has additional flexibility that is controlled by a small set of parameters.

We propose to consider a basis that is linear in some of the parameters but has several other parameters that allow greater flexibility. In other words, we consider bases that are linear with respect to some of the terms (related to the fast time scale), and nonlinear with respect to the rest (related to the slow time scale). The idea is that one does not lose much from such an approach in general if it fails, but in many cases it is possible to obtain better fitness and thus a better performance, due to this additional flexibility. Mathematically,

$$\tilde{J}(x, r, s) = \phi\left(x, s\right)^\top r, \quad s \in \mathbb{R}^{K_s}, \tag{14}$$

where r is a linear parameter related to the fast time scale, and s is the nonlinear parameter related to the slow time scale. In the view of (4), we note that from now on the matrix Φ depends on s, i.e., $\Phi \equiv \Phi_s$, and in matrix form we have $\tilde{J} = \Phi_s r$, but for the ease of exposition we drop the dependency on s.

An example for an adaptive base is the *trigonometric adaptive base*, e.g., $\{\sin(xs/d)\}_{d=1}^{K_r} \bigcup \{\cos(xs/d)\}_{d=1}^{K_r}$. In this basis, the linear dimension is $2K_r$ but it is controlled by the scalar s. Another example is the *RBF*[3] *adaptive base*, i.e., $\{\exp\{-(x-s_d)^\top A(x-s_d)\}_{d=1}^{K_r}$, where the basis functions centers are determined by the nonlinear parameter s_d. We note that in contrast to previous works using

[3] Radial Basis Functions; see [15].

RBFs, in the current RBF formulation the linear and nonlinear parameters of the RBFs changes on different time scales. The following assumption is needed for proving later results.

Assumption 6. *The columns of the matrix Φ are linearly independent, $K_r < N$, and $\Phi r \neq e$, where e is a vector of 1's. Moreover, the functions $\phi(x,s)$ and $\partial\phi(x,s)/\partial s_i$ for $1 \leq i \leq K_s$ are Liphschitz in s with a coefficient L_ϕ, and bounded with coefficient B_ϕ.*

Notation comment: for ease of exposition, we drop the dependency on x_n, e.g., $\phi_n \equiv \phi(x_n, s_n)$, $g_n \equiv g(x_n)$. Denote $\phi \triangleq \phi(x,s)$, $\phi' \triangleq \phi(y,s)$ (where as in Section 2.1, y is the state followed the state x), $\phi'_n \triangleq \phi(x_{n+1}, s_n)$, $d_n \triangleq d(x_n, x_{n+1})$, and $d \triangleq d(x,y)$. Thus, $d = g - \eta + \phi'^\top r - \phi^\top r$ and $d_n = g_n - \eta_n + \phi_n'^\top r_n - \phi_n^\top r_n$.

3.2 Minimum Mean Square Error and TD

Assume a basis parameterized as in (14). The *mean square error* (MSE) is defined as

$$\text{MSE} = \frac{1}{2}\text{E}\left[\left(\tilde{J}(x) - J(x)\right)^2\right].$$

The gradients with respect to r and s are

$$\nabla_r \text{MSE} = \frac{1}{2}\text{E}\left[\left(\tilde{J}(x) - J(x)\right)\phi\right] \approx \text{E}\left[d\phi\right], \tag{15}$$

$$\frac{\partial \text{MSE}}{\partial s_i} = \text{E}\left[\left(\tilde{J}(x) - J(x)\right)\frac{\partial \tilde{J}(x)}{\partial s_i}\right] \approx \text{E}\left[d\frac{\partial \phi^\top}{\partial s_i}r\right], \tag{16}$$

where in the approximation we use the bootstrapping method (see [15] for a disussion) in order to get the well known TD algorithm (i.e., substituting $J \approx T\tilde{J}$). We use SA in order to solve the stochastic equations (15) and (16), which together with Theorem 2 is the basis for the following algorithm. For technical reasons, we add an requirement that the iterates for θ and s are bounded, which practically is not constraining[4].

Algorithm 7. *Adaptive basis TD (ABTD).*

$$\eta_{n+1} = \eta_n + \alpha_n^{(3)}\left(g_n - \eta_n\right), \tag{17}$$

$$r_{n+1} = r_n + \alpha_n^{(3)} d_n \phi_n, \tag{18}$$

$$\theta_{n+1} = H_P^{(\theta)}\left[\theta_n + \alpha_n^{(2)}\psi_n d_n\right], \tag{19}$$

$$s_{i,n+1} = H_P^{(s)}\left[s_{i,n} + \alpha_n^{(1)} d_n \frac{\partial \phi_n^\top}{\partial s_i} r_n\right], \quad i = 1, \ldots, K_s, \tag{20}$$

where $H_P^{(\theta)}$ and $H_P^{(s)}$ are projection operators into a non-empty open constraints set whenever $\theta_n \notin H_p$ and $s \notin H_s$, respectively, and the step size series $\{\alpha_n^{(i)}\}$ for $i = 1,2,3$ satisfy Assumption 3.

[4] See Kushner and Yin [10] for a discussion on constrained SA.

We note that this algorithm is an AC algorithm with three time scales: the usual two time scales, i.e., choosing $\{\alpha_n^{(1)}\}_{n=1}^{\infty} \equiv 0$ yields Algorithm 1 of [5], and the third iterates is added for the basis adaptation, which is the slowest.

3.3 Minimum Square Bellman Error

The Square Bellman Error (BE) is defined as

$$\mathrm{BE} = \frac{1}{2}\mathrm{E}\left[\left(T\tilde{J}(x) - \tilde{J}(x)\right)^2\right].$$

The gradients with respect to r and s are

$$\nabla_r \mathrm{BE} = \mathrm{E}\left[d\left(\phi' - \phi\right)\right], \quad \frac{\partial \mathrm{BE}}{\partial s_i} = \mathrm{E}\left[d\left(\frac{\partial \phi'^{\top}}{\partial s_i} - \frac{\partial \phi^{\top}}{\partial s_i}\right) r\right].$$

Based on this we have the following SA algorithm, that is similar to Algorithm 7 except for the iterates for r_n and s_n.

Algorithm 8. - *Adaptive Basis for Bellman Error (ABBE). Consider the iterates for η and θ in Algorithm 7. The iterates for r and s_i are*

$$\begin{aligned} r_{n+1} &= r_n - \alpha_n^{(3)} d_n\left(\phi_n' - \phi_n\right), \\ s_{i,n+1} &= H_P^{(s)}\left[s_{i,n} - \alpha_n^{(1)} d_n \left(\frac{\partial \phi_n'}{\partial s_i} - \frac{\partial \phi_n}{\partial s_i}\right)^{\top} r_n\right], \qquad i = 1, \ldots, K_s. \end{aligned}$$

3.4 Minimum Square Projected Bellman Error

The Projected Bellman Error (PBE) is defined as

$$\mathrm{PBE} = \mathrm{E}\left[\left(\Pi T\tilde{J}(x) - \tilde{J}(x)\right)^2\right] = \mathrm{E}\left[d\phi\right]'\left(\mathrm{E}\left[\phi\phi'\right]\right)^{-1}\mathrm{E}\left[d\phi\right],$$

where the projection operator is defined in (5) and where the second equality was proved by Sutton et al. [16], Section 4. We note that the projection operator is independent of r but dependent on the basis parameter s. Define $w = \left(\mathrm{E}\left[\phi\phi'\right]\right)^{-1}\mathrm{E}\left[d\phi\right]$. Thus, w is the solution to the equation $\left(\mathrm{E}\left[\phi\phi'\right]\right) w = \mathrm{E}\left[d\phi\right]$, which yields $\mathrm{PBE} = w'\mathrm{E}\left[d\phi\right]$. Define similarly to Section 6.3.3 of [4] $Ar + b \triangleq \mathrm{E}\left[d\phi\right]$, where $A = \mathrm{E}[\phi(\phi' - \phi)^{\top}]$ and $b = \mathrm{E}[\phi(g - \eta)]$. Define $A^{(i)}$ to be the i-th column of A. For later use, we give here the gradient of w with respect to r and s in implicit form

$$\begin{aligned} \left(\mathrm{E}\left[\phi\phi^{\top}\right]\right)\frac{\partial}{\partial r_i} w &= A^{(i)}, \\ \mathrm{E}\left[\phi\phi^{\top}\right]\frac{\partial}{\partial s_i} w + \frac{\partial}{\partial s_i}\mathrm{E}\left[\phi\phi^{\top}\right] w &= \frac{\partial A}{\partial s_i} r + \frac{\partial b}{\partial s_i}. \end{aligned}$$

Denote by A_n, $A^s_{i,n}$,$b^s_{i,n}$,w_n, $w^r_{i,n}$, and $w^s_{i,n}$ the estimators at time n of A, $\partial A/\partial s_i$, $\partial b/\partial s_i$, w, $\partial w/\partial r_i$, and $\partial w/\partial s_i$, respectively. Define $A^{(i)}_n$ to be the i-th column of A_n. Thus, the SA iterations for these estimators are

$$\begin{aligned}
A_{n+1} &= A_n + \alpha^{(4)}_n \left(\phi_n \left(\phi_n - \phi_{n+1}\right)^\top - A_n\right), \\
A^s_{i,n+1} &= A^s_{i,n} + \alpha^{(4)}_n \left(\frac{\partial \phi_n}{\partial s_i}\left(\phi_n - \phi_{n+1}\right)^\top + \phi_n \frac{\partial}{\partial s_i}\left(\phi_n - \phi_{n+1}\right)^\top - A^s_{i,n}\right), \\
b^s_{i,n+1} &= b^s_{i,n} + \alpha^{(4)}_n \left(g\frac{\partial \phi_n}{\partial s_i} - b^s_{i,n}\right), \\
w_{n+1} &= w_n + \alpha^{(4)}_n \left(\phi_n d_n - \phi_n \phi_n^\top w_n\right), \\
w^r_{i,n+1} &= w^r_{i,n} + \alpha^{(4)}_n \left(A^{(i)}_n - \phi_n \phi_n^\top w^r_{i,n}\right), \\
w^s_{i,n+1} &= w^s_{i,n} + \alpha^{(4)}_n \left(A^s_{i,n} r_n + b^s_{i,n} - \left(\frac{\partial}{\partial s_i}\left(\phi_n \phi_n^\top\right)\right) w_n - \phi_n \phi_n^\top w^s_{i,n}\right).
\end{aligned}$$

where $\left\{\alpha^{(4)}_n\right\}$ satisfies Assumption 3. Next, we compute the gradient of the objective function PBE with respect to r and s and suggest a gradient descent algorithm to find the optimal value. Thus,

$$\frac{\partial \mathrm{PBE}}{\partial r_i} = \mathrm{E}\left[d\phi\right]^\top \frac{\partial}{\partial r_i} w^\top + w^\top \frac{\partial}{\partial r_i}\mathrm{E}\left[d\phi\right], \quad \frac{\partial \mathrm{PBE}}{\partial s_i} = \frac{\partial w^\top}{\partial s_i}\mathrm{E}\left[d\phi\right] + w^\top \frac{\partial \mathrm{E}\left[d\phi\right]}{\partial s_i}.$$

The following algorithm gives the SA iterates for r and s, where the iterates for η and θ are the same as in Algorithms 7 and 8 and therefore omitted. This algorithm has four time scales: estimators, critic, actor, and basis adaptation correspond to the step sizes $\{\alpha^{(4)}_n\}$, $\{\alpha^{(3)}_n\}$, $\{\alpha^{(2)}_n\}$, and $\{\alpha^{(1)}_n\}$, respectively.

Algorithm 9. - *Adaptive Basis for PBE (ABPBE). Consider the iterates for η and θ in Algorithm 7. The iterates for r and s are*

$$\begin{aligned}
r_{i,n+1} &= r_{i,n} - \alpha^{(3)}_n \left(d_n \phi_n^\top w^r_{i,n} + w_n^\top A^{(i)}_n r_{i,n} r_n\right), \\
s_{i,n+1} &= s_{i,n} - \alpha^{(1)}_n \left(d_n \phi_n^\top w^s_{i,n} + \left(A^s_{i,n} r_n + b^s_{i,n}\right)^\top w_n\right), \quad i = 1, \ldots, K_s.
\end{aligned}$$

4 Analysis

In this section we prove the convergence of the previous section Algorithm 7 and 8. We omit the convergence proof of Algorithm 9 that is similar to the convergence proof of Algorithm 8.

4.1 Convergence of ABTD

We begin by stating a theorem regarding the ABTD convergence. Due to space limitations, we give only a proof sketch based on the convergence proof of Theorem 2 of Bhatnagar et al. [5]. The self-contained proof under more general conditions is left for the long version of this work.

Theorem 10. *Consider Algorithm 7 and suppose Assumption 1, 3, and 6, hold. Then, the iterates* (17)-(20) *of Algorithm 7 converge w.p. 1 to a point that locally maximizes* η *and solves the equation* $\mathrm{E}[d\nabla_s\phi^\top r] = 0$.

Proof. (Sketch) There are three time-scales in (17)-(20), therefore, we wish to use Theorem 5, i.e., we need to prove that the requirements of Assumption 4 are valid w.r.t. to all iterations, i.e., η_n, r_n, θ_n, and s_n.
Requirement 1-4 w.r.t. iterates η_n, r_n, θ_n. Bhatnagar et al. proved in [5] that (17)-(19) converge for a specific s. Assumption 6 implies that the requirements 1-4 of Assumption 4 are valid regarding the iterates of η_n, r_n and θ_n uniformly for all $s \in \mathrm{I\!R}^{K_s}$. Therefore, it is sufficient to prove that on top of (17)-(19) iterate (20) also converges, i.e., that requirements 1-4 of Assumption 4 are valid w.r.t. s_n.
Requirement 1 w.r.t. iterate s_n. Define the σ-algebra $\mathcal{F}_n \triangleq \sigma(\eta_k, r_k, \theta_k, s_k : k \le n)$, and define $F_n^{(\eta)} \triangleq \mathrm{E}[g_n - \eta_n|\mathcal{F}_n]$, $F_n^{(r)} \triangleq \mathrm{E}[d_n\phi_n|\mathcal{F}_n]$, $F_n^{(\theta)} \triangleq H_P^{(\theta)} \mathrm{E}[\psi_n d_n|\mathcal{F}_n]$, $F_n^{(s_i)} \triangleq H_P^{(s)}\mathrm{E}[d_n\frac{\partial\phi_n^\top}{\partial s_i}r_n|\mathcal{F}_n]$, and $M_{n+1}^{(s_i)} \triangleq H_P^{(s)}[(d_n\frac{\partial\phi_n^\top}{\partial s_i}r_n) - F_n^{(s_i)}]$. Thus, (20) can be expressed as

$$s_{i,n+1} = s_{i,n} + \alpha_n^{(1)}\left(F_n^{(s_i)} + M_{n+1}^{(s_i)}\right). \tag{21}$$

Trivially, using Assumption 6, $F_n^{(r)}$, $F_n^{(\theta)}$, and $F_n^{(s)}$ are Liphschitz, with respect to s, with coefficients B_ϕ^2, L_ϕ, and L_ϕ, respectively. Also, $F_n^{(s_i)}$ is Liphschitz with respect to η, r, and θ with coefficients 1, B_ϕ, and 1, respectively. Thus, requirement 1 of Assumption 4 is satisfied.
Requirements 2 and 3 w.r.t. iterate s_n. By construction, the iterate s_n is bounded. Requirement 3 of Assumption 4 is valid using the boundedness of the martingale difference noise $M_{n+1}^{(s_i)}$ that implies, using the martingale convergence theorem [4], that the martingale $\sum_n \alpha_n^{(3)} M_{n+1}^{(s_i)}$ converges.
Requirement 4 w.r.t. iterate s_n. Using the result of Bhatnagar et al. [5], the fast time scales converge w.r.t. the slow time scale. Thus, Requirement 4 is valid based on the fact that the iterates (17)-(19) converge. □

4.2 Convergence of Adaptive Basis for Bellman Error

We begin by stating the theorem followed by its proof.

Theorem 11. *Consider Algorithm 8 and suppose that Assumption 1, 3, and 6, hold. Then Algorithm 8 converge w.p. 1 to a point that locally maximizes* η *and locally minimizes* $\mathrm{E}[d^2]$.

Proof. (Sketch) To use Theorem 5 we need to check that Assumption 4 is valid. Define the σ-algebra $\mathcal{F}_n \triangleq \sigma(\eta_k, r_k, \theta_k, s_k : k \le n)$, and define $F_n^{(\eta)} \triangleq \mathrm{E}[g_n - \eta_n|\mathcal{F}_n]$, $M_{n+1}^{(\eta)} \triangleq (g_n - \eta_n) - F_n^{(\eta)}$, $F_n^{(r)} \triangleq -\mathrm{E}[d_n(\phi_{n+1} - \phi_n)|\mathcal{F}_n]$, $M_{n+1}^{(r)} \triangleq -(d_n(\phi_{n+1} - \phi_n)) - F_n^{(r)}$, $F_n^{(\theta)} \triangleq \mathrm{E}[\psi_n d_n|\mathcal{F}_n]$, $M_{n+1}^{(\theta)} \triangleq (\psi_n d_n) - F_n^{(\theta)}$, $F_n^{(s_i)} \triangleq -\mathrm{E}[d_n(\frac{\partial\phi_{n+1}^\top}{\partial s_i}r_n - \frac{\partial\phi_n^\top}{\partial s_i}r_n)|\mathcal{F}_n]$, and $M_{n+1}^{(s_i)} \triangleq -(d_n(\frac{\partial\phi_{n+1}^\top}{\partial s_i}r_n) - \frac{\partial\phi_n^\top}{\partial s_i}r_n)) - F_n^{(s_i)}$.

On the fast time scale (which is related to $a_n^{(3)}$), as in Theorem 10, η_n converges to $\mathrm{E}[g(x)]$. On the same time scale we need to show that the iterate for r_n converges. Using the above definitions, we can write the iteration r_n as

$$r_{n+1} = r_n + \alpha_n^{(3)} \left(F_n^{(r)} + M_{n+1}^{(r)} \right). \tag{22}$$

We use Theorem 2.2 of Borkar and Meyn [7] to achieve this. Briefly, this theorem states that given an iteration as (22), this iteration is bounded w.p.1 if

(A1) The process $F_n^{(r)}$ is Lipschitz, the function $F_\infty(\sigma) \triangleq \lim_{\sigma\to\infty} F^{(r)}(\sigma r)/r$ is Lipschitz, and $F_\infty(\sigma)$ is asymptotically stable in the origin.

(A2) The sequence $M_{n+1}^{(r)}$ is a martingale difference noise and for some C_0

$$\mathrm{E}\left[(M_{n+1}^{(r)})^2 | \mathcal{F}_n \right] \le C_0(1 + \|r_n\|^2).$$

Trivially, the function $F_n^{(r)}$ is Lipschitz continuous, and we have

$$\lim_{\sigma\to\infty} F^{(r)}(\sigma r)/r = -\mathrm{E}\left[(\phi' - \phi)(\phi' - \phi)^\top | \right] r.$$

Thus, it is easy to show, using Assumption 6, that the ODE $\dot{r} = F_\infty^{(r)}$ has a unique global asymptotically stable point at the origin and (A1) is valid. For (A2) we have

$$\begin{aligned} \mathrm{E}\left[\left\| M(n+1)^{(r)} \right\|^2 \middle| \mathcal{F}_n \right] &\le \mathrm{E}\left[\| d_n \left(\phi_n' - \phi_n \right) \|^2 \middle| \mathcal{F}_n \right] \\ &\le 2 \left(B_g + B_\eta + 4B_\phi^2 r_n \right)^2 \triangleq K''(1 + \|r_n\|^2), \end{aligned}$$

where the first inequality results from the inequality $\mathrm{E}[(x - \mathrm{E}[x])^2] \le \mathrm{E}[x^2]$, and the second inequality follows from the uniform boundedness of the involved variables. We note that the related ODE for this iteration is given by $\dot{r} = F^{(r)}$, and the related Lyapunov function is given by $\mathrm{E}[d^2]$. Next, we need show that under the convergence of the fast time scales for η_n and r_n, the slower iterate for θ converges. The proof of this is identical to that of Theorem 2 of [5] and is therefore omitted. We are left with proving that if the fast time scales converge, i.e., the iterates η_n, r_n, and θ_n, then the iterate $s_n^{(i)}$ converge as well. The proof follows similar lines as of the proof for $s_n^{(i)}$ in the proof of Theorem 10, whereas here the iterate s_n converge to the stable point of the ODE $\dot{s} = \nabla_s \mathrm{E}[d(x, y)^2]$. □

5 Simulations

In this section we report empirical results applying the algorithms on two types of problems: GARNET problems [1] and the mountain car problem.

5.1 Garnet problems

The GARNET[5] problems [1,5] are a class of randomly constructed finite MDPs serving as a test-bench for RL algorithms. A GARNET problem is characterized by four parameters and is denoted by GARNET(X, U, B, σ). The parameter X is the number of states, U is the number of actions, B is the branching factor, and σ is the variance of each transition reward. When constructing such a problem, we generate for each state a reward, distributed according to $\mathcal{N}(0, 1)$. For each state-action the reward is distributed according to $\mathcal{N}(g(x), \sigma^2)$. The transition matrix for each action is composed of B non-zero terms. We consider the same GARNET problems as those simulated by [5]. For the critic's feature vector, we use the basis functions $\phi(x, s) = \cos\left(\frac{x}{d}s + \varrho_{x,d}\right)$, where $x = 1, \ldots, N$, $1 \le d \le K_r$, $s \in \mathbb{R}^1$, and $\varrho_{x,d}$ are i.i.d. uniform random phases. Note that only one parameter in this simulation controls the basis functions. The actor's feature vectors are of size $K_a \times |U|$, and are constructed as

$$\xi(x, u) \triangleq (\overbrace{0, \ldots, 0}^{K_a \times (u-1)}, \phi(x, s(t = 0)), \overbrace{0, \ldots, 0}^{K_a \times (|U|-u)}).$$

The policy function is $\mu(u|x, \theta) = e^{\theta^\top \xi(x,u)} / \sum_{u' \in U} e^{\theta^\top \xi(x,u')}$. Bhatnagar et al. [5] reported simulation results for two GARNET problems: GARNET$(30, 4, 2, 0.1)$ and GARNET$(100, 10, 3, 0.1)$. We based our simulations on these results where the time steps are identical to those of [5]. The GARNET$(30, 4, 2, 0.1)$ problem (Fig. 1 left pane) was simulated for $K_r = 4$ (two lower graphs) and $K_r = 12$ (two upper graphs), where each graph is an average of 100 repeats. The GARNET$(100, 10, 3, 0.1)$ problem (Fig. 1 right pane) was simulated for $K_r = 4$ (two lower graphs) and $K_r = 12$ (two upper graphs), where each graph is an average of 100 repeats. We can see that in such problems there is an evident advantage to an adaptive base, which can achieve additional fitness to the problem, and thus even for low dimensional problems the adaptation may be crucial.

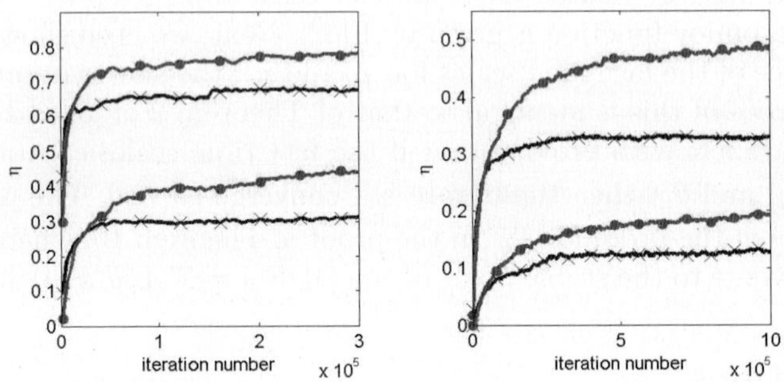

Fig. 1. Results for GARNET$(30, 4, 2, 0.1)$ (left pane) and GARNET$(100, 10, 3, 0.1)$ (right pane) where circled graphs are for adaptive bases. In each graph the lower two graphs are for $K_r = 4$ and the upper graphs are for $K_r = 12$. See the text for detail.

[5] Brevity for Generic Average Reward Non-stationary Environment Test-bench.

5.2 The Mountain Car

The mountain car task (see [15] for details) is a physical problem where a car is positioned randomly between two mountains (see Fig. 2 left pane) and needs to climb the right mountain, but the engine of the car does not support such a straight climb. Thus, the car needs to accumulate sufficient gradational energy, by applying back and forth actions, in order to succeed.

We applied the adaptive basis TD algorithm on this problem. We chose the critic basis functions to be radial basis functions (RBF) (see [15]), where the value function is represented by $\sum_{i=1}^{M} r_i \exp\{-(p-s_i^{(p)})^2/s_{p,i}^2 - (v-s_i^{(v)})^2/s_{v,i}^2\}$. The centers of the RBFs are parameterized by $(s_i^{(p)}, s_i^{(v)})_{i=1}^{M}$ while the variance is represented by $(s_{p,i}^2, s_{v,i}^2)_{i=1}^{M}$. In the right pane of Fig. 2 we present simulation results for 4 cases: SARSA (blue dash) which is based on the implementation of example 8.2 of [15], AC (red dash-dot) with 64 basis functions uniformly distributed on the parameter space, ABTD with 64 basis functions (magenta dotted) where both the location and the variance of the basis functions can adapt, ABAC with 16 basis functions (black solid) with the same adaptation. We see that the adaptive basis gives a significant advantage in performance. Moreover, we see that even with a small number of parameters, the performance is not affected. In the middle pane, the dynamics of a realization of the basis functions is presented where the dots and circles are the initial positions and final positions of the basis functions, respectively. The circle sizes are proportional to the basis functions standard deviations, i.e., $(s_{p,i}, s_{v,i})_{i=1}^{M}$.

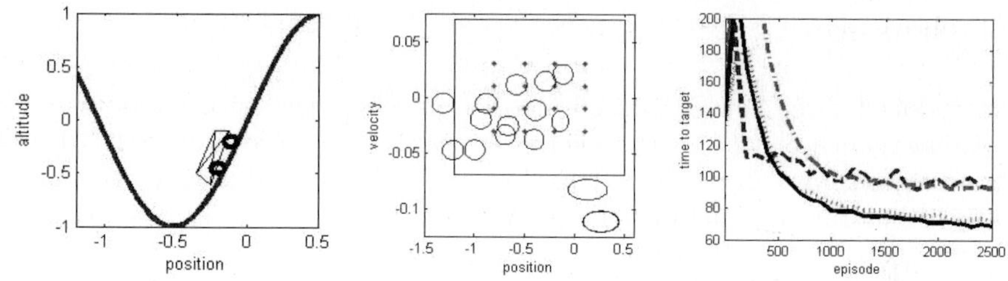

Fig. 2. (Left pane) illustration of the mountain car task. **(Middle pane)** Realization of ABTD with 16 basis functions where the red dots are the basis functions initial position and the circles are their final position. The radii are proportional to the variance. The rectangle represents the bounded parameter set of the car. **(Right pane)** Simulation results for the mountain car problem with solutions of SARSA (blue dash), AC (red dash-dot), AB-AC with 64 basis functions (magenta dotted), and AB-AC with 16 basis functions (black solid).

5.3 The Performance of Multiple Time Scales vs. Single Time Scale

In this section we discuss the differences in performance between the MTS algorithm to the STS algorithms. Unlike mistakenly thought, neither MTS algorithms nor STS algorithms have advantage in terms of convergence. This

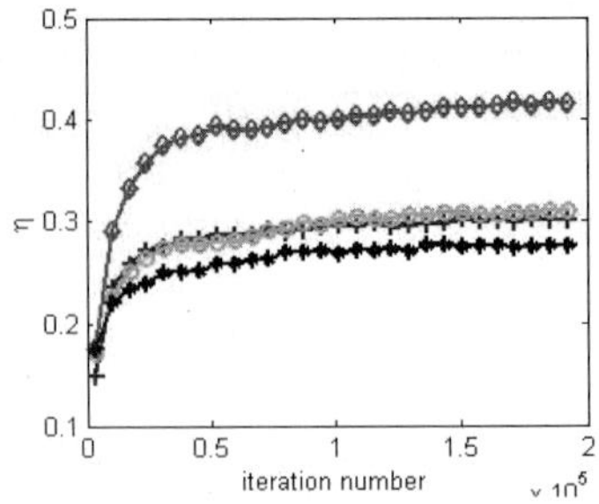

Fig. 3. Results for GARNET(30, 5, 5, 0.1) for $K_r = 8$. The upper diamond red graph is MTS ABTD algorithm, the circled green graph is STS ABTD acting on slow time scale, the blue crossed line is MTS static basis AC algorithm as in [5], and the black stared line is STS ABTD acting on fast time scale. Each graph is average of 100 simulation runnings.

difference comes from the fact that both methods perform the gradient algorithm differently, thus, they may result different trajectories. In Fig. 3 we can see a case on a GARNET(30,5,5,0.1) where the MTS ABTD algorithm (upper red diamond graph) has an advantage over STS ABTD algorithms or MTS static basis AC algorithm as in [5] (rest of the graphs). We note that this is not always the case and it depends on the problem parameters or the initial conditions.

6 Discussion

We introduced three new AC based algorithms where the critic's basis is adaptive. Convergence proofs, in the average reward case, were provided. We note that the algorithms can be easily transformed to discounted reward. When considering other target functions, more AC algorithms with adaptive basis can be devised, e.g., considering the objective function $||\mathrm{E}[d\phi]||^2$ yields $\mathrm{A}^\top$TD and GTD(0) algorithms [17]. Also, mixing the different algorithm introduced in here, can yield new algorithms with some desired properties. For example. we can devise an algorithm where the linear part is updated similar to (18) and the non-linear part is updated similar to (21). Convergence of such algorithms will follow the same lines of proof as introduced here.

An interesting question in the adaptive bases context is the following. How well do adaptive base algorithms cope with increased dimensionality when the intrinsic dimensionality stays small?

To conclude, the advantage of adaptive bases is evident: they relieve the domain expert from the task of carefully designing the basis. Instead, he may choose a flexible base, where one use algorithms as introduced here to adapt the base to the problem at hand. From a methodological point of view, the method we introduced in this paper demonstrates how to easily transform an existing RL algorithm to an adaptive base algorithm. The analysis of the original problem is used to show convergence of the faster time scale and the slow time scale is used for modifying the basis, analogously to "code reuse" concept in software engineering.

References

1. Archibald, T., McKinnon, K., Thomas, L.: On the Generation of Markov Decision Processes. Journal of the Operational Research Society 46, 354–361 (1995)
2. Bradtke, S.J., Barto, A.G.: Linear least-squares algorithms for temporal difference learning. Machine Learning 22, 33–57 (1996)
3. Bertsekas, D.: Dynamic programming and optimal control, 3rd edn. Athena Scientific, Belmont (2007)
4. Bertsekas, D., Tsitsiklis, J.: Neuro-dynamic programming. Athena Scinetific, Belmont (1996)
5. Bhatnagar, S., Sutton, R., Ghavamzadeh, M., Lee, M.: Natural actor–critic algorithms. Technical report Univ. of Alberta (2007)
6. Borkar, V.: Stochastic approximation with two time scales. Systems & Control Letters 29, 291–294 (1997)
7. Borkar, V., Meyn, S.: The ODE method for convergence of stochastic approximation and reinforcement learning. SIAM Journal on Cont. and Optim. 38, 447–469 (2000)
8. Busoniu, L., Ernst, D., De Schutter, B., Babuska, R.: Cross-Entropy Optimization of Control Policies With Adaptive Basis Functions. IEEE Transactions on Systems, Man, and Cybernetics, Part B: Cybernetics (99), 1–14 (2010)
9. Girgin, S., Preux, P.: Basis expansion in natural actor critic methods. In: Girgin, S., Loth, M., Munos, R., Preux, P., Ryabko, D. (eds.) EWRL 2008. LNCS (LNAI), vol. 5323, pp. 110–123. Springer, Heidelberg (2008)
10. Kushner, H., Yin, G.: Stochastic approximation and recursive algorithms and applications. Springer, Heidelberg (2003)
11. Leslie, D., Collins, E.: Convergent multiple-timescales reinforcement learning algorithms in normal form games. The Annals of App. Prob. 13, 1231–1251 (2003)
12. Menache, I., Mannor, S., Shimkin, N.: Basis function adaptation in temporal difference reinforcement learning. Annals of Operations Research 134, 215–238 (2006)
13. Mokkadem, A., Pelletier, M.: Convergence rate and averaging of nonlinear two-time-scale stochastic approximation algorithms. Annals of Applied Prob. 16, 1671
14. Schoknecht, R.: Optimality of reinforcement learning algorithms with linear function approximation. In: Proceedings of Neural Information Processing and Systems, pp. 1555–1562 (2002)
15. Sutton, R.S., Barto, A.G.: Reinforcement Learning - an Introduction. MIT Press, Cambridge (1998)
16. Sutton, R.S., Maei, H.R., Precup, D., Bhatnagar, S., Silver, D., Szepesvári, C., Wiewiora, E.: Fast gradient-descent methods for temporal-difference learning with linear function approximation. In: Proceedings of the 26th Annual International Conference on Machine Learning (2009)
17. Sutton, R.S., Szepesvari, C., Maei, H.R.: A convergent $o(n)$ temporal-difference algorithm for off-policy learning with linear function approximation. In: Advances in Neural Information Processing Systems, vol. 21, pp. 1609–1616 (2009b)
18. Yu, H., Bertsekas, D.: Basis function adaptation methods for cost approximation in MDP. In: Proc. of IEEE International Symposium on Adaptive Dynamic Programming and Reinforcement Learning, Nashville, TN (2009)

Constructing Nonlinear Discriminants from Multiple Data Views

Tom Diethe[1], David Roi Hardoon[2], and John Shawe-Taylor[1]

[1] Department of Computer Science, University College London
[2] Data Mining Department, Institute for Infocomm Research, A*Star Singapore
{t.diethe,j.shawe-taylor}@cs.ucl.ac.uk, drhardoon@i2r.a-star.edu.sg

Abstract. There are many situations in which we have more than one view of a single data source, or in which we have multiple sources of data that are aligned. We would like to be able to build classifiers which incorporate these to enhance classification performance. Kernel Fisher Discriminant Analysis (KFDA) can be formulated as a convex optimisation problem, which we extend to the Multiview setting (MFDA) and introduce a sparse version (SMFDA). We show that our formulations are justified from both probabilistic and learning theory perspectives. We then extend the optimisation problem to account for directions unique to each view (PMFDA). We show experimental validation on a toy dataset, and then give experimental results on a brain imaging dataset and part of the PASCAL 2007 VOC challenge dataset.

Keywords: Fisher Discriminant Analysis, Convex Optimisation, Multiview Learning, Kernel methods.

1 Introduction

We consider related but subtly differing settings within the domain of supervised learning. In Multi-View Learning (MVL), we have multiple views of the same underlying semantic object, which may be derived from different sensors, or different sensing techniques. In Multi-Source Learning (MSL), we have multiple sources of data which come from different sources but whose label space is aligned. Finally, in Multiple Kernel Learning (MKL), we have multiple kernels built from different feature mappings of the same data source. In general, any algorithm built to solve any of the three problems will also solve the others, but this may not be in the most optimal or desirable manner. For example, MKL algorithms do not make any attempt to integrate the sources of information from each view, and work by simply placing weights over the kernels [1]. Anecdotally, it seems that in many practical situations in which the number of kernels is small, the performance of MKL algorithms can actually be worse than simply choosing the best kernel through a heuristic method such as cross-validation (CV)[1].

[1] Amongst others, this topic was discussed at the NIPS 2009 Workshop "Understanding Multiple Kernel Learning Methods".

J.L. Balcázar et al. (Eds.): ECML PKDD 2010, Part I, LNAI 6321, pp. 328–343, 2010.

In the MVL or MSL paradigm, we are assuming that the number of views or sources is typically small (*i.e.* $2 \rightarrow 10$), and hence another viewpoint is needed in which the sources are combined more usefully. The basic idea of MVL is to introduce one function per view which only uses the features from that view, and then jointly optimize these functions such that learning is enhanced. In MVL, we are also usually interested in having weight vectors and loadings for each of the views, which we do not have when we concatenate features (or equivalently sum kernel matrices), or take convex combinations of kernels as in the MKL setting. Without loss of generality, we will assume that we are in the MVL setting for the rest of the paper.

Canonical Correlation Analysis (CCA) and Kernel Canonical Correlation Analysis (KCCA) [7] attempt to integrate two sources of information by maximising the correlations between projections of each view. They are unsupervised techniques, and as such are not ideally suited to a classification setting. A common way of performing classification on two-view data using KCCA is to use the projected data from one of the views as input to a standard classification algorithm, such as a Support Vector Machine (SVM). However, as with Principal Components Analysis (PCA), the subspace that is learnt through such unsupervised methods may not always align well with the label space.

SVM-2K [5] was an attempt to take this to its logical conclusion by combining this two stage learning into a single optimisation. The algorithm introduces the constraint of similarity between two 1-dimensional projections which identify two distinct SVMs in the two feature spaces. However SVM-2K requires extra parameters (the C-parameter for each SVM, and another mixing parameter, along with any kernel parameters) that the methods presented here will not require. In addition, it is not easy to see how the SVM-2K formulation can be generalised to more than two views. There has been one related approach that tries to find the optimum combination of Fisher classifiers [8] using the MKL architecture [1]. In its initial form this problem is non-convex, although the authors do recast the problem in terms of a semi-definite programme (SDP), at the expensive of an increase in the problem scale. In addition, the MKL architecture means that the output of the algorithm is a single weight vector for the convex combination of kernels. The formulation presented here has some similarities to that of [8], except cast here in the MVL framework and also providing additional modelling flexibility.

2 Preliminaries

We first review the convex formulation of Kernel Fisher Discriminant Analysis (KFDA) in the form given by [13]. Let $(\mathbf{x}, y) \sim S$ be an input-output pair from an m-sample S with $\mathbf{x} \in \mathbb{R}^n$ and $y \in \{-1, +1\}$. Let $\mathbf{X} = (\mathbf{x}_1, \ldots, \mathbf{x}_m)'$ be the input vectors stored in matrix $\mathbf{X}$ as row vectors, and $\mathbf{y} = (y_1, \ldots, \mathbf{y}_m)'$ be a vector of outputs, where $'$ denote the transpose of vectors or matrices. For simplicity we always assume that the examples are already projected into the kernel defined feature space $\mathcal{F}$, so that the kernel matrix $\mathbf{K}$ has entries $\mathbf{K}[i, j] = \langle \mathbf{x}_i, \mathbf{x}_j \rangle$. The explicit feature mapping is defined as $\phi : \mathbf{x} \rightarrow \phi(\mathbf{x}) \in \mathcal{F}$. Furthermore we define $\mathbf{1} \in \mathbb{R}^m$ as the vector of all ones and $\mathbf{I} \in \mathbb{R}^{m \times m}$ the $m-$dimensional identity matrix.

To proceed, we can use the fact that KFDA minimises the variance of the data along the projection whilst maximising the separation of the classes. If we characterise the variance within a vector of slack variables $\boldsymbol{\xi} \in \mathbb{R}^n$, we can directly minimise the variance as follows,

$$\begin{aligned} \min_{\boldsymbol{\alpha},\boldsymbol{\xi}} \quad & \|\boldsymbol{\xi}\|^2 + \mu \boldsymbol{\alpha}' \mathbf{K} \boldsymbol{\alpha} \\ \text{s.t.} \quad & \mathbf{K}\boldsymbol{\alpha} + \mathbf{1}b = \mathbf{y} + \boldsymbol{\xi} \\ & \boldsymbol{\xi}'\mathbf{e}^c = 0 \;\; \text{for} \;\; c = -1, +1, \qquad \text{where } \mathbf{e}_i^c = \begin{cases} 1 & \text{if } y_i = c \\ 0 & \text{otherwise.} \end{cases} \end{aligned} \tag{1}$$

3 Convex Multiview Fisher Discriminant Analysis

Here the convex formulation for KFDA given above will be extended to multiple views. Given p "views" of the same data source, or alternatively p aligned data sources, to form an $m-$sample S with input output $p+1$ tuples $(\mathbf{x}_{(1)}, \mathbf{x}_{(2)}, \ldots, \mathbf{x}_{(p)}, y)$. It is assumed that each view has already been projected into a feature space $\mathcal{F}_d$, so that the kernel matrix $\mathbf{K}_d$ for that view has entries $\mathbf{K}_d[i,j] = \langle \mathbf{x}_{(d)i}, \mathbf{x}_{(d)j} \rangle$. The explicit feature mapping for a each view is defined as $\phi_d : \mathbf{x}_{(d)} \rightarrow \phi_d(\mathbf{x}_{(d)}) \in \mathcal{F}_d$. Given matrices of inputs $\mathbf{X}_d = [\mathbf{x}_{(d)1}, \ldots, \mathbf{x}_{(d)m}]'$, the formulation (1) is extended to find p dual weight vectors $\boldsymbol{\alpha}_d$, $d = 1, \ldots, p$. The concatenation of these weight vectors will be denoted by $\tilde{\boldsymbol{\alpha}} = [\boldsymbol{\alpha}_1', \ldots, \boldsymbol{\alpha}_p']'$. The convex form of Multiview Fisher Discriminant Analysis (MFDA) is given in equation (2) below. The goal is now to minimise the variance of the data along the projection whilst maximising the distance between the average outputs for each class over all of the views.

$$\begin{aligned} \min_{\boldsymbol{\alpha}_d, b, \boldsymbol{\xi}} \quad & \mathcal{L}(\boldsymbol{\xi}) + \mu \mathcal{P}(\tilde{\boldsymbol{\alpha}}), \\ \text{s.t.} \quad & \sum_{d=1}^{p} (\mathbf{K}_d \boldsymbol{\alpha}_d + \mathbf{1}b_d) = \mathbf{y} + \boldsymbol{\xi}, \qquad d = 1, \ldots, p \\ & \boldsymbol{\xi}'\mathbf{e}^c = 0 \;\; \text{for} \;\; c = 1, 2, \end{aligned} \tag{2}$$

where $\mathcal{L}(\cdot)$ and $\mathcal{P}(\cdot)$ are the loss function and regulisation function respectively, as follows,

$$\mathcal{L}(\boldsymbol{\xi}) = \|\boldsymbol{\xi}\|_2^2, \tag{3}$$

$$\mathcal{P}(\tilde{\boldsymbol{\alpha}}) = \sum_{d=1}^{p} (\boldsymbol{\alpha}_d' \mathbf{K}_d \boldsymbol{\alpha}_d). \tag{4}$$

The first constraint in (2) ensures that the average loss between the output and its class label is minimised. The second constraint ensures that the average output for each class is each label. The classification function on a set of examples $\mathbf{x}_{(d),i}$ from views $d = 1, \ldots, p$ now becomes,

$$f(\mathbf{x}_{(d),i}) = \operatorname{sgn}\left(\sum_{d=1}^{p} f(\mathbf{x}_{(d)i})\right) = \operatorname{sgn}\left(\sum_{d=1}^{p} \mathbf{K}_d[:, i]'\boldsymbol{\alpha}_d + b_d\right). \tag{5}$$

Clearly (2) collapses to (1) for $p = 1$. Observe that the solutions given will, in the linearly separable case, be equivalent to summing kernels. Meaning that viewed in the primal form, the result is the standard criterion in the space defined by the concatenation of the features, and the norm of the full weight vector is given by (4). However this formulation leads to two main advantages. Firstly, it provides a flexible framework that allows for different noise models and regularisation functions. Secondly, explicit weight vectors are available for each view, which allows the calculation of implicit weightings over the views (see Section 3.2 below). In the non-linearly separable case, the equivalence breaks down, as the optimisation ties the views together through the shared slack variables.

Further intuition on the operation of the algorithm is as follows. Given two views $\mathbf{x}_{(1)}$ and $\mathbf{x}_{(2)}$, and using the standard ℓ_2 loss function, MFDA is trying to minimise the summed errors committed: $\left\|f_1(\mathbf{x}_{(1)}) + f(\mathbf{x}_{(2)}) - \mathbf{y}\right\|_2^2$. So if some slack is added to one of the examples, *e.g.* $\mathbf{x}_{(1)i}$, then the algorithm will try to push the corresponding example $\mathbf{x}_{(2)i}$ the other way to try to minimise the overall slack. This can be seen as "view disagreement" which means that the algorithm tries to use information from both views to aid the classification. However of course the algorithm can "give up" and allow the slack to be big for that example, meaning that $\mathbf{x}_{(1)}$ and $\mathbf{x}_{(2)}$ can be pushed the same way.

It is actually possible to state the problem as the reverse - saying that normally in MVL the goal is to search for view agreement, which would (for example) be minimising $\left\|f(\mathbf{x}_{(1)}) - f(\mathbf{x}_{(2)})\right\|_2^2$ (ignoring the labels). This is one particular form of the so-called "Co-Training" problem, which in order to work requires that each of the views are *sufficient* for classification, and methods that use this break down when there is significant view disagreement. A recent paper tried to get around this by learning separate classifiers and then looking for view agreement/disagreement between them, before combining them into a final classifier (a form of bootstrapping)[3]. MFDA should have an advantage over this as it is directly optimising the combined classifier. However, we also provide an alternative 'Private' method with separate slacks for each view as well as the overall slacks (see Section 3.4 to follow). Essentially, if there is a "trouble" point in view $\mathbf{x}_{(1)}$, but not in view $\mathbf{x}_{(2)}$, the disagreement can be soaked up by the private slack, allowing the two views to move into agreement with zero shared slack.

3.1 Probabilistic Interpretation

Following the analysis of [13], it is possible to view the KFDA algorithm from a probabilistic point of view. It is known that Fisher Discriminant Analysis (FDA) is Bayes optimal for two Gaussian distributions with equal covariance in the input space. The data may not fall naturally into this model, but it may be the case that for certain feature spaces (*e.g.* the space defined by the Radial Basis Function (RBF) kernel), the examples projected into a manifold in this space may be well approximated by Gaussian distributions with diagonal covariance[2]. In this case KFDA would be Bayes optimal in the feature space.

Consider data generated according to a Gaussian noise model, $y_i = \operatorname{sgn}(\mathbf{x}_i\mathbf{w} + n_i)$ where n is assumed to be an *independently and identically distributed* (i.i.d.) random variable (noise) with mean 0 and variance σ^2. If one considers KFDA as regression on to the labels, then a Gaussian noise model with known variance σ would result in the following expression for the likelihood: $\Pr(\mathbf{y}|\boldsymbol{\alpha}) = \exp(-\|\boldsymbol{\xi}\|_2^2)$. If a prior over the weights with hyperparameters μ is used, the log of the posterior is simply $\log(\Pr(\mathbf{y}|\boldsymbol{\alpha})\Pr(\boldsymbol{\alpha}|\mu)) = -\|\boldsymbol{\xi}\|_2^2 - \log(\Pr(\boldsymbol{\alpha}|\mu))$. The choice of prior then becomes equivalent to the choice of regularisation function, which will be discussed in Section 3.3. When viewed in this way the outputs produced by KFDA can be interpreted as probabilities, which in turn makes it possible to assign confidence to the final classifications.

This view of KFDA also motivates the Multiview extension of the algorithm. We can extend and combine the graphical interpretations of [2] and [6] using the above definitions as seen in Figure 1. Note that explicit mixing weights $\boldsymbol{\beta}$ paramaterised by ρ are shown (dotted). Note that due to the optimisation (which constrains the functions over each feature space with the shared slack variable) and the fact that we have separate $\boldsymbol{\alpha}$ vectors for each view, we are able to drop the mixing weights $\boldsymbol{\beta}$ from our formulation. Under the assumption that the kernels are normalised, we can calculate these weights *post-hoc* as will be shown in Section 3.2. Taking the approach of Naïve Bayes Probabilistic Label Fusion (NBF) [9], the first step is to assume conditional independence between classifiers given a class label. Suppose the set of labels $\mathbf{s} = \{s_1, \ldots, s_p\}$ are given from p classifiers for a given point $\mathbf{x}_i$. Denoting $\Pr(s_d)$ as the probability that classifier D_d labels an example $\mathbf{x}_i$ in class $\omega_c \in \Omega$, (in this case $\Omega = \{-1, +1\}$), then the likelihood of the classifiers given a label is,

$$\Pr(\mathbf{s}|\omega_c) = \Pr(s_1, \ldots, s_p|\omega_c) = \prod_{d=1}^{p} \Pr(s_d|\omega_c). \tag{6}$$

The posterior probability needed to label $\mathbf{x}_i$ is then given by,

$$\Pr(\omega_c|\mathbf{s}) = \frac{\Pr(\omega_c)\Pr(\mathbf{s}|\omega_c)}{\Pr(\mathbf{s})} = \frac{1}{Z}\Pr(\omega_c)\prod_{d=1}^{p} \Pr(s_d|\omega_c), \tag{7}$$

[2] After (empirical) whitening has been performed on the data. It may also be necessary to restrict to the main eigenvalues as the eigenvectors corresponding to smaller eigenvalues will start to be very random. In the space spanned by the top eigenvectors the data will then have diagonal covariance.

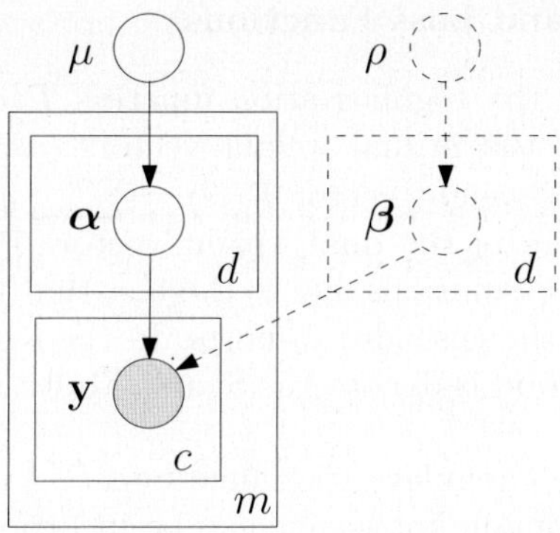

Fig. 1. Plates diagram showing the hieararhical Bayesian interpretation of MFDA. $\boldsymbol{\beta}$ are the hypothetical mixing parameters with prior weights ρ if an explicit mixing was used - in the case of MFDA these are fixed and hence can be removed, but can be calculated post-hoc.

where Z is a normalisation constant. Assume a uniform prior over labels, the log posterior is then given by,

$$\log(\Pr(\omega_c|\mathbf{s})) \propto \sum_{d=1}^{p} \log(\Pr(s_d|\omega_c)). \tag{8}$$

This implies that by directly optimising this sum, we are optimising the NBF over KFDA classifiers, which is precisely the motivation for both the objective function and the classification function for MFDA, both of which will be described in the next Section. At first glance it seems that this conditional independence assumption could be problematic, as this assumption is seldom true. However, Kuncheva made the point that despite this NBF is experimentally observed to be surprisingly accurate and efficient [9]. However, it does open the door to further possibilities for combining KFDA classifiers, but this is outside the scope of the present work.

3.2 Implicit Weighting

In order to determine the importance of each of the views after training, it is possible to calculate the implicit weighting of each view simply through the weighted sum of the absolute values of the classification functions. This is justified by the intuition made in Section 3.1 that the outputs of each classifier can be interpreted as probabilities, with the assumption that each kernel is normalised as per [16], *i.e.* $\text{trace}(\mathbf{K}_d) = m,\ d = 1, \dots, p$. This in turn means that the overall confidence of the classifier can be calculated as the sum of the log probabilities that the function $f(\mathbf{x}_{(d)i})$ for classifier d on example i give the class label ω_c.

$$\beta_d \approx \frac{1}{Z} \sum_{c \in \Omega} \log(\Pr(s_d|\omega_c)) = \frac{\sum_{i=1}^{m} |\mathbf{K}_d[:, i]'\boldsymbol{\alpha}_d + b_d|}{\sum_{i=1}^{m} \sum_{d=1}^{p} |\mathbf{K}_d[:, i]'\boldsymbol{\alpha}_d + b_d|}. \tag{9}$$

3.3 Regularisation and Loss Functions

The natural choices for the regularisation function $\mathcal{P}(\tilde{\boldsymbol{\alpha}})$ would either be the sum of the ℓ_2-norms of the primal weight vectors (as in (4)), or the sum of the ℓ_2-norms of the dual weight vector $\mathcal{P}(\tilde{\boldsymbol{\alpha}}) = \sum_{d=1}^{p} \|\boldsymbol{\alpha}_d\|_2^2$. Potentially more interesting is the ℓ_1-norm of the dual weight vector, $\mathcal{P}(\tilde{\boldsymbol{\alpha}}) = \sum_{d=1}^{p} \|\boldsymbol{\alpha}_d\|_1$, as this choice leads to sparse solutions due to the fact that the ℓ_1-norm can be seen as an approximation to the (pseudo) ℓ_0-norm. In the rest of the chapter the ℓ_1-norm regularisation method is denoted as Sparse Multiview Fisher Discriminant Analysis (SMFDA).

In some situations these regularisation functions $\mathcal{P}(\cdot)$ may be too simplistic, in which case additional domain knowledge can be incorporated into the function. For example, there is reason to believe *a-priori* that most of the views are likely not to be useful, but the individual weights in that view are, then $\mathcal{P}(\tilde{\boldsymbol{\alpha}}) = \|\mathbf{A}\|_{2,1}$ could be used where $\mathbf{A} = [\boldsymbol{\alpha}_1, \ldots, \boldsymbol{\alpha}_p]$ is $\tilde{\boldsymbol{\alpha}}$ reshaped as a matrix of weights and the block (r, p)-norm of $\mathbf{A}$ is defined as $\|\mathbf{A}\|_{r,p} = (\sum_{i=1}^{m} \|\boldsymbol{\alpha}_i\|_p^r)^{1/p}$. Another example would be a situation it may be desirable to impose sparsity on some views but not others. For two views, this would simply be $\mathcal{P}(\tilde{\boldsymbol{\alpha}}) = \|\boldsymbol{\alpha}_1\|_2^2 + \|\boldsymbol{\alpha}_2\|_1$ in order to promote sparsity in the second view but not the first. One could also promote sparsity in the primal version of one view by passing in the explicit features for that view (if available) and penalising $\mathbf{X}'_d \boldsymbol{\alpha}_d$. In this way any mixture of linear with nonlinear features and primal with dual sparsity can be combined across the views, all in a single optimisation framework. One can also pre-specify the weights of views by parameterising them, if one has a strong prior belief that a view will be more or less useful, but it in general it is not necessary or helpful to do this.

Following [13] the assumption of a Gaussian noise model can also be removed, resulting in different loss functions on the slacks $\boldsymbol{\xi}$. For example, if a Laplacian noise model is chosen $\|\boldsymbol{\xi}\|_2^2$ can be replaced with $\|\boldsymbol{\xi}\|_1$ in the objective function. The advantage of this is if the ℓ_1-norm regulariser from above is chosen, the resulting optimisation is a linear programme, which can be solved efficiently using methods such as column generation. From a modelling perspective, it may be advantageous to choose a noise model that is robust to outliers, such as Huber's Robust loss, which can easily be used in the framework presented here.

3.4 Incorporating Private Directions

The above formulations seek to find the projection that is maximally discriminative averaged across views. However these problems are very tightly constrained, and optimisation may be difficult in situations where one or more of the views is not informative of the labels (*i.e.* is essentially noise). This leads to considering the allowance of some extra slack $\boldsymbol{\zeta}_d$ that is private to each view, which is similar to the approach taken by [11] to probabilistic latent space modelling. This leads to the following formulation which we term Private Multiview Fisher Discriminant Analysis (PMFDA),

$$\begin{aligned}
\min_{\boldsymbol{\alpha}_d, b, \boldsymbol{\xi}, \boldsymbol{\zeta}_d} \quad & \mathcal{L}(\boldsymbol{\xi}, \tilde{\boldsymbol{\zeta}}, \tau) + \mu \mathcal{P}(\boldsymbol{\alpha}_d), & d = 1, \ldots, p \\
\text{s.t.} \quad & \mathbf{K}_d \boldsymbol{\alpha}_d + \mathbf{1} b = \mathbf{y} + \boldsymbol{\xi} + \boldsymbol{\zeta}_d & d = 1, \ldots, p \\
& \mathbf{1}_i' \boldsymbol{\xi} = 0 & i = 1, 2, \qquad (10)
\end{aligned}$$

with $\tilde{\boldsymbol{\zeta}} = [\boldsymbol{\zeta}_1', \ldots, \boldsymbol{\zeta}_p']'$. The regularisation function $\mathcal{P}(\cdot)$ is as before (4), and the loss function is updated to incorporate $\boldsymbol{\zeta}_d$ as follows,

$$\mathcal{L}(\boldsymbol{\xi}, \tilde{\boldsymbol{\zeta}}, \tau) = \|\boldsymbol{\xi}\|_2^2 + \tau \sum_{d=1}^{p} \|\boldsymbol{\zeta}_d\|_2^2. \qquad (11)$$

Note the extra parameter τ which enables the tuning of the relative importance of private or shared slacks. If $\tau = 1$ the penalties of the private slack for an example i are proportional to $\boldsymbol{\xi}_i/p$, which means that the more views that are added, the less each view is allowed to dominate. In the experiments conducted here this was simply set heuristically to 0.1 to allow a small amount of leeway for each view.

3.5 Generalisation Error Bound for MFDA

We now construct a generalisation error bound for MFDA by applying the following results from [15] and [10] and extending to the Multiview case. The first bounds the difference between the empirical and true means (Theorem 3 in [15]).

Theorem 1 (Bound on the true and empirical means). *Let S_d be a view of a sample of m points drawn independently according to a probability distribution P_d, where R_d is the radius of the ball in the feature space $\mathcal{F}_d$ containing the support of the distribution. Consider the mean vector $\boldsymbol{\mu}_d$ and the empirical estimate $\hat{\boldsymbol{\mu}}_d$ defined as*

$$\begin{aligned}
\boldsymbol{\mu}_d &= \mathbb{E}_{P_d}\left[\phi(\mathbf{x}_d)\right], \\
\hat{\boldsymbol{\mu}}_d &= \hat{\mathbb{E}}_{\mathbf{x}_d}\left[\phi(\mathbf{x}_d)\right] = \frac{1}{p} \sum_{d=1}^{p} \phi(\mathbf{x}_d). \qquad (12)
\end{aligned}$$

Then with probability at least $1 - \delta$ over the choice of S_d, we have

$$\|\hat{\boldsymbol{\mu}}_d - \mathbb{E}_{\mathbf{x}_d}[\phi(\mathbf{x}_d)]\| \leq \frac{R_d}{\sqrt{m}} \left(2 + \sqrt{2 \ln \frac{1}{\delta}}\right). \qquad (13)$$

Consider the covariance matrix $\boldsymbol{\Sigma}_d$ and the empirical estimate $\hat{\boldsymbol{\Sigma}}_d$ defined as

$$\begin{aligned}
\boldsymbol{\Sigma}_d &= \mathbb{E}\left[(\phi(\mathbf{x}_d) - \boldsymbol{\mu}_d)(\phi(\mathbf{x}_d) - \boldsymbol{\mu}_d)'\right], \\
\hat{\boldsymbol{\Sigma}}_d &= \hat{\mathbb{E}}\left[(\phi(\mathbf{x}_d) - \hat{\boldsymbol{\mu}}_d)(\phi(\mathbf{x}_d) - \hat{\boldsymbol{\mu}}_d)'\right]. \qquad (14)
\end{aligned}$$

The following corollary bounds the difference between the empirical and true covariance (Corollary 6 in [15]).

Corollary 1 (Bound on the true and empirical covariances). *Let S_d be an m sample from P_d as above, where R_d is as defined above. Provided $m \geq (2+\sqrt{2\ln 2/\delta})^2$, we have*

$$\left\|\hat{\boldsymbol{\Sigma}}_d - \boldsymbol{\Sigma}_d\right\|_F \leq \frac{2R_d^2}{\sqrt{m}}\left(2+\sqrt{2\ln\frac{2}{\delta}}\right), \tag{15}$$

The following Lemma is connected with the classification algorithm "Robust Minimax Classification" developed by [10], adapted here for MFDA.

Lemma 1. *Let $\boldsymbol{\mu}_d$ be the mean of a distribution and $\boldsymbol{\Sigma}_d$ its covariance matrix, $\mathbf{w}_d \neq 0$, b given, such that $\mathbf{w}_d'\boldsymbol{\mu}_d + b \leq 0$ and $\Delta \in [0,1)$, then if*

$$-\left(\mathbf{w}_d'\boldsymbol{\mu}_d + b\right) \geq \kappa(\Delta)\sqrt{\mathbf{w}_d'\boldsymbol{\Sigma}_d\mathbf{w}_d},$$

where $\kappa(\Delta) = \sqrt{\frac{\Delta}{1-\Delta}}$, then

$$\Pr\left(\mathbf{w}_d'\phi(\mathbf{x}_d) + b \leq 0\right) \geq \Delta$$

In order to provide a true error bound we must bound the difference between this estimate and the value that would have been obtained had the true mean and covariance been used.

Theorem 2 (Main). *Let S_d be a view of a sample of m points drawn from P_d as above, where R_d is the radius of the ball in the feature space $\mathcal{F}_d$ containing the support of the distribution. Let $\hat{\boldsymbol{\mu}}_d$ ($\boldsymbol{\mu}_d$) be the empirical (true) mean of a sample of m points from the view S_d, $\hat{\boldsymbol{\Sigma}}_d$ ($\boldsymbol{\Sigma}_d$) its empirical (true) covariance matrix, $\mathbf{w}_d \neq \mathbf{0}$, $\|\mathbf{w}\|_2 = 1$, and b given, such that $\mathbf{w}_d'\boldsymbol{\mu}_d + b \leq 0$ and $\Delta \in [0,1)$. Then with probability $1-\delta$ over the draw of the random sample, if*

$$-\left(\mathbf{w}_d'\hat{\boldsymbol{\mu}}_d + b\right) \geq \kappa(\Delta)\sqrt{\mathbf{w}_d'\hat{\boldsymbol{\Sigma}}_d\mathbf{w}_d} \qquad d = 1,\ldots,p,$$

then

$$\Pr\left(\left(\mathbf{w}_d'\phi_d(\mathbf{x}_d) + b\right) > 0\right) < 1-\Delta,$$

where

$$\Delta = \frac{\left(\mathbf{w}_d'\hat{\boldsymbol{\mu}}_d + b - A_d\right)^2}{\mathbf{w}_d'\hat{\boldsymbol{\Sigma}}_d\mathbf{w}_d + B_d + \left(\mathbf{w}_d'\hat{\boldsymbol{\mu}}_d + b - A_d\right)^2},$$

such that $\|\hat{\boldsymbol{\mu}}_d - \boldsymbol{\mu}_d\| \leq A_d$ where $A_d = \frac{R_d}{\sqrt{m}}\left(2+\sqrt{2\ln\frac{2m}{\delta}}\right)$, and $\left\|\hat{\boldsymbol{\Sigma}}_d - \boldsymbol{\Sigma}_d\right\|_F \leq B_d$ where $B_d = \frac{2R_d^2}{\sqrt{m}}\left(2+\sqrt{2\ln\frac{4m}{\delta}}\right)$.

Proof. (sketch). First we re-arrange $\mathbf{w}'_d\boldsymbol{\mu}_d + b \geq \kappa(\Delta)\sqrt{\mathbf{w}'_d\boldsymbol{\Sigma}_d\mathbf{w}_d}$ from Lemma 1 for each view in terms of $\kappa(\Delta)$:

$$\kappa(\Delta) = \frac{\mathbf{w}'_d\boldsymbol{\mu}_d + b}{\sqrt{\mathbf{w}'_d\boldsymbol{\Sigma}\mathbf{w}_d}}. \tag{16}$$

These quantities are in terms of the true means and covariances. In order to achieve an upper bound we need the following sample compressed results for the true and empirical means (Theorem 1) and covariances (Corollary 1):

$$\|\hat{\boldsymbol{\mu}}_d - \mathbb{E}_{\mathbf{x}_d}[\hat{\mu}_d(\mathbf{x}_d)]\| \leq A_d = \frac{R_d}{\sqrt{m}}\left(2 + \sqrt{2\ln\frac{2m}{\delta}}\right),$$

$$\left\|\hat{\boldsymbol{\Sigma}}_d - \boldsymbol{\Sigma}_d\right\|_F \leq B_d = \frac{2R_d^2}{\sqrt{m}}\left(2 + \sqrt{2\ln\frac{4m}{\delta}}\right).$$

Given equation (16) we can use the empirical quantities for the means and covariances in place of the true quantities. However, in order to derive a genuine upper bound we also need to take into account the upper bounds between the empirical and true means. Including these in the expression above for $\kappa(\Delta)$ by replacing δ with $\delta/2$, to derive a lower bound, we get:

$$\kappa(\Delta) = \frac{\mathbf{w}'_d\hat{\boldsymbol{\mu}}_{dS_d} + b - A_d}{\sqrt{\mathbf{w}'_d\hat{\boldsymbol{\Sigma}}_d\mathbf{w}_d + B_d}}.$$

Finally, making the substitution $\kappa(\Delta) = \sqrt{\frac{\Delta}{1-\Delta}}$ and solving for Δ yields the result. □

The following Proposition upper bounds the generalisation error of Multiview Fisher Discriminant Analysis (MFDA).

Proposition 1. *Let $\mathbf{w}_d$, b, be the (normalised) weight vector and associated threshold returned by the Multiview Fisher Discriminant Analysis (MFDA) when presented with a view of the training set S_d. Furthermore, let $\hat{\boldsymbol{\Sigma}}_d^+$ ($\hat{\boldsymbol{\Sigma}}_d^-$) be the empirical covariance matrices associated with the positive (negative) examples of the m training samples from S_d projected using $\mathbf{w}_d$. Then with probability at least $1-\delta$ over the draw of all the views of the random training set S_d, $d = 1, \ldots, p$ of m training examples, the generalisation error $\mathcal{R}$ is bounded by*

$$\mathcal{R} \leq \max(1 - \Delta^+, 1 - \Delta^-)$$

where Δ^j, $j = +, -$ such that

$$\Delta^j = \frac{j\left(\left(\sum_{d=1}^p(\mathbf{w}'_d\hat{\boldsymbol{\mu}}_{S_d}^j + b) - C^j\right)^2\right)}{\left(\sum_{d=1}^p \mathbf{w}'_d\hat{\boldsymbol{\Sigma}}_d^j\mathbf{w}_d\right) + D^j + \left(j(\sum_{d=1}^p \mathbf{w}'_d\hat{\boldsymbol{\mu}}_{S_d}^j + b) - C^j\right)^2},$$

where $C^j = \frac{\sum_{d=1}^p R_d}{\sqrt{m^j}}\left(2 + \sqrt{2\ln\frac{4mp}{\delta}}\right)$, $D^j = \frac{2\sum_{d=1}^p R_d^2}{\sqrt{m^j}}\left(2 + \sqrt{2\ln\frac{8mp}{\delta}}\right)$.

Proof. For the negative part of the proof we require $\mathbf{w}'_d\hat{\boldsymbol{\mu}}_d^- + b \geq \kappa(\Delta)\sqrt{\mathbf{w}'_d\hat{\boldsymbol{\Sigma}}_d^-\mathbf{w}_d}$ which is a straight forward application of Theorem 2 with δ replaced with $\delta/2$. For the positive part, observe that we require $\mathbf{w}'_d\hat{\boldsymbol{\mu}}_d^+ - b \geq \kappa(\Delta)\sqrt{\mathbf{w}'_d\hat{\boldsymbol{\Sigma}}_d^+\mathbf{w}_d}$, hence, a further application of Theorem 2 with δ replaced by $\delta/2$ suffices. Finally, we take a union bound over the p views such that m is replaced by mp. □

3.6 Experiments: Toy Data

In order to show that MVL methods can be beneficial, and demonstrate the validity of the outlined methods, experiments were first conducted with simulated toy data. A data source S was created by taking two 1−dimensional Gaussian distributions (S^+, S^-) which were well separated, which was then split into 100 train and 50 test points. The source S was embedded into 2−dimensional views through complementary linear projections (ϕ_1, ϕ_2) to give new "views" $\mathbf{X}_1, \mathbf{X}_2$. Differing levels of independent "measurement noise" were added to each view (n_1, n_2), and identical "system noise" was added to both views (n_S). A third view was constructed of pure noise to simulate a faulty sensor $(\mathbf{X}_3)$. The labels $\mathbf{y}$ were calculated as the sign of the original data source.

$$
\begin{aligned}
&S = \{S^+, S^-\} && \text{(source)}\\
&S^+ \sim \mathcal{N}(5,1), S^- \sim \mathcal{N}(-5,1) &&\\
&\mathbf{y} = \text{sgn}(S) && \text{(labels)}\\
&\phi_1 = [1,-1], \phi_2 = [-1,1] && \text{(projections)}\\
&n_1 \sim \mathcal{N}(0,5)^2, n_2 \sim \mathcal{N}(0,3)^2 && \text{(meas. noise)}\\
&n_S \sim \mathcal{N}(0,2)^2 && \text{(system noise)}\\
&\mathbf{X}_1 = \phi'_1 S + n_1 + n_S && \text{(view 1)}\\
&\mathbf{X}_2 = \phi'_2 S + n_2 + n_S && \text{(view 2)}\\
&\mathbf{X}_3 = n_S && \text{(view 3)}
\end{aligned}
$$

$\mathbf{X}_1$ and $\mathbf{X}_2$ are noisy views of the same signal, with correlated noise, which can be a typical problem in multivariate signal processing (*e.g.* sensors in close proximity). Linear kernels were used for each view. A small value for the regularisation parameter $\mu = 10^{-3}$ was chosen heuristically for all the experiments. Table 1 gives an overview of the results on the toy dataset. Comparisons were made against: KFDA on each of the views (denoted as $f(1)$, $f(2)$ and $f(3)$ respectively); summing the classification functions of these ($fsum$); summing the kernels of each view ($ksum$); followed by MFDA, PMFDA and SMFDA. Note that an unweighted sum of kernels is equivalent to concatenating the features before creating a single kernel. The table shows the test error over 10 random repeats of the experiment in first column, followed by the implicit weightings for each of the algorithms calculated via (9). Note that the $ksum$ method returns single m−dimensional weight vector, and unless a kernel with an explicit feature space is used it is not possible to recalculate the implicit weightings over the features. In this case, since linear kernels are used the weightings have been calculated. For the three methods outlined in this paper (MFDA, PMFDA, SMFDA),

Table 1. Test errors over ten runs on the toy dataset. Methods described in the text. $W(\cdot)$ refers to the implicit weightings given by each algorithm for each of the views. Note that the weightings closely match the actual SNR.

Method	**Test error**	$W(1)$	$W(2)$	$W(3)$
$f(a)$	0.19	1.00	0.00	0.00
$f(b)$	0.10	0.00	1.00	0.00
$f(c)$	0.49	0.00	0.00	1.00
$fsum$	0.39	0.33	0.33	0.33
$ksum$	0.04	0.29	0.66	0.05
MFDA	0.04	0.29	0.66	0.05
PMFDA	0.04	0.29	0.66	0.05
SMFDA	0.04	0.29	0.66	0.05
Actual		0.35	0.65	0.00

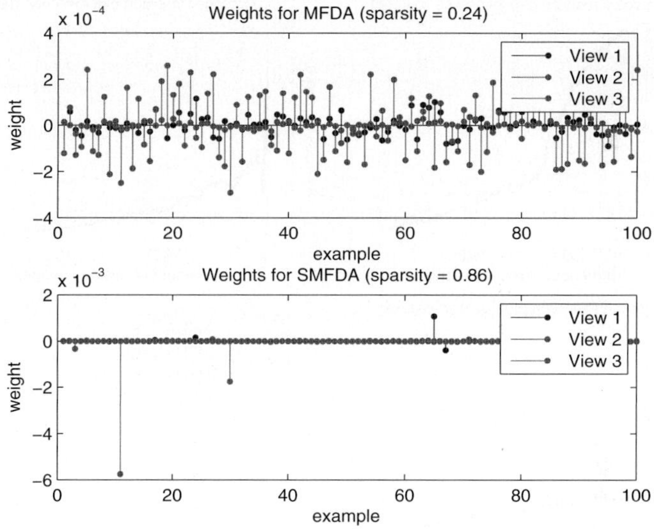

Fig. 2. Weights given by MFDA and SMFDA on the toy dataset. Notice that many of the weights for SMFDA are close to zero, indicating sparse solutions. Also notice that most of the weights for view 3 (pure noise) are close to zero.

as expected the performance is roughly equivalent to the $ksum$ method. The last row in the table (actual) is the empirical Signal to Noise Ratio (SNR) calculated as $SNR_d = \sum(\mathbf{X}_d'\mathbf{X}_d)/\text{var}(S - \mathbf{X}_d)$ for view d, which as can be seen is closely matched by the weightings given.

The sparsity of SMFDA can be seen in figure 2. The sparsity level quoted in the figure is the proportion of the weights below 10^{-5}.

4 Experiments

4.1 VOC 2007 DATASET

The sets of features ("views") used can be found in [17], with an extra feature extraction method known as Scale Invariant Feature Transformation (SIFT) [12]. RBF kernels were constructed for each of these feature sets, the RBF width parameter was set using a heuristic method[3]. The Pattern Analysis, Statistical Modelling and Computational Learning (PASCAL) Visual Object Classes (VOC) 2007 challenge database was used which contains 9963 images, each with at least 1 object. The number of objects in each image ranges from 1 to 20, with, for instance, objects of people, sheep, horses, cats, dogs etc. For a complete list of the objects, and description of the data set see the VOC 2007 challenge website[4].

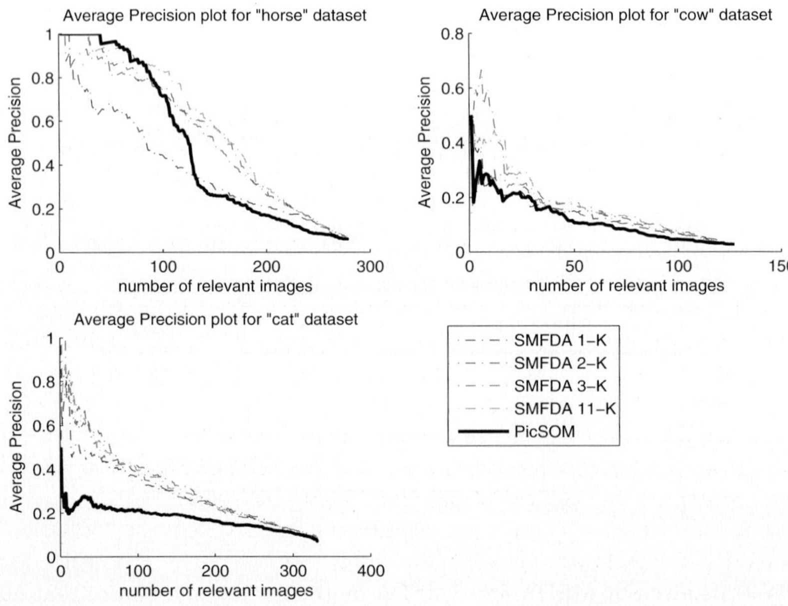

Fig. 3. Average precision recall curves for 3 VOC 2007 datasets for SMFDA plotted against PicSOM results

Figure 3 shows Recall-Precision curves for SMFDA with 1, 2, 3 or 11 kernels and PicSOM [17], and Table 2 shows the balanced error rate (the average of the errors on each class) and overall average precision for the PicSOM, KFDA

[3] For each setting of the width parameter, histograms of the kernel values were created. The chosen kernel was the one whose histogram peak was closest to 0.5 (*i.e.* furthest from 0 and 1).

[4] http://www.pascal-network.org/challenges/VOC/voc2007/workshop/index.html

using cross-validation to choose the best single kernel, and SMFDA. For the purposes of training, a random subset of 200 irrelevant images was used rather than the full training set. Results for three of the object classes (cat, cow, dog) are presented. The results show that, in general, adding more kernels into the optimisation can assist in recall performance. For each object class, the subsets of kernels (*i.e.* 1,2, or 3) were chosen by the weights given by SMFDA on the 11 kernels. The best single kernel (based on SIFT features) performs well alone, yet the improvement in some cases is quite marked. Results are competitive with the PicSOM algorithm, which uses all 11 feature extraction methods, and all of the irrelevant images.

Table 2. Balanced Error Rate (BER) and Average Precision (AP) for four of the VOC challenge datasets, for four different methods: PicSOM, KFDA with cross validation (KFDA CV), and SMFDA

Dataset →	**Cat**		**Cow**		**Horse**	
Method ↓	BER	AP	BER	AP	BER	AP
PicSOM	n/a	0.18	n/a	0.12	n/a	0.48
KFDA CV	**0.26**	**0.36**	0.32	0.14	0.22	0.51
SMFDA	**0.26**	**0.36**	**0.27**	**0.15**	**0.19**	**0.58**

4.2 Neuroimaging Dataset

This section describes analysis of functional Magnetic Resonance Imaging (fMRI) data[5] that was acquired from 16 subjects who viewed image stimuli from two categories (pleasant (+ve) and unpleasant (-ve)). The images were presented in 6 blocks of 42 images (7 volumes) per category. The image stimuli are represented using SIFT features [12], and conventional pre-processing was applied to the fMRI data with linear kernels. A leave-subject-out paradigm was used where 15 subjects are combined for training and a single subject is withheld for testing. This gave a total of $42 \times 2 \times 15 = 1260$ training and $42 \times 2 = 84$ testing fMRI volumes and paired image stimuli. In the following experiment, the following comparisons were made: An SVM on the fMRI data (single view); KCCA on the fMRI + Image Stimuli (two views) followed with an SVM trained on the fMRI data projected into the learnt KCCA semantic space; MFDA on the fMRI + Image Stimuli (two views). The results are given in Table 3 where it can be observed that on average MFDA performs better than both the SVM (which is a single view approach), and the KCCA/SVM which similarly to MFDA incorporates two views into the learning process. In this case the label space is clearly not well aligned with the KCCA projections, whereas a supervised method such as MFDA is able to find this alignment.

[5] Data donated by Mourão-Miranda *et. al.* [14].

Table 3. In the following table the leave-one-out errors for each subject are presented. The following methods are compared: SVM on the fMRI data alone; KCCA analysis on the two views fMRI and Image Stimuli followed by an SVM on the projected fMRI data; the proposed MFDA on the two views fMRI+Image.

Sub.	SVM	KCCA/SVM	MFDA
1	0.1310	0.1667	**0.1071**
2	0.1905	0.2739	**0.1429**
3	0.2024	**0.1786**	0.1905
4	0.1667	0.2125	**0.1548**
5	**0.1905**	0.2977	0.2024
6	0.1667	0.1548	**0.1429**
7	**0.1786**	0.2262	0.1905
8	0.2381	0.2858	**0.2143**
9	0.3096	0.3334	**0.2619**
10	0.2977	0.3096	**0.2262**
11	**0.1191**	0.1786	0.1429
12	0.1786	0.2262	**0.1667**
13	0.2500	0.2381	**0.0714**
14	0.4405	0.4405	**0.2619**
15	**0.2500**	0.2977	0.2738
16	**0.1429**	0.1905	0.1860
Mean:	0.2158±0.08	0.2508±0.08	**0.1860±0.06**

5 Conclusions

KFDA can be formulated as a convex optimisation problem, which we extended to the Multiview setting MFDA using justifications from a probabilistic point of view. We also provide a generalisation error bound. A sparse version SMFDA was then introduced, and the optimisation problem further extended to account for directions unique to each view PMFDA. Experimental validation was shown on a toy dataset, followed by experimental results on part of the PASCAL 2007 VOC challenge dataset and a fMRI dataset, showing that the method is competitive with state-of-the-art methods whilst providing additional benefits.

Mika *et. al.* [13] demonstrate that their convex formulation of KFDA can easily be extended to both multi-class problems and regression problems, simply by updating the final two constraints. The same is also true of MFDA and its derivatives, which enhances its flexibility. The possibility of replacing the Naïve Bayes Fusion method for combining classifiers is another interesting avenue for research.

Finally, for the special case of SMFDA there is the possibility of using a stagewise optimisation procedure similar to the Least Angle Regression Solver (LARS) [4] which would have the benefit of computing the full regularisation path.

References

1. Bach, F.R., Lanckriet, G.R.G., Jordan, M.I.: Multiple kernel learning, conic duality, and the smo algorithm. In: ICML 2004: Proceedings of the twenty-first international conference on Machine learning, p. 6. ACM Press, New York (2004)
2. Centeno, T.P., Lawrence, N.D.: Optimising kernel parameters and regularisation coefficients for non-linear discriminant analysis. Journal of Machine Learning Research 7, 455–491 (2006)
3. Christoudias, C.M., Urtasun, R., Darrell, T.: Multi-view learning in the presence of view disagreement. In: Proceedings of Conference on Uncertainty in Artificial Intelligence, UAI (2008)
4. Efron, B., Hastie, T., Johnstone, L., Tibshirani, R.: Least angle regression. Annals of Statistics 32, 407–499 (2002)
5. Farquhar, J., Hardoon, D., Meng, H., Shawe-Taylor, J., Szedmak, S.: Two view learning: SVM-2k, theory and practice. In: Weiss, Y., Schölkopf, B., Platt, J. (eds.) Advances in Neural Information Processing Systems 18, pp. 355–362. MIT Press, Cambridge (2006)
6. Girolami, M., Rogers, S.: Hierarchic bayesian models for kernel learning. In: ICML, pp. 241–248 (2005)
7. Hardoon, D., Szedmak, S., Shawe-Taylor, J.: Canonical correlation analysis: An overview with application to learning methods. Neural Computation 16(12), 2639–2664 (2004)
8. Kim, S.J., Magnani, A., Boyd, S.: Optimal kernel selection in kernel fisher discriminant analysis. In: ICML 2006: Proceedings of the 23rd international conference on Machine learning. pp. 465–472. ACM, New York (2006)
9. Kuncheva, L.I.: Combining Pattern Classifiers: Methods and Algorithms. Wiley Interscience, Hoboken (2004)
10. Lanckriet, G.R., Ghaoui, L.E., Bhattacharyya, C., Jordan, M.I.: A robust minimax approach to classification. J. Mach. Learn. Res. 3, 555–582 (2003)
11. Leen, G., Fyfe, C.: Learning shared and separate features of two related data sets using GPLVMs. Tech. rep., Presented at the NIPS 2008 workshop Learning from Multiple Sources (2008)
12. Lowe, D.G.: Object recognition from local scale-invariant features. In: Proceedings of the 7th IEEE International Conference on Computer vision, Kerkyra, Greece, pp. 1150–1157 (1999)
13. Mika, S., Rätsch, G., Müller, K.R.: A mathematical programming approach to the kernel Fisher algorithm. In: Leen, T., Dietterich, T., Tresp, V. (eds.) Advances in Neural Information Processing Systems, vol. 13, pp. 591–597 (2001)
14. Mourão-Miranda, J., Reynaud, E., McGlone, F., Calvert, G., Brammer, M.: The impact of temporal compression and space selection on svm analysis of single-subject and multi-subject fmri data. NeuroImage 33(4), 1055–1065 (2006)
15. Shawe-Taylor, J., Cristianini, N.: Estimating the moments of a random vector. In: Proceedings of GRETSI 2003 Conference, vol. 1, pp. 47–52 (2003)
16. Shawe-Taylor, J., Cristianini, N.: Kernel Methods for Pattern Analysis. Cambridge University Press, Cambridge (2004)
17. Viitaniemi, V., Laaksonen, J.: Techniques for image classification, object detection and object segmentation applied to VOC challenge 2007. Tech. rep., Department of Information and Computer Science, Helsinki University of Technology, TKK (2008)

Learning Algorithms for Link Prediction Based on Chance Constraints

Janardhan Rao Doppa[1], Jun Yu[1], Prasad Tadepalli[1], and Lise Getoor[2]

[1] School of EECS, Oregon State University, Corvallis, OR 97330 USA
{doppa,yuju,tadepall}@eecs.orst.edu
[2] Computer Science Dept., University of Maryland, College Park, MD 20742 USA
getoor@cs.umd.edu

Abstract. In this paper, we consider the link prediction problem, where we are given a partial snapshot of a network at some time and the goal is to predict the additional links formed at a later time. The accuracy of current prediction methods is quite low due to the extreme class skew and the large number of potential links. Here, we describe learning algorithms based on chance constrained programs and show that they exhibit all the properties needed for a good link predictor, namely, they allow *preferential bias* to positive or negative class; handle *skewness* in the data; and *scale* to large networks. Our experimental results on three real-world domains—co-authorship networks, biological networks and citation networks—show significant performance improvement over baseline algorithms. We conclude by briefly describing some promising future directions based on this work.

1 Introduction

Network analysis, performed in domains including social networks, biological networks, transaction networks, and the web, has received a lot of interest in recent years. These networks evolve over time and it is a challenging task to understand the dynamics that drives their evolution. Link prediction is an important research direction within this area. The goal here is to predict the potential future interaction between two nodes, given a partial view of the current state of the network.

This problem occurs in several domains. In many cases, we are interested in the links that are likely to form in the future. For example, in citation networks describing collaboration among scientists, we want to predict which pairs of authors are likely to collaborate in future; in social networks, we would want to predict new friendships; in query graphs, we want to predict the related queries in the context of web search and in biological networks we want to predict which proteins are likely to interact. On the other hand, we may be interested in anomalous links; for example, in financial transaction networks, the unlikely transactions might indicate fraud, and on the web, they might indicate spam.

There is a large literature on link prediction. Early approaches to this problem are based on defining a measure for analyzing the *proximity* of nodes in the

J.L. Balcázar et al. (Eds.): ECML PKDD 2010, Part I, LNAI 6321, pp. 344–360, 2010.

network [1,19,14]. For example, shortest path, common neighbors, Katz measure, Adamic-adar etc., all fall under this category. More recently, Sarkar et al. [22] gave a theoretical justification of these link prediction heuristics. Liben-Nowell and Klienberg studied the usefulness of all these topological features by experimenting on bibliographic datasets [14]. It was found that, no one measure is superior in all cases. Statistical relational models were also tried with some success [7,8,24,20]. Recently, the link prediction problem is studied in the supervised learning framework by treating it as an instance of binary classification [9,11,4,25,27]. These methods use the topological and semantic measures defined between nodes as features for learning classifiers. Given a snapshot of the social network at time t for training, they consider all the links present at time t as positive examples and consider a large sample of absent links (pair of nodes which are not connected) at time t as negative examples. The learned classifiers performed consistently across all the datasets unlike heuristic methods which were inconsistent, although the accuracy of prediction is still very low. There are several reasons for this low prediction accuracy. One of the main reasons is the huge class skew associated with link prediction. In large networks, it's not uncommon for the prior link probability on the order of 10^{-4} or less, which makes the prediction problem very hard, resulting in poor performance. In addition, as networks evolve over time, the negative links grow quadratically whereas positive links grow only linearly with new nodes. Further, in some cases we are more concerned with *link formation*, the problem of predicting new positive links, and in other cases we are more interested in *anomalous link detection* [21], the problem of detecting unlikely links. In general, we need the following properties for a good link predictor: allow *preferential bias* to the appropriate class; handle *skewness* in the data; *scale* to large networks.

Chance-constraints and Second-Order Cone Programs(SOCPs) [15] are a special class of convex optimization problems that have become very popular lately, due to the efficiency with which they can be solved using fast interior point methods. They are used in a variety of settings such as feature selection [3], dealing with missing features [23], classification and ordinal regression algorithms that scale to large datasets [18], and formulations to deal with unbalanced data [17,10]. In this work, we give a scalable cost-sensitive formulation based on chance-constraints which satisfies all the requirements needed for learning a good link predictor mentioned above and show how it can be used for link prediction to significantly improve performance. The chance constraints can be converted into deterministic ones using Chebyschev-Cantelli inequality, resulting in a SOCP. The complexity of SOCPs is moderately higher than linear programs and they can be solved using general purpose SOCP solvers like SeDuMi[1] or YALMIP[2].

The main contributions of this paper include: 1. We identify important requirements of the link prediction task and propose a new cost-sensitive formulation based on chance constraints satisfying all the requirements. We describe its connections to other frameworks like biased classification and uneven margin

[1] http://sedumi.ie.lehigh.edu

[2] http://users.isy.liu.se/johanl/yalmip/

algorithms. 2. We perform a detailed evaluation on multiple datasets from three real-world domains–co-authorship networks, biological networks and citation networks– to investigate the effectiveness of our methods. We show significant improvement in link prediction accuracy.

2 Cost-Sensitive Learning for Link Prediction

In this work, we consider the link prediction problem as an instance of binary classification. We are given training data $\mathcal{D} = \{(x_1, y_1), (x_2, y_2), \cdots, (x_n, y_n)\}$ where, each $x_i \in \Re^n$ is a feature vector defined between two nodes and $y_i \in \{-1, +1\}$ is the corresponding label that stands for the presence or absence of an edge between the two nodes. In our case, we have a huge class imbalance problem, i.e., the number of negative examples $\gg$ the number of positive examples. There are two different ways of addressing the class imbalance problem. In the first approach, it is turned into a balanced problem either by over-sampling the minority class or under-sampling the majority class. However, both these sampling methods have their drawbacks. By doing under-sampling, we lose some information and over-sampling introduces noise into the data. In the second approach, class imbalance problem is addressed by reducing it to a cost-sensitive learning problem where misclassification costs are unknown. Then, the ratio of misclassification costs is varied to find out the best decision function based on the validation set. In this work, we will follow the second approach which is considered to be more principled. In particular we are interested in a cost-sensitive formulation in the max-margin framework. We require a solution which is scalable to large data sets; this is very important for the link prediction task. For now, we work with only linear decision functions of the form $f(x) = w^T x - b$. However, all the formulations described in this work can be kernelized to construct non-linear classifiers.

Cost-Sensitive Learning Problem: In the traditional binary classification problem, all misclassifications are considered to be of the same cost, i.e., $C_{12} = C_{21}$ where, C_{12} is the misclassification cost of predicting a data point of class 1 as class 2 and C_{21} the misclassification cost of predicting a data point of class 2 as class 1. However, this assumption is not true for many real-world applications like medical domains e.g., predicting whether a patient has breast cancer or not. In these problems, some mistakes are considered more costly than others and are studied under *cost-sensitive* framework. In a cost-sensitive learning problem, we are given a set of training examples, along with the misclassification costs. The goal of learning is to find a hypothesis that minimizes the expected cost of misclassification.

3 Clustering-Based Cost-Sensitive Formulation

In this formulation, we assume that class conditional densities of positive and negative points can be modeled as mixture models with component distributions. Let k_1 and k_2 denote the number of components in the mixture model for positive and negative class respectively and say $k = k_1 + k_2$. We can cluster the positive and negative points separately, and estimate the first and second order moments (μ, Σ) of all the clusters. Given these second order moments, our goal is to find a discriminating hyperplane $w^T x - b = 0$, which separates these positive and negative clusters in such a way that it minimizes the expected cost of misclassification. To this end, consider the following formulation:

$$\begin{aligned} \min_{w,b,\eta_i} \quad & \frac{1}{2}\|w\|_2^2 + \mathcal{C}_{reg}\left\{\mathcal{C}_{12}\sum_{i=1}^{k_1}\eta_i + \mathcal{C}_{21}\sum_{i=k_1+1}^{k}\eta_i\right\} \\ \text{s.t. } & Pr(X_i \in \mathcal{H}_2) \leq \eta_i, : \forall i = 1, \cdots, k_1 \\ & Pr(X_i \in \mathcal{H}_1) \leq \eta_i : \forall i = k_1+1, \cdots, k \\ & 0 \leq \eta_i \leq 1 : \forall i = 1, \cdots, k \end{aligned} \tag{1}$$

Here $X_i, \forall i = 1, \cdots, k_1$ and $X_i, \forall i = k_1 + 1, \cdots, k$ are random variables corresponding to the components of the mixture models for positive and negative classes respectively; $\mathcal{H}_1$ and $\mathcal{H}_2$ denote the positive and negative half spaces i.e., $\mathcal{H}_1(w, b) = \{x|w^T x - b \geq 0\}$ and $\mathcal{H}_2(w, b) = \{x|w^T x - b \leq 0\}$; η_i stands for the probability with which any point drawn from a mixture component lies on the wrong side of the hyperplane. The objective function consists of two terms: the first term $\frac{1}{2}\|w\|_2^2$ is the standard squared-norm regularizer and second term $\mathcal{C}_{12}\sum_{i=1}^{k_1}\eta_i + \mathcal{C}_{21}\sum_{i=k_1+1}^{k}\eta_j$ is the empirical expected cost of misclassification. $\mathcal{C}_{reg}$ is the regularization parameter that controls the trade off between empirical error and generalization error.

The above probabilistic constraints can be written as deterministic constraints using multivariate Chebyshev-Cantelli inequality [10].

3.1 Conversion of Chance-Constraint to Second-Order Cone Constraint

This conversion can be done in several different ways [12,10]. We present the variant based on a multi-variate generalization of Chebyschev-Cantelli inequality [16] which is stated below.

Theorem 1. *Let Z be a random variable whose second order moments are (μ, σ^2). Then for any $t > 0$,*

$$Pr(Z - \mu \geq t) \leq \tfrac{\sigma^2}{\sigma^2 + t^2}$$

We can use the above theorem to do this conversion. Let X be an n-dimensional random variable with second order moments (μ, Σ). By applying the above theorem to random variable $-w^T x$, $w \in \Re^n$ and with $t = w^T\mu - b$, we get

$$Pr(-w^T X \geq -b) \leq \frac{w^T \Sigma w}{w^T \Sigma w + (w^T \mu - b)^2} \tag{2}$$

Now, satisfying the constraint $Pr(w^T X - b \geq 0) \geq \eta$ is same as satisfying $Pr(-w^T X \geq -b) \leq 1 - \eta$. By applying Theorem 1, we can satisfy $Pr(-w^T X \geq -b) \leq 1 - \eta$ if:

$$\frac{w^T \Sigma w}{w^T \Sigma w + (w^T \mu - b)^2} \leq 1 - \eta \tag{3}$$

Re-arranging the terms in the above inequality gives us:

$$w^T \mu - b \geq \kappa \sqrt{w^T \Sigma w} \tag{4}$$

where, $\kappa = \sqrt{\frac{\eta}{1-\eta}}$.

3.2 Separable Case

By using the above conversion with $X = X_i$ and $\eta = 1 - \eta_i$ and re-writing it in the standard SOCP form, we get the following formulation:

$$\begin{aligned} \min_{w,b,\eta_i} \quad & \mathcal{C}_{12} \sum_{i=1}^{k_1} \eta_i + \mathcal{C}_{21} \sum_{i=k_1+1}^{k} \eta_j \\ \text{s.t. } & w^T \mu_i - b \geq 1 + \kappa_i \sqrt{w^T \Sigma_i w} : \forall i = 1, \cdots, k_1 \\ & b - w^T \mu_i \geq 1 + \kappa_i \sqrt{w^T \Sigma_i w} : \forall i = k_1 + 1, \cdots, k \\ & 0 \leq \eta_i \leq 1 : \forall i = 1, \cdots, k \\ & W \geq \|w\|_2 \end{aligned} \tag{5}$$

where, $\kappa_i = \sqrt{\frac{1-\eta_i}{\eta_i}}$; W is a user-defined parameter which plays similar role as $\mathcal{C}_{reg}$ in the previous formulation. The geometric interpretation of the above constraints is that of finding a hyperplane which separates the positive and negative ellipsoids whose centers are at μ_i, shapes determined by Σ_i, and the sizes of the ellipsoids, i.e., κ_i (see Figure 1) to be classified correctly in order to minimize the expected cost of misclassification.

3.3 Non-separable Case

In the above formulation, if the means of the clusters are not separable, then the optimization problem is infeasible. For example, in the worst case say η_i is 1 for some of the non-separable ellipsoids; but even in this worst case the constraints require the means μ_i of these ellipsoids to lie on the correct side of the hyperplane, i.e., $w^T \mu_i - b \geq 1$ and $w^T \mu_i - b \geq -1$. To avoid this problem, we can introduce slack variables ξ_i as in soft-margin SVM formulation and fix the values of η_1 and η_2, the false-positive and false-negative probabilities, to very small values (say 0.1). Note that, η_1 and η_2 are shared by all the clusters of positive and negative classes respectively. In this case the objective function will

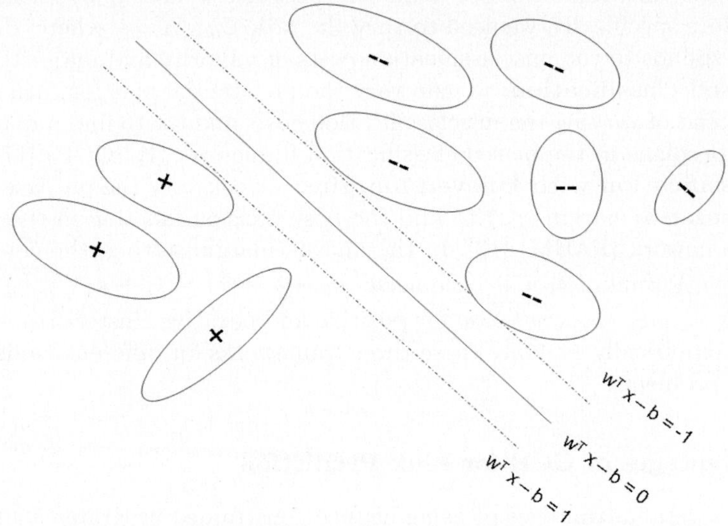

Fig. 1. Geometric interpretation of SOCP formulation

be replaced with slack variables ξ_i instead of η_i in the separable case and leads to the following formulation:

$$
\begin{aligned}
\min_{w,b,\eta_i} \quad & \mathcal{C}_{12} \sum_{i=1}^{k_1} \xi_i + \mathcal{C}_{21} \sum_{i=k_1+1}^{k} \xi_j \\
\text{s.t. } & w^T \mu_i - b \geq 1 - \xi_i + \kappa_1 \sqrt{w^T \Sigma_i w} : \forall i = 1, \cdots, k_1 \\
& b - w^T \mu_i \geq 1 - \xi_i + \kappa_2 \sqrt{w^T \Sigma_i w} : \forall i = k_1 + 1, \cdots, k \\
& \xi_i \geq 0 : \forall i = 1, \cdots, k \\
& W \geq \|w\|_2
\end{aligned}
\tag{6}
$$

We can see that cost-sensitive SVM is now a special case of this formulation when we consider each data point as a cluster, i.e., the covariance matrix is null. By solving the above SOCP problem using standard SOCP solvers like SeDuMi, we get the optimum values of w and b, and a new data point x can be classified as $\text{sign}(w^T x - b)$.

3.4 Unbalanced Data

In the case of skewed class distribution, one class will have more representative data points (majority class) when compared to the other class (minority class). We can handle the unbalanced problem in three different ways.

1) Cost-Sensitive classification: we can transform the unbalanced problem into a cost-sensitive learning problem where costs are unknown and by varying the costs based on a validation set to find the best discriminating hyperplane(CS-SOCP). More specifically, we need to vary the ratio $\mathcal{C}_{min}/\mathcal{C}_{maj}$ where, $\mathcal{C}_{min}$ and $\mathcal{C}_{maj}$ corresponds to the misclassification costs of minority and majority class.

2) Biased classification: we can vary the preferential bias for each class η_1 and η_2 instead of varying the misclassification costs and try to find a maximum-margin hyperplane in the biased classification framework (B-SOCP) [17].

3) Classification with Uneven margins: we can vary the positive margin (τ_+) and negative margin (τ_-) to find the best decision function in the Uneven Margin framework (PAUM) [13]. In the uneven margin setting, the constraints in the above formulation will become $w^T\mu_i - b \geq \tau_+ - \xi_i + \kappa_1\sqrt{w^T\Sigma_i w}$ and $b - w^T\mu_i \geq \tau_- - \xi_i + \kappa_2\sqrt{w^T\Sigma_i w}$ for positive and negative clusters respectively.

We will empirically evaluate these three frameworks for different kinds of link prediction problems.

3.5 Advantages of CCP for Link Prediction

There are several advantages of using chance-constrained programs for the link prediction.

Scalability: The SOCP formulation based on chance constraints is scalable to large datasets because the number of constraints in this formulation is linear in the number of clusters, whereas the number of constraints in the SVM formulation (QP problem) is linear in the number of data points. In addition, there are very efficient interior point algorithms for solving SOCP.

Missing Links: As described before, we consider a large sample of node pairs which are not connected at time t as negative examples. However, some of these negative examples may be noisy, i.e., the link may exist, but was simply not observed at time t. In the case of SVMs the gemoetric interpretation of dual is that of finding the distance between two convex hulls corresponding to the positive and negative points respectively [2]. Conversely, the interpretation of dual for SOCP formulation is that of finding distance between two convex hulls corresponding to the positive and negative ellipsoids. In the presence of noisy labels, The SVM solution is much more sensitive to noisy labels than the solution with ellipsoids.

Missing features: Chance-constrained programs can naturally handle missing features [23]. The key idea here is to use chance constraints to deal with uncertainty in the missing values based on the second order moments. The Gaussian assumption allows us to use EM to impute the missing values. The resulting formulation again leads to an SOCP problem.

Applications: We can use this framework for several applications like recommendations, collaborative filtering, online advertisement and marketing, and anomalous link discovery in financial networks, terrorist networks, power grids and disease transmission networks.

4 Experimental Results and Discussion

In this section, we describe our experimental setup, description of datasets, features used for learning the classifier, evaluation methodology, followed by our results and discussion.

4.1 Datasets

We use three different kinds of real-world domains namely co-authorship networks, biological networks, and citation networks for evaluating our learning algorithms.

Co-authorship networks. In co-authorship networks, we want to predict which pair of authors are likely to collaborate in future. We use two different co-authorship networks:

1) DBLP dataset: we use a dataset which was generated using DBLP collection of computer science articles[3], and contains all the papers from the proceedings of 28 conferences related to machine learning, data mining and databases from 1997 to 2006.

2) Genetics dataset: The *genetics* dataset contains articles published in 14 journals related to genetics and molecular biology from 1996 to 2005. The genetics dataset was generated from the popular PubMed database[4].

For each dataset we have the data for 10 years. We consider the data from first 9 years for training and the data from the 10th year for testing. We consider all the links formed in the 9th year as positive training examples and among all the negative links (those links that are not formed in the first 9 years), we randomly collect a large sample and label them as negative training examples. Note that the features of these training examples are constructed based on the first 8 years of data. Similarly for the test set, we consider all the links that are formed during the 10th year as positive examples and collect a sample of all the negative links as negative examples. Also the features of these testing examples are constructed based on the first 9 years of data.

Biological networks. We use two biological networks, a protein-protein interaction network[5] and a metabolic network[6]. The details are described below:

1) Metabolic network: This network contains 668 nodes and 2782 links. In the metabolic network, proteins are represented as nodes, and an edge indicates that the two proteins are enzymes that catalyze successive reactions between them. This dataset has several features for each protein based on gene expression, localization, phylogenetic profiles and chemical compatibility along with some kernel features as well.

2) Protein-protein interaction network: This network contains 2617 nodes and 8844 edges. Each protein is described by a 76 dimensional feature

[3] http://dblp.uni-trier.de/
[4] http://www.ncbi.nlm.nih.gov/entrez
[5] http://noble.gs.washington.edu/proj/maxent/
[6] http://web.kuicr.kyoto-u.ac.jp/supp/yoshi/ismb05/

vector, where each feature indicates whether the protein is related to a particular function or not.

Since we do not have temporal information for either of these networks, we will choose a random two thirds of the data for training and the remaining one third for testing.

Citation networks. For the citation prediction task, we used the KDD Cup 2003[7] dataset which contains the citation network for both training and testing periods separately. Also for each paper we have all the information including the title, authors, abstract and textual content of the paper. We consider two different kinds of prediction tasks.

1) Complete bibliography prediction: Given a new paper we want to predict the complete bibliography of the paper, i.e., all those papers in the training which will be cited by this paper. In this task, we connect this new paper to all the previous papers written by its authors before the prediction for constructing features.

2) Partial bibliography prediction: In this task, given a new paper and its partial bibliography, we want to predict the remaining entries.

We sample roughly 10 times the number of positive links from the pool of absent links resulting in a positive to negative class ratio of 1:10. The exact number of positive and negative examples used for different link prediction tasks are shown in Table 1.

Table 1. Details of training and testing data for different link prediction tasks

Prediction Task	TRAINING		TESTING	
	# positives	# negatives	# positives	# negatives
DBLP	1404	14040	1021	10210
Genetics	2422	24220	3017	30170
Metabolic network	618	6180	928	9280
Protein network	1966	19660	2948	29480
Complete citation	3000	30000	3000	30000
Partial citation	3000	30000	3000	30000

4.2 Feature Description

We use two different kinds of features between two nodes, namely *content features* and *structural features*.

The *content feature function* $\phi_{cont} : \Re^d \times \Re^d \mapsto \Re^n$ is defined based on the attributes of the two nodes. For example, in the case of co-authorship networks the features of each author corresponds to occurrences of a particular word in the author's papers. The content feature function could be the kronecker product of these binary vectors. Similarly for friend recommendation problems in social

[7] http://www.cs.cornell.edu/projects/kddcup/datasets.html

networks, content features will be defined based on the user profiles – geographic location, college/university, work place, hobbies/interests, browsing/navigation history on the network etc. In the case of protein-protein networks, content features can be defined as the Kronecker product of the features of each protein. Therefore, weights on each of these kronecker features will tell us how likely those proteins will interact.

The *structural feature function* $\phi_{struct} : G_{n_1,n_2} \mapsto \Re^m$ is defined over the local subgraph around the two nodes n_1 and n_2. One can also call them relational features, e.g., approximate Katz measure which is calculated on the ensemble of paths between two nodes (say upto depth 4), number of common neighbors, social connectivity which shows how connected these two nodes are with the other nodes in their neighborhood etc., which are meaningful for each network. For example, in the citation prediction task the network between papers and authors is very complex, i.e., links are between one paper and another–*paper1 cites paper2*, and between an author and a paper–*author1 writes paper1*. Therefore, these complex multi-way relationships could be used to define relational features which will be useful for our link prediction task.

4.3 Evaluation

We use the precision and recall metrics from Information Retrieval context for evaluation, and compare the chance-constraints based algorithms, namely cost-sensitive SOCP (CS-SOCP), biased SOCP (B-SOCP) against cost-sensitive SVMs[8] (CS-SVM) and perceptron with uneven margins (PAUM) [13]. We rank all the test examples according to the margin of the classifiers and calculate precision and recall from *top-k* by varying the value of k. Here, precision is defined as the percentage of true-positive links that are predicted correctly among the *top-k* and recall is defined as the percentage of true-positive links that are predicted correctly out of the total true-positive links. Note that, majority of the applications of link prediction algorithms are in recommendation systems like movie recommendations in Netflix, music recommendation engines like *last.fm*, friends suggestions in social networks etc. Therefore, link prediction algorithms should be evaluated based on the quality of the top-k recommendations produced by them. According to the above definitions of precision and recall, precision need not always monotonically decrease with k. We report the precision and recall curves by varying the value of k along the x-axis. We also report the AUC values calculated for top 20% of the total testing links (see Table 2).

We use $k_1 = k_2 = 100$ clusters for all clustering-based SOCP formulations and k-means++[9], a variant of k-means algorithm which is fast and proven to be near optimal for clustering in our experiments. We observe that the number of clusters will not make much difference in the results as long as they are not too small a number of clusters. As the number of clusters increases SOCP based algorithms will tend to move closer towards their SVM counterparts. Note that, SOCP and

[8] LIBSVM with -wi option to specify costs.
[9] http://en.wikipedia.org/wiki/K-means++

Table 2. AUC values for top 20% of the total testing links for different learning algorithms

	CS-SOCP	B-SOCP	CS-SVM	PAUM
DBLP	**0.4019**	0.3707	0.3186	0.0682
Genetics	**0.2314**	0.1981	0.1526	0.0638
Metabolic	0.619	0.6183	**0.6447**	0.0816
Protein	0.2754	**0.2786**	0.2471	0.1274
Complete citation	**0.3684**	0.3186	0.3252	0.3586
Partial citation	0.4994	0.469	**0.5356**	0.3607

Table 3. Training and classification time results

	Training time		Classification time	
	CS-SVM	CS-SOCP	CS-SVM	CS-SOCP
DBLP	39.68s	0.69s	7.64s	0.46s
Genetics	3m 34s	9s	1m 44s	27s
Metabolic	15.1s	4.89s	7.31s	4.29s
Protein	42m 23s	56.64s	1m 53s	19.64s
Complete citation	3m 17.6s	8.92s	1m 16.6s	13.92s
Partial citation	5m 19.5s	10.21s	1m 3.3s	13.78s

SVM based algorithms are exactly the same when we consider each data point as a cluster, i.e., the covariance matrix is null. We use diagonal covariance for our SOCP experiments. We report the best results for CS-SVM and CS-SOCP by varying the ratio $\mathcal{C}_+/\mathcal{C}_-$ on validation set. Similarly, we give the best results for B-SOCP by varying η_1 and η_2 . For PAUM, we pick the best values for τ_- from $\{-1.5, -1, -0.5, 0, 0.1, 0.5, 1\}$ and for τ_+ from $\{-1, -0.5, 0, 0.1, 0.5, 1, 2, 5, 10, 50\}$ based on the validation set. We run PAUM for a maximum of 1000 iterations or until convergence.

4.4 Results and Discussion

The precision and recall curves for all the 6 datasets are shown in Figures 2,3 and 4. As we can see, both CS-SOCP and B-SOCP outperform CS-SVM in precision and recall for majority of the datasets namely, DBLP, genetics, complete and partial bibliographic prediction tasks. Particularly, they achieve significantly higher recall on the complete bibliography prediction task (72.4% and 66.16% compared to 52.53% of CS-SVM) and partial bibliographic prediction task (82.96% and 75.13% compared to 70.13% of CS-SVM). Similarly, if we look at the AUC values in Table 2, SOCP based algorithms significantly outperform the other algorithms on 4 out of 6 prediction tasks, including the protein-protein interaction network which is a very high-dimensional dataset. We conjecture that noisy labels for the missing links (explained in the previous section) might have contributed to the bad performance of CS-SVM in both these tasks. Also note that the prediction accuracies significantly improve in the case of partial prediction

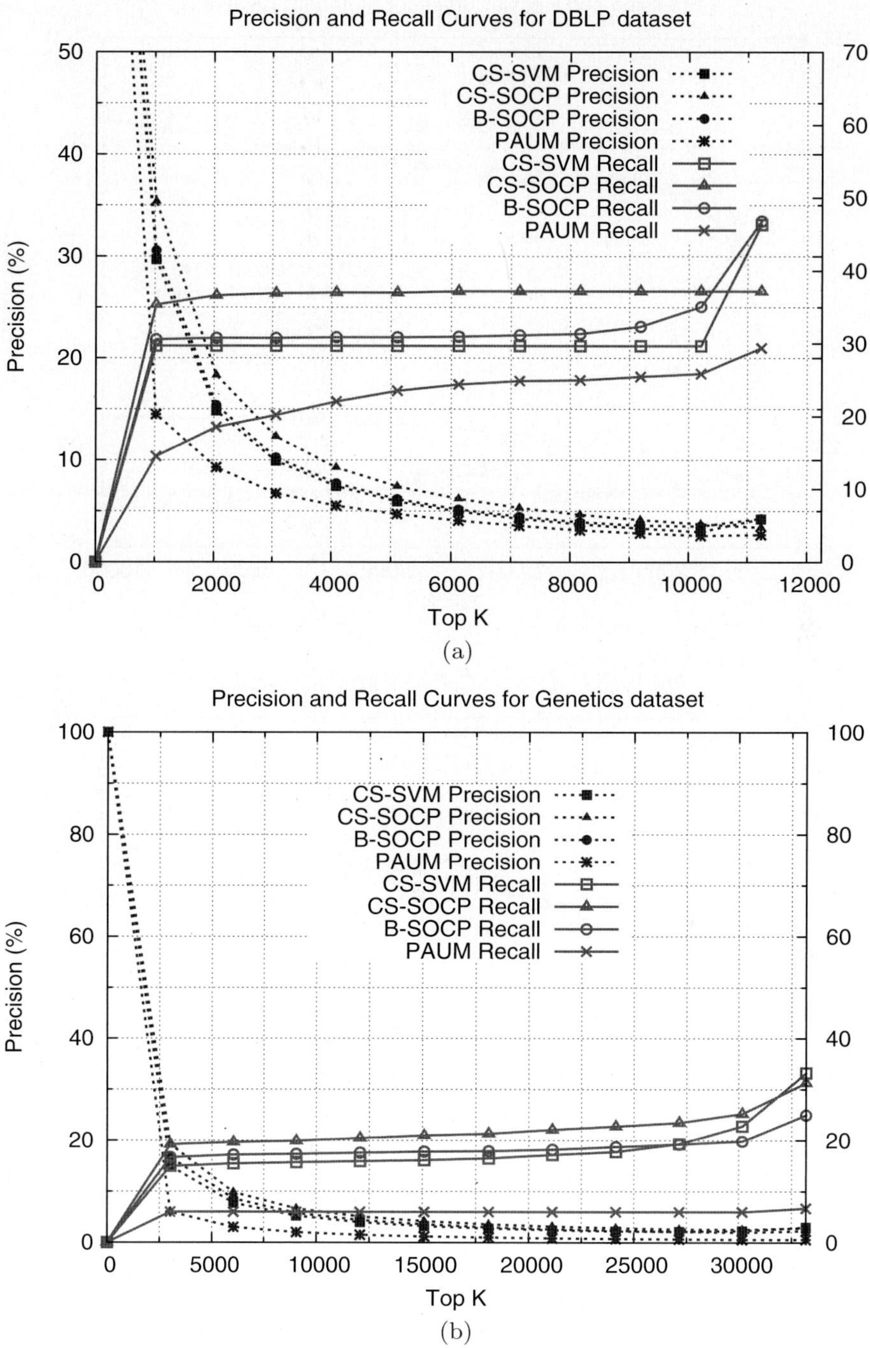

Fig. 2. Precision and Recall curves for Co-authorship networks (a) DBLP (b) Genetics

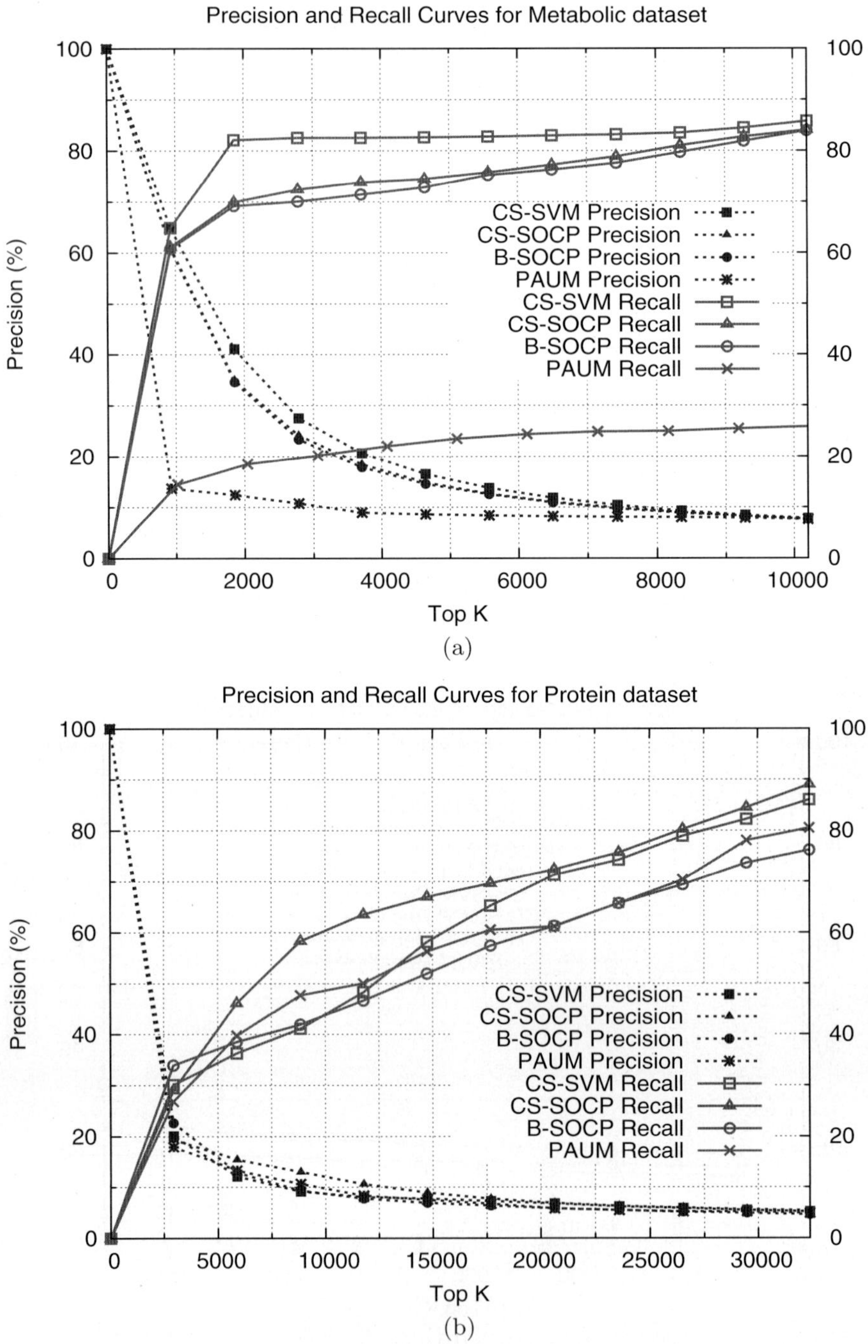

Fig. 3. Precision and Recall curves for Biological networks (a) Metabolic netowork (b) Protein-protein interaction network

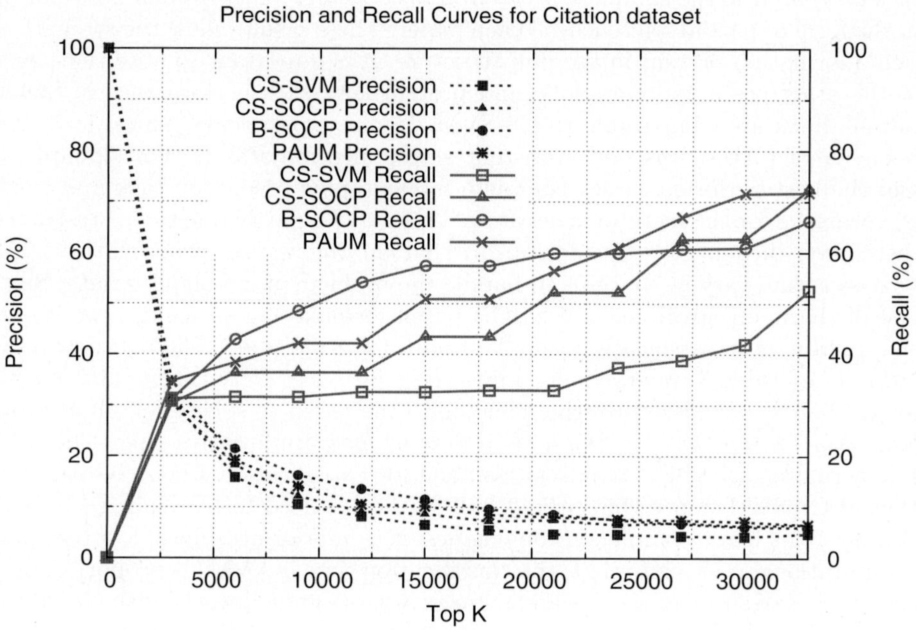

(a)

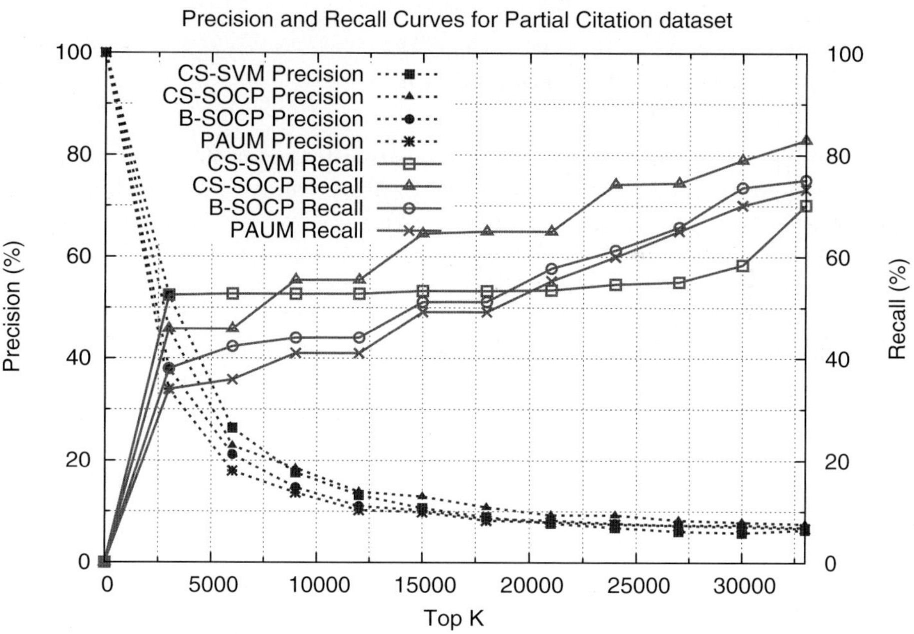

(b)

Fig. 4. Precision and Recall curves for citation networks (a) Complete bibliography prediction task (b) Partial bibliography prediction task

task compared to the complete prediction task because of additional information in the form of partial references of each paper. These results show the strength of rich information present in the link structure. It is important to note that, even in the other cases like metabolic and protein networks, performance of SOCP formulations are comparable to CS-SVM. In our experiments, we noticed that behavior of PAUM was not consistent across all the datasets. For example, it had the best performance for complete bibliographic prediction task and worst performance for the metabolic network. This may be partly due to our restricted search over the margin space. It appears that varying costs or probabilities might be easier than varying margins to handle the problem of unbalanced data. Since one of the main advantages of SOCP based formulations is scaling, we report the training and classification time[10] of both CS-SVM and CS-SOCP for all the datasets in Table 3 (m stands for mins and s for secs). Note that, training time for CS-SOCP includes clustering time and time taken to solve the SOCP problem. We can see that CS-SOCP is orders of magnitude faster than CS-SVM. Furthermore, CS-SOCP requires less time for classification when compared to that of CS-SVM. Since the learned link predictors need to be deployed in real-time systems like recommendation engines, it is important to have low test time computational cost. Note that, the classification time in SVMs is proportional to the number of support vectors and support vectors grow linearly with size of the data. On the other hand, the number of support vectors in CS-SOCP is bounded by the number of clusters k and does not depend on the size of the data.

5 Conclusions and Future Work

In this work, we proposed a new cost-sensitive formulation based on chance constraints and described its connections to other frameworks like biased classification and uneven margin algorithms. We showed how learning algorithms based on chance-constraints can be used to solve different kinds of link prediction problems and showed empirical evidence with experiments on several real-world datasets. It is interesting to note that we could formulate link-prediction as a complex structured prediction problem with exponential number of constraints. The manner in which the absent links are sampled to be used as negative examples for our classification problem, is roughly equivalent to randomly sampling the constraints for the structured prediction problem [6,5]. We believe that this is a very fruitful direction towards solving some of these hard problems. Wick et al. use similar ideas for their SampleRank algorithm and got some impressive results [26]. In future, we would like to extend the current framework to a relational setting similar to Taskar's work [24]. However, formulating it as relational or structured prediction poses an enormous inference problem, especially in large networks. One possible approach is to take a middle path between complete independence and arbitrary relational structure.

---

[10] All experiments were run on a machine with 2GB RAM and 2.16 GHz Intel dual core processor.

Acknowledgements

This material is based upon work supported by the Defense Advanced Research Projects Agency (DARPA) under Contract No. FA8750-09-C-0179. Any opinions, findings and conclusions or recommendations expressed in this material are those of the author(s) and do not necessarily reflect the views of the DARPA, or the Air Force Research Laboratory (AFRL). This work is also partially funded by NSF Grant No. 0746930. We would like to thank Saketha Nath for useful discussions which helped our understanding on chance-constrained programs.

References

1. Adamic, L., Adar, E.: Friends and neighbors on the web. Social Networks 25, 211–230 (2001)
2. Bennett, K.P., Bredensteiner, E.J.: Duality and geometry in svm classifiers. In: International Conf. on Machine Learning (ICML), pp. 57–64 (2000)
3. Bhattacharyya, C.: Second order cone programming formulations for feature selection. Journal of Machine Learning Research (JMLR) 5, 1417–1433 (2004)
4. Bilgic, M., Namata, G.M., Getoor, L.: Combining collective classification and link prediction. In: Workshop on Mining Graphs and Complex Structures at the IEEE International Conference on Data Mining(ICDM) (2007)
5. Calafiore, G., Campi, M.: Uncertain convex programs: Randomized solutions and confidence levels. Mathematical Programming 102, 25–46 (2005)
6. Farias, D.P.D., Roy, B.V.: On constraint sampling in the linear programming approach to approximate dynamic programming. Mathematics of Operations Research 29(3), 462–478 (2001)
7. Getoor, L., Friedman, N., Koller, D., Taskar, B.: Learning probabilistic models of relational structure. In: International Conference on Machine Learning(ICML) (2001)
8. Getoor, L., Friedman, N., Koller, D., Taskar, B.: Learning probabilistic models of link structure. Journal of Machine Learning Research 3, 679–707 (2002)
9. Hasan, M., Chaoji, V., Salem, S., Zaki, M.: Link prediction using supervised learning. In: SDM Workshop on Link Analysis Counter-terrorism and Security (2006)
10. Jagarlapudi, S.N.: Learning Algorithms using Chance-Constrained Programming. Ph.d dissertation, Computer Science and Automation, IISc Bangalore (2007)
11. Kashima, H., Abe, N.: A parameterized probabilistic model of network evolution for supervised link prediction. In: International Conference on Data Mining (ICDM) (2006)
12. Lanckriet, G., Ghaoui, L.E., Bhattacharya, C., Jordan, M.: Minimax probability machine. In: Annual Conference on Neural Information Processing Systems (NIPS) (2001)
13. Li, Y., Zaragoza, H., Herbrich, R., Shawe-Taylor, J., Kandola, J.S.: The perceptron algorithm with uneven margins. In: International Conference on Machine Learning (ICML), pp. 379–386 (2002)
14. Libennowell, D., Kleinberg, J.: The link prediction problem for social networks. In: International Conference on Knowledge Management(CIKM) (2003)
15. Lobo, M.S., Vandenberghe, L., Boyd, S., Lebret, H.: Applications of second-order cone programming. Linear Algebra and its Applications 238, 193–228 (1998)

16. Marshall, A., Olkin, I.: Multivariate chebyshev inequalities. Annals of Mathematical Statistics 31, 1001–1014 (1960)
17. Nath, J.S., Bhattacharyya, C.: Maximum margin classifiers with specified false positive and false negative error rates. In: SIAM International Conference on Data Mining (SDM) (2007)
18. Nath, J.S., Bhattacharyya, C., Murty, M.N.: Clustering based large margin classification: a scalable approach using socp formulation. In: International Conference on Knowledge Discovery and Data Mining(KDD) (2006)
19. Newman, M.: Clustering and preferential attachment in growing networks. Physical Review Letters, 64 (2001)
20. Popescul, A., Popescul, R., Ungar, L.: Statistical relational learning for link prediction. In: IJCAI workshop on Learning Statistical Models for Relational Data (2003)
21. Rattngon, M., Jensen, D.: The case for anomalous link discovery. SIGKDD Explorations 7(2), 41–47 (2005)
22. Sarkar, P., Chakrabarti, D., Moore, A.: Theoretical justification of popular link prediction heuristics. In: International Conference on Learning Theory (COLT), pp. 295–307 (2010)
23. Shivaswamy, P.K., Bhattacharyya, C., Smola, A.J.: Second order cone programming approaches for handling missing and uncertain data. Journal of Machine Learning Research (JMLR) 7, 1283–1314 (2006)
24. Taskar, B., Fai Wong, M., Abbeel, P., Koller, D.: Link prediction in relational data. In: Annual Conference on Neural Information Processing Systems (NIPS) (2003)
25. Wang, C., Satuluri, V., Parthasarathy, S.: Local probabilistic models for link prediction. In: International Conference on Data Mining (ICDM) (2007)
26. Wick, M., Rohanimanesh, K., Culotta, A., McCallum, A.: Samplerank: Learning preferences from atomic gradients. In: Neural Information Processing Systems (NIPS) Workshop on Advances in Ranking (2009)
27. Zheleva, E., Getoor, L., Golbeck, J., Kuter, U.: Using friendship ties and family circles for link prediction. In: 2nd ACM SIGKDD Workshop on Social Network Mining and Analysis (SNA-KDD) (2008)

Sparse Unsupervised Dimensionality Reduction Algorithms

Wenjun Dou, Guang Dai, Congfu Xu, and Zhihua Zhang

College of Computer Science and Technology
Zhejiang University
Hangzhou, Zhejiang 310027, China
{xucongfu, zhzhang}@cs.zju.edu.cn

Abstract. Principal component analysis (PCA) and its dual—principal coordinate analysis (PCO)—are widely applied to unsupervised dimensionality reduction. In this paper, we show that PCA and PCO can be carried out under regression frameworks. Thus, it is convenient to incorporate sparse techniques into the regression frameworks. In particular, we propose a sparse PCA model and a sparse PCO model. The former is to find sparse principal components, while the latter directly calculates sparse principal coordinates in a low-dimensional space. Our models can be solved by simple and efficient iterative procedures. Finally, we discuss the relationship of our models with other existing sparse PCA methods and illustrate empirical comparisons for these sparse unsupervised dimensionality reduction methods. The experimental results are encouraging.

1 Introduction

Unsupervised dimensionality reduction methods are widely used in many applications such as image processing, microarray data analysis, information retrieval, etc. PCA [13] and PCO (or the classical multidimension scaling) [9,16] are two classical unsupervised techniques for dimensionality reduction. PCA aims to find the principal components (PCs) with the largest variance, while PCO directly calculates the coordinate configurations in the dimension-reduced space.

However, it is sometimes difficult to interpret the results with PCA, because each principal component is a linear combination of all the original variables and its loadings are typically nonzero. Many approaches have been developed to deal with this drawback of PCA. Recently, a sparse approach has been introduced. Roughly speaking, the approach is to impose some sparsity constraints such as lasso [19] and elastic net [22] to loadings, then some of loadings are naturally zero. There are mainly two families of sparse PCA methods in the literature. The first one uses the maximum-variance property of principal components, such as SCoTLASS [14], DSPCA [2], sPCA-rSVD [17], SOCA [20], sPCA-DC [18], etc. The other family is based on regression-type problems such as sparse PCA [23].

In this paper we develop a new sparse PCA model to achieve sparseness. Our model is built on the notion of optimal scoring, which was originally used to carry out the Fisher discriminant analysis [11]. Recently, Clemmensen [1] proposed a sparse discriminant analysis method by optimal scoring. Zhang and Dai [21] showed that some

J.L. Balcázar et al. (Eds.): ECML PKDD 2010, Part I, LNAI 6321, pp. 361–376, 2010.

unsupervised learning methods, such as spectral clustering and PCA, can be cast into an optimal scoring framework. This work immediately motivates our sparse PCA model. An advantage of our model over the other sparse PCA methods is that our model can achieve more sparseness when the total explained variance of principal components is approximately same. Moreover, since our sparse PCA is derived from the optimal scoring, it is more appropriately applied to discriminant analysis problems.

When applying these sparse PCA methods, we are able to obtain sparse principal loadings. However, the coordinate configurations in the dimension-reduced space are not necessarily sparse. In practical applications, it would be sometimes interesting to find sparse principal coordinates. As we know, however, there is no work about this theme. In this paper we present a sparse PCO model. In particular, we exploit the Eckart-Young theorem [3], because the theorem shows a dual relationship between the conventional PCA model and the conventional PCO model. Moreover, we note that it also provides a regression framework to perform PCO. Introducing the elastic net penalty for principal coordinates into this framework, we devise our sparse PCO.

The rest of this paper is organized as follows. Sections 2 and 3 present our sparse PCA and PCO models respectively. Section 4 discusses the relationship of our sparse models with other existing sparse PCA methods. In Section 5, we conduct our experimental evaluations. Finally, we give our conclusions in Section 6.

2 Sparse PCA

Optimal scoring was first introduced by Hastie [11] to formulate the Fisher linear discriminant analysis as a multiple linear regression problem. In recent work, Zhang and Dai [21] extended the concept of optimal scoring to unsupervised learning problems and developed a framework for unsupervised clustering. Based on the optimal scoring framework, we now develop our sparse principal component analysis model, namely *sparse PCA via optimal scoring* (sPCA-OS).

First of all, we list some notations. Throughout this paper, $\mathbf{I}_m$ denotes the $m \times m$ identity matrix, $\mathbf{1}_m$ the $m \times 1$ of ones, and $\mathbf{0}$ the zero matrix or vector whose dimensionality is dependent upon the context.

For an $n \times m$ matrix $\mathbf{A} = [a_{ij}]$, let $\text{vec}(\mathbf{A}) = (a_{11}, \ldots, a_{n1}, a_{12}, \ldots, a_{nm})^T$ be the $nm \times 1$ vector, $\|\mathbf{A}\|_F = \|\text{vec}(\mathbf{A})\|_2 = \sqrt{\sum_{i,j} a_{ij}^2}$ be the Frobenius norm of $\mathbf{A}$ or the 2-norm of $\text{vec}(\mathbf{A})$, and $\|\mathbf{A}\|_1 = \sum_{i,j} |a_{ij}|$ be the 1-norm of $\text{vec}(\mathbf{A})$. In addition, $\mathbf{A} \otimes \mathbf{B} = [a_{ij}\mathbf{B}]$ represents the Kronecker product of $\mathbf{A}$ and $\mathbf{B}$, and

$$\mathcal{O}^{n \times m} = \{\mathbf{A} : \mathbf{A} \in \mathbb{R}^{n \times m} \text{ and } \mathbf{A}^T\mathbf{A} = \mathbf{I}_m\}.$$

2.1 Optimal Scoring for PCA

We are given an $n \times p$ data matrix $\mathbf{X} = [\mathbf{x}_{ij}]$, where n is the number of observations and p is the number of variables. Let $\mathbf{Y} = [\mathbf{y}_1, \ldots, \mathbf{y}_q]$ be an $n \times q$ sample scoring matrix such that $\mathbf{1}_n^T\mathbf{Y} = \mathbf{0}$ and $\mathbf{Y}^T\mathbf{Y} = \mathbf{I}_q$ and $\mathbf{W} = [\mathbf{w}_1, \ldots, \mathbf{w}_q]$ be a $p \times q$ weight matrix. For dimensionality reduction problems, q represents the number of PCs (or loadings) and must be less than p.

Without loss of generality, suppose that $\mathbf{X}$ is centered; that is, $\mathbf{1}_n^T\mathbf{X} = \mathbf{0}$. The framework of optimal scoring for unsupervised learning is then defined by

$$\min_{\mathbf{Y},\mathbf{W}} \Big\{ f(\mathbf{Y},\mathbf{W}) \equiv \frac{1}{2}\|\mathbf{Y}-\mathbf{X}\mathbf{W}\|_F^2 + \frac{\delta^2}{2}\mathrm{tr}(\mathbf{W}^T\mathbf{W}) \Big\}$$

subject to $\mathbf{1}_n^T\mathbf{Y} = \mathbf{0}$ and $\mathbf{Y}^T\mathbf{Y} = \mathbf{I}_q$. Zhang and Dai [21] proved that the minimum of f is obtained when $\mathbf{Y}$ is the $n \times q$ matrix of the top orthonormal eigenvectors of $\mathbf{X}(\mathbf{X}^T\mathbf{X} + \delta^2\mathbf{I}_p)^{-1}\mathbf{X}^T$ and $\mathbf{W} = (\mathbf{X}^T\mathbf{X} + \delta^2\mathbf{I}_p)^{-1}\mathbf{X}^T\mathbf{Y}$. Obviously, $\mathbf{W}$ can be treated as a non-orthogonal matrix of loadings and then $\mathbf{X}\mathbf{W}$ is the low-dimensional principal coordinate matrix of $\mathbf{X}$.

Let $r = \min\{n, p\}$, we make the full singular value decomposition (SVD) [7] of $\mathbf{X}$ as $\mathbf{X} = \mathbf{U}\mathbf{D}\mathbf{V}^T$, where $\mathbf{U} = [\mathbf{u}_1, \ldots, \mathbf{u}_n]$ $(n \times n)$ and $\mathbf{V} = [\mathbf{v}_1, \ldots, \mathbf{v}_p]$ $(p \times p)$ are orthogonal matrices, and $\mathbf{D} = \mathrm{diag}(d_1, \ldots, d_r)$ $(n \times p)$ is a diagonal matrix with $d_1 \geq d_2 \geq \cdots \geq d_r \geq 0$. We then obtain the minimizers of $f(\mathbf{Y},\mathbf{W})$ as $\mathbf{Y} = [\mathbf{u}_1, \ldots, \mathbf{u}_q]$ and $\mathbf{W} = \mathbf{V}_1(\mathbf{D}_1^2 + \delta^2\mathbf{I}_q)^{-1}\mathbf{D}_1$, where $\mathbf{V}_1 = [\mathbf{v}_1, \ldots, \mathbf{v}_q]$ and $\mathbf{D}_1 = \mathrm{diag}(d_1, \ldots, d_q)$. This immediately leads us to the following theorem.

Theorem 1. *Assume that* $\mathbf{Y}$ *is an* $n \times q$ *optimal scoring matrix and* $\mathbf{W}$ *is a* $p \times q$ *loading matrix. Consider*

$$(\hat{\mathbf{Y}}, \hat{\mathbf{W}}) = \operatorname*{argmin}_{\mathbf{Y},\mathbf{W}} \ f(\mathbf{Y},\mathbf{W}) \tag{1}$$

under the constraints $\mathbf{1}_n^T\mathbf{Y} = 0$ *and* $\mathbf{Y}^T\mathbf{Y} = \mathbf{I}_q$. *Then* $\hat{\mathbf{w}}_j \propto \mathbf{v}_j$ *for* $j = 1, \ldots, q$.

Zou [23] proposed a regression approach for solving PCA. Theorem 1 shows that we can develop an alternative regression formulation of PCA.

2.2 Sparse PCA via Optimal Scoring

It is natural to impose a sparse penalty to the loading matrix $\mathbf{W}$. Particularly, we exploit the elastic net in our sPCA-OS method. That is, we consider the following optimization problem:

$$\begin{aligned} &\operatorname*{argmin}_{\mathbf{Y},\mathbf{W}} \frac{1}{2}\|\mathbf{Y}-\mathbf{X}\mathbf{W}\|_F^2 + \frac{\delta^2}{2}\|\mathbf{W}\|_F^2 + \sum_{j=1}^{q}\lambda_j\|\mathbf{w}_j\|_1 \\ &\text{s.t.}\ \ \mathbf{1}_n^T\mathbf{Y} = \mathbf{0} \text{ and } \mathbf{Y}^T\mathbf{Y} = \mathbf{I}_q, \end{aligned} \tag{2}$$

where δ is applied to all the q components and the λ_j is used to let each loading $\mathbf{w}_j$ has different degree of sparseness.

Problem (2) can be solved by an iternative procedure. First, with fixed $\mathbf{Y}$, the optimization problem (2) is converted into the conventional elastic net problem [22]; namely,

$$\min_{\mathbf{W}} \frac{1}{2}\sum_{j=1}^{q}\|\mathbf{y}_j - \mathbf{X}\mathbf{w}_j\|_2^2 + \frac{\delta^2}{2}\sum_{j=1}^{q}\|\mathbf{w}_j\|_2^2 + \sum_{j=1}^{q}\lambda_j\|\mathbf{w}_j\|_1.$$

This problem can be decomposed into q separable optimization problems; that is, for $j = 1, \ldots, q$, we have

$$\min_{\mathbf{w}_j} \frac{1}{2}\|\mathbf{y}_j - \mathbf{X}\mathbf{w}_j\|_2^2 + \frac{\delta^2}{2}\|\mathbf{w}_j\|_2^2 + \lambda_j\|\mathbf{w}_j\|_1. \tag{3}$$

Second, with fixed $\mathbf{W}$, we can ignore the penalty terms in the optimization problem (2) and it becomes a Procrustes problem [5] as follows:

$$\begin{aligned} &\underset{\mathbf{Y}}{\operatorname{argmin}} \ \frac{1}{2}\sum_{j=1}^{q}\|\mathbf{y}_j - \mathbf{X}\mathbf{w}_j\|_2^2 \\ &\text{s.t. } \mathbf{1}_n^T\mathbf{Y} = \mathbf{0} \text{ and } \mathbf{Y}^T\mathbf{Y} = \mathbf{I}_q. \end{aligned} \tag{4}$$

This problem is easily solved in a closed form (see, e.g., Gower and Dijksterhuis [10]). Let the thin SVD [7] of $\mathbf{XW}$ be $\boldsymbol{\Psi}\boldsymbol{\Delta}\boldsymbol{\Phi}^T$ where $\boldsymbol{\Psi}$ ($n \times q$) and $\boldsymbol{\Phi}$ ($q \times q$) satisfy $\boldsymbol{\Psi}^T\boldsymbol{\Psi} = \mathbf{I}_q$ and $\boldsymbol{\Phi}^T\boldsymbol{\Phi} = \mathbf{I}_q$ and $\boldsymbol{\Delta}$ ($q \times q$) is the diagonal matrix with nonnegative entries. Then $\mathbf{XW} = \boldsymbol{\Psi}\boldsymbol{\Delta}\boldsymbol{\Phi}^T$ is a solution of (4). The procedure for solving our sPCA-OS is summarized in Algorithm 1.

Algorithm 1. SparsePCA via Optimal Scoring(sPCA-OS)

1: Initialize a $\mathbf{Y}$ subject to $\mathbf{1}_n^T\mathbf{Y} = \mathbf{0}$ and $\mathbf{Y}^T\mathbf{Y} = \mathbf{I}_q$.
2: With a fixed $\mathbf{Y} = [\mathbf{y}_1, \ldots, \mathbf{y}_q]$, solve the elastic net problems for $j = 1, \ldots, q$:

$$\underset{\mathbf{w}_j}{\operatorname{argmin}} \ \frac{1}{2}\|\mathbf{y}_j - \mathbf{X}\mathbf{w}_j\|_2^2 + \frac{\delta^2}{2}\|\mathbf{w}_j\|_2^2 + \lambda_j\|\mathbf{w}_j\|_1.$$

3: With a fixed $\mathbf{W} = [\mathbf{w}_1, \ldots, \mathbf{w}_q]$, perform the thin SVD of $\mathbf{XW}$ as $\mathbf{XW} = \boldsymbol{\Psi}\boldsymbol{\Delta}\boldsymbol{\Phi}^T$ and update $\mathbf{Y}$ by $\mathbf{Y} = \boldsymbol{\Psi}\boldsymbol{\Phi}^T$.
4: Repeat Steps 2 and 3 until convergence.

3 Sparse PCO

Although $\mathbf{W}$ obtained via Algorithm 1 is sparse, the coordinate matrix $\mathbf{Z} = \mathbf{XW}$ is not necessarily sparse. However, it would be also interesting in the situation that $\mathbf{Z}$ is sparse. We thus attempt to develop a sparse PCO algorithm in which the coordinate matrix is sparse. First of all, we present the following theorem.

Theorem 2. *Let the full SVD of* $\mathbf{X}$ *be* $\mathbf{X} = \mathbf{UDV}^T$ *where* $\mathbf{U} \in \mathcal{O}^{n\times n}$, $\mathbf{V} = [\mathbf{v}_1, \ldots, \mathbf{v}_p] \in \mathcal{O}^{p\times p}$ *and* $\mathbf{D} = \mathsf{diag}(d_1, \ldots, d_r)$ *with* $r = \min\{n, p\}$ *and* $d_1 \geq d_2 \geq \ldots \geq d_r \geq 0$. *Assume that* g *is defined by*

$$g(\mathbf{A}, \mathbf{Z}) = \|\mathbf{X} - \mathbf{Z}\mathbf{A}^T\|_F^2 + \gamma\|\mathbf{Z}\|_F^2 \tag{5}$$

where $\mathbf{A} \in \mathcal{O}^{p\times q}$, $\mathbf{Z} \in \mathbb{R}^{n\times q}$ *and* $\gamma \geq 0$. *Then the minimum of* g *is obtained when* $\mathbf{A} = [\mathbf{v}_1, \ldots, \mathbf{v}_q]$ *and* $\mathbf{Z} = \frac{1}{1+\gamma}\mathbf{XA}$.

The proof of this theorem is given in Appendix A. When $\gamma = 0$, Theorem 2 degenerates to the Eckart-Young theorem [3,15].

Theorem 2 shows that $\mathbf{Z}$ is just the coordinate matrix up to the constant $\frac{1}{1+\gamma}$. In order to make $\mathbf{Z}$ sparse, we impose the elastic net penalty on it. Accordingly, we have our sparse PCO model which is defined by the following optimization problem:

$$\min_{\mathbf{A}\in\mathcal{O}^{p\times q},\mathbf{Z}\in\mathbb{R}^{n\times q}} \frac{1}{2}\|\mathbf{X}-\mathbf{Z}\mathbf{A}^T\|_F^2 + \frac{\gamma_1}{2}\|\mathbf{Z}\|_F^2 + \gamma_2\|\mathbf{Z}\|_1, \tag{6}$$

where $\gamma_1 > 0$ and $\gamma_2 > 0$ are regularization parameters. We also resort to an alternatively iterative procedure to solve the problem (6).

With a fixed $\mathbf{Z}$, the optimization problem (6) becomes

$$\min_{\mathbf{A}\in\mathcal{O}^{p\times q}} \frac{1}{2}\|\mathbf{X}-\mathbf{Z}\mathbf{A}^T\|_F^2,$$

which is a reduced rank Procrustes problem illustrated by Zou [23]. Suppose that the thin SVD of $\mathbf{X}^T\mathbf{Z}$ is $\mathbf{X}^T\mathbf{Z} = \boldsymbol{\Psi}\boldsymbol{\Delta}\boldsymbol{\Phi}^T$. Then $\mathbf{A} = \boldsymbol{\Psi}\boldsymbol{\Phi}^T$ is the solution of this Procrustes problem.

With a fixed $\mathbf{A}$, the optimization problem (6) degenerates to

$$\min_{\boldsymbol{\beta}} \frac{1}{2}\|\boldsymbol{\alpha}-(\mathbf{A}\otimes\mathbf{I}_n)\boldsymbol{\beta}\|_2^2 + \frac{\gamma_1}{2}\|\boldsymbol{\beta}\|_2^2 + \gamma_2\|\boldsymbol{\beta}\|_1, \tag{7}$$

where $\boldsymbol{\alpha} = \mathsf{vec}(\mathbf{X})$ and $\boldsymbol{\beta} = \mathsf{vec}(\mathbf{Z})$. Since $(\mathbf{A}\otimes\mathbf{I}_n)^T(\mathbf{A}\otimes\mathbf{I}_n) = (\mathbf{A}^T\mathbf{A})\otimes\mathbf{I}_n = \mathbf{I}_{qn}$, this problem can be directly solved via the soft thresholding algorithm [19,22].

In summary, we have our SPCO algorithm which is given in Algorithm 2.

Algorithm 2. Sparse PCO (SPCO)

1: Give $\mathbf{X}$ and initialize $\mathbf{A}$.

2: Fix $\mathbf{A}$ and solve the elastic net problem w.r.t. $\boldsymbol{\beta}$.

$$\min_{\boldsymbol{\beta}} \frac{1}{2}\|\boldsymbol{\alpha}-(\mathbf{A}\otimes\mathbf{I}_n)\boldsymbol{\beta}\|_2^2 + \frac{\gamma_1}{2}\|\boldsymbol{\beta}\|_2^2 + \gamma_2\|\boldsymbol{\beta}\|_1.$$

3: Fix $\boldsymbol{\beta}$ and perform the thin SVD of $\mathbf{X}^T\mathbf{Z}$ as $\mathbf{X}^T\mathbf{Z} = \boldsymbol{\Psi}\boldsymbol{\Delta}\boldsymbol{\Phi}^T$ and update $\mathbf{A}$ by $\mathbf{A} = \boldsymbol{\Psi}\boldsymbol{\Phi}^T$.

4: Repeat Step 2 and 3 until convergence.

4 Related Work

In the literature [20], the authors illustrated the relationship among SCoTLASS [14], sPCA-rSVD [17], the SPCA method of Zou *et al.* [23] (called sPCA-ZHT), and the SPCA method of Witten *et al.* [20] (called SPCA-WTH). We further discuss the relationship of our sPCA-OS and SPCO with these existing sparse PCA methods.

First, the SPCA method of Zou *et al.* [23] (SPCA-ZHT) is defined as

$$\min_{\mathbf{W}\in\mathbb{R}^{p\times q}} \frac{1}{2}\|\mathbf{X}-\mathbf{X}\mathbf{W}\mathbf{A}^T\|_F^2 + \frac{\gamma_1}{2}\|\mathbf{W}\|_F^2 + \sum_{j=1}^{q}\gamma_{2,j}\|\mathbf{w}_j\|_1,$$
$$\text{s.t. } \mathbf{A}^T\mathbf{A} = \mathbf{I}_q.$$

Comparing this model with our SPCO model in (6), a connection between these two models is immediately obtained via letting $\mathbf{Z} = \mathbf{X}\mathbf{W}$. In our SPCO, we devise a different penalty term

$$\frac{\gamma_1}{2}\text{tr}(\mathbf{W}^T\mathbf{X}^T\mathbf{X}\mathbf{W}) + \gamma_2\|\mathbf{X}\mathbf{W}\|_1,$$

which can be regarded as a weighted norm of $\mathbf{W}$ with respect to $\mathbf{X}$.

Since $\mathbf{A}$ $(p\times q)$ is orthogonal, there exists a $p\times(p-q)$ matrix $\mathbf{A}_0$ such that $\mathbf{A}_0^T\mathbf{A}_0 = \mathbf{I}_{p-q}$ and $\mathbf{A}^T\mathbf{A}_0 = \mathbf{0}$. Thus,

$$\|\mathbf{X}-\mathbf{X}\mathbf{W}\mathbf{A}^T\|_F^2 = \|\mathbf{X}\mathbf{A}_0\|_F^2 + \|\mathbf{X}\mathbf{A}-\mathbf{X}\mathbf{W}\|_F^2.$$

This implies that SPCA-ZHT is equivalent to

$$\min_{\mathbf{W}\in\mathbb{R}^{p\times q}} \frac{1}{2}\|\mathbf{X}\mathbf{A}-\mathbf{X}\mathbf{W}\|_F^2 + \frac{\gamma_1}{2}\|\mathbf{W}\|_F^2 + \sum_{j=1}^{q}\gamma_{2,j}\|\mathbf{w}_j\|_1,$$
$$\text{s.t. } \mathbf{A}^T\mathbf{A} = \mathbf{I}_q,$$

which with (2) together shows an interesting connection between SPCA-ZHT and our sPCA-OS by setting $\mathbf{Y} = \mathbf{X}\mathbf{A}$. However, in our model we employ the constraint $\mathbf{A}^T\mathbf{X}^T\mathbf{X}\mathbf{A} = \mathbf{I}_q$.

Second, when $q = 1$, the sPCA-rSVD model of Shen & Huang [17] is defined as

$$\min_{\mathbf{A}\in\mathbb{R}^{p\times q}} \frac{1}{2}\|\mathbf{X}-\mathbf{Z}\mathbf{A}^T\|_F^2 + \frac{\gamma_1}{2}\|\mathbf{A}\|_F^2 + \gamma_2\|\mathbf{A}\|_1,$$
$$\text{s.t. } \mathbf{Z}^T\mathbf{Z} = \mathbf{I}_q.$$

We see that there is a duality between sPCA-rSVD and SPCO, in which the roles of $\mathbf{Z}$ and $\mathbf{A}$ are exchanged.

We now study the relationship of sPCA-OS with SPCA-WTH [20] and SCoTLASS [14]. Assume $q = 1$, we write (2) for sPCA-OS as

$$\min_{\mathbf{y},\mathbf{w}} f(\mathbf{y},\mathbf{w}) = \|\mathbf{y}-\mathbf{X}\mathbf{w}\|_2^2,$$
$$\text{s.t. } \|\mathbf{w}\|_2^2 \le c_1,\ \|\mathbf{w}\|_1 \le c_2,\ \|\mathbf{y}\|_2^2 = 1,$$

which can be used to associate sPCA-OS with SPCA-WTH and SCoTLASS, because SPCA-WTH is based on the following problem

$$\max_{\mathbf{u},\mathbf{v}} \mathbf{u}^T\mathbf{X}\mathbf{v},$$
$$\text{s.t. } \|\mathbf{v}\|_2^2 \le 1,\ \|\mathbf{v}\|_1 \le c_2, \|\mathbf{u}\|_2^2 \le 1,$$

while SCoTLASS is based on the problem:

$$\max_{\mathbf{w}} \ \mathbf{w}^T \mathbf{X}^T \mathbf{X} \mathbf{w},$$
$$\text{s.t. } \|\mathbf{w}\|_2^2 \leq c_1, \ \|\mathbf{w}\|_1 \leq c_2.$$

5 Experiments

In this section we compare our sPCA-OS and SPCO algorithms with two closely related sparse PCA methods, i.e., SPCA-ZHT [23] and SPCA-WTH [20]. The experiments are conducted on the `pitprops` dataset, two synthetic datasets, six UCI datasets and one gene microarray dataset. Following the setting in [17,23], we also employ *cumulative percentage of explained variance* (CPEV) as an evaluation criterion.

5.1 Evaluations on the `Pitprops` Dataset

The `pitprops` dataset was first put forward by Jeffers [12] for difficulty of interpreting PCs. The dataset consists of 13 variables and 180 observations, and has become a standard example illustrating the potential difficulty of interpreting principal components. Jeffers [12] suggested explaining the first six components. Thus, we also select the first six PCs to analyze SPCA-ZHT, SPCA-WTH and sPCA-OS. Similar to [23], the corresponding regularized parameters are identified on the basis of that each sparse approximation explains almost the same amount of variance as the ordinary PC does. Tables 1-3 show the experimental results with these sparse PCA methods.

It is seen from Tables 1-3 that with regard to CPEV, the sPCA-OS method should be competitive with the other two sparse PCA methods, because the values of CPEV for SPCA-ZHT, SPCA-WTH and sPCA-OS are 75.76%, 75.22% and 75.47%, respectively. Moreover, sPCA-OS and SPCA-ZHT have higher sparseness than SPCA-WTH on the whole, specially as the number of principal components increases. In addition, Figure 1 shows the corresponding variation of variance with respect to the different principal components. It is also worth mentioning that unlike SPCA-WTH, the variances of sPCA-OS and SPCA-ZHT strictly monotonously decrease.

5.2 Evaluations on Two Synthetic Datasets

We also conduct our comparison on the synthetic dataset employed in [23]. In particular, three hidden factors are first created as follows:

$$v_1 \sim \mathcal{N}(0, 290), \ v_2 \sim \mathcal{N}(0, 300), \ v_3 = -0.3v_1 + 0.925v_2 + \epsilon, \ \epsilon \sim \mathcal{N}(0, 1),$$

where v_1, v_2 and ϵ are mutually independent. Then, 10 observed variables $x_1, \ldots, x_{10}$ are generated by:

$$x_i = v_j + \epsilon_i^j, \ \ \epsilon_i^j \sim \mathcal{N}(0, 1),$$

where $j = 1$ corresponds to $i = 1, 2, 3, 4$, $j = 2$ corresponds to $i = 5, 6, 7, 8$, and $j = 3$ corresponds to $i = 9, 10$, and the ϵ_i^j are independent of each other. From the

Table 1. The first six PCs obtained by SPCA-ZHT on the `pitprops` dataset

Variable	$PC1$	$PC2$	$PC3$	$PC4$	$PC5$	$PC6$
topdiam	-0.4796	0	0	0	0	0
length	-0.4689	0	0	0	0	0
moist	0	0.7769	0	0	0	0
testsg	0	0.6289	0	0	0	0
ovensg	0.1903	0	0.6551	0	0	0
ringtop	0	0	0.6048	0	0	0
ringbut	-0.2802	0	0.4528	0	0	0
bowmax	-0.3401	-0.0312	0	0	0	0
bowdist	-0.4148	0	0	0	0	0
whorls	-0.3844	0	0	0	0	0
clear	0	0	0	-1	0	0
knots	0	0	0	0	-1	0
diaknots	0	0	0	0	0	1
CPEV(%)	28.06	42.06	55.16	62.61	69.45	75.76

Table 2. The first six PCs obtained by SPCA-WTH on the `pitprops` dataset

Variable	$PC1$	$PC2$	$PC3$	$PC4$	$PC5$	$PC6$
topdiam	0	-0.6974	0	0	0	0
length	0	-0.6988	0	0	0	0
moist	0	0	0.6825	0	0	0
testsg	0	0	0.6967	0	0	0
ovensg	0	0.1494	0	0	0.2810	0.4391
ringtop	0	0	0	0	0.4995	0
ringbut	-0.8099	0	0	0	0	0
bowmax	0	0	0	-0.2208	0	0
bowdist	0	-0.0544	0	-0.2752	0	-0.8740
whorls	-0.5214	0	0	0	0	-0.0021
clear	0	0	0	0	-0.8195	0.1054
knots	0	0	0	0.8152	0	-0.1795
diaknots	0.2687	0	0	0	0	0
CPEV(%)	14.49	30.89	46.37	57.99	68.12	75.22

formulation above, it is interesting to note that the three hidden factors have about the same variances, and the variables (x_1, x_2, x_3, x_4) are independent of the variables (x_5, x_6, x_7, x_8). Moreover, as the mixed roles, the variables (x_9, x_{10}) have closed relationship with the variables (x_5, x_6, x_7, x_8).

We implement classical PCA, SPCA-ZHT, SPCA-WTH and sPCA-OS on the 500 synthetic points here. Table 4 summarizes the comparison results. Obviously, unlike PCA, the three sparse PCA algorithms have the correct sparse representations

Table 3. The first six PCs obtained by sPCA-OS on the `pitprops` dataset

Variable	$PC1$	$PC2$	$PC3$	$PC4$	$PC5$	$PC6$
topdiam	0.6304	0	0	0	0	0
length	0.6092	0	0	0	0	0
moist	0	0	0	0.6894	0	0
testsg	0	0	0	0.7244	0	0
ovensg	-0.4739	0	0	0	0	0
ringtop	0	0	0	0	0	0.8517
ringbut	0	0	0	0	0	0.5240
bowmax	0	0	0	0	0.5033	0
bowdist	0	0.7976	0	0	0	0
whorls	0	0.0599	-0.4074	0	0	0
clear	0	0	0.9135	0	0	0
knots	0	0	0	0	0	0
diaknots	0.0832	-0.6002	0	0	0	0
CPEV(%)	23.42	40.60	53.69	62.76	70.25	75.47

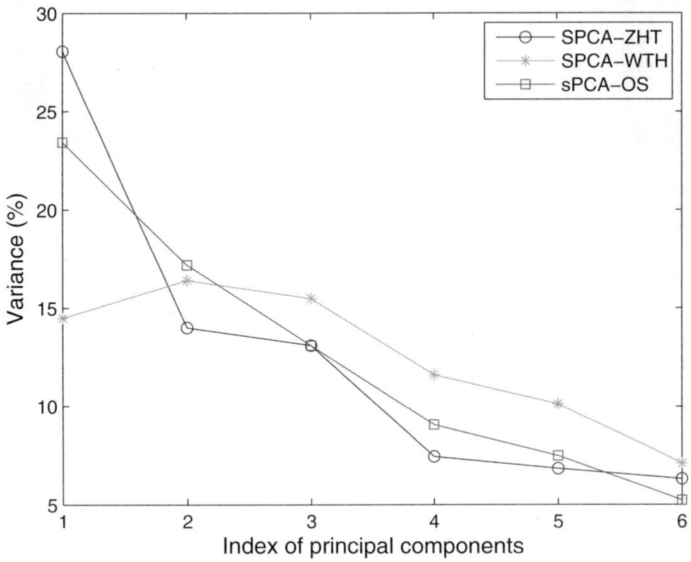

Fig. 1. Variation of variance (%) corresponding to the principal components on the `pitprops` dataset

recovering the same hidden factors. Moreover, the variance of nonzero loadings of sPCA-OS is less than those of the other two sparse PCA. This shows that our sPCA-OS should be more robust than SPCA-ZHT and SPCA-WTH. This agrees with that our sPCA-OS based on the optimal scoring can capture the underlying discriminative property

Table 4. Results of the simulation example: loadings and variance

	PCA		SPCA-ZHT		SPCA-WTH		sPCA-OS	
	$PC1$	$PC2$	$PC1$	$PC2$	$PC1$	$PC2$	$PC1$	$PC2$
X_1	-0.0901	-0.4767	0	0.3945	-0.0024	0	0	-0.4982
X_2	-0.0875	-0.5107	0	0.6348	-0.7270	0	0	-0.5028
X_3	-0.0886	-0.4719	0	0.4789	-0.4704	0	0	-0.5007
X_4	-0.0856	-0.4801	0	0.4604	-0.5001	0	0	-0.4983
X_5	0.4134	-0.0840	0.3616	0	0	-0.1826	0.5292	0
X_6	0.3948	-0.1266	0.3831	0	0	-0.8578	0.5328	0
X_7	0.3991	-0.1442	0.4218	0	0	-0.4115	0.4538	0
X_8	0.4047	-0.1173	0.7380	0	0	-0.2481	0.4798	0
X_9	0.3996	0.0270	0	0	0	0	0	0
X_{10}	0.3996	0.0221	0	0	0	0	0	0
CPEV(%)	61.03	94.01	60.86	75.27	36.50	72.71	61.59	79.66

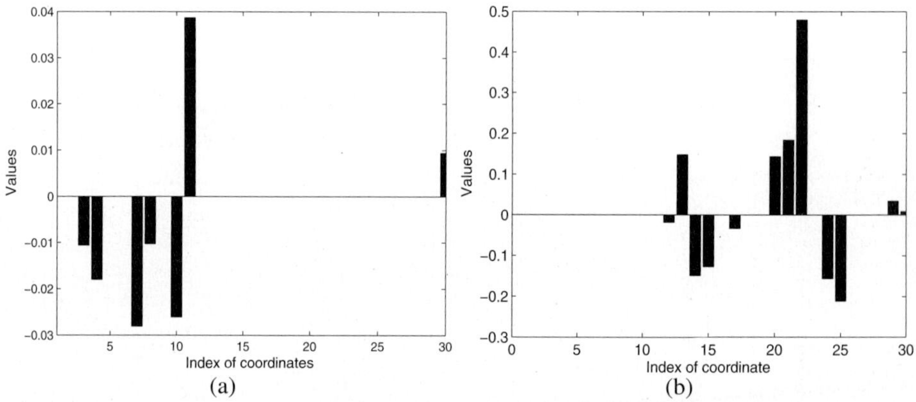

Fig. 2. Sparse results of SPCO on the synthetic dataset: (a) principal coordinate on the first principal component; (b) principal coordinate on the second principal component

in datasets. In addition, our CPEV is higher than the others, when all the sparse PCA algorithms keep the same number of nonzero loadings.

In order to reveal the effectiveness of SPCO, we also generate another synthetic dataset via the following Gaussian distributions:

$$G_1 \sim \mathcal{N}(10 * \mathbf{1}_{50},\ \mathbf{I}_{50}),\ G_2 \sim \mathcal{N}(-10 * \mathbf{1}_{50},\ 9 * \mathbf{I}_{50}),\ G_3 \sim \mathcal{N}(\mathbf{0},\ 17 * \mathbf{I}_{50}).$$

Obviously, G_2 is more close to G_3 in comparison with G_1. This synthetic dataset consists of 30 points with 50 variables, and each 10 points corresponds to one distribution. Without loss of generality, the points $(\mathbf{x}_{(i-1)*10+1}, \cdots, \mathbf{x}_{i*10})$ are sampled from G_i, $i = 1, 2, 3$. The results of implementing SPCO are shown in Figure 2, where the i-th coordinate correspond to the i-th point. Figure 2 (a) depicts that the nonzero coordinates correspond to the first 10 points, while the coordinates for the other points are zeros.

Figure 2 (b) depicts that the coordinates for the first 10 points are zeros, but the nonzero coordinates are associated with the other points. In essence, it also tells us an appealing property that SPCO can effectively capture the underlying relationship among data by strengthening zero or nonzero coordinates. We also carry out the similar experiments with the three sparse PCA algorithms. Unfortunately, the resulting principal coordinates are not sparse so that the property mentioned above does not remain.

5.3 Evaluations on Classification

In this experiment, we conduct the comparison of the four sparse dimensionality reduction algorithms in classification problems. We first implement sPCA-OS, SPCO, SPCA-ZHT and SPCA-WTH for dimensionality reduction. Then, we perform classification respectively on the dimensionality-reduced data matrices by simply applying nearest neighbor classifier.

Our experiments are implemented on six UCI datasets, the details of which are summarized in Table 5.

In order to make comparison fair, we keep CPEV as equal as possible when we carry out the four dimensionality reduction methods. This can be done by adjusting the regularization parameters. Note that for SPCO, there is no obvious effect on the explained variance when varying the regularization parameter λ. For each dataset, we randomly sample 90% of the instances for training and the remaining 10% for test. This procedure is repeated 20 times for each dataset, and the evaluation criteria are reported in the classification accuracy rate($\%$) and corresponding standard deviation.

The classification results are listed in Table 6. We find that when the explained variance obtained from the four sparse dimensionality reduction methods are nearly equal, our sPCA-OS and SPCO outperform SPCA-ZHT and SPCA-WTH on the whole. It also successfully confirms that as shown in [21], our sPCA-OS based on optimal scoring can effectively detect the underlying discriminative information among data, and that SPCO can also effectively detect the underlying distribution among data.

It should be also worth mentioning here that the sparsity of the loadings obtained by sPCA-OS, SPCA-ZHT, SPCA-WTH are different when keeping approximately same explained variance. In particular, Figure 3 depicts the sparsity of loadings obtained by sPCA-OS, SPCA-ZHT and SPCA-WTH on the six UCI datasets. Obviously, the similar conclusion for the `pitprops` dataset can be followed here.

Table 5. The summaries of datasets. (c—the number of classes, p—the number of the variables and n—the number of instances.).

Datasets	c	p	n
`dermatology`	6	34	358
`segmentation`	7	18	2310
`glass`	6	9	214
`letter`	10	16	1978
`pageblocks`	5	10	5473
`pendigits`	10	16	7494

Table 6. Classification results using the four sparse methods on the different datasets. ("CPEV" for "the cumulative percentage of explained variance", "acc" for "the accuracy of classification" and "std" for "the standard deviation".).

Dataset	CPEV	acc($\pm$ std)	Algorithm
dermatology	34.67	80.54 ($\pm$3.39)	SPCA-ZHT
	34.70	80.02 ($\pm$3.46)	SPCA-WTH
	34.64	82.43($\pm$5.48)	sPCA-OS
	35.97	83.56($\pm$5.48)	SPCO
segmentation	70.70	66.52($\pm$2.25)	SPCA-ZHT
	73.00	70.80($\pm$1.53)	SPCA-WTH
	73.35	77.80($\pm$1.53)	sPCA-OS
	71.04	71.13($\pm$0.94)	SPCO
glass	82.14	45.78($\pm$5.95)	SPCA-ZHT
	82.10	45.81($\pm$4.95)	SPCA-WTH
	82.01	46.87($\pm$6.42)	sPCA-OS
	82.69	47.56($\pm$6.00)	SPCO
letter	81.34	60.96($\pm$2.74)	SPCA-ZHT
	81.30	63.04($\pm$2.74)	SPCA-WTH
	81.67	64.26($\pm$1.78)	sPCA-OS
	80.80	63.12($\pm$2.93)	SPCO
pageblocks	67.83	93.83($\pm$0.48)	SPCA-ZHT
	67.90	94.03($\pm$0.31)	SPCA-WTH
	67.90	93.80($\pm$0.45)	sPCA-OS
	67.62	94.35($\pm$0.43)	SPCO
pendigits	82.86	93.02($\pm$0.76)	SPCA-ZHT
	82.50	93.48($\pm$0.57)	SPCA-WTH
	82.80	93.18($\pm$0.58)	sPCA-OS
	82.63	93.67($\pm$0.61)	SPCO

5.4 Application in Gene Microarray

The SRBCT microarray dataset [8] has 2308 genes (variables) and 63 samples (observations). We implement SPCO on this dataset to explore sparse representation of the original data in a dimensionality-reduced space. Here q is the number of principal coordinates and it is also the dimensionality of the low-dimensional space.

Similarly, a new evaluation criterion, i.e., *cumulative percentage of zero coordinates* (CPZC) is defined to measure SPCO, and it is the ratio of the number of zero coordinates to the total number of coordinates in the dimensionality-reduced data matrix.

A sequence of principal coordinates are taken from $q = 3$ to $q = 63$. As shown in Figure 4, the CPEV of the principal component $\mathbf{A}$ increases sharply while the CPZC of data matrix $\mathbf{Z}$ in the low-dimensional space dramatically declines, when q is set from 3 to 10. When $q > 20$, however, the obtained CPZC is nearly same while the CPEV still increases steadily. Thus, in practice, we can choose a suitable q through the trade-off between CPEV and CPZC.

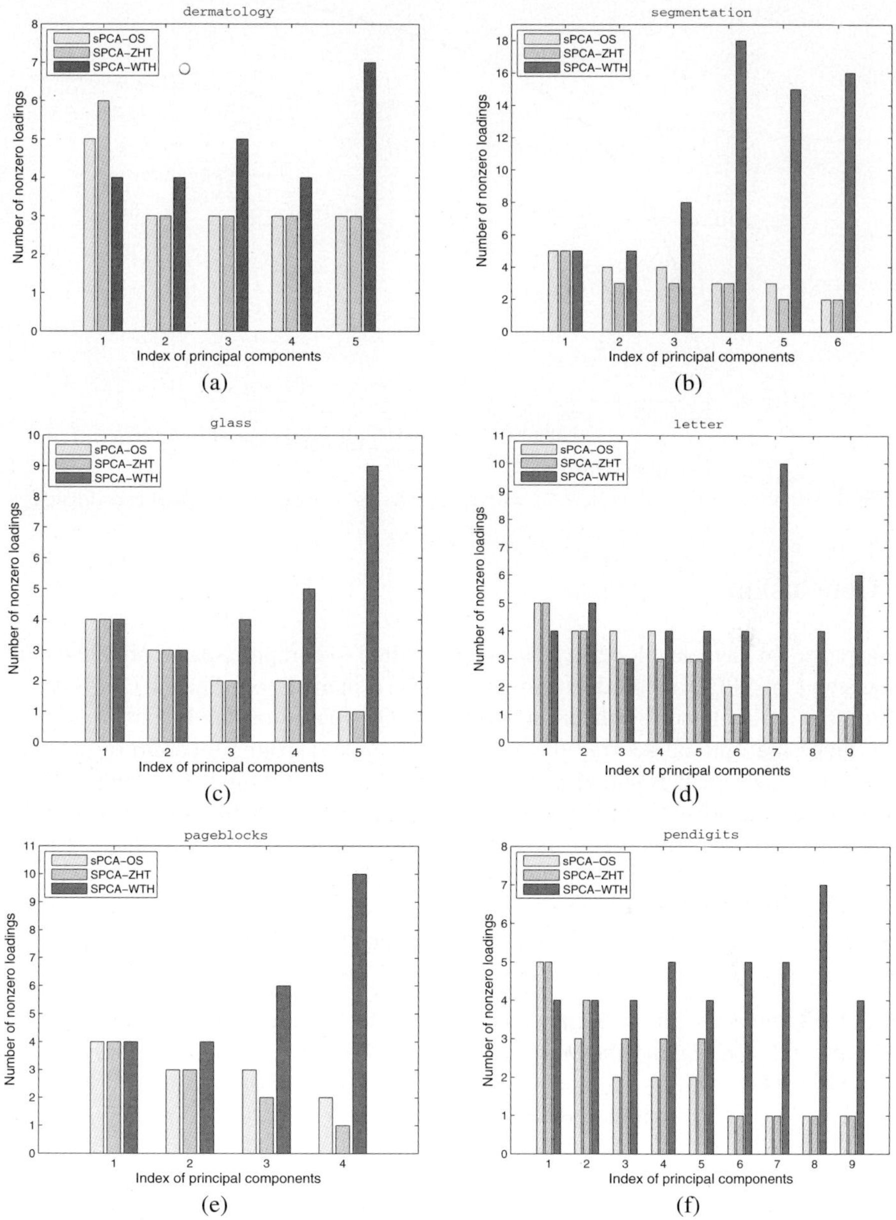

Fig. 3. The sparsity of loadings obtained by sPCA-OS, SPCA-ZHT and SPCA-WTH on the six UCI datasets

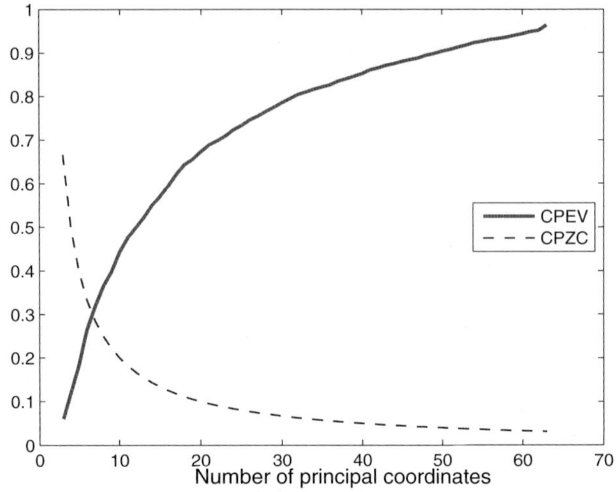

Fig. 4. Variation of CPEV and CPZC with respect to the number of principal coordinates

6 Conclusion

In this paper we have developed a new SPCA method to compute sparse principal components and an SPCO method to compute sparse principal coordinates. Our SPCA is built on the optimal scoring theorem, while SPCO is based on the Eckart-Young theorem. Since the optimal scoring theorem and the Eckart-Young theorem respectively provide the theoretical foundation to carry out the conventional PCA and PCO by regression-type problems, our SPCA and SPCO can be efficiently solved by existing algorithms for sparse regression models such that in [4,6,19,22]. There are a lot of treatments for sparse PCA in the literature. To our knowledge, however, we have first provided an attempt for sparse PCO. In our future work, we will dig out the potential application of our sparse PCO method in multivariate data analysis.

Acknowledgments. This work is supported by the Natural Science Foundations of China (No. 60970081) and the 973 Program of China (No. 2010CB327903). Zhihua Zhang also acknowledges support from Doctoral Program of Specialized Research Fund of Chinese Universities (No. J20091608) and from the Fundamental Research Funds for the Central Universities.

References

1. Clemmensen, L., Hastie, T., Erbøll, B.: Sparse discriminant analysis. Technical report (June 2008)
2. d'Aspremont, A., El Ghaoui, L., Jordan, M.I., Lanckriet, G.R.G.: A direct formulation for sparse PCA using semidefinite programming. SIAM Review 49(3), 434–448 (2007)

3. Eckart, C., Young, G.: The approximation of one matrix by another of lower rank. Psychometrika 1, 211–218 (1936)
4. Efron, B., Hastie, T., Johnstone, I., Tibshirani, R.: Least angle regression (with discussions). The Annals of Statistics 32(2), 407–499 (2004)
5. Elden, L., Park, H.: A procrustes problem on the stiefel manifold. Numerische Mathematik (1999)
6. Friedman, J.H., Hastie, T., Hoefling, H., Tibshirani, R.: Pathwise coordinate optimization. The Annals of Applied Statistics 2(1), 302–332 (2007)
7. Golub, G.H., Van Loan, C.F.: Matrix Computations, 3rd edn. The Johns Hopkins University Press, Baltimore (1996)
8. Golub, T., Slonim, D., Tamayo, P., Huard, C., Gaasenbeek, M., Mesirov, J., Coller, H., Loh, M., Downing, J., Caligiuri, M.: Molecular classification of cancer: class discovery and class prediction by gene expression monitoring. Science 286, 531–536 (1999)
9. Gower, J.C.: Some distance properties of latent root and vector methods used in multivariate data analysis. Biometrika 53, 315–328 (1966)
10. Gower, J.C., Dijksterhuis, G.B.: Procrustes Problems. Oxford University Press, Oxford (2004)
11. Hastie, T., Tibshirani, R., Buja, A.: Flexible discriminant analysis by optimal scoring. Journal of the American Statistical Association 89(428), 1255–1270 (1994)
12. Jeffers, J.: Two case studies in the application of principal compnent. Appl. Statist. 16, 225–236 (1967)
13. Jolliffe, I.T.: Principal component analysis, 2nd edn. Springer, New York (2002)
14. Jolliffe, I.T., Trendafilov, N.T., Uddin, M.: A modified principal component technique based on the lasso. Journal of Computational and Graphical Statistics 12, 531 (2003)
15. Magnus, J.R., Neudecker, H.: Matrix Calculus with Applications in Statistics and Econometric. John Wiley & Sons, New York (1999) (revised edn.)
16. Mardia, K.V., Kent, J.T., Bibby, J.M.: Multivariate Analysis. Academic Press, New York (1979)
17. Shen, H., Huang, J.: Sparse principal component analysis via regularized low rank matrix approximation. Journal of Multivariate Analysis 99, 1015–1034 (2008)
18. Sriperumbudur, B.K., Torres, D., Lanckriet, G.R.G.: Sparse eigen methods by d.c. programming. In: ICML (2007)
19. Tibshirani, R.: Regression shrinkage and selection via the lasso. Journal of the Royal Statistical Society, Series B 58, 267–288 (1996)
20. Witten, M., Tibshirani, R., Hastie, T.: A penalized matrix decomposition, with applications to sparse principal components and canonical correlation analysis. Biostatisitics 10(3), 515–534 (2009)
21. Zhang, Z., Dai, G.: Optimal scoring for unsupervised learning. In: Advances in Neural Information Processing Systems, vol. 23 (2009)
22. Zou, H., Hastie, T.: Regularization and variable selection via the elastic net. Journal of the Royal Statistical Society, Series B 67, 301–320 (2005)
23. Zou, H., Hastie, T., Tibshirani, R.: Sparse principal component analysis. Journal of Computational and Graphical Statistics 15, 265–286 (2006)

A The Proof of Theorem 2

Consider the Lagrange function

$$L(\mathbf{Z}, \mathbf{A}, \mathbf{C}) = \frac{1}{2}\Big[\|\mathbf{X}-\mathbf{Z}\mathbf{A}^T\|_F^2 + \gamma\|\mathbf{Z}\|_F^2 - \mathrm{tr}(\mathbf{C}(\mathbf{A}^T\mathbf{A}-\mathbf{I}_q))\Big],$$

where $\mathbf{C}$ is a $q\times q$ symmetric matrix of Lagrange multipliers. The first-order conditions are

$$\begin{aligned} \frac{\partial L}{\partial \mathbf{Z}} &= (\mathbf{Z}\mathbf{A}^T - \mathbf{X})\mathbf{A} + \gamma\mathbf{Z} = \mathbf{0}, \\ \frac{\partial L}{\partial \mathbf{A}} &= (\mathbf{A}\mathbf{Z}^T - \mathbf{X}^T)\mathbf{Z} - \mathbf{A}\mathbf{C} = \mathbf{0}, \\ \frac{\partial L}{\partial \mathbf{C}} &= \mathbf{A}^T\mathbf{A} - \mathbf{I}_q = \mathbf{0}, \end{aligned} \tag{8}$$

which yield

$$\mathbf{Z} = \frac{1}{1+\gamma}\mathbf{X}\mathbf{A}.$$

Premultiplying both sides of (8) by $\mathbf{A}^T$ then gives

$$\mathbf{C} = \mathbf{Z}^T\mathbf{Z} - \mathbf{Z}^T\mathbf{X}\mathbf{A} = -\gamma\mathbf{Z}^T\mathbf{Z}.$$

Hence, (8) can be rewritten as

$$(\mathbf{X}^T\mathbf{X})\mathbf{A} = \mathbf{A}(\mathbf{A}^T\mathbf{X}^T\mathbf{X}\mathbf{A}).$$

Suppose that the full SVD of $\mathbf{A}^T\mathbf{X}^T\mathbf{X}\mathbf{A}$ is $\mathbf{A}^T\mathbf{X}^T\mathbf{X}\mathbf{A} = \mathbf{G}\boldsymbol{\Lambda}\mathbf{G}^T$ where $\mathbf{G}$ is an orthogonal $q\times q$ matrix and $\boldsymbol{\Lambda}$ is a diagonal $q\times q$ matrix with the eigenvalues (singular values) of $\mathbf{A}^T\mathbf{X}^T\mathbf{X}\mathbf{A}$ on its diagonal. Then we have

$$(\mathbf{X}^T\mathbf{X})\mathbf{A}\mathbf{G} = \mathbf{A}\mathbf{G}\boldsymbol{\Lambda},$$

which implies that $\mathbf{A}\mathbf{G}$ is the orthogonal eigenvector matrix of $\mathbf{X}^T\mathbf{X}$ and the diagonal entries of $\boldsymbol{\Lambda}$ are its corresponding eigenvalues.

Now given $\mathbf{Z} = \frac{1}{1+\gamma}\mathbf{X}\mathbf{A}$, we have

$$\begin{aligned} g(\mathbf{A}, \mathbf{Z}) &= \mathrm{tr}(\mathbf{X}^T\mathbf{X}) - \frac{1}{1+\gamma}\mathrm{tr}(\mathbf{A}^T\mathbf{X}^T\mathbf{X}\mathbf{A}) \\ &= \mathrm{tr}(\mathbf{X}^T\mathbf{X}) - \frac{1}{1+\gamma}\mathrm{tr}(\boldsymbol{\Lambda}). \end{aligned}$$

In order to minimize g, we should maximize $\mathrm{tr}(\boldsymbol{\Lambda})$. We thus take $\boldsymbol{\Lambda} = \mathrm{diag}(d_1, \ldots, d_q)$. Moreover, we can also let $\mathbf{A} = [\mathbf{v}_1, \ldots, \mathbf{v}_q]$.

Asking Generalized Queries to Ambiguous Oracle

Jun Du and Charles X. Ling

Department of Computer Science
The University of Western Ontario, London, Ontario, N6A 5B7, Canada
{jdu42,cling}@csd.uwo.ca

Abstract. Asking generalized queries (by regarding some features as don't-care) in active learning has been proposed and studied recently. As each generalized query is equivalent to a set of specific ones, the answers from the oracle can usually provide more information thus speeding up the learning effectively. However, as the answers to the generalized queries might be uncertain, previous studies often assume that the oracle is capable of providing (accurate) probabilistic answers. This assumption, however, is often too stringent in real-world situations. In this paper, we make a more realistic assumption that the oracle can only provide (non-probabilistic) ambiguous answers, similar to the setting in *multiple-instance learning*. That is, the generalized query is labeled positive if at least one of the corresponding specific queries is positive, and is labeled negative otherwise. We therefore propose an algorithm to construct the generalized queries and improve the learning model with such ambiguous answers in active learning. Empirical study shows that, the proposed algorithm can significantly speed up the learning process, and outperform active learning with either specific queries or inaccurately answered generalized queries.

1 Introduction

Active learning, as an effective paradigm to speed up the learning process, has been intensively studied in recent years. In most traditional active learning studies, the learner usually regards the specific examples directly as queries, and requests the corresponding labels from the oracle. For instance, given a prostatitis patient data set, the learner usually presents the entire patient example, such as {ID = 7354288, name = John, age = 50, gender = male, weight = 230, blood-type = AB, fever = yes, painful-urination = yes, $\cdots$} (with all the features), to the oracle, and requests the corresponding label whether this patient has prostatitis or not. However, in this case, many features (such as ID, name, blood-type, and so on) might be *irrelevant* to prostatitis diagnosis. Not only could queries like this confuse the oracle, but each answer responded from the oracle is also applicable for only one specific example.

In many real-world active learning applications, the oracles are often human experts, thus they are usually capable of answering more general queries. For

J.L. Balcázar et al. (Eds.): ECML PKDD 2010, Part I, LNAI 6321, pp. 377–392, 2010.

instance, given the same patient data set, the learner could ask a generalized query, such as "are men between 45 and 55 with fever and painful-urination likely to have prostatitis?", where only four features (gender, age, fever and painful-urination) are provided, and all the rest are considered as don't-care. Not only are such generalized queries more natural and relevant, each generalized query can often represent a set of specific examples, thus the answer for the query is also applicable to all these examples. This allows the active learner to improve learning more effectively.

However, when asking such generalized queries in active learning, the answers are often uncertain. For instance, the answer could be "yes, with 80% probability" that men between 45 and 55 with fever and painful-urination are likely to have prostatitis. In the ideal situation, the oracle is expected to provide such accurate probabilistic answers for the generalized queries, in order to improve the learning model accordingly. This requirement, however, is usually too stringent in reality. Instead, generalized queries are often responded with ambiguous answers in real-world situations. For instance, if the query is "are women (or boys under 10) likely to have prostatitis", the specialist (oracle) would always respond "No", indicating none of such people would have this disease. However, if the query is "are men between 45 and 55 likely to have prostatitis", the answer would often be "Yes", indicating some of such people indeed have this disease, but the accurate proportion (probability) is unknown. Clearly, such generalized queries and ambiguous answers commonly occur in our daily life, thus it is reasonable and desired to study them together.

In this paper, therefore, we make a more realistic assumption that the oracle can only provide ambiguous (non-probabilistic) answers to the generalized queries. That is: the oracle labels the query as negative *if (and only if)* all the examples represented by this query are negative; otherwise the query is always labeled as positive. The similar setting of ambiguous answers has been extensively studied in *multiple-instance learning* [1], and applied to many real-world applications, such as drug activity prediction [1], content-based image retrieval [2] and text categorization [3]; see Section 2 for more details.

In active learning scenario, such setting of ambiguous answers is reasonable yet flexible. On one hand, even though the oracle is still required to answer generalized queries, the answer only needs to be "Yes (positive)" or "No (negative)", which is more natural and applicable in real-world situations. On the other hand, such ambiguous answers can also be regarded as a more general form of the specific (accurate) answers. Specifically, when some features are discovered as don't-care and the query is generalized, the answer is indeed ambiguous ("Yes" indicates at least one specific example is positive, whereas "No" indicates all corresponding examples are negative). However, when no don't-care feature is discovered and the query turns to be a specific one, such "Yes-No" response naturally becomes the *accurate* answer to the specific query. Therefore, such setting of ambiguous answers is more flexible than the regular setting in active learning, yet still applicable in many real-world situations.

In this paper, by assuming that the oracle is capable of providing such ambiguous answers to the generalized queries, we propose a novel method to, first construct the generalized queries, and then update the learning model according to the ambiguous answers. Empirical study on UCI ([4]) data sets shows that, the proposed method can significantly speed up the learning process, and outperform active learning with either specific queries or inaccurately answered generalized queries.

The rest of the paper is organized as follows. Section 2 reviews previous works on active learning and multiple-instance learning. Section 3 describes our algorithm to ask generalized queries and improve the learning model with ambiguous answers. In Section 4, empirical study is conducted on real-world UCI data sets to verify the superiority of the proposed method. Section 5 presents conclusions.

2 Related Work

Most previous studies of active learning can be categorized into two types: the pool-based active learning and the membership query.[1] In the pool-based active learning, a pool of unlabeled examples is given, among which the learner can choose the examples and request the corresponding labels [7]. Briefly speaking, the pool-based active leaner first evaluates each example in the pool, to decide which one can maximumly improve the performance of the current model. Then the learner acquires its label from the oracle, and update the learning model accordingly. The process repeats, until some stop criterion is met. On the other hand, active learning with membership queries (or direct query construction) can directly construct examples (without the need of the pool) and request the corresponding labels from the oracle [8,9]. Previous research shows that, in most situations, both of these two types of active learning can significantly reduce the number of labeled examples needed, compared with labeling examples randomly.

Without the restriction of the pool, membership query can always construct the *optimal* queries to improve the performance of the current model. However, in reality, these *optimal* queries might be unrealistic thus hard to be answered by the oracle (such as, handwritten character recognition, text classification, and so on). Therefore, the pool-based active learning is extensively studied in recently years. In this paper, the proposed learning algorithm can be adapt to both of these two types, and we will use pool-based setting for illustration.

The essence of active learning lies in measuring the "goodness" of the unlabeled examples. Many criteria have been proposed in the literature. *Uncertainty sampling* [7] considers the most uncertain example as most valuable, and has been extensively studied and widely used in many previous studies [10,11,6,12,13]. *Query-by-committee* (QBC) [14] is a more theory-based approach, and considers the example that minimizes the version space as optimal. In addition, other criteria, such as *variance reduction* [15], *Fisher information ratio* [16], and *estimated error reduction* [17], are also elaborately designed and well accepted.

[1] Stream-based active learning [5] is considered as another type in some literatures. In essence, it could be viewed as an online version of the pool-based active learning [6].

Most previous works of active learning assume that the oracle could only answer specific queries, with all features provided. [18,19] consider a more natural situation that the oracle is capable of answering generalized queries, and propose a novel algorithm to ask such queries and improve the learning model. However, as the answers for such generalized queries are often uncertain, it is also assumed in [18,19] that the oracle could provide (accurate) probabilistic answers to those queries. This assumption, however, is too stringent in many real-world situations.

On the other hand, a more relaxed assumption has been studied in *multiple-instance learning* ([1]), where examples are given in bags and oracle is only required to provide one ambiguous answer for each bag. More specifically, given a bag of unlabeled examples, the oracle will respond negative *if (and only if)* these examples are all negative, and respond positive otherwise. In this setting, it is more likely for the learner to be responded with a positive answer; and more importantly, such positive answer only indicates that at least one example in the given bag is positive, but the true label of each specific example is still unknown. Many algorithms, such as *diverse density* [20], *citation kNN* [21], *multiple-decision tree* [22] and *multiple-instance logistic regression* [23], have been developed to tackle such ambiguous answers, so as to predict labels of the future unseen bags. Despite of the ambiguity of the answers, multiple-instance learning has been applied to many real-world applications, such as drug activity prediction [1], content-based image retrieval [2], and text categorization [3].

In this paper, we apply such ambiguous oracle to active learning with generalized queries. More specifically, in active learning, the learner always tends to construct generalized queries and request the corresponding labels from the oracle. However, the oracle is only capable of respond with ambiguous answers. That is, given a generalized query, the oracle will respond negative *if (and only if)* the examples represented by the query are all negative, and respond positive otherwise. Such setting of ambiguous oracle relaxes the stringent assumptions in the previous studies of active learning with generalized queries, and is applicable to more real-world situations.

It is also worth noting that, our study in this paper is not a simple combination of active learning and multiple instances learning.[2] Instead of aiming to improve the predictive performance on the *unseen bags of examples* in multiple-instance learning, in this paper, we still attempt to improve the predictive performance on the *unseen specific examples* (as in traditional supervised setting). In addition, we also consider generalized queries in active learning scenario, thus the problem we are attempting to solve is more complex and difficult.

3 Algorithm

In this section, we design active learning algorithm to ask generalized queries and further improve the predictive performance based on the responded ambiguous answers. Specifically, we use *logistic regression* as the base active learner, due to

[2] [24] categorizes multiple-instance active learning into four scenarios, and develops a novel algorithm to deal with one of those formulations.

its good performance in probability estimation and the convenience of designing objective function (see Section 3.1 for details).

The learning process can be roughly broken into the following two steps in each iteration:

- **Step 1:** Based on the current training and unlabeled data sets, the learner constructs a generalized query (according to certain objective function).
- **Step 2:** After obtaining the ambiguous answer of the generalized query, the learner updates the learning model (according to certain objective function).

In the above two steps, objective functions are required for both constructing the generalized queries and updating the learning model. Therefore, in the rest of this section, we first design a *universal* objective function for both of the above two steps; and then present the implementation details for each of them.

3.1 Objective Function

In each learning iteration, when constructing the generalized query, the optimal query is expected to be the one that yields the best performance of the learning model; likewise, when updating the learning model, the optimal model parameters are also the ones that yield the best predictive performance. Therefore, we can design one universal objective function to evaluate and optimize both the generalized queries and the model parameters.

In the current setting, the desired objective function needs to suit all of the following requirements: 1) logistic regression; 2) active learning; 3) generalized queries; and 4) ambiguous answers.

In the traditional supervised learning, *maximum likelihood* is commonly used to train a logistic regression classifier. Thus, it could also be considered as the most primitive objective function to find the optimal queries and model parameters, as follows:

$$< q, \mathbf{w} >_{opt} = \arg\max_{q,\mathbf{w}} \sum_{(\mathbf{x}_i, y_i) \in D} \log p(y_i | \mathbf{x}_i; q, \mathbf{w}) \quad (1)$$

where, $< q, \mathbf{w} >_{opt}$ denotes the tuple of optimal query q and model parameter $\mathbf{w}$, $\mathbf{x}_i$ and y_i denote the ith example in the given training data set D. This primitive objective function satisfies the requirement of logistic regression.

In active learning, however, the labeled training data set is usually small, thus the classifier trained via maximum likelihood alone (Equation 1) might be unreliable. In addition to the training data, we are often given a large amount of unlabeled data in active learning (for the pool-based setting), and this set of data can also help to evaluate the generalized queries and the model parameters. Intuitively, if the query (q) can indeed improve the performance of the learning model, the updated model would also be more confident in predicting all the unlabeled data; likewise, if the parameter ($\mathbf{w}$) can indeed yield a high-performance model, the model would also predict unlabeled data more confidently. Therefore, in addition to the maximum likelihood on the labeled training data, the predictive certainty (calculated by *entropy*) on the unlabeled data can also be considered as an additional measurement. This yields a more sophisticated objective function, as follows:

$$
\begin{aligned}
< q, \mathbf{w} >_{opt} &= \arg\max_{q,\mathbf{w}} \sum_{(\mathbf{x}_i,y_i)\in D} \log p(y_i|\mathbf{x}_i; q, \mathbf{w}) - \alpha \sum_{\mathbf{x}_j\in U, y\in\{0,1\}} H(p(y|\mathbf{x}_j; q, \mathbf{w})) \\
&= \arg\max_{q,\mathbf{w}} \sum_{(\mathbf{x}_i,y_i)\in D} \log p(y_i|\mathbf{x}_i; q, \mathbf{w}) \\
&\quad + \alpha \sum_{\mathbf{x}_j\in U} \sum_{y\in\{0,1\}} p(y|\mathbf{x}_j; q, \mathbf{w}) \log p(y|\mathbf{x}_j; q, \mathbf{w})
\end{aligned}
\quad (2)
$$

where $\mathbf{x}_j$ denotes the jth unlabeled example in the given unlabeled data set U, and α represents a trade-off parameter to balance the influence of the labeled and unlabeled data. This objective function satisfies the requirement of active learning.[3]

In our current setting, however, instead of requesting the labels for specific examples, the active learner always attempts to ask generalized queries. Moreover, these generalized queries are often responded with ambiguous answers by the oracle. Equation 2 therefore cannot suit this requirement. Instead, under current conditions, in each learning iteration, there always exist three data (query) sets: the initial training data set D, the query set Q which contains all the previous queries asked by the learner (one in each iteration), and the current unlabeled data set U. Therefore, in the tth learning iteration, the query q_t and the model parameter $\mathbf{w}_t$ can be optimized by: maximizing the likelihood with respect to both the initial training data D and the query set Q, and minimizing the predictive uncertainty with respect to the unlabeled data U, as follows:

$$
\begin{aligned}
< q_t, \mathbf{w}_t >_{opt} = \arg\max_{q,\mathbf{w}} \quad & \alpha_1 \sum_{(\mathbf{x}_i,y_i)\in D} \log p(y_i|\mathbf{x}_i; q_t, \mathbf{w}_t) \\
& + \alpha_2 \sum_{(q_k,y_k)\in Q} \log p(y_k|q_k; q_t, \mathbf{w}_t) \\
& + \alpha_3 \sum_{\mathbf{x}_j\in U} \sum_{y\in\{0,1\}} p(y|\mathbf{x}_j; q_t, \mathbf{w}_t) \log p(y|\mathbf{x}_j; q_t, \mathbf{w}_t)
\end{aligned}
\quad (3)
$$

where all the notations are the same as in the previous objective functions, and α_1, α_2 and α_3 represent the trade-off parameters to balance the influence of the three data (query) sets.

This objective function suits all the requirements in our current setting, and is applied to both query searching and model updating in each learning iteration, as follows.

- In the query searching step, the objective function is applied to find the optimal query (q_{opt}). Specifically, given one candidate query (and the estimated label, see Section 3.2 for details), Equation 3 can be regarded as a univariate

[3] The similar objective function has been applied in semi-supervised learning [25] and batch mode active learning [26]; and these previous studies have shown its applicability in real-world applications.

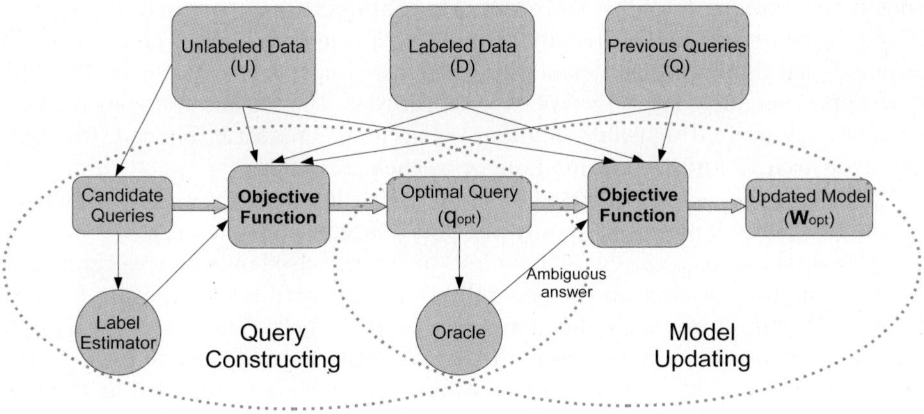

Fig. 1. The framework of the proposed algorithm

function of the model parameter $\mathbf{w}$ (denoted by $f(\mathbf{w})$). Thus, gradient decent can be directly applied to find the optimized $f(\mathbf{w})$. Thereafter, among all the candidate queries, the one that yields the maximum $f(\mathbf{w})$ is chosen and regarded as the optimal query (q_{opt}).

- In the model updating step, the objective function is applied to find the optimal model parameter ($\mathbf{w}_{opt}$). Specifically, given the optimal query (q_{opt}) and the true label provided by the oracle, Equation 3 is optimized in a similar way, and the optimal model parameter ($\mathbf{w}_{opt}$) can be determined.

This entire process is also illustrated in Figure 1. However, there are still some issues to be solved during this learning process, such as, how to search the candidate queries, how to estimate the posterior probability for each specific example ($p(y|\mathbf{x})$) and each generalized query ($p(y|q)$), how to set the trade-off parameters (α_1, α_2 and α_3) and so on. We will describe the details and answer these questions in the following sections.

3.2 Constructing Generalized Queries

In each active learning iteration, constructing the optimal generalized query can be implemented by searching the best one in the query space (according to Equation 3). However, two issues need to be solved at this stage: how to search in the query space, and how to estimate the labels for the candidate queries. We will provide solutions for these two problems in this subsection.

In most traditional active learning studies, each unlabeled example is directly regarded as a candidate query. Thus, in each iteration, the query space simply contains all the current unlabeled examples, and exhaustive search is usually applied directly. However, when asking generalized queries, each unlabeled example can generate a set of candidate generalized queries, due to the existence of the don't-care features. For instance, given a specific example with d features, there exist $\binom{d}{1}$ generalized queries with one don't-care feature, $\binom{d}{2}$ generalized queries with two don't-care features, and so on. Thus, altogether 2^d corresponding generalized queries could be constructed from each specific example. Therefore, given an unlabeled data set with l examples, the entire query space would be $2^d l$. This query space is quite large (grows exponentially to the feature dimension), thus exhaustively evaluating every candidate is no longer realistic. Instead, we apply greedy search to find the optimal query in each iteration.

Specifically, for each unlabeled example (with d features), we first construct all the generalized queries with only one don't-care feature (i.e., $\binom{d}{1} = d$ queries), and choose the best as the current candidate. Then, based only on this candidate, we continue to construct all the generalized queries with two don't-care features (i.e., $\binom{d-1}{1} = d-1$ queries), and again only keep the best. The process repeats to greedily increase the number of don't-care features in the query, until no better query can be generated. The last generalized query thus is regarded as the best for the current unlabeled example. We conduct the same procedure on all the unlabeled examples, thus we can find the optimal generalized query based on the whole unlabeled set.

With such greedy search strategy, the computation complexity of searching is thus $O(d^2)$ with respect to the feature dimension d. This indicates an exponential improvement over the complexity of the original exhaustive search ($\Theta(2^d)$). Note that, it is true that such local greedy search cannot guarantee finding the *true* optimal generalized query in the entire query space, but the empirical study (see Section 4) will show it still works effectively in most cases.

With greedy search, all these candidate queries are expected to be evaluated by Equation 3, such that the optimal one could be determined. Note that, if the true labels for these candidate queries are known, the evaluation process can be implemented in exactly the same way as we will describe in Section 3.3. However, all the candidate queries are not yet labeled at the current stage, thus the objective function cannot be directly applied. Here, we use a simple strategy to estimate the label probabilities of these queries, and then evaluate them accordingly.

Specifically, given a specific query (with no don't-care feature), we simply assume that it is equally likely to be labeled positive or negative, thus we evaluate the query by regarding its label as 0.5 positive and 0.5 negative. However, in terms of the generalized queries with don't-care features, as they are expected to be responded with ambiguous answers (negative if all the examples represented are negative, and positive otherwise), we also attempt to estimate the probability of such ambiguous answer. More specifically, we suppose that each generalized query with n don't-care features can be represented by 2^n specific examples,[4] and each of these specific example is still equally likely to be positive or negative. Thus the probability of such generalize query being negative would be calculate as 0.5^{2^n} (i.e., the probability of all the examples represented being negative), and the probability of being positive would consequently be $(1 - 0.5^{2^n})$. Therefore,

[4] See Section 3.3 for details why and how each generalized query can be represented by this many examples.

for each candidate query, we apply this simple strategy to estimate its label probability and make the evaluation accordingly.

To summarize, in each learning iteration, we use greedy search to select candidate queries from the entire query space, and evaluate these candidates (with the estimated labels) according to Equation 3. The optimal query thus could be discovered.

3.3 Updating Learning Model

After discovering the optimal generalized query in each iteration, the active learner requests the corresponding label from the oracle, and then updates the learning model (i.e., optimize the model parameter $\mathbf{w}$ according to Equation 3). However, as the active learner always tends to ask generalized queries, and is always responded with ambiguous answers, the objective function is difficult to be directly specified.

Specifically, with logistic regression, the posterior probability for each example $\mathbf{x}$ can be specified as

$$p(y = 1|\mathbf{x}) = \sigma(\mathbf{w}^T \mathbf{x}) = \frac{1}{1 + e^{-\mathbf{w}^T \mathbf{x}}} \tag{4}$$

Therefore, in Equation 3, the part to maximize the log likelihood on the original training data D (i.e., the first term in Equation 3) is easy to calculate, so is the part to minimize the predictive uncertainty (entropy) on the unlabeled data U (i.e., the last term in Equation 3). However, it is difficult to specify the posterior probability for all the previous queries (i.e., $p(y|q)$, as in the middle term of Equation 3), due to the generalization of these queries and the ambiguity of the answers. We will solve this issue in this subsection.

The basic idea to estimate the posterior probability for each query ($p(y|q)$) works as follows:

- We first specify each generalized query with a set of representative examples, the posterior probability for each of these examples thus can be presented as in Equation 4.
- Then, we combine the probabilities of all these representative examples together, to form the probability for the corresponding generalized query.

More specifically, as generalized queries contain don't-care features, each generalized query can often represent a set of specific examples. For instance, if "temperature" is the don't-care feature in a generalized query, this query represents infinite examples with *any* temperature values (while keeping other features unchanged). Therefore, it seems difficult to find appropriate representative examples to specify the generalized queries.

However, in our current setting, the classifier is always a linear seperator (due to logistic regression) and the labels of the generalized queries are negative if (and only if) *all* the corresponding examples are negative (due to ambiguous oracle). Under these conditions, we can have an intuition that, given any generalized query with one don't-care feature, if (and only if) the corresponding example with the maximum value (for the don't-care feature) and the one with the minimum value are both negative, the generalized query will be labeled negative. For instances, we suppose "temperature" is the only don't-care feature, and its valid range is $[94F, 108F]$. Thus, a generalized query {age = 65, gender = male, temperature = ∗, ⋯} will definitely be labeled as negative, if (and only if) the two specific examples, {age = 65, gender = male, temperature = 94F, ⋯} and {age = 65, gender = male, temperature = 108F, ⋯}, are both negative. This intuition can be illustrated in Figure 2.

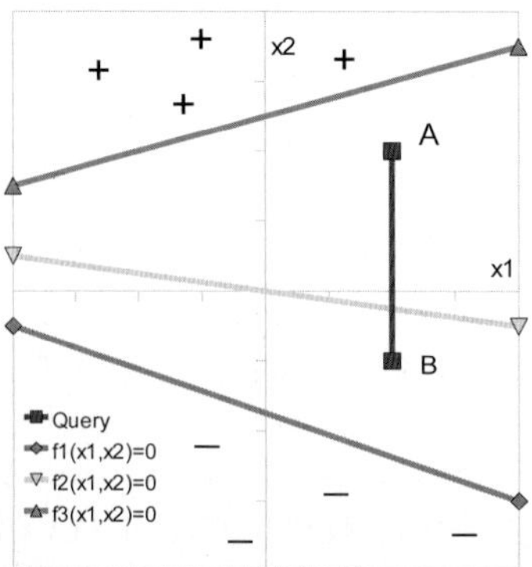

This figure illustrates the label of a generalized query $\{x1 = 2, x2 = *\}$ (where the valid range of $x2$ is $[-1, 2]$) with respect to three linear separators. Line "Query" denotes all the examples represented by the generalized query, where "A" $\{x1 = 2, x2 = 2\}$ and "B" $\{x1 = 2, x2 = -1\}$ represent two specific examples with the maximum and minimum values for the don't-care feature $x2$. Lines "f1(x1,y1)=0", "f2(x1,x2)=0" and "f3(x1,x2)=0" represent three linear separators in the given 2-D space, and all of them can only provide ambiguous answers for the queries. We can clearly see that, "f3(x1,x2)" can simultaneously label both "A" and "B" as negative[a], it consequently labels the query as negative; whereas both "f1(x1,x2)" and "f2(x1,x2)" always label at least one of "A" and "B" as positive, they label the query as positive consequently.

[a] We suppose all the examples *above* the linear separators are positive, and the ones *below* are negative.

Fig. 2. An illustration for representing generalized queries with specific examples

This illustration indicates that, given a generalized query with one don't-care feature, as long as the labels of the two specific examples (with maximum and minimum values for the don't-care feature) are known, the label for query can be easily determined.[5] Therefore, we can simply represent such generalized query by these two specific examples in the learning process. Furthermore, we can extend the conclusion that, given a generalized query with two don't-care features, four examples with the combinations of the min and max values for the two features could be used to represent the query; and further, a generalized query with n don't-care features would be specified by 2^n examples.

Now, we are able to specify any generalized query with a set of representative examples, and the posterior probability for these specific examples can

[5] Note that, in active learning, the minimum and maximum values of any feature can be reliably estimated from both the labeled and unlabeled data.

also be presented through Equation 4. Next, we will combine all these probabilities together, to form the posterior probability for the corresponding generalized query.

As we have introduced in Section 1, in the current setting, the ambiguous oracle provides negative answer only when all the corresponding specific examples are negative, and provides positive answer otherwise. This mechanism is similar to the labeling process in multi-instance learning (MIL). Therefore, we can simply apply the existing combining functions in MIL, to form the probabilities of the generalized queries. More specifically, we will apply the noise-or function [20]

$$p(y = 1|q) = 1 - \prod_{k=1}^{n} (1 - p(y = 1|\mathbf{x}_k)) \tag{5}$$

to form the probability of the generalized query, where q denotes the generalized query and $(\mathbf{x}_1, \cdots, \mathbf{x}_n)$ denotes the corresponding n representative examples. The probability for any generalized query therefore can be specified.[6]

To formalize, given the optimal generalized query (and the true ambiguous label) in each iteration, by combining Equations 3, 4 and 5, the optimal model parameter $\mathbf{w}$ can be determined by minimizing the following error function:

$$\begin{aligned} E(\mathbf{w}) = &-\alpha_1 \sum_{(\mathbf{x},y)\in D} \{y_i \log \sigma(\mathbf{w}^T\mathbf{x}_i) + (1 - y_i) \log(1 - \sigma(\mathbf{w}^T\mathbf{x}_i))\} \\ &-\alpha_2 \sum_{(q_k,y_k)\in Q} \{y_k \log(1 - \prod_{\mathbf{x}_{ki}\in q_k} (1 - \sigma(\mathbf{w}^T\mathbf{x}_{ki})) + (1 - y_k) \sum_{\mathbf{x}_{ki}\in q_k} \log(1 - \sigma(\mathbf{w}^T\mathbf{x}_{ki}))\} \\ &-\alpha_3 \sum_{\mathbf{x}_j\in U} \{\sigma(\mathbf{w}^T\mathbf{x}_j) \log \sigma(\mathbf{w}^T\mathbf{x}_j) + (1 - \sigma(\mathbf{w}^T\mathbf{x}_j)) \log(1 - \sigma(\mathbf{w}^T\mathbf{x}_j))\} \end{aligned} \tag{6}$$

where $\sigma(\mathbf{w}^T\mathbf{x}_i)$ can be further specified by Equation 4, and $\mathbf{x}_{ik}$ denotes the representative example for query q_k. In addition, gradient decent is applied to implement optimization, and the gradient of the error function with respect to $\mathbf{w}$ can be calculated:

$$\begin{aligned} \nabla E(\mathbf{w}) = &-\alpha_1 \sum_{(\mathbf{x}_i,y_i)\in D} \{(y_i - \sigma(\mathbf{w}^T\mathbf{x}_i))\mathbf{x}_i\} \\ &-\alpha_2 \sum_{(q_k,y_k)\in Q} \{\frac{(1 - \prod_{\mathbf{x}_{ki}\in q_k} (1 - \sigma(\mathbf{w}^T\mathbf{x}_{ki})) - y_k) \sum_{\mathbf{x}_{ki}\in q_k} \sigma(\mathbf{w}^T\mathbf{x}_{ki})\mathbf{x}_{ki}}{1 - \prod_{\mathbf{x}_{ki}\in q_k} (1 - \sigma(\mathbf{w}^T\mathbf{x}_{ik}))}\} \\ &-\alpha_3 \sum_{\mathbf{x}_j\in U} \{(\log \frac{\sigma(\mathbf{w}^T\mathbf{x}_j)}{1 - \sigma(\mathbf{w}^T\mathbf{x}_j)})\sigma(\mathbf{w}^T\mathbf{x}_j)(1 - \sigma(\mathbf{w}^T\mathbf{x}_j))\mathbf{x}_j\} \end{aligned} \tag{7}$$

Consequently, in each learning iteration, given the optimal generalized query and the ambiguous answer, the optimal model parameter can be obtained, and the learning model can be updated.

[6] Note that, other combining functions, such as the softmax function [23] $p(y = 1|q) = \frac{\sum_{k=1}^{n} p(y=1|\mathbf{x}_k)e^{p(y=1|\mathbf{x}_k)}}{\sum_{k=1}^{n} e^{p(y=1|\mathbf{x}_k)}}$, can also be applied to form the probability of the generalized queries.

In the next section, we will conduct empirical study to discuss the setting for the trade-off parameters α_1, α_2 and α_3, and compare the proposed algorithm with the traditional ones.

4 Empirical Study

In this section, we empirically study the performance of the proposed algorithm with generalized queries and ambiguous answers, and compare it with the existing active learning algorithms on seven real-world data sets from UCI Machine Learning Repository [4].

4.1 Experimental Configurations

We conduct experiments with two settings of the trade-off parameters (α_1, α_2 and α_3) for Equation 3. Specifically, we first consider a uniform parameter setting, that is, $\alpha_1 = \alpha_2 = \alpha_3$ (the corresponding algorithm is denoted by "AL-GQA(u)"). We can notice from Equation 3 that, all the three terms (i.e., log likelihood on D, log likelihood on Q, and predictive entropy on U) are specified by the summation of the examples (queries) in each corresponding set. It therefore indicates that, with uniform trade-off parameters, those three terms are implicitly weighed by the number of examples (queries) in the corresponding set. For instance, in the initial learning iterations, D and Q usually contain fewer examples (queries), which consequently yields lower implicit weights on the first and second terms (i.e., log likelihood on D and Q); on the other hand, U usually contains a large amount of examples, thus the third term (i.e., predictive entropy on U) will be weighted higher. To compensate for this effect, we consider another non-uniform parameter setting: $\alpha_1 = 1/|D|$, $\alpha_2 = 1/|Q|$[7], and $\alpha_3 = 1/|U|$, where $|D|$, $|Q|$ and $|U|$ denote the size of the corresponding data (query) sets (the algorithm is denoted by "AL-GQA(n)").

We also conduct experiments on active learning with specific queries (*uncertainty sampling* [7]; denoted by "AL-US") and active learning with inaccurately answered generalized queries [18] (denoted by "AL-GQN") for comparison. More specifically, "AL-US" always asks one specific example in each learning iteration, and the answer to the example is always accurate; whereas "AL-GQN" tends to ask generalized queries, and is always responded with inaccurate probabilistic answers (with up to 30% noise). In contrast, the proposed algorithms "AL-GQA(u)" and "AL-GQA(n)" also tend to ask generalized queries in each iteration, but are always responded with ambiguous (non-probabilistic) answers.

All of the seven UCI data sets have numeric features, binary class and no missing values. Information on these data sets is tabulated in Table 1. Each whole data set is first split randomly into three disjoint subsets: the training set, the unlabeled set, and the test set. The test set is always 25% of the whole data set. To make sure that active learning can possibly show improvement when the

[7] When the query set Q is empty (i.e., $|Q| = 0$), we directly set $\alpha_2 = 0$, indicating that the empty query set plays no role in this situation.

Table 1. The seven UCI data sets used in the experiments

Dataset	No. of Attributes	No. of Examples	Class Distribution	Training Size
breast-w	9	699	458/241	1/20
diabetes	8	768	500/268	1/10
heart-statlog	13	270	150/120	1/10
hepatitis	19	155	32/123	1/5
ionosphere	33	351	126/225	1/20
sonar	60	208	97/111	1/5
spambase	57	4601	1813/2788	1/100

unlabeled data are labeled and included into the training set, we choose a small training set for each data set such that the "maximum reduction" of the error rate[8] is large enough (greater than 10%). The training sizes of the seven UCI data sets range from 1/100 to 1/5 of the whole sets, also listed in Table 1. The unlabeled set is the whole set taking away the test set and the training set.

As for all the UCI data sets, we have neither true target functions nor human oracles to answer the generalized queries, we simulate the target functions by constructing learning models on the entire data sets in the experiments. The simulated target function provides ambiguous answer to each generalized query. The experiment is repeated 10 times on each data set (i.e., each data set is randomly split 10 times), and the experimental results are recorded for comparison.

4.2 Experimental Results

Based on the seven tested UCI data sets, Figure 3 plots the learning curves of the four algorithms, and presents the summary of t-test (the paired two-tailed t-test with a 95% confidence level) for comparison (where each entry, $w/t/l$, means that the algorithm in the corresponding row wins on w data sets, ties on t data sets, and loses on l data sets, compared with the algorithm in the corresponding column). We therefore can make some clear observations from these experimental results:

- On most tested data sets, both "AL-GQA(u)" and "AL-GQA(n)" perform significantly better than "AL-US". This demonstrates the superiority of active learning with generalized queries and ambiguous answers to the traditional specific-query based learning.
- On most tested data sets, both "AL-GQA(u)" and "AL-GQA(n)" perform significantly better than "AL-GQN". This indicates that, compared with inaccurate probabilistic answers, ambiguous answers are often more effective in speeding up active learning with generalized queries.

[8] The "maximum reduction" of the error rate is the error rate on the initial training set D alone (without any benefit of the unlabeled examples) subtracting the error rate on D plus all the unlabeled data in U with correct labels. The "maximum reduction" roughly reflects the upper bound on error reduction that active learning can achieve.

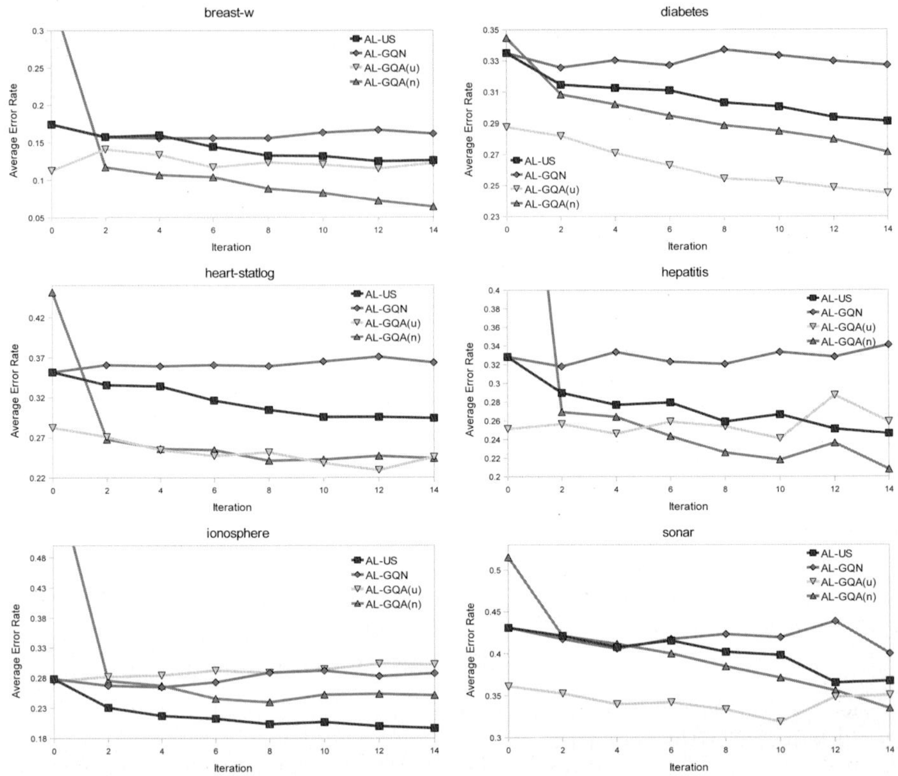

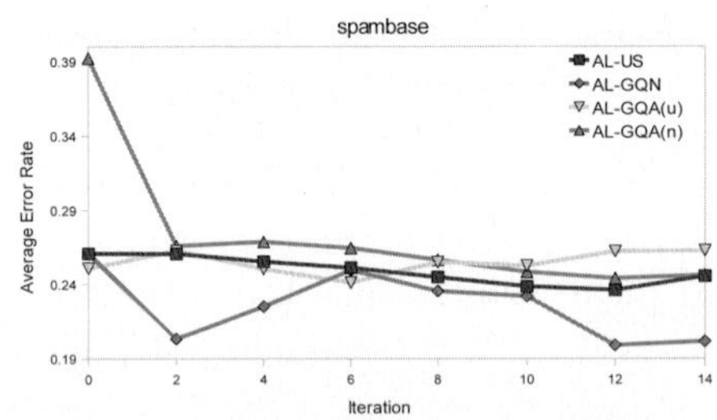

	AL-US	AL-GQN
AL-GQA(u)	4/2/1	5/1/1
AL-GQA(n)	5/1/1	6/0/1

Summary of t-test on the seven UCI data sets

Fig. 3. Learning curves and summary of t-test of the four algorithms on seven UCI data sets

– "AL-GQA(u)" and "AL-GQA(n)" work comparably well on most tested data sets. Note that, "AL-GQA(u)" usually starts from lower error rates in the initial iteration (without asking any queries), and then keeps improving the predictive performance; whereas, "AL-GQA(n)" often has rather high error rates in the initial iteration, but the predictive performance can be promptly improved once it starts asking queries.

To conclude, the experimental results clearly demonstrates the advantage of generalized queries and ambiguous answers: such type of active learning is superior in speeding up the learning process, compared with active learning with either specific queries or inaccurately answered generalized queries.

5 Conclusions

In this paper, we assume that the active learner can ask generalized queries, and the oracle is capable of respond with ambiguous answers (i.e., positive if at least one corresponding specific example is positive, and negative otherwise). After demonstrating the wide applicability of this setting, we develop a novel algorithm to ask generalized queries and update learning model with ambiguous answers in active learning. Empirical study on UCI data sets demonstrates the superiority of such type of active learning, and shows that the proposed algorithms can significantly speed up the learning process, compared with active learning with either specific queries or inaccurately answered generalized queries.

References

1. Dietterich, T.G., Lathrop, R.H., Perez, T.L.: Solving the multiple instance problem with axis-parallel rectangles. Artificial Intelligence 89(1-2), 31–71 (1997)
2. Zhang, Q., Goldman, S.A., Yu, W., Fritts, J.: Content-based image retrieval using multiple-instance learning. In: ICML 2002: Proceedings of the Nineteenth International Conference on Machine Learning, pp. 682–689. Morgan Kaufmann Inc., San Francisco (2002)
3. Andrews, S., Tsochantaridis, I., Hofmann, T.: Support vector machines for multiple-instance learning. In: Advances in Neural Information Processing Systems 15, vol. 15, pp. 561–568 (2003)
4. Asuncion, A., Newman, D.J.: UCI machine learning repository (2007)
5. Cohn, D.A., Atlas, L., Ladner, R.E.: Improving generalization with active learning. Machine Learning 15(2), 201–221 (1994)
6. Baram, Y., El-Yaniv, R., Luz, K.: Online choice of active learning algorithms. Journal of Machine Learning Research 5, 255–291 (2004)
7. Lewis, D.D., Catlett, J.: Heterogeneous uncertainty sampling for supervised learning. In: Cohen, W.W., Hirsh, H. (eds.) Proceedings of ICML 1994, 11th International Conference on Machine Learning, pp. 148–156. Morgan Kaufmann Publishers, San Francisco (1994)
8. Angluin, D.: Queries and concept learning. Machine Learning 2(4), 319–342 (1988)
9. Ling, C.X., Du, J.: Active learning with direct query construction. In: KDD 2008: Proceeding of the 14th ACM SIGKDD International Conference on Knowledge Discovery and Data Mining, pp. 480–487. ACM, New York (2008)
10. Tong, S., Koller, D.: Support vector machine active learning with applications to text classification. Journal of Machine Learning Research 2, 45–66 (2002)
11. Saar-Tsechansky, M., Provost, F.: Active sampling for class probability estimation and ranking. Machine Learning 54(2), 153–178 (2004)
12. Lindenbaum, M., Markovitch, S., Rusakov, D.: Selective sampling for nearest neighbor classifiers. Machine Learning 54(2), 125–152 (2004)
13. Margineantu, D.D.: Active cost-sensitive learning. In: The Nineteenth International Joint Conference on Artificial Intelligence, Edinburgh, Scotland (2005)

14. Seung, H.S., Opper, M., Sompolinsky, H.: Query by committee. In: COLT 1992: Proceedings of the Fifth Annual Workshop on Computational Learning Theory, pp. 287–294. ACM, New York (1992)
15. Cohn, D.A., Ghahramani, Z., Jordan, M.I.: Active learning with statistical models. Journal of Artificial Intelligence Research 4, 129–145 (1996)
16. Zhang, T., Oles, F.J.: A probability analysis on the value of unlabeled data for classification problems. In: Proc. 17th International Conf. on Machine Learning, pp. 1191–1198 (2000)
17. Roy, N., Mccallum, A.: Toward optimal active learning through sampling estimation of error reduction. In: Proc. 18th International Conf. on Machine Learning, pp. 441–448. Morgan Kaufmann, San Francisco (2001)
18. Du, J., Ling, C.X.: Active learning with generalized queries. In: Proceedings of the 9th IEEE International Conference on Data Mining, pp. 120–128 (2009)
19. Du, J., Ling, C.X.: Asking generalized queries to domain experts to improve learning. IEEE Transactions on Knowledge and Data Engineering 22(6), 812–825 (2010)
20. Maron, O., Lozano-Pérez, T.: A framework for multiple-instance learning. In: Advances in Neural Information Processing Systems, vol. 10, pp. 570–576 (1998)
21. Wang, J., Zucker, J.D.: Solving the multiple-instance problem: A lazy learning approach. In: ICML 2000: Proceedings of the Seventeenth International Conference on Machine Learning, pp. 1119–1126. Morgan Kaufmann Inc., San Francisco (2000)
22. Zucker, J.-D., Chevaleyre, Y.: Solving multiple-instance and multiple-part learning problems with decision trees and decision rules. application to the mutagenesis problem. In: Proceedings of the 14th Biennial Conference of the Canadian Society on Computational Studies of Intelligence: Advances in Artificial Intelligence, pp. 204–214. Springer, Heidelberg (2001)
23. Ray, S., Craven, M.: Supervised versus multiple instance learning: an empirical comparison. In: De Raedt, L., Wrobel, S., De Raedt, L., Wrobel, S. (eds.) ICML. ACM International Conference Proceeding Series, vol. 119, pp. 697–704. ACM, New York (2005)
24. Settles, B., Craven, M., Ray, S.: Multiple-instance active learning. In: Advances in Neural Information Processing Systems, vol. 20, pp. 1289–1296 (2008)
25. Grandvalet, Y., Bengio, Y.: Semi-supervised learning by entropy minimization. In: Advances in Neural Information Processing Systems, vol. 17, pp. 529–536 (2005)
26. Guo, Y., Schuurmans, D.: Discriminative batch mode active learning. In: Advances in Neural Information Processing Systems, vol. 20, pp. 593–600 (2008)

Analysis of Large Multi-modal Social Networks: Patterns and a Generator

Nan Du[1], Hao Wang[1], and Christos Faloutsos[2]

[1] Nokia Research Center, Beijing
{daniel.du,hao.ui.wang}@nokia.com
[2] Carnegie Mellon University, Pittsburgh
christos@cs.cmu.edu

Abstract. On-line social networking sites often involve multiple relations simultaneously. While people can build an explicit social network by adding each other as friends, they can also form several implicit social networks through their daily interactions like commenting on people's posts, or tagging people's photos. So given a real social networking system which changes over time, what can we say about people's social behaviors ? Do their daily interactions follow any pattern ? The majority of earlier work mainly mimics the patterns and properties of a single type of network. Here, we model the formation and co-evolution of multi-modal networks emerging from different social relations such as "*who-adds-whom-as-friend*" and "*who-comments-on-whose-post*" simultaneously. The contributions are the following : (a) we propose a new approach called *EigenNetwork Analysis* for analyzing time-evolving networks, and use it to discover temporal patterns with people's social interactions; (b) we report inherent correlation between *friendship* and *co-occurrence* in on-line settings; (c) we design the first multi-modal graph generator *xSocial*[1] that is capable of producing multiple weighted time-evolving networks, which match most of the observed patterns so far. Our study was performed on two real datasets (Nokia FriendView and Flickr) with 100,000 and 50,000,000 records respectively, each of which corresponds to a different social service, and spans up to two years of activity.

Keywords: Social Network Analysis, Graph Generator, Multi-modal Networks.

1 Introduction

Research of real world complex networks, like social networks [24], biological networks[11], topology [15] of WWW and Internet raises many significant and important problems. What patterns do the human-to-human interactions follow in large-scale social networks ? How can we use such patterns to facilitate existing applications, such as anomaly detection [1] [18] and collective classification [8], and make further innovations?

[1] http://research.nokia.com/people/hao_ui_wang/index.html

J.L. Balcázar et al. (Eds.): ECML PKDD 2010, Part I, LNAI 6321, pp. 393–408, 2010.

As a result of the widespread adoption of Web 2.0 technology, social networking sites or services(SNS) are becoming ubiquitous and penetrate into every corner of people's daily lives. In such systems, people often belong to multiple social networks because of different person-to-person interactions. For example, in Nokia FriendView(`http://betalabs.nokia.com/apps/nokia-friend-view`), Flickr(`www.flickr.com`), Facebook(`www.facebook.com`), eBay(`www.ebay.com`), LinkedIn(`www.linkedin.com`), and Twitter(`www.twitter.com`), they all provide the basic function that enables people to add each other as friends through their content and conversations, which contributes to the emergence of our first type of social network, namely, the "*friend network*" or the "*buddy network*".

In addition, they also allow people to participate in specific activities. In FriendView, we can comment on the posts written by our colleagues. In Flickr, we can tag the photos uploaded by our friends. In eBay, we can rate the products sold by our partners. As a consequence, interactions with people centered around content form another type of social network called the "*comment network*" or the "*participation network*" from such activities as "commenting-on-posts", "tagging-photos" and "rating-products". Therefore, these two types of social networks describe different facets of the same social networking system. For each of them, recent research has reported fascinating patterns, like [26] or lognormal [7] or Double Pareto LogNormal (DPLN) distribution [25] for the degree, as well as small and shrinking diameter [20].

In this paper, we are interested in answering the following questions:

- Do human social interactions and behaviors follow any temporal pattern ? Is there any regularity inherent in the daily activities of individuals and groups? Can we use such patterns to make predictions of their future behaviors ?
- Given a real social networking site, is there any correlation between the *buddy network* and the *participation network* ? For instance, can we infer the friendship between two people in *buddy network* according to the discrete observations of their co-occurrence in the *participation network* ?
- How can we produce an intuitive generator that will mimic the behaviors, and correlations of these networks within a real social networking site simultaneously ? Most existing generators try to mimic the skewed distribution of degree or weight of only a single network, and thus fail to incorporate the possible correlations with other networks. Here, we want a multi-modal graph generator, which should describe the way in which the different social networks discussed above could co-evolve over time through the local interactions and activities between individuals.

Answering these questions can have many practical applications. First, identifying meaningful patterns hidden in human activities contributes to classifying people into different groups according to the similarity of their social behaviors, based on which we can have a deep insight about the composition and evolution of the network they belong to. Discovering new patterns also helps to discard unrealistic graph models. Second, knowing the correlation between different social relations is good for us to design better systems that further expand the range of human interactions by offering particular friend or product recommendations according to specific user context. Finally, intuitive graph models are also vital for simulation studies of routing algorithms when it is hard or even impossible to collect large real data, for understanding how the macro and global patterns of networks can emerge through the micro and local interactions among people over time, and for compressing and summarizing the real networks by model parameters.

The paper is then organized as follows. Section 2 reviews related work. Section 3 presents our observed patterns. Section 4 describes the *xSocial* model in detail. Section 5 gives the conclusion.

2 Related Work

In this section, we mainly survey the various discovered properties of real world networks, and several well-known graph generators.

2.1 Network Patterns

Many interesting patterns that real graphs follow have been discovered in recent work like the power-law distribution of the number of messages(photos), power law comment distribution, power law interval distribution[16], power law degree distribution[26], power law edge-weight distribution[24], power law node-weight distribution[24], Snapshot Power Law(SPL)[22], Clique Participation Law(CPL) [13], Clique-Degree Power Law(CDPL)[13], Triangle Weight Law[13], Eigenvalue Power Law(EPL)[2], shrinking diameter[20], and oscillating connected component (GCC&NLCC)[22]. These patterns are important for us to understand the static and temporal properties of real world networks, to identify authorities and subgroups, as well as to refine routing algorithms and recommendations. Moreover, they are also vital for eliminating unrealistic graph generators and guiding us to design better ones, because ideally a graph model should be able to mimic all these patterns as many as possible.

2.2 Graph Generators

Generally, the graph generators of recent literature can be mainly classified as *emergent* graph models, and *generative* graph models. The basic principle of *emergent* graph models is that the macro network properties should *emerge* from the micro interactions of nodes over time. This type of models include Erdös-Rényi(ER) model [14], small-world model [27], BA model [6], Copy model [9], Random Multiplication Model [9], Forest Fire model [20], 'butterfly' model [22], and 'RTG' model [2]. [See [5] and [9] for a detailed review and discussion]. Recently, Goetz [16] also provides models to mimic the evolving and spreading mechanism of blog systems. Moreover, research from the fields of economics and game theory also brought utility-based models [17][4][12][13] where each node tries to optimize a specific utility function, and the network structure can arise from the collective strategic activities of all the nodes. *Generative* graph models often assume a global mathematic rule and perform iterations of such rule recursively until the generated networks meet several properties of real networks. Such models include kronecker multiplication model [19] and tensor model [3].

In summary, the majority of earlier graph generators often focused on modeling some main properties of only one single network. For example, [27][6][20][22] are limited in trying to model unweighted networks, and cannot be generalized to weighted networks. Goetz[16] describes the evolving process of blogs, but fail to incorporate the weights. Although RTG[2] can generate weighted graphs, it still only focused on one single network. As to the generative models, they usually cannot mimic the micro mechanism of node and edge addition, which makes it hard for us to understand the inherent natural process of real networks. In contrast, our work not only considers to mimic most of the known patterns, such as generating weighted networks from local nodes' interactions, but also focuses on co-evolution of different networks simultaneously.

3 Tools and Observations

In this section, we seek to find patterns inherent in large-scale on-line social networking sites. We first give a preliminary description of Nokia FriendView and Flickr datasets, and then we present the proposed *EigenNetwork* analysis method, and the discovered *CoParticipation Friendship Correlation* pattern.

3.1 Data Description

The datasets that we have analyzed include the interaction records from Nokia FriendView, and Flickr. Nokia FriendView is a location-enhanced experimental microblogging application and service operated by Nokia Beta Labs from the beginning of November 2008 to the end of September 2009 when the service was finished. It allows users to post messages about their status and activities from GPS-enabled Nokia S60 phones or from the web. Any two users can add each other to their buddy list through email request and confirmation. The users can also comment on the status messages posted by the buddies in their social network. As a result, we use three different types of record, $< usrID, msgID, postTime, length >$, $< usrID, buddyID, addTime >$, $< userID, msgID, commentTime, length >$, to describe these actions respectively.

Here, the edge weight of *buddy network* is the total number of comment times between them. For the dataset, there are 34,980 users, 20,873 buddy links, 62,736 status messages, and 22,251 comments [10]. The unique feature of this dataset is that it has recorded a complete evolving process of a social networking site from the very beginning to the end, over the course of 11 months. The detailed records enable us to have a deep insight about the way that people interact with each other. In the Flickr dataset (where people can upload photos, add contacts, and comment on or tag photos), we use similar tuples as Friend View to describe the data which includes about 542,105 users, 46,668,661 contact links, 101,520,484 photos, and 8,999,983 comments from 2005 to 2007. Because these datasets belong to different services, have different scales, and were collected from different time, the diversity of our data can thus be guaranteed. Notice we only use the encrypted user id in this study, and restrict our interest only in the statistical findings within the data.

3.2 EigenNetwork Analysis

While the activities and interactions where each of us is involved every day appear nearly random, intuition tells that there also seems to be some regular recurrence of patterns, especially when we take the temporal, spatial, and social context into consideration. For instance, we may check several emails, and see some news after arriving at the office in the morning. Then we might chat with our friends through instant messaging during the working hours, and in the evening, we might write blogs, make comments, upload photos, or even play on-line games. Since a social network is inherently the collection of people and their interactions, analyzing the temporal behaviors of individuals and subgroups can help us to have a deep insight about the overall composition of the entire network.

We formulate our approach as follows. Given graph $\mathcal{G}$, for $\forall e_{ij} \in \mathcal{E}(\mathcal{G})$, we characterize the temporal activity of all the edges by a two-dimensional $E \times D$ binary matrix $\mathcal{M}$, where $E = |\mathcal{E}(\mathcal{G})|$, and D is the total number of days that graph $\mathcal{G}$ has been in study.

$$\mathcal{M}(p,q) = \begin{bmatrix} 0 & 0 & 1 & 0 & \ldots \\ 0 & 1 & 0 & 0 & \ldots \\ 1 & 0 & 1 & 1 & \ldots \\ \ldots & \ldots & \ldots & \ldots & \ldots \end{bmatrix} \tag{1}$$

Therefore, the pth row represents the behavior of a particular edge e_{ij} spanning the D days. On a specific day q, if node v_i and v_j has at least one interaction with each other, then $\mathcal{M}(p,q) = 1$; otherwise $\mathcal{M}(p,q) = 0$. We then do Singular Value Decomposition(SVD) on matrix $\mathcal{M}$ and it is factorized as

$$\mathcal{M} = U \times \Sigma \times V^T \tag{2}$$

where the columns of D-by-K matrix V form a set of orthonormal *input* basis vectors for $\mathcal{M}$, the columns of E-by-K matrix U form a set of corresponding orthonormal *output* basis vectors, and the diagonal values in K-by-K matrix Σ are the singular values arranged in the descending order by which each corresponding *input* is multiplied to give a corresponding *output*.

By intuition, the *SVD* on matrix $\mathcal{M}$ implicitly decomposes the E edges into K groups. Each column (or singular vector) i of the E-by-K matrix U describes the extent to which each edge of $\mathcal{G}$ participates in the ith group. Every column j of the D-by-K matrix V shows the extent to which the jth group is active on each day. The nonnegative real numbers on the diagonal of the K-by-K matrix Σ indicates the strength of each group. For each singular value s_i, the *energy* of s_i is defined as s_i^2, so we keep the first few strongest singular values whose sum covers 80-90 percentile of the total energy. Here, we build matrix $\mathcal{M}$ for

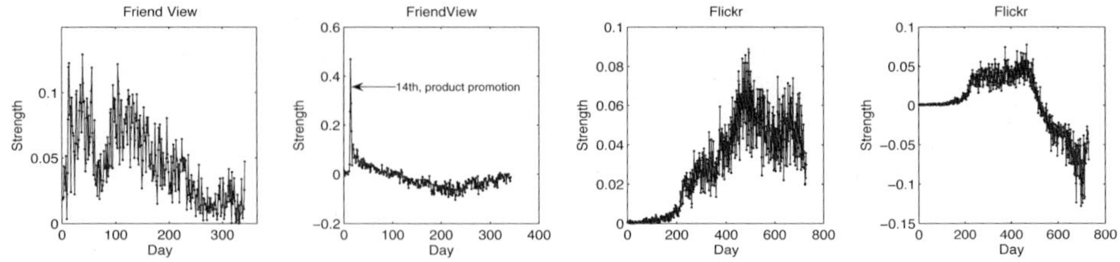

(a) 1st vector (b) 2nd vector (c) 1st vector (d) 2nd vector

Fig. 1. The 1st and 2nd singular vector of matrix V that describe the corresponding daily activities of the 1st and 2nd subgraph consisting of the selected edges in the *participation network* (formed by the *comment* relation) of FriendView (a-b), and Flickr (c-d) respectively

the *participation network* which emerges from the comment interactions among users in FriendView and Flickr respectively. $\mathcal{M}(p,q) = 1$ means that for the pth edge e_{ij}, at least one of the two nodes (v_i and v_j) commented on the messages or photos posted by the other one on the qth day.

Figure 1 shows the top two singular vectors of the matrix V from FriendView and Flickr. In Figure 1(a-b), we have two groups of edges that show different patterns of behavior. The first group of Figure 1(a) has basically a periodic pattern, while the second group of Figure 1(b) appears more bursty, where the spike occurs on the 14th day. Based on the complete records of FriendView, it was discovered that the 14th day was just during the week that Nokia did lots of advertising work to promote FriendView by calling for more open beta testers. For Flickr, both of the two groups shown in Figure 1(c-d) behave periodically. There is a clear trend of overall growth in the amplitude with some oscillation. We guess this may be caused by the quickly increased popularity and fast development of Flickr as more and more users joined in the system after the year 2006.

Figure 2 further presents the evolving process of the subgraph G_x^1 and G_x^2 consisting of the selected edges that actively participate in the 1st and 2nd singular vector of matrix U. Being active means that we only keep the set of edges whose sum of the energy (which is the square of the corresponding value) covers 80-90 percentile of the total energy. In Figure 2, the evolving pattern of G_x^1 and G_x^2 are clearly different. Subgraph G_x^1 contains a size-4 clique (complete graph) where each blue-square node has connections with each other. This clique remains stable in topology and in total number of activities over the whole period, except for G_2^1 where five edges shown in red had significantly increased number of activities, and for G_3^1 where the the number of their activities dropped back. For $\forall e_{ij} \in \mathcal{E}(\mathcal{G}_x^1), x > 1$, *red* color of e_{ij} indicates that its weight (which is the total number of times that node v_i and v_j interact with each other in the xth month) is significantly higher than its previous value in graph $\mathcal{G}_{x-1}^1$, and *green* color means the reverse. We made further investigations into the egocentric subgraph of around such 4 blue-square nodes in the entire network. Their average *degree*, and *node betweenness* [24] are 39 and 0.42 respectively. Because *degree*,

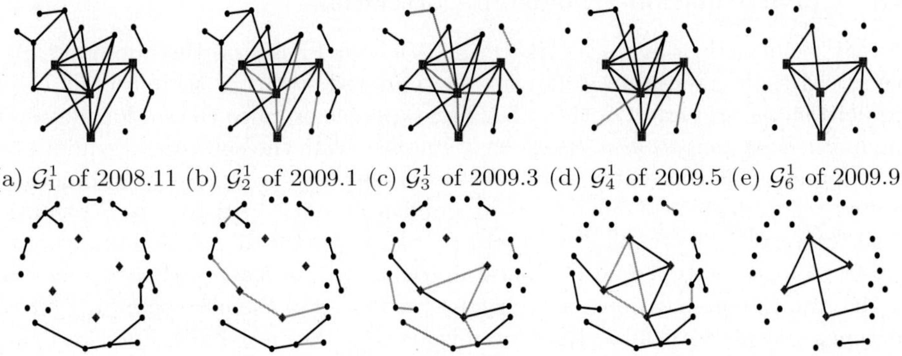

(a) $\mathcal{G}_1^1$ of 2008.11 (b) $\mathcal{G}_2^1$ of 2009.1 (c) $\mathcal{G}_3^1$ of 2009.3 (d) $\mathcal{G}_4^1$ of 2009.5 (e) $\mathcal{G}_6^1$ of 2009.9

(f) $\mathcal{G}_1^2$ of 2008.11 (g) $\mathcal{G}_2^2$ of 2009.1 (h) $\mathcal{G}_3^2$ of 2009.3 (i) $\mathcal{G}_4^2$ of 2009.5 (j) $\mathcal{G}_6^2$ of 2009.9

Fig. 2. The evolving process of the subgraph G_x^1 and G_x^2 consisting of the selected edges belonging to the 1st(top row) and 2nd(bottom row) singular vector of matrix U in the *participation network* of FriendView. $\forall G_x^1$ (G_x^2) where $x > 1$, *red* indicates that the weight (representing the number of times that two users comment on each other's messages) is at least an order of magnitude higher than its previous value in G_{x-1}^1 (G_{x-1}^2), *green* means the reverse, and *black* shows the same level.

and *node betweenness* are two popular measures to quantify a node's authority or centrality in a social network, the subgraph formed from these active edges in the 1st singular vector of matrix U actually represents the central part or the core of FriendView's *participation network*.

We see that in November, 2008 and January, 2009, there are two significant increases in the number of interactions as most edges in the subgraph are red compared with the previous graph, which also coincides with the two spikes in Figure 1 (a). Moreover, because the open beta testing for FriendView actually finished in September, 2009, in Figure 2, the subgraph becomes sparse, when the interactions between users dropped gradually, and also conforms with the decreasing trend in Figure 1(a). In contrast, the subgraph G_x^2 is loosely connected. In the beginning, it only consisted of several separated edges. Notice in Figure 1(b), there is a bursty in the first month when Nokia did a lot of publicity work. As a result, there were many separated short-term interactions at that time.

Therefore, because subgraphs formed by the selected edges from the singular vectors of matrix U (which are also the eigenvectors of $\mathcal{M} \times \mathcal{M}^T$) hold different local temporal patterns, and represent different compositions of the overall network, they are defined as the *EigenNetworks*, and our methodology is thus called *EigenNetwork* analysis.

Observation 1. *EigenNetwork. The EigenNetworks can reveal local compositions of real world social networks, and hold different temporal patterns over time.*

3.3 CoParticipation-Friendship Correlation

In real social networking sites like FriendView or Flickr, on the one hand, people spend their daytime in following the updated status of their friends in the explicit *buddy network*. On the other hand, people are also the major players in the implicit *participation network* that emerges from the activities we adopt. As a consequence, is there any correlation between these two types of interaction ? Will the reoccurrence of one particular implicit activity contribute to a formation of the corresponding explicit interaction ? More specifically, can we quantify the extent to which two people will become friends in the *buddy network* according to the discrete observations of their co-occurrences in the corresponding *participation network* ? An underlying premise is that the probability for two people to become friends increases with the number of activities in which they have engaged together.

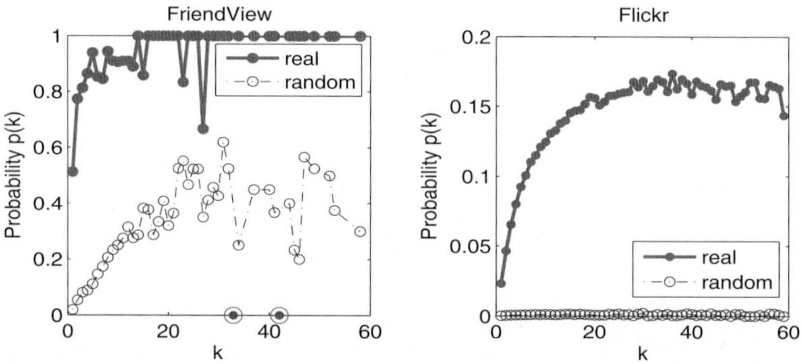

(a) Friend View (b) Flickr

Fig. 3. The probability $P(k)$ of being friends as a function of the number of co-commented messages k in Friend View (a), and photos in Flickr (b) respectively. For each k, *red* curve indicates the actual probability of being friends, and *blue* curve shows the expected value in random graphs. The outliers are marked by *red* circles.

Figure 3 shows this basic relationship in red color for FriendView and Flickr respectively, that is, the probability $P(k)$ of two people to become friends as a function of the total number of times that they have participated in k common activities. $P(k)$ is calculated as follows. We first find all tuples $< i, j, k >$ such that node v_i and v_j have k participated activities in common. Then $P(k)$ is the fraction of such tuples for a given k that node v_i and v_j are also friends in the *buddy network*. We see that when k is roughly small ($k < 30$), $P(k)$ has a strictly monotonic increase as k increases. However, as k becomes large, the marginal effect diminishes as k increases.

Moreover, we would also like to evaluate how this empirical correlation compares to the corresponding result if comments were produced randomly with the same background distribution in real datasets. Specifically, for each node v_i, while we keep the number of posts(photos) on which she would comment the same as that in real dataset, we let her randomly choose among the posts(photos) this time. We then use $P_0(k)$ to denote the expected probability for a pair of nodes to become friends. If $P(k) > P_0(k)$, we say that the correlation is over-represented in the data compared to chance;on the other hand, if $P(k) < P_0(k)$, then this correlation is underrepresented. To quantify the significance of $P(k)$ being over-or-underrepresented, we use the *surprise*[21] $S(k)$ which is defined as

$$S(k) = \Delta(k) \times (P(k) - P_0(k))/\sqrt{\Delta(k) \times P(k) \times (1 - P(k))} \tag{3}$$

where $\Delta(k)$ is the total number of tuples that have k common posts(photos). $S(k)$ indicates the number of standard deviations by which the actual number of pairs being friends deviates from the expected number in random graphs. In Figure 3, the plots in *blue* dash-line describe $P_0(k)$ vs. k in random graphs for FriendView and Flickr respectively. According to the Cental Limit Theorem, the distribution of each $S(k)$ conforms approximately to a standard normal distribution, and it is expected on the order of tens to already be significant ($S(k) = 6$ gives a p-value of 10^{-8} approximately)[21]. However, in our datasets, we have found that the average $S(k)$ for $k < 60$ and $P(k) \neq 0$ is 54.0 and 371.6 for FriendView and Flickr respectively, which is much larger and means that this correlation is statistically significant.

There are also some outliers(marked by red circles). The existence of such outliers means that although two people have engaged in many common activities together, they are still not friends yet. We guess this might be caused by users' ignorance or unawareness of each other. These types of users may only care about the messages or photos themselves by ignoring other people's comments at all. As a result, one possible application of the correlation shown in Figure 3 may be to help us with better recommendation systems, especially for the situation where k is large, because as k continuously increases , the number of pairs who have k activities together decreases significantly, which makes it easier to give specific recommendations.

Observation 2. **CoParticipation-Friendship** Correlation(CPF). *Given a real social networking site, the probability $P(k)$ of being friends for any pairwise persons increases with their k activities in common. Although the marginal effect diminishes as k increases, the effect remains significant.*

4 xSocial Model

Next, we present our *xSocial* model where the "x" means that it is a multi-modal graph generator that mimics real social networking sites to produce the *buddy network* and the *participation network* simultaneously. The guiding principle is that based on our understanding of existing patterns, we will devise a set of simple rules that each user would follow, and the entire social network will arise and evolve through the local interactions between individuals over time. Notice that this is actually a very challenging task, because the majority of prior work mostly focused on modeling only a single network. Our work is different as all the synthetic networks generated by *xSocial* should follow both the old and the new patterns mentioned in the previous section.

4.1 Model Description

xSocial model consists of the following four essential components:

- It is designed by using agent-based modeling approach. We have a set $\mathcal{A}$ of n distinct agents, each of which has a *preference* value f_i.
- At each time, every agent performs three independent actions(*write a message*, *add a friend* and *comment on a message*) guided by the one-dimensional random walk mechanism.
- An agent chooses his friends either by their popularity or by the number of messages on which they have commented together, which is determined by his *preference* f_i.
- An agent can also follow the updated status of his friends by putting comments on the corresponding newly written messages.

Preference Value. For any agent $a_i \in \mathcal{A}$, $f_i \in (0,1)$ represents two different behaviors of people while they are using on-line social networking services. f_i approaching to 1 means the agent likes to follow those active agents who have already written many messages, and continuously put new messages, while f_i close to 0 indicates that he is interested in and pays more attention to the comments put by other agents.

Random Walk. In every step, each agent does a random walk on a line, and then chooses to write a message, or add a friend, or comment on a message whenever the walk returns to the origin (at state 0). We use three integers: S_w, S_a, and S_c to represent the state of the corresponding action respectively. The initial state of S_w, S_a, and S_c is 0. For each variable, there are two types of transition. An agent a_i adds or subtracts 1 from the variable's current state with probability p_i and $1 - p_i$ respectively. The agent a_i performs the corresponding action whenever S_w, S_a, or S_c returns to 0 again. Newman [23] shows that the inter-posting times follow a power-law distribution with exponent -1.5. Intuitively, the random walk reflects how frequently an agent uses the social networking service. On the one hand, when the probability p_i approaches to 0 or 1, the agents may just use the system in the very beginning, and never come back, just like that most users only register an account for curiosity in the beginning, but almost seldom use the service later. On the other hand, when p_i is near 0.5, the corresponding agents are relatively active users who can successively use the service, although they may be distracted by some random events to the nearby state around the origin.

Add a friend. For $\forall a_i \in \mathcal{A}$, once $a_i.S_a$ hits zero, he decides to expand his *buddy network* by exploring more friends. With probability f_i, a_i trusts word-of-mouth and chooses an agent a_j proportionally with the number of messages she has written, because the user who has published many messages will naturally attract attention of others so that she can expand the number of her followers.

Once she has more followers, she would probably like to publish even more messages. In the opposite case, with probability $1 - f_i$, a_i picks an agent a_k proportionally with the number of messages on which they have put comments together.

Make a comment. For $\forall a_i \in \mathcal{A}$, when a_i decides to comment on some other messages at the moment $a_i.S_c = 0$, she prefers the candidates newly written by her friends, because for most social networking services, we can often receive a notification once any one of our friends has updated her status. Therefore, a_i chooses the message proportionally with $\frac{\#comments+1}{age+1}$, where $\#comments$ is the number of existing comments on such message, and age is the number of times since its publication. As a result, the newly written messages which already have many comments will be chosen with very high probability.

Algorithm 1. *xSocial* Model

```
Input: A, T, time ← 0
while time < T do
    foreach a_i ∈ A do
        with probability p_i, add a_i.S_w, a_i.S_a, and a_i.S_c by 1;otherwise, subtract them all
        by 1;
        if a_i.S_w = 0 then a_i writes a message;
        if a_i.S_a = 0 then
            if SampleUniform(0, 1) < f_i then
                the probability of a_i choosing a_j is P(a_i → a_j) ∝ #messages(a_j);
                (#messages is the number of a_j's messages);
            else
                the probability of a_i choosing a_j is P(a_i → a_j) ∝ #cocomments(a_i, a_j);
                (#cocomments is the number of messages commented together);
        if a_i.S_c = 0 then
            the probability of a_i choosing a message o_j is P(a_i → o_j) ∝ (#comments+1)/(age+1);
    time ← time + 1;
```

In summary, all these four major components in our *xSocial* model include very simple rules without assuming any prior sophisticated distributions or constraints. However, as we will show in the next section, both the *buddy network* and the *participation network* generated by this simple model can still match most patterns found on the real datasets. Pseudocode for *xSocial* is shown in algorithm 1.

4.2 Model Analysis

The *xSocial* model incorporates the interlinked evolving process of *buddy network* and *participation network* together, which is much more challenging to model jointly than separately. On the one hand, because the majority of existing graph generators mostly considers modeling a single type of network, there is no natural model to compare with our model. On the other hand, since the *xSocial* model also uses random walk to determine when to put a message or

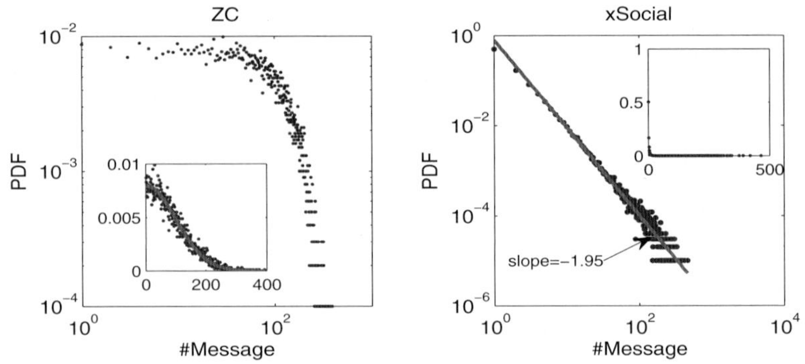

Fig. 4. PDF of #Message. Left : ZC model gives a folded normal distribution $p(x) \propto N_f(0,T)$ where T is the number of times that random walk repeats; Right: xSocial model produces a power-law distribution $p(x) \propto x^{-1.95}$.

photo, for this point, we can at least make a comparison to the *ZC* model [16]. However, as we will show as follows, even *ZC* model still cannot give the correct distribution of the number of messages that users have posted. In Figure 4, *ZC* model actually gives a folded normal distribution for people's posting behavior with 0 mean and T deviation where T is the number of times that random walk repeats. In contrast, our *xSocial* model matches the power law exponents well : -1.95 vs. -2.07 in Figure 6(m) and 6(q).

Because normal distribution and power-law distribution corresponds to two extreme cases : a *homogeneous* network, vs. a *heterogeneous* network, we believe that this difference arises as a result of the different random walk behaviors. In *ZC* model, the probability for an agent to change his state is all set to 0.5, then each agent has equal opportunity to cross zero (make a post), although they may be distracted by some random events to the nearby state. However, this only models the behavior of active users who frequently use the system although they can be away from his computer for some random distractions. However, a real social networking site not only includes active users, but also involve lots of inactive users. These users just register an account for curiosity in the beginning, but seldom come back and use the system later. As a result, the probability of a random walk to change state in *xSocial* is designed to be different for each agent, which essentially enhances the system's heterogeneity. Because such heterogeneity significantly increases the model complexity, rigorous mathematical proofs are our current ongoing work.

4.3 Model Validation

How accurate is our model? A model is considered to be good if it is able to produce patterns and properties similar to those found in real world networks as many as possible. We simulated the model 15,000 times with 100,000 agents. For

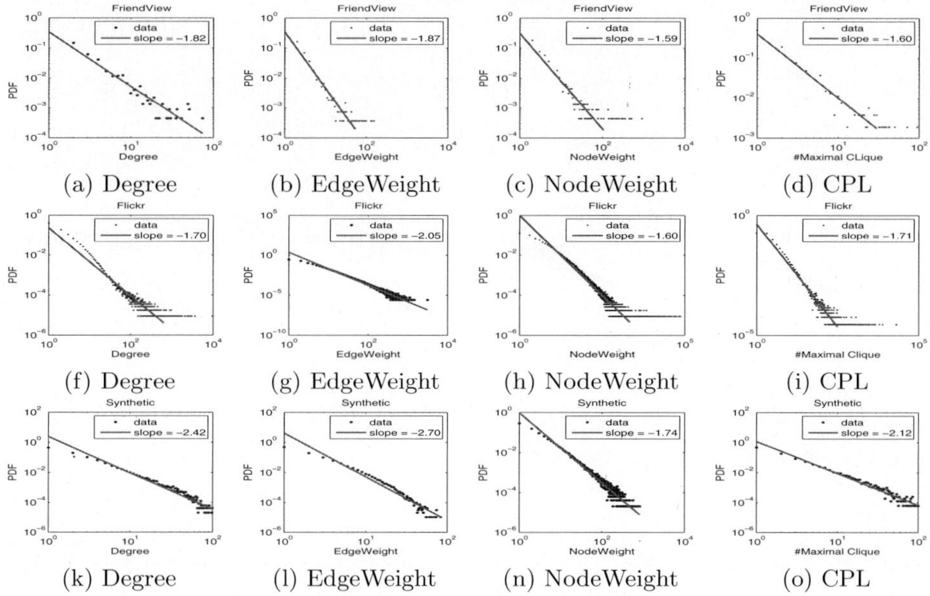

(a) Degree (b) EdgeWeight (c) NodeWeight (d) CPL

(f) Degree (g) EdgeWeight (h) NodeWeight (i) CPL

(k) Degree (l) EdgeWeight (n) NodeWeight (o) CPL

Fig. 5. Comparison of the weighted buddy network, between the two real graphs (top two rows) for FriendView and Flickr, and our synthetic graph (bottom row)

each agent a_i, as f_i and p_i are independently and uniformly chosen at random from 0 to 1, there is inherently no user-predefined parameters for *xSocial* to set.

We check with Edge-Weight Distribution[24], Node-Weight Distribution[24], as well as TWL[13] for weighted network, and Degree Distribution[26], SPL[22], CPL[13], CDPL[13], GCC&NLCC[22], as well as EPL[2] for unweighted network. For each agent, we check with the distribution of the number of written posts, and the CoParticipation Friendship Correlation(CPF). For each post, we check with the distribution of the number of received comments.

Figure 5 and 6 show the related old and new patterns of the *buddy network* for Flickr and FriendView as well as for *xSocial* results, respectively. Figure 7 further compares the *participation network* (formed by the comment relation) between the real graph and the synthetic graph. Because we have similar observations on Flickr for the basic known patterns, here, we have to only present the results on FriendView in Figure 6(m)$\sim$(p) and in Figure 7 due to the limit of space.

According to [22], there often exists a gelling point corresponding to the spike in the early stage of the growing network in Figure 6(s) and in Figure 7(h), after which several small connected components start to fuse into a giant connected component, and the 2nd and 3rd connected component remain constant size with small oscillations.

We see that the gelling point in Figure 6(o) and in Figure 7(d) is not as clear as in Figure 6(s) and in Figure 6(h), probably because the data was collected as a snapshot by month, and the data at the gelling point could be absorbed by

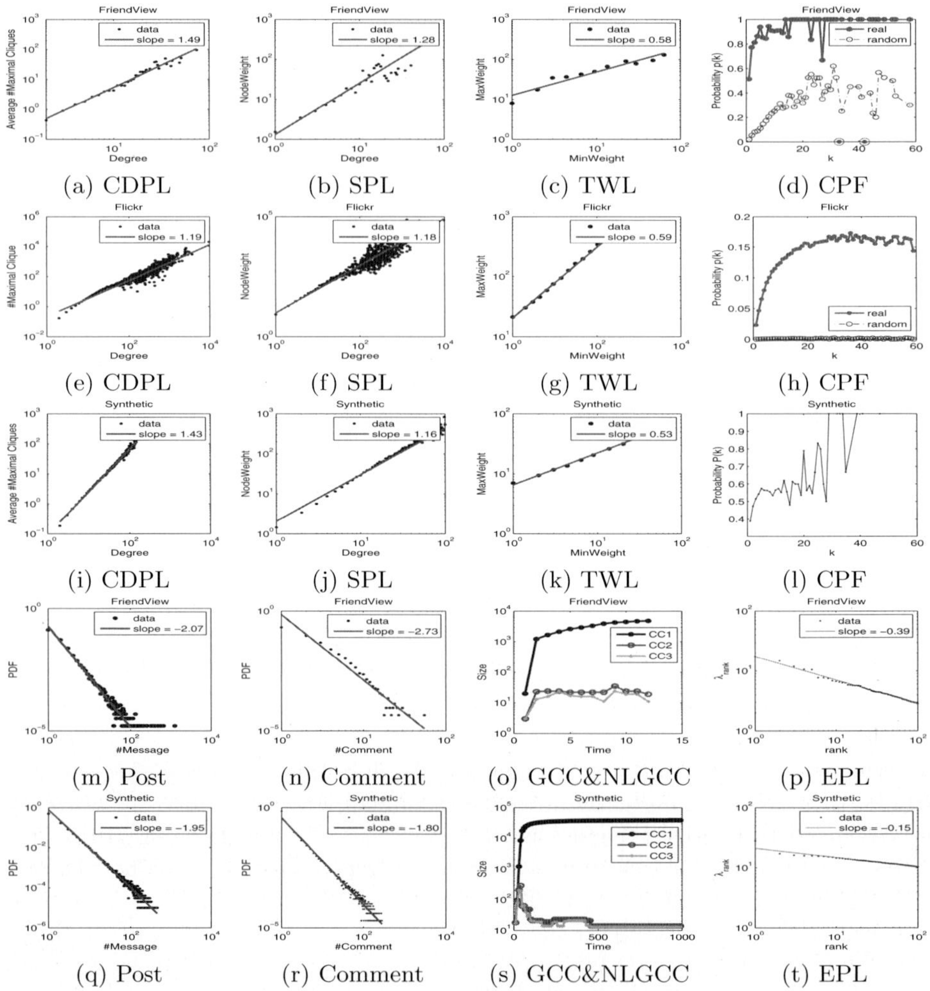

Fig. 6. Comparison of the weighted buddy network cont'd, between the two real graphs from (a) to (h), and (m) to (p) for FriendView and Flickr, and our synthetic graphs from (i) to (l), and (q) to (t)

the later networks. By contrast, the synthetic networks are more fine-grained. However, after the gelling point, the size of the connected components keeps constant, showing that the synthetic networks coincide with the real networks. The effective diameter of the weighted *buddy network* and *participation network* of FirendView is approximately 9 and 10, while *xSocial* gives 9.7 and 9.5 respectively. In all cases, *xSocial* can give skewed distributions for both the *buddy network* and *participation network* which are remarkably close to the real ones.

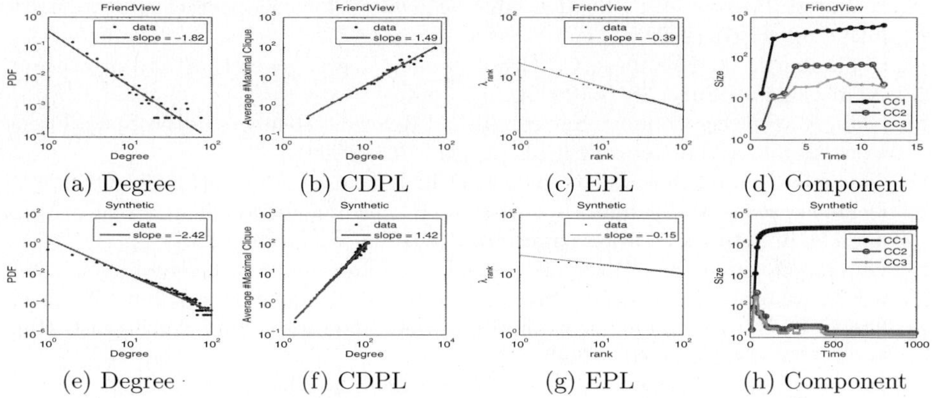

(a) Degree (b) CDPL (c) EPL (d) Component

(e) Degree (f) CDPL (g) EPL (h) Component

Fig. 7. Qualitative comparison of the comment network, between the real graph (top row), and our synthetic graph (bottom row)

5 Conclusion

We study multi-modal networks formed by *friend* and *comment* relations in two different datasets which have over 50 million records and span the course of 2 years. The main contributions are: (a) we proposed the *EigenNetwork* approach to analyzing time-evolving networks, and revealed that there exists temporal regularity with people's on-line social interactions; (b) we discovered inherent correlations between friendship and occurrence in on-line social networking settings; (c) we design the first multi-modal graph generator *xSocial* that stands out from the rest, because it does not include any user predefined parameters, it only uses local information, and it is capable of describing the co-evolving process of multiple weighted social networks that match the old and new patterns observed so far.

References

1. Aggarwal, C.C., Yu, P.S.: Outlier detection with uncertain data. In: SDM, pp. 483–493 (2008)
2. Akoglu, L., Faloutsos, C.: Rtg: A recursive realistic graph generator using random typing. In: PKDD, pp. 13–28 (2009)
3. Akoglu, L., McGlohon, M., Faloutsos, C.: Rtm: Laws and a recursive generator for weighted time-evolving graphs. In: ICDM, pp. 701–706 (2008)
4. Albers, S., Eilts, S., Even-Dar, E., Mansour, Y., Roditty, L.: On nash equilibria for a network creation game. In: SODA, pp. 89–98 (2006)
5. Albert, R., Barabasi, A.-L.: Statistical mechanics of complex networks. Reviews of Modern Physics (2002)
6. Barabasi, A.L., Albert, R.: Emergence of scaling in random networks. Science 286(5439), 509–512 (1999)
7. Bi, Z., Faloutsos, C., Korn, F.: The "DGX" distribution for mining massive, skewed data. In: KDD (August 2001), Runner up for Best Paper Award

8. Bilgic, M., Getoor, L.: Effective label acquisition for collective classification. In: KDD, pp. 43–51 (2008)
9. Chakrabarti, D., Faloutsos, C.: Graph mining: Laws, generators, and algorithms. ACM Comput. Surv. 38(1) (2006)
10. Chin, A.: Finding Cohesive Subgroups and Relevant Members in the Nokia Friend View Mobile Social Network. CSE (4), 278–283 (2009)
11. Chung, F., Lu, L., Dewey, T.G., Galas, D.J.: J. Comput. Biol.10(5), 677–687 (2003)
12. Demaine, E.D., Hajiaghayi, M., Mahini, H., Zadimoghaddam, M.: The price of anarchy in network creation games. In: PODC, pp. 292–298 (2007)
13. Du, N., Faloutsos, C., Wang, B., Akoglu, L.: Large human communication networks: patterns and a utility-driven generator. In: KDD 2009, pp. 269–278 (2009)
14. Erdos, P., Renyi, A.: On the evolution of random graphs. Publ. Math. Inst. Hungary. Acad. Sci. 5, 17–61 (1960)
15. Faloutsos, M., Faloutsos, P., Faloutsos, C.: On power-law relationships of the internet topology. In: SIGCOMM, pp. 251–262 (August-September 1999)
16. Goetz, M., Leskovec, J., Mcglohon, M., Faloutsos, C.: Modeling blog dynamics. In: ICWSM 2009 (2009)
17. Laoutaris, N., Poplawski, L.J., Rajaraman, R., Sundaram, R., Teng, S.-H.: Bounded budget connection (bbc) games or how to make friends and influence people, on a budget. CoRR (2008)
18. Lee, J.-G., Han, J., Li, X.: Trajectory outlier detection: A partition-and-detect framework. In: ICDE 2008, pp. 140–149 (2008)
19. Leskovec, J., Chakrabarti, D., Kleinberg, J.M., Faloutsos, C.: Realistic, mathematically tractable graph generation and evolution, using kronecker multiplication. In: Jorge, A.M., Torgo, L., Brazdil, P.B., Camacho, R., Gama, J. (eds.) PKDD 2005. LNCS (LNAI), vol. 3721, pp. 133–145. Springer, Heidelberg (2005)
20. Leskovec, J., Kleinberg, J., Faloutsos, C.: Graphs over time: densification laws, shrinking diameters and possible explanations. In: KDD 2005, pp. 177–187 (2005)
21. Leskovec, J., Hunttenlocher, D., Kleinberg, J.: Signed Networks in Social Media. In: CHI, pp. 1361–1370 (2010)
22. McGlohon, M., Akoglu, L., Faloutsos, C.: Weighted graphs and disconnected components: patterns and a generator. In: KDD 2008, pp. 524–532 (2008)
23. Newman, M.E.J.: Power laws, pareto distributions and zipf's law. Contemporary Physics 46, 323 (2005)
24. Onnela, J.-P., Saramäki, J., Hyvönen, J., Szabó, G., de Menezes, A.M., Kaski, K., Barabási, A.-L., Kertész, J.: Analysis of a large-scale weighted network of one-to-one human communication. New J. Phys. 9(6) (June 2007)
25. Reed, W., Jorgensen, M.: The double pareto-lognormal distribution - a new parametric model for size distribution (2004)
26. Watts, D.: Small Worlds:The Dynamics of Networks between Order and Randomness. Princeton University Press, Princeton (1999)
27. Watts, D.J., Strogatz, S.H.: Collective dynamics of 'small-world' networks. Nature 393(6684), 440–442 (1998)

A Cluster-Level Semi-supervision Model for Interactive Clustering

Avinava Dubey[1], Indrajit Bhattacharya[2,⋆], and Shantanu Godbole[1]

[1] IBM Research - India
{kumdubey,shantanugodbole}@in.ibm.com
[2] Indian Institute of Science
indrajit@csa.iisc.ernet.in

Abstract. Semi-supervised clustering models, that incorporate user provided constraints to yield meaningful clusters, have recently become a popular area of research. In this paper, we propose a cluster-level semi-supervision model for inter-active clustering. Prototype based clustering algorithms typically alternate between updating cluster descriptions and assignment of data items to clusters. In our model, the user provides semi-supervision directly for these two steps. Assignment feedback re-assigns data items among existing clusters, while cluster description feedback helps to position existing cluster centers more meaningfully. We argue that providing such supervision is more natural for exploratory data mining, where the user discovers and interprets clusters as the algorithm progresses, in comparison to the pair-wise instance level supervision model, particularly for high dimensional data such as document collection. We show how such feedback can be interpreted as constraints and incorporated within the k-means clustering framework. Using experimental results on multiple real-world datasets, we show that this framework improves clustering performance significantly beyond traditional k-means. Interestingly, when given the same number of feedbacks from the user, the proposed framework significantly outperforms the pair-wise supervision model.

1 Introduction

While clustering has been one of the most effective tools for exploratory data mining for decades, it is widely accepted that the clusters generated without any supervision often do not lead to meaningful insights for the user. Accordingly, there has been a lot of interest in recent years in developing semi-supervised clustering models that can accommodate supervision from the user to guide the clustering process[4]. In the most popular model for semi-supervised clustering, the user provides must-link and cannot-link constraints over pairs of data instances [17]. It has been shown that such constraints can significantly improve clustering performance beyond that of unsupervised models.

An interesting feature of this instance-level model is that the constraints are independent of each other, and also of all other instances in the dataset. However, it is straight forward to imagine example datasets, as in Figure 1, where the same pair of instances may be 'must-linked' or 'cannot-linked' depending on the other instances present in the

⋆ Work done while at IBM Research - India.

J.L. Balcázar et al. (Eds.): ECML PKDD 2010, Part I, LNAI 6321, pp. 409–424, 2010.

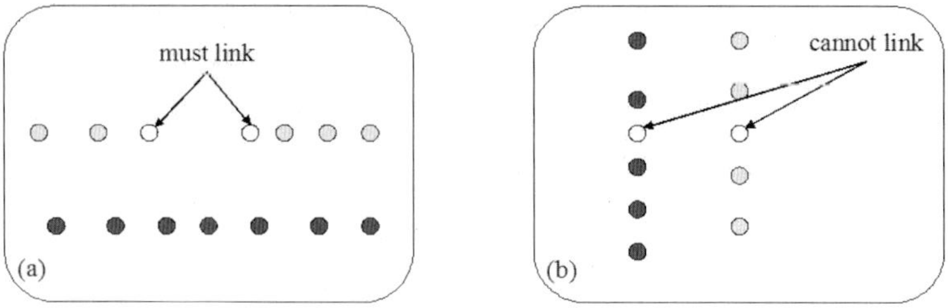

Fig. 1. Cluster examples involving the same pair of instances, shown in white. Based on the other instances in the dataset, in (a) they are must-linked to belong to the same cluster, while in (b) they are cannot-linked.

dataset. Therefore, such independent constraints required by the instance-level model can only be provided when the human supervisor can visualize the space of all the data instances and then decide on the desired shape of the clusters, as in most illustrative examples for instance-level supervision [4]. Such visualization of data items is difficult, if not impossible, in high dimensions, for example when clustering a collection of text documents. A pair of documents, one on the English soccer league and the other on the Spanish soccer league, would belong to the same cluster if the document collection is on different types of sports, but not in a different collection that discusses different European soccer leagues. Thus providing constraints on this pair of documents is not possible using independent pair-wise constraints without visualizing or understanding the document collection in its entirety.

In this paper, as an alternative we propose an interactive cluster-level semi-supervision framework for clustering, where such conditional constraints can be provided by the human supervisor. Prototype or model based clustering algorithms typically iterate over two steps — assignment of data points to clusters, and adjustment of clusters to minimize distortion. In our model, the user provides two different types of feedback, aimed directly at supervising these two different steps, while the algorithm executes. Using *assignment feedback*, the user moves a data point from one of the current clusters to another. Using *cluster description feedback*, the user modifies the feature vector of any current cluster to make it more meaningful. Such an interactive framework is particularly useful for exploring large high-dimensional data sets when the clusters are not known in advance. The current set of clusters provides the user with a summarized, aggregated view of the entire dataset. Conditioned on this current set of clusters, and also enabled by the summary that it provides, he then re-assigns and re-adjusts this clustering as he thinks appropriate. The algorithm learns from this feedback, and from other feedbacks provided in earlier stages, to re-cluster the dataset, which the user can again inspect and critique. The iterative process continues until the user is satisfied with the clustering. The basic idea of interaction for clustering [6,12], interpreted as pair-wise constraints, was proposed as early as 1999, but to the best of our knowledge, there has not been any follow-up work around cluster-level supervision to address large dimensionality of the data.

We show how both these types of feedback can be interpreted as constraints and incorporated within the k-means formulation, and the assignment and update steps can be modified to minimize constraint violation while maintaining low distortion error. Though we focus on the k-means objective function for this paper, we believe that a similar semi-supervision framework can be built around other model-based clustering objective functions as well.

The rest of the paper is organized as follows. In Section 2, we illustrate and motivate the two types of feedback using examples, followed by a review of related work in Section 3. We formalize the problem in Section 4, followed by the inter-active clustering algorithm in Section 5. Possible models for the supervisor and the issue of convergence are discussed in Section 6, and then our experimental evaluation is described in Section 7. We end with concluding remarks and possible future directions in Section 8.

2 Cluster-Level Supervision: An Example

In this section, we present an illustrative example of cluster-level supervision. Though the example is for document collections, we believe that the framework also offers similar advantages for other high dimensional domains.

Consider a very large collection of postings on a vehicle related mailing list. An analyst wishes to understand the issues being discussed, and, in order to do so, decides to partition the posts into k clusters, using the first stage of the algorithm, which is completely unsupervised. Note that he does not have any idea of possible issues ahead of time, apart from that they all relate to vehicles. On inspecting the cluster descriptions — the top words in decreasing order of weight (or importance) for the cluster — he finds that one cluster (c_1) is about {Yamaha, Honda, car, bike, GM}, while another (c_2) is about {parts, power-steering, door, power-windows}. Using his domain knowledge, he understands that the c_1 is about *Bikes & Cars*, while c_2 is about *Car Parts*.

In the first scenario, he likes these cluster descriptions, and goes on to inspect some of the individual posts contained in them. He notices that some posts 'truly' relating to c_1 (*Bikes & Cars*) — for example a few mentioning 'the best part about Honda' — have been assigned incorrectly by the algorithm to the *Car Parts* cluster c_2. He corrects this by appropriately re-assigning these posts to c_1. We call this an *assignment feedback* from the user, where he moves a data instance from one existing cluster to another. The clustering algorithm learns from this feedback — to add 'part' to the description of the first cluster c_1 as well, possibly with a small weight — so that other similar posts get correctly assigned.

In the second scenario, the user decides that he would rather prefer cluster c_1 to be about *Bikes* and cluster c_2 to be about *Cars & Car Parts*. To achieve this, he adjusts the description of the clusters, by changing c_1's description to {Yamaha, Honda, bike} and c_2's description to {car, GM, power-steering, door, power-windows, Honda}. We call this a *cluster description feedback*, where the user directly modifies the features in the cluster description according to his preference. Observe that it would not have been possible for the user to create these new cluster descriptions, without knowing the summaries provided by the existing cluster descriptions. Again, the algorithm learns from this feedback to correctly reassign the posts to appropriate clusters. We will assume in our formulation that the user provides a new weight vector for the cluster description, but in practice the user need not specify weights explicitly. For example, in a working system, he may simply rank order the top few features, or click and drag a weight curve over them. However, the details of the user interface is outside the scope of this paper, and we intend to investigate this in future work.

In general, the user is expected to provide both these types of feedback to the algorithm inter-actively, as new clusters emerge and documents get assigned to them. At any stage, the algorithm considers all feedback provided so far, even those at earlier stages, to recluster the documents.

We note that it is possible to provide cluster description feedback indirectly through assignment feedback, and vice versa. Re-assignment of points to centroids will automatically result from cluster description changes and similarly, many assignment constraints will lead to movement of the centroid in the next iteration. However we will see how using them directly to achieve the intended effect saves the user considerable time and effort.

3 Related Work

There has been a lot of research over the last decade on clustering with constraints [4]. The most popular approach provides pair-wise instance-level supervision, which is either used to learn distance metrics [19,16,2] or to provide additional constraints for the clustering algorithm [17,18,5,9,10,11]. Other work on cluster-level constraints looks to control size and balancing of clusters [9,1], or to find alternative clusters [14]. In contrast, we look to provide on-line supervision on descriptions and data assignments for existing clusters.

The idea of assignment feedback is closely related to active learning [7,8]. However, active learning assumes the classes in the data to be known apriori, while in our framework the user aims to simultaneously discover the clusters and the assignments to them in the spirit of exploratory data mining.

The concept of active supervision for clustering has been explored [15,6,3,12] but mostly for pair-wise constraints. Cohn et. al.[6] proposed the idea of inter-active clustering as early as 1999 in an unpublished manuscript. Though they suggest the possible use of cluster-item assignment feedback, the proposed model only incorporates pair-wise constraints. Similarly, desJardins et. al.[12] explore how user interaction with clusters, visualized in a two-dimensional space, can be interpreted as pair-wise constraints for clustering. To the best of our knowledge, the idea of using assignment feedback in conjunction with description feedback on current clusters to address high-dimensionality of the data is novel in the literature.

Many clustering objective functions more sophisticated than k-means distortion have been proposed. For example, co-clustering[13] looks to cluster the features and the items simultaneously. Note that this paper does not propose a new clustering objective function. Using k-means as an illustrative example, we have shown how cluster-level inter-action can be used to improve unsupervised clustering. Such interaction can be built on top of co-clustering and other proto-type based clustering models as well.

4 Problem Formulation

In the traditional k-means problem, we have a set of n data points $\{x_1, \ldots, x_n\}$, each drawn from domain $\mathcal{X}$. A clustering of these data points is defined by k clusters $\{c_1, \ldots, c_k\}$ with corresponding centers $\{\mu_1, \ldots, \mu_k\}$, and an assignment $\delta(c_i, x_j)$ of data points to clusters. We will consider each data point and the cluster center to be defined as weight vectors over the features of $\mathcal{X}$. Given such a clustering C, we can measure the distortion error for C as the summed distance of data points from their corresponding cluster centers:

$$E^x(C, \delta) = \sum_i \sum_j (\mu_i - x_j)^2 \delta(\mu_i, x_j) \tag{1}$$

Typically, given the set of data points, the goal of k-means clustering is to find the k cluster centers and an assignment of data points to clusters such that the total distortion error is minimized. In the rest of the formulation, we will not distinguish between clusters and centers. For example, we will use the notation μ to refer to both a center and its corresponding cluster.

In our version of the problem, we additionally have two different types of feedback provided by the user.

We have a set F^a of l assignment feedbacks, provided by the user possibly over different stages of the inter-active procedure. The ith assignment feedback f_i^a can be represented as $\{x_i^a, \boldsymbol{\mu}_i^a, \mu_i^a\}$, indicating that data point x_i^a is assigned by the user to a specific cluster μ_i^a from the set of current clusters $\boldsymbol{\mu}_i^a$.

We also have a set F^d of m cluster description feedbacks obtained from the user, again over various stages of the inter-active process. For the ith such feedback f_i^d, we assume that the user observes the top t features of a cluster ordered by weight (o_i^d) and provides his preferred feature vector (p_i^d) for it as feedback. We call t the observed description length for cluster description feedback. Accordingly, we represent f_i^d as $\{o_i^d, p_i^d\}$, where o_i^d is an ordered set of features and p_i^d is a weight vector over all features.

Though each feedback is provided at a specific stage of the interaction for a specific set of clusters C, it is desirable to make use of it at later stages when the current set of clusters is C', depending on how different C' is from C. Therefore, at any stage of the clustering, we consider *all* feedbacks that have been provided up to that stage.

In the presence of these two sets of feedbacks from the user, our reformulated clustering goal is to conform with these feedbacks as much as possible, while still maintaining low distortion error. In order to capture this in our objective function, we associate one constraint for each feedback and a penalty that the clustering algorithm has to pay for violating that constraint.

Let us first consider an assignment feedback f_i^a. The most specific constraint that arises from it is that every time the current set of clusters exactly matches $\boldsymbol{\mu}_i^a$, the data point x_i^a always has to be assigned to the specific cluster μ_i^a from among them. The penalty for violating this constraint is the distance between μ_i^a and the cluster $\delta(x_i^a)$ to which x_i^a is assigned instead. (Note that, without ambiguity, we have overloaded the symbol δ to use $\delta(x_i^a)$ as a function that returns a specific cluster.) However, this very specific interpretation would render this constraint irrelevant at later stages of the clustering when the current set of clusters is even slightly different from $\boldsymbol{\mu}_i^a$. To get around this, we define the relevance $R_i^a(C)$ of an assignment constraint f_i^a, given a current set of clusters C, as the 'similarity' of C with the set of clusters $\boldsymbol{\mu}_i^a$ specified in the feedback. For a 'similar' set of clusters C, there is no penalty when x_i^a is assigned to the cluster $N_i^a(C)$ which is 'nearest' to μ_i^a in the current cluster set C. If, however, x_i^a is assigned to some cluster $\delta(x_i^a)$ which is different from $N_i^a(C)$, then the clustering algorithm has to pay a penalty equal to the distance between $\delta(x_i^a)$ and $N_i^a(C)$. The total assignment error takes into account both the penalty and the current relevance of the constraint. The higher the relevance, the higher is the assignment error for violating the constraint.

In summary, the clustering error associated with an assignment feedback f_i^a is captured as

$$E_i^a(C,\delta) = (\delta(x_i^a) - N_i^a(C))^2 \times R_i^a(C) \tag{2}$$

The total error for the entire set of assignment feedbacks F^a is obtained by summing over the errors for the individual feedbacks: $E^a(C,\delta) = \sum_{i=1}^{l} E_i^a(C,\delta)$.

The relevance of the current set of clusters C to the feedback clusters $\boldsymbol{\mu}_i^a$ is measured using the best mapping $M_i^a(C)$ between the two sets of clusters. The weakness of such a mapping can be measured by the summed distances between mapped clusters from the two sets. The relevance is then defined using an exponential function as

$$R_i^a(C) = \exp(-(\sum_{\mu \in C}(\mu - M_i^a(\mu))^2) \tag{3}$$

Let us now consider a cluster description feedback $f_i^d \in F^d$. The most specific constraint that can be associated with f_i^d is that for any cluster μ from the current set of clusters C, if the observed description of μ is the same as the description o_i^d in the feedback, then the center μ of the cluster should match the user preferred weight vector p_i^d. Recall that the observed description is the ordered set top(μ,t) of top t features of μ. In case the current and preferred weight vectors (μ and p_i^d) over features do not match, the clustering algorithm has to pay a penalty equal to the distance between the two weight vectors.

As for the assignment feedback, the specificity of this interpretation would make f_i^d irrelevant for most clusters at later stages, where the top feature sequence top(μ,t) does not exactly match o_i^d. We deal with this, as before, by introducing a relevance measure $R_i^d(\mu)$ for each cluster description feedback f_i^d and any cluster μ. The higher the relevance of the feedback for any cluster, the higher is the description error for not conforming with it. The relevance $R_i^d(\mu)$ of a description feedback may be measured using various similarity measures defined for ordered sets:

$$R_i^d(\mu) = \text{RankSim}(\text{top}(\mu,t), o_i^d) \tag{4}$$

For simplicity, we presently define RankSim(s_1, s_2) to be 1 if the *unordered* sets corresponding to s_1 and s_2 match, and 0 otherwise. We are investigating better measures than reward subset matches, such as Jaccard Similarity.

In summary, the clustering error associated with each cluster description feedback f_i^d is given as:

$$E_i^d(C, \delta) = \lambda_i \sum_{\mu \in C} (\mu - p_i^d)^2 \times R_i^d(\mu) \tag{5}$$

where λ_i is the strength of the ith cluster description feedback. To appreciate the use of λ_i, observe that for assignment feedback, the algorithm can typically satisfy the user by assigning the feedback point to the user specified cluster. However, for description feedback, the points currently assigned to a cluster also affect the position of its new center in conjunction with the user specified description. λ_i is used to specify the relative importance of the user's feedback and the assigned points. We further elaborate on the role of λ_i in Section 5.

The total error for the entire set of cluster description feedbacks F^d is obtained by summing over the errors for the individual feedbacks: $E^d(C, \delta) = \sum_{i=1}^{m} E_i^d(C, \delta)$.

Finally, the total clustering error is the sum of the errors due to distortion, assignment constraints and cluster description constraints:

$$E(C, \delta) = E^x(C, \delta) + E^a(C, \delta) + E^d(C, \delta) \tag{6}$$

Our goal is to find the optimal combination of clusters and assignments that minimize this total error.

5 Interactive Clustering Algorithm

In this section, we look at an algorithm that iteratively alternates between interacting with the user to acquire feedback and minimizing the total error in Equation (6) considering all the feedback obtained from the user so far over all stages of the algorithm. Algorithms for proto-type or model based clustering typically follow an iterative alternating optimization style, where each step consists of two sub-steps - prototype update and re-assignment. In the following subsections, we describe how these two steps can be modified to handle user feedback, and how the user interacts with the algorithm to provide feedback.

Cluster Update: In the cluster update sub-step, the existing clusters C are updated based on the current assignment δ of data points to clusters, and the current relevance $R^a(C)$ and $R^d(C)$ of the feedbacks F^a and F^d. Unfortunately, the different clusters cannot be updated independently as for the traditional k-means algorithm. This is because the feedbacks introduce dependencies across clusters. As a result, we cyclically update each of the k clusters keeping the other $k-1$ clusters fixed. This procedure is repeated until all the clusters stabilize. When the other $k-1$ clusters are held fixed, along with the assignments and the feedback relevances, updating cluster μ_i to minimize total error becomes a quadratic optimization problem. Solving it leads to the following update step:

$$\mu = \frac{1}{Z} \times \sum_{x} x\delta(x, \mu) + \sum_{i=1}^{l} R_i^a(C)[\delta(x_i^a, \mu)N_i^a(C) +$$

$$\sum_{\mu'} I(\mu', N_i^a(C))\delta(x_i^a, \mu')\mu'] + \sum_{i=1}^{m} \lambda_i R_i^u(\mu) p_i^d$$

where $I()$ is the indicator function, and Z is an appropriate normalization term.

The update rule may be interpreted as follows.

The first term shows the traditional movement of the cluster towards the centroid of the data points currently assigned to it. The second and third terms demonstrate the dependence on the other current centers brought about by the assignment constraints. An assignment feedback f_i^a is relevant for cluster μ either if the feedback data point x_i^a is currently assigned to this cluster, or if μ is the currently preferred cluster $N_i^a(C)$ for feedback f_i^a. In the first case, the cluster μ moves towards that current cluster $N_i^a(C)$ which is the currently preferred cluster for the feedback f_i^a. This is reflected by the second term. In the other case, cluster μ tries to move closer to that cluster μ' to which the feedback point x_i^a is currently assigned. This is reflected by the third term. Both of these movements are influenced by the current relevance $R_i^a(C)$ of the assignment feedback in question.

The effect of the cluster description feedbacks is captured by the last term. For any description feedback f_i^d that is relevant for this cluster, the cluster moves closer to the preferred description p_i^d in the feedback. As before, this movement is also tempered by the relevance $R_i^u(\mu)$ of the feedback for this cluster.

Finally, the updated position of the cluster is the net effect of the influence of all the relevant assignment and description constraints, as well as all of the data points currently assigned to this cluster. Observe that in the update rule, the user preferred description p_i^d for a description feedback behaves similarly to any other data point assigned to the cluster, and would have minimal effect in determining its new position without the weighting term λ_i.

Once all of the k clusters have been iteratively updated and have stabilized, the relevance R^a and R^d of the assignment and description constraints is recalculated based on the updated cluster positions, according to Equation (3) and Equation (4) respectively.

Point Reassignment: In the re-assignment step, the assignment δ of the data points is recalculated based on the updated cluster positions and the current relevance of the constraints. The contribution to clustering error by assigning a data point x to an existing cluster μ can be calculated by considering the distance from the cluster, and, for any assignment feedback f_i^a specified on the point x, the distance of μ from the currently preferred cluster $N_i^a(C)$ for the feedback, and its current relevance $R_i^a(C)$:

$$(\mu - x)^2 + \sum_{i=1}^{l} R_i^a(C) I(x, x_i^a)(\mu - N_i^a(C))^2$$

where $I()$ is again the indicator function. The point is then assigned to that cluster μ among the k current clusters for which this assignment error is minimized. Observe that cluster description feedbacks do not influence the assignment of data points. Also observe that, unlike cluster updates, the reassignment of each data point can still be done independently of the other data points, as in the traditional k-means algorithm.

```
    Algo InteractiveCluster
    Params: Set of data points X, Int k

1.  Initialize F^a and F^d to empty set

    % Initialize clusters
2.  Initialize k clusters in C
3.  Iterate n times or until convergence
4.      Assign each data point in X to nearest cluster
5.      Recompute k clusters from assigned data points

    % Start inter-active k-means
6.  Iterate until user is satisfied with C
7.      Acquire new feedback and add to F^a and F^d
8.      Iterate n times or until convergence
9.          Iteratively update each of k clusters in C
                based on relevance R^a, R^d and assignment δ
10.         Re-calculate relevance R^a, R^d, based on updated clusters C
11.         Re-calcluate assignment of data points in X
                based on updated clusters C and relevance R^a

12. Return k clusters C and assignment δ
```

Fig. 2. High level pseudo-code describing the cluster-level inter-active k-means (CLIKM) algorithm

User Interaction: At each stage, after minimizing total error in Equation (6) considering all the feedback obtained so far, the algorithm returns the new set of clusters for inspection. The user browses over the new cluster descriptions and assignments and provides fresh feedback on them. While it may be possible for the user to inspect all cluster descriptions, or at least the ones that have changed significantly since his previous inspection, it is extremely unlikely that he can inspect cluster assignments of all data items. We will assume that he can provide provide only $n_f = n_a + n_d$ feedbacks at each interaction stage, where n_a is the number of assignment feedbacks and n_d is the number of description feedbacks. We assume $n_d = k$, which means he inspects all clusters, but $n_a \ll n$, where n is the number of data points. Currently, we assume the the user randomly selects n_a data points for inspecting and providing feedback. However, it is possible to do better than this, as in the case of active learning [7]. While we have done initial work on actively selecting the data points for presenting to the user for feedback, this is largely a subject of future research.

The overall cluster-level interactive k-means algorithm (**CLIKM**) is shown in Figure 2. The algorithm starts by creating an initial set of clusters in steps 2-5, based only on distortion error. Then at every step, it updates the clusters (step 9), recalculates the relevance of all feedbacks (step 10) and then re-assigns the data points based on the updated clusters and the relevance of the all constraints acquired so far (step 11). These steps are repeated until convergence based on the current set of feedbacks. The algorithm terminates when the user is satisfied with the current clustering. Otherwise, the algorithm acquires more feedback from the user (step 7) and reclusters the data points based on the feedback set, which now additionally includes the most recently received feedbacks.

6 A Supervisor Model

Large-scale evaluations with interactive algorithms with human supervisors require a lot of time and effort. In this section, as a substitute, we describe a parameterized supervisor model for our framework, based on gold-standard cluster labels on data points.

The challenge is that cluster-level supervision is conditioned on the *current* set of clusters, which may be different from the true clusters in the gold-standard. We assume that the supervisor is able to construct a correspondence between the true clusters T in the gold standard and the current clusters C available at any stage of the interactive process. This correspondence is found using a maximum weighted matching between the true clusters and the current clusters in bipartite graph, where the edge weight between a true cluster t and a current cluster c is the number of data points from t in c.

As the first supervisor parameter, we control the supervisor's knowledge about the exact description of a true cluster t using a parameter $p \in [0, 1]$. When averaging over documents in a true cluster t to construct its description, any specific document is included in the average computation with probability p, so that the user only has partial knowledge of t's description for $p < 1.0$.

The second supervisor parameter is a recognition threshold r for true clusters from computed clusters. For exploratory data mining, the supervisor often becomes aware of clusters existing in the data as they gradually emerge during the clustering process. We assume that the supervisor recognizes a true cluster t from the current cluster c only if c has ambiguity (measured as entropy over true clusters) below threshold r, and if t is the majority true cluster within c. At any stage of the clustering algorithm, the supervisor has a set $T_r \subseteq T$ of recognized true clusters, and is able map current classes only to these true clusters.

Now, when asked to provide assignment feedback for a data point x given current clusters C and true clusters T, the supervisor first retrieves the true cluster t for x, and then returns the corresponding current cluster. On the other hand, when asked for description feedback on a current cluster $c \in C$, the supervisor first retrieves the corresponding true cluster t, and then returns its inexact description based on his knowledge level p. Note that the supervisor can provide feedback only if the relevant true cluster t belongs in his recognized set of clusters T_r.

In the experimental evaluation of our interactive clustering algorithm in the next section, we consider supervisors with different recognition levels, as well as different levels of knowledge.

Convergence. An important issue that naturally arises for any interactive data mining task is that of convergence. While emphasizing that a detailed investigation of supervisor behavior and convergence is beyond the scope of this paper, here we briefly discuss conditions under which convergence can be guaranteed in the context of our supervisor model.

At a high level, the interactive clustering process converges if and only if the supervisor provides a *consistent sequence of feedbacks*. Here we provide a very strict definition of consistency. A sequence of feedbacks is consistent if all of the following conditions hold. First, the resulting sequence of recognized cluster sets is monotonically non-decreasing, i.e., a cluster once recognized by the supervisor cannot be decided as unrecognizable later. Secondly, for two cluster description feedbacks $f_i^d \equiv \{o_i^d, p_i^d\}$ and $f_j^d \equiv \{o_j^d, p_j^d\}$ provided at two different stages of CLIKM, if they are provided on the same recognized cluster ($o_i^d = o_j^d$), then the preferred descriptions also are the same ($p_i^d = p_j^d$). Thirdly, for two assignment feedbacks $f_i^a \equiv \{x_i^a, \boldsymbol{\mu}_i^a, \mu_i^a\}$ and $f_j^a \equiv \{x_j^a, \boldsymbol{\mu}_j^a, \mu_j^a\}$ provided at different stages, if they are provided for the same data point ($x_i^a = x_j^a$) given the same current set of clusters ($\boldsymbol{\mu}_i^a = \boldsymbol{\mu}_j^a$), then the preferred cluster also has to be the same ($\mu_i^a = \mu_j^a$). Under the above conditions, the interactive process is guaranteed to convergence.

We think that it is possible to relax the first two conditions, but note that third is an absolute requirement. Also, the number of iterations to convergence for a consistent feedback sequence depends on multiple factors, such as the rate of growth of the recognized cluster set, and for multiple assignment feedbacks on the same data point, the similarity between their set of clusters.

7 Experiments

In this section, we experimentally evaluate the effectiveness of our proposed interactive clustering framework. We considered two benchmark real-world text categorization datasets from two different domains, Twenty Newsgroups[1] and Reuters-21578[2]. The goal of the clustering task is to produce clusters that correspond to the gold-standard categories. Generally, this is not expected of unsupervised clustering. We investigate how accurately and efficiently the gold standard categories can be discovered when provided with semi-supervision. We also investigate the relative impact of assignment and cluster description feedback, and the effect of different supervisor parameters on our cluster-level interactive framework. We next describe our datasets, baselines and evaluation metrics before discussing experimental results.

Datasets: For our first dataset (8NG), we selected eight 20 Newsgroup classes (all of $rec.*$ and $sci.*$) having 1000 documents each. Our second dataset (R10) was created by selecting the top 10 categories from the Reuters-21578 corpus, and including all train/test documents, resulting in a collection of 9118 documents. We pre-processed all documents in a standard way using word stemming, and pruning stop-words and infrequent words (occurring less than 5 times in the dataset). We have made the actual processed datasets and their descriptions available online[3].

---

[1] http://people.csail.mit.edu/jrennie/20Newsgroups/

[2] http://www.daviddlewis.com/resources/testcollections/reuters21578/

[3] http://www.godbole.net/shantanu/work/ecml10iclust.html

Evaluation Metric: To evaluate a set of clusters against a gold-standard, we check the correctness of clustering decisions over all pairs of data points. We report the standard F1 measure and Adjusted Rand Index (ARI) over the pairwise clustering decisions. The F1 measure is the harmonic mean of precision and recall over pairwise decisions. The Adjusted Rand Index is defined as $2(ab-cd)/((a+d)(d+b)+(a+c)(c+b))$ where a is the number of true positive pairs, b is the number of true negative pairs, c is the number of false positive pairs and d is the number of false negative pairs. We also evaluated using Normalized Mutual Information and found the trends to be similar to that with F1 and ARI.

Baselines: As the first baseline for our cluster-level interactive k-means algorithm (**CLIKM**), we consider completely unsupervised k-means (**KM**). We also compare against pair-wise constrained clustering with instance-level must-link and cannot-link constraints (**PCC**) [3]. For this comparison, we incrementally provide pair-wise constraints to PCC and cluster-level constraints to CLIKM and compare their performance for the same number of provided constraints. Since PCC infers additional constraints from the provided ones using transitivity of must-link constraints, the actual number of constraints considered by PCC is much larger than the provided number. In contrast, the actual and provided number of constraints is the same for CLIKM. While doing this comparison, it needs to be borne in mind that the nature of the two constraints are quite different from each other. First, since CLIKM constraints are conditioned on the current set of clusters, they cannot be converted to an equivalent set of independent pair-wise constraints over data points. Secondly, the supervisor effort required to provide these two types of constraints will also be quite different, and can only be measured using extensive user studies. Keeping in mind these differences, we study the relative performance of PCC and CLIKM when given an equal — though not necessarily equivalent — number of constraints.

Since all of these algorithms only find local optima, when comparing any two, we provide them with the same initialization. All the reported plots are averaged over 10 runs.

Parameter Settings: As default parameters, we set observed cluster description length $t = 10$, supervisor recognition threshold $r = 0.95$ and knowledge level $p = 0.25$. The strength of description feedback λ is set to be the average cluster size (n/k), so that the importance of the supervisor's feedback is roughly the same as that of the points assigned to the cluster. We set the number of feedbacks at each step $n_f = 200$ for both CLIKM and PCC. (Trends are similar with $n_f = 100$ and 200) For CLIKM, description feedback is provided for all current clusters ($n_d = k$), so that assignment feedback is provided for $n_a = 200 - k$ random data points. For PCC, must-link or cannot-link feedback is provided for n_f random pairs of data points. Recall that the actual number of constraints considered by PCC is much larger than n_f, since PCC infers more constraints using transitivity of must-link constraints.

Experiment 1 - CLIKM vs Baselines: In our first experiment, we compare CLIKM with point assignment and cluster description feedback, PCC with pair-wise item-level feedback, and KM on two datasets. The results are shown in Figure 3, where clustering performance is plotted against the cumulative number feedbacks provided to PCC and CLIKM. The trends are similar for F1 and ARI as the evaluation measure. Expectedly,

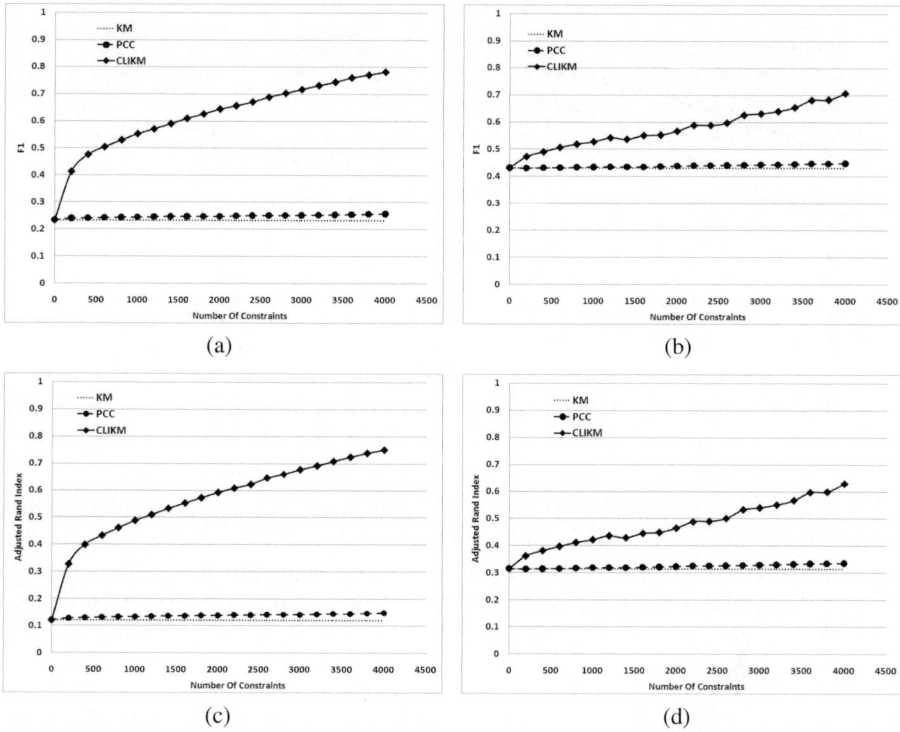

Fig. 3. Performance (F1 and ARI) of CLIKM and PCC vs number of feedbacks from the user against KM as baseline on 8NG (a,c) and R10 (b,d)

with increasing number of feedbacks, the performance CLIKM improves significantly over unsupervised KM. Performance improves most sharply at the beginning, so that after a few hundred feedbacks F1 increases from 0.22 to 0.4 for 8NG and from 0.4 to 0.47 for R10, and increases steadily, but at a slower rate, after that. This is because the user is able to recognize all true clusters during the very first interaction with the default recognition threshold $r = 0.95$. Also observe that performance of CLIKM drops slightly at a couple of places for R10 in Figure 3(b,d). This is due to the supervisor's inexact knowledge when providing description feedback ($p = 0.25$). We investigate the effect of the user's knowledge and recognition ability in greater detail later in the section.

Interestingly, the rate of performance improvement for CLIKM is significantly higher than PCC for both datasets. One potential reason for this is that the space of constraints is quadratic in the number of data items for PCC. In comparison, CLIKM needs at most a linear number of constraints for each clustering iteration, and the number of iterations is usually a constant. As a result, the user can drive the clustering towards his desired state with significantly fewer feedbacks using CLIKM.

In the rest of our experiments, we report only F1 numbers. All trends are similar with ARI.

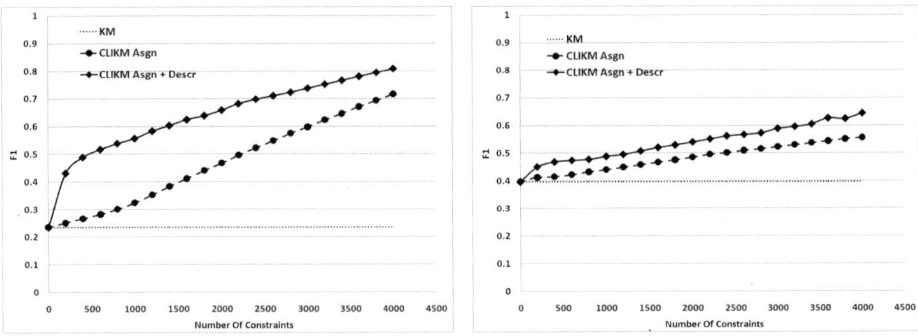

(a) (b)

Fig. 4. Effect of assignment and cluster description feedback on CLIKM for (a) 8NG and (b) R10

Experiment 2 - Ablation Study: In our second experiment, we perform an ablation study to evaluate the impact of assignment and cluster description feedback separately. The results are shown in Figure 4. The plots clearly shows that improvement brought about by cluster description feedback on top of that from assignment feedback. It demonstrates that cluster description feedback enables the user to guide the clustering much faster than when empowered only with assignment feedback.

Experiment 3 - Varying Supervisor Parameters: The success of interactive clustering depends a lot on the user's ability to recognize desired clusters, his knowledge about the correct description of these clusters and the strength of description feedback that he sets. We evaluate this over the next two experiments.

First, in Table 1, we record the effect of providing description feedback to CLIKM with different combinations of user knowledge p and strength of description feedback λ, after it has already received 3000 assignment feedbacks. The first trend is that performance improves with supervisor knowledge when λ is fixed, over all 3 columns. Recall that $\lambda = n/k$ corresponds to equally weighting user's cluster description feedback and the data points currently assigned to a cluster, so that the first two columns correspond to weighting the feedback 10 times and 2 times lower than the data respectively. The rows provide a more interesting insight. For reasonable supervisor knowledge, performance improves with higher λ. However, when supervisor knowledge is weak (first two rows), increasing λ hurts performance. This suggests that when the supervisor is not confident

Table 1. Clustering performance (F1) after 3000 feedbacks for varying user knowledge (p) and user confidence (λ) on R10

	$\lambda = n/10k$	$\lambda = n/2k$	$\lambda = n/k$
p=0.001	0.486	0.471	0.410
p=0.01	0.487	0.497	0.487
p=0.10	0.489	0.502	0.522
p=0.25	0.490	0.510	0.511

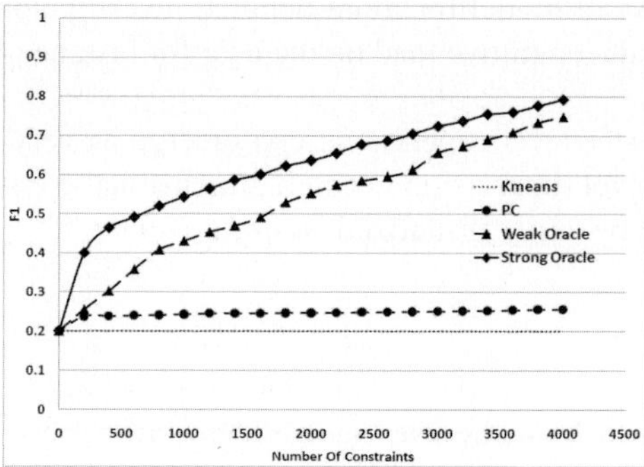

Fig. 5. F1 using strong ($r = 0.95$) and weak supervisor ($r = 0.9$) for 8NG

about his knowledge of the clusters, he should allow the data to influence the clustering more than his supervision.

Finally, we explore the impact of recognition threshold r of the supervisor. In Figure 5, we compare CLIKM performance with the default strong supervisor ($r = 0.95$) against that with a weak supervisor ($r = 0.9$). For the strong supervisor, performance improves significantly right at the beginning when all the true clusters are recognized and description feedback is provided for them. For the weak supervisor, only 50% of the true clusters are recognized at the first stage, and the rest at various later stages of the inter-active process, as marked by the jumps in performance. The gap between the two curves closes steadily as the inter-active process progresses.

Summary of Experiments: In summary, our experiments demonstrate that cluster-level semi-supervision leads to significant and steady improvements in clustering accuracy in comparison with unsupervised k-means. Improvements persist over varying levels of supervisor knowledge and cluster recognition ability. The rate of improvement is several times faster compared to that with an equal number pair-wise instance-level constraints.

8 Conclusions

In this paper we have proposed a novel semi-supervised model for interactive clustering, where the user provides two different types of feedback that align naturally with the update and assignment steps of prototype based clustering. Taking k-means as an example, we have shown how such feedback can be incorporated within prototype based objective functions as additional constraints to be satisfied. We have demonstrated the effectiveness of our model for clustering two real-life benchmark text datasets. Interestingly, our experiments show that performance increases significantly faster using this cluster-level semi-supervision compared to pair-wise instance-level supervision.

The proposed model can be further improved, for example by considering the source cluster in addition to the destination cluster for assignment feedback, and generalizing cluster description feedback with better measures of rank similarity. We also intend to improve on our supervisor model for more detailed investigation of convergence, and also better understand the cognitive load on the user for instance-level and cluster-level semi-supervision.

We believe that cluster-level supervision can emerge as very promising alternative to pair-wise instance-level supervision for high dimensional domains such as document collections that cannot be easily visualized.

References

1. Banerjee, A., Ghosh, J.: Scalable clustering algorithms with balancing constraints. Data Mining and Knowledge Discovery 13(3) (2006)
2. Bar-Hillel, A., Hertz, T., Shental, N., Weinshall, D.: Learning distance functions using equivalence relations. In: Proc. of ICML (2003)
3. Basu, S., Banjeree, A., Mooney, E.: Active semi-supervision for pairwise constrained clustering. In: Proc. of SDM (2004)
4. Basu, S., Davidson, I., Wagstaff, K.: Constrained clustering: Advances in algorithms, theory, and applications. Chapman and Hall/CRC Data Mining and Knowledge Discovery Series (2008)
5. Basu, S., Banerjee, A., Mooney, R.: Semi-supervised clustering by seeding. In: Proc. of ICML (2002)
6. Cohn, D., Caruana, R., McCallum, A.: Semi-supervised clustering with user feedback. Tech. rep., TR2003-1892, Cornell University (2003)
7. Cohn, D., Atlas, L., Ladner, R.: Improving generalization with active learning. Machine Learning 15(2) (1994)
8. Cohn, D., Ghahramani, Z., Jordan, M.: Active learning with statistical models. Journal of Artificial Intelligence Research 4(1) (1996)
9. Davidson, I., Ravi, S.: Clustering with constraints: Feasibility issues and the k-means algorithm. In: Proc. of SDM (2005)
10. Davidson, I., Ravi, S.: Identifying and generating easy sets of constraints for clustering. In: Proc. of AAAI (2006)
11. Davidson, I., Ravi, S.: Intractability and clustering with constraints. In: Proc. of ICML (2007)
12. desJardins, M., MacGlashan, J., Ferraioli, J.: Interactive visual clustering. In: Proc. of IUI (2007)
13. Dhillon, I., Mallela, S., Modha, D.: Information-theoretic co-clustering. In: Proc. of SIGKDD (2003)
14. Gondek, D., Hofmann, T.: Non-redundant data clustering. In: Proc. of ICDM (2004)
15. Hofmann, T., Buhmann, J.: Active data clustering. In: Proc. of NIPS (1998)
16. Klein, D., Kamvar, S., Manning, C.: From instance-level constraints to space-level constraints: Making the most of prior knowledge in data clustering. In: Proc. of ICML (2002)
17. Wagstaff, K., Cardie, C.: Clustering with instance-level constraints. In: Proc. of ICML (2000)
18. Wagstaff, K., Cardie, C., Rogers, S., Schrödl, S.: Constrained k-means clustering with background knowledge. In: Proc. of ICML (2001)
19. Xing, E., Ng, A., Jordan, M., Russell, S.: Distance metric learning, with application to clustering with side-information. In: Proc. of NIPS (2002)

Software-Defect Localisation by Mining Dataflow-Enabled Call Graphs

Frank Eichinger, Klaus Krogmann, Roland Klug, and Klemens Böhm

Karlsruhe Institute of Technology (KIT), Germany
{eichinger,krogmann,klemens.boehm}@kit.edu, klugr@ipd.uka.de

Abstract. Defect localisation is essential in software engineering and is an important task in domain-specific data mining. Existing techniques building on call-graph mining can localise different kinds of defects. However, these techniques focus on defects that affect the controlflow and are agnostic regarding the dataflow. In this paper, we introduce dataflow-enabled call graphs that incorporate abstractions of the dataflow. Building on these graphs, we present an approach for defect localisation. The creation of the graphs and the defect localisation are essentially data mining problems, making use of discretisation, frequent subgraph mining and feature selection. We demonstrate the defect-localisation qualities of our approach with a study on defects introduced into Weka. As a result, defect localisation now works much better, and a developer has to investigate on average only 1.5 out of 30 methods to fix a defect.

1 Introduction

Software quality is a huge concern in industry and in the software-engineering community. Software is rarely free from defects, and finding them is difficult. Especially when projects are large, several developers make changes in the source code, and a developer works with code somebody else has written, localising defects is tedious. (Semi-)Automatic tools for defect localisation are desirable. Clearly, such tools should be able to deal with many different kinds of defects.

In the past years, a number of studies has investigated defect localisation with *graph-mining techniques* [3,5,7,8,18]. They build on the analysis of *dynamic call graphs* (see [6] for an overview). One such graph is a concise representation of a programme execution and reflects the method-invocation structure. The localisation techniques do frequent subgraph mining with call graphs of correct and of failing executions. The rationale for defect localisation is that patterns occurring more frequently in graphs of failing executions contain methods which are more likely to be defective. Techniques that incorporate the analysis of call frequencies have proven to be more accurate and discover more types of defects than techniques without this feature [7]. See Figure 1 for a simple call graph representing a programme execution – each node stands for a method, edges represent method calls, and edge weights represent method-call frequencies.

An important characteristic of the existing call-graph-based techniques is that they merely analyse the call-graph structure and the call frequencies. They can

J.L. Balcázar et al. (Eds.): ECML PKDD 2010, Part I, LNAI 6321, pp. 425–441, 2010.

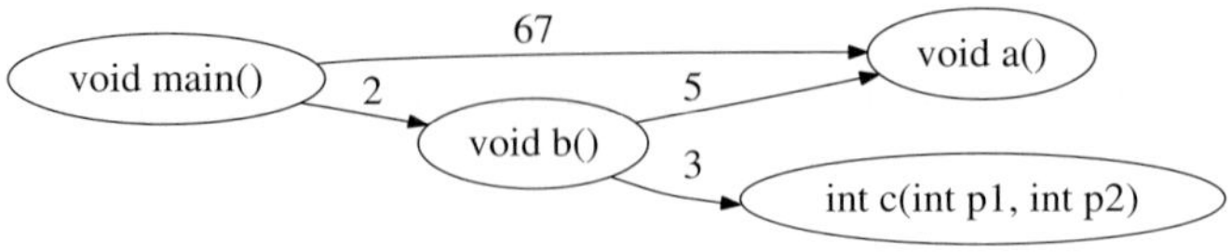

Fig. 1. Example call graph with call frequencies (not dataflow enabled)

only localise defects which affect the call graph of a programme execution (simplified, the controlflow). While this is an important class of defects, Cheng et al. [3] point out that the current techniques are agnostic regarding defects that influence the dataflow. We refer to such defects as *dataflow-affecting bugs*. They influence the data exchanged between methods. For example, think of a method which wrongly calculates some value, and which needs to be localised. A call-graph-based technique can only recognise such a defect if the value affects a control statement. Although this happens frequently, it might occur in methods which are actually defect-free, leading to erroneous localisations.

Han and Gao identify software engineering and defect localisation as a major area in domain-specific data mining [9]. They point out that the integration of domain knowledge (in our case, the exact specification of call graphs) and dedicated analysis techniques are crucial for the success of data mining in any domain. In this paper, we present a call-graph-based technique which localises both dataflow-affecting and call-graph-affecting bugs. The specification of the underlying graphs is not obvious: On the one hand, a call graph is a compact representation of an execution. On the other hand, dataflow-related information refers to values of many method calls within one execution. This information needs to be available at a level of detail which allows to locate defects. To illustrate the difficulties, an edge in a call graph typically represents thousands to millions of method calls. Annotating each edge with the method-call parameters and method-return values of all invocations corresponding to it incurs huge annotations and is not practical. In this paper we propose *dataflow-enabled call graphs* (*DEC graphs*) which incorporate concise numeric dataflow information.

DEC graphs are augmentations of call graphs with abstractions of method-call parameters and of method-return values. To obtain DEC graphs, we treat different data types differently. In particular, we discretise numerical parameter and return values. Figure 2 is a DEC graph corresponding to Figure 1. The call from method `b` to method `c` is attributed with a tuple of integers, containing the total number of calls and the numbers of calls with parameter and return values falling into different intervals. (We explain the details later.) When the DEC graphs are assembled, we do frequent subgraph mining with the graphs, not considering the dataflow abstractions for the moment. We then analyse the tuples of integers assigned to the edges with a feature-selection algorithm in the different subgraphs mined separately. Finally, we derive a likelihood of defectiveness for every method in the programme considered. These likelihoods form a ranking of suspicious methods which can guide the manual debugging process. We demonstrate the appropriateness and precision of our DEC-graph-based

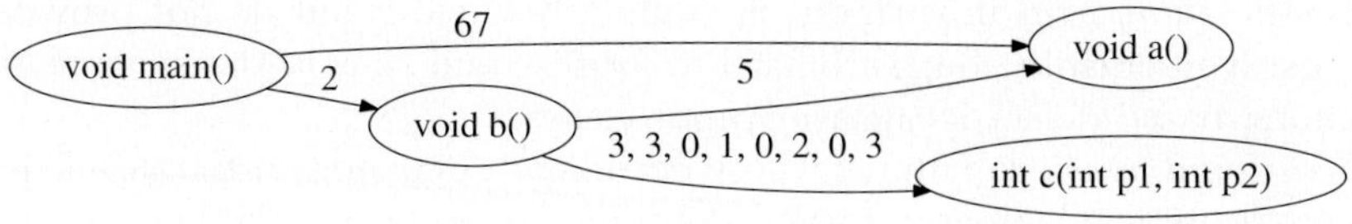

Fig. 2. Example dataflow-enabled call graph (DEC graph)

approach for the localisation of defects. In a case study we evaluate the approach using defects introduced into the Weka machine-learning suite [23].

All in all, our new technique for defect localisation features contributions at different stages of the analysis process and in the application domain:

Dataflow-Enabled Call Graphs. We introduce *DEC graphs* as sketched before, featuring dataflow abstractions. We describe an efficient implementation of their generation for Java programmes. To the best of our knowledge, this is the first study considering the dataflow within call graphs for defect localisation.

A Defect-Localisation Approach for Dataflow-Affecting Bugs. We present a defect localisation technique for DEC graphs. It is a case of weighted graph mining, which ultimately identifies defective methods.

Results in Software Engineering. We demonstrate the usefulness of our approach by means of defect-localisation experiments. Localisation works better with DEC graphs, as compared to graphs that are not dataflow enabled. Some defects can only be localised well with DEC graphs. Further, we describe and evaluate extensions of our approach which improve the defect localisation.

Paper Outline: Section 2 introduces the fundamentals of call-graph-based defect localisation. Sections 3 and 4 introduce DEC graphs and explain how we use them for defect localisation. Section 5 contains the experimental evaluation, Section 6 discusses related work, and Section 7 concludes.

2 Fundamentals of Call-Graph-Based Defect Localisation

In this section we first introduce our notion of bugs, followed by fundamentals on call-graph-based defect localisation.

Types of Bugs in Software. In the field of debugging, one usually avoids the terms *bug* and *fault*, but distinguishes between *defects*, *infections* and *failures* [26]. In this frequently-cited classification, *defects* are the places in the source code which cause a problem, an *infection* is an incorrect programme state (usually triggered by a defect), and *failures* are an observable incorrect programme behaviour (e.g., a user experiences wrong calculations). In the context of our study, we use a further differentiation, introduced in [6]:

- *Crashing bugs* lead to an unexpected termination of a programme. Examples include null-pointer exceptions and divisions by zero, which in various programming languages are easy to localise with the help of a stack trace.

Non-crashing bugs in turn display failing behaviour but do not provide any hints. Non-crashing bugs are hard to localise and thus in the focus of graph-mining-based defect-localisation approaches.
- *Occasional bugs* are failures whose occurrence depends on the input data of a programme. Compared to *non-occasional bugs*, occasional bugs are harder to find since they require more test cases to be reproduced.
- *Structure-affecting bugs* are infections which change the *structure* of a call graph. This is, when comparing graphs from correct and failing executions, certain graph substructures might or might not occur in either of the two variants. Such infections typically occur when `if` statements have defects or contain infected variables. In contrast, *call-frequency-affecting bugs* are infections which change the call *frequency* of a certain substructure in failing executions, rather than completely missing or adding structures. Such infections typically occur when loops are involved. Both structure-affecting and call-frequency-affecting bugs are also called *call-graph-affecting bugs*.

As any call-graph-based technique, we focus on *non-crashing* and *occasional bugs*. Opposed to other techniques, we consider *dataflow-affecting bugs* besides *call-graph-affecting bugs*. *Dataflow-affecting bugs* manifest themself by infected method-call parameters or return values, i.e., a method returns a wrong value. Dataflow-affecting bugs might affect the structure of a call graph and/or the call frequencies as a side effect. In this case, existing techniques can locate the defects in principle. However, infected behaviour often appears in other methods then those with the actual defect. This might disturb defect localisation. In such cases, our dataflow-enabled technique can deliver more precise localisations.

Localising Call-Graph-Affecting Bugs. In the past years, a number of call-graph-based techniques for defect localisation have been proposed [3,5,7,8,18]. Their basic idea is to mine for patterns in the call graph which are characteristic for failing executions. Then, they derive some defectiveness likelihood for the methods. We now briefly review the different call graph variants used, as well as the corresponding defect-localisation techniques. [6] is a detailed survey.

Existing techniques focus on structure-affecting bugs [3,5,7,18] and call-frequency-affecting bugs [7,8]. The graphs in [3,5,18] incorporate temporal information, the ones in [7,8] do not. In [6,7] we explain that the increased graph size when temporal information is incorporated leads to scalability issues when it comes to the mining. [3,8,18] build on call graphs where exactly one node represents a method, while [5,7] allow for more than one node. This promises more precise results, as more detailed contexts of method calls can be included in the analysis. At the same time, the size of the graphs increases, which also tends to increase runtime [6]. In contrast to the other representations, the graphs in [7,8] are weighted. Edge weights represent the number of corresponding method calls.

Besides different graph representations, the different approaches derive defectiveness likelihoods in different ways. [18] builds on graph classification with subgraph patterns. The authors first mine frequent subgraph patterns with a variant of CloseGraph [25] before they use them to train a support-vector machine (SVM). The authors consider the difference in accuracy between two

SVMs – one built with graph patterns including a certain method and one without them – as an evidence for defective methods. [3] builds on the same graphs, but relies on discriminative subgraph mining with the LEAP algorithm [24]. This directly pinpoints suspicious subgraph-structures and thus methods that are possibly defective. [5] derives defect likelihoods from support values of subgraph patterns in call graphs. Finally, [7] combines the idea from [5] to use support-based structural likelihoods with the analysis of call frequencies. While [5] incorporates only basic frequency-related information, [7,8] analyse call frequencies by means of a feature-selection algorithm. It does so in a step subsequent to graph mining with the CloseGraph algorithm [25]. This analysis allows to localise call-frequency-affecting and structure-affecting bugs. In [8] we successfully investigate the usage of call graphs for the localisation of defects in multithreaded programmes.

In this paper, we borrow concepts from both graph representations and localisation techniques from previous work. However, our graphs and our technique are different as they are tailored for the localisation of dataflow-affecting bugs – which is new in this study – in addition to call-graph-affecting ones. The shape of our graphs is similar to those in [3,8,18], but without temporal information. Furthermore, we generalise the concept of edge weights and their analysis [7,8] by introducing tuples of weights incorporating dataflow abstractions.

3 Dataflow-Enabled Call Graphs (DEC Graphs)

In this section we introduce and specify *dataflow-enabled call graphs* (*DEC graphs*) and explain how we obtain them. These graphs and their analysis (described in the following section) are the core of our approach to localise dataflow-affecting bugs. The basic idea of DEC graphs is to extend edges in call graphs with tuples which are abstractions of method parameters and return values. Obtaining these abstractions is a data-mining problem by itself: Huge amounts of values from method-call monitoring need to be condensed to enable a later analysis and ultimately the localisation of defects. We address this problem by means of discretisation. In the following, we first explain how we derive programme traces from programme executions. Next, we describe the structure of our call graphs. We then explain the dataflow abstractions and explain why they are useful for defect localisation. Finally, we say how we obtain the graphs from programme traces and give a concrete example.

Derivation of Programme Traces. We employ the aspect-oriented programming language AspectJ [13] to weave tracing functionality into Java programmes. By defining so-called pointcuts in AspectJ, we instrument method calls. At each call, we insert logging statements which save caller-callee relations. For each invocation, we log call frequency and data values (parameters and return values) that occur at runtime. Finally, we use this data to build call graphs.

When logging data values, we log primitive data types as they are, capture arrays and collections by their size, and reduce strings to their length. Such an abstraction from concrete dataflow has before successfully been used in the area of software performance prediction, e.g. [14]. Certainly, these simplifications can be severe, but logging the full data would result in overly large amounts of data. Our evaluation (Section 5) primarily studies primitive data types. A systematic evaluation of arrays, collections and strings as well as techniques for complex data types is subject to future work.

Call-Graph Structure. Based on the experience from previous studies [6], we decide to make use of a *total-reduction* variant of call graphs. This allows for better scalability of mining algorithms for large software projects and for an on-the-fly generation. In such graphs, every individual method which is called during a programme execution forms a single node (see Figure 1 for an example).

Dataflow Abstractions. As mentioned before, we use discretisation in order to find an abstraction of method parameters and return values based on the values monitored. Discretisation gives us a number of intervals for every parameter and for the return value (we discuss respective techniques in the following). We then count the number of method invocations falling into the intervals determined and attribute these counts to the edges.

Definition 1. *An* edge-weight tuple *in a* dataflow-enabled call graph (DEC graph) *consists of the counts of method calls falling into the respective intervals:*

$$(t, p_1^{i_1}, p_1^{i_2}, ..., p_1^{i_{n_1}}, p_2^{i_1}, p_2^{i_2}, ..., p_2^{i_{n_2}}, ..., p_m^{i_1}, p_m^{i_2}, ..., p_m^{i_{n_m}}, r^{i_1}, r^{i_2}, ..., r^{i_{n_r}})$$

where t is the total number of calls, $p_1, p_2, ..., p_m$ are the method-call parameters, r is the method-return value and $i_1, i_2, ..., i_{n_x}$ (n_x denotes the number of intervals of parameter/return value x) are the intervals of the parameters/return values.

The idea is that values referring to an infection tend to fall into different intervals than values which are not infected. For example, infected values might always be lower than correct values. Alternatively, infected values might be outliers which do not fall into the intervals of correct values as well. In order to be suited for defect localisation, intervals must respect correct and failing programme executions as well as distributions of values. Generally, it might be counter-productive to divide a value range like integer into intervals of equal size. Groups of close-by values of the same class might fall into different intervals, which would complicate defect localisation.

Derivation of Dataflow-Enabled Call Graphs (DEC Graphs). The CAIM (class-attribute interdependence maximisation) algorithm [15] suits our requirements for intelligent discretisation: It (1) discretises single numerical attributes, (2) takes classes associated with tuples into account (i.e., *correct* and *failing* executions in our scenario) and (3) automatically determines a (possibly) minimal number of intervals. Internally, the algorithm maximises the attribute-class interdependence. Comparative experiments by the CAIM inventors have demonstrated a high accuracy in classification settings.

In concrete terms, we let CAIM find intervals for every method parameter and return value of every method call corresponding to a certain edge. We do so for all edges in all call graphs belonging to the programme executions considered.

We then assemble the edge-weight tuples as described in Definition 1. Example 1 illustrates the discretisation. As we are faced with millions of method calls from hundreds to thousands of programme executions, frequently consisting of duplicate values, we pre-aggregate values during the execution. To avoid scalability problems, we then utilise a proprietary implementation of CAIM which is able to handle large amounts of data in pre-aggregated form. Note that the dataflow abstractions in DEC graphs can only be derived for a set of executions, as discretisation for a single execution is not meaningful.

Example 1. We consider the call of method `c` from method `b` in Figure 1 (Exec. 1 in Table 1) and three further programme executions (Exec. 2–4) invoking the same method with a frequency of one to three. Method `c` has two parameters `p1`, `p2` and returns value `r`. A discretisation of `p1`, `p2` and `r` based on the example values given in Table 1(a) leads to two intervals of `p1` and `r` ($p_1^{i_1}, p_1^{i_2}$ and r^{i_1}, r^{i_2}) and three for `p2` ($p_2^{i_1}, p_2^{i_2}, p_2^{i_3}$). See Table 1(b) for the exact intervals. The occurrences of elements of edge-weight tuples can then be counted easily – see Table 1(c), the discretised version of Table 1(a). The edge-weight tuple of `b` → `c` in Exec. 1 then is as displayed in Figure 2, referring to $(t, p_1^{i_1}, p_1^{i_2}, p_2^{i_1}, p_2^{i_2}, p_2^{i_3}, r^{i_1}, r^{i_2})$.

Table 1. Example discretisation for the call of `int c(int p1, int p2)` from `b`

(a) Example call data.

Exec.	p1	p2	r	*class*
1	2	43	12	*correct*
1	1	44	11	*correct*
1	3	4	9	*correct*
2	12	33	8	*failing*
3	23	27	6	*failing*
3	15	28	5	*failing*
3	16	23	7	*failing*
4	6	2	10	*correct*
4	11	47	13	*correct*

(b) Intervals generated.

Value	Intervals
p1	i_1 : [1, 11.5]
	i_2 : (11.5, 23]
p2	i_1 : [2, 13.5]
	i_2 : (13.5, 38]
	i_3 : (38, 47]
r	i_1 : [5, 8.5]
	i_2 : (8.5, 13]

(c) Discretised data.

Exec.	p1	p2	r
1	i_1	i_3	i_2
1	i_1	i_3	i_2
1	i_1	i_1	i_2
2	i_2	i_2	i_1
3	i_2	i_2	i_1
3	i_2	i_2	i_1
3	i_2	i_2	i_1
4	i_1	i_1	i_2
4	i_1	i_3	i_2

4 Localising Dataflow-Affecting Bugs

We now explain how to derive defect localisations from DEC graphs. We first give an overview, then describe subgraph mining (Section 4.1) and the actual defect localisation (Section 4.2) and two extensions (Sections 4.3 and 4.4).

Overview. Algorithm 1 works with a set of traces T of programme executions. At first, it assigns a class (*correct*, *failing*) to every trace $t \in T$ (Line 3), using a test oracle. Then the procedure generates DEC graphs from every trace t (Line 4). Next, the procedure derives frequent subgraphs of these graphs which are used as contexts where defects are located (Line 6). The last step calculates

Algorithm 1. Procedure of defect localisation with DEC graphs.

Input: a set of programme traces $t \in T$
Output: a method ranking based on each method's likelihood to be defective $P(m)$
1: $G = \emptyset$ // initialise a set of DEC graphs
2: **for all** traces $t \in T$ **do**
3: check if t was a correct execution and assign a *class* $\in \{$*correct*, *failing*$\}$ to t
4: $G = G \cup \{derive\_dataflow\text{-}enabled\_call\_graph(t)\}$
5: **end for**
6: $SG = frequent\_subgraph\_mining(G)$
7: calculate $P(m)$ for all methods m; based on SG

a likelihood of containing a defect for every method m (Line 7). This facilitates a ranking of the methods, which can be given to software developers. They would then review the suspicious methods manually, starting with the one which is most likely to be defective.

4.1 Frequent Subgraph Mining

As shown in Line 6 in Algorithm 1, we use frequent subgraph mining to derive subgraphs which are frequent within the call graphs considered. This particular step mines the pure graph structure only and ignores the edge-weight-tuples for the moment. The subgraphs obtained serve as different contexts, and further analyses are done for every subgraph context separately. This aims at a higher precision than an analysis without such contexts. For example, a failure might occur when method a is called from method b, only when method c is called as well. Then, the defect might be localised only in call graphs containing all methods mentioned, but not in graphs without method c.

We rely on the ParSeMiS implementation [21] of CloseGraph [25] for frequent subgraph mining. CloseGraph has successfully been used in related studies [7,18]. In a set of graphs G, it discovers subgraphs with a user-defined minimum support. For this value, we use $\min(|G_{\text{corr}}|, |G_{\text{fail}}|)/2$, where G_{corr} and G_{fail} are the sets of call graphs of correct and failing executions, respectively ($G = G_{\text{corr}} \cup G_{\text{fail}}$). This ensures not to miss any structure which occurs in less then half of all executions belonging to the smaller class. Preliminary experiments have shown that this minimum support allows for both short runtimes and good results.

4.2 Entropy-Based Defect Localisation

Next, we calculate the likelihood that a method contains a defect (Line 7 in Algorithm 1). The rationale is to identify methods which call other methods with discriminative parameter values or which have return values discriminating well between correct and failing executions. As mentioned before, we analyse every edge-weight tuple in the DEC graphs in the context of every subgraph mined. This aims at a high probability to actually reveal a defect, as every tuple is typically investigated in many different contexts. We assemble a table which contains the elements of the tuples of all edges in all subgraphs as columns

Table 2. Example feature table. g_1 refers to Exec. 1 from Example 1 (Figure 2)

	sg_1									sg_2		
Exec.	main → b	b → c								main → a	···	*class*
	t	t	$\frac{p_1^{i1}}{t}$	$\frac{p_1^{i2}}{t}$	$\frac{p_2^{i1}}{t}$	$\frac{p_2^{i2}}{t}$	$\frac{p_2^{i3}}{t}$	$\frac{r^{i1}}{t}$	$\frac{r^{i2}}{t}$	t		
g_1	2	3	1.00	0.00	0.33	0.00	0.67	0.00	1.00	67	···	*correct*
⋮	⋮	⋮	⋮	⋮	⋮	⋮	⋮	⋮	⋮	⋮	⋱	⋮
g_n	2	9	1.00	0.00	0.33	0.00	0.67	0.67	0.33	0	···	*failing*

and all programme executions (represented by their DEC graphs) as rows. The table cells contain the tuple values: the total call frequencies t and normalised interval frequencies. More precisely, we divide every interval frequency by the corresponding t in order to obtain the ratio of calls falling into each interval.

Table 2 is an example table which assumes that two subgraphs were found in the previous graph mining step, sg_1 (`main` → `b` → `c`) and sg_2 (`main` → `a`). The first column lists the call graphs $g \in G$. The second column corresponds to sg_1 and edge `main` → `b` with the total call frequency t. The following eight columns correspond to the second edge in this subgraph. Besides the total call frequency t, these columns represent intervals and are derived from the frequencies of parameter and return values. The very last column contains the class *correct* or *failing*. If a certain subgraph is not contained in a call graph, the corresponding cells have value 0, like g_n, which does not contain sg_2.

After assembling the table, we employ the *information-gain-ratio feature-selection* algorithm (*GainRatio*, [22]) in its Weka implementation [23] to calculate the discriminativeness of the columns and thus of the different edge-weight-tuple values. The *GainRatio* is a measure from information theory and builds – similar to information gain – on *entropy*. Values of *GainRatio* are in the interval $[0, 1]$. High values indicate a table column affected by a defect. Previous work [7,8] has shown that entropy-based measures are well-suited for defect localisation.

So far, we have derived defect likelihoods for every column in the table. However, we are interested in likelihoods for methods m, and every method corresponds to more than one column in general. This is due to the fact that a method can call several other methods and might itself be invoked from various other methods, in the context of different subgraphs. Furthermore, methods might have several parameters and a return value, each with possibly several intervals. To obtain method likelihood $P(m)$, we assign every column containing a total frequency t or a parameter-interval frequency p^{i} to the calling method and every return-value-interval frequency r^{i} to the callee method. We then calculate $P(m)$ as the maximum of the *GainRatio* values of the columns assigned to method m. By doing so, we identify the defect likelihood of a method by its most suspicious invocation and the most suspicious element of its tuple. Other invocations are less important, as they might not be related to a defect. The call context of a likely defective method and suspicious data values are supplementary information which we report to software developers to ease debugging.

Example 2 illustrates how our technique is able to localise *dataflow-affecting bugs* based on the ratios of executions falling into the different intervals of the method parameters and return values. Furthermore, it localises *call-frequency-affecting bugs* based on the call frequencies in the edge-weight tuples. In addition, our technique is able to localise most *structure-affecting bugs* as well: (1) The call structure is implicitly contained in the feature tables (e.g., Table 2) – value 0 indicates subgraphs not supported by an execution. (2) Such defects are frequently caused by control statements (e.g., `if`) evaluating previously wrongly calculated values. Our analysis based on dataflow can detect such situations more directly.

Example 2. The graphs g_1 and g_n in Table 2 display very similar values, but refer to a correct and a failing execution. Assume that method `c` contains a defect which occasionally leads to a wrongly calculated return value. This is reflected in the columns $\frac{r^{i_1}}{t}$ and $\frac{r^{i_2}}{t}$ of $\mathtt{b} \rightarrow \mathtt{c}$ in sg_1. The *GainRatio* measure will recognise fluctuating values in these columns, leading to a high ranking of method `c`.

4.3 Follow-Up-Infection Detection

Call graphs of failing executions frequently contain infection-like patterns which are caused by a preceding infection (not a defect). We call such patterns *follow-up infections.* We now describe an extension (for Line 7 in Algorithm 1) which detects certain follow-up infections and enhances the method ranking.

A follow-up infection occurs when a defective method m_1 calls another method m_2 which in turn calls method m_3. m_1 can often be localised because of the call frequency associated with edge $m_1 \rightarrow m_2$. However, $m_2 \rightarrow m_3$ might have a call frequency proportional to $m_1 \rightarrow m_2$. Thus, m_2 will have the same *GainRatio* as m_1. We make use of this observation and remove methods within the same subgraph belonging to $m_2 \rightarrow m_3$ from the ranking when the following conditions hold: (1) $\mathit{GainRatio}(m_1 \rightarrow m_2) = \mathit{GainRatio}(m_2 \rightarrow m_3)$ (we consider the *GainRatio* values from columns belonging to total call frequencies and parameters), and (2) $m_1 \rightarrow m_2 \rightarrow m_3$ is not part of a cycle within any $g \in G$. (2) is necessary as the origin of an infection cannot be determined within a cycle. However, our detection is a heuristic which is helpful in practice (see Section 5). In the presence of noise, this follow-up-infection detection might not work, and – pathologically – two edges might have the same *GainRatio* value by chance.

4.4 Improvements for Structure-Affecting Bugs

The subgraphs mined in Line 6 in Algorithm 1 can be used for an enhanced localisation of structure-affecting bugs. There are two kinds of such bugs: (1) those which lead to additional structures and (2) those leading to missing structures. To deal with both of them, we use the support *supp* of every subgraph *sg* in G_{corr} and G_{fail} separately to define two intermediate rankings. The rationale is that methods in subgraphs having a high support in either correct or failing executions are more likely to be defective. We again use the maximum:

$$P_{\mathrm{corr}}(m) := \max_{m \in sg \in SG} \mathit{supp}(sg, G_{\mathrm{corr}}); \; P_{\mathrm{fail}}(m) := \max_{m \in sg \in SG} \mathit{supp}(sg, G_{\mathrm{fail}})$$

With these two values, we define a structural score as follows:

$$P_{\text{struct}}(m) = |P_{\text{corr}}(m) - P_{\text{fail}}(m)|$$

To integrate P_{struct} into our *GainRatio*-based method ranking $P(m)$ (in Line 7 in Algorithm 1), we calculate the average:

$$P_{\text{comb}}(m) = \frac{P(m) + P_{\text{struct}}(m)}{2}$$

5 Experimental Evaluation

To investigate the defect-localisation capabilities of our approach, we use the Weka [23] machine-learning suite, manually add a number of defects to it, instrument the code and execute it using test-input data. Finally, we compare the defect ranking returned by our approach with the de-facto defect locations. Overall, we carry out six experiments[1]:

(E1) Application of the new approach featuring DEC graphs,
(E2) —— with follow-up-infection detection,
(E3) —— with follow-up-infection detection and structural ranking,
(E4) the same approach with call graphs that are not dataflow enabled,
(E5) —— with follow-up-infection detection, and
(E6) —— with follow-up-infection detection and structural ranking.

Experimental Setting. Weka is a data-intensive open-source application with a total of 19,938 methods and 301k lines of code (LOC). We introduce five different kinds of defects. They are of the same types as the defects in related evaluations, e.g., the *Siemens programmes* [10], which are often used [3,5,18] to evaluate defect localisation techniques for C programmes. Yet, a single Siemens programme comprises at most 566 LOC, which makes them unrealistically small and makes defect localisation less challenging.

The defect types introduced to Weka are typical programming mistakes, are non-crashing, occasional and dataflow-affecting and/or call-graph-affecting:

- *Variable assignment.* The assigned value of a variable differs from the correct value. An example for such a defect is `counter = a + b` where the correct code is `counter = a`.
- *Off-by-one.* This defect often happens when accessing an array or a collection. For example, `coll.get(i)` is accessed instead of `coll.get(i + 1)`.
- *Return value.* In this case, only the return statement of a method has a defect. For example, `return 0` is used instead of `return bestValue`.
- *Loop iterations.* This kind of defect affects the number of executions of a loop. For example, a `for` loop uses the wrong counter variable or misses an iteration: `for(int i = 0; i < max; i++)` instead of `for(int i = 0; i <= max; i++)`.

[1] (E4–6) essentially are a comparison to [7] ("total reduction"). We use the same localisation technique as with the DEC graphs for a fair comparison.

– *Branching condition.* This kind of defect covers wrong comparisons like `a > b` instead of `a < b`. Furthermore, the Boolean expressions `and`, `or`, `true` and `false` can be easily misplaced in branching conditions.

In total, we evaluate ten separate defects (Defect 1–10) as well as six combinations of two of these defects (Defects 11–16).[2] The defects introduced are 4x variable assignment, 3x return value, 1x off-by-one, 1x loop condition and 1x branch condition. Variable-assignment defects include array manipulations, string operations and inline variable assignments (e.g., `doSth(op(a) + b)`). Return values are manipulated by a value offset, returning a constant instead of a variable and by returning a wrong constant. We introduce some kinds of defects repeatedly to cover different characteristics of each defect. Defects 11–16 mimic typical situations where a programme contains more than one defect.

We introduce all defects in `weka.classifiers.trees.DecisionStump`. This class is the implementation of a decision-tree algorithm which comprises 18 methods. We emphasise that we instrument all 19,938 methods of Weka, and all of them are potential subjects to defect locations. A typical execution of `DecisionStump` involves a total of 30 methods.

We execute each defective version of Weka with 90 sets of sampled data from the UCI machine-learning repository [1] and classify correct and failing executions of the programme. To this end, we first execute a correct reference version of Weka with all 90 UCI data sets. After that, we execute the defective versions with the same data. We then interpret any deviation in the output of the two versions as a failure. The number of correct executions is in the same range as the number of failing ones. They differ by a factor of 2.7 on average and by 5.3 in the worst case.

Experimental Results. We present the results – the ranking position which pinpoints the actual defect – of the six experiments for all sixteen defects in Table 3. This position quantifies the number of methods a software developer has to review in order to find the defect (smaller numbers are preferred). We compare the experimental results pairwise between DEC graphs (E1–3) and non-DEC graphs (E4–6), as indicated by the arcs. A grey-coloured cell means worse results, non-coloured cells mean same or improved results. Bold-face rankings indicate same or improved results compared to the preceding row (separately for DEC/non-DEC graphs). In programmes with more than one defect (i.e., Defects 11–16), we present numbers corresponding to the defect ranked best. This reflects that a developer would first fix one defect, before applying our technique again. Sometimes two or more methods have the same defect likelihood. In this case, we use the worst ranking position for all methods with the same likelihood. This is in line with the methodology of related studies [11].

The experiments clearly show the improved defect-localisation capabilities of the new approach based on DEC graphs. Even without extensions (E1), a top ranking is obtained in 15 out of 16 cases. We consider a method ranked top

[2] We provide the defective programme versions online: `http://www.ipd.kit.edu/~eichi/papers/eichinger10software-defect/`

Table 3. Defect-localisation results. (E2/3/5/6) incl. follow-up, (E3/6) incl. struct

Experiment \ Defect	1	2	3	4	5	6	7	8	9	10	11	12	13	14	15	16	∅
(E1) DEC graphs	3	3	1	3	2	2	12	3	1	1	2	2	1	2	1	3	2.3
(E2) DEC graphs	**2**	**2**	**1**	**2**	**2**	**2**	**9**	**2**	**1**	**1**	**2**	**2**	**1**	**2**	**1**	**2**	**1.9**
(E3) DEC graphs	**1**	**1**	**1**	**2**	6	**1**	**1**	**1**	3	5	**1**	**1**	**1**	**1**	**1**	**1**	**1.6**
(E4) Non-DEC graphs	1	1	11	13	10	3	13	10	9	6	3	8	1	3	8	10	6.1
(E5) Non-DEC graphs	**1**	**1**	**4**	**5**	**4**	**2**	**7**	**3**	**5**	**4**	**2**	**4**	**1**	**2**	**3**	**3**	**2.8**
(E6) Non-DEC graphs	**1**	**1**	**1**	**2**	7	**1**	**2**	**1**	8	6	**1**	**1**	**1**	**1**	**1**	**1**	**2.0**

when a developer has to investigate only 3 methods out of the 30 ones actually executed. With non-DEG graphs (E4), only 6 defects are ranked top. In only 5 out of 48 measurement points, compared to 26 out of 48 ones, the DEC-graph-based approach is worse than the reference. DEC graphs have reached a top ranking in 44 cases, whereas non-DEC graphs had a top ranking in only 28 cases. When directly comparing DEC graphs (E1) with non-DEC graphs (E4) without extensions, the defect localisation was better in 13 out of 16 cases. Furthermore, looking at the average values ('∅'), the number of methods to be investigated could be reduced by more than half.

Using the follow-up detection (E2/5), the ranking could be improved in all cases or has generated results of the same quality compared to the respective initial approach. This is remarkable, as the follow-up-infection detection is a heuristic approach. The use of both the follow-up and structural extension (E3/6) results in further improvements. For DEC graphs (E3) in comparison to (E2), the extension improves the ranking in 9 cases and lowers the ranking in 3 cases, i.e., better overall results. For non-DEC graphs (E6) in comparison to (E5), the picture is similar: 10 improved cases and 3 worse ones.

Analysing the localisation capabilities per defect class results in an inhomogeneous picture when looking at (E1–3). For the variable-assignment Defects 1, 3, 5, 6, localisation is mostly fine; the off-by-one Defect 2 is well located; return-value Defects 4, 8, 9 can be located with both approaches. Only for Defect 9 we achieve a further improvement compared to non-DEC graphs. The structural extension is misleading for the localisation of branch-condition Defect 10, while it enables the identification of the loop-iteration Defect 7 (a structure-affecting bug; the changed loop condition hinders the loop from being executed and thus other methods from being called).

Regarding the Weka versions with two defects (Defects 11–16), defect localisation always works better on average than for versions with only one defect (E1–6). Our explanation is that defect localisation has a higher chance to be correct when two methods have a defect.

Overall, the experiments show a large improvement of the ranking with the new approach. In combination with follow-up detections and the structural ranking (E3), results are best. Using the structural ranking leads to a slightly worse ranking for some defects. The experiments also show that only 1.6 out of the

19,938 methods of Weka (of which 30 methods are actually executed) must be investigated on average in order to find a defect (E3). The results promise a strong reduction of time spent on defect localisation in software-engineering projects.

Improved Experimental Results using Static Analysis. Despite the good results achieved so far, we investigate further improvements. One starting point is the handling of methods with the same defect likelihood. As in related studies [11], we currently use the worst ranking position for all methods which have the same defect likelihood. A second static ranking criterion helps distinguishing methods with the same defect likelihood: We sort such methods decreasingly by their size in lines of code (LOC)[3]. Research has shown that the size in LOC frequently correlates with the defectiveness likelihood [20]. Applied to our experiments, we can observe an improvement of the average ranking position as follows: 2.3 to 1.9 (E1), 1.9 to 1.7 (E2), 1.6 to 1.5 (E3), 6.1 to 3.6 (E4), 2.8 to 2.6 (E5) and 2.0 to 1.9 (E6). Although the additional static ranking criterion leads to improvements in all experiments, the non-DEC graphs (E4–6) benefit from the improved ranking to a larger extent. As feature selection for non-DEC graphs considers fewer columns, the defect likelihood of methods has fewer different values than for DEC graphs, and this more frequently leads to equal rankings. However, even after the combination with static analysis, defect localisation with DEC graphs is always better on average than with non-DEC graphs. The same observations as described in the preceding paragraphs hold.

6 Related Work

Defect-localisation techniques are *static* or *dynamic*. Dynamic techniques rely on the analysis of programme runs (like our approach) while static techniques do not require any execution and investigate the source code only.

Static Analysis. *Mining software repositories* maps post-release failures from a bug database to defects in static-source code. For example, [20] derives standard code metrics and builds regression models which then predict possible post-release failures. Such approaches require a large collection of defects and extensive version-history data.

FindBugs [2] is an approach complementary to ours. Its static-code analyses for Java generally cannot localise most dataflow-related defects. Regarding the defects in our evaluation (Section 5), FindBugs recognises none of them. Instead, it aims at defects like possible null-pointer accesses due to missing initialisation.

Dynamic Analysis. Approaches in this category are based on instrumentation, like our approach. Such approaches tend to either have a large memory footprint or do not capture all defects due to selective logging of executions.

Statistical defect localisation is a family of dynamic techniques which utilise pattern detection on data values monitored during execution. Liblit et al. [16]

[3] Here we use the sum of non-blank and non-comment LOC inside method bodies.

analyse monitored data values using regression techniques to identify defective code. Compared to our work, Liblit et al. record only three intervals for return values of methods. We use a variable number of dynamically identified intervals for data characterisations. The approach reduces its footprint by collecting only small samples of executed programmes. A similar approach by Liu et al. [17] focuses on controlflow and instruments variables in condition statements. It then calculates a ranking which yields high values when the evaluation of these statements differs significantly in correct and failing executions. Opposed to our approach, none of these approaches takes structural properties of call graphs into account. Hence, structure-affecting bugs can be detected less easily.

Analysis of execution traces is the basis for call-graph-based methods. Tarantula [11,12] is a technique using tracing and visualisation. To localise defects, it utilises a ranking of programme components which are executed more often in failing programme executions. Though this technique is rather simple, it produces good defect-localisation results. Our technique comprises the method-invocation structure and dataflow information, which is not covered by [11,12].

Masri [19] performs a dynamic information-flow analysis to localise defects in source code. Specifically, sub-paths of information flow of correct and failing executions are compared, to rank defect positions. Information-flow sub-paths comprise frequency, source and target types (e.g., branch, statement), and the length of the information-flow path executed. Opposed to [19], our approach deals with abstractions of data values in the dataflow analysis and not only relies on the relation of correct and failing executions for defect localisation.

7 Conclusions and Future Work

Defect localisation is essential in software engineering, but very time-consuming. (Semi-)Automated localisation therefore is desirable. We have presented an approach based on newly introduced dataflow-enabled call graphs (DEC graphs). It outperforms existing techniques. It targets at defects which affect the dataflow or the controlflow of a programme. Both technical contributions of our approach, the generation and the analysis of DEC graphs, rely on data-mining techniques. Our approach generates DEC graphs by means of meaningful discretisation and derives defect localisations with a weighted graph-mining approach.

Our experiments have shown that the approach may significantly reduce the time required to localise defects in software. On average, only 1.5 out of the 30 methods executed in the case-study system must be investigated to fix a defect.

Future work will extend the approach: (1) Currently, global variables are not handled. Static code analysis might help to identify global variables that are read within a method. They can then be treated like method-call parameters. (2) We plan to investigate an integration with ideas from Masri's approach [19]. Furthermore, as mentioned in Section 3, we will investigate non-primitive data types and 'real defects', originating from open-source software projects [4].

References

1. Asuncion, A., Newman, D.J.: UC Irvine Machine-Learning Repository, http://archive.ics.uci.edu/ml/
2. Ayewah, N., Hovemeyer, D., Morgenthaler, J.D., Penix, J., Pugh, W.: Using Static Analysis to Find Bugs. IEEE Softw. 25(5), 22–29 (2008)
3. Cheng, H., Lo, D., Zhou, Y., Wang, X., Yan, X.: Identifying Bug Signatures Using Discriminative Graph Mining. In: Proc. Int. Symposium on Software Testing and Analysis, ISSTA (2009)
4. Dallmeier, V., Zimmermann, T.: Extraction of Bug Localization Benchmarks from History. In: Proc. Int. Conf. on Automated Software Engineering, ASE (2007)
5. Di Fatta, G., Leue, S., Stegantova, E.: Discriminative Pattern Mining in Software Fault Detection. In: Proc. Int. Workshop on Software Quality Assurance (2006)
6. Eichinger, F., Böhm, K.: Software-Bug Localization with Graph Mining. In: Aggarwal, C.C., Wang, H. (eds.) Managing and Mining Graph Data. Springer, Heidelberg (2010)
7. Eichinger, F., Böhm, K., Huber, M.: Mining Edge-Weighted Call Graphs to Localise Software Bugs. In: Daelemans, W., Goethals, B., Morik, K. (eds.) ECML PKDD 2008, Part I. LNCS (LNAI), vol. 5211, pp. 333–348. Springer, Heidelberg (2008)
8. Eichinger, F., Pankratius, V., Große, P.W.L., Böhm, K.: Localizing Defects in Multithreaded Programs by Mining Dynamic Call Graphs. In: Proc. Testing: Academic and Industrial Conference – Practice and Research Techniques (2010)
9. Han, J., Gao, J.: Research Challenges for Data Mining in Science and Engineering. In: Kargupta, H., Han, J., Yu, P.S., Motwani, R., Kumar, V. (eds.) Next Generation of Data Mining. Chapman & Hall/CRC (2008)
10. Hutchins, M., Foster, H., Goradia, T., Ostrand, T.: Experiments on the Effectiveness of Dataflow- and Controlflow-Based Test Adequacy Criteria. In: Proc. Int. Conf. on Software Engineering, ICSE (1994)
11. Jones, J.A., Harrold, M.J.: Empirical Evaluation of the Tarantula Automatic Fault-Localization Technique. In: Proc. Int. Conf. on Automated Software Engineering, ASE (2005)
12. Jones, J.A., Harrold, M.J., Stasko, J.: Visualization of Test Information to Assist Fault Localization. In: Proc. Int. Conf. on Software Engineering, ICSE (2002)
13. Kiczales, G., Hilsdale, E., Hugunin, J., Kersten, M., Palm, J., Griswold, W.G.: An Overview of AspectJ. In: Knudsen, J.L. (ed.) ECOOP 2001. LNCS, vol. 2072, p. 327. Springer, Heidelberg (2001)
14. Krogmann, K., Kuperberg, M., Reussner, R.: Using Genetic Search for Reverse Engineering of Parametric Behaviour Models for Performance Prediction. IEEE Trans. Softw. Eng. (accepted for publication, to appear 2010)
15. Kurgan, L.A., Cios, K.J.: CAIM Discretization Algorithm. IEEE Trans. Knowl. Data Eng. 16(2), 145–153 (2004)
16. Liblit, B., Aiken, A., Zheng, A.X., Jordan, M.I.: Bug Isolation via Remote Program Sampling. SIGPLAN Not. 38(5), 141–154 (2003)
17. Liu, C., Yan, X., Fei, L., Han, J., Midkiff, S.P.: SOBER: Statistical Model-Based Bug Localization. SIGSOFT Softw. Eng. Notes 30(5), 286–295 (2005)
18. Liu, C., Yan, X., Yu, H., Han, J., Yu, P.S.: Mining Behavior Graphs for "Backtrace" of Noncrashing Bugs. In: Proc. SDM (2005)
19. Masri, W.: Fault Localization Based on Information Flow Coverage. Softw. Test., Verif. Reliab. 20(2), 121–147 (2009)

20. Nagappan, N., Ball, T., Zeller, A.: Mining Metrics to Predict Component Failures. In: Proc. Int. Conf. on Software Engineering, ICSE (2006)
21. Philippsen, M., et al.: ParSeMiS: The Parallel and Sequential Mining Suite, `http://www2.informatik.uni-erlangen.de/EN/research/ParSeMiS/`
22. Quinlan, J.R.: C4.5: Programs for Machine Learning. Morgan Kaufmann, San Francisco (1993)
23. Witten, I.H., Frank, E.: Data Mining: Practical Machine Learning Tools and Techniques with Java Implementations. Morgan Kaufmann, San Francisco (2005)
24. Yan, X., Cheng, H., Han, J., Yu, P.S.: Mining Significant Graph Patterns by Leap Search. In: Proc. SIGMOD (2008)
25. Yan, X., Han, J.: CloseGraph: Mining Closed Frequent Graph Patterns. In: Proc. KDD (2003)
26. Zeller, A.: Why Programs Fail: A Guide to Systematic Debugging. Morgan Kaufmann, San Francisco (2009)

Induction of Concepts in Web Ontologies through Terminological Decision Trees

Nicola Fanizzi, Claudia d'Amato, and Floriana Esposito

Dipartimento di Informatica
Università degli studi di Bari "Aldo Moro"
Campus Universitario – Via Orabona 4, 70125 Bari, Italy
{fanizzi,claudia.damato,esposito}@di.uniba.it

Abstract. A new framework for the induction of logical decision trees is presented. Differently from the original setting, tests at the tree nodes are expressed with Description Logic concepts. This has a number of advantages: expressive terminological languages are endowed with full negation, thus allowing for a more natural division of the individuals at each test node; these logics support the standard ontology languages for representing knowledge bases in the Semantic Web. A top-down method for inducing *terminological decision trees* is proposed as an adaptation of well-known tree-induction methods. This offers an alternative way for learning in Description logics as concept descriptions can be associated to the terminological trees. A new version of the System TermiTIS, implementing the methods, is experimentally evaluated on ontologies from popular repositories.

1 Introduction

Among the other facets, the Semantic Web (SW) is a Web of Data[1]. Countless structured data sets are distributed all over the Web and containing all kinds of information. They are the property of companies or institutions that tend to make them broadly accessible. Next generation knowledge bases are envisioned to be based on shared ontologies and a number of distributed repositories that contain resources which are annotated using the concepts and properties defined in terms of such ontologies expressed through such standardized representations.

The collection of SW technologies (RDF/S, OWL, etc.) provides an environment where applications can query that data, draw inferences using vocabularies, etc. They are ultimately based on Description Logics (DLs). These languages constitute particular fragments of Predicate Logic (with extensions to higher order logics for the most expressive ones). They differ from the clausal representations for having a variable-free syntax and especially the *open-world semantics* [1] which makes them particularly well suited for Web-scale distributed scenarios.

These considerations justify the growing interest in investigating machine learning and knowledge discovery methods that cope with these formalisms.

[1] See http://www.w3.org/standards/semanticweb/data

J.L. Balcázar et al. (Eds.): ECML PKDD 2010, Part I, LNAI 6321, pp. 442–457, 2010.

Early works on learning in DLs essentially focused on the learnability of terminological languages, like CLASSIC and its successors [2], that are the ancestors of the current DL languages. More recently, DL concept learning approaches based on refinement operators have been investigated [3]. In [4] a coverage algorithm is presented whose downward operator exploits the notion of *counterfactuals*. This has been exploited also for other related tasks, such as conceptual clustering [5]. Since the algorithm works on examples expressed as *most specific concepts* (MSCs) [1], it tends to provide correct yet excessively complex concepts definitions. To avoid this problem, other top-down refinement algorithm, based on new downward operators, utilizes a syntactic heuristic to guide the search towards correct definitions of limited complexity [6]. A similar approach is followed in DL-FOIL [7], which adapts the well-known learning method to the different representation.

The induction of decision trees is among the most well-known machine learning techniques [8], also in its more recent extensions that are able to work with more expressive logical representations in clausal form [9]. In this work, the general framework is extended to to cope with yet more different logical representations as those designed for formal Web ontologies. We adopt an expressive DL language for representing the tests at the tree nodes of a logical decision tree. This allows to express different concepts w.r.t. the original clausal representation and is naturally compliant with the open-world semantics which is required by the evolving distributed environments for SW applications.

The tree-induction algorithm adopts a classical top-down *divide-and-conquer* strategy [10] which differs from previous DL concept learning methods based on sequential covering or heuristic search, with the use of refinement operators for DL concept descriptions [4, 7, 6]. Once a terminological tree is induced, similarly to the logical decision trees, a definition of the target concepts can be drawn exploiting the nodes in the tree structure. The algorithm has also a useful side-effect: the suggestion of new intermediate concepts which may have no definition in the current ontology. From a technical viewpoint, the likely chance that an instance a might not be assigned[2] to a given concept C (nor to its complement) requires a different setting for the learning problem [7], that is similar to learning with unknown class attributes [11], with a special treatment of the cases of uncertain classification mentioned above.

The resulting system, TERMITIS (TERMI*nological* T*ree* I*nduction* S*ystem*), ver. 1.2, was applied, for comparative purposes, to ontologies that have been considered in previous experiments with other DL learning systems [4, 7]. As they are real ontologies artificial learning problems were crafted by randomly building concept descriptions and determining the respective training and test sets. This also demonstrates the usage of the method as a means for performing approximations of concepts across ontologies (even when they are described with different languages [1]). Standard performance indices require that the class-membership can be determined for each test instance. This is not possible when the open-world semantics is assumed (as discussed before), then we resort to

[2] For the open-world semantics, $\neg C(a)$ does not necessarily follow from $\not\models C(a)$.

different indices, measuring the alignment of the classification decided by the inductive model with the one found deductively by a DL reasoner [7]. This allows measuring the amount of unlabeled instances that may be ascribed to the newly induced concepts (or to their complements), which may constitute a real added value brought by inductive methods to DL reasoning, although this sort of *abductive* conclusions would have to be validated by a domain expert.

The paper is organized as follows. After the next section introducing the representation, in Sect. 3 the DL learning problem is formalized and discussed. Sect. 4, presents the terminological tree model and the algorithms for growing them and for deriving concept descriptions. In Sect. 5 the experiments proving the effectiveness of the approach are reported. Finally, possible developments are discussed in Sect. 6.

2 Description Logics: Syntax and Semantics

In this section we shortly recall syntax and semantics of the DL representation. For brevity, we cannot report syntax and semantics of the various constructors, which can be easily be found in the reference manual [1]. In turn, the DL concept descriptions are straightforwardly mapped onto XML serializations of the standard ontology languages.

Roughly, the terminological formalisms are concept-centric: they distinguish *concepts* from *relations* that are used to describe restrictions on concepts. In a DL language, primitive *concepts* $N_C = \{C, D, \ldots\}$ are interpreted as subsets of a domain of objects (resources) and primitive *roles* $N_R = \{R, S, \ldots\}$ are interpreted as binary relations on such a domain (properties). *Individuals* represent the objects through names chosen from the set $N_I = \{a, b, \ldots\}$. Complex concept descriptions are built using atomic concepts and primitive roles by means of specific constructors. The meaning of the descriptions is defined by an *interpretation* $\mathcal{I} = (\Delta^{\mathcal{I}}, \cdot^{\mathcal{I}})$, where $\Delta^{\mathcal{I}}$ is the *domain* of the interpretation and the functor $\cdot^{\mathcal{I}}$ stands for the *interpretation function*, mapping the intension of concepts and roles to their extension (respectively, a subset of the domain and a relation defined on such domain).

The *top* concept $\top$ is interpreted as the whole domain $\Delta^{\mathcal{I}}$, while the *bottom* concept $\bot$ corresponds to $\emptyset$. Complex descriptions can be built in $\mathcal{ALC}$ using the following constructors[3]. The language supports *full negation*: given any concept description C, denoted $\neg C$, it amounts to $\Delta^{\mathcal{I}} \setminus C^{\mathcal{I}}$. The *conjunction* of concepts, denoted with $C_1 \sqcap C_2$, yields an extension $C_1^{\mathcal{I}} \cap C_2^{\mathcal{I}}$ and, dually, concept *disjunction*, denoted with $C_1 \sqcup C_2$, yields $C_1^{\mathcal{I}} \cup C_2^{\mathcal{I}}$. Finally, there are two restrictions on roles: the *existential restriction*, denoted with $\exists R.C$, and interpreted as the set $\{x \in \Delta^{\mathcal{I}} \mid \exists y \in \Delta^{\mathcal{I}} : (x, y) \in R^{\mathcal{I}} \wedge y \in C^{\mathcal{I}}\}$ and the *value restriction*, denoted with $\forall R.C$, whose extension is $\{x \in \Delta^{\mathcal{I}} \mid \forall y \in \Delta^{\mathcal{I}} : (x, y) \in R^{\mathcal{I}} \rightarrow y \in C^{\mathcal{I}}\}$.

[3] In fact, $\mathcal{ALC}$ corresponds to the fragment of first-order logic obtained by restricting the syntax to formulae containing two variables. $\mathcal{ALC}$ has a modal logic counterpart, namely the multi-modal version of the logic $\mathbf{K_m}$ [1].

Further constructors extend the expressiveness of the $\mathcal{ALC}$ language. We are interested in the logic that constitutes the counterpart of OWL-DL ontology language, namely $\mathcal{SHOIQ}(D)$, that extends $\mathcal{ALC}$ with transitive roles, role hierarchies, individual classes, inverse roles and qualified number restrictions. Besides, concrete domains[4] (**D**) with their specific semantics can be dealt with.

The set-theoretic notion of *subsumption* between concepts (or roles) can be given in terms of the interpretations:

Definition 2.1 (subsumption). *Given two concept descriptions C and D, C is subsumed by D, denoted by $C \sqsubseteq D$, iff for every interpretation $\mathcal{I}$ of $\mathcal{T}$ it holds that $C^\mathcal{I} \subseteq D^\mathcal{I}$. Hence, $C \equiv D$ amounts to $C \sqsubseteq D$ and $D \sqsubseteq C$.*

This notion may be easily extended to the case of role descriptions (for languages admitting role-hierarchies).

A *knowledge base* $\mathcal{K} = \langle \mathcal{T}, \mathcal{A} \rangle$ contains two components: a TBox $\mathcal{T}$ and an ABox $\mathcal{A}$. $\mathcal{T}$ is a set of terminological axioms $C \sqsubseteq D$, yet we will consider only definitions $A \equiv D$, where $A \in N_C$ is a concept name (atomic) and D is a concept description given in terms of the language constructors, meaning $A^\mathcal{I} = D^\mathcal{I}$. The ABox $\mathcal{A}$ contains extensional assertions (ground facts) on concepts and roles, e.g. $C(a)$ and $R(a, b)$, meaning, respectively, that $a^\mathcal{I} \in C^\mathcal{I}$ and $(a^\mathcal{I}, b^\mathcal{I}) \in R^\mathcal{I}$. Note that the *unique names assumption* is not necessarily made by default[5].

An interpretation satisfying all the axioms in the knowledge base is said to be a *model* for it. Hence, the usual notions of satisfiability, consistency, etc. apply also for these logics. reasoning is generally performed by resorting to tableau algorithms [1].

Example 2.1 (kinship). Given primitive concepts like `Male` and primitive roles like `hasChild`, an example of TBox (in the proposed language) is:

$$\begin{aligned}\mathcal{T} = \{\ & \mathtt{Female} \equiv \neg\mathtt{Male},\\ & \mathtt{Father} \equiv \mathtt{Male} \sqcap \exists \mathtt{hasChild}.\top,\\ & \mathtt{Mother} \equiv \mathtt{Female} \sqcap \exists \mathtt{hasChild}.\top,\\ & \mathtt{Parent} \equiv \mathtt{Mother} \sqcup \mathtt{Father}\ \}\end{aligned}$$

The concept `Father` translates the sentence: *"a father is a male individual that has some individual as his child"* ($\top$ denotes the most general concept). It is easy to see that $\mathtt{Father} \sqsubseteq \mathtt{Parent}$, yet $\mathtt{Parent} \not\sqsubseteq \mathtt{Father}$.

Now, if we define a new concept:
$\mathtt{FatherWithoutSons} \equiv \mathtt{Father} \sqcap \forall\mathtt{hasChild}.(\neg\mathtt{Male})$
then $\mathtt{FatherWithoutSons} \sqsubseteq \mathtt{Father}$ yet $\mathtt{Father} \not\sqsubseteq \mathtt{FatherWithoutSons}$.

ABox assertions are ground facts like:
`Father(OEDIPUS)`, `Male(OEDIPUS)`, `hasChild.Male(JOCASTA,OEDIPUS)`, $\exists$`hasChild.Female(OEDIPUS)`, $\neg\forall$`hasChild.Male(JOCASTA)` and so on. □

4 Concrete domains include basic data types, such as numerical types, strings, etc., but also more complex types, such as tuples of the relational calculus, spatial regions, or time intervals.

5 Different individual names (that ultimately correspond to URIs in RDF/OWL) may be mapped onto the same object (resource), if not explicitly forbidden.

The most important inference service from the inductive point of view is *instance checking* [1], that amounts to ascertain class-membership assertions: $\mathcal{K} \models C(a)$, where $\mathcal{K}$ is the knowledge base a is an individual name and C is a concept definition given in terms of the concepts accounted for in $\mathcal{K}$. An important difference with other FOL fragments is the *open-world assumption* (OWA) which makes it more difficult to answer class-membership queries. Thus it may happen that an object that cannot be proved to belong to a certain concept is not necessarily a counterexample for that concept. That would only be interpreted[6] as a case of insufficient (incomplete) knowledge for that assertion.

Example 2.2 (Oedipus' family). Given a TBox $\mathcal{T}$ containing the kinship concept definitions of the previous example and also the concept
$\texttt{MotherWithNoDaughter} \equiv \texttt{Mother} \sqcap \forall \texttt{hasChild}.\neg\texttt{Female}$
and the following ABox (using another atomic concept $\mathsf{Parricide}$):

$$\begin{aligned}\mathcal{A} = \{ & \texttt{Female(JOCASTA)}, \texttt{Female(POLYNEIKES)}, \\ & \texttt{Male(OEDIPUS)}, \texttt{Male(THERSANDROS)}, \\ & \texttt{hasChild(JOCASTA, OEDIPUS)}, \\ & \texttt{hasChild(JOCASTA, POLYNEIKES)}, \\ & \texttt{hasChild(OEDIPUS, POLYNEIKES)}, \\ & \texttt{hasChild(POLYNEIKES, THERSANDROS)}, \\ & \texttt{Parricide(OEDIPUS)}, \neg\texttt{Parricide(THERSANDROS)} \}\end{aligned}$$

one may infer the truth for assertions such as Parent(OEDIPUS), but not for MotherWithNoDaughter(POLYNEIKES) because it may well be that a daughter of POLYNEIKES is merely not known.

In order to better appreciate the difference of ABox reasoning w.r.t. query answering let us consider the classic reasoning problem [1]:
given the query $(\texttt{hasChild}.(\texttt{Parricide} \sqcap \texttt{hasChild}.\neg\texttt{Parricide}))(\texttt{JOCASTA})$? (*does Jocasta have a child that is a parricide and that, in turn, has a child that is not a parricide ?*), a query answering system under a closed-world semantics[7] would return a negative answer (false) because it cannot prove the assertion Parricide(POLYNEIKES). Conversely, by reasoning on the possible models of the ABox, one may divide these interpretations into two classes: one contains those satisfying Parricide(POLYNEIKES) and the other with the models of its negation. In both cases JOCASTA can be recognized as an instance of the query concept, hence the answer to the query is true. □

This assumption is perfectly compatible with the typical scenario related to the Semantic Web, where new resources may continuously be made available across the Web, hence a complete knowledge cannot be assumed at any time.

Another useful inference service provided by the DL reasoners is concept *retrieval*: given a certain concept C, retrieve all the individuals that belong to it. Formally: $\{a \in \mathsf{Ind}(\mathcal{A}) \mid \mathcal{K} \models C(a)\}$.

[6] Models could be constructed for both the membership and non-membership case.
[7] Query answering on databases amounts to *finite model checking* [1] (i.e. evaluation of a formula in a fixed finite model).

3 Learning Problems in Description Logics

For the basic terminology on DL languages and terminologies see [1]. Given a DL knowledge base $\mathcal{K} = (\mathcal{T}, \mathcal{A})$ as the source for the background knowledge, suppose that an expert provides proper ABox assertions to deem some individuals as examples (or counter-examples) of some new target concept for which one wants to learn a definition in the form of a DL concept description:

Definition 3.1 (learning problem). *Let $\mathcal{K} = (\mathcal{T}, \mathcal{A})$ be a DL knowledge base.* ***Given***

- *a (new) target concept name C;*
- *a set of positive and negative examples for C:*
  $\mathcal{S}_C^+(\mathcal{A}) = \{a \in \mathsf{Ind}(\mathcal{A}) \mid \mathcal{K} \models C(a)\}$ *and*
  $\mathcal{S}_C^-(\mathcal{A}) = \{b \in \mathsf{Ind}(\mathcal{A}) \mid \mathcal{K} \models \neg C(b)\}$

Build *a concept description D so that the following relations are satisfied by the definition $C \equiv D$:*

- $\mathcal{K} \models C(a) \quad \forall a \in \mathcal{S}_C^+(\mathcal{A})$ *and*
- $\mathcal{K} \not\models C(b) \quad \forall b \in \mathcal{S}_C^-(\mathcal{A})$

Note that in this supervised concept learning task $\mathcal{S}_C^+(\mathcal{A})$ and $\mathcal{S}_C^-(\mathcal{A})$ (i.e. sets of individuals) represent, resp., the sets of positive and negative examples, whereas D is the hypothesis to be induced. In the original setting for logic decision trees [9] multiple *disjoint* concepts are to be learned. In the case of DLs disjointness must be explicitly stated as an axiom in the knowledge base. Note that this is different from settings where negative examples are such that $\mathcal{K} \not\models C(a)$, that may be fulfilled by a single interpretation satisfying $\neg C(a)$.

The example on the domain of *cars* employed in [9] to illustrate the task of learning logical decision trees can be transposed as follows:

Example 3.1 (car checking). An engineer must check a set of cars. Each car is made of several parts that may need be replaced. This can be done by either their manufacturer or by the engineer himself. In the former case the car is an instance of the `SendBack` concept, in the latter it belongs to the `Fix` concept. In case no worn parts are detected, the car is `Ok`. The TBox $\mathcal{T}$ includes the following background knowledge:

$$\{\, \texttt{Gear} \sqsubseteq \texttt{Replaceable},\ \texttt{Chain} \sqsubseteq \texttt{Replaceable},\ \texttt{Engine} \sqsubseteq \neg\texttt{Replaceable},\ \texttt{Wheel} \sqsubseteq \neg\texttt{Replaceable} \,\} \subseteq \mathcal{T}$$

Besides, since the target concepts are meant to be disjoint, the following axioms must be added to $\mathcal{T}$:

$$\{\texttt{SendBack} \sqsubseteq \neg(\texttt{Fix} \sqcup \texttt{Ok}),\ \texttt{Fix} \sqsubseteq \neg(\texttt{Ok} \sqcup \texttt{SendBack}),\ \texttt{Ok} \sqsubseteq \neg(\texttt{SendBack} \sqcup \texttt{Fix})\}$$

The original examples can be encoded as the following set of assertions:

$$\begin{aligned}\mathcal{A}' = \{ & \mathtt{Machine(M_1), hasPart(M_1, G_1), Gear(G_1), Worn(G_1),} \\ & \mathtt{hasPart(M_1, C_1), Chain(C_1), Worn(C_1),} \\ & \mathtt{Machine(M_2), hasPart(M_2, E_2), Engine(E_2), Worn(E_2),} \\ & \mathtt{hasPart(M_2, C_2), Chain(C_2), Worn(C_2),} \\ & \mathtt{Machine(M_3), hasPart(M_3, W_2), Wheel(W_3), Worn(W_3),} \\ & \mathtt{Machine(M_4)} \} \subseteq \mathcal{A}\end{aligned}$$

Then given this knowledge base and the example sets $\mathcal{S}_C^+(\mathcal{A}) = \{\mathtt{M_1}, \mathtt{M_3}\}$ and $\mathcal{S}_C^-(\mathcal{A}) = \{\mathtt{M_2}, \mathtt{M_4}\}$, a good definition for $C = \mathtt{SendBack}$ may be:

$$\mathtt{SendBack} \equiv \mathtt{Machine} \sqcap \exists \mathtt{hasPart.}(\mathtt{Worn} \sqcap \neg \mathtt{Replaceable})$$

A more general setting may be conceived to manage the case of refinement problems, in which a definition for the target concept is already available but it may be defective w.r.t. to some positive or negative examples [7, 6]. Note that this task differs also from settings where the aim is building a classifier, rather than DL concept definitions, through parametric and non-parametric statistical methods [12, 13, 14]. These related settings will not be further discussed.

4 Terminological Decision Trees and Their Induction

First-order logical decision trees (FOLDTs) are defined [9] as binary decision trees in which

1. the nodes contain tests in the form of conjunctions of literals;
2. left and right branches stand, resp., for the truth-value (resp. true and false) determined by the test evaluation;
3. different nodes may share variables, yet a variable that is introduced in a certain node must not occur in the right branch of that node.

Terminological decision trees (TDTs) extend the original definition, allowing DL concept descriptions as (variable-free) node tests. Fig. 1 shows a TDT denoting also the definition of the SendBack concept proposed in Ex. 3.1.

4.1 Classification

The TDTs can be used for classifying individuals. Fig. 2 shows the related classification procedure. It uses other functions: LEAF() to determine whether a node is a leaf of the argument tree, ROOT() which returns the root node of the input tree, and INODE() which retrieves the test concept and the left and right subtrees branching from a given internal node. Given an individual a, starting from the root node, the algorithm checks the class-membership w.r.t. the test concept D_i in the current node, i.e. $\mathcal{K} \models D_i(a)$, sorting a to the left branch if the test is successful while the right branch is chosen if $\mathcal{K} \models \neg D_i(a)$. Eventually the classification is found as a leaf-node concept.

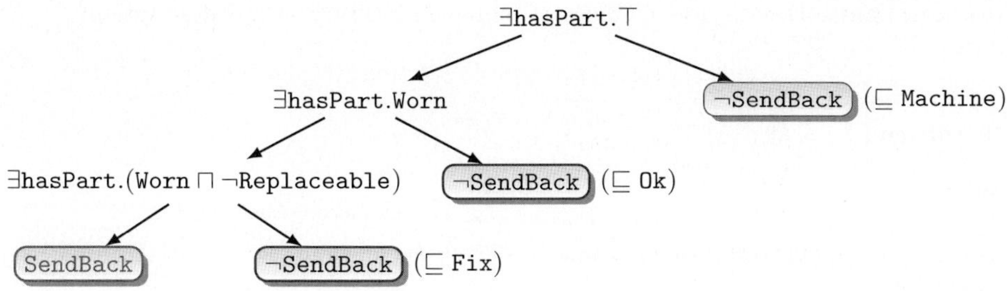

Fig. 1. A TDT whose leftmost path corresponds to the DL concept definition SendBack ≡ ∃hasPart.(Worn ⊓ ¬Replaceable). Other definitions can be associated to the paths to leaves labeled with ¬SendBack that are related to other (disjoint) concepts.

function CLASSIFY(a: individual, T: TDT, $\mathcal{K}$: KB): concept;
begin

1. $N \leftarrow$ ROOT(T);
2. **while** ¬LEAF(N, T) **do**
   (a) $(D, T_{\text{left}}, T_{\text{right}}) \leftarrow$ INODE(N);
   (b) **if** $\mathcal{K} \models D(a)$ **then** $N \leftarrow$ ROOT(T_{left})
   (c) **elseif** $\mathcal{K} \models \neg D(a)$ **then** $N \leftarrow$ ROOT(T_{right})
   (d) **else return** $\top$
3. $(D, \cdot, \cdot) \leftarrow$ INODE(N);
4. **return** D;

end

Fig. 2. Classification with TDTs

Note that the open-world semantics may cause unknown answers (failure of both left and right branch tests) that can be avoided by considering a weaker (default) right-branch test: $\mathcal{K} \not\models D_i(a)$. This differs from the FOLDTs where the test actually consists of several conjunctions that occur in the path from the root to the current node.

4.2 From Terminological Decision Trees to Concept Descriptions

Note that each node in a path may be used to build a concept description through specializations. This can be given 1) by adding a conjunctive concept description, 2) by refining a sub-description in the scope of an existential, universal or number restriction or 3) by narrowing a number restriction (which may be allowed by the underlying language, e.g. $\mathcal{ALN}$ or $\mathcal{ALCQ}$). No special care[8] is to be devoted to negated atoms and their variables. The underlying tableau reasoning algorithm for reasoning with $\mathcal{ALC}$ is indeed sound and complete (see [1], Ch. 3).

[8] We are considering expressive (and decidable) DL languages like $\mathcal{ALCQ}$, that are endowed with full negation, hence the situation is perfectly symmetric.

function DERIVEDEFINITION(C: concept name, T: TDT): concept description;
begin

1. $S \leftarrow$ ASSOCIATE$(C, T, \top)$;
2. **return** $\bigsqcup_{D \in S} D$;

end

function ASSOCIATE(C: concept name; T: TDT; D_c: current concept description): set of descriptions;
begin

1. $N \leftarrow$ ROOT(T);
2. $(D_n, T_{\text{left}}, T_{\text{right}}) \leftarrow$ INODE(N);
3. **if** LEAF(N, T)
   then
   (a) **if** $D_n = C$ **then**
   return $\{D_c\}$;
   else
   return $\emptyset$;
   else
   (a) $S_{\text{left}} \leftarrow$ ASSOCIATE$(C, T_{\text{left}}, D_c \sqcap D_n)$;
   (b) $S_{\text{right}} \leftarrow$ ASSOCIATE$(C, T_{\text{right}}, D_c \sqcap \neg D_n)$;
   (c) **return** $S_{\text{left}} \cup S_{\text{right}}$;

end

Fig. 3. Mapping a TDT onto a DL concept description

For each target concept name C it is possible to derive a *single* concept definition from a TDT. The algorithm (see Fig. 3) follows all the paths leading to success nodes i.e. leaves labeled with C (the heads of the clauses in the original setting) collecting the intermediate test concepts (formerly, the body literals). In this way, each path yields a different conjunctive concept description. that represents a different version of the target concept in conjunctive form $D_i = D_1^i \sqcap \cdots \sqcap D_l^i$. The final single description for the target concept is obtained as the disjunctive description built with concepts from this finite set $S = \{D_i\}_{i=1}^M$. Hence, the final definition is $C \equiv \bigsqcup_{i=1}^M D_i$. As an example, looking back at the TDT depicted in Figure 1, a concept definition that may be extracted is

$$\texttt{Ok} \equiv \exists\texttt{hasPart}.\top \sqcap \neg\exists\texttt{hasPart.Worn} \equiv \exists\texttt{hasPart}.\top \sqcap \forall\texttt{hasPart}.\neg\texttt{Worn}$$

i.e. something that has exclusively parts which are not worn.

Like in the original logic tree induction setting, also internal nodes may be utilized to induce new intermediate concepts.

4.3 Induction of TDTs

The subsumption relationship $\sqsubseteq$ induces a partial order on the space of DL concept descriptions. Then, the learning task can be cast as a search for a solution of the problem in the partially ordered space. In such a setting, suitable operators to traverse the search space are required [4, 6].

While existing DL concept induction algorithms generally adopt a separate-and-conquer covering strategy, the TDT-learning algorithm adopts a divide-and-conquer strategy [10]. It also tries to cope with the limitations of the other learning systems, namely approximation and language-dependence. Indeed, since the early works [2], instances are required to be transposed to the concept level before the learning can start. This is accomplished by resorting to the computation, for each training individual, of the related MSC the individual belongs to [1], which need not exist, especially for expressive DLs, and thus has to be approximated. Even in an approximated version, the MSCs turn out to be extremely specific descriptions which affects both the efficiency of learning and the effectiveness of the learned descriptions as this specificity easily leads to overfitting the data [4].

The algorithms implemented by DL-Learner [6] partly mitigate these disadvantages being based on stochastic search using refinement operators and a heuristic computed on the grounds of the covered individuals (and a syntactic notion of concept length). Generate-and-test strategies may fall short when considering growing search spaces determined by more expressive languages. This drawback is hardly avoidable and it has been tackled by allowing more interaction with the knowledge engineer which can be presented with partial solutions and then decide to stop further refinements.

Our TDT-induction algorithm adapts the classic schema implemented by C4.5 [8] and Tilde [9]. A sketch of the main routine is reported in Fig. 4. It reflects the standard tree induction algorithms with the addition of the treatment of unlabeled training individuals. The three initial conditional statements take care of the base cases of the recursion, namely:

1. no individuals got sorted to the current subtree root then the resulting leaf is decided on the grounds of the prior probabilities of positive and negative instances (resp. Pr_+ and Pr_-);
2. no negative individual yet a sufficient rate (w.r.t. the threshold θ) of positive ones got sorted to the current node, then the leaf is labeled accordingly;
3. dual case w.r.t. to the previous one.

The second half of the algorithm (randomly) generates a set *Specs* of (satisfiable) candidate descriptions (calling generateNewConcepts), that can specialize the current description D when added as a conjunction. Then, the best one (D_{best}) is selected in terms of an improvement of the purity of the subsets of individuals resulting from a split based on the test description. The (im)purity measure is based on the entropic *information gain* [8] or on the *Gini index* which was finally preferred. In the DL setting the problem is made more complex by the presence of instances which cannot be labelled as positive or negative (see [7]) whose contributions are considered as proportional to the prior distribution of positive and negative examples.

Once the best description D_{best} has been selected (calling selectBestConcept), is is installed as the current subtree root and the sets of individuals sorted

function INDUCETDTREE(C: concept name; D: current description;
    Ps, Ns, Us: set of (positive, negative, unlabeled) training individuals): TDT;
const θ; {purity threshold}
begin
Initialize new TDT T;
if $|Ps| = 0$ **and** $|Ns| = 0$
then {*pure* leaf node}
    if $Pr_+ \geq Pr_-$
    then T.root $\leftarrow C$
    else T.root $\leftarrow \neg C$;
    return T;
if $|Ns| = 0$ **and** $|Ps|/(|Ps| + |Ns| + |Us|) > \theta$ **then**
    begin T.root $\leftarrow C$; **return** T; **end**
if $|Ps| = 0$ **and** $|Ns|/(|Ps| + |Ns| + |Us|) > \theta$ **then**
    begin T.root $\leftarrow \neg C$; **return** T; **end**
$Specs \leftarrow$ GENERATENEWCONCEPTS(D, Ps, Ns);
$D_{best} \leftarrow$ SELECTBESTCONCEPT($Specs$, Ps, Ns, Us);
$((P^l, N^l, U^l), (P^r, N^r, U^r)) \leftarrow$ SPLIT(D_{best}, Ps, Ns, Us);
T.root $\leftarrow D_{best}$;
T.left $\leftarrow$ INDUCETDTREE($C, D \sqcap D_{best}, P^l, N^l, U^l$);
T.right $\leftarrow$ INDUCETDTREE($C, D \sqcap \neg D_{best}, P^r, N^r, U^r$);
return T;
end

Fig. 4. The main routine for inducing terminological decision trees

to this node are subdivided according to their classification w.r.t. such a concept. Note that unlabeled individuals must be sorted to both subtrees. Finally the recursive calls for the construction of the subtrees are made, passing the proper sets of individuals and the concept descriptions $D \sqcap D_{best}$ and $D \sqcap \neg D_{best}$ related to either path.

5 Experimental Evaluation

5.1 Experimental Setting

To test the new algorithm on real ontologies, the resulting system TERMITIS was applied to a number of individual classification problems solved by inducing TDTs w.r.t. random query concepts. To this purpose, a number of ontologies represented in OWL concerning different domains have been selected[9], namely: FINITESTATEMACHINES (FSM) concerning finite state machines, NEWTESTAMENTNAMES (NTN) accounting for characters and places mentioned in that

[9] The ontologies can be found in standard repositories: the Protégé library (`http://protege.stanford.edu/plugins/owl/owl-library`) and TONES (`http://owl.cs.manchester.ac.uk/repository`).

Table 1. Facts concerning the ontologies employed in the experiments

ontology	DL language	#concepts	#object properties	#datatype properties	#individuals
FSM	$\mathcal{SOF}(\mathcal{D})$	20	10	7	37
MDM0.73	$\mathcal{ALCHOF}(\mathcal{D})$	196	22	3	112
WINES	$\mathcal{ALCOF}(\mathcal{D})$	75	12	1	161
BIOPAX	$\mathcal{ALCIF}(\mathcal{D})$	74	70	40	323
HDISEASE	$\mathcal{ALCIF}(\mathcal{D})$	1498	10	15	639
NTN	$\mathcal{SHIF}(\mathcal{D})$	47	27	8	676
FINANCIAL	$\mathcal{ALCIF}$	60	16	0	1000

book, the BioPax glycolysis ontology (BioPax) describing the glycolysis pathway from the EcoCyc database, the WINE ontology from a project describing Wines and Food, the FINANCIAL ontology, built for eBanking applications and two medical ontologies MDM0.73 and HDISEASE. Tab. 1 summarizes important details concerning these ontologies, in terms of the numbers of concepts, object and datatype properties and individuals. The sizes of the ontologies are to be measured in terms of (thousands of) RDF triples.

Artificial learning problems were created generating 50 random queries per ontology by composition of 2 through 8 concepts built by means of the language constructors: complement ($\neg$), intersection ($\sqcap$), union ($\sqcup$), universal ($\forall$) or existential ($\exists$) restrictions. The general experiment design adopted a *.632 bootstrap* strategy. A standard reasoner[10] was employed to decide on the real class-membership (and non-membership) w.r.t. the query concepts.

The performance was evaluated comparing the classification of the test individuals w.r.t. the query concepts performed using both the induced trees and the deductive instance-checking provided by the reasoner. The prior distribution of positive and negative instances were computed for each ontology. The default setting of the threshold ($\theta = .05$) was considered.

5.2 Outcomes

Due to the open-world semantics, it may happen that the membership of an individual w.r.t. a query concept cannot be determined by a reasoner (see Sect. 2). Then a three-way classification problems were considered and the induced model was evaluated using the following indices also for allowing a comparison to the outcomes of experiments with another DL concept learning system like DL-FOIL [7]. Essentially they measure the correspondence between the deductive and inductive classification for the test instances w.r.t. the query concepts provided, resp., using a DL reasoner and the TDT algorithm (see Fig. 2):

– *match rate*, i.e. number of cases of individuals that got the same classification with both modes;

[10] PELLET ver. 2 (available at `http://clarkparsia.com/pellet`).

Table 2. Results: average values ± standard deviations

ontology	match rate	commission rate	omission rate	induction rate
FSM	96.68±01.98	00.99±01.35	00.02±00.18	02.31±00.51
MDM0.73	93.96±05.44	00.39±00.61	03.50±04.16	02.15±01.47
Wines	74.36±25.63	00.67±04.63	12.46±14.28	12.13±23,49
BioPax	96.51±06.03	01.30±05.72	02.19±00.51	00.00±00,00
hDisease	78.60±39.79	00.02±00.10	01.54±06.01	19.82±39.17
NTN	91.65±15.89	00.01±00.09	00.36±01.58	07.98±14.60
Financial	96.21±10.48	02.14±10.07	00.16±00.55	01.49±00.16

- *omission error rate*, amount of individuals for which the membership w.r.t. the given query could not be determined using the TDT, while they can be proven to belong to the query concept or to its complement;
- *commission error rate*, amount of individuals found to belong to the query concept according to the induced TDT, while they can be proven to belong to its complement and vice-versa;
- *induction rate*, amount of individuals found to belong to the query concept or its complement according to the TDT, while either case is not logically derivable from the knowledge base.

Tab. 2 reports the outcomes in terms of these indices. Preliminarily, we found that the search procedure was accurate enough: it made few critical mistakes, especially when the considered concepts are known to have many examples (and counterexamples) in the ontology. However, it is important to note that, in each experiment, the commission error was limited but not absent, as in the experiments with other classification methods. The cases of queries for which this measure was high are due to the limited amount of examples available (concepts with narrow extensions). Even few mistakes provoked high error rates. This is also due to the absence of axioms stating explicitly the disjointness between some concepts. Also the omission error rates are quite low. They are comparable with the amount of inductive conclusions that could be drawn with the induced definitions. Again these figures may vary as a consequence of the presence / absence of knowledge about the disjunction of (sibling) concepts in the subsumption hierarchies. In an ontology population perspective, the cases of induction are interesting because they suggest new assertions which cannot be logically derived by using a deductive reasoner yet they might be used to complete a knowledge base, e.g. after being validated by an ontology engineer and/or a domain expert. Better results were obtained on the same task with different inductive methods. Yet, like with DL-Foil we have the added value of having an intensional definition of the target concepts.

The elapsed time was very limited: about 0.5 hour for a whole experiment on a medium sized ontology (in terms of number of individuals) including the time consumed by the reasoner for the deductive instance checks.

BioPax
induced concept:
```
(Or (And physicalEntity protein) dataSource)
```
original concept:
```
(Or (And (And dataSource externalReferenceUtilityClass)
        (ForAll ORGANISM (ForAll CONTROLLED phys icalInteraction)))
     protein)
```

NTN
induced conceot:
```
(Or EvilSupernaturalBeing (Not God))
```
original concept:
```
(Not God)
```

Financial
induced concept:
```
(Or (Not Finished) NotPaidFinishedLoan Weekly)
```
original concept:
```
(Or LoanPayment (Not NoProblemsFinishedLoan))
```

Fig. 5. Examples of induced description for given target concepts employed for the generation of training examples

A comparison with previous works is difficult because some of them consider only binary problems so that standard accuracy measures can be applied. However, as the unknown membership is inherent in the semantics of the representation, it appears to be fairer to consider ternary problems distinguishing counter-examples from those of uncertain membership. As we used the same measures employed for the evaluation of DL-Foil [7], some comparison can be made with that system (although a 10 fold cross-validation design was adopted there). For all the three ontologies employed in the former experiment (BioPax, NTN and Financial) the performance was improved (from 75% to over 90% match rates while the commission errors, formerly up to 16-19% are reduced to a few percentage points). Also in terms of variance, there is a significant reduction of the standard deviation, a sign that the new method is also more stable.

5.3 Qualitative Evaluation

For some of the ontologies, we report in Figure 5 some examples of the concept descriptions (in a Lisp-like syntax) that were learned during the experiments and compare them to the original query concept that generated the examples and counterexamples.

The same concepts have been processed by DL-Foil with the same results, while former systems (such as YinYang [4]) tended to output accurate yet unnecessarily long descriptions. Of course for a correct qualitative interpretation of the value of these concepts some familiarity is assumed with the domain of the ontologies. However we notice that the induced concepts generally tend to be slightly more complex than the original descriptions. This is not so when learning from MSCs [2, 4]. Moreover, they are correct w.r.t. the individuals occurring in the ontology: the extensions of the original and induced concepts, measured in terms of the known instances (retrieval set), generally overlap.

6 Conclusions and Outlook

Introducing terminological decision trees, we investigated new methods (based on logical decision trees) for learning concepts in expressive DLs representations that support the standard Web ontology languages. In the TERMITIS system, a top-down tree induction algorithm was implemented that is an adaptation of standard tree induction methods to the issues related to the different representation. Indeed, as with the predecessor DL-FOIL, it essentially makes use of a different gain function which takes into account the open-world semantics. This requires a different setting and a special treatment of the unlabeled individuals which is similar to the semi-supervised learning problem [11] in a FOL context.

The presented experimental evaluation, applying the TERMITIS system to the task of individual classification real ontologies (some had been already used for experimenting DL-FOIL) using the same performance indices, measuring the alignment of the classifications decided by the induced TDTs with those derived deductively by a DL reasoner. This allowed measuring the amount of unlabeled instances that may be ascribed to the newly induced concepts (or to their negations), which constituted a real added value brought by the inductive method. The experiments made on various ontologies showed that the method is quite effective, and its performance depends on the number (and distribution) of the available training individuals. Besides, the procedure appears also robust to noise since commission errors were limited.

Actually this sort of *abductive* assertions produced by the inductive method should be evaluated by domain experts (e.g. those who supported the construction of the ontology). Namely, validated assertions may be employed in the task of ontology population. This also allows for a more focused diagnosis of the ontologies as it may elicit specific parts that would require some amendment.

We plan to extend this work in various directions. First of all the underlying DL language is being extended. The method can already manage KBs represented in more expressive languages than $\mathcal{ALCQ}$ but use the concepts therein as atoms and building new ones exclusively thorough $\mathcal{ALCQ}$ concept constructors. More impurity indices have to be explored especially to better take into account the uncertainty related to the unlabeled individuals.

Finally, the presented method may be the basis for alternative hierarchical clustering algorithms where clusters would be formed grouping individuals on the grounds of the invented subconcept instead of their similarity, as this may be hardly defined with such complex representations.

References

[1] Baader, F., Calvanese, D., McGuinness, D., Nardi, D., Patel-Schneider, P. (eds.): The Description Logic Handbook. Cambridge University Press, Cambridge (2003)
[2] Cohen, W., Hirsh, H.: Learning the CLASSIC description logic. In: Torasso, P., et al. (eds.) Proceedings of the 4th International Conference on the Principles of Knowledge Representation and Reasoning, pp. 121–133. Morgan Kaufmann, San Francisco (1994)
[3] Badea, L., Nienhuys-Cheng, S.H.: A refinement operator for description logics. In: Cussens, J., Frisch, A.M. (eds.) ILP 2000. LNCS (LNAI), vol. 1866, pp. 40–59. Springer, Heidelberg (2000)
[4] Iannone, L., Palmisano, I., Fanizzi, N.: An algorithm based on counterfactuals for concept learning in the semantic web. Applied Intelligence 26, 139–159 (2007)
[5] Fanizzi, N., Iannone, L., Palmisano, I., Semeraro, G.: Concept formation in expressive description logics. In: Boulicaut, J.-F., Esposito, F., Giannotti, F., Pedreschi, D. (eds.) ECML 2004. LNCS (LNAI), vol. 3201, pp. 99–113. Springer, Heidelberg (2004)
[6] Lehmann, J., Hitzler, P.: Concept learning in description logics using refinement operators. Machine Learning 78, 203–250 (2010)
[7] Fanizzi, N., d'Amato, C., Esposito, F.: DL-Foil: Concept learning in Description Logics. In: Železný, F., Lavrač, N. (eds.) ILP 2008. LNCS (LNAI), vol. 5194, pp. 107–121. Springer, Heidelberg (2008)
[8] Quinlan, R.: Induction of decision trees. Machine Learning 1, 81–106 (1986)
[9] Blockeel, H., De Raedt, L.: Top-down induction of first-order logical decision trees. Artificial Intelligence 101, 285–297 (1998)
[10] Boström, H.: Covering vs. divide-and-conquer for top-down induction of logic programs. In: Proceedings of the 14th International Joint Conference on Artificial Intelligence, IJCAI 1995, pp. 1194–1200. Morgan Kaufmann, San Francisco (1995)
[11] Goldman, S.A., Kwek, S., Scott, S.D.: Learning from examples with unspecified attribute values. Information and Computation 180, 82–100 (2003)
[12] d'Amato, C., Fanizzi, N., Esposito, F.: Query answering and ontology population: An inductive approach. In: Bechhofer, S., Hauswirth, M., Hoffmann, J., Koubarakis, M. (eds.) ESWC 2008. LNCS, vol. 5021, pp. 288–302. Springer, Heidelberg (2008)
[13] Fanizzi, N., d'Amato, C., Esposito, F.: Statistical learning for inductive query answering on OWL ontologies. In: Sheth, A.P., Staab, S., Dean, M., Paolucci, M., Maynard, D., Finin, T., Thirunarayan, K. (eds.) ISWC 2008. LNCS, vol. 5318, pp. 195–212. Springer, Heidelberg (2008)
[14] Luna, J.O., Cozman, F.G.: An algorithm for learning with probabilistic description logics. In: Bobillo, F., et al. (eds.) Proceedings of the 5th International Workshop on Uncertainty Reasoning for the Semantic Web, URSW2009. CEUR Workshop Proceedings, vol. 527, pp. 63–74. CEUR-WS.org (2009)

Classification with Sums of Separable Functions

Jochen Garcke

MATHEON and Technische Universität Berlin
Institut für Mathematik, MA 3-3
Straße des 17. Juni 136, 10623 Berlin, Germany
garcke@math.tu-berlin.de

Abstract. We present a novel approach for classification using a discretised function representation which is independent of the data locations. We construct the classifier as a sum of separable functions, extending the paradigm of separated representations. Such a representation can also be viewed as a low rank tensor product approximation. The central learning algorithm is linear in both the number of data points and the number of variables, and thus is suitable for large data sets in high dimensions. We show that our method achieves competitive results on several benchmark data sets which gives evidence for the utility of these representations.

1 Introduction

We consider the basic binary classification problem, where one starts from a set of labelled data,

$$D = \left\{(\underline{x}_j; y_j)\right\}_{j=1}^{N} = \left\{(x_1^j, \cdots, x_d^j; y_j)\right\}_{j=1}^{N}, \tag{1}$$

with $y_i \in \{-1, 1\}$ labelling the two classes and $\underline{x}$ a d-dimensional feature vector. There exist numerous algorithms for classification (see e.g. [1,2]). Function based methods for classifying data construct a function $g(\underline{x})$ such that the sign of $g(\underline{x}_j)$ matches y_j for the given data, and the sign of $g(\underline{x})$ correctly predicts y for other $\underline{x}$. Since the data may contain errors, or may simply not provide enough information, one cannot expect to completely satisfy this goal, so one tries to minimise the classification error rates.

An approach using nonlinear functions in high dimensions typically has to address the curse of dimensionality, where the complexity of the function representation, which is the number of unknowns, typically grows exponentially with the dimensions. For example, support vector machines are based on a data centred function representation using kernels. One of the reasons for the success of this approach is that here the dimensionality turns up in the complexity of the kernel computation but not in the complexity of the function representation which essentially only depends on the number of data. One can view other approaches and how they address the curse of dimensionality in a similar fashion, e.g. for neural networks the dimension turns up in the perceptron.

J.L. Balcázar et al. (Eds.): ECML PKDD 2010, Part I, LNAI 6321, pp. 458–473, 2010.

In the last years numerical approaches using function representations based on a sparse tensor product approach were successfully used in high dimensions [3,4]. These approaches are based, in an abstract fashion, on the approximation by a sum of separable functions,

$$g(\underline{x}) = \sum_{l=1}^{r} s_l \prod_{i=1}^{d} g_i^l(x_i)\,, \tag{2}$$

also known as low rank tensor decomposition or sparse tensor product approximation. The number of terms r is called the *(separation) rank.* Note that the coefficients s_l are solely for later (computational) convenience, so that one can scale the individual functions to $\|g_i^l\| = 1$ in some suitable function norm, e.g. the maximum norm. Since the s_l depend on the g_i^l they are not explicitly learned, but are derived values. If one applies a log to such a function with $r = 1$ the resulting approach is well known in the statistical literature as additive models.

Many methods are based on this formulation but differ in how they use it. Sparse grid methods are based on a multi-scale tensor product basis where the g_i^l are combined from a set of orthogonal functions. Here basis functions of small importance, justified by decay estimates exploiting hierarchical properties, are omitted [4]. In the statistics literature, representations of the form (2) appear under the names "parallel factorisation" or "canonical decomposition", see [5] for a review and further references. They are used primarily to analyse data on a grid, typically in $d = 3$. Since the goal is to interpret data, constraints on g_i^l, such as positivity when one is interested in probabilities, are often imposed. Similarly, since they only describe data on a grid, a general function is not built.

Following [6] we use sums of separable functions of the form (2) but without constraints such as orthogonality or positivity on the g_i^l. The resulting nonlinear approximation method is called a *separated representation* [3]. The functions g_i^l may be constrained to a *subspace*, but are not restricted to come from a particular *basis set.* This extra freedom allows one to find good approximations with surprisingly small r, and reveals a much richer structure than one would believe beforehand. Although there are at present no useful theorems on the size r needed for a general class of functions, there are examples where removing constraints produces expansions that are *exponentially* more efficient than one would expect a priori, i.e. $r = d$ instead of 2^d or $r = \log d$ instead of d. These example are discussed in detail in [3], we will sketch a few here as illustrations.

First, as a simple example, note that in the separated representation one can have a two-term representation

$$\prod_{i=1}^{d} \phi_i(x_i) + \prod_{i=1}^{d} (\phi_i(x_i) + \phi_{i+d}(x_i)) \tag{3}$$

where $\{\phi_j\}_{j=1}^{2d}$ form an orthonormal set. To represent the same function as (3) while requiring all factors to come from a master orthogonal set would force one to multiply out the second term and thus obtain a representation with 2^d terms. Thus a function that would have $r = 2^d$ in an orthogonal basis may be reduced to $r = 2$. Second, consider the additive model $\sum_{i=1}^{d} \phi_i(x_i)$, and note that it is equal to

$$\lim_{h \to 0} \frac{1}{2h} \left(\prod_{i=1}^{d} (1 + h\phi_i(x_i)) - \prod_{i=1}^{d} (1 - h\phi_i(x_i)) \right) . \tag{4}$$

Thus we can approximate a function that naively would have $r = d$ using only $r = 2$ by choosing h small enough for a given approximation error ϵ. This formula provides an example of converting addition to multiplication; it is connected to exponentiation, since one could use $\exp(\pm h\phi_i(x_i))$ instead of $1 \pm h\phi_i(x_i)$. Third, note that Gaussians are separable, since

$$\exp(-c\|\underline{x} - \underline{z}\|^2) = \prod_{i=1}^{d} \exp(-c(x_i - z_i)^2) . \tag{5}$$

By expanding a radial function in Gaussians, one can obtain a separated representation for it.

In a data mining context one can view such a representation as a sum of different influences. Consider a case where several effects are responsible for the overall distinction into two classes. Each rank in (2) now plays the role of one such effect. This is a somewhat simplified view, as only products are considered as summands in our approach and the interaction between different attributes will not be just of product nature. On the other hand our algorithm optimises the non-linear separated formulation in all ranks at the same time and has therefore interaction between the different "effects". In the end the goal is not to compute summands describing effects, but to achieve the best compressed representation. As mentioned above, addition can be converted to multiplication (4) and additive representations of certain effects can therefore be represented more efficiently. Also note again, that for $r = 1$ (2) falls back to an additive model. Overall, there is some underlying structure present in machine learning applications which our approach exploits at least implicitly.

The representation (2) provides a rich class of functions from which to construct approximations, while also allowing algorithms that scale linearly in both N and d. It is especially appropriate for the case where the dimension d is large, but the underlying function that generated the data is fairly "simple". One can of course construct functions where the representation (2) would fail, in the sense that r must be very large. It appears, however, that such bad functions do not appear naturally. In some sense such bad functions are excluded by the very nature of the problem we consider. For example, consider a "complicated" function, whose discretisation would require a grid with M points in each direction and, thus, $N = M^d$ samples. Since M^d is impossibly large for even moderate values of M and d, we must assume $N \ll M^d$ and, therefore, the data cannot describe such a function to begin with.

In [6] it was shown that such representations are effective as regression functions. The goals of this paper are to present algorithms to construct classifiers of the form (2), and to give numerical evidence that such representations are worth using as classifiers as well, which from a function point of view possess quite different properties than regression functions.

To adapt the method to the classification problem we replace the least-squares error by other more suited loss functions. We use log-likelihood estimation, which in the sense of probability estimation is more appropriate and statistically sound for classification than a least squares approach [1,2], and the hinge loss used for support vector machines in a smooth variant introduced in [7].

These loss functions make a different solution strategy necessary. Instead of an alternating least squares procedure, which at its core involves the solution of a linear equation system "living" in one dimension x_i, we now have a non-quadratic function to minimise using non-linear minimisation algorithms. We investigate both an alternating minimising procedure and a global minimisation which optimises in all dimensions simultaneously. For both we need to formulate a suitable regularised problem. This is necessary to avoid overfitting, but also for numerical stabilisation. We extend the approach from [6] and also investigate the use of $\|\nabla g(\underline{x})\|^2$ as a simple regularisation term to enforce smoothness.

In the following Section 2 we describe the regularised minimisation problem while Section 3 contains the employed algorithms. After presenting numerical results in Section 4 we conclude with an outlook.

2 Definition of the Problem

2.1 Loss Function

The representation of a function by sums of separable functions was studied recently for regression [6]. We now adapt this approach for loss functions preferred in classification problems, the overall aim is to minimise the expected classification error on test data. First is the negative log likelihood [1,2]

$$\frac{1}{N}\sum_{j=1}^{N} L_{LL}(y_j, g(\underline{x}_j)) = \frac{1}{N}\sum_{j=1}^{N} \log\left(1+\exp\left(-y_j g(\underline{x}_j)\right)\right) . \tag{6}$$

Note that we use the encoding $y_i \in \{-1, 1\}$ for the classes instead of the encoding $y_i \in \{0, 1\}$ often employed in the log likelihood approach.

As an alternative we also adapt the hinge loss used for support vector machines. Since this function is not differentiable at $yg(\underline{x}) = 1$ one would need a general descent method. Instead we use a smooth approximation to it proposed by [7], which is inspired by the Huber loss,

$$L_{HH}(y,t) = \begin{cases} 0 & \text{if } yt > 1+h \\ \frac{(1+h-yt)^2}{4h} & \text{if } |1-yt| \le h \\ 1-yt & \text{if } yt < 1-h \end{cases} \tag{7}$$

where h is a parameter to be chosen, typically between 0.01 and 0.5; for $h = 0$ one obtains the hinge loss. Therefore one minimises in this case

$$\frac{1}{N}\sum_{j=1}^{N} L_{HH}(y_j, g(\underline{x}_j)). \tag{8}$$

Note that we are not actually minimising the hinge loss, but from a machine learning point of view there is, besides tractability, no reason to prefer the hinge loss over a smoothed version [7]. In other words, if one does not gain algorithmic advantages using the hinge loss, like for support vector machines in the dual optimisation, one can use a huberised version and expect comparable results.

In the following we describe our approach for the negative log likelihood (LL) loss function (6), but one can replace it at every step with the huberised hinge (HH) loss function (8).

2.2 Basis in One Dimension

Up to now we have not specified the representation for the one-dimensional functions g_i^l. For the above, and most of the following description, it is enough to assume that we are given a function space of (finite) dimension M_l in which to search for g_i^l. For example, one could choose polynomials of some degree, splines, or piecewise linear functions. This space may be different for each term l in the sum, each attribute i, and in general also for each (l, i) pair. We next choose some basis $\{\phi_k^l\}_{k=1}^{M_l}$ for this function space, but we emphasise that the main results are independent of the particular choice. The function g_i^l will be represented by the vector of its M_l coefficients $c_i^l(k)$ in its expansion into $\{\phi_k^l\}$:

$$g_i^l(x_i) = \sum_k^{M_l} c_i^l(k)\phi_k^l(x_i). \tag{9}$$

In our numerical experiments in Section 4 we use a multi-scale basis of tent functions on the interval $[0, 1]$, as was used e.g. in [6,8]. On level 0 this consists of the functions 1 and x. On level 1 we additionally include the tent function of support 1 centred at 1/2, i.e. the line segments from $(0, 0)$ to $(1/2, 1)$ and then to $(1, 0)$. Level 2 adds two tent functions of width 1/2, centred at 1/4 and 3/4, etc. This function space consists of piecewise linear functions.

We will solve for the values of $c_i^l(k)$ for all i, l and k, so those are the free parameters with respect to which we minimise the error.

2.3 Avoiding Over-Fitting

There are two ways in which over-fitting can occur here. The first is when r is too large. Since r is the main complexity parameter, it is natural to take a parametric approach and choose r very low. As with all parametric methods, various more-or-less justified tests, or simple cross-validation, can be used to choose the appropriate r.

The second way over-fitting can occur is when there is over-fitting in the one-dimensional functions g_i^l. There are two natural ways to avoid over-fitting within this framework. One is again to use a parametric approach, and choose M small. Note that the discrete function space and the choice of its resolution can also be viewed in the context of regularisation by projection [9,10].

The other way is to use a nonparametric approach and incorporate regularisation to encourage smoothness, as we describe next. We will use two different strategies for regularisation, one in regard to the full function g, and one in regard to the one-dimensional factors g_i^l. Furthermore, regularisation is usually also beneficial for the stability of the employed numerical solution strategy.

Global Regularisation. One possibility is to add a weighted regularisation term to the loss function (6) or (8) which results in a new functional for the minimisation

$$\frac{1}{N}\sum_{j=1}^{N} L(y_j, g(\underline{x}_j)) + \lambda\|\mathcal{S}(g(\underline{x}))\|^2. \tag{10}$$

The particular choice of $\mathcal{S}$ depends on the chosen discrete function space, or, viewed in the kernel context, the particular choice of $\mathcal{S}$ corresponds, under certain conditions, to a reproducing Kernel Hilbert space (RKHS) and therefore defines a function space (which is then discretised). The regularisation parameter λ has to be chosen suitably as usual. Since we employ multi-scale linear functions as our basis for $g(\underline{x})$ we can only use first derivatives in $\mathcal{S}$. Therefore we use $\|\nabla g(\underline{x})\|^2$ as a simple regularisation term. Although this does not define a RKHS, it was shown to be a reasonable choice in [11,12].

Regularisation of One Factor. We will see in section 3.1 that for the alternating minimisation procedure the problem collapses to one-dimensional subproblems in coordinate direction x_i. Here one can use the global regularisation term from the last section. But one can also encourage smoothness of the function $g(\underline{x})$ just in regard to direction x_i and assume here for the minimisation that the other components have no influence on the smoothness of the function. Furthermore one enforces regularisation separately for each summand $g_i^l(x_i)$ in (2) and not combined. Note that we present the resulting form of regularisation in the following to completely formulate the problem setup. That this choice of the regularisation term has in particular numerical advantages will be clearer after the study of section 3.1, where the alternating minimisation procedure for the one-dimensional problems is explained.

The one-dimensional function $g_i^l(x_i)$ from (2) is using the basis functions ϕ_k^l – we assume the same basis in all dimensions here – and will be represented by M_l coefficients $c_i^l(k)$ according to (9). One now chooses a list of penalty weights γ_k^l and adds to the one dimension problem (13)

$$\lambda\sum_{l} s_l^2 \sum_{k} \gamma_k^l |c_i^l(k)|^2 . \tag{11}$$

This approach was taken in [6] with a particular choice of weights. We use a slight modification and choose the weights for one $g_i^l(x)$ by $\|\mathcal{S}(g_i^l(x))\|^2$. Due to the orthogonality of the basis we have $\int \phi_k^{l\,\prime}(x)\phi_{\tilde{k}}^{l\,\prime}(x)dx = \delta_{k\tilde{k}}C(k)$, where the constant $C(k)$ depends on the size of the support of ϕ_k^l. Due to the choice of $\|\mathcal{S}(\phi_k(x))\|^2$ basis functions with small support will be penalised stronger than those with large support. This will penalise large local variance much stronger than a change of the function over larger intervals.

Let us remark, that in the least-squares case the minimisation problem can be ill-posed [13], but if each coefficient is penalised by some value larger than zero the problem becomes well-posed; see [3] for discussion on controlling condition number in this way. We conjecture that the situation is similar when minimising log likelihood or the huberised hinge loss.

2.4 Sums of Separable Functions in the Learning Theory Context

This approach fits into the framework of Sobolev spaces, these were studied in the learning theory context for example in [14]. This then gives us an infinite function space, with bounds on its properties for learning, in which a discrete approximation takes place, in our case by the use of sums of separable functions.

Such an approximation of an element from a function space by a linear combination of functions from a given dictionary is much less studied in learning theory. From the perspective of approximation theory in [15] bounds for convergence rates for the problem of approximating a given function f from a Hilbert space H by means of greedy algorithms are given and applied to the statistical learning theory context.

In regard to approximation properties of our approach, some results are given in [3] which show how other approaches can be formulated in the form of (2), and how with increasing number of ranks r and increasing resolution M one can approximate a function from a Sobolev space of certain smoothness arbitrarily close. But the convergence order for these somewhat related approaches grows exponentially in d.

To give theoretical results for our approach a characterisation of functions with low separation rank (or tensors with a low rank decomposition) is needed, but currently there is no characterisation of this kind. The examples above and in [3] as well as the successful use of the related "parallel factorisation" or "canonical decomposition" in statistics show that there are surprising mechanisms that allow low separation rank. At this stage the lack of a complete theory should not prevent a study of sums of separable functions for learning.

3 Minimisation Procedures

There are many algorithms for solving least-squares problems using representations like (2) described in the literature, see [5,16] for surveys. It is a non-trivial problem and there is active research to improve convergence and reduce the dependence on the starting guess.

These algorithms can be classified in three main groups: alternating algorithms, which update only a subset of the unknowns at each step; derivative-based methods, seeking an update for all the parameters simultaneously by successive approximations; and direct (non-iterative) methods. The latter cannot be applied in our setting.

3.1 Alternating Minimisation Procedure

For the least squares error this approach is well-known as alternating least-squares (ALS). The idea of partitioning the space of unknowns and solving the optimisation alternatingly in these partitions is known under several other names like coordinate descent and goes back at least to [17]. In the following we will call it alternating minimisation procedure (AMP).

Collapse to One-Dimensional Subproblems. We now assume that an initial guess g of the form (2) is given, with some choice of representation for g_i^l. We fix the components in all directions but one, and so collapse to an one-dimensional problem. For simplicity we describe the case for direction $i = 1$, and so fix g_i^l for $i > 1$. We define the (fixed) partial products from the remaining directions by

$$p_j^l = s_l \prod_{i=2}^{d} g_i^l(x_i^j), \qquad l = 1, \ldots, r, \quad j = 1, \ldots, N. \tag{12}$$

The loss (6) then reduces to

$$\frac{1}{N} \sum_{j=1}^{N} \log \left(1 + \exp \left(-y_j \sum_{l=1}^{r} p_j^l g_1^l(x_1^j) \right) \right). \tag{13}$$

To minimise (13) we must solve a one-dimensional non-linear problem involving r one-dimensional functions g_1^l, each described by M_l coefficients. As a minimiser one has the choice under several algorithms. We did experiments with the quasi-Newton method BFGS, a non-linear CG-method, and a trust-region method (see e.g. [18]). One could numerically estimate the needed derivatives (and hessian for the trust-region method) of the loss function. But e.g. the derivative for the log-likelihood with respect to $c_1^l(k)$ can be given explicitly as

$$\frac{1}{N} \sum_{j=1}^{N} -y_j s_l \phi_k^l(x_1^j) \frac{\exp\left(-y_j \sum_{l=1}^{r} p_j^l g_1^l(x_1^j)\right)}{1 + \exp\left(-y_j \sum_{l=1}^{r} p_j^l g_1^l(x_1^j)\right)}, \tag{14}$$

and the derivative of the regularisation terms is straightforward.

The minimisation finishes once a suitable stopping criteria for the employed non-linear solver is fulfilled, e.g. the objective function does decrease smaller than a given threshold. Since we have an outer iteration the stopping criteria for the one-dimensional minimisation can be coarser than typically used. We then re-normalise g_1^l and incorporate the norm into s_l, this is not strictly necessary, we need not normalise at all; we do so only to prevent over/under-flows.

Before we describe the full algorithm including the iteration over the dimensions we now consider the computational cost bounds; exemplary for the BFGS-method. Here we assume $M_l = M$ for all l and that the cost to evaluate ϕ_k^l is $\mathcal{O}(1)$, which is the case for our choice of basis functions. Therefore the cost to evaluate a single g_i^l at a single point is $\mathcal{O}(M)$. The computation count for one iteration of BFGS is $\mathcal{O}(r^2 M^2)$ plus the costs for the evaluation of the loss function and the gradient for the iteration update and in particular the line search [18]. Given the p_j^l, it costs $\mathcal{O}(rMN)$ to compute the loss function (13). To evaluate (14) we compute the fraction

$$\frac{\exp\left(-y_j \sum_{l=1}^{r} p_j^l g_1^l(x_1^j)\right)}{1+\exp\left(-y_j \sum_{l=1}^{r} p_j^l g_1^l(x_1^j)\right)} \tag{15}$$

once for each j, again in $\mathcal{O}(rMN)$. Using that (now fixed) value we compute the derivative with respect to a single $c_1^l(k)$ in $\mathcal{O}(N)$. Since we have rM different $c_1^l(k)$, the total complexity for the computation of the derivatives for the loss part is $\mathcal{O}(rMN)$.

The two regularisation alternatives have different costs. To compute the global regularisation for the regularised loss (10) one needs $\mathcal{O}(r^2M^2)$ operations. The prime of the regularisation needs $\mathcal{O}(r^2M)$ each time, it simplifies for the employed multi-scale linear basis. The contribution of the other dimensions to the regularisation term in (10) can be computed once at the beginning of the minimisation and needs $\mathcal{O}(d^2r^2M^2)$ If we denote the number of BFGS iterations by S and take it all together the cost to minimise the one-dimensional problem with the global regularisation is

$$\mathcal{O}((r^2M^2 + rMN)S + d^2r^2M^2). \tag{16}$$

The simpler regularisation term (11) only needs $\mathcal{O}(rM)$ operations for the evaluation of the regularised loss function and its prime. Which gives a total cost of

$$\mathcal{O}((r^2M^2 + rMN)S). \tag{17}$$

Alternating Improvement. If we can solve the one-dimensional subproblems, then we can iteratively solve such problems to reduce the loss (6). The alternating strategy [3,5,16] is to loop through the directions $i = 1, \ldots, d$. One then repeats this alternating process and monitors the change in the loss (6), or the regularised loss (10), to detect convergence. It is certainly possible to hit local minima. Even when we approach the true minima, we have no reason to expect any better than linear convergence.

To account for the computational cost to set up these problems we assume that the number of AMP iterations is K. The cost to compute all p_j^l for a single i is then $\mathcal{O}(rdMN)$. It would appear that we have cost $\mathcal{O}(rd^2MNK)$ in the outer loop through the d directions. However, when we switch from, say, $i = 1$ to $i = 2$, we can simply update p_j^l by multiplying it by $g_1^l(x_1^j)/g_2^l(x_2^j)$, at cost $\mathcal{O}(rMN)$. The total cost for the update in the AMP formulation (without the cost for solving the one-dimensional subproblems) is thus

$$\mathcal{O}(drMNK)\,. \tag{18}$$

If we incorporate this algorithm into the overall method and account for the total cost we get

$$\mathcal{O}(K(dr^2M^2 + drMN)S). \tag{19}$$

for the local regularisation from Section 2.3. The cost is linear in both d and N, and so the method is feasible for large data sets in high dimensions.

Using the global regularisation from Section 2.3 we observe a complexity of

$$\mathcal{O}(K[(dr^2M^2 + drMN)S + d^3r^2M^2]). \tag{20}$$

Again linear in N, but in parts cubic in d, although the inner non-linear solver is linear in d. Nevertheless, the complexity still suggests the method for large data sets in high dimensions.

The computational complexity for a non-linear CG-method or the trust-region method in regard to N and d are similar. Only the evaluation of the regularised loss functions, their prime and hessian (for the trust-region method) depends on these and there we have a linear scaling in N and d for the regularisation (11) and linear in N and cubic in d for the regularisation (10).

The number of iterations needed in the non-linear solvers is the remaining important computational aspect. These will depend implicitly on the complexity of the function and therefore on the number of data. At this point we have not investigated the non-linear solvers in detail but use well tested and publically available implementations. Line-search procedures adopted to the problem and suitable pre-conditioners for the non-linear CG approach are needed for a fully efficient scheme. But for now we focus on the investigation of the accuracy and representation power of our approach for classification problems. Therefore it is enough to solve the non-linear problems to a sufficient degree in reasonable time.

Also note that we expect that after a few steps of the alternating procedure we will have good starting values for the non-linear minimisation.

3.2 Global Minimisation Procedure

Alternating algorithms are in particular attractive for the least square loss because at their inner core a linear equation system needs to be solved. We here use other loss functions and therefore have to use a non-linear minimisation procedure anyway, therefore one can consider treating the full problem (10) directly, as it is also often used for least squares minimisation [5,16]. In the following we will call it global minimisation procedure (GMP).

The amount of data and the dimensionality now have a different influence on the computational complexity. Again we focus on the BFGS-algorithm. The number of unknowns of our representation is drM, therefore the cost for one BFGS-iteration is of the order $\mathcal{O}(d^2r^2M^2)$ plus the cost for evaluating the regularised loss function (10) and its prime. It costs $\mathcal{O}(drMN)$ to evaluate the loss function and $\mathcal{O}(d^2r^2M^2)$ for the regularisation term. The same holds for one partial derivative of which there are drM. In total we have for the complexity of the algorithm

$$\mathcal{O}((d^2r^2M^2N + d^2r^3M^3)S), \tag{21}$$

where S is the number of BFGS-iterations.

Although the global minimisation procedure has the larger order of computational complexity it can be competitive if the number of iterations S is small.

But for this often special care has to taken in the line search procedure. Furthermore, implementing such an algorithm is often more cumbersome, especially in regard to the needed derivative. As we will see in the numerical results using standard implementations of non-linear solver does not achieve the wanted accuracy. Here further investigation of the procedures and their adaptation to the particular loss function and problem setup are necessary, but out of scope of this paper.

4 Numerical Results

In this section we give numerical results for several benchmark problems. Our goal is to demonstrate that the representation (2) is powerful enough to build good classifiers.

We compare against data used in the benchmark study [19], where the classification methods support vector machines with RBF-kernel (svm), classification trees, linear discriminant analysis, quadratic discriminant analysis, neural networks, generalised linear models (glm), multinomial logit models, nearest neighbours (nn), learning vector quantisation (lvq), flexible discriminant analysis, mixture discriminant analysis, bagging, double bagging (dbagg), random forests (rForst), and multiple additive regression trees were compared empirically[1].

As in [19], we measure the classification performance using the prediction error. Ten-fold cross-validation was performed ten-times; we report the means and medians of the test set error rates of all 100 runs, whereas the standard deviation and inter-quartile range are computed with regard to the ten-fold results. For comparison we give the best result from the benchmark study and note the rank of our approach in comparison to the other methods used.

The separation rank r, the discretisation level of the multi-scale basis (which correlates with the basis size M) and the size of the regularisation parameter were selected similar to [19]: we split the training data 2:1, train on the first two thirds and evaluate on the last third to select good parameters. With these we learn on all training data and evaluate on the as-yet-unseen test data. Note that depending on the problem we used up to level 4 of the multi-scale basis and rank $r = 7$, although often $r \leq 4$ was sufficient.

It was observed that the test error is relatively unaffected by the value of h in the huberised hinge loss, as long as it is not too large when it resembles more the L_2 loss [7]. In our experiments we observed this behaviour as well and use a fixed $h = 0.05$.

Pre-processing of the data consists of omitting missing values, like in [19], and scaling all data to $[0, 1]^d$. We concentrate in this paper on data sets with metric attributes. Therefore we use the five synthetic data sets and five real ones, including one with some categorical variables which were transformed into binary attributes. For binary attributes i we only use a linear function, i.e. $M_i = 2$.

[1] The data is available from (`http://www.ci.tuwien.ac.at/~meyer/benchdata/`).

4.1 Alternating and Global Minimisation

First we remark on the empirical behaviour of the two different minimisation procedures. We consider the circle in a square (CIRCLE) and the two noisy spirals (SPIRALS) data sets from [19], both are two dimensional. One might expect that in this low dimensional case there would not be too much difference in the behaviour of the two minimisation procedures. But already here the global optimisation procedure does not cope with the problem and produces worse results. This does not depend on the employed non-linear solver.

To be more precise. Using the same minimisation procedure in a standard implementation using standard line search procedures the alternating minimisation procedure achieves better results than the global minimisation procedure. This observations does not depend on the non-linear solver nor which public implementation was used. We expect that with a detailed investigation of the minimisation problem and an adaption of the line search procedure the global minimisation will perform better. There is active research in this regard in the least-squares context [5,16,20].

In any case, not only does the alternating minimisation procedure show the better order of computational complexity, it also achieves better correctness rates as we can see in Table 1.

Somewhat promising are further results using a few iterations of the alternating procedure to compute a good starting point for the global minimiser. For the spiral data set we achieve in this way a median of 0.1 (0.075) and mean of 0.22 (0.13) using the log-likelihood as a loss function. Although the huberised hinge loss does not benefit from such an approach for this data set; nor do we observe such an improvement for the circle data set.

For higher dimensional data sets the global minimisation performs, as one would expect after these results, even worse. We therefore abstain form giving detailed results for the global minimisation procedure in the following.

We also use the two dimensional spiral data set to illustrate the approach. In Figure 1 we show the components $g_i^l(x_i)$ of the solution for one instance of the data set and the resulting classifier. One might not expect that with just the sum of four product functions such a complicated classifier can be obtained.

Table 1. Results on low dimensional synthetic data sets for both loss functions and (A)lternating and (G)lobabl minimisation procedures. We give the mean (with standard deviation) and median (with inter-quartile range).

	CIRCLE					SPIRALS			
	LL - A	LL - G	HH - A	HH - G		LL - A	LL - G	HH - A	HH - G
mean	**2.22**	2.66	2.45	3.65	mean	**0.25**	0.53	1.03	2.79
	(0.46)	(0.39)	(0.47)	(0.47)		(0.10)	(0.28)	(0.22)	(0.57)
median	**1.95**	2.35	2.30	3.45	median	**0.20**	**0.20**	0.75	2.30
	(0.61)	(0.60)	(0.80)	(0.49)		(0.16)	(0.44)	(0.29)	(0.93)

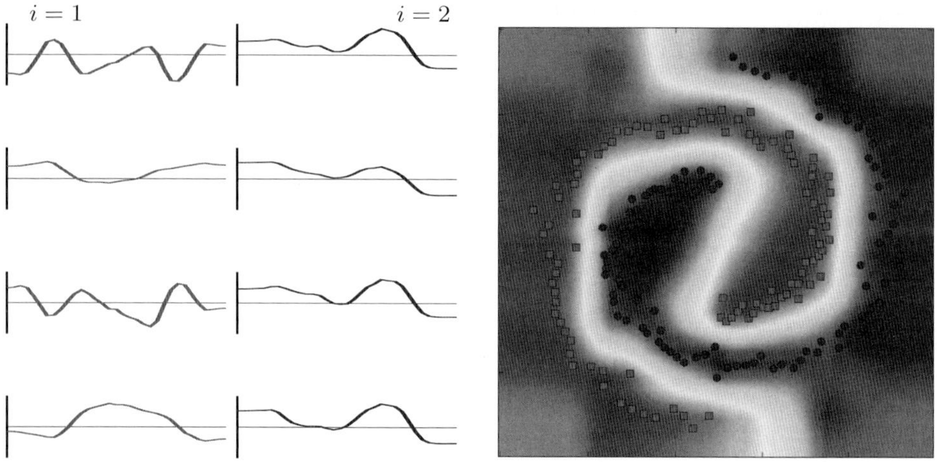

Fig. 1. Classifier with $r = 4$ using the multi-scale basis with level five produced using the negative log likelihood error on one instance of the spiral data set. Left: The function of the form (2) with $r = 4$ using the multi-scale basis with level five. Each subplot shows a $g_i^l(x_i)$. The magnitude s_l has been distributed. Right: Value of the classifier over the two-dimensional domain.

4.2 Results on Benchmark Data

We give the results of our experiments in Table 2 for the synthetic data sets and Table 3 for the real ones. Our current code is a somewhat experimental python implementation, whose purpose is to produce the approximation results that are the main point of this paper. For one run on a data set the computational time varied between a few seconds and a few hundred seconds, depending on the data set, the rank, the degree and the regularisation parameter. In python, loops with numerical computations are known to be inefficient and the algorithm consists of a fair amount of loops over the data or the basis functions. We expect that a proper implementation would decrease the runtime significantly. Therefore we abstain for now from giving more detailed numbers for the computational times, but will followup once a scaleable implementation is available.

Overall the log likelihood estimation performs somewhat better than the huberised hinge loss, the latter might benefit from the h in its definition (7) chosen depending on the data. The run time difference between these two loss functions in our experiments did not appear to be significant.

The simple local regularisation method performs slightly better than the global regularisation. For the twonorm, bupa liver and credit data set it results in smaller misclassification rates, while for the spirals and threenorm data set the global regularisation method is better. In particular for some of the real data the minimisation procedure had problems to converge for the global regularisation; again, better adapted non-linear solution strategies might improve

Table 2. Results on synthetic data from the study [19]. We give the mean (with standard deviation) and median (with inter-quartile range) for the best results from [19], our approach, and our rank in comparison to all 17 approaches used. We use log likelihood estimation (LL), the huberised hinge loss (HH) and both forms of regularisation.

data set		best	other	LL-Reg. (11) SumSep	rank	HH-Reg. (11) SumSep	rank	LL-Reg. (10) SumSep	rank	HH-Reg. (10) SumSep	rank
CIRCLE	mean	2.66	svm	2.26 (0.44)	1	2.39 (0.44)	1	**2.22** (0.46)	1	2.45 (0.47)	1
	median	2.50 (0.49)	svm	2.00 (0.41)	1	2.10 (0.62)	1	**1.95** (0.61)	1	2.30 (0.80)	1
SPIRALS	mean	0.17	nn	1.04 (0.18)	3	1.19 (0.22)	3	**0.25** (0.10)	2	1.03 (0.22)	3
	median	0.10 (0.07)	nn	0.90 (0.22)	3	1.10 (0.31)	3	**0.20** (0.16)	2	0.75 (0.29)	3
TWONORM	mean	2.82	svm	**3.61** (0.34)	5	4.08 (0.34)	5	12.04 (7.85)	17	6.37 (1.34)	12
	median	2.70 (0.20)	svm	**3.40** (0.40)	5	3.85 (0.54)	5	5.50 (8.93)	10	5.90 (0.59)	11
THREENORM	mean	14.17	lvq	18.94 (0.98)	9	19.21 (0.60)	9	**14.45** (0.40)	2	15.62 (0.69)	2
	median	13.70 (0.77)	lvq	18.95 (0.98)	8	19.10 (0.50)	9	**14.40** (0.79)	2	15.40 (1.22)	2
RINGNORM	mean	3.58	svm	5.44 (0.58)	2	5.92 (2.64)	2	**4.85** (0.31)	2	5.43 (0.32)	2
	median	2.90 (0.70)	svm	4.80 (0.75)	2	4.90 (0.61)	2	**4.70** (0.33)	2	5.30 (0.31)	2

the results. The good performance of the simpler local regularisation is an indication that the limitation to a discrete function representation has a large effect in the avoidance of overfitting, known as regularisation by projection in other fields [9,10].

In comparison to the 17 other methods our approach achieves very competitive results, here we look at the median as is done in [19]. For all data sets at least one variant of our approach is in the top five and for eight of the data sets we are in the top three. For six data sets, the majority, at least one version of our procedure achieved better results than a support vector machine, which was the best method in the referenced study [19].

5 Outlook

We described a new classification algorithm using sums of separable functions to represent the classifier. Numerical evidence shows that typical data sets can be efficiently described by this approach, which is the main message of this paper.

There are several extensions and generalisations possible. Foremost, instead of the standard and publically available non-linear minimisation algorithms a

Table 3. Results on real data from the study [19]. We give the mean (with standard deviation) and median (with inter-quartile range) for the best results from [19], our approach, and our rank in comparison to all 17 approaches used. We use log likelihood estimation (LL), the huberised hinge loss (HH) and both forms of regularisation.

data set		best	other	LL-Reg. (11) SumSep	rank	HH-Reg. (11) SumSep	rank	LL-Reg. (10) SumSep	rank	HH-Reg. (10) SumSep	rank
	mean	2.28	rForst	3.30	5	**3.21**	4	3.33	5	3.66	6
CANCER				(0.24)		(0.26)		(0.22)		(0.49)	
	median	1.49	rForst	2.94	4	**2.92**	3	**2.92**	3	2.94	4
		(0.16)		(0.25)		(0.33)		(0.34)		(0.71)	
	mean	27.04	rForst	26.63	1	**26.58**	1	31.74	8	30.66	7
LIVER				(1.48)		(0.98)		(1.50)		(1.86)	
	median	27.02	rForst	**25.71**	1	**25.71**	1	30.56	6	30.56	6
		(2.15)		(2.52)		(1.48)		(2.80)		(3.24)	
	mean	22.65	dbagg	**23.70**	9	24.73	10	27.11	11	26.43	11
CREDIT				(0.67)		(0.87)		(1.40)		(1.10)	
	median	22.77	5 appr.	**22.77**	1	24.25	10	26.87	13	26.73	12
		(0.20)		(0.57)		(0.87)		(2.40)		(0.72)	
	mean	5.93	svm	**8.01**	4	9.03	6	16.11	15	11.49	8
IONOSPHERE				(0.84)		(0.63)		(1.67)		(2.23)	
	median	5.71	2 appr.	**8.57**	4	**8.57**	4	**8.57**	4	**8.57**	4
		(0.70)		(1.35)		(0.95)		(3.21)		(4.32)	
	mean	22.37	2 appr.	23.37	5	23.35	5	**23.08**	5	24.10	9
DIABETIS				(0.59)		(0.52)		(0.71)		(0.73)	
	median	22.08	4 appr.	**22.08**	1	23.23	5	22.72	5	23.38	6
		(0.26)		(0.65)		(0.80)		(0.94)		(0.41)	

method more specifically tuned to the problem could be used, e.g. similar to [20,21]. Second, the approach can easily be extended for categorical attributes x_j. One can use a basis of vectors rather than functions, and index their coordinates by the categories of such an attribute. Third, if one formulates the loss (13) using something besides negative log likelihood or the hinge loss used in support vector machines one obtains a similar nonlinear optimisation problem which can be treated analogously. Fourth, note that in [6] it was shown how to extend the regression algorithm to vector-valued regression functions. By letting the vector represent the probabilities of the data point being in the different classes, one obtains a multi-class classifier that produces probabilities rather than a decision on the class. Suitable multi-class loss functions for support vector machines or penalised likelihood estimation can be found in [22]. Finally, one can use different one-dimensional spaces for different attributes and for different l. For example, if one assumes an almost normal distribution underlying the data, one might use a suitable Gaussian for the $l = 0$ term, but another basis for other l to approximate necessary adjustments.

Acknowledgements. The author cordially thanks Martin Mohlenkamp (Ohio University) for helpful discussions and suggestions.

References

1. Bishop, C.M.: Pattern recognition and machine learning. Springer, Heidelberg (2006)
2. Hastie, T., Tibshirani, R., Friedman, J.: The Elements of Statistical Learning. Springer, Heidelberg (2001)
3. Beylkin, G., Mohlenkamp, M.J.: Algorithms for numerical analysis in high dimensions. SIAM J. Sci. Comput. 26(6), 2133–2159 (2005)
4. Bungartz, H.J., Griebel, M.: Sparse grids. Acta Numer. 13, 147–269 (2004)
5. Kolda, T.G., Bader, B.W.: Tensor decompositions and applications. SIAM Review 51(3), 455–500 (2009)
6. Beylkin, G., Garcke, J., Mohlenkamp, M.J.: Multivariate regression and machine learning with sums of separable functions. SIAM Journal on Scientific Computing 31(3), 1840–1857 (2009)
7. Chapelle, O.: Training a support vector machine in the primal. Neural Computation 19(5), 1155–1178 (2007)
8. Garcke, J., Griebel, M., Thess, M.: Data mining with sparse grids. Computing 67(3), 225–253 (2001)
9. Engl, H., Hanke, M., Neubauer, A.: Regularization of Inverse Problems. Kluwer, Dordrecht (1996)
10. Natterer, F.: Regularisierung schlecht gestellter Probleme durch Projektionsverfahren. Numer. Math. 28, 329–341 (1977)
11. Garcke, J.: Regression with the optimised combination technique. In: Cohen, W., Moore, A. (eds.) Proceedings of the 23rd ICML 2006, pp. 321–328. ACM Press, New York (2006)
12. Garcke, J., Hegland, M.: Fitting multidimensional data using gradient penalties and the sparse grid combination technique. Computing 84(1-2), 1–25 (2009)
13. de Silva, V., Lim, L.H.: Tensor rank and the ill-posedness of the best low-rank approximation problem. SIAM J. Matrix Anal. Appl. 30(3), 1084–1127 (2008)
14. Cucker, F., Smale, S.: On the mathematical foundations of learning. Bulletin of the AMS 39(1), 1–49 (2001)
15. Barron, A.R., Cohen, A., Dahmen, W., Devore, R.A.: Approximation and learning by greedy algorithms. Ann. Stat. 36(1), 64–94 (2008)
16. Tomasi, G., Bro, R.: A comparison of algorithms for fitting the parafac model. Computational Statistics and Data Analysis 50(7), 1700–1734 (2006)
17. Yates, F.: The analysis of replicated experiments when the field results are incomplete. Emp. J. Exp. Agric. 1, 129–142 (1933)
18. Nocedal, J., Wright, S.J.: Numerical optimization, 2nd edn. Springer, Heidelberg (2006)
19. Meyer, D., Leisch, F., Hornik, K.: The support vector machine under test. Neurocomputing 55, 169–186 (2003)
20. Espig, M.: Effiziente Bestapproximation mittels Summen von Elementartensoren in hohen Dimensionen. PhD thesis, Universität Leipzig (2008)
21. Espig, M., Hackbusch, W., Rohwedder, T., Schneider, R.: Variational calculus with sums of elementary tensors of fixed rank. Numerische Mathematik (submitted 2010), Preprint 52/2009, MPI for Mathematics in the Sciences
22. Wahba, G.: Soft and hard classification by reproducing kernel Hilbert space methods.. Proc. Natl. Acad. Sci. USA 99(26), 16524–16530 (2002)

Feature Selection for Reinforcement Learning: Evaluating Implicit State-Reward Dependency via Conditional Mutual Information

Hirotaka Hachiya and Masashi Sugiyama

Tokyo Institute of Technology, Tokyo, 152-8552, Japan
hachiya@sg.cs.titech.ac.jp, sugi@cs.titech.ac.jp

Abstract. Model-free reinforcement learning (RL) is a machine learning approach to decision making in unknown environments. However, real-world RL tasks often involve high-dimensional state spaces, and then standard RL methods do not perform well. In this paper, we propose a new feature selection framework for coping with high dimensionality. Our proposed framework adopts *conditional mutual information* between return and state-feature sequences as a feature selection criterion, allowing the evaluation of implicit state-reward dependency. The conditional mutual information is approximated by a least-squares method, which results in a computationally efficient feature selection procedure. The usefulness of the proposed method is demonstrated on grid-world navigation problems.

1 Introduction

Optimal decision making in unknown environment is a challenging task in the machine learning community. *Reinforcement learning* (RL) is a popular framework for this purpose, and has been actively studied. In RL, a *policy* (the decision rule of an agent) is determined so that *return* (the sum of discounted rewards the agent will receive) is maximized. So far, various RL approaches such as *policy iteration* [1,2] and *policy search* [3,4,5] have been explored and demonstrated to be promising in small- to medium-sized problems.

However, when the dimensionality of the state space is high, existing RL approaches tends to perform poorly. Unfortunately, this critically limits the range of applicability of RL in practice since real-world RL tasks such as robot control often involve high-dimensional state spaces. To cope with high dimensionality of the state space, choosing a subset of relevant features from the high-dimensional state variables, i.e., *feature selection*, is highly useful.

For example, let us consider developing a security guard robot that can deal with a variety of tasks such as navigation, patrol, and intruder detection. For this purpose, the robot is equipped with various types of sensors such as position, orientation, distance, vision, sound, smell, and temperature sensors. However, when a security guard robot is engaged in a particular task such as navigation, all the sensors may not be needed. Since sensors necessary for solving a task are

J.L. Balcázar et al. (Eds.): ECML PKDD 2010, Part I, LNAI 6321, pp. 474–489, 2010.

different depending on the tasks, it is not possible to choose the subset of sensors in advance. Thus, adaptive feature selection based on currently available data samples is indispensable in this scenario.

Various feature selection strategies have been explored so far, which can be categorized into the *wrapper* and *filter* approaches [6]. In the wrapper approach, features are selected depending on the subsequent learning process such as least-squares fitting of the *value function* [7,8,9]. A supervised dimensionality reduction method called *neighborhood component analysis* [10] was applied to feature selection in RL [7], while the decomposition of value function approximation error into reward prediction error and transition prediction error was utilized for feature selection in [9]. These wrapper methods would be useful for specific RL frameworks such as policy iteration, but they may not be directly employed in other frameworks such as policy search.

On the other hand, in the filter approach, features are selected independently of subsequent learning processes [11,12]. More specifically, a subset of features is chosen in an *information-theoretic* way that the remaining subset of features is statistically independent of the outcome (typically, the rewards). Such an information-theoretic approach is versatile as preprocessing of high-dimensional data.

A supervised dimensionality reduction method called *kernel dimension reduction* (KDR) [13] was applied to feature selection in RL [11]. Based on the *Markov* property of RL problems, their method evaluates the conditional independence between the entire state features and a state subset which directly influences rewards at the next time-step. However, since RL deals with sequential decision making problems, there can exist an *implicit* dependency between state features and *rewards* through the process of sequential decision making. This is illustrated using a simple example in Figure 1. A feature $s^{(2)}$ influences $s^{(1)}$ at the next time-step which have a direct effect on rewards. Thus, features $s^{(1)}$ and $s^{(2)}$ can be selected by KDR. However, $s^{(3)}$ cannot be selected by KDR since there is no dependency between $s^{(1)}$ and $s^{(3)}$ in a single time-step, although it actually influences rewards in two time-steps through $s^{(2)}$ and $s^{(1)}$.

The implicit dependency in the sequential process can be detected in principle by recursively evaluating the dependency between states and rewards [12]. However, such a recursive approach is computationally demanding particularly when there exist *cascaded* dependency relations. For example, in Figure 1, two recursions are needed to find the relevant features $\{s^{(1)}, s^{(2)}, s^{(3)}\}$. First, $s^{(2)}$ is selected due to its dependency to $s^{(1)}$, and then $s^{(3)}$ is chosen because of its dependency to $s^{(2)}$. In addition, an assumption that the model of *factored* Markov decision processes is available as a dynamic Baysian network was imposed in [12], which may not be realistic.

In order to overcome the drawbacks of existing approaches, we introduce a new framework of filter-type feature selection for RL. More specifically, we propose to directly evaluate the independence between return and state-feature sequences using the *conditional mutual information* [14]. In order to efficiently approximate the conditional mutual information from samples, we utilize a least-squares

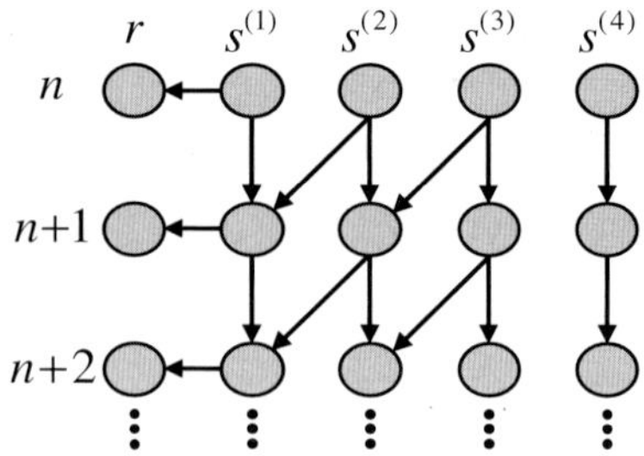

Fig. 1. An example of implicit dependency between state features and rewards. Each row represents the time-step $(n, n+1, n+2, \ldots)$, and columns represent reward(r) and state features ($s^{(1)}$, $s^{(2)}$, $s^{(3)}$, $s^{(4)}$). Arrows indicate the dependency between two variables; for example, an arrow from $s_n^{(2)}$ to $s_{n+1}^{(1)}$ exists if $s_{n+1}^{(1)}$ depends on $s_n^{(2)}$. In this example, a state feature $s^{(3)}$ does not have direct influence on the next reward r_n, but has indirect influence on r_{n+2} through $s_{n+1}^{(2)}$ and $s_{n+2}^{(1)}$.

(un-conditional) mutual information estimator which was proved to possess the optimal convergence rate [15].

The rest of this paper is organized as follows. In Section 2, we mathematically formulate the problem of RL. In Section 3, we describe our proposed feature selection procedure. Experimental results are reported in Section 4, demonstrating the effectiveness of the proposed method in grid-world navigation. Finally, we conclude in Section 5 by summarizing our contributions and describing future work.

2 Formulation of RL

In this section, we formulate the RL problem as a Markov decision process (MDP).

2.1 Markov Decision Process

Let us consider an MDP specified by $(\mathcal{S}, \mathcal{A}, p_{\mathrm{T}}, p_{\mathrm{I}}, R, \gamma)$, where $\mathcal{S}$ ($\in \mathbb{R}^v$) is a set of v-dimensional states, $\mathcal{A}$ ($\in \mathbb{R}$) is a set of one-dimensional actions, $p_{\mathrm{T}}(\boldsymbol{s}'|\boldsymbol{s}, a)$ (≥ 0) is the transition probability-density from state $\boldsymbol{s}$ to next state $\boldsymbol{s}'$ when action a is taken, $p_{\mathrm{I}}(\boldsymbol{s})$ (≥ 0) is the probability density of initial states, $R(\boldsymbol{s}, a, \boldsymbol{s}')$ ($\in \mathbb{R}$) is an immediate reward for transition from $\boldsymbol{s}$ to $\boldsymbol{s}'$ by taking action a, and γ ($\in (0, 1]$) is the discount factor for future rewards. By following initial probability p_{I}, transition probability p_{T}, and policy π, an MDP generates a sequence of states, actions, and rewards as

$$\boldsymbol{s}_1, a_1, r_1, \boldsymbol{s}_2, a_2, r_2, \boldsymbol{s}_3, a_3, r_3, \ldots,$$

where the subscript indicates the time step. Let $p^{\pi}(\boldsymbol{s}|n)$ be the probability density of state $\boldsymbol{s}$ at time step n:

$$p^\pi(\boldsymbol{s}|n=1) = p_{\mathrm{I}}(\boldsymbol{s}),$$
$$p^\pi(\boldsymbol{s}|n=2) = \iint p_{\mathrm{I}}(\boldsymbol{s}_1)\pi(a_1|\boldsymbol{s}_1)p_{\mathrm{T}}(\boldsymbol{s}|\boldsymbol{s}_1,a_1)\mathrm{d}\boldsymbol{s}_1\mathrm{d}a_1,$$
$$p^\pi(\boldsymbol{s}|n=3) = \iiiint p_{\mathrm{I}}(\boldsymbol{s}_1)\pi(a_1|\boldsymbol{s}_1)p_{\mathrm{T}}(\boldsymbol{s}_2|\boldsymbol{s}_1,a_1) \times \pi(a_2|\boldsymbol{s}_2)p_{\mathrm{T}}(\boldsymbol{s}|\boldsymbol{s}_2,a_2)\mathrm{d}\boldsymbol{s}_1\mathrm{d}a_1\mathrm{d}\boldsymbol{s}_2\mathrm{d}a_2,$$
$$\vdots$$

2.2 Optimal Policy

Let η_n ($\in \mathbb{R}$) be the *return* which is the sum of discounted rewards the agent will receive when starting from the n-th time step:

$$\eta_n \equiv \sum_{n'=n}^{\infty} \gamma^{n'-n} R(\boldsymbol{s}_{n'}, a_{n'}, \boldsymbol{s}_{n'+1}).$$

Let $p^\pi(\eta|\boldsymbol{s})$ be the probability density of return η when starting from a state $\boldsymbol{s}$ and then following a policy π. Let $V^\pi(\boldsymbol{s})$ be the *expected return*:

$$V^\pi(\boldsymbol{s}) \equiv \int \eta p^\pi(\eta|\boldsymbol{s})\mathrm{d}\eta.$$

The goal of RL is to learn the optimal policy π^* that maximizes the expected return $V^\pi(\boldsymbol{s})$:

$$\pi^*(\cdot|\boldsymbol{s}) \equiv \operatorname*{arg\,max}_{\pi(\cdot|\boldsymbol{s})} V^\pi(\boldsymbol{s}). \tag{1}$$

2.3 Data Samples

We suppose that a dataset consisting of M episodes of N steps is available. The agent initially starts from a randomly selected state $\boldsymbol{s}_1$ following the initial-state probability density $p_{\mathrm{I}}(\boldsymbol{s})$, and chooses an action based on a policy $\pi(a_n|\boldsymbol{s}_n)$. Then the agent makes a transition following $p_{\mathrm{T}}(\boldsymbol{s}_{n+1}|\boldsymbol{s}_n,a_n)$, and receives a reward r_n ($= R(\boldsymbol{s}_n, a_n, \boldsymbol{s}_{n+1})$). This is repeated for N steps—thus the training data $\mathcal{D}^\pi$ is expressed as

$$\mathcal{D}^\pi \equiv \{d_n^\pi\}_{n=1}^N, \tag{2}$$

where each time-step data d_n^π consists of M sets of 3-tuple elements observed at each time step n as

$$d_n^\pi \equiv \{(\boldsymbol{s}_{m,n}^\pi, a_{m,n}^\pi, r_{m,n}^\pi)\}_{m=1}^M. \tag{3}$$

Let $\eta_{m,n}^\pi$ be the return in the m-th episode defined by

$$\eta_{m,n}^\pi \equiv \sum_{n'=n}^{N} \gamma^{n'-n} r_{m,n'}^\pi. \tag{4}$$

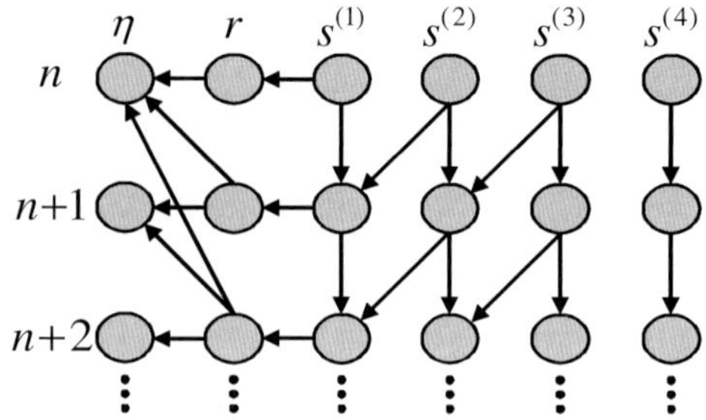

Fig. 2. An example of dependency between state features and returns. State features $s^{(1)}$, $s^{(2)}$ and $s^{(3)}$ influence returns at the same time-step, e.g., from $s_n^{(3)}$ to η_n.

3 Feature Selection via Conditional Mutual Information

In this section, we describe our proposed feature selection method.

Let η_n and $\boldsymbol{s}_n = (s_n^{(1)}, s_n^{(2)}, \ldots, s_n^{(v)})$ be the return and the state features at the n-th time step. For u $(\leq v)$ being the number of features we want to select, our goal is to find a 'subset' $\boldsymbol{z}_n = (z_n^{(1)}, z_n^{(2)}, \ldots, z_n^{(u)})^\top$ of the state features $\boldsymbol{s}_n$ such that

$$\eta_n \perp \boldsymbol{s}_n \mid \boldsymbol{z}_n, \quad \forall n = 1, 2, \ldots, N. \tag{5}$$

This means that, for all time steps, the return η_n is conditionally independent of the entire state features $\boldsymbol{s}_n$ given the subset $\boldsymbol{z}_n$.

The criterion (5) allows us to capture an indirect dependency from state features to rewards r since returns η contain all subsequent rewards. This is illustrated using a simple example in Figure 2. A state feature $s^{(3)}$ does not influence a reward r at the next time-step, but it does affect a return η at the same time-step since the return is the sum of discounted subsequent rewards, i.e., $\eta_n = r_n + \gamma r_{n+1} + \gamma^2 r_{n+2} + \cdots$.

3.1 Conditional Mutual Information

Mutual information (MI) is a popular measure of independence between random variables [14]. Here, we use a variant of MI based on the squared-loss [15] defined by

$$I^\pi(\eta; \mathrm{z}|\mathrm{n} = n) \equiv \iint \left(\frac{p^\pi(\eta, \boldsymbol{z}|n)}{p^\pi(\eta|n)p^\pi(\boldsymbol{z}|n)} - 1 \right)^2 p^\pi(\eta|n)p^\pi(\boldsymbol{z}|n)\mathrm{d}\eta\mathrm{d}\boldsymbol{z}, \tag{6}$$

where $I^\pi(\eta; \mathrm{z}|\mathrm{n} = n)$ denotes MI between return η and features $\boldsymbol{z}$ at the n-th time step when following a policy π. $p^\pi(\eta, \boldsymbol{z}|n)$ denotes the joint density of η and $\boldsymbol{z}$ at the n-th time step, and $p^\pi(\eta|n)$ and $p^\pi(\boldsymbol{z}|n)$ denote the marginal densities of return η and features $\boldsymbol{z}$ at the n-th time step, respectively. $I^\pi(\eta; \mathrm{z}|\mathrm{n} = n)$ is non-negative and is equal to zero if and only if

$$p^\pi(\eta, \boldsymbol{z}|n) = p^\pi(\eta|n)p^\pi(\boldsymbol{z}|n),$$

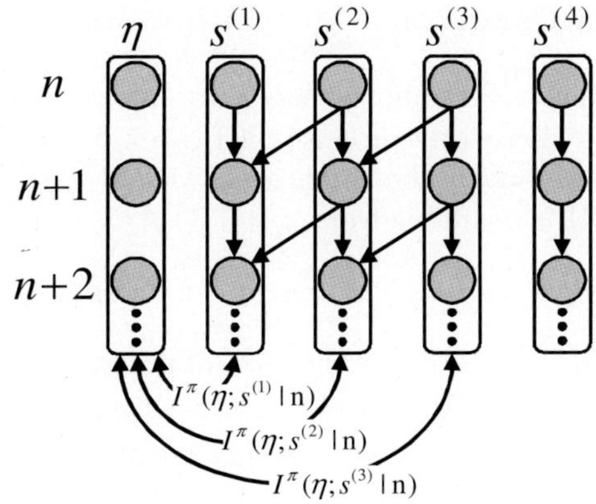

Fig. 3. An example of dependency between returns and elements of states. State elements $s^{(1)}$, $s^{(2)}$, and $s^{(3)}$ directly influence the return η.

i.e., η and $\boldsymbol{z}$ are conditionally independent of each other given n.

We propose to use *conditional MI* $I^{\pi}(\eta; \mathrm{z}|\mathrm{n})$ as our feature selection criterion, which is defined as the average of MI $I^{\pi}(\eta; \mathrm{z}|\mathrm{n} = n)$ over time steps $n = 1, 2, \ldots, N$ [14]:

$$I^{\pi}(\eta; \mathrm{z}|\mathrm{n}) = \frac{1}{N} \sum_{n=1}^{N} I^{\pi}(\eta; \mathrm{z}|\mathrm{n} = n).$$

The conditional MI between returns and state features can be seen as a measure of dependency between returns and state-feature sequences as illustrated in Figure 3.

The rationale behind the use of conditional MI for feature selection relies on the following lemma (its proof is provided in Appendix).

Lemma 1.

$$\begin{aligned} I^{\pi}(\eta; \mathrm{s}|\mathrm{n}) - I^{\pi}(\eta; \mathrm{z}|\mathrm{n}) = & \frac{1}{N} \sum_{n=1}^{N} \iint \frac{p^{\pi}(\eta, \boldsymbol{z}|n)^2}{p^{\pi}(\eta|n) p^{\pi}(\boldsymbol{z}|n)^2} \\ & \times \left(\frac{p^{\pi}(\eta, \boldsymbol{s}|\boldsymbol{z}, n)}{p^{\pi}(\boldsymbol{s}|\boldsymbol{z}, n) p^{\pi}(\eta|\boldsymbol{z}, n)} - 1 \right)^2 p^{\pi}(\boldsymbol{s}|n) \mathrm{d}\boldsymbol{s} \mathrm{d}\eta \\ \geq & \, 0. \end{aligned}$$

This lemma implies that $I^{\pi}(\eta; \mathrm{s}|\mathrm{n}) \geq I^{\pi}(\eta; \mathrm{z}|\mathrm{n})$ and the equality holds if and only if

$$p^{\pi}(\eta, \boldsymbol{s}|\boldsymbol{z}, n) = p^{\pi}(\eta|\boldsymbol{z}, n) p^{\pi}(\boldsymbol{s}|\boldsymbol{z}, n), \ \forall n = 1, 2, \ldots, N.$$

This is equivalent to Eq.(5), and thus Eq.(5) can be attained by maximizing $I^{\pi}(\eta; \mathrm{z}|\mathrm{n})$ with respect to $\boldsymbol{z}$.

3.2 Estimation of Conditional Mutual Information

Since $I^\pi(\eta; \mathrm{z}|\mathrm{n})$ is not accessible, it needs to be estimated from data samples. Here, we employ a recently-proposed MI estimator called *least-squares MI* (LSMI) [15] for approximating the conditional MI $I^\pi(\eta; \mathrm{z}|\mathrm{n})$. A MATLAB® implementation of LSMI is available from

`http://sugiyama-www.cs.titech.ac.jp/~sugi/software/LSMI/`

The basic idea of LSMI is to estimate the ratio of probability densities $w_n(\eta, \boldsymbol{z}) \equiv \frac{p^\pi(\eta,\boldsymbol{z}|n)}{p^\pi(\eta|n)p^\pi(\boldsymbol{z}|n)}$ contained in MI without going through density estimation of $p^\pi(\eta, \boldsymbol{z}|n)$, $p^\pi(\eta|n)$, and $p^\pi(\boldsymbol{z}|n)$. Since density estimation is known to be a hard task [16], avoiding density estimation and directly estimating their ratio would be preferable [17,18].

The density ratio function $w_n(\eta, \boldsymbol{z})$ is approximated by the following linear model:

$$\widehat{w}_n(\eta, \boldsymbol{z}) \equiv \boldsymbol{\alpha}_n^\top \boldsymbol{\psi}_n(\eta, \boldsymbol{z}),$$

where $\boldsymbol{\alpha}_n = (\alpha_{n,1}, \alpha_{n,2}, \ldots, \alpha_{n,B})^\top$ are parameters to be learned, B is the number of parameters, and

$$\boldsymbol{\psi}_n(\eta, \boldsymbol{z}) = (\psi_{n,1}(\eta, \boldsymbol{z}), \psi_{n,2}(\eta, \boldsymbol{z}), \ldots, \psi_{n,B}(\eta, \boldsymbol{z}))^\top$$

are basis functions such that

$$\psi_{n,b}(\eta, \boldsymbol{z}) \geq 0, \;\; \forall b, \;\; \forall(\eta, \boldsymbol{z}).$$

The parameter $\boldsymbol{\alpha}_n$ is determined so that the following squared error J_0 is minimized:

$$\begin{aligned} J_0(\boldsymbol{\alpha}_n) &\equiv \frac{1}{2} \iint \left(\widehat{w}_n(\eta, \boldsymbol{z}) - w_n(\eta, \boldsymbol{z})\right)^2 p^\pi(\eta|n)p^\pi(\boldsymbol{z}|n)\mathrm{d}\eta\mathrm{d}\boldsymbol{z} \\ &= \frac{1}{2} \iint \widehat{w}_n^2(\eta, \boldsymbol{z})p^\pi(\eta|n)p^\pi(\boldsymbol{z}|n)\mathrm{d}\eta\mathrm{d}\boldsymbol{z} \\ &\quad - \iint \widehat{w}_n(\eta, \boldsymbol{z})p^\pi(\eta, \boldsymbol{z}|n)\mathrm{d}\eta\mathrm{d}\boldsymbol{z} + C, \end{aligned}$$

where

$$C \equiv \frac{1}{2} \iint w_n^2(\eta, \boldsymbol{z})p^\pi(\eta, \boldsymbol{z}|n)\mathrm{d}\eta\mathrm{d}\boldsymbol{z}$$

is a constant and thus can be safely ignored. Let us denote the first two terms by J:

$$\begin{aligned} J(\boldsymbol{\alpha}_n) &\equiv J_0(\boldsymbol{\alpha}_n) - C \\ &= \frac{1}{2}\boldsymbol{\alpha}_n^\top \boldsymbol{H}_n \boldsymbol{\alpha}_n - \boldsymbol{h}_n^\top \boldsymbol{\alpha}_n, \end{aligned} \tag{7}$$

where

$$\boldsymbol{H}_n \equiv \iint \boldsymbol{\psi}_n(\eta, \boldsymbol{z})\boldsymbol{\psi}_n(\eta, \boldsymbol{z})^\top p^\pi(\eta|n)p^\pi(\boldsymbol{z}|n)\mathrm{d}\eta\mathrm{d}\boldsymbol{z},$$

$$\boldsymbol{h}_n \equiv \iint \boldsymbol{\psi}_n(\eta, \boldsymbol{z})p^\pi(\eta, \boldsymbol{z}|n)\mathrm{d}\eta\mathrm{d}\boldsymbol{z}.$$

The expectations in $\boldsymbol{H}_n$ and $\boldsymbol{h}_n$ are approximated by the empirical averages using a one-step data sample d_n^π (see Eq.(3)).

$$\widehat{\boldsymbol{H}}_n \equiv \frac{1}{M^2}\sum_{m,m'=1}^{M} \boldsymbol{\psi}_n(\eta_{m,n}, \boldsymbol{z}_{m',n})\boldsymbol{\psi}_n(\eta_{m,n}, \boldsymbol{z}_{m',n})^\top,$$

$$\widehat{\boldsymbol{h}}_n \equiv \frac{1}{M}\sum_{m=1}^{M} \boldsymbol{\psi}_n(\eta_{m,n}, \boldsymbol{z}_{m,n}).$$

Then the following optimization problem is obtained:

$$\widehat{\boldsymbol{\alpha}}_n \equiv \mathop{\arg\min}_{\boldsymbol{\alpha}_n \in \mathbb{R}^B} \left[\frac{1}{2}\boldsymbol{\alpha}_n^\top \widehat{\boldsymbol{H}}_n \boldsymbol{\alpha}_n - \widehat{\boldsymbol{h}}_n^\top \boldsymbol{\alpha}_n + \frac{\lambda}{2}\boldsymbol{\alpha}_n^\top \boldsymbol{\alpha}_n\right], \tag{8}$$

where a regularization term $\lambda\boldsymbol{\alpha}_n^\top\boldsymbol{\alpha}_n/2$ is included. Differentiating the above objective function with respect to $\boldsymbol{\alpha}_n$ and equating it to zero, the solution can be obtained analytically as

$$\widehat{\boldsymbol{\alpha}}_n = (\widehat{\boldsymbol{H}}_n + \lambda \boldsymbol{I}_B)^{-1}\widehat{\boldsymbol{h}}_n,$$

where I_B denotes the B-dimensional identity matrix.

Using a density ratio estimator $\widehat{\boldsymbol{\alpha}}_n^\top \boldsymbol{\psi}_n(\eta, \boldsymbol{z})$, we can construct a conditional MI estimator between return and state-feature sequences as

$$\widehat{I}^\pi(\eta; \mathrm{z}|\mathrm{n}) \equiv \frac{1}{N}\sum_{n=1}^{N} \widehat{I}^\pi(\eta_n; \mathrm{z}_n), \tag{9}$$

where $\widehat{I}^\pi(\eta_n; \mathrm{z}_n)$ is an MI estimator between returns and state features at the n-th time step given as

$$\widehat{I}^\pi(\eta_n; \mathrm{z}_n) \equiv \frac{1}{M^2}\sum_{m,m'=1}^{M} \left(\widehat{\boldsymbol{\alpha}}_n^\top \boldsymbol{\psi}_n(\eta_{m,n}, \boldsymbol{z}_{m',n}) - 1\right)^2. \tag{10}$$

3.3 Feature Selection Algorithm

Finally, we describe how features are selected based on the conditional MI estimator $\widehat{I}^\pi(\eta; \mathrm{z}|\mathrm{n})$.

Forward selection and *backward elimination* would be two major strategies of feature selection [6,19]. Here we employ forward selection since it was computationally more efficient and performed well in our preliminary experiments.

Algorithm 1: FORWARDSELECTION($u, \mathcal{D}^\pi$)

$//u$: Number of features we want to choose
$//\mathcal{D}^\pi$: Data samples collected following π
$//v$: Number of all features
$//\mathcal{I}$: Remaining feature indices
$//\mathcal{J}$: Chosen feature indices
$\mathcal{I} \leftarrow \{1, 2, \ldots, v\}$
$\mathcal{J} \leftarrow \{\}$
for $u' = 1, 2, \ldots, u$
  // Find the feature that maximizes the conditional mutual information
  $k \leftarrow \arg\max_{i \in \mathcal{I}} \sum_{n=1}^{N} \mathrm{LSMI}(d_n^\pi, \{s_n^{(j)}\}_{j \in \mathcal{J}} \cup s_n^{(i)})$
  $\mathcal{I} \leftarrow \mathcal{I} \backslash k$ // Remove k from $\mathcal{I}$
  $\mathcal{J} \leftarrow \mathcal{J} \cup k$ // Add k to $\mathcal{J}$
return $(\mathcal{J})$

Fig. 4. A pseudo code of the proposed feature selection algorithm with forward selection. By the LSMI function, MI between return and state features is computed using the n-th time step data d_n^π.

Let $\mathcal{J}$ be the set of chosen feature indices. The forward selection algorithm starts from the empty feature-index set $\mathcal{J} = \{\}$. The index of the most relevant feature, together with features whose indices are included in $\mathcal{J}$, is sequentially added to $\mathcal{J}$ at each iteration. The relevance of each state feature $s^{(i)}$ is evaluated using the conditional MI estimator $\widehat{I}^\pi(\eta; \mathrm{z}|\mathrm{n})$ described in Section 3.2. This forward selection process is repeated u times, where u is the number of features we want to choose.

A pseudo code of the proposed feature selection algorithm with forward selection is described in Figure 4.

4 Numerical Experiments

In this section, we evaluate the performance of our proposed feature selection method on a grid-world navigation problem illustrated in Figure 5. The two-dimensional maze consists of walls (black cells) and target states (light-gray cells). The goal of the task is to navigate an agent to the target location by avoiding the walls.

4.1 Setup

The state space $\mathcal{S}$ consists of 14-dimensional discrete features $\boldsymbol{s} = (s^{(1)}, s^{(2)}, \ldots, s^{(14)})^\top$, where $s^{(1)}, s^{(2)} \in \{1, 2, \ldots, 10\}$ are the horizontal and vertical positions of the agent, respectively. $s^{(3)} \in \{0, 1, 2, \ldots, 20\}$ is the remaining battery level; it is initially set to 20 (fully charged) and is decreased by 1 at

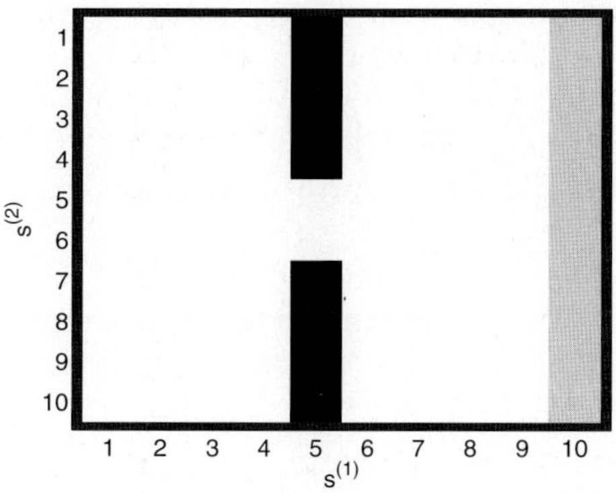

Fig. 5. A grid-world navigation problem. An agent is placed randomly in the grid-world, and can move around in the white area based on the four actions: moving up, down, left, or right at every time step. Black boxes in the middle represent walls through which the agent cannot go, and target location to which we want to guide the agent is the light-gray area at $s^{(1)} = 10$.

every agent's movement. The rest of features $s^{(4)}, s^{(5)}, s^{(6)}, \ldots, s^{(14)}$ corresponds to noise, each of which independently follows the Gaussian distribution with different mean:

$$\frac{1}{\sigma_{\text{noise}}\sqrt{2\pi}} \exp\left(-\frac{(s^{(i)} - \nu_i)^2}{2\sigma_{\text{noise}}^2}\right), \quad \forall i = 4, 5, 6, \ldots, 14.$$

We set $\nu_i = i - 3$ and $\sigma_{\text{noise}} = 1$, and round the value of $s^{(i)}$ down to the nearest integer for discretization. These additional dimensions of the state space may be regarded as information brought by irrelevant sensors such as sound, smell, and temperature sensors.

The action space $\mathcal{A}$ consists of four discrete actions, each of which corresponds to the direction of the agent's move: up, down, left, and right. For instance, if the agent chooses the 'right'-action, $s^{(1)}$ is incremented unless there is an wall and the battery level ($s^{(3)}$) is zero.

The reward +2 is given when the agent visits the target location; otherwise the reward is zero:

$$R(\boldsymbol{s}, a, \boldsymbol{s}') = \begin{cases} 2 & \text{if } s'^{(1)} = 10, \\ 0 & \text{otherwise.} \end{cases}$$

The discount factor is set to $\gamma = 0.95$.

Data samples $\mathcal{D}^\pi$ consisting M episodes with $N = 20$ steps are collected. The initial position of the agent is set to

$$(\boldsymbol{s}_1^{(1)}, \boldsymbol{s}_1^{(2)}) = (1, \beta),$$

where β is randomly chosen from $\{1, 2, \ldots, 10\}$. Then, the agent follows a stochastic policy $\pi(a|\boldsymbol{s})$ defined by

$$\pi(a|\boldsymbol{s}) = \begin{cases} 0.7 & \text{if } a = a^*, \\ 0.1 & \text{otherwise.} \end{cases}$$

where a^* is 'down' when $s^{(1)} = 4$ and $1 \leq s^{(2)} \leq 4$, a^* is 'up' when $s^{(1)} = 4$ and $7 \leq s^{(2)} \leq 10$, and a^* is 'right' in other states. We compute the conditional MI estimator $\widehat{I}^\pi(\eta; \mathrm{z}|\mathrm{n})$ from the dataset $\mathcal{D}^\pi$ (see Section 3.2). Gaussian kernels are used as basis functions:

$$\psi_{n,b}(\eta, \boldsymbol{z}) \equiv \exp\left(-\frac{\|(\eta, \boldsymbol{z}^\top)^\top - \boldsymbol{\mu}_{n,b}\|^2}{2\sigma_n^2}\right),$$

where $\boldsymbol{\mu}_{n,b}$ and σ_n are the mean and standard deviation of the Gaussian kernel, respectively. We set $B = M$, i.e., the number B of basis functions is equal to the number M of episodes. The mean $\boldsymbol{\mu}_{n,b}$ is selected from the dataset d_n^π as $\boldsymbol{\mu}_{n,b} = (\eta_{n,b}^\pi, \boldsymbol{z}_{n,b}^{\pi\ \top})^\top$. The standard deviation σ_n as well as the regularization parameter λ (see Eq.(8)) is determined by cross-validation with respect to J (see Eq.(7)) [15].

We compare the performance of our proposed method with the KDR-based method [11]. The feature selection criterion used by the KDR-based method is the conditional independence between states $\boldsymbol{s}$ and its subset $\boldsymbol{s}^r$ which directly influences rewards:

$$\boldsymbol{s}_{n+1}^r \perp \boldsymbol{s}_n \mid \boldsymbol{z}_n, \quad \forall n = 1, 2, \ldots, N,$$

where $\boldsymbol{s}^r = \{s^{(1)}\}$ in the current navigation problem. This criterion is evaluated using the *conditional cross-covariance* operator in a Gaussian reproducing kernel Hilbert space.

Similarly to our proposed method, we implement the KDR-based method based on the forward selection strategy: starting from the empty set $\mathcal{J} = \{\}$, the index of the state feature $s^{(i)}$, which, together with features whose indices are included in $\mathcal{J}$, attains the above conditional independence the most is added to $\mathcal{J}$ at every iteration. Following the suggestion in [13], we fix the width of the Gaussian kernel to the median of the distance between all the data samples, and fix the regularization parameter to 0.1.

To illustrate how the feature selection methods work in our grid-world navigation task, we run the forward selection algorithms for 14 iterations to rank all the state features. Let us consider the following two cases: the "without-gravel" and "with-gravel" scenarios. In the "without-gravel" scenario, state features $s^{(1)}$ and $s^{(2)}$ should have higher ranks because the horizontal position of the agent $s^{(1)}$ determines the reward directly and its vertical position $s^{(2)}$ is necessary to avoid walls in the middle of the maze.

On the other hand, in the "with-gravel" scenario, gravel exists in some grids and the state feature $s^{(4)}$ detects the existence of gravel; when $s^{(4)} = 2$, there

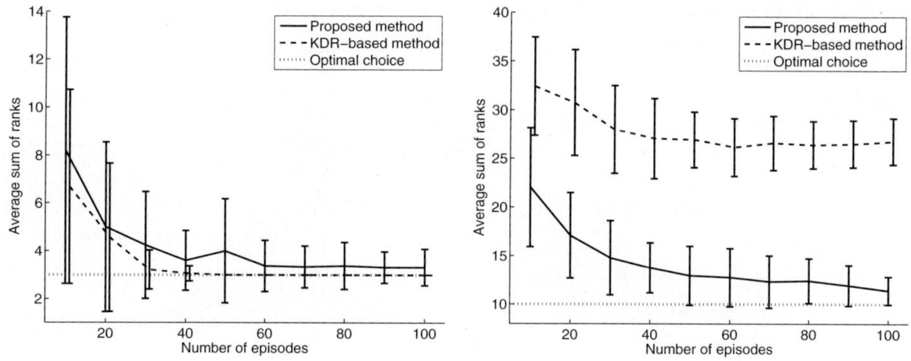

(a) "Without-gravel" scenario (b) "With-gravel" scenario

Fig. 6. Feature selection performance in the grid-world navigation task. The graphs depict the sum of ranks of relevant features averaged over 50 trials as a function of the number M of episodes. The optimal values are 3 (=1 + 2) and 10 (= 1 + 2 + 3 + 4) in the "without-gravel" and "with-gravel" scenarios, respectively.

exists gravel in the right grid of the agent. The agent can avoid the gravel area by moving to the left when $s^{(4)} = 2$; otherwise, it gets into the gravel area which continues for 4 time steps (this is indicated by $s^{(4)} = 3$). When the agent is in the gravel area, the battery level $s^{(3)}$ is decreased by 3 at each step. Then, the agent can not be able to reach the target place due to lack of battery (recall that the battery level is decreased by 1 at each step outside the gravel area). This indicates that the gravel-feature $s^{(4)}$ indirectly influences rewards after several time steps through the battery level $s^{(3)}$ and the horizontal position $s^{(1)}$. Therefore, in this case, in addition to $s^{(1)}$ and $s^{(2)}$, $s^{(3)}$ and $s^{(4)}$ should also have higher ranks to avoid the gravel area.

4.2 Results

Figure 6(a) depicts the sum of ranks of the features $s^{(1)}$ and $s^{(2)}$ averaged over 50 trials as a function of the number M of episodes for the "without-gravel" scenario. Since the features $s^{(1)}$ and $s^{(2)}$ should be ranked first and second, the optimal value is 3 (= 1+2). The graph overall shows that the performance of both the proposed and KDR-based methods improves as the number M of episodes increases. The KDR-based method converges to the optimal value (= 3) at 40 episodes, while a small error remains in the proposed method. This difference is caused by the fact that the battery level $s^{(3)}$ is occasionally ranked higher than the position $s^{(1)}$ and $s^{(2)}$ in the proposed method. Since the battery level is also somewhat relevant to returns, the result of the proposed method would be reasonable.

Figure 6(b) depicts the sum of ranks of the features $s^{(1)}$, $s^{(2)}$, $s^{(3)}$, and $s^{(4)}$ averaged over 50 trials as a function of the number M of episodes for the "with-gravel" scenario. Unlike the case of the "without-gravel" scenario, the performance

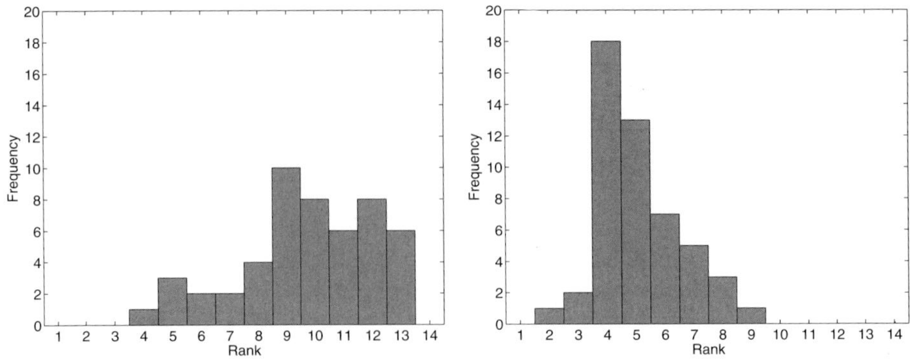

(a) KDR-based method (b) Proposed method

Fig. 7. Histograms of the rank of the feature $s^{(4)}$ in the "with-gravel" scenario. The number M of episodes is fixed to 100 and the number of trials is 50.

improvement of the KDR-based method is very slow and is saturated after 60 episodes. This happened because evaluating the one-step dependency from the gravel feature $s^{(4)}$ to the horizontal position $s^{(1)}$ cannot find the relevance of $s^{(4)}$ to subsequent rewards.

Figure 7(a) depicts the histogram of the rank of the gravel-feature $s^{(4)}$ in the "with-gravel" scenario when the number M of episodes is fixed to 100. The graph shows that the KDR-based method ranks $s^{(4)}$ in lower positions, particularly in the range between 8 and 13. This implies that the feature $s^{(4)}$ is less frequently selected by the KDR-based method and then the gravel cannot be avoided properly.

On the other hand, the performance of the proposed method improves as the number M of episodes increases, and approaches the optimal value (= 10) (see Figure 6(b)). Figure 7(b)) shows that the proposed method ranks $s^{(4)}$ in higher positions, particularly around 4. This was achieved because the proposed method evaluates the dependency between returns η and $s^{(4)}$.

Overall, the proposed method was shown to be a promising feature selection method in RL.

5 Conclusions

In real-world reinforcement learning problems, selecting a subset of relevant attributes from the high-dimensional state variable is considerably important since standard reinforcement learning methods do not perform well with high-dimensional state spaces. An existing feature selection approach relies on the conditional independence between state and its subset which directly influences rewards. However, this is not appropriate when there is an indirect dependency between states and rewards, which is often the case in practical RL scenarios.

To overcome this limitation, we proposed a new framework of feature selection by considering the dependency between return and state-feature sequences. Our framework adopts conditional mutual information as the dependency measure, and it is approximated using least-squares estimation. The effectiveness of the proposed method was shown through experiments.

Acknowledgment

The authors were supported by the FIRST program.

Appendix: Proof of Lemma 1

Here, we give a proof of Lemma 1.

From the definition of conditional mutual information, we have

$$
\begin{aligned}
&I^{\pi}(\eta;\mathrm{s}|\mathrm{n}) - I^{\pi}(\eta;\mathrm{z}|\mathrm{n}) \\
&= \frac{1}{N}\sum_{n=1}^{N}\int p^{\pi}(\eta|n)\left\{\int\left(\frac{p^{\pi}(\eta,\boldsymbol{s}|n)}{p^{\pi}(\eta|n)p^{\pi}(\boldsymbol{s}|n)}-1\right)^{2}p^{\pi}(\boldsymbol{s}|n)\mathrm{d}\boldsymbol{s}\right. \\
&\qquad\qquad \left.-\int\left(\frac{p^{\pi}(\eta,\boldsymbol{z}|n)}{p^{\pi}(\eta|n)p^{\pi}(\boldsymbol{z}|n)}-1\right)^{2}p^{\pi}(\boldsymbol{z}|n)\mathrm{d}\boldsymbol{z}\right\}\mathrm{d}\eta \\
&= \frac{1}{N}\sum_{n=1}^{N}\int p^{\pi}(\eta|n)\left\{\int\left(\frac{p^{\pi}(\eta,\boldsymbol{s}|n)}{p^{\pi}(\eta|n)p^{\pi}(\boldsymbol{s}|n)}\right)^{2}p^{\pi}(\boldsymbol{s}|n)\mathrm{d}\boldsymbol{s} - 2\int\frac{p^{\pi}(\eta,\boldsymbol{s}|n)}{p^{\pi}(\eta|n)}\mathrm{d}\boldsymbol{s}\right. \\
&\qquad\qquad \left.-\int\left(\frac{p^{\pi}(\eta,\boldsymbol{z}|n)}{p^{\pi}(\eta|n)p^{\pi}(\boldsymbol{z}|n)}\right)^{2}p^{\pi}(\boldsymbol{z}|n)\mathrm{d}\boldsymbol{z} + 2\int\frac{p^{\pi}(\eta,\boldsymbol{z}|n)}{p^{\pi}(\eta|n)}\mathrm{d}\boldsymbol{z}\right\}\mathrm{d}\eta \\
&= \frac{1}{N}\sum_{n=1}^{N}\int p^{\pi}(\eta|n)\left\{\int\left(\frac{p^{\pi}(\eta,\boldsymbol{s}|n)}{p^{\pi}(\eta|n)p^{\pi}(\boldsymbol{s}|n)}\right)^{2}p^{\pi}(\boldsymbol{s}|n)\mathrm{d}\boldsymbol{s}\right. \\
&\qquad\qquad \left.-\int\left(\frac{p^{\pi}(\eta,\boldsymbol{z}|n)}{p^{\pi}(\eta|n)p^{\pi}(\boldsymbol{z}|n)}\right)^{2}p^{\pi}(\boldsymbol{z}|n)\mathrm{d}\boldsymbol{z}\right\}\mathrm{d}\eta \\
&= \frac{1}{N}\sum_{n=1}^{N}\int p^{\pi}(\eta|n)\int\left(\frac{p^{\pi}(\eta,\boldsymbol{s}|n)}{p^{\pi}(\eta|n)p^{\pi}(\boldsymbol{s}|n)}-\frac{p^{\pi}(\eta,\boldsymbol{z}|n)}{p^{\pi}(\eta|n)p^{\pi}(\boldsymbol{z}|n)}\right)^{2}p^{\pi}(\boldsymbol{s}|n)\mathrm{d}\boldsymbol{s}\mathrm{d}\eta.
\end{aligned}
$$

To obtain the second equality above, we used

$$
\int\frac{p^{\pi}(\eta,\boldsymbol{s}|n)}{p^{\pi}(\eta|n)}\mathrm{d}\boldsymbol{s} = \iint\frac{p^{\pi}(\eta,\boldsymbol{z},\bar{\boldsymbol{z}}|n)}{p^{\pi}(\eta|n)}\mathrm{d}\boldsymbol{z}\mathrm{d}\bar{\boldsymbol{z}} = \int\frac{p^{\pi}(\eta,\boldsymbol{z}|n)}{p^{\pi}(\eta|n)}\mathrm{d}\boldsymbol{z},
$$

where $\bar{\boldsymbol{z}}$ is the complement of $\boldsymbol{z}$. To obtain the third equality, we used

$$
\begin{aligned}
&\int\left(\frac{p^{\pi}(\eta,\boldsymbol{z}|n)}{p^{\pi}(\eta|n)p^{\pi}(\boldsymbol{z}|n)}\right)^{2}p^{\pi}(\boldsymbol{z}|n)\mathrm{d}\boldsymbol{z} \\
&\qquad = \int\frac{p^{\pi}(\eta,\boldsymbol{s}|n)p^{\pi}(\eta,\boldsymbol{z}|n)}{p^{\pi}(\eta|n)^{2}p^{\pi}(\boldsymbol{s}|n)p^{\pi}(\boldsymbol{z}|n)}p^{\pi}(\boldsymbol{s}|n)\mathrm{d}\boldsymbol{s}.
\end{aligned}
$$

Since

$$p^\pi(\boldsymbol{s}|\boldsymbol{z},n)p^\pi(\boldsymbol{z}|n) = p^\pi(\boldsymbol{s},\boldsymbol{z}|n) = p^\pi(\boldsymbol{s}|n),$$
$$\frac{p^\pi(\eta,\boldsymbol{z}|n)}{p^\pi(\eta|\boldsymbol{z},n)p^\pi(\boldsymbol{z}|n)} = 1,$$
$$\frac{p^\pi(\eta,\boldsymbol{s}|n)}{p^\pi(\eta|n)p^\pi(\boldsymbol{s}|n)} = \frac{p^\pi(\eta,\boldsymbol{s}|n)p^\pi(\eta,\boldsymbol{z}|n)}{p^\pi(\boldsymbol{s}|\boldsymbol{z},n)p^\pi(\eta|\boldsymbol{z},n)p^\pi(\eta|n)p^\pi(\boldsymbol{z}|n)},$$

we have

$$\begin{aligned}
&I^\pi(\eta;\mathrm{s}|\mathrm{n}) - I^\pi(\eta;\mathrm{z}|\mathrm{n}) \\
&= \frac{1}{N}\sum_{n=1}^{N}\int p^\pi(\eta|n)\int \frac{p^\pi(\eta,\boldsymbol{z}|n)^2}{p^\pi(\eta|n)^2 p^\pi(\boldsymbol{z}|n)^2} \\
&\qquad\qquad \times\left(\frac{p^\pi(\eta,\boldsymbol{s}|\boldsymbol{z},n)}{p^\pi(\boldsymbol{s}|\boldsymbol{z},n)p^\pi(\eta|\boldsymbol{z},n)} - 1\right)^2 p^\pi(\boldsymbol{s}|n)\mathrm{d}\boldsymbol{s}\mathrm{d}\eta, \\
&= \frac{1}{N}\sum_{n=1}^{N}\iint \frac{p^\pi(\eta,\boldsymbol{z}|n)^2}{p^\pi(\eta|n)p^\pi(\boldsymbol{z}|n)^2}\left(\frac{p^\pi(\eta,\boldsymbol{s}|\boldsymbol{z},n)}{p^\pi(\boldsymbol{s}|\boldsymbol{z},n)p^\pi(\eta|\boldsymbol{z},n)} - 1\right)^2 p^\pi(\boldsymbol{s}|n)\mathrm{d}\boldsymbol{s}\mathrm{d}\eta,
\end{aligned}$$

which concludes the proof. (Q.E.D.)

References

1. Sutton, R.S., Barto, G.A.: Reinforcement Learning: An Introduction. MIT Press, Cambridge (1998)
2. Lagoudakis, M.G., Parr, R.: Least-squares policy iteration. Journal of Machine Learning Research 4, 1107–1149 (2003)
3. Williams, R.J.: Simple statistical gradient-following algorithms for connectionist reinforcement learning. Machine Learning 8, 229–256 (1992)
4. Dayan, P., Hinton, G.E.: Using expectation-maximization for reinforcement learning. Neural Computation 9(2), 271–278 (1997)
5. Kakade, S.: A natural policy gradient. Advances in Neural Information Processing Systems 14, 1531–1538 (2002)
6. Guyon, I., Elisseeff, A.: An introduction to variable and feature selection. Journal of Machine Learning Research 3, 1157–1182 (2003)
7. Keller, P.W., Mannor, S., Precup, D.: Automatic basis function construction for approximate dynamic programming and reinforcement learning. In: Proceedings of the 23rd International Conference on Machine learning, pp. 449–456 (2006)
8. Parr, R., Painter, C.W., Li, L., Littman, L.M.: Analyzing feature generation for value-function approximation. In: Proceedings of the 24th International Conference on Machine Learning, pp. 737–744 (2007)
9. Parr, R., Li, L., Taylor, G., Painter, C.W., Littman, L.M.: An analysis of linear models, linear value-function approximation, and feature selection for reinforcement learning. In: Proceedings of the 25th International Conference on Machine Learning, pp. 752–759 (2008)

10. Goldberger, J., Roweis, S., Hinton, G., Salakhutdinov, R.: Neighbourhood components analysis. In: Saul, L.K., Weiss, Y., Bottou, L. (eds.) Advances in Neural Information Processing Systems, vol. 17, pp. 513–520. MIT Press, Cambridge (2005)
11. Morimoto, J., Hyon, S., Atkeson, C.G., Cheng, G.: Low-dimensional feature extraction for humaniod locomotion using kernel dimension reduction. In: Proceedings of 2007 IEEE International Conference on Robotics and Automation, pp. 2711–2716 (2008)
12. Kroon, M., Whiteson, S.: Automatic feature selection for model-based reinforcement learning in factored mdps. In: Proceedings of the 2009 International Conference on Machine Learning and Applications, pp. 324–330 (2009)
13. Fukumizu, K., Bach, F.R., Jordan, M.I.: Kernel dimension reduction in regression. Annals of Statistics 37(4), 1871–1905 (2009)
14. MacKay, D.J.C.: Information Theory, Inference, and Learning Algorithms. Cambridge University Press, Cambridge (2003)
15. Suzuki, T., Sugiyama, M., Kanamori, T., Sese, J.: Mutual information estimation reveals global associations between stimuli and biological processes. BMC Bioinformatics 10(1), S52 (2009)
16. Vapnik, V.N.: Statistical Learning Theory. Wiley, New York (1998)
17. Kanamori, T., Hido, S., Sugiyama, M.: A least-squares approach to direct importance estimation. Journal of Machine Learning Research 10, 1391–1445 (2009)
18. Kanamori, T., Suzuki, T., Sugiyama, M.: Condition number analysis of kernel-based density ratio estimation. Technical report, arXiv (2009), `http://www.citebase.org/abstract?id=oai:arXiv.org:0912.2800`
19. Song, L., Smola, A., Gretton, A., Borgwardt, K., Bedo, J.: Supervised feature selection via dependence estimation. In: Proceedings of the 24th International Conference on Machine Learning, pp. 823–830 (2007)

Bagging for Biclustering: Application to Microarray Data

Blaise Hanczar and Mohamed Nadif

LIPADE, University Paris Descartes, 45 rue des saint-pres, 75006 Paris, France
blaise.hanczar@parisdescartes.fr

Abstract. One of the major tools of transcriptomics is the biclustering that simultaneously constructs a partition of both examples and genes. Several methods have been proposed for microarray data analysis that enables to identify groups of genes with similar expression pro?les only under a subset of examples. We propose to improve the quality of these biclustering methods by adapting the approach of bagging to biclustering problems. The principle consists in generating a set of biclusters and aggregating the results. Our method has been tested with success on artificial and real datasets.

1 Introduction

The capacity of microarray to measure simultaneously the expression of a whole genome under different experimental condition, is of great interest for biologists. Clustering techniques are one of the major tools to analyse these data. They allow the discovery of groups of genes that share a similar expression profile over all experimental conditions. We assume that genes that share similar expression profiles, have close biological functions. The clustering of gene expression over homogeneous conditions is therefore a relevant tool for functional analyses [2]. With the classic methods of clustering, like k-means, hierarchical clustering or self organizing maps, it is assume that genes in the same cluster have a similar behavior over all conditions. However, when the experimental conditions are heterogeneous, it may be more appropriate to form clusters of genes over only subset of conditions. In this case, biclustering methods are more adapted. For example, when a set of genes participates in a cellular process that is active only in a subset of conditions, or when a gene is implied in multiple pathways that may or not be co-active under a subset of conditions. Biclustering methods allow the identification of relevant groups of genes and conditions that cannot be identified by classic clustering techniques. These kinds of methods consist in simultaneous clustering on rows and columns, to reorganize the data set into homogeneous blocks. It is an old approach but over the last years it has attracted many authors, (see for instance; [12,13]).

In this paper, we try to improve the performance of biclustering algorithms by using the ensemble approach. The principle of ensemble methods is to construct a set of models, then to aggregate them into a single model, generally by using generally a voting scheme. It is well-known that these methods often perform better than a single model (see for instance; [9]). Ensemble methods first appeared in supervised learning problems. A combination of classifiers is more accurate than single classifiers [18]. A pioneer method, boosting, whose most popular algorithm adaboost, was developed

J.L. Balcázar et al. (Eds.): ECML PKDD 2010, Part I, LNAI 6321, pp. 490–505, 2010.

mainly by Shapire [22]. The principle is to assign a weight to each training example, then several classifiers are learned iteratively and between each learning step the weight of examples is adjusted depending on the classifier results. The final classifier is a weighted vote of classifiers constructed during the procedure. Boosting has been used with success on several microarray datasets [7,17]. The other type of popular ensemble methods, proposed by Breiman, is bagging [3]. The principle is to create a set a classifiers based on bootstrap samples of the original data. This approach, and especially the random forest, is also efficient for microarray based classification [8]. In the last years, several works have shown that ensemble methods can also be used in unsupervised learning. The principle of boosting is exploited by Frossyniotis et al. [11] in order to provide a consistent partitioning of the data. The boost-clustering approach creates, at each iteration, a new training set using weighted random sampling from original data, and a simple clustering algorithm is applied to provide new clusters. Dudoit and Fridlyand [10] used bagging to improve the accuracy of clustering in reducing the variability of PAM (Partitioning Around Medoids) results [15]. Their method has been applied to leukemia and melanoma datasets and permitted to differentiate the different subtypes of tissues. Strehl [23] have proposed an approach to combine multiple partitioning obtained from different sources into a single one. They introduced three heuristics to solve this problem: 1) a hypergraph partitioning algorithm that approximates the maximum mutual information objective function with a constrained minimum cut, 2) a cluster-based similarity partitioning algorithm that establishes a distance between elements based on the individual clustering, 3) a meta-clustering algorithm where groups of clusters (meta-clusters) are identified and consolidated.

Since ensemble methods allow the improvement of the performance of supervised classification and clustering, it is reasonable to think that they can also be used to tackle the biclustering problem. In this paper, we propose in this paper a bagging approach for the biclustering of microarray data. Although the problem of ensemble biclustering shares some traits with classic biclustering, there are two major issues which are specific to ensemble methods. The first one is the generation of a collection of biclusters. This raises the question of how to generate different biclusters and what is the the source of diversity ? We have chosen the bagging approach to generate biclusters from bootstrapped datasets. The second issue is about the aggregation function. How to combine the different biclusters and resolve the label correspondence problem? Based on the works of Sterhl and Ghosh [23], we propose a solution that consists in creating *meta-clusters* of biclusters. Finally, for each gene and example, we compute the probabilities of their belonging to each meta-cluster. We test our method both on artificial and real data. The results on artificial data show that ensemble methods allow to improve the accuracy of biclustering. The results obtained on real data show that ensemble biclusters give a lower residue than classic biclusters. Moreover the ensemble biclusters are also biologically more relevant with respect to the prior knowledge of data.

2 State of the Art

Consider the data matrix $X = \{E, G\}$ where $E = E_1, \ldots, E_N$ is a set of N examples represented by M-dimensional vectors and $G = G_1, \ldots, G_M$ is a set of M genes. A bicluster B is a submatrix of X defined by a subset of examples and a subset of genes;

$$B = \{(E^B, G^B); E^B \subseteq E, G^B \subseteq G\}.$$

A biclustering operator Φ_K is a function that delivers one or several biclusters $B_0 = \emptyset, B_1, \ldots, B_K$ given (E_i, G_j).

Note that a submatrix is considered as a bicluster if it presents a particular pattern. There is no definition of what these patterns are. The choice of considering a submatrix as a bicluster, is subjective and depends on the context. However there are some basic patterns that can be used to identify a bicluster. They are called constant, additive and multiplicative models. In constant models, all values in a bicluster are equal. In additive and multiplicative models, there is an additive and multiplicative factor between rows and columns respectively. Biclusters can also be identified by a mixture of these three models. We also consider different bicluster structures. Biclusters can overlap on the rows and/or columns, or present a tree or checkerboard structure. This diversity in the nature of biclusters accounts for the fact that no biclustering algorithm can identify all types of biclusters.

Several biclustering algorithms have been developed and applied to microarray analysis. Cheng and Church [6] were the first to propose an algorithm for this task. They consider that biclusters follow an additive model and use a greedy iterative search to minimizing the mean square residue. This algorithm identifies biclusters one by one. They applied their method to *yeast cell cycle data* and identified several biologically relevant biclusters. Lazzeroni and Owen [16] have proposed the popular plaid model. They assume that biclusters are organized in layers and follow a given statistical model incorporating additive two way ANOVA models. The search approach is iterative: Once $K-1$ layers (biclusters) have been identified, the Kth bicluster that minimizes a merit function depending on all layers is selected. They also applied their method to yeast data and found that genes in same biclusters share biological functions. Kluger et al. [14] used a spectral approach for biclustering assuming that the data matrix contains a checkerboard structure after normalization. This structure is identified by a singular value decomposition. They applied their method to *Lymphoma and Leukemia* datasets which contained different subtypes of cancer. On both datasets, conditions of the same subtype have been grouped together into the same biclusters. Tanay et al. [24] have developed *SAMBA*, an approach based on the graph theory coupled with statistical modeling of the data. SAMBA, applied to a lymphoma dataset, produces biclusters representing new concrete biological associations. Cheng et al. [5] have proposed the $pCluster$ method that has the advantage it can identify both additive and multiplicative biclusters in presence of overlap. They validated their method on yeast cell-cycle dataset using Gene Ontology annotations. Prelic et al. [21] made a comparative study of different biclustering methods for gene expression. They used a very simple divide and conquer the *Bimax* algorithm as a reference to investigate the usefulness of different biclustering algorithms. They concluded that Bimax produces results similar to those of more complex methods. Abdullah et al. [1] proposed a graph-drawing-based biclustering technique based on the crossing minimization paradigm. These algorithms are the most popular ones used in bioinformatics but this list is not exhaustive. Madeira and Oliveira [19] have published a good survey of biclustering methods for biological data analysis and enumerated more than 15 used in this context. Note also a more recent review of biclustering methods in data mining [4].

3 Bagged Biclustering

The principle of bagged biclustering consists in applying a biclustering method on multiple bootstrapped datasets and aggregate the results. Our method can be divided into 3 steps. The first one is the construction of a collection of biclusters. These biclusters are generated by the multiple application of a classic biclustering algorithm on bootstrapped data. Then, a distance matrix between the obtained biclusters is computed based on their similarity. This distance is used to make a hierarchical clustering of the biclusters. From the resulting dendrogram, K meta-clusters of biclusters, are extracted. In the last step, we compute for each element (examples and genes) the probabilities of its belonging to each meta-cluster. If the probability is higher than a given threshold, the element is assigned to the meta-cluster. At the end of the procedure, the K meta-clusters contain a set of examples and genes that represent our final biclusters. Note that the number of biclusters K is a parameter to be fixed before the computation. In the next section, we describe in detail the three steps of our process.

3.1 Bicluster Collection Generation

The aim of this part is to generate a high number of different biclusters. To do so, we generate bootstrap samples of the original data. A bootstrap sample is a random drawing with replacement of the same size as the original data. It contains on average 68% of the original elements. In the case of biclustering we should sample the data in both dimensions, genes and examples. If we do that, the bootstrap sample will contain 63% of the original examples and 63% of the original genes, which means that the bootstrapped data will contain only 46% of the original matrix. To keep the same proportion as in the classic bootstrap, we decide to perform the bootstrap sample on only one dimension. Since microarray data contain much many genes than examples, the bootstrap sample will be done on genes. From the original data $X = \{E, G\}$, R bootstrapped datasets are generated $\{X^b = \{E, G^b\}, b = 1, ..., R\}$ where G^b is a bootstrap sample of G assumed an iid sample.

On each of the R bootstrapped datasets, a biclustering algorithm,with the same parameters, is applied to produce K biclusters. We obtain a collection of KR biclusters noted B^b that are used to identify meta-clusters.

3.2 Metacluster Identification

The objective is to identify K meta-clusters merging the similar biclusters. The idea is that if two biclusters, generated from different bootstrapped data, are similar, it is likely that they represent the same bicluster. All bootstrapped biclusters representing the same bicluster should be grouped into a meta-cluster. The notion of similarity between two biclusters depends on the number of elements (genes and examples) they have in common. We use the Jaccard index to evaluate this similarity:

$$Sim(B_k, B_\ell) = \frac{|B_k \cap B_\ell|}{|B_k \cup B_\ell|} = \frac{|B_k \cap B_\ell|_E + |B_k \cap B_\ell|_G}{|B_k \cup B_\ell|_E + |B_k \cup B_\ell|_G}$$

where $|.|_C$ corresponds to the cardinality computed on the set C. From this similarity, we can define the dissimilarity between B_k and B_ℓ by $d(B_k, B_\ell) = 1 - Sim(B_k, B_\ell)$. This distance belongs to $[0, 1]$, 0 indicating that two biclusters are identical and 1 that they have no element in common. From the distance matrix, a hierarchical clustering of the bicluster is constructed using the average linkage. From the obtained dendrogram we can identify K meta-clusters in cutting the dendrogram. Each meta-cluster M_g; $g = 1, \ldots, K$, is a set of $\{B_1^b, ..., B_{K_g}^b\}$ where K_g is the cardinality of M_g. Note that we do not consider trivial meta-clusters containing a few biclusters. Before deducing these meta-clusters in the third step, we compute the probability of (E_i, G_j) belonging to the meta-clusters $M_1, \ldots, M_K$. It is denoted $p_g(E_i, G_j)$ and can be estimated by the proportion of biclusters of M_g containing (E_i, G_j). Note that the parameter K is the same than the one used in bicluster collection generation step.

3.3 Bicluster Computation

The last step consists in computing the final biclusters of the original data. Each meta-cluster is assigned to a bicluster. Then each element can be assigned to biclusters depending on computed probabilities $p_g(E_i, G_j)$. We have to distinguish two cases. In the first one we consider that there is no overlapping between the biclusters, i.e. (E_i, G_j) belongs to at most one bicluster. If there are no probabilities superior to a given threshold t then (E_i, G_j) is assigned to no bicluster. If there is at least one probability superior to t then (E_i, G_j) is assigned to the bicluster $\hat{B}$ belonging to the meta-cluster M_g and maximizing the probability $p_g(E_i, G_j)$. Then $\Phi_K(E_i, G_j)$ is equal to

$$\begin{cases} B_0 = \emptyset & \text{if } p_g(E_i, G_j) < t \forall g \\ \hat{B} = \arg\max_{M_g} p_g(E_i, G_j) & \text{otherwise} \end{cases}$$

In the second case, overlapping between two biclusters is possible, i.e. an element can belong to several biclusters. If there are no probabilities superior to the threshold t then the element is assigned to no bicluster. Otherwise, all biclusters whose corresponding probabilities are higher than the threshold t are assigned to the element.

$$\Phi_K(E_i, G_j) = \begin{cases} B_0 = \emptyset & \text{if } p_g(E_i, G_j) < t \forall g \\ B_k \in M_g & \text{such as } p_g(E_i, G_j) \geq t \end{cases}$$

Depending on the overlapping choice, we will use the first or the second case. Note that we can mix these two cases if we want different overlapping choices for examples and genes. For instance, a current choice in microarray analysis is to allow the overlapping on the examples and not on the genes because of the disproportion between the number of genes and examples. The function of the second case is used on examples and the function of the first case on genes. The threshold t has a high influence on the bicluster computation. The choice of its value will be discussed later.

4 Experiments on Artificial Data

In our simulation study, we evaluate the performance of bagged biclustering and compare it to single biclustering. We performed our experiments on artificial and real datasets with five biclustering algorithms: Bimax [21], Cheng & Church [6], plaid model [16], spectral biclustering [14] and Xmotifs [20]. This part concerns the results on artificial data.

4.1 Generation of Artificial Datasets

The first part of our experiments is based on artificial data. These data cannot be considered as a reliable representation of real microarray data. Nevertheless they can be used to measure the performance and behavior of the new methods in conventional cases. The artificial data is a matrix with random values in which we included several biclusters. The dataset contains M genes, N examples and K biclusters. In our simulations $M = 200$, $N = 100$ and $K = 2, 4, 6$. The size of each bicluster is randomly chosen between 10 examples by 20 genes to 20 examples by 40 genes. The partition on the genes is defined by the classification $M \times K$ matrix $\mathbf{z}$ defined by $z_{ik} = 1$ if the gene i belongs to the bicluster B_k and 0 otherwise. In the same way, we consider the classification $N \times K$ matrix $\mathbf{w} = (w_{jk})$ representing the partition of the examples. The partitions of biclusters are defined randomly. Note that from $\mathbf{z}$ and $\mathbf{w}$, we can define different cases of bicluster overlapping. The simplest one corresponds to no overlapping structure. The second case consists in considering the presence of the overlapping of genes. In the same way, we define the overlapping of examples. The last and most complex case is total overlapping where genes and examples can belong to several biclusters. An overlapping between biclusters can occur only in this last case. Note also that a gene or example can belong to no bicluster. The total overlapping structure is the most general case, so we consider only this situation in our experiments.

There is no general definition of what a bicluster is. The different algorithms used search different models of biclusters. In our experiments we use five different biclustering algorithms. These algorithms do not identify biclusters of the same nature. For each of them, we define a bicluster with the same model as the one used in the original paper where the algorithm was published.

- **Plaid model.** We consider the more general model. The values of X_{ij} belonging to a bicluster B_k (layer) depend on 4 parameters: a constant μ_0 describing the background layer, μ_k the average of B_k, α_{ik} and β_{jk} allowing to identify respectively a subset of genes and examples having identical responses. The values of X_{ij} are then represented as:$X_{ij} = \mu_0 + \sum_{k=1}^{K} z_{ik} w_{jk}(\mu_k + \alpha_{ik} + \beta_{jk})$. In our experiments, all values belonging to biclusters are generated according to a uniform distribution $U[0, 5]$. The values outside a bicluster, are generated according a uniform distribution $U[-10, 10]$.
- **CC model.** For the Cheng and Church model, The values of X_{ij} belonging to a bicluster B_k depend on 3 parameters: μ_k the average of B_k, μ_{ik} and μ_{jk} are respectively the means of E_i and G_j belonging to the bicluster B_k. The values of X_{ij} are then represented as:$X_{ij} = \sum_{k=1}^{K} z_{ik} w_{jk}(\mu_{ik} + \mu_{jk} - \mu_k)$. The values are generated in the same manner that for the Plaid model.
- **Xmotifs.** we consider binary data and biclusters that depend on two parameters $\alpha > 0$ and $\beta < 1$ defined by the three following rules. A bicluster must contain a number of examples superior to αN. All values in a bicluster are equal to (0 or

1). If a gene does not belong to a bicluster, there is a maximum βN values that are equal.
- **Spectral biclustering.** It is a special case since the data follow a checkboard structure. Each value belongs to a bicluster. The examples are partitioned into two clusters and genes into $K/2$ clusters. Each bicluster is defined by its means value μ_k. In our simulations, all values of a bicluster are chosen equal to their mean value.
- **Bimax**. The model used for Bimax is the simplest. The dataset is considered binary, all values in a bicluster are equal to 1 and all values out of any bicluster are 0. Then, biclusters are sub-matrices containing only 1s. Each X_{ij} value is then defined by $X_{ij} = z_{ik} \times w_{jk}$.

Once the biclusters are defined and included in the dataset, we add a Gaussian noise $N(0, \sigma^2)$. The variance permits to control the difficulty of the biclustering task.

4.2 Study Design

We evaluate the performance of each biclustering method by comparing the biclusters obtained by different algorithms and true biclusters. We compute the error of biclustering by estimating the total number of misclassified values. Since this estimation should not depend on the labellings of the biclusters, we enumerate all possible relabellings and retain the label that gives the smallest error noted e. The study design used in our experiment is the following:

1. Generate an artificial dataset X with K biclusters
2. Apply a biclustering algorithm on X to identify K biclusters
3. Compare the true biclusters to the biclusters obtained by the five algorithms and compute the biclustering error noted e^{single}.
4. Apply the associated bagged biclustering of the five algorithms on X to identify K biclusters
5. Compare the true biclusters to the obtained biclusters and compute the biclustering error noted e^{bagged}.
6. Iterate steps 1-5 200 times and compute the means of e^{single} and e^{bagged}.

4.3 Results on Artificial Data

The choice of the decision threshold t is crucial. A too small value of t may imply that an element, belonging to a bicluster, will be assigned to no bicluster. The obtained biclusters will be small and several genes will miss to be assigned. But we will be highly confident that all identified genes and examples are really assigned to biclusters. On the other hand, a too high value of t leads to assign an element to biclusters not containing this element. The obtained biclusters will be large and tend to contain all elements actually belonging to the biclusters. The risk is that they will also contain false positives, i.e. elements wrongly assigned. The dependence of t on these different situations implies its difficult choice. Here, in our experiments the assessing of t is performed empirically, giving a good tradeoff between false and true positives.

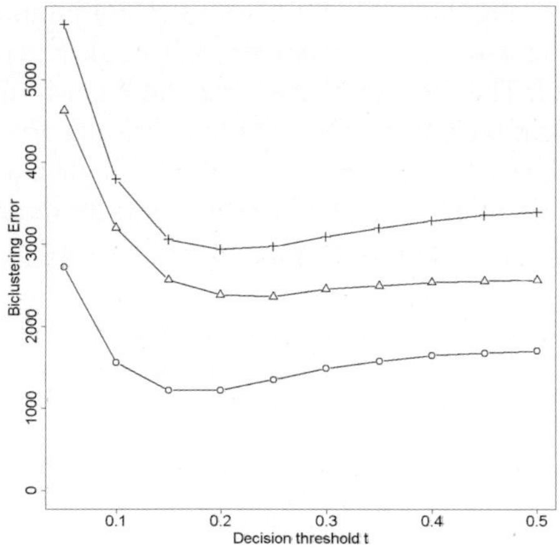

Fig. 1. $\circ : K = 2$, $\triangle : K = 4$, $+ : K = 6$. Biclustering error of the Bimax algorithm in function of the value of the decision threshold t. The dot line represents results for problems with 2 biclusters, the triangle line with 4 biclusters and the cross line with 6 biclusters.

Figure 1 shows the biclustering error in function of the decision threshold t with the Bimax algorithm. The three lines, dot, triangle and cross, represent respectively the results on dataset containing 2, 4 and 6 biclusters. The behavior of the three curves is the same: they are strongly decreasing from $t = 0$ to $t \approx 0.2$ then the biclustering error increases slowly. For all curves, the minimum error is around $t = 0.2$. We use this threshold in our experiments. We employ the same procedure to assess the threshold for the other algorithms.

We performed different experiments giving the same results for different values of $K = 2, 4, 6$. Figure 2 and 3 show the biclustering error with the five algorithms: Bimax, Cheng & Church, plaid model, spectral biclustering and Xmotifs. For each algorithm, three graphics represent the results on datasets containing 2,4 and 6 biclusters. In each graphic, the error biclustering is computed in function of the quantity of noise introduced in the dataset. The dot curve represents the error of the single biclustering algorithm, the triangle curve the bagged biclustering. We see that in all cases the error is naturally increasing with the noise. For Bimax, we see that at $\sigma = 0.4$ the error of single biclustering "jumps" from 0 to 1700. The error of the bagged biclustering increases slowly to join the error of the single biclustering at $\sigma = 1.8$. We can interpret this as follows: for $\sigma \leq 0.4$ the biclustering problem is easy, the performance of biclustering is maximal. For $\sigma \geq 1.8$ the datasets are so noisy that all biclusterings are meaningless, we have checked that the error is the same as a random biclustering. These graphics show that bagged Bimax gives a better biclustering than single Bimax. We find the same type of results in the major part of our experiments. The Cheng & Church algorithm and Bimax have the same behavior. Error of single CC increases faster than bagged CC. In plaid model results, we see that the error of the bagged plaid model is lower than that of the single plaid model for $\sigma < 1$. For higher level of noise, $\sigma \geq 1$, the two algorithms give the same performance as "random" biclustering. The results of

Spectral biclustering are singular since the error does not jump for a given value of σ but is regularly increasing with σ. The error curves of single and bagged spectral biclustering are almost parallel. The error of bagged spectral biclustering is always below the error of the single spectral biclustering. For Xmotifs results, the error curve of bagged Xmotifs always begins below the curve of single Xmotifs, then joins it for higher levels of noise. The results, including various biclustering algorithms and different number of biclusters (not reported here), show that bagged biclustering gives better results than single biclustering.

In these experiments on artificial data, we assume the knowledge of the true number of biclusters, i.e. the number of meta-clusters, but in real data problems this number is generally unknown. Here we show that a measure of clustering goodness, like the *within-group sum of squares* criterion noted commonly W, also named *within-group inertia*, can be used on the bagged bicluster dendogram to find an appropriated number of meta-clusters. In our simulations we perform the agglomerative hierarchical clustering on the biclusters and we try several cuts of the tree in order to obtain different numbers of meta-clusters. For each cut, we compute W of the obtained partition of bagged biclusters. The W criterion, depending on a partition, decreases with the number of meta-clusters, and so a scree plot with one or several elbows may used to propose a cut of the dendrogram. For instance, the scree plots of Figure 4 express W computed in function of the number of meta-clusters with the CC algorithm. From each scree plot, we can retain the first elbow corresponding respectively to 2, 4 and 6 meta-clusters (represented with a dotted line in the graphics). Note that we have also tested the influence of the number of bootstrap iterations R on our method. Our simulations show that, when R is large ($R > 100$), this parameter has no effect on the results of bagged biclustering.

5 Experiments on Real Data

We tested our approach on real microarray datasets. The evaluation is more subjective than on artificial data since we cannot know the true biclusters. We rely on statistical measures and the biological context of the datasets to evaluate the goodness of the biclusters. In the following, we have applied the plaid and CC models and evaluate the quality of a bicluster B_k by computing the Mean Square Residue (MSR) defined by

$$\frac{1}{\#B_k}\sum_{i,j} z_{ik}w_{jk}(X_{ij}-\mu_{ik}-\mu_{jk}+\mu_k)^2.$$

The symbol # denotes the cardinality. We have computed the MSR for the biclusters in single and bagged contexts. The partition of genes, extracted from biclusters, should be coherent with known genetic information contained in public databases like KEGG. We expect to find genes belonging to the same biological pathways in a bicluster. We have applied single biclustering algorithms and bagged biclustering to four microarray datasets. These datasets are available online[1] and their characteristics can be found in table 1.

Firstly, the Bimax, Xmotifs (after binarization process) and spectral algorithms, are not commented in the following. Using both single and bagged versions they did not

[1] http://algorithmics.molgen.mpg.de/Static/Supplements/CompCancer/datasets.htm

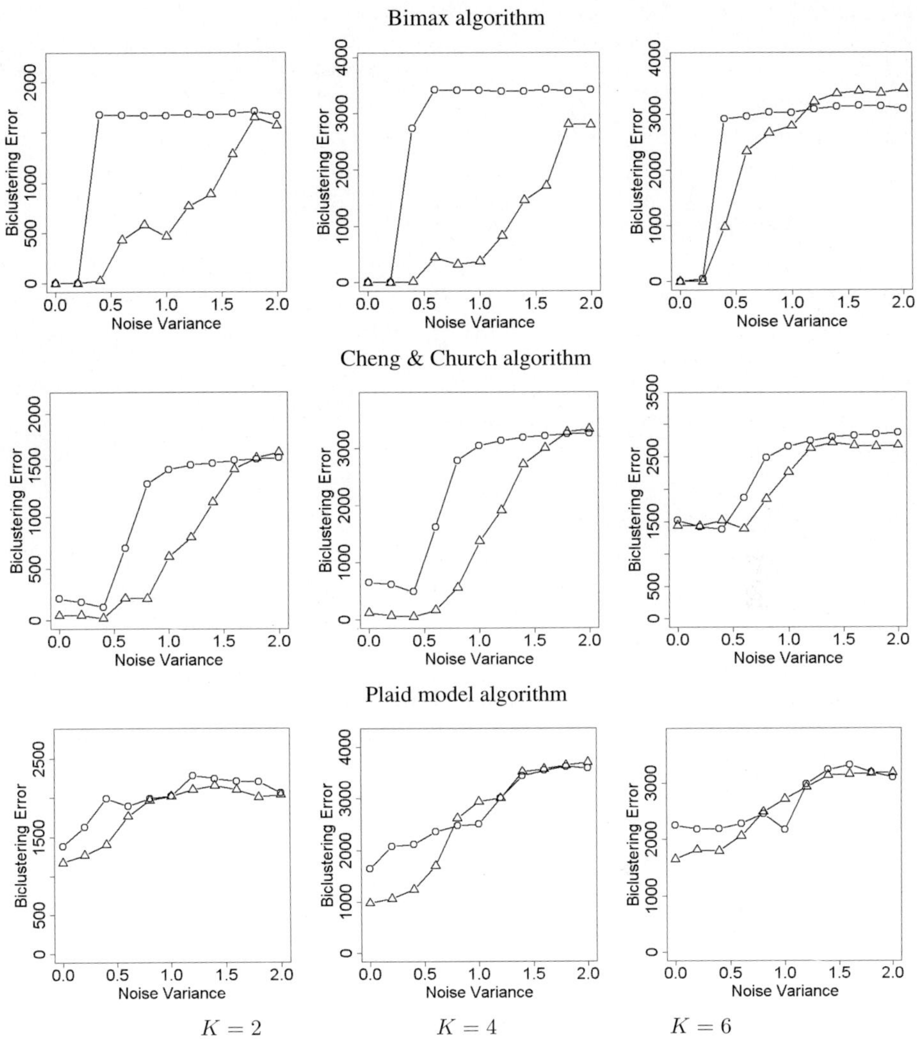

Fig. 2. Biclustering error in function of noise variance on artificial data. The three graphic lines correspond respectively to results with Bimax, CC algorithm and plaid model. The columns correspond respectively to results with 2, 4 and 6 biclusters. Dot lines and triangle lines correspond respectively to single and bagged biclustering.

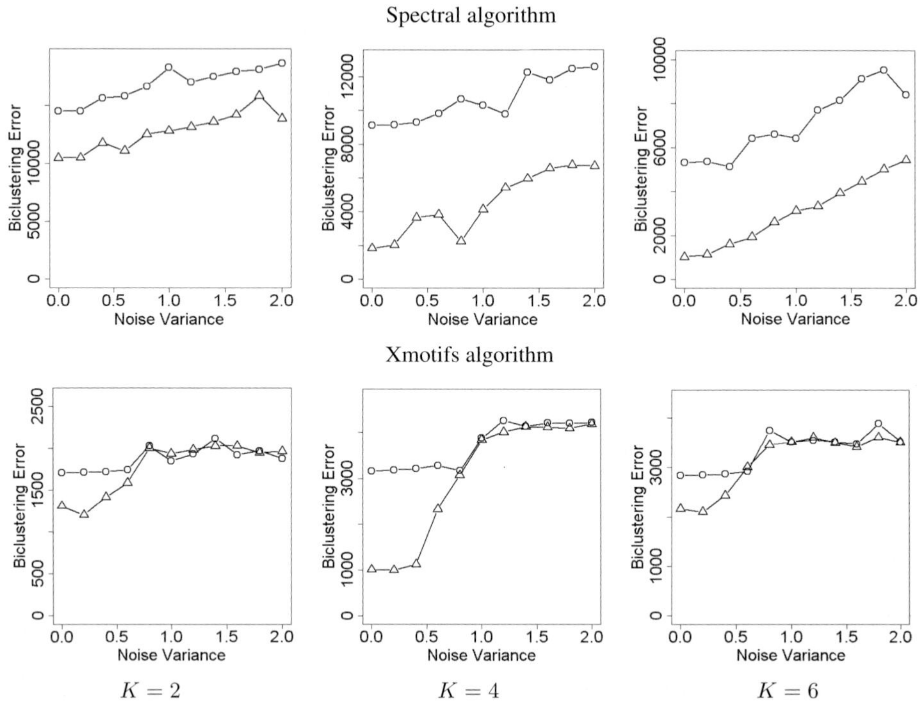

Fig. 3. Biclustering error in function of noise variance on artificial data. The two graphic lines correspond respectively to results with Spectral biclustering and Xmotifs. The columns correspond respectively to results with 2, 4 and 6 biclusters. Dot lines and triangle lines correspond respectively to single and bagged biclustering.

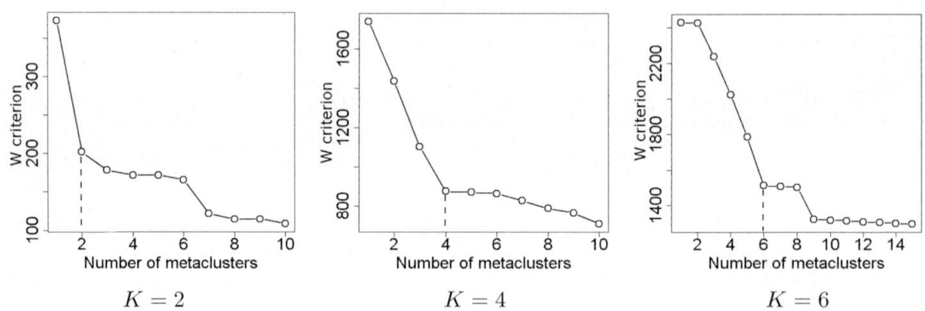

Fig. 4. The within-group sum of squares in function of the number of meta-clusters with the CC algorithm

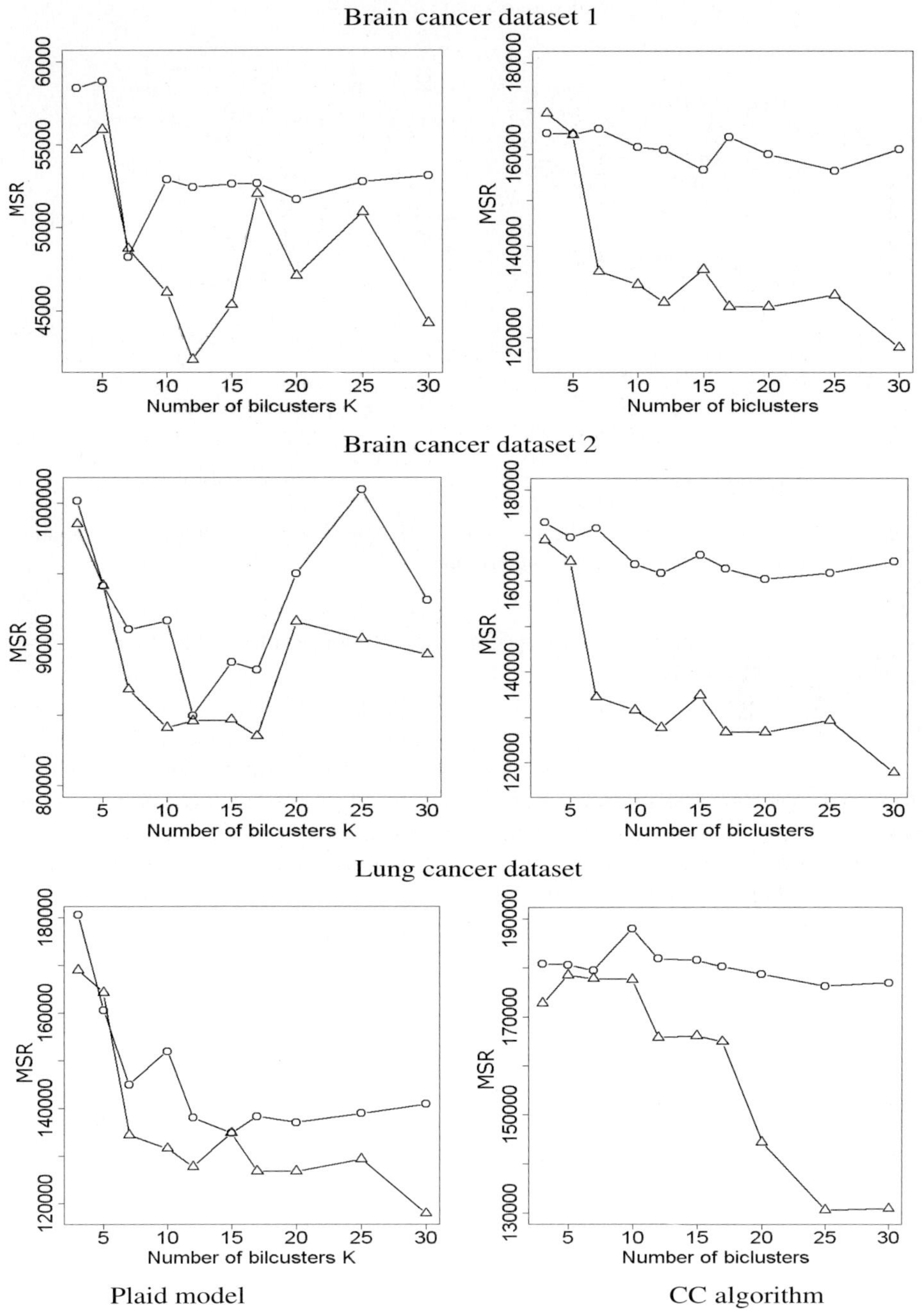

Fig. 5. MSR in function of the number of biclusters K. Dot line correspond to MSR of single biclustering and triangle line of bagged biclustering.

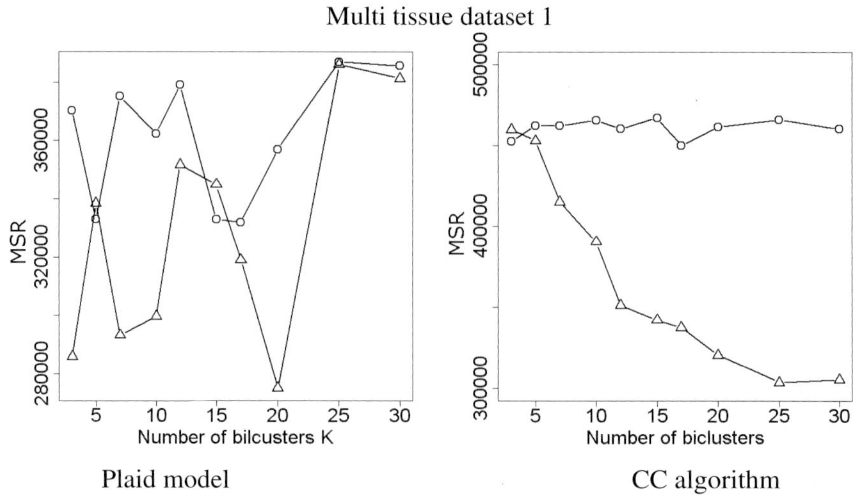

Fig. 6. MSR in function of the number of biclusters K. Dot line correspond to MSR of single biclustering and triangle line of bagged biclustering.

find any biclusters from the 4 datasets. The type of biclusters sought by these methods, are not suitable for our datasets.

Applying single and bagged biclustering with the plaid model and the CC models on the four microarray datasets, we compute the MSR of results produced by each method according different values of K. The number of bootstrap iterations for bagged biclustering is $R = 100$. The figures 5 and 6 show the MSR of the single (dot lines) and bagged (triangle lines) biclustering in function of the number of biclusters K. First column represents results with plaid model and second column with CC algorithm. The results with the plaid model exhibit a lot of variability, but the MSR of bagged biclustering is lower than single biclustering. In the four datasets there is a peak phenomenon, MSR is firstly decreasing then increasing with the number of biclusters. The minimum is around $K = 12$. For lung cancer dataset, MSR is strongly decreasing until $K = 10$ then is stable. In multi tissue dataset, the results are very variable, it is hard to see the general behavior of MSR in function of K. The important point is that MSR of bagged biclustering is below MSR of single biclustering. The only cases where the MSR of the two approaches are equal is where MSR are at its maximum. That means for an optimal number of biclusters, bagged biclustering produces better results than single biclustering. The results with CC algorithm are clearer. All five graphics present the same behavior and we note that the performance increases with the number of biclusters. The MSR of single biclustering remains stable whereas the MSR of bagged biclustering decreases strongly and becomes stable for high number of biclusters. In all datasets, bagged biclustering is much better than single biclustering for $K > 7$.

We also compared single and bagged biclustering based on the coherence of the obtained gene partition. If two genes are in the same bicluster, we can assume that these two genes are biologically related. A good tool to check if there is an identified relation between two genes, is the pathway database of the Kyoto Encyclopedia of Genes

Table 1. Characteristics of the datasets and number of over-represented pathways

			Plaid model		CC algorithm	
Datasets	#ex.	#genes	Single	Bagged	Single	Bagged
Brain cancer 1	50	1377	8	18	19	25
Brain cancer 2	42	1379	10	26	3	14
Lung cancer	203	1543	12	20	15	19
Multi tissue	190	1363	4	15	26	41

and Genomes (KEGG). These pathways represent molecular interaction and reaction networks for metabolism, various cellular processes, and human diseases. All genes in the same pathway are considered biologically related. The number of over-represented pathways is computed for each bicluster. We use the hypergeometric test to check if a path is over-represented in a bicluster. Given pathway a PW_i, let p_i be the probability that a gene belongs to the pathway PW_i and $S_i = [p_i M]$ ($[.]$ denotes the integer part) the number of genes belonging to PW_i. The probability of obtaining k genes belonging to the pathway PW_i from a random selection of A genes follows a hypergeometric law:

$$P(k, S_i, M, A) = \frac{C_{S_i}^{k} C_{M-S_i}^{A-k}}{C_M^A}.$$

Given a bicluster $B = \{E, G\}$, let X_i be the number of genes belonging to the pathway PW_i, the probability of obtaining at least X_i genes belonging to PW_i by a random selection is defined by:

$$Pr(k \geq X_i) = \sum_{x \geq X_i} P(x, S_i, M, |G|).$$

We compute this probability for all pathways and the Holm correction is applied to address the problem of multiple comparisons. Finally all pathways with a corrected p-value inferior to 0.05 is considered over-represented. We assume that the number of over-represented pathways represents the biological information captured by the genes of the biclusters. A good biclustering should provide high number of over-represented pathways. We use this measure to compare the results of the different algorithms. The chosen number of biclusters is the one that minimizes the MSR. In the table 1, we report the number of over-expressed pathways in datasets and it appears clearly that bagged biclusters contains much more over-expressed pathways and can be consider more biologically relevant than single biclusters.

6 Conclusion

In this paper we have introduced the concept of ensemble methods for biclustering in the context of microarray data. The proposed bagged biclustering generates a collection of biclusters based on bootstrap samples of the original data. Then a distance matrix between biclusters is computed based on the Jaccard index. From these distances we construct a dendrogram and identify meta-clusters. The probability to belong to each meta-cluster is computed for each element (E_i, G_j). Finally the final biclusters are built based on these probabilities. The performance of our method has been tested on both artificial and real microarray data. On artificial data, we have shown that ensemble method enables to strongly decrease the biclustering error compared to classic methods whatever the used biclustering algorithm, the number of biclusters and the chosen noise level. On real data, bagged biclustering provides biclusters more relevant than single biclustering according to their MSR value. In addition, we have noted that bagged biclusters capture more biologic information since they contain more overexpressed pathways. The use of our approach is a new powerful tool for microarray analysis and should allow biologists to identify new relevant patterns in gene expression data.

Aknowledgments

This research was supported by the CLasSel ANR project.

References

1. Abdullah, A., Hussain, A.: A new biclustering technique based on crossing minimization. Neurocomputing 69(16-18), 1882–1896 (2006)
2. Alizadeh, A.: Distinct types of diffuse large b-cell lymphoma identified by gene expression profiling. Nature 403, 503–511 (2000)
3. Breiman, L.: Bagging predictors. Machine Learning 24, 123–140 (1996)
4. Busygin, S., Prokopyev, O., Pardalos, P.: Biclustering in data mining. Computers and Operations Research 35(9), 2964–2987 (2008)
5. Cheng, K.O., Law, N.F., Siu, W.C., Liew, A.W.: Identification of coherent patterns in gene expression data using an efficient biclustering algorithm and parallel coordinate visualization. BMC Bioinformatics 9, 210 (2008)
6. Cheng, Y., Church, G.M.: Biclustering of expression data. In: Proc. Int. Conf. Intell. Syst. Mol. Biol., vol. 8, pp. 93–103 (2000)
7. Dettling, M., Bühlmann, P.: Boosting for tumor classification with gene expression data. Bioinformatics 19(9), 1061–1069 (2003)
8. Diaz-Uriarte, R., Alvarez de Andres, S.: Gene selection and classification of microarray data using random forest. BMC Bioinformatics 7(3) (2006)
9. Dietterich, T.G.: Ensemble methods in machine learning. In: Kittler, J., Roli, F. (eds.) MCS 2000. LNCS, vol. 1857, pp. 1–15. Springer, Heidelberg (2000)
10. Dudoit, S., Fridlyand, J.: Bagging to improve the accuracy of a clustering procedure. Bioinformatics 19(9), 1090–1099 (2003)
11. Frossyniotis, D., Likas, A., Stafylopatis, A.: A clustering method based on boosting. Pattern Recognition Letters 25, 641–654 (2004)
12. Govaert, G., Nadif, M.: Clustering with block mixture models. Pattern Recognition 36, 463–473 (2003)
13. Govaert, G., Nadif, M.: Block clustering with Bernoulli mixture models: Comparison of different approaches. Computational Statistics and Data Analysis 52, 3233–3245 (2008)
14. Kluger, Y., Basri, R., Chang, J.T., Gerstein, M.: Spectral biclustering of microarray data: coclustering genes and conditions. Genome Res. 13(4), 703–716 (2003)
15. van der Laan, M., Pollard, K., Bryan, J.: A new partitioning around medoids algorithm. Journal of Statistical Computation and Simulation 73(8), 575–584 (2003)

16. Lazzeroni, L., Owen, A.: Plaid models for gene expression data. Tech. rep., Stanford University (2000)
17. Long, P., Long, P.M., Vega, V.B.: Boosting and microarray data. Machine Learning 1-2(52), 31–44 (2003)
18. Maclin, R.: An empirical evaluation of bagging and boosting. In: Proceedings of the Fourteenth National Conference on Artificial Intelligence, pp. 546–551. AAAI Press, Menlo Park (1997)
19. Madeira, S.C., Oliveira, A.L.: Biclustering algorithms for biological data analysis: a survey. IEEE/ACM Transactions on Computational Biology and Bioinformatics 1(1), 24–45 (2004)
20. Murali, T., Kasif, S.: Extracting conserved gene expression motifs from gene expression data. Pacific Symposium on Biocomputing 8, 77–88 (2003)
21. Prelic, A., Bleuler, S., Zimmermann, P., Wille, A., Buhlmann, P., Gruissem, W., Hennig, L., Thiele, L., Zitzler, E.: A systematic comparison and evaluation of biclustering methods for gene expression data. Bioinformatics 22(9), 1122–1129 (2006)
22. Schapire, R.: The boosting approach to machine learning: An overview. In: Nonlinear Estimation and Classification. Springer, Heidelberg (2003)
23. Strehl, A., Ghosh, J.: Cluster ensembles - a knowledge reuse framework for combining multiple partitions. Journal of Machine Learning Research 3, 583–617 (2002)
24. Tanay, A., Sharan, R., Shamir, R.: Discovering statistically significant biclusters in gene expression data. Bioinformatics 18(Suppl. 1), 136–144 (2002)

Hub Gene Selection Methods for the Reconstruction of Transcription Networks

José Miguel Hernández-Lobato[1] and Tjeerd M.H. Dijkstra[2]

[1] Computer Science Department, Universidad Autónoma de Madrid,
Francisco Tomás y Valiente, 11, 28049, Madrid, Spain
josemiguel.hernandez@uam.es

[2] Institute for Computing and Information Sciences,
Radboud University Nijmegen, The Netherlands
t.dijkstra@science.ru.nl

Abstract. Transcription control networks have a scale-free topological structure: While most genes are involved in a reduced number of links, a few hubs or key regulators are connected to a significantly large number of nodes. Several methods have been developed for the reconstruction of these networks from gene expression data, e.g. ARACNE. However, few of them take into account the scale-free structure of transcription networks. In this paper, we focus on the hubs that commonly appear in scale-free networks. First, three feature selection methods are proposed for the identification of those genes that are likely to be hubs and second, we introduce an improvement in ARACNE so that this technique can take into account the list of hub genes generated by the feature selection methods. Experiments with synthetic gene expression data validate the accuracy of the feature selection methods in the task of identifying hub genes. When ARACNE is combined with the output of these methods, we achieve up to a 62% improvement in performance over the original reconstruction algorithm. Finally, the best method for identifying hub genes is validated on a set of expression profiles from yeast.

Keywords: Transcription network, ARACNE, Automatic relevance determination, Group Lasso, Maximum relevance minimum redundancy, Scale-free, Hub.

1 Introduction

High-throughput molecular technologies like DNA microarrays allow researchers to simultaneously measure the concentration of thousands of known molecules (e.g. mRNAs) in a cellular population. This technological breakthrough has encouraged a systems biology approach to the analysis of complex cellular processes [1,2]. The new paradigm focuses on the interactions between molecules, usually represented by a network, and how these interactions give rise to cellular behavior. In particular, functional properties are not in the molecules themselves but they appear as a result of their coordinated interactions. A challenging task is

J.L. Balcázar et al. (Eds.): ECML PKDD 2010, Part I, LNAI 6321, pp. 506–521, 2010.

then to reconstruct the network of molecule interactions from high-throughput experimental data [3,4,5].

Transcription networks are a particular class of interaction networks in which each node represents a different gene and each link corresponds to an interaction between two genes at the transcription level. The network captures the regulatory role of certain genes whose final products, e.g. transcription factors, govern the transcription rate of other genes. Transcription control networks display two typical characteristics. First, they have a sparse connectivity matrix [6] and second, they have a scale-free topological organization [7,4]. In particular, while most genes are involved in only a reduced number of interactions, a few hubs or key regulators are connected to a significantly large number of nodes. These features can be observed in the transcription network displayed in Fig. 1.

Under the assumption that mRNA concentration is predictive of regulatory molecule activity, several machine learning methods have been proposed for the reconstruction of transcription networks from steady-state gene expression data. Some examples are given by Bayesian network methods [8], mutual information approaches [4,5] and linear regression techniques [9,10]. While most of these methods enforce sparsity in the resulting interaction map, few of them actually exploit the scale-free topological structure of transcription control networks. An exception is the method described in [11] which is based on a scale-free prior within a framework of Gaussian graphical models. However, this latter method has a very high computational cost and is not feasible for large-scale problems. As a more efficient approach in terms of computational time, we concentrate on the hubs that commonly appear in transcription networks.

This paper is based on the assumption that reverse engineering methods can be significantly improved by predicting beforehand those genes that are more likely to be hubs or key regulators in the transcription network. For this task, we introduce a multiple linear regression model which describes the interactions between genes and transcriptional regulators. The model includes a coefficient matrix which is expected to be columnwise sparse. A set of hub genes can then be selected by identifying those columns of the regression matrix which have many non-zero entries. In practice, this is equivalent to selecting those features that are more relevant for solving the multiple regression problem. Three feature selection methods are proposed for this task: The group Lasso regularization method [12], the automatic relevance determination (ARD) approach [13] and the maximum relevance minimum redundancy (MRMR) criterion [14]. The performance of the different methods for identifying hub genes is evaluated in a series of experiments with simulated gene expression data. The program SynTREN is used in the data generation process [15]. These experiments indicate that the best performance is obtained by the hub gene selection method based on the ARD technique.

In a second step, we show how the list of hub genes generated by the feature selection methods can be used to significantly improve the results of ARACNE, a state-of-the-art technique for reverse engineering transcription networks [4,16]. Another series of experiments with simulated gene expression data indicate that a combination of ARACNE with the ARD feature selection method yields a 62%

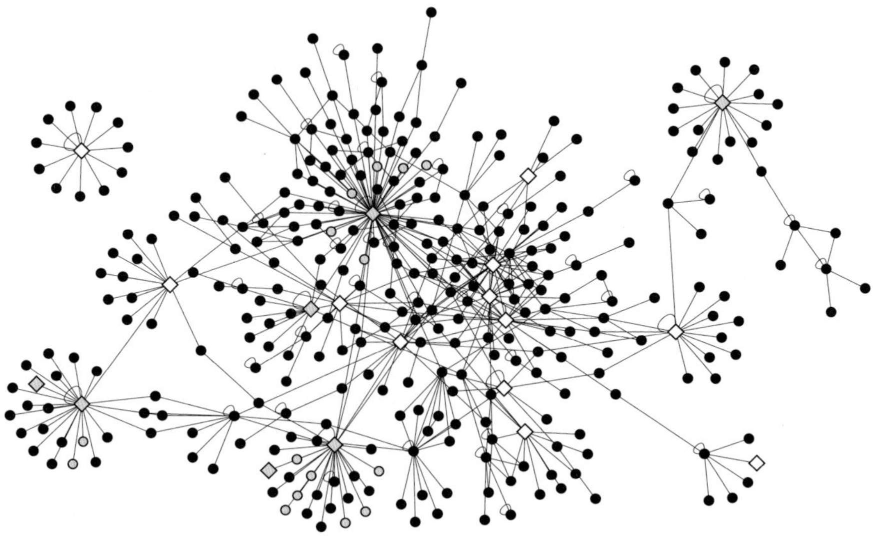

Fig. 1. Approximation of the transcriptional regulatory network of *Escherichia coli*. Nodes represent genes and edges indicate a direct transcriptional interaction between two genes. Diamond shaped nodes correspond to those genes selected by ARD, most of them are key regulators or hubs in the network. Light-gray nodes are those genes selected by the group Lasso approach, most of them are connected to the same hub node and their expression levels are therefore highly correlated. The visualization of the network was implemented using the software Cytoscape [18].

improvement in performance over the results of the original algorithm. Finally, the ARD method for hub gene identification is run on a set of expression profiles from *Saccharomyces cerevisiae* collected under different experimental conditions [17]. This dataset contains 247 expression measurements for a total of 5520 genes. An analysis of the the top ten genes identified by the ARD technique reveals a large number of global transcriptional regulators.

2 A Linear Model of Transcription Control

First of all, we describe a linear regression model for steady-state gene expression data. Transcription control in biological systems is a dynamic process typically characterized by a set of differential equations. Michaelis Menten interaction kinetics and the Hill equation [19,20] model the rate of production of a transcript X by the enzyme RNA polymerase P when activator A is required for transcription:

$$\frac{d\,[X]}{dt} = V_m \cdot \frac{[A]^{\alpha}}{[A]^{\alpha} + K_A} \cdot \frac{[P]}{[P] + K_P} - \delta\,[X]\,, \quad (1)$$

where δ is the degradation rate of transcript X, V_m is the maximum rate of synthesis, K_P and K_A are activation thresholds, α is the Hill coefficient for cooperative binding and $[\cdot]$ stands for *"concentration of"*. When competitive inhibition occurs between the enzyme RNA polymerase P and the repressor R, the rate of transcription is given by

$$\frac{d[X]}{dt} = V_m \cdot \frac{[P]}{[P] + K_P\left(1 + [R]^\beta / K_R\right)} - \delta [X] \,, \tag{2}$$

where K_R is a repressor threshold and the parameter β is similar to α. When steady-state conditions hold and molecule concentrations are far from saturating, we obtain

$$[X] = \frac{V_m [A]^\alpha [P]}{\delta K_A K_P} \,, \qquad [X] = \frac{V_m K_R [P]}{\delta K_P [R]^\beta} \,, \tag{3}$$

for the activation and repression cases, respectively. Taking logarithms, assuming that activation and repression are both possible at the same time and considering that mRNA concentration is predictive of final product concentration, we end up with a linear model of transcript log-concentration

$$\log [X] = \sum_{i=1}^{m} b_i \log [X_i] + \text{constant} \,, \tag{4}$$

where the b_i are real coefficients and the X_i are the transcripts of the activators, the repressors and the enzyme RNA polymerase. When X is a self-regulating gene, $\log [X]$ is included in the right-hand side of (4) with associated coefficient b_{m+1}. This autoregulatory term can be eliminated by algebraic simplification. For this, b_i is replaced by $b_i' = b_i/(1 - b_{m+1})$ where $i = 1, \ldots, m$ and b_{m+1} is set to zero.

The previous linear model is easily extended to account for all the transcripts present in a biological system. Let $\mathbf{X}$ denote a $d \times n$ gene log-expression matrix where each row corresponds to a different gene and each column represents an independent sample obtained under steady-state conditions. Additionally, let us assume that the rows of $\mathbf{X}$ have been centered so that they have zero mean. Then, if the log-expression measurements are contaminated with additive Gaussian noise, (4) suggests that $\mathbf{X}$ should approximately satisfy

$$\mathbf{X} = \mathbf{B}\mathbf{X} + \sigma \mathbf{E} \,, \tag{5}$$

where $\mathbf{B}$ is a $d \times d$ matrix of regression coefficients that links each gene to its transcriptional regulators, σ is a positive constant that is determined by the level of noise in $\mathbf{X}$ and $\mathbf{E}$ is a $d \times n$ matrix whose elements are i.i.d. random variables that follow a standard Gaussian distribution. In this linear model, the diagonal elements of $\mathbf{B}$ are all zero since any autoregulatory term in (5) can be eliminated using the process described in the previous paragraph.

Note that the coefficient matrix $\mathbf{B}$ encodes the connectivity of the underlying transcription network. In particular, let b_{ij} be the element in the i-th row and j-th column of $\mathbf{B}$, then $b_{ij} \neq 0$ whenever gene j is a transcriptional regulator of gene i and $b_{ij} = 0$ otherwise. Consequently, $\mathbf{B}$ is constrained by the topological structure of the network. Transcription networks are sparsely connected [6], with most genes having only a few connections. However, a few hubs or key regulators are connected to a significantly large number of genes [21]. Fig. 1 displays a sample transcription network with these characteristics. Consequently, we may expect $\mathbf{B}$ to be a sparse matrix whose rows typically have a few coefficients different from zero and these non-zero coefficients are often clustered on a few of the columns of $\mathbf{B}$, where each of these columns corresponds to a different hub gene or global regulator. We may say that $\mathbf{B}$ is a columnwise sparse matrix. Note that the column of $\mathbf{B}$ that corresponds to RNA polymerase is the one with more non-zero elements since this enzyme is involved in the production of all the transcripts in any biological system.

Given $\mathbf{X}$, the previous paragraph suggests a procedure for identifying those genes that are more likely to be hubs in the transcription network. The method consists in first, solving the multiple regression problem represented by (5) under the assumption that $\mathbf{B}$ is columnwise sparse and second, identifying the columns of this regression matrix with more non-zero coefficients. This is equivalent to selecting the rows of $\mathbf{X}$ that are more relevant for solving the multiple regression problem, i.e., we have to identify a reduced subset of genes whose expression pattern accounts for the expression pattern of all the other genes in the network.

3 Hub Gene Selection Methods

Let k be an approximation of the number of hub genes in the transcription network. We propose three linear methods for the identification of the k rows of $\mathbf{X}$ that are more relevant for solving the multiple regression task. The proposed techniques are based on extensions of standard sparse linear models [22,13] and standard feature selection methods [14] to the multiple regression case.

3.1 Automatic Relevance Determination

In this section, we extend the automatic relevance determination (ARD) framework [13] to enforce columnwise sparsity of matrix $\mathbf{B}$ in (5). The first step is to assign a Gaussian prior to the rows of $\mathbf{B}$. Let $\mathbf{b}_i = \left(b_i^1, \ldots, b_i^{i-1}, b_i^{i+1}, \ldots, b_i^d\right)$ be the i-th row in this matrix, then its prior distribution is

$$\mathcal{P}\left(\mathbf{b}_i|\mathbf{a}\right) = (2\pi)^{-(d-1)/2} \prod_{j \neq i} a_j^{1/2} \exp\left[-a_j (b_i^j)^2/2\right] , \tag{6}$$

where $\mathbf{a} = (a_1, \ldots, a_d)$ is a d-dimensional vector of hyperparameters. The prior for the i-th column of $\mathbf{B}$ is then a zero-mean multivariate Gaussian distribution with independent components that have an inverse variance given by a_i. The ARD method works by optimizing the evidence of the resulting Bayesian model with respect to $\mathbf{a} = (a_1, \ldots, a_d)$. During this process, some of the a_i take an infinite value and consequently, the resulting posterior for the corresponding rows of $\mathbf{B}$ is a point mass at zero. However, a direct optimization of the evidence function under the ARD framework is very likely to get stuck into a sub-optimal solution. To avoid this problem, we propose a greedy optimization method in which all the a_i are initially set to infinite and only k of these parameters are optimized sequentially in a greedy manner.

The logarithm of the evidence of the resulting Bayesian model is

$$\mathcal{L}(\mathbf{a}) = -\frac{1}{2}\sum_{i=1}^{d} \log |\mathbf{C}_{-i}| - \frac{1}{2}\sum_{i=1}^{d} \mathbf{x}_i^{\mathrm{t}} \mathbf{C}_{-i}^{-1} \mathbf{x}_i + \text{constant}\,, \tag{7}$$

where $\mathbf{x}_i$ represents the transpose of the i-th row of $\mathbf{X}$ and

$$\mathbf{C} = \sigma^2 \mathbf{I} + \sum_{i=1}^{d} a_i^{-1} \mathbf{x}_i \mathbf{x}_i^{\mathrm{t}}\,, \qquad \mathbf{C}_{-i} = \sigma^2 \mathbf{I} + \sum_{j \neq i}^{d} a_j^{-1} \mathbf{x}_j \mathbf{x}_j^{\mathrm{t}}\,. \tag{8}$$

Following [23], we separate out the contribution of each a_j on the total log-evidence

$$\begin{aligned} \mathcal{L}(\mathbf{a}) &= \mathcal{L}(\mathbf{a}_{-j}) + \ell(a_j) \\ &= \mathcal{L}(\mathbf{a}_{-j}) + \sum_{i \neq j} \frac{1}{2} \left[\log a_j - \log(a_j + s_{ji}) + \frac{q_{ji}^2}{a_j + s_{ji}} \right], \end{aligned} \tag{9}$$

where $s_{ji} = \mathbf{x}_j^{\mathrm{t}} \mathbf{C}_{-ij}^{-1} \mathbf{x}_j$ and $q_{ji} = \mathbf{x}_j^{\mathrm{t}} \mathbf{C}_{-ij}^{-1} \mathbf{x}_i$. For the computation of s_{ji} and q_{ji} we use

$$s_{ji} = \frac{a_j S_{ji}}{a_j - S_{ji}}\,, \qquad q_{ji} = \frac{a_j Q_{ji}}{a_j - S_{ji}}\,, \tag{10}$$

$$S_{ji} = D_{jj} + \frac{D_{ji}^2}{a_i - D_{ii}}\,, \qquad Q_{ji} = \frac{a_i D_{ji}}{a_i - D_{ii}}\,, \tag{11}$$

where $S_{ji} = \mathbf{x}_j^{\mathrm{t}} \mathbf{C}_{-i}^{-1} \mathbf{x}_j$, $Q_{ji} = \mathbf{x}_j^{\mathrm{t}} \mathbf{C}_{-i}^{-1} \mathbf{x}_i$ and $D_{ji} = \mathbf{x}_j^{\mathrm{t}} \mathbf{C}^{-1} \mathbf{x}_i$. Note that when $a_i = +\infty$, it is satisfied that $S_{ji} = D_{jj}$ and $Q_{ji} = D_{ji}$. Thus, it is only necessary to compute D_{ii} when $a_i < +\infty$ and with (10), (11) and matrix $\mathbf{C}^{-1}$ we can compute all the s_{ji} and q_{ji} very efficiently.

A greedy process allows us to select the k most relevant rows of $\mathbf{X}$. First, all the components of $\mathbf{a}$ are initialized to $+\infty$ and a copy of matrix $\mathbf{C}^{-1}$ is kept in memory with initial value $\sigma^{-2}\mathbf{I}$. For $j = 1, \ldots, d$ such that $a_j = +\infty$, we compute the increment in the log-evidence that can be achieved by optimizing $\mathcal{L}(\mathbf{a})$ with respect to a_j when $\mathbf{a}_{-j} = (a_1, \ldots, a_{j-1}, a_{j+1}, \ldots, a_d)$ is kept fixed. The a_j that yields the highest increment is then updated so that $\mathcal{L}(\mathbf{a})$ is maximized. All the optimizations are performed using a combination of golden section search and successive parabolic interpolation. The main loop of the process is repeated until k components of $\mathbf{a}$ are finite. For a particular j, we compute D_{ji} for all i as the product $\mathbf{X}\mathbf{C}^{-1}\mathbf{x}_j$. Once an a_j becomes finite, we efficiently update $\mathbf{C}^{-1}$ using

$$\mathbf{C}_{\text{new}}^{-1} = \mathbf{C}_{\text{old}}^{-1} - \frac{\mathbf{C}_{\text{old}}^{-1} \mathbf{x}_j \mathbf{x}_j^{\mathrm{t}} \mathbf{C}_{\text{old}}^{-1}}{a_j^{\text{new}} + \mathbf{x}_j^{\mathrm{t}} \mathbf{C}_{\text{old}}^{-1} \mathbf{x}_j}\,. \tag{12}$$

```
Input: A gene log-expression matrix X and a positive integer k.
Output: A value for the regularization parameter M.

   bottomM ← 0.
   topM ← 100.
   converged ← FALSE.
   while not (converged)
        newM ← (bottomM + topM) / 2.
        numberOfNonZeroColumns ← groupLasso(X, newM).
        if (numberOfNonZeroColumns < k)
             bottomM ← newM.
        else if (numberOfNonZeroColumns > k)
             topM ← newM.
        else converged ← TRUE.
   M ← newM.
   return M.
```

Fig. 2. Pseudocode of the binary search process for finding M in the group Lasso method. The figure displays the steps that are followed to obtain a value for M in (13) so that the minimizer of this expression has exactly k non-zero columns. The function groupLasso returns the number of non-zero columns in the minimizer of (13) when $\mathbf{X}$ and M are equal to the arguments passed to this function.

To compute D_{ii} for all i such that $a_i < +\infty$, we keep in memory a matrix $\mathbf{F}$ whose components are $\mathbf{x}_i^{\text{t}}\mathbf{C}^{-1}\mathbf{x}_j$ for all a_i and a_j finite. When an a_j becomes finite, $\mathbf{F}$ is efficiently updated by using (12) and appending a new row and a new column to the matrix. The required D_{ii} values are then easily obtained from the main diagonal of $\mathbf{F}$. Once this greedy process is finished, the k hub genes selected by the ARD method are represented by the k components of $\mathbf{a}$ that are finite. The final computational cost of this technique is $\mathcal{O}(kd^2n)$.

3.2 Group Lasso

An extension of the Lasso regularization method [22] denoted group Lasso [12] is useful to obtain groupwise sparse linear models. Given (5), we can obtain a columnwise sparse estimate of the coefficient matrix $\mathbf{B}$ as the minimizer of

$$\| \mathbf{X} - \mathbf{BX} \|_{\text{F}}^2 \quad \text{subject to} \quad \sum_{i=1}^{d} \| \mathbf{b}_i \|_2 \leq M \quad \text{and} \quad \text{diag}(\mathbf{B}) = \mathbf{0}\,, \tag{13}$$

where $\| \cdot \|_{\text{F}}$ and $\| \cdot \|_2$ stand for the Frobenius and ℓ_2 norms, respectively, $\mathbf{b}_i$ is the i-th column of matrix $\mathbf{B}$ and M is a positive regularization parameter. Since $\mathbf{B}$ is constrained by the sum of ℓ_2 norms, some of its non-relevant columns will shrink to zero during the optimization process [12]. In this regularization method, there is a direct relationship between M and the number of non-zero columns in the resulting estimate of $\mathbf{B}$. The larger M is, the more non-zero columns in the coefficient matrix estimate. Consequently, to select the k most relevant rows of $\mathbf{X}$, we perform a binary search on M so that the corresponding minimizer of (13) has exactly k non-zero columns. This process is illustrated by the pseudocode displayed in Fig. 2. The optimization problem involved by (13) is efficiently solved using the gradient projection method described in [24]. The final computational cost of the group Lasso method is $\mathcal{O}(d^3)$. This is comparatively much more expensive than the cost of the ARD approach when $d \gg n$, which is often the case in practical applications.

3.3 Maximum Relevance Minimum Redundancy

The maximum relevance minimum redundancy (MRMR) criterion for feature selection [14] is generalized in this section to the particular case of the multiple regression problem displayed in (5). Let $\mathbf{x}_i = (x_{i1}, \ldots, x_{in})$ be the i-th row of $\mathbf{X}$ and let $I(\mathbf{x}_i, \mathbf{x}_j)$ denote the empirical mutual information for genes i and j. Now, suppose that S is a set that contains those genes already selected by the method. The next gene to be selected is a gene j that is not yet in S and that maximizes

$$\frac{1}{d}\sum_{i=1}^{d} I(\mathbf{x}_i, \mathbf{x}_j) - \frac{1}{|S|}\sum_{i \in S} I(\mathbf{x}_i, \mathbf{x}_j)\,. \tag{14}$$

The redundancy term is here the same as in standard MRMR. However, the relevance term is an average of the relevance terms for each of the d individual regressions in (5). For the computation of the empirical mutual information, $I(\mathbf{x}_i, \mathbf{x}_j)$, we assume that the marginal distribution of the pairs $\{x_{ik}, x_{jk}\}_{k=1}^{n}$ is bivariate Gaussian. In consequence, $I(\mathbf{x}_i, \mathbf{x}_j) = -0.5 \log(1 - \hat{\rho}_{ij}^2)$, where $\hat{\rho}_{ij}$ is the empirical correlation of the vectors $\mathbf{x}_i$ and $\mathbf{x}_j$. The final computational cost of the MRMR method is $\mathcal{O}(d^2 n)$. Hence, MRMR is the most efficient of the three feature selection methods analyzed in this paper.

4 Evaluation of the Hub Gene Selection Methods

The performance of the ARD, group Lasso and MRMR methods for identifying hub genes is evaluated in a series of experiments with simulated steady-state gene expression data. The program SynTREN [15] is used to generate synthetic gene expression data from three different transcription control networks. Two of the networks are from the organism *Escherichia coli* and they have 423 and 1330 genes, respectively [25,26]. The last network is from *Saccharomyces cerevisiae* and has 690 genes [27]. 50 gene expression matrices with 250 measurements are generated from each network using the following configuration for SynTREN: All the noise levels are fixed to 0.3, transition functions are linear, the number of burn in-cycles is fixed to 1000, complex interactions are deactivated and the number of correlated externals is 0.

To reduce the negative impact of outliers, we map each row of each gene log-expression matrix to the standard Gaussian distribution. For this task, an approximation of the probability integral transform is used in combination with the standard Gaussian quantile function:

$$x_{ij}^{\text{map}} = \Phi^{-1}\left[F_n(x_{ij})\frac{n}{n+1}\right] \quad \text{for} \quad j = 1, \ldots, n\,, \tag{15}$$

where x_{ij} is the j-th element in the i-th row of a $d \times n$ gene log-expression matrix, Φ^{-1} is the standard Gaussian quantile function, F_n is the empirical cumulative distribution of the elements in the i-th row of the log-expression matrix and the factor $n/(n+1)$ guarantees that the argument passed to Φ^{-1} is lower than one. The resulting data matrix is the input to the hub gene selection methods.

The ARD, group Lasso and MRMR methods are executed on each of the gene expression matrices generated by the program SynTREN (after applying the previous data preprocessing method). The parameter σ in the ARD approach is initialized to 1 and k, the number of hub genes that each method must select, is fixed to a reduced fraction of the total number of nodes in the underlying transcription network, e.g. 5%. Hub genes are by definition highly connected nodes in the network. Hence, as a measure of quality, we compute the average connectivity of the genes selected by each method. For comparative purposes, we also include in the analysis an extra method that only returns a list of k genes selected randomly. The highest possible average connectivities that a method can obtain are 17.27 and 37.31 for the small and large E. coli networks and 22.51 for the yeast network.

Table 1 summarizes the results of each feature selection method. The three proposed methods perform significantly better than the random approach, with ARD obtaining the best accuracy. For small networks, the results of MRMR and ARD are more or less similar. However, ARD is much better than MRMR when analyzing large transcription networks. Additionally, the Group Lasso method performs rather poorly when compared with the MRMR and ARD approaches. The reason for this poor performance is the propensity of the group Lasso method for selecting genes whose expression patterns are very correlated. By contrast, ARD and MRMR do not have this problem since these methods penalize the selection of redundant features [23,14].

Finally, Fig. 1 displays a fraction of the transcription control network from *Escherichia coli* with 423 genes. Those nodes with a diamond shape correspond to the genes selected by the ARD method when trained on one of the synthetic 423×250 gene expression matrix generated by SynTREN. The figure clearly shows that most of the genes selected by ARD are hubs or key regulators in the network. Light-gray nodes are those genes selected by the group Lasso method. In this method, most of the selected genes are connected to the same hub or key regulator and in consequence, their expression patterns are highly correlated. The nodes selected by MRMR, although not shown, are in this case very similar to those selected by ARD.

Table 1. Average connectivity and corresponding standard deviation for each method

Network	Genes	Random	Lasso	MRMR	ARD
Small E. coli	423	2.54±0.77	7.23±0.40	13.31±0.77	14.01±0.71
Yeast	690	3.14±0.91	9.46±0.51	13.91±1.27	16.51±0.74
Large E. coli	1330	4.22±1.94	8.56±0.72	14.70±2.48	23.48±1.62

5 Combining ARACNE with the Output of a Hub Gene Selection Method

In this section we describe how to improve the performance of ARACNE using the output of the previous hub gene selection methods. First, we introduce the original ARACNE algorithm. ARACNE is a state-of-the-art method for reverse engineering transcription networks using only steady-state gene expression data [4,16]. ARACNE belongs to a class of network reconstruction algorithms known as mutual information networks [28]. In these methods, a connection between two nodes is created if the corresponding transcript concentrations exhibit large mutual information. The first step of ARACNE is to compute the empirical mutual information for any two transcripts i and j, i.e. $I(\mathbf{x}_i, \mathbf{x}_j)$. The connection weight w_{ij} for any two genes i and j is initialized as $w_{ij} = I(\mathbf{x}_i, \mathbf{x}_j)$. After this, in a pruning step, ARACNE employs the data processing inequality to eliminate indirect interactions between genes. This inequality states that when genes i and j interact indirectly through gene k, it should be satisfied

$$I(\mathbf{x}_i, \mathbf{x}_j) \leq \min\left\{I(\mathbf{x}_i, \mathbf{x}_k), I(\mathbf{x}_j, \mathbf{x}_k)\right\} . \tag{16}$$

For each triplet of transcripts, ARACNE applies inequality (16) and sets w_{ij} to zero when (16) holds for some k, otherwise w_{ik} is not changed. The last step of ARACNE is to link each pair of genes i and j when w_{ij} is higher than threshold $\theta > 0$. Note that ARACNE enforces sparsity in the reconstructed network first, by making use of the data processing inequality and second, by means of the threshold θ. However, ARACNE does not take into account the scale-free topological structure of transcription control networks.

Prediction errors in ARACNE are mainly due to the large variance displayed by the mutual information estimates $I(\mathbf{x}_i, \mathbf{x}_j)$ for any two genes i and j. This excessive variance has its origin in the reduced number of measurements and the high level of noise in gene expression data. The performance of ARACNE can be significantly reduced by these random fluctuations in the estimation of the mutual information. In particular, ARACNE may fail to eliminate some of the indirect gene interactions. This occurs when genes i and j interact through gene k indirectly but fluctuations in the mutual information estimation process make $I(\mathbf{x}_i, \mathbf{x}_j)$ larger than $I(\mathbf{x}_i, \mathbf{x}_k)$ or $I(\mathbf{x}_j, \mathbf{x}_k)$. In a similar manner, ARACNE may mistakenly remove a direct interaction between genes i and j when these fluctuations in the estimation process make $I(\mathbf{x}_i, \mathbf{x}_j)$ lower than $I(\mathbf{x}_i, \mathbf{x}_k)$ and $I(\mathbf{x}_j, \mathbf{x}_k)$ for some gene k.

5.1 Modification of the Original ARACNE Method

Prior knowledge about the particular topology of the network can be very useful to reduce the negative impact of random fluctuations in the estimation of the mutual information. In particular, when the underlying network is scale-free any arbitrary node is more likely to be connected to a hub node than to any other ordinary node. This means that, on average, $I(\mathbf{x}_i, \mathbf{x}_j)$ should be larger when either i or j is a hub gene than when none of them is. Using this idea, we propose to improve the performance of ARACNE as follows. After ARACNE has computed the empirical mutual information for all genes, we update the mutual information estimates as

$$I_{\text{new}}(\mathbf{x}_i, \mathbf{x}_j) = \begin{cases} I_{\text{old}}(\mathbf{x}_i, \mathbf{x}_j) + \Delta_{ij} & \text{if } i \in S_H \text{ or } j \in S_H \\ I_{\text{old}}(\mathbf{x}_i, \mathbf{x}_j) & \text{otherwise} \end{cases} \quad (17)$$

for $i = 1, \ldots, d$, $j = 1, \ldots, d$ and $j \neq i$, where S_H is a set of hub genes and Δ_{ij} is a positive number that scales with the level of noise in the estimation of the mutual information for transcripts i and j. Here, S_H has been generated previously by either the ARD, group Lasso, or MRM methods. The purpose of Δ_{ij} is to generate an increment in the mutual information estimate between genes i and j when one of these genes is a hub. However, this increment has to be adjusted with respect to the level of noise in $I_{\text{old}}(\mathbf{x}_i, \mathbf{x}_j)$. In particular, when the number of available gene expression measurements increases, the amount of noise in $I_{\text{old}}(\mathbf{x}_i, \mathbf{x}_j)$ decreases and Δ_{ij} should decrease too. Otherwise, $I_{\text{new}}(\mathbf{x}_i, \mathbf{x}_j)$ would be asymptotically biased and the performance of the new ARACNE algorithm would not improve as more data samples are available.

If we assume that the elements of $\mathbf{x}_i$ and $\mathbf{x}_j$ have a bivariate Gaussian marginal distribution, then the first step of ARACNE will set

$$I(\mathbf{x}_i, \mathbf{x}_j) = -0.5 \log(1 - \hat{\rho}_{ij}^2) \,, \quad (18)$$

where $\hat{\rho}_{ij}$ is the empirical correlation between $\mathbf{x}_i$ and $\mathbf{x}_j$. In this case, we can fix a probabilistic upper bound on the noise in the estimation of $I(\mathbf{x}_i, \mathbf{x}_j)$ using the delta method, namely

$$\Delta_{ij} = n^{-1/2} |\hat{\rho}_{ij}| \Phi^{-1} (1 - \gamma) \,, \quad (19)$$

where we have approximated the sampling distribution of $\hat{\rho}_{ij}$ by a Gaussian with variance $n^{-1}(1 - \hat{\rho}_{ij}^2)^2$ and mean $\hat{\rho}_{ij}$, n is the number of available expression measurements, Φ^{-1} is the standard Gaussian quantile function and γ is a small positive number that determines the probability of the actual noise being greater than the upper bound.

The update rule described by (17) and (19) generates an increment in the mutual information estimates when one of the two transcripts is a hub gene. This increment is proportional to the uncertainty in the estimation of the mutual information so that we can achieve a balance between the connectivity structure given by the data and our prior assumption that the genes in S_H should be highly connected. The update rule guarantees that direct connections where one

Table 2. Average AUC-PR and corresponding standard deviation for each method

Network	Genes	Standard ARACNE	New ARACNE with Random	Lasso	MRMR	ARD
Small E. coli	423	0.41±0.03	0.28±0.05	0.41±0.04	0.57±0.04	0.56±0.03
Yeast	690	0.30±0.02	0.16±0.03	0.30±0.02	0.35±0.03	0.39±0.02
Large E. coli	1330	0.16±0.01	0.06±0.01	0.15±0.01	0.18±0.03	0.26±0.02

of the two transcripts is a hub gene are less likely to be mistakenly removed during the pruning step of ARACNE. Furthermore, indirect links between two transcripts regulated by a common hub gene are also more likely to be pruned out during this step. Consequently, the new ARACNE algorithm should have an improved performance over the original version of the method.

6 Evaluation of the New ARACNE Method

To evaluate the accuracy of the new ARACNE method based on a list of hub genes, we perform a series of experiments with synthetic gene expression data. The standard ARACNE algorithm is compared with four versions of the modified method, each one corresponding to a different technique for obtaining the set S_H of hub genes: ARD, group Lasso, MRMR and the random approach. The data used in these experiments are the 150 gene expression matrices generated for the experiments of Sect. 4. Again, the data are transformed using the mapping given by (15). The parameters of the gene selection methods are the same as in Sect. 4. Mutual information in ARACNE is computed as in MRMR and γ is fixed to d^{-1}, where d is the number of genes in the network under analysis.

The performance of each reverse engineering technique is evaluated using the area under the Precision-Recall curve (AUC-PR) which is obtained by altering threshold θ [29]. The precision and the recall rates are computed by comparing the edges of the true network with those edges inferred by the method under analysis. Precision-Recall curves are preferred as a measure of quality over other alternatives like Receiver Operating Characteristics (ROC) curves. The reason for this is that the target class distribution in the network reconstruction problem (edge, no edge) is very skewed and in this case Precision-Recall curves are more suitable for evaluating the performance of a method [29].

Table 2 displays the average and the standard deviation of the AUC-PR measures obtained by each method in the experiments. The random approach decreases the performance of the standard ARACNE algorithm and the group Lasso approach does not seem to provide any relevant improvement. However, the MRMR and ARD feature selection methods give a significant boost to the performance of the original ARACNE algorithm, with ARD obtaining the best results: up to a 62% improvement for the *E. coli* network with 1330 genes.

Finally, Fig. 3 displays the Precision-Recall curves obtained by the different network reconstruction methods when they are executed on a particular gene expression matrix sampled from each of the three transcription control networks:

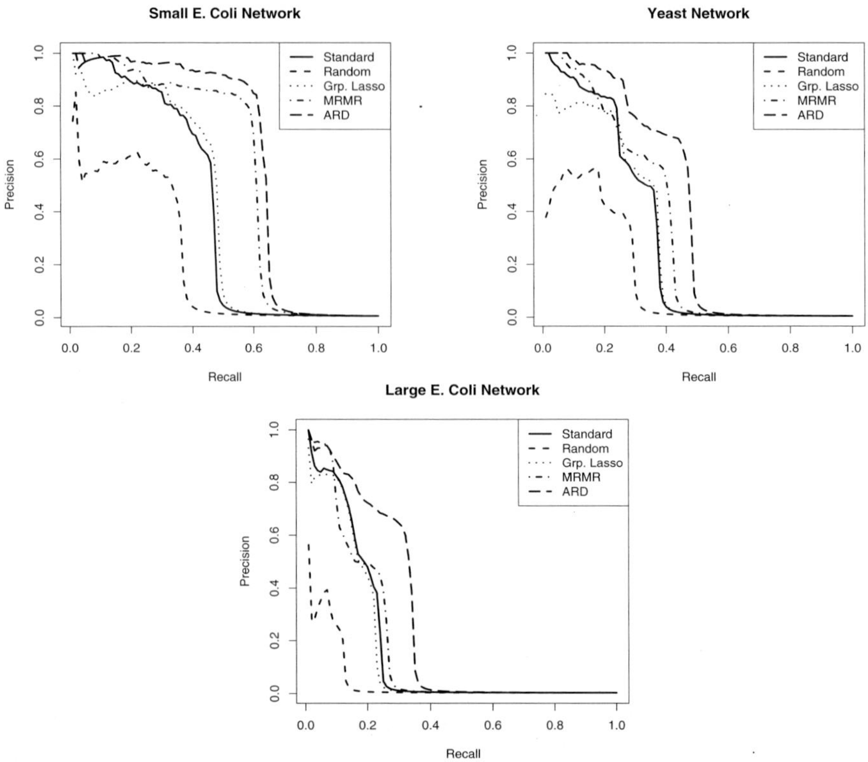

Fig. 3. Precision-Recall curves for the network reconstruction methods when they are run on a specific gene expression matrix sampled from: The small E. coli network (top left), the yeast network (top right) and the large E. coli network (bottom). The best performing method is the ARACNE algorithm which employs the list of hub genes obtained by the ARD feature selection approach.

The small E. coli network, the yeast network and the large E. coli network. The best performing method is the ARACNE algorithm that employs the list of hub genes obtained by the ARD feature selection approach.

7 Experiments with Real Microarray Data

We evaluate the ability of the ARD method for selecting key regulators on a set of expression profiles from *Saccharomyces cerevisiae*. This dataset contains 247 expression measurements for a total of 5520 genes and it is publicly available at the Many Microbe Microarrays Database [17]. Before running the analysis, the gene expression matrix is transformed using (15) and σ is fixed to 1.

Table 3 displays the top ten genes selected by the ARD technique and their description at the *Saccharomyces* Genome Database. From the list of ten genes, YOR224C, YBR289W and YBR160W are involved in transcription or in the

Table 3. Top ten genes selected by ARD on the yeast dataset

Rank	Gene	Description
1	YOR224C	RNA polymerase subunit.
2	YPL013C	Mitochondrial ribosomal protein.
3	YGL245W	Glutamyl-tRNA synthetase.
4	YPL012W	Ribosomal RNA processing.
5	YER125W	Ubiquitin-protein ligase.
6	YER092W	Associates with the INO80 chromatin remodeling complex
7	YBR289W	Subunit of the SWI/SNF chromatin remodeling complex.
8	YBR272C	Involved in DNA mismatch repair.
9	YBR160W	Catalytic subunit of the main cell cycle kinase.
10	YOR215C	Unknown function.

regulation of transcription. YER092W is likely to be a regulator of transcription and YPL013C, YGL245W and YPL012W could have some type of role in post-transcriptional regulation. It is significant that the first gene selected by the ARD method is a component of the RNA polymerase enzyme. In particular, the linear model of transcription that is described in Sect. 2 precisely states that the transcript concentration of RNA polymerase is the most informative feature for solving the multiple regression problem given by (5). Since only 7 genes are described at the *Saccharomyces* Genome Database as RNA polymerase subunits, the probability of obtaining this result by chance is approximately 10^{-4}. Finally, note that YER125W is involved in the degradation of RNA polymerase [30] and also regulates many cellular processes, including transcription.

8 Conclusions

Several machine learning methods are available in the literature for the reverse engineering of transcription networks using only steady-state gene expression data. Most of these methods enforce sparsity in the resulting interaction map. However, few of them take into account the scale-free topology of transcription networks. In this paper, we have focused on the hubs that commonly appear in networks with a scale-free topological structure. Reverse engineering methods can be significantly improved by predicting beforehand those genes that are likely to be hubs or key regulators in the underlying network. For this task, we have introduced a multiple linear regression model which describes the interactions between genes and transcriptional regulators. The model includes a coefficient matrix which is expected to be columnwise sparse. A set of hub genes can then be selected by identifying those columns of the regression matrix which have many non-zero entries. In practice, this is equivalent to selecting those features that are more relevant for solving the multiple regression problem. Three feature selection methods have been proposed for this task. The group Lasso method [12], the automatic relevance determination (ARD) approach [13] and the maximum relevance minimum redundancy (MRMR) criterion [14]. A series of experiments with simulated gene expression data validate the capacity of these methods for identifying hub genes. The best results are obtained by the feature selection method based on the ARD approach.

Additionally, the performance of ARACNE, a popular method for reverse engineering transcription networks, has been improved by taking into account the list of hub genes generated by the aforementioned feature selection methods. Experiments with simulated gene expression data indicate that a combination of ARACNE and the best feature selection method, ARD, yields up to a 62% improvement in performance with respect to the original algorithm.

Finally, the best feature selection method, ARD, is validated using expression profiles from yeast. This method obtains from an original set with thousands of genes a final set of ten genes that includes many global mRNA regulators.

Acknowledgements

This study was supported under project TIN2007-66862-C02-02 by the Spanish *Dirección General de Investigación* and by a Dutch NWO Computational Life Science grant awarded to Tjeerd M. H. Dijkstra (RUMPHI 600.635.100.08N28).

References

1. Kitano, H.: Systems biology: a brief overview. Science 295(5560), 1662–1664 (2002)
2. Stolovitzky, G., Califano, A.: Systems biology: Making sense of oceans of biological data. The New York Academy of Sciences Update Magazine, 20–23 (March/April 2006)
3. Tong, et al.: Global mapping of the yeast genetic interaction network. Science 303(5659), 808–813 (2004)
4. Basso, K., Margolin, A.A., Stolovitzky, G., Klein, U., Dalla-Favera, R., Califano, A.: Reverse engineering of regulatory networks in human B cells. Nature Genetics 37, 382–390 (2005)
5. Faith, J.J., Hayete, B., Thaden, J.T., Mogno, I., Wierzbowski, J., Cottarel, G., Kasif, S., Collins, J.J., Gardner, T.S.: Large-scale mapping and validation of escherichia coli transcriptional regulation from a compendium of expression profiles. PLoS Biology 5, 54–66 (2007)
6. Thieffry, D., Huerta, A.M., Pérez-Rueda, E., Collado-Vides, J.: From specific gene regulation to genomic networks: a global analysis of transcriptional regulation in escherichia coli. BioEssays 20(5), 433–440 (1998)
7. Albert, R.: Scale-free networks in cell biology. Journal of Cell Science 118(21), 4947–4957 (2005)
8. Friedman, N.: Inferring cellular networks using probabilistic graphical models. Science 303(5659), 799–805 (2004)
9. Yeung, M.K.S., Tegnér, J., Collins, J.J.: Reverse engineering gene networks using singular value decomposition and robust regression. Proceedings of the National Academy of Sciences of the United States of America 99(9), 6163–6168 (2002)
10. Gardner, T., di Bernardo, D., Lorenz, D., Collins, J.J.: Inferring genetic networks and identifying compound mode of action via expression profiling. Science 301(5629), 102–105 (2003)

11. Sheridan, P., Kamimura, T., Shimodaira, H.: On scale-free prior distributions and their applicability in large-scale network inference with gaussian graphical models. In: Complex Sciences, pp. 110–117. Springer, Heidelberg (2009)
12. Yuan, M., Lin, Y.: Model selection and estimation in regression with grouped variables. Journal of the Royal Statistical Society, Series B 68, 49–67 (2006)
13. Tipping, M.E.: Sparse bayesian learning and the relevance vector machine. Journal of Machine Learning Research 1, 211–244 (2001)
14. Peng, H., Long, F., Ding, C.: Feature selection based on mutual information: Criteria of max-dependency, max-relevance, and min-redundancy. IEEE Transactions on Pattern Analysis and Machine Intelligence 27(8), 1226–1238 (2005)
15. den Bulcke, T.V., Leemput, K.V., Naudts, B., van Remortel, P., Ma, H., Verschoren, A., Moor, B.D., Marchal, K.: Syntren: a generator of synthetic gene expression data for design and analysis of structure learning algorithms. BMC Bioinformatics 7(1), 43 (2006)
16. Margolin, A., Nemenman, I., Basso, K., Wiggins, C., Stolovitzky, G., Favera, R., Califano, A.: Aracne: An algorithm for the reconstruction of gene regulatory networks in a mammalian cellular context. BMC Bioinformatics 7(Suppl. 1), S7 (2006)
17. Faith, J.J., Driscoll, M.E., Fusaro, V.A., Cosgrove, E.J., Hayete, B., Juhn, F.S., Schneider, S.J., Gardner, T.S.: Many microbe microarrays database: uniformly normalized affymetrix compendia with structured experimental metadata. Nucleic Acids Research 36, D866–D870 (2008)
18. Shannon, P., Markiel, A., Ozier, O., Baliga, N.S., Wang, J.T., Ramage, D., Amin, N., Schwikowski, B., Ideker, T.: Cytoscape: a software environment for integrated models of biomolecular interaction networks. Genome Research 13, 2498–2504 (2003)
19. Gardner, T.S., Faith, J.J.: Reverse-engineering transcription control networks. Physics of Life Reviews 2(1), 65–88 (2005)
20. Alon, U.: An introduction to systems biology. CRC Press, Boca Raton (2006)
21. Barabási, A.L., Oltvai, Z.N.: Network biology: understanding the cell's functional organization. Nature Reviews Genetics 5(2), 101–113 (2004)
22. Tibshirani, R.: Regression shrinkage and selection via the lasso. Journal of the Royal Statistical Society. Series B (Methodological) 58(1), 267–288 (1996)
23. Bishop, C.: Pattern Recognition and Machine Learning. Springer, Heidelberg (2006)
24. Kim, Y., Kim, J., Kim, Y.: Blockwise sparse regression. Statistica Sinica 16, 375–390 (2006)
25. Shen-Orr, S.S., Milo, R., Mangan, S., Alon, U.: Network motifs in the transcriptional regulation network of escherichia coli. Nature Genetics 32, 64–68 (2002)
26. Ma, H.W., Kumar, B., Ditges, U., Gunzer, F., Buer, J., Zeng, A.P.: An extended transcriptional regulatory network of escherichia coli and analysis of its hierarchical structure and network motifs. Nucleic Acids Research 32(22), 6643–6649 (2004)
27. Guelzim, N., Bottani, S., Bourgine, P., Képès, F.: Topological and causal structure of the yeast transcriptional regulatory network. Nature Genetics 31, 60–63 (2002)
28. Meyer, P.E., Lafitte, F., Bontempi, G.: minet: A r/bioconductor package for inferring large transcriptional networks using mutual information. BMC Bioinformatics 9(1), 461 (2008)
29. Davis, J., Goadrich, M.: The relationship between precision-recall and roc curves. In: ICML 2006, pp. 223–240 (2006)
30. Huibregtse, J.M., Yang, J.C., Beaudenon, S.L.: The large subunit of RNA polymerase II is a substrate of the Rsp5 ubiquitin-proteinligase. Proceedings of the National Academy of Sciences of the United States of America 94(8), 3656–3661 (1997)

Expectation Propagation for Bayesian Multi-task Feature Selection

Daniel Hernández-Lobato[1], José Miguel Hernández-Lobato[2], Thibault Helleputte[1], and Pierre Dupont[1]

[1] Machine Learning Group, ICTEAM institute, Université catholique de Louvain, Place Sainte Barbe 2, B-1348 Louvain-la-Neuve, Belgium
{daniel.hernandez-lobato, thibault.helleputte, pierre.dupont}@uclouvain.be

[2] Computer Science Department, Universidad Autónoma de Madrid, C/ Francisco Tomás y Valiente, 11 28049 Madrid, Spain
josemiguel.hernandez@uam.es

Abstract. In this paper we propose a Bayesian model for multi-task feature selection. This model is based on a generalized *spike and slab* sparse prior distribution that enforces the selection of a common subset of features across several tasks. Since exact Bayesian inference in this model is intractable, approximate inference is performed through expectation propagation (EP). EP approximates the posterior distribution of the model using a parametric probability distribution. This posterior approximation is particularly useful to identify relevant features for prediction. We focus on problems for which the number of features d is significantly larger than the number of instances for each task. We propose an efficient parametrization of the EP algorithm that offers a computational complexity linear in d. Experiments on several multi-task datasets show that the proposed model outperforms baseline approaches for single-task learning or data pooling across all tasks, as well as two state-of-the-art multi-task learning approaches. Additional experiments confirm the stability of the proposed feature selection with respect to various sub-samplings of the training data.

Keywords: Multi-task learning, feature selection, expectation propagation, approximate Bayesian inference.

1 Introduction

The automatic induction of a predictor for a dependent variable y given a feature vector $\mathbf{x}$ can be a difficult task when the number of training instances is very small and the number of explanatory variables is large. Examples of learning applications with these characteristics include, among others, the classification of microarray data [1] or the analysis of high-dimensional images [2]. Under these circumstances, an underlying linear model is often considered, possibly in an expanded feature space. A potential way of improving the robustness of these models is to assume that only a small subset of the original features are relevant

J.L. Balcázar et al. (Eds.): ECML PKDD 2010, Part I, LNAI 6321, pp. 522–537, 2010.

for prediction [3]. That is, the underlying linear model is assumed to be sparse with many coefficients being equal to zero. The identification of the relevant features is typically implemented by optimizing an objective function penalized by a sparsity enforcing regularizer. Such a regularizer drives to zero some of the coefficients during the optimization process. A common choice is the ℓ_1 norm of the vector of model coefficients [4]. Within a Bayesian framework, sparsity can be favored by considering sparse prior distributions for the coefficients of the linear model. Examples of such priors include the Student's t distribution, the Laplace [2] and the *spike and slab* [5]. Among these, the *spike and slab* is the only prior that can assign a non-zero probability to solutions with many coefficients being equal to zero. Under this prior it is furthermore possible to specify intuitively the fraction of coefficients that are *a priori* different from zero. The *spike and slab* prior also provides a selective shrinkage of the model coefficients [6]: for high sparsity levels, most of the coefficients are close to zero while a few of them have significantly larger values. By contrast, other priors shrink towards zero *all* the coefficients when the sparsity level is increased. Since exact Bayesian inference is typically intractable under sparse priors, approximate algorithms have to be used in practice. An alternative Bayesian approach for feature selection is automatic relevance determination (ARD) [7]. However, ARD does not consider uncertainty in the feature selection process nor does it provide a posterior probability of using each feature for prediction

We address here prediction problems for which the number of instances is typically small. In such cases, it may be beneficial for the induction to rely on several distinct but related tasks. Microarray datasets offer examples of such tasks for which the common objective is typically to discriminate between normal and tumor samples, while tissue and RNA extraction protocols may differ across tasks. Specifically, the multi-task approach proposed by Obozinski *et al.* assumes that the distinct tasks share a reduced set of common relevant features [8]. They propose to solve an optimization problem that minimizes a logistic loss function combined with a term that penalizes the ℓ_1 norm of the vector of ℓ_2 norms of the feature-specific coefficient vectors across the different tasks. Such a mixed norm regularization drives to zero the same coefficients of the task-specific vectors during the optimization process. This favors the selection of a common set of features to describe each task. The amount of sparsity is determined in this model by a hyper-parameter λ which has to be tuned by cross validation. In particular, [8] gives an efficient path-following algorithm to find the potential values of λ.

In a different work, Evgeniou and Pontil consider that the hyperplanes of the distinct tasks are the linear combination of a common hyperplane and a task-specific hyperplane [9]. They specifically propose to minimize a hinge loss function that is penalized by two terms. The first term is proportional to the hyper-parameter λ_1 and penalizes the squared values of the ℓ_2 norms of the task-specific hyperplanes. The second term is proportional to the hyper-parameter λ_2 and penalizes the squared value of the ℓ_2 norm of the common hyperplane. These two parameters are tuned by cross-validation and their ratio determines the contribution of the common hyperplane to each task. If λ_1/λ_2 is high, the task-specific hyperplanes are penalized more and all the hyperplanes of the different tasks tend to be equal to the common hyperplane. By contrast, when λ_1/λ_2 is low, the common hyperplane is penalized more and the models for the different tasks tend to be equal to task-specific hyperplanes.

In the present work, we propose a Bayesian model for multi-task feature selection based on a generalization of the *spike and slab* prior distribution. This generalized prior, detailed in Sect. 2, enforces the selection of a common subset of features to describe each different task. Exact Bayesian inference in this model is infeasible and approximate techniques have to be used in practice. We consider here the expectation propagation (EP) algorithm [10], which is briefly reviewed in Sect. 3. EP approximates the posterior distribution of the model using a parametric distribution that belongs to the exponential family. We detail in Sect. 4 a specific posterior approximation for our multi-task Bayesian model. We also introduce an efficient parametrization that guarantees that EP has a time complexity that can be made linear in the number of features d, under the assumption that d is significantly larger than the number of instances for each task. EP also approximates the posterior probability of using each feature for prediction. These probabilities are particularly useful to identify relevant features. Finally, experiments reported in Sect. 5 are conducted on a collection of multi-task learning problems. They show that our Bayesian model is competitive with respect to baseline approaches of single-task learning and data pooling across all tasks, and with respect to the multi-task learning methods [8,9] mentioned above. Additional experiments detail the stability of the various feature selection methods under different sub-samplings of the training data.

2 Bayesian Multi-task Feature Selection

Following Obozinski *et al.* [8], we assume that the different tasks share a small subset of common relevant features. To identify these features we define a Bayesian model relying on a set of linear functions that discriminate between two class labels. A set of $k = 1, \ldots, K$ learning tasks are assumed to be available, each one consisting of n_k d-dimensional input samples $\mathbf{X}_k = \{\mathbf{x}_{k1}, \ldots, \mathbf{x}_{kn_k}\}$ and the corresponding class labels $\mathbf{y}_k = \{y_{k1}, \ldots, y_{kn_k}\}$, where $y_{ki} \in \{-1, 1\}, \forall i, k$. We further assume that $n_k \ll d, \forall k$. Given $\mathbf{X}_k$ and $\mathbf{y}_k$, let us consider the following labeling rule:

$$y_{ki} = \begin{cases} 1 & \text{if } \mathbf{w}_k^T \mathbf{x}_{ki} + \epsilon_{ki} \geq 0 \\ -1 & \text{otherwise,} \end{cases} \tag{1}$$

where ^{T} denotes vector transpose, $\mathbf{w}_k$ are the model coefficients for task k, and a noise term ϵ_{ki} is assumed to follow a standard Gaussian distribution. The likelihood for $\mathbf{w}_k$ is hence defined as:

$$\mathcal{P}(\mathbf{y}_k|\mathbf{w}_k, \mathbf{X}_k) = \prod_{i=1}^{n_k} \mathcal{P}(y_{ki}|\mathbf{w}_k, \mathbf{x}_{ki}) = \prod_{i=1}^{n_k} \Phi(y_{ki}\mathbf{w}_k^T \mathbf{x}_{ki}) , \tag{2}$$

where $\Phi(\cdot)$ denotes the cumulative probability function of a standard Gaussian distribution. For notational convenience, we will remove the explicit conditional dependence on $\mathbf{X}_k$ in the subsequent expressions. We consider a bias term by extending each vector $\mathbf{x}_{ki}$ with a constant component equal to one. Each input sample will also be assumed to have been multiplied by its corresponding label y_{ki} and the result will be simply denoted by $\mathbf{x}_{ki}$.

The prior distribution for the model coefficients $\mathbf{w}_k$ is a generalization of the *spike and slab* sparse prior distribution [5]. In particular, we assume that all components of the vectors $\mathbf{w}_k$ are independent. Binary latent variables γ_j, with $j = 1, \ldots, d+1$ are introduced to indicate whether the j-th feature is used for classification *in each* of the different tasks ($\gamma_j = 1$) or not ($\gamma_j = 0$)[1]. Given the vector $\boldsymbol{\gamma}$, the prior for each $\mathbf{w}_k$ is

$$\mathcal{P}(\mathbf{w}_k|\boldsymbol{\gamma}) = \prod_{j=1}^{d+1} \mathcal{N}(w_{kj}|0, \sigma_{1j}^2)^{\gamma_j} \mathcal{N}(w_{kj}|0, \sigma_{0j}^2)^{1-\gamma_j} , \tag{3}$$

where $\mathcal{N}(w_{kj}|0, \sigma_{1j}^2)$ denotes a Gaussian density with zero mean and variance equal to σ_{1j}^2. In a single task setting the prior (3) reduces to the standard spike and slab prior. We set the *spikes* to be deltas centered at the origin, *i.e.* $\sigma_{0j}^2 \to 0$, $\forall j$, to enforce sparsity among the components of each vector $\mathbf{w}_k$. The variances of the *slabs*, σ_{1j}^2, are set equal to one for $j = 1, \ldots, d$. The variance of the *slab* corresponding to the bias term, $\sigma_{1(d+1)}^2$, is set to a significantly larger value (*e.g* 10). Since we further assume that each feature can be used *a priori* independently for classification, the prior on $\boldsymbol{\gamma}$ is simply defined as a multivariate Bernoulli distribution:

$$\mathcal{P}(\boldsymbol{\gamma}) = \prod_{j=1}^{d+1} p_{0j}^{\gamma_j} (1 - p_{0j})^{1-\gamma_j} , \tag{4}$$

where p_{0j} specifies the prior probability for $\gamma_j = 1$. The prior probability corresponding to the bias component $p_{0(d+1)}$ is set equal to one in this model.

According to the above definitions, and given $\mathbf{X}_k$ and $\mathbf{y}_k$ for $k = 1, \ldots, K$, we can use the Bayes' theorem to make inference about each $\mathbf{w}_k$ and $\boldsymbol{\gamma}$. Let $\mathbf{W} = \{\mathbf{w}_1, \ldots, \mathbf{w}_K\}$ and $\mathbf{Y} = \{\mathbf{y}_1, \ldots, \mathbf{y}_K\}$ be two matrices summarizing the model coefficients and the class labels of the different tasks, respectively. The posterior for $\mathbf{W}$ and $\boldsymbol{\gamma}$ is

$$\mathcal{P}(\mathbf{W}, \boldsymbol{\gamma}|\mathbf{Y}) = \frac{\mathcal{P}(\mathbf{Y}|\mathbf{W})\mathcal{P}(\mathbf{W}|\boldsymbol{\gamma})\mathcal{P}(\boldsymbol{\gamma})}{\mathcal{P}(\mathbf{Y})} , \tag{5}$$

where $\mathcal{P}(\mathbf{Y}|\mathbf{W}) = \prod_{k=1}^K \mathcal{P}(\mathbf{y}_k|\mathbf{w}_k)$, $\mathcal{P}(\mathbf{W}|\boldsymbol{\gamma}) = \prod_{k=1}^K \mathcal{P}(\mathbf{w}_k|\boldsymbol{\gamma})$ and $\mathcal{P}(\mathbf{Y})$ is just a normalization constant, known as the model evidence, which can be used to perform model selection under a Bayesian framework [11].

[1] In the Bayesian model described in the present work a given feature need not be strictly required for each task since only the *prior* distribution relies on the assumption of features used by either all tasks or none.

In this model the class label $y_k^{\text{new}} \in \{-1, 1\}$ of a new unlabeled instance $\mathbf{x}_k^{\text{new}}$ corresponding to the task k is computed using the predictive distribution:

$$\mathcal{P}(y_k^{\text{new}}|\mathbf{x}_k^{\text{new}}, \mathbf{Y}) = \int \sum_{\gamma} \mathcal{P}(y_k^{\text{new}}|\mathbf{x}_k^{\text{new}}, \mathbf{w}_k)\mathcal{P}(\mathbf{W}, \gamma|\mathbf{Y})\, d\mathbf{W}\,. \tag{6}$$

This probabilistic output is useful to quantify the uncertainty in the prediction.

Finally, those features with the highest contribution in *all* the classification tasks can be identified using the posterior distribution for γ:

$$\mathcal{P}(\gamma|\mathbf{Y}) = \int \mathcal{P}(\mathbf{W}, \gamma|\mathbf{Y})\, d\mathbf{W}\,. \tag{7}$$

Unfortunately, the exact computation of (5), (6) and (7) is too expensive for typical learning problems and have to be approximated. We rely here on expectation propagation [10], a fast algorithm for approximate Bayesian inference. This algorithm is described in the next section.

3 Expectation Propagation

In the Bayesian model for multi-task feature selection described in Sect. 2, the joint probability of $\mathbf{W}$, γ and $\mathbf{Y}$, *i.e.* the numerator in the right hand side of (5), can be written as a product of several terms t_i

$$\mathcal{P}(\mathbf{W}, \gamma, \mathbf{Y}) = \mathcal{P}(\mathbf{Y}|\mathbf{W})\mathcal{P}(\mathbf{W}|\gamma)\mathcal{P}(\gamma) = \prod_i t_i(\mathbf{W}, \gamma)\,, \tag{8}$$

where the first $n = n_1 + \ldots + n_K$ terms correspond to $\mathcal{P}(\mathbf{Y}|\mathbf{W})$, the next $K(d+1)$ terms correspond to $\mathcal{P}(\mathbf{W}|\gamma)$ and the last term corresponds to $\mathcal{P}(\gamma)$. Expectation propagation (EP) approximates each term t_i in (8) by a corresponding simpler term $\tilde{t}_i$. These approximate terms are restricted to have the form of a parametric probability distribution that belongs to the exponential family. They however do not need to integrate to one. Once normalized with respect to $\mathbf{W}$ and γ, (8) becomes the posterior distribution for the model parameters. Similarly, when normalized with respect to $\mathbf{W}$ and γ, the product of the approximate terms $\tilde{t}_i$ becomes the posterior approximation:

$$\mathcal{Q}(\mathbf{W}, \gamma) = \frac{1}{Z} \prod_i \tilde{t}_i(\mathbf{W}, \gamma) \approx \mathcal{P}(\mathbf{W}, \gamma|\mathbf{Y})\,, \tag{9}$$

where the normalization constant Z approximates $\mathcal{P}(\mathbf{Y})$, the model evidence. Because of the closure property of the exponential family, $\mathcal{Q}$ has the same parametric form as the approximate terms $\tilde{t}_i$. In practice, the form of $\mathcal{Q}$ is selected first and the approximate terms $\tilde{t}_i$ are constrained by this form. EP iteratively updates each approximate term $\tilde{t}_i$, until the convergence of the posterior approximation $\mathcal{Q}$, in such a way that $\tilde{t}_i \prod_{j \neq i} \tilde{t}_j$ is as close as possible to $t_i \prod_{j \neq i} \tilde{t}_j$. Closeness is defined here in terms of the Kullback-Leibler (KL) divergence. This procedure guarantees that each approximate term $\tilde{t}_i$ is similar to the corresponding exact term t_i in regions of high posterior probability, as defined by the product of the other approximate terms [10]. The different steps of the EP algorithm are:

1. Initialize all $\tilde{t}_i$ and $\mathcal{Q}$ to be uniform.
2. Repeat until $\mathcal{Q}$ converges:
   (a) Select a $\tilde{t}_i$ to refine and compute $\mathcal{Q}^{\setminus i} \propto \frac{\mathcal{Q}}{\tilde{t}_i}$.
   (b) Update $\tilde{t}_i$ so that $\mathrm{KL}(t_i\mathcal{Q}^{\setminus i}||\tilde{t}_i\mathcal{Q}^{\setminus i})$ is minimized.
   (c) Compute an updated posterior approximation $\mathcal{Q}^{\mathrm{new}} \propto \tilde{t}_i\mathcal{Q}^{\setminus i}$.

Whenever needed for model selection, an approximate model evidence can also be computed by integrating the product of all $\tilde{t}_i$'s.

When $\mathcal{Q}$ is assumed to be Gaussian, the first step of the EP algorithm is implemented by setting the mean and the variance of all the approximate terms and the posterior approximation $\mathcal{Q}$ equal to zero and $+\infty$ respectively. In step 2-(a), $\mathcal{Q}^{\setminus i}$ has the same form as $\mathcal{Q}$ because of the closure property of the exponential family. The optimization problem of step 2-(b) is convex and it can be efficiently solved by matching the sufficient statistics between $t_i\mathcal{Q}^{\setminus i}$ and $\tilde{t}_i\mathcal{Q}^{\setminus i}$ [11]. The EP algorithm is not guaranteed to converge although extensive empirical evidence shows that most of the times it converges to a fixed point solution [10]. Oscillations without convergence can be prevented by using *damped* updates [12]. Finally, the EP algorithm has shown an excellent performance in terms of its computational cost versus its approximation accuracy when compared to other approximate inference methods such as the Laplace approximation, variational inference or Markov chain Monte Carlo sampling [10].

4 The Posterior Approximation

We propose to approximate the posterior (5) by the following parametric distribution belonging to the exponential family:

$$\mathcal{Q}(\mathbf{W},\gamma) = \prod_{k=1}^{K} \mathcal{N}(\mathbf{w}_k|\mathbf{m}_k,\mathbf{V}_k) \prod_{j=1}^{d+1} p_j^{\gamma_j}(1-p_j)^{1-\gamma_j}\,, \tag{10}$$

where $\mathcal{N}(\cdot|\mathbf{m}_k,\mathbf{V}_k)$ denotes a multivariate Gaussian distribution with mean vector $\mathbf{m}_k$ and covariance matrix $\mathbf{V}_k$. In (10), $\mathbf{m}_k$, $\mathbf{V}_k$ and $\mathbf{p} = (p_1,\ldots,p_{d+1})^T$ are free parameters that determine the posterior approximation. We denote t_{ki} the exact term of the true posterior associated to the likelihood of the i-th training instance of the k-th learning task:

$$t_{ki}(\mathbf{w}_k) = \Phi(\mathbf{w}_k^T\mathbf{x}_{ki})\,, \tag{11}$$

and $\tilde{t}_{ki}$ its associated approximate term. The definition of $\mathcal{Q}$ given in (10) constrains the form of $\tilde{t}_{ki}$ to be:

$$\tilde{t}_{ki}(\mathbf{w}_k) = \tilde{s}_{ki}\exp\left\{-\frac{1}{2\tilde{v}_{ki}}\left(\mathbf{w}_k^T\mathbf{x}_{ki} - \tilde{m}_{ki}\right)^2\right\}\,, \tag{12}$$

where $\tilde{s}_{ki}$, $\tilde{v}_{ki}$ and $\tilde{m}_{ki}$ are free parameters. Both the exact term t_{ki} and its approximation $\tilde{t}_{ki}$ do not depend on $\boldsymbol{\gamma}$. This is known as the locality property of EP [13]. Additionally, $\tilde{t}_{ki}$ can be written as a univariate Gaussian distribution since each likelihood term only constrains the corresponding vector $\mathbf{w}_k$ through the direction of $\mathbf{x}_{ki}$ [10]. Similarly, we denote t_{kj} the exact term corresponding to the prior for the j-th component of $\mathbf{w}_k$:

$$t_{kj}(\mathbf{w}_k, \boldsymbol{\gamma}) = \mathcal{N}(w_{kj}|0, \sigma_{1j}^2)^{\gamma_j} \mathcal{N}(w_{kj}|0, \sigma_{0j}^2)^{1-\gamma_j}\,, \tag{13}$$

and $\tilde{t}_{kj}$ its approximation. From (10), it follows that

$$\tilde{t}_{kj}(\mathbf{w}_k, \boldsymbol{\gamma}) = \tilde{s}_{kj} \tilde{p}_{kj}^{\gamma_j} (1 - \tilde{p}_{kj})^{1-\gamma_j} \exp\left\{ -\frac{1}{2\tilde{\nu}_{kj}} (w_{kj} - \tilde{\mu}_{kj})^2 \right\}, \tag{14}$$

where $\tilde{s}_{kj}$, $\tilde{p}_{kj}$, $\tilde{\nu}_{ki}$ and $\tilde{\mu}_{ki}$ are free parameters. Again, $\tilde{t}_{kj}$ only depends on one component of $\mathbf{w}_k$ and $\boldsymbol{\gamma}$ because of the locality property of EP. Finally, the exact term corresponding to the prior for $\boldsymbol{\gamma}$ is given by (4). This term can be estimated exactly, *i.e.* $\tilde{t}(\boldsymbol{\gamma}) = t(\boldsymbol{\gamma}) = \mathcal{P}(\boldsymbol{\gamma})$. Since $\tilde{t}(\boldsymbol{\gamma})$ has no free parameters, it does not require updating.

From the definition of $\mathcal{Q}$ as the normalized product of all the approximate terms, the parameters of the posterior approximation are:

$$\mathbf{V}_k = \left(\mathbf{X}_k \boldsymbol{\Lambda}_k \mathbf{X}_k^T + \boldsymbol{\Delta}_k \right)^{-1}, \qquad \mathbf{m}_k = \mathbf{V}_k \left(\mathbf{X}_k \boldsymbol{\eta}_k + \boldsymbol{\Delta}_k \tilde{\boldsymbol{\mu}}_k \right),$$
$$p_j = \frac{\prod_{k=1}^K \tilde{p}_{kj} p_{0j}}{\prod_{k=1}^K \tilde{p}_{kj} p_{0j} + \prod_{k=1}^K (1 - \tilde{p}_{kj})(1 - p_{0j})}, \qquad \text{for } j = 1, \ldots, d+1\,, \tag{15}$$

where we have defined $\boldsymbol{\Lambda}_k = \text{diag}(\tilde{v}_{k1}^{-1}, \ldots, \tilde{v}_{kn_k}^{-1})$, $\boldsymbol{\Delta}_k = \text{diag}(\tilde{\nu}_{k1}^{-1}, \ldots, \tilde{\nu}_{k(d+1)}^{-1})$, $\boldsymbol{\eta}_k = (\tilde{m}_{k1}/\tilde{v}_{k1}, \ldots, \tilde{m}_{kn_k}/\tilde{v}_{kn_k})^T$, $\tilde{\boldsymbol{\mu}}_k = (\tilde{\mu}_{k1}, \ldots, \tilde{\mu}_{k(d+1)})^T$ and $\text{diag}(\cdot)$ denotes a diagonal matrix.

4.1 Efficient EP Update Scheme

The EP algorithm iteratively updates the posterior approximation $\mathcal{Q}$, which includes the computation of K covariance matrices of size $(d+1)\times(d+1)$. Thus, a straightforward implementation would have a time complexity in $\mathcal{O}(Kd^2)$ for each EP iteration. Since, in our context $d \gg n_k$ $\forall k$, we introduce an alternative parametrization of the posterior approximation which provides a more efficient updating scheme. This parametrization is similar to the one that arises in kernel classifiers when the EP algorithm is written in terms of inner products [10]. Specifically, instead of explicitly storing the parameters $\mathbf{m}_k$ and $\mathbf{V}_k$ of $\mathcal{Q}$, we define and store

$$\mathbf{A}_k = \mathbf{X}_k^T \mathbf{V}_k \mathbf{X}_k\,, \qquad \mathbf{B}_k = \mathbf{X}_k^T \mathbf{V}_k \boldsymbol{\Delta}_k\,, \qquad \mathbf{h}_k = \mathbf{X}_k^T \mathbf{m}_k\,, \tag{16}$$

where $\mathbf{A}_k \in \mathbb{R}^{n_k, n_k}$, $\mathbf{B}_k \in \mathbb{R}^{n_k, d}$, and $\mathbf{h} \in \mathbb{R}^{n_k}$. These new parameters are updated after each EP iteration.

The first step of the EP algorithm would typically set the approximate terms to be uniform. Some iterations of the EP algorithm may however be saved if the approximate terms are initialized in such a way that the Gaussian part of $\mathcal{Q}$ has the same mean and variance as the exact prior for $\mathbf{W}$:

$$\begin{aligned} &\tilde{v}_{ki} = +\infty\,, && \tilde{m}_{ki} = 0\,, && h_{ki} = 0\,, && p_j = p_{0j}\,, \\ &\tilde{\nu}_{kj} = \sigma^2_{1j} p_{0j}\,, && \tilde{\mu}_{kj} = 0\,, && \mathbf{B}_k = \mathbf{X}_k^T\,, && \mathbf{A}_k = \mathbf{X}_k^T \mathbf{D} \mathbf{X}_k\,, \quad \forall k,i,j\,, \end{aligned} \tag{17}$$

where $\mathbf{D} = \mathrm{diag}\left(\sigma^2_{11} p_{01}, \ldots, \sigma^2_{1(d+1)} p_{0(d+1)}\right)$ and $\mathcal{Q}$ has been initialized to the normalized product of all approximate terms. The constants $\tilde{s}_{ki}$ and $\tilde{s}_{kj}$ of the approximate terms can be set to any arbitrary positive value.

Let $\mathcal{Q}^{\backslash ki}$ denote the posterior approximation that results from removing from $\mathcal{Q}$ the approximate likelihood term $\tilde{t}_{ki}$. Furthermore, let $\mathbf{m}_k^{\backslash ki}$ and $\mathbf{V}_k^{\backslash ki}$ denote the parameters of the resulting Gaussian approximation to the posterior of $\mathbf{w}_k$. If we define

$$h_{ki}^{\backslash ki} = \mathbf{x}_{ki}^T \mathbf{m}_k^{\backslash ki}\,, \qquad \lambda_{ki}^{\backslash ki} = \mathbf{x}_{ki}^T \mathbf{V}_k^{\backslash ki} \mathbf{x}_{ki}\,, \tag{18}$$

the step 2-(a) of the EP algorithm for these approximate terms becomes:

$$h_{ki}^{\backslash ki} = h_{ki} + \frac{(\mathbf{A}_k)_{ii}}{\tilde{v}_{ki} - (\mathbf{A}_k)_{ii}} (h_{ki} - \tilde{m}_{ki})\,, \quad \lambda_{ki}^{\backslash ki} = (\mathbf{A}_k)_{ii} + \frac{(\mathbf{A}_k)_{ii}^2}{\tilde{v}_{ki} - (\mathbf{A}_k)_{ii}}\,. \tag{19}$$

Part of an updated posterior distribution $\mathcal{Q}^{\mathrm{new}}$ may be computed:

$$h_{ki}^{\mathrm{new}} = \mathbf{x}_{ki}^T \mathbf{m}_k^{\mathrm{new}} = h_{ki}^{\backslash ki} + \alpha_{ki} \lambda_{ki}^{\backslash ki}\,, \tag{20}$$

where

$$\alpha_{ki} = \frac{\mathcal{N}(u_{ki}|0,1)}{Z_{ki}} \frac{1}{\sqrt{\lambda_{ki}^{\backslash ki} + 1}}\,, \qquad u_{ki} = \frac{h_{ki}^{\backslash ki}}{\sqrt{\lambda_{ki}^{\backslash ki} + 1}}\,, \qquad Z_{ki} = \Phi(u_{ki})\,. \tag{21}$$

The updated approximate term $\tilde{t}_{ki}$ follows from

$$\begin{aligned} \tilde{v}_{ki} &= \lambda_{ki}^{\backslash ki} \left(\frac{1}{\alpha_{ki}(h_{ki}^{\mathrm{new}} + \alpha_{ki})} - 1 \right)\,, \qquad \tilde{m}_{ki} = h_{ki}^{\mathrm{new}} + \alpha_{ki} \tilde{v}_{ki}\,, \\ \tilde{s}_{ki} &= Z_{ki} \sqrt{1 + \lambda_{ki}^{\backslash ki} / \tilde{v}_{ki}} \exp\left\{ \frac{\alpha_{ki}}{2} \frac{\lambda_{ki}^{\backslash ki} + 1}{h_{ki}^{\mathrm{new}} + \alpha_{ki}} \right\}\,. \end{aligned} \tag{22}$$

Finally, $\mathcal{Q}^{\mathrm{new}}$ results from updating the matrices $\mathbf{A}_k$ and $\mathbf{B}_k$ and the vector $\mathbf{h}_k$ using the Woodbury formula:

$$\begin{aligned} \mathbf{A}_k^{\mathrm{new}} &= \mathbf{A}_k - \frac{(\mathbf{A}_k)_{\cdot,i} (\mathbf{A}_k)_{i,\cdot}}{c_k^{-1} + (\mathbf{A}_k)_{ii}}\,, \qquad \mathbf{B}_k^{\mathrm{new}} = \mathbf{B}_k - \frac{(\mathbf{A}_k)_{\cdot,i} (\mathbf{B}_k)_{i,\cdot}}{c_k^{-1} + (\mathbf{A}_k)_{ii}}\,, \\ \mathbf{h}_k^{\mathrm{new}} &= \mathbf{A}_k^{\mathrm{new}} \boldsymbol{\eta}_k + \mathbf{B}_k^{\mathrm{new}} \tilde{\boldsymbol{\mu}}_k\,, \end{aligned} \tag{23}$$

where $c_k = 1/\tilde{v}_{ki}^{\text{new}} - 1/\tilde{v}_{ki}^{\text{old}}$. The computational complexity of these updates is in $\mathcal{O}(n_k d)$.

The approximate terms $\tilde{t}_{kj}$ corresponding to the prior for $\mathbf{W}$ are updated in parallel for each task k, as in [14]. For all $j = 1, \ldots, d+1$, $\tilde{t}_{kj}$ is removed from $\mathcal{Q}$ and the posterior approximation $\mathcal{Q}^{\setminus kj}$ is computed . Given each $\mathcal{Q}^{\setminus kj}$, the corresponding approximate terms are updated simultaneously. The updates for such a parallel scheme are simpler as they only require the marginals of the approximate posterior for $\mathbf{w}_k$. Once the parallel updates have been performed, the posterior approximation for $\mathbf{w}_k$ needs to be recomputed as the normalized product of the approximate terms.

Given $\mathcal{Q}$, the cost of computing the marginals of the posterior approximation for $\mathbf{w}_k$ is in $\mathcal{O}(n_k^2 d)$ when $d \gg n_k$. Let $\mathbf{v}_k = (v_{k1}, \ldots, v_{k(d+1)})^T$ be a vector summarizing the variance of each marginal and $\mathbf{m}_k$ a vector summarizing the corresponding means:

$$\mathbf{v}_k = \tilde{\boldsymbol{\nu}}_k - \tilde{\boldsymbol{\nu}}_k \circ ((\mathbf{X}_k \mathbf{L}_k \circ \mathbf{X}_k)\,\mathbf{1}) \circ \tilde{\boldsymbol{\nu}}_k\,, \quad \mathbf{m}_k = \boldsymbol{\zeta} - \tilde{\boldsymbol{\nu}}_k \circ \left(\mathbf{X}_k \mathbf{L}_k \mathbf{X}_k^T \boldsymbol{\zeta}\right), \tag{24}$$

where $\circ$ indicates the Hadamard element-wise product, $\tilde{\boldsymbol{\nu}}_k = (\tilde{\nu}_{k1}, \ldots, \tilde{\nu}_{k(d+1)})^T$, $\mathbf{1}$ is a vector of ones $\mathbf{L}_k = \boldsymbol{\Lambda}_k - \boldsymbol{\Lambda}_k \mathbf{A}_k \boldsymbol{\Lambda}_k$ and $\boldsymbol{\zeta} = \tilde{\boldsymbol{\nu}}_k \circ (\mathbf{X}_k \boldsymbol{\eta}_k + \boldsymbol{\Delta}_k \tilde{\boldsymbol{\mu}}_k)$. $\mathcal{Q}^{\setminus kj}$ is computed simultaneously for each approximate term $\tilde{t}_{kj}$ from (24). Let $v_{kj}^{\setminus kj}$ and $m_{kj}^{\setminus kj}$ denote the variance and the mean of the posterior distribution of w_{kj} under each $\mathcal{Q}^{\setminus kj}$. Let $p_j^{\setminus kj}$ denote the parameter that determines the posterior probability of $\gamma_j = 1$ in $\mathcal{Q}^{\setminus kj}$. For each $\mathcal{Q}^{\setminus kj}$, these parameters are defined as:

$$\begin{aligned} p_j^{\setminus kj} &= \frac{p_j/\tilde{p}_{kj}}{p_j/\tilde{p}_{kj} + (1-p_j)/(1-\tilde{p}_{kj})}\,, \quad \forall j\,, \quad v_{kj}^{\setminus kj} = \left(v_{kj}^{-1} - \tilde{\nu}_{kj}^{-1}\right)^{-1}\,, \quad \forall j\,, \\ m_{kj}^{\setminus kj} &= v_{kj}^{\setminus kj}\left(v_{kj}^{-1} m_{kj} - \tilde{\nu}_{kj}^{-1}\tilde{\mu}_{kj}\right)^{-1}\,, \quad \forall j\,. \end{aligned} \tag{25}$$

The corresponding approximate terms $\tilde{t}_{kj}$ are

$$\begin{aligned} \tilde{\nu}_{kj} &= c_3^{-1} - v_{ki}^{\setminus ki}\,, \quad \forall j\,, \quad \tilde{s}_{kj} = (\mathcal{G}_1 + \mathcal{G}_0)\sqrt{1 + v_{ki}^{\setminus ki}/\tilde{\nu}_{kj}}\exp\left\{\frac{1}{2}\frac{c_1^2}{c_3}\right\}\,, \quad \forall j\,, \\ \tilde{p}_{kj} &= \frac{\mathcal{G}_1}{\mathcal{G}_1 + \mathcal{G}_0}\,, \quad \forall j\,, \qquad \tilde{\mu}_{kj} = m_{kj}^{\setminus kj} - c_1\left(\tilde{\nu}_{kj} + v_{ki}^{\setminus ki}\right)\,, \quad \forall j\,, \end{aligned} \tag{26}$$

where

$$\begin{aligned} c_1 &= \rho_{kj} a_1 + (1-\rho_{kj}) a_0\,, & c_2 &= \rho_{kj}\left[a_1^2 - \frac{a_1}{m_{kj}^{\setminus kj}}\right] + (1-\rho_{kj})\left[a_0^2 - \frac{a_0}{m_{kj}^{\setminus kj}}\right]\,, \\ c_3 &= c_1^2 - c_2\,, & Z_{kj} &= p_j^{\setminus kj}\mathcal{G}_1 + (1-p_j^{\setminus kj})\mathcal{G}_0\,, \\ \mathcal{G}_0 &= \mathcal{N}(0|m_{kj}^{\setminus kj}, v_{kj}^{\setminus kj} + \sigma_{0j}^2)\,, & \mathcal{G}_1 &= \mathcal{N}(0|m_{kj}^{\setminus kj}, v_{kj}^{\setminus kj} + \sigma_{1j}^2)\,, \\ a_0 &= m_{kj}^{\setminus kj}/\left(v_{kj}^{\setminus kj} + \sigma_{0j}^2\right)\,, & a_1 &= m_{kj}^{\setminus kj}/\left(v_{kj}^{\setminus kj} + \sigma_{1j}^2\right)\,, \\ \rho_{kj} &= p_j^{\setminus kj}\mathcal{G}_1/Z_{kj}\,. \end{aligned} \tag{27}$$

The parallel updates for the corresponding posterior distribution $\mathcal{Q}^{\text{new}}$ are

$$m_{kj}^{\text{new}} = m_{kj}^{\setminus kj} - c_1 v_{kj}^{\setminus kj}, \quad \forall j, \qquad v_{kj}^{\text{new}} = v_{kj}^{\setminus kj}\left(1 - c_3 v_{kj}^{\setminus kj}\right), \quad \forall j,$$

$$p_{kj}^{\text{new}} = p_j^{\setminus kj} + \frac{\mathcal{G}_1 - \mathcal{G}_0}{Z_{kj}} p_j^{\setminus kj}(1 - p_j^{\setminus kj}), \quad \forall j. \tag{28}$$

Finally, $\mathcal{Q}^{\text{new}}$ results from updating the matrices $\mathbf{A}_k$ and $\mathbf{B}_k$:

$$\mathbf{A}_k^{\text{new}} = \mathbf{M}_k - \mathbf{M}_k \left(\mathbf{M}_k + \boldsymbol{\Lambda}_k^{-1}\right)^{-1} \mathbf{M}_k, \tag{29}$$

$$\mathbf{B}_k^{\text{new}} = \mathbf{X}_k^T - \mathbf{M}_k \left(\mathbf{M}_k + \boldsymbol{\Lambda}_k^{-1}\right)^{-1} \mathbf{X}_k^T, \tag{30}$$

where $\mathbf{M}_k = \mathbf{X}_k^T \boldsymbol{\Delta}_k^{-1} \mathbf{X}_k \in \mathbb{R}^{n_k, n_k}$. The computational complexity of these updates is in $\mathcal{O}(n_k^2 d)$ since $d \gg n_k$. The vector $\mathbf{h}_k$ is updated as in (23).

4.2 Predictive Distribution and Feature Selection

The predictive distribution (6) can be approximated using $\mathcal{Q}$ as an estimate of the exact posterior:

$$\mathcal{P}(y_k^{\text{new}} | \mathbf{x}_k^{\text{new}}, \mathbf{Y}) \approx \Phi\left(\frac{\mathbf{m}_k^T \mathbf{x}_k^{\text{new}}}{\sqrt{(\mathbf{x}_k^{\text{new}})^T \mathbf{V}_k \mathbf{x}_k^{\text{new}} + 1}}\right), \tag{31}$$

where $\mathbf{m}_k$ can be obtained as in (24) and

$$(\mathbf{x}_k^{\text{new}})^T \mathbf{V}_k \mathbf{x}_k^{\text{new}} = (\mathbf{x}_k^{\text{new}})^T \boldsymbol{\Delta}_k^{-1} \mathbf{x}_k^{\text{new}} - (\mathbf{x}_k^{\text{new}})^T \boldsymbol{\Delta}_k^{-1} \mathbf{X}_k \mathbf{L}_k \mathbf{X}_k^T \boldsymbol{\Delta}_k^{-1} \mathbf{x}_k^{\text{new}}.$$

Since $\mathbf{m}_k$ is already computed once the model is estimated, the cost of evaluating (31) is in $\mathcal{O}(n_k d)$.

The EP approximation of (7) is used to identify the most relevant features:

$$\mathcal{P}(\boldsymbol{\gamma} | \mathbf{Y}) \approx \prod_{j=1}^{d+1} p_j^{\gamma_j} (1 - p_j)^{1 - \gamma_j}, \tag{32}$$

where p_j estimates the posterior probability of using attribute j for prediction in *all* the tasks.

Assuming a constant number of EP iterations until convergence, a reasonable assumption in practice, the time complexity of the EP algorithm is in $\mathcal{O}(\sum_{k=1}^K n_k^2 d)$. This complexity, linear in d, is good since $d \gg n_k, \forall k$. Finally, our actual EP implementation also relies on *damped* updates [12], as they seem to improve the overall convergence.

5 Experiments

We detail here several multi-task experiments to assess the performance of the Bayesian multi-task feature selection (BMFS) introduced in Sect. 2. Comparative results are reported with single-task learning (one classifier estimated independently on each task) or data pooling across all tasks (one global classifier). In both cases, those individual classifiers are estimated through the same EP procedure as in BMFS but with the original single-task spike and slab prior [5]. We also present the performances of linear models regularized with a mixed norm [8] (the ℓ_1/ℓ_2 method) and the regularized multi-task learning (RMLT) method [9][2].

5.1 Arabic Digits

A first batch of experiments is carried out using the dataset of Arabic handwritten digits MADBase [15]. This dataset contains 70,000 images of size 28×28 pixels of each Arabic digit written by 700 different writers. There are 10 images of each digit (0 to 9) written by the same writer. Fig. 1 displays some examples of the images contained in this dataset. We consider binary classification tasks to discriminate each digit from the digit 0 for a particular writer, which is arguably a difficult problem [15]. For each digit i versus digit 0 with $i \neq 0$, we extract the 800 available images corresponding to 40 different writers (writers 601 to 640). We thus consider 40 tasks (one per writer) with 20 samples per task. The data for each task are randomly split 100 times into training and test sets, with 17 and 3 instances respectively, and the average prediction accuracy is reported.

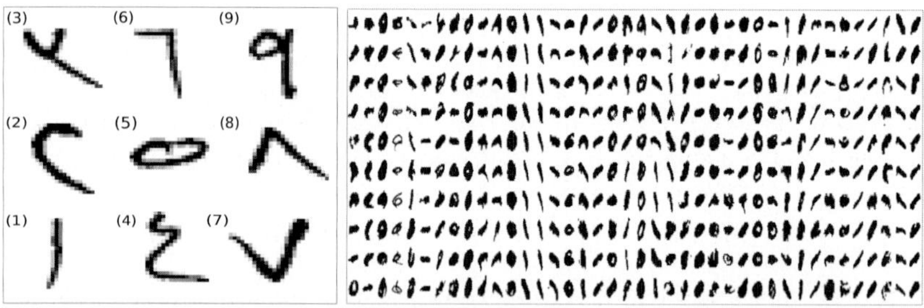

Fig. 1. Arabic digits from 1 to 9 (left). The Arabic digit 0 written ten times by forty different persons, with a strongly writer-dependent style (right).

In BMFS, single-task learning and pooling we set $p_{0j} = 5\%$ for $j = 1, \ldots, d$ to model our prior belief that an accurate classification may only depend on a few pixels (approximately 40). In the ℓ_1/ℓ_2 method, λ is chosen using the algorithm described in [8] by minimizing the cross-validation error[3]. Similarly,

[2] As for BMFS, we consider a bias term in those multi-task methods by extending each input sample with a constant component equal to one. Additionally, in the ℓ_1/ℓ_2 method it is straightforward to not regularize the corresponding coefficient.

[3] The parameters of this algorithm are $\epsilon = 0.02$, $\lambda_{\max} = \lambda_0/500$ and $\xi = \min(10^{-3}, 0.01\lambda)$, as suggested in [8].

Table 1. Prediction error in % (and standard deviation) of each method on each binary problem

Problem	BMFS	ℓ_1/ℓ_2	RMTL	Single-Task	Pooling
0 vs 1	**0.5±0.4**	1.3±0.7	4.0±2.1	3.2±1.2	2.1±1.1
0 vs 2	**0.6±0.4**	1.0±0.7	4.7±1.5	6.6±1.7	3.5±2.1
0 vs 3	**1.6±0.7**	2.3±0.8	4.5±1.8	5.0±1.4	4.5±1.8
0 vs 4	**1.3±0.7**	**1.3±0.7**	4.9±1.7	5.0±1.4	2.1±0.9
0 vs 5	**1.0±0.6**	1.6±0.8	7.3±2.2	9.2±1.9	8.0±3.4
0 vs 6	**0.4±0.4**	1.1±0.6	3.5±1.8	3.2±1.2	3.0±6.1
0 vs 7	**0.5±0.4**	1.2±0.7	3.5±1.4	6.1±1.6	3.5±1.6
0 vs 8	**1.4±0.7**	2.2±1.0	4.2±2.2	6.0±1.7	4.5±1.9
0 vs 9	**1.2±0.7**	1.4±0.7	5.0±1.5	3.8±1.3	3.5±1.4
Average	**1.0±0.2**	1.5±0.2	4.6±0.5	5.3±0.4	3.8±0.9

Fig. 2. Feature importance in gray scale as computed by EP for the Bayesian multi-task approach (left), single-task learning (middle) and pooling (right). The multi-task approach is the most confident method about the feature importance.

in the RMTL method λ_1 and λ_2 are chosen with a grid search over ten values from 0.01 to 100 by cross-validation. The data are also normalized to have zero mean and unit standard deviation on the training set.

Table 1 displays the prediction error of each method for each binary problem of the form digit i vs digit 0 with $i \neq 0$ averaged over the 40 different tasks. The error of the best method for each binary problem is high-lighted in bold face. BMFS obtains the best prediction error in all the problems investigated, outperforming the ℓ_1/ℓ_2 and the RMTL methods, except for the problem 0 vs 4, where the ℓ_1/ℓ_2 method obtains similar results. The Bayesian approach also outperforms single-task learning or data pooling in all cases. A Friedman rank test (p-value $= 6.1 \cdot 10^{-81}$) and a Nemenyi post-hoc test (p-value $= 7.6 \cdot 10^{-5}$) confirm that there is statistical evidence supporting a performance difference in favor of BMFS when the average prediction error across the nine problems is considered [16].

Fig. 2 shows the estimate of the relative importance of each feature (pixel) as computed by EP in (32) for BMFS, single-task learning (we display the results for

Table 2. Characteristics of the three prostate cancer microarray datasets

Name	Normal	Tumor	Features	Platform	Ref.
Singh	50	52	12,626	HGU95Av2	[18]
Stuart	50	38	12,626	HGU95Av2	[19]
Welsh	9	25	12,625	HGU95A	[20]

the first writer) and pooling. The figure displays the values of p_j for $j = 1, \ldots, d$ using a gray scale from white ($p_j = 0$) to black ($p_j = 1$). These results correspond to a fixed train / test partition of the binary classification problem 0 vs 8. The gray background color indicates a value for p_j equal to 5%, *i.e.* the prior value for $\gamma_j = 1$. The figure shows that BMFS allows us to be the most confident about the relative importance of each feature. With single-task learning all pixels are very close to the prior value, which is likely related to the reduced number of training instances for each task. Pooling improves over single-task learning but is still much less informative than BMFS. The smaller confidence obtained with pooling is likely related to the estimation of a single hyperplane to describe all the learning tasks. Similar observations can be made from the different binary problems considered and the different train/test partitions.

5.2 Microarray Data

A second batch of experiments is carried out using three microarray prostate cancer datasets described in Table 2, where each dataset is identified by the first author of the corresponding reference. The Welsh dataset is made of gene expression values. The Singh and Stuart original data are made of laser intensity images from each microarray. The RMA preprocesing method [17] is used to produce gene expression values from these images. The microarray technology is common for Singh and Stuart but is slightly older for Welsh. We restrict our attention to the $12,600$ features they have in common. For each dataset there is one learning task to discriminate between normal and tumor samples. We randomly split 100 times the data for each task into 2/3 (training) and 1/3 (test). In BMFS, we set $p_{0j} = 50/d$, for $j = 1, \ldots, d$, to model our prior belief that only a few genes (50) may be relevant for classification . The data and the hyper-parameters of the other multi-task methods are respectively normalized and set as described in Sect. 5.1.

Table 3 reports balanced classification rates (BCR) for each method, averaged over data partitions. This evaluation metric is the arithmetic mean of the accuracy within each class (tumor, normal). It is preferred over standard accuracy to assess prediction performance from highly unbalanced classes [21]. It is also the arithmetic mean between specificity and sensitivity, commonly used in the medical domain. The table shows that BMFS obtains the best performance for Welsh, the ℓ_1/ℓ_2 model is optimal for Singh and the RMTL method is optimal for Stuart. BMFS is overall the best method according to average results

Table 3. BCR in % (and standard deviation) of each method on each prostate cancer classification task

Task	BMFS	ℓ_1/ℓ_2	RMTL	Single-Task	Pooling
Singh	90.3±4.5	**90.4±4.6**	89.6±5.1	89.4±4.5	89.4±5.0
Stuart	78.6±6.3	76.9±5.5	**80.0±5.7**	77.7±6.3	79.8±6.0
Welsh	**97.3±6.0**	94.0±10.9	93.4±6.8	95.6±8.4	93.4±6.8
Average	**88.7±2.9**	87.0±3.9	87.6±3.3	87.5±3.2	87.5±3.3

over the three tasks. A Friedman rank test (p-value $= 8.4 \cdot 10^{-5}$) and a Nemenyi post-hoc test (p-value $= 3.5 \cdot 10^{-3}$) confirm that these differences are statistically significant.

Finally, we compare the different methods in terms of their stability for identifying relevant features when the training data are slightly modified. The Kuncheva stability index [22] measures to which extent T sets of m selected features share common elements. Let $\mathcal{S}_i^m$ denote the set of the top m features identified by a method from the i-th train/test partition. The Kuncheva index over the several data partitions $\mathcal{A}_m = \{\mathcal{S}_i^m : i = 1, \ldots, T\}$ is defined as

$$\mathcal{I}(\mathcal{A}_m) = \frac{2}{T(T-1)} \sum_{i=1}^{T-1} \sum_{j=i+1}^{T} \frac{|\mathcal{S}_i^m \cap \mathcal{S}_j^m| - \frac{m^2}{d}}{m - \frac{m^2}{d}}, \quad (33)$$

where $T = 100$ is the number of training sets, d is the total number of features and m^2/d is the expected value of $|\mathcal{S}_i^m \cap \mathcal{S}_j^m|$ by chance. The index satisfies $-1 < \mathcal{I}(\mathcal{A}_m) \leq 1$ and the closer to one, the larger the number of common features in the different sets. A value of the index near zero indicates commonly selected features at a chance level.

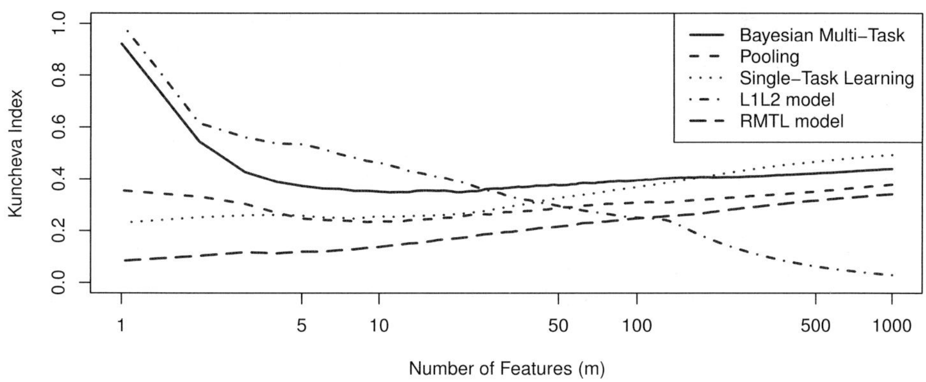

Fig. 3. Stability (Kuncheva index) of the feature ranking implemented by each method

Fig. 3 displays the stability of each method as a function of m. For BMFS, pooling and single-task learning, features are ranked according to the vector $\mathbf{p}$. For the ℓ_1/ℓ_2 method, features are ranked according to the order in which they are included in the active set of the path-following algorithm of [8]. For the RMTL approach features are ranked according to the sum of the corresponding squared coefficients of each hyperplane, *i.e.* $\sum_{k=1}^{K} w_{kj}^2$ for feature j. This value is used as an estimate of the relative feature importance. For the single-task method, we display the value of the stability index averaged over the three learning tasks. The figure shows that the ℓ_1/ℓ_2 method offers the most stable selection for small subsets of features, followed by BMFS, whose selection is more stable than the ones of single-task learning and data pooling.

6 Conclusion

We propose a novel Bayesian model for multi-task feature selection. This model is based on a generalized *spike and slab* sparse prior distribution that enforces the selection of a common subset of features to describe each task. Since exact Bayesian inference is intractable for this model, approximate inference is performed through expectation propagation (EP). We propose an original parametrization of the EP procedure which offers a linear complexity in the number of features. Practical experiments on multi-task digit recognition and microarray data classification illustrate the benefits of the proposed approach, as compared to simple baselines and state-of-the-art multi-task approaches, in terms of predictive performance and stability of the feature selection.

Acknowledgment

T. Helleputte is funded by a FRIA grant (1.E091.07). Computations were run on the INGRID cluster of the Center for Intensive Computation and Mass Storage (Louvain). J. M. Hernández-Lobato is funded by the Spanish MCyT (prj. TIN2007-66862-C02).

References

1. Dudoit, S., Fridlyand, J.: Classification in microarray experiments. In: Statistical Analysis of Gene Expression Microarray Data, pp. 93–158. Chapman and Hall/CRC Press (2003)
2. Seeger, M., Nickisch, H., Pohmann, R., Schölkopf, B.: Optimization of k-space trajectories for compressed sensing by Bayesian experimental design. Magnetic Resonance in Medicine 63(1), 116–126 (2009)
3. Johnstone, I., Titterington, D.: Statistical challenges of high-dimensional data. Philosophical Transactions of the Royal Society A: Mathematical, Physical and Engineering Sciences 367(1906), 4237 (2009)
4. Tibshirani, R.: Regression shrinkage and selection via the lasso. Journal of the Royal Statistical Society. Series B (Methodological) 58(1), 267–288 (1996)

5. George, E.I., McCulloch, R.E.: Approaches for Bayesian variable selection. Statistica Sinica 7(2), 339–373 (1997)
6. Ishwaran, H., Rao, J.: Spike and slab variable selection: frequentist and Bayesian strategies. The Annals of Statistics 33(2), 730–773 (2005)
7. Tipping, M.E.: Sparse Bayesian learning and the relevance vector machine. Journal of Machine Learning Research 1, 211–244 (2001)
8. Obozinski, G., Taskar, B., Jordan, M.: Joint covariate selection and joint subspace selection for multiple classification problems. Statistics and Computing, 1–22 (2009)
9. Evgeniou, T., Pontil, M.: Regularized multi–task learning. In: Proceedings of the Tenth ACM SIGKDD International Conference on Knowledge Discovery and Data Mining, pp. 109–117. ACM, New York (2004)
10. Minka, T.: A Family of Algorithms for approximate Bayesian Inference. PhD thesis, Massachusetts Institute of Technology (2001)
11. Bishop, C.M.: Pattern Recognition and Machine Learning (Information Science and Statistics). Springer, Heidelberg (August 2006)
12. Minka, T., Lafferty, J.: Expectation-propagation for the generative aspect model. In: Darwiche, A., Friedman, N. (eds.) Proceedings of the 18th Conference on Uncertainty in Artificial Intelligence, pp. 352–359. Morgan Kaufmann, San Francisco (2002)
13. Seeger, M.: Notes on Minka's expectation propagation for Gaussian process classification. Technical report, University of Edinburgh (2002)
14. Gerven, M.V., Cseke, B., Oostenveld, R., Heskes, T.: Bayesian source localization with the multivariate Laplace prior. In: Bengio, Y., Schuurmans, D., Lafferty, J., Williams, C.K.I., Culotta, A. (eds.) Advances in Neural Information Processing Systems, vol. 22, pp. 1901–1909 (2009)
15. Abdleazeem, S., El-Sherif, E.: Arabic handwritten digit recognition. International Journal on Document Analysis and Recognition 11(3), 127–141 (2008)
16. Demšar, J.: Statistical comparisons of classifiers over multiple data sets. Journal of Machine Learning Research 7, 1–30 (2006)
17. Irizarry, R., Hobbs, B., Collin, F., Beazer-Barclay, Y., Antonellis, K., Scherf, U., Speed, T.: Exploration, normalization, and summaries of high density oligonucleotide array probe level data. Biostatistics 4(2), 249 (2003)
18. Singh, D., Febbo, P.G., Ross, K., Jackson, D.G., Manola, J., Ladd, C., Tamayo, P., Renshaw, A.A., D'Amico, A.V., Richie, J.P., Lander, E.S., Loda, M., Kantoff, P.W., Golub, T.R., Sellers, W.R.: Gene expression correlates of clinical prostate cancer behavior. Cancer Cell 1, 203–209 (2002)
19. Stuart, R., Wachsman, W., Berry, C., Wang-Rodriguez, J., Wasserman, L., Klacansky, I., Masys, D., Arden, K., Goodison, S., McClelland, M., et al.: In silico dissection of cell-type-associated patterns of gene expression in prostate cancer. Proceedings of the National Academy of Sciences 101(2), 615 (2004)
20. Welsh, J., Sapinoso, L., Su, A., Kern, S., Wang-Rodriguez, J., Moskaluk, C., Frierson Jr., H., Hampton, G.: Analysis of gene expression identifies candidate markers and pharmacological targets in prostate cancer. Cancer Research 61(16), 5974 (2001)
21. Helleputte, T., Dupont, P.: Feature selection by transfer learning with linear regularized models. In: Buntine, W., Grobelnik, M., Mladenić, D., Shawe-Taylor, J. (eds.) ECML PKDD 2009. LNCS, vol. 5781, pp. 533–547. Springer, Heidelberg (2009)
22. Kuncheva, L.I.: A stability index for feature selection. In: Proceedings of the 25th IASTED International Multi-Conference on Artificial Intelligence and Applications, Anaheim, CA, USA, pp. 390–395. ACTA Press (2007)

Graphical Multi-way Models

Ilkka Huopaniemi[1,⋆], Tommi Suvitaival[1], Matej Orešič[2], and Samuel Kaski[1]

[1] Aalto University School of Science and Technology, Department of Information and Computer Science, Helsinki Institute for Information Technology HIIT, P.O. Box 15400, FI-00076 Aalto, Finland
[2] VTT Technical Research Centre of Finland, P.O. Box 1000, FIN-02044 VTT, Espoo, Finland
{ilkka.huopaniemi,tommi.suvitaival,samuel.kaski}@tkk.fi, matej.oresic@vtt.fi
http://www.cis.hut.fi/projects/mi

Abstract. Multivariate multi-way ANOVA-type models are the default tools for analyzing experimental data with multiple independent covariates. However, formulating standard multi-way models is not possible when the data comes from different sources or in cases where some covariates have (partly) unknown structure, such as time with unknown alignment. The "small n, large p", large dimensionality p with small number of samples n, settings bring further problems to the standard multivariate methods. We extend our recent graphical multi-way model to three general setups, with timely applications in biomedicine: (i) multi-view learning with paired samples, (ii) one covariate is time with unknown alignment, and (iii) multi-view learning without paired samples.

Keywords: ANOVA, Bayesian latent variable modeling, data integration, multi-view learning, multi-way learning.

1 Introduction

Multivariate multi-way ANOVA-type methods are the default tool for analyzing data with multiple covariates. A prototypical example in biomedical data analysis is studying the effects of disease and treatment in populations of biological measurements. Formulating the data analysis as a linear model makes it possible to ask if the covariates ("ways", disease and treatment), or more interestingly, their interactions have an effect on the data.

In the two-way case, to explain the covariate-related variation in one data source, say $\mathbf{x}$, the following linear model is usually assumed:

$$\mathbf{x}_j|_{(a,b)} = \boldsymbol{\mu}^x + \boldsymbol{\alpha}_a^x + \boldsymbol{\beta}_b^x + (\boldsymbol{\alpha\beta})_{ab}^x + \epsilon_j. \quad (1)$$

⋆ I.H., T.S and S.K belong to the Adaptive Informatics Research Centre and Helsinki Institute for Information Technology HIIT. The work was funded by Tekes MASI program and by Tekes Multibio project. I.H. is funded by the Graduate School of Computer Science and Engineering. S.K is partially supported by EU FP7 NoE PASCAL2, ICT 216886.

J.L. Balcázar et al. (Eds.): ECML PKDD 2010, Part I, LNAI 6321, pp. 538–553, 2010.

Here $\mathbf{x}_j$ is a continuous-valued data vector, observation number j, and the a and b ($a = 0, \ldots A$ and $b = 0, \ldots B$) are the two independent covariates, such as disease and treatment. The $\boldsymbol{\alpha}_a^x$ and $\boldsymbol{\beta}_b^x$ are parameter vectors describing the covariate-specific effects, called main effects. The $(\boldsymbol{\alpha\beta})_{ab}^x$ denotes the interaction effect; the apparently complicated notation is standard, it simply means a parameter vector. In the biomedical example this interaction is the most interesting parameter, describing if the treatment has disease-specific effects (cures the disease). These effects model the variation from the baseline level (called grand mean) $\boldsymbol{\mu}^x$. The ϵ_j is a noise term. The traditional methods for finding and testing the statistical significance of the effects of the covariates on the data are Analysis of Variance (ANOVA) [4] and its multivariate generalization (MANOVA).

A recurring problem in modern data analyses, especially in biomedical experiments, is that the number of samples n is small and dimensionality p is large. The "small n, large p" has recently gained increasing attention in the machine learning community, whereas only a few methods for multi-way modelling have been reported. The currently popular approaches, multi-task learning and multi-label prediction that attempt to share statistical strength between related tasks help if tasks are assumed related, but are not targeted for studying the effects of multiple independent covariates in the data.

It is evident that with small sample-sizes, harsh dimension reduction is needed and the modelling should be done in a low-dimensional latent factor space, say $\mathbf{x}^{lat}$. In addition to trivial approaches such as a prior PCA dimension reduction, two approaches exist for multivariate multi-way analysis in the case of "small n, large p"-conditions. The first, intended for modelling the effects of multiple covariates, is Sparse factor regression [10,14].

The second, hierarchical generative modelling approach [6] forms factors by assuming the variables are grouped, and the variation of the latent variables is generated by the external covariates $p(\mathbf{x}^{lat}|a, b)$, in the spirit of linear models.

We will now extend multi-way modelling to three novel tasks which cannot be solved by standard ANOVA-models or our earlier [6], and not easily by supervised regression/classification either. The associated machine learning problems are illustrated in Figure 1.

We first consider multi-way analysis when the data comes from different sources ("views") with different domains (has unmatched data spaces). A typical biological example is using two or more measurement techniques or having measurements from several tissues of each individual, the underlying experiment having a multi-way experimental setup. We consider the "view" as an additional "way" (or covariate) in the multi-way analysis. However, since different views have different domains, a standard multi-way model is not applicable. The model we present extends multi-view learning with paired samples into multi-way cases, which has plenty of applications in modern molecular biological experiments in terms of integration of multiple data sources. This first extension has already been described in [5];. we include it for completeness.

We then extend multi-way learning into cases where, for one of the "ways", the covariates have partly unknown structure. We concentrate on time, having an unknown alignment; an intuitive application is learning of unknown alignments with Hidden Markov Models (HMM) having linear chains. We consider "time with unknown alignment" as one covariate in a multi-way model. An example considered in this paper is having time-series measurements with unknown alignment from both healthy and diseased populations. The modelling task is to find, based on data, the effect of time, the effect of the other covariate(s) (disease) and, most interestingly, their interaction. The time alignment is learned at the same time.

As a third extension we consider integrating multiple views in different domains when even the samples are not paired. This almost impossible task becomes weakly possible if the experiments are similar in the sense of having a similar covariate design. An example of "multi-view learning without paired samples" is having a similar healthy-diseased time-series dataset with unknown alignments from two species, man and mouse, with different variable-spaces. We assume and search for some shared covariate-related behavior in the datasets. We propose "view without paired samples" as a covariate for an extended multi-way analysis. This makes it possible to evaluate statistical significance of shared covariate-related behavior, in contrast to having only view-specific effects (interaction effect of "view" and other covariates). We choose the generative approach [6] to extend multi-way modelling to the novel cases, because its hierarchical structuring of the effects acting on latent variables makes the extensions reasonable. The new modelling elements, the generative model of Canonical Correlation Analysis (CCA) [1,7], a standard method for multi-view learning with paired samples and unmatched dimensions, and the HMM-model, for time-dependent covariates turn out to be fully compatible with the generative multi-way modelling approach.

We will call the different types of covariates as follows: Covariates which can in principle be studied with existing ANOVA-type methods are **standard covariates**; examples include disease, treatment, gender. "Time with unknown alignment" is a special case of a **covariate with unknown structure**. "View" is a **view-covariate** in the case of paired (co-occurring) samples, and "view without paired samples" is a **view-covariate** in the case of no pairing between the samples.

The key point why we need to distinguish **view-covariates** from the standard covariates is that it actually does not make sense to define main effects for the view-covariates at all, since the domains of the views are different. However, it is sensible to define interaction effects *b*etween a view-covariate and a standard covariate (including "time with unknown alignment"). This allows us to rigorously decompose standard covariate effects into shared and view-specific effects. Furthermore, this decomposition actually forms the connection between the different views, allowing the multi-way problem to be formulated in the first place.

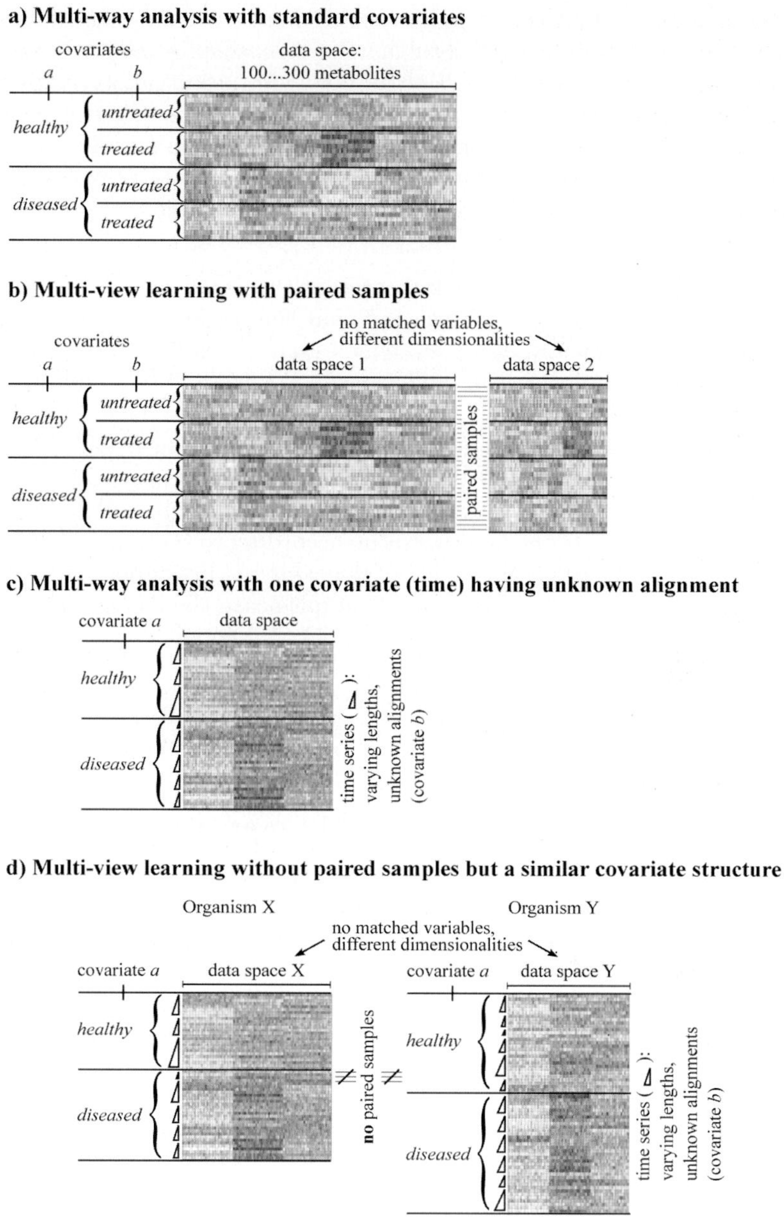

Fig. 1. Illustration of the four data analysis tasks in this paper. (a) Standard ANOVA setup, but with large dimensionality (metabolites) compared to number of samples (rows). (b) Extension to multi-view learning with paired samples. (c) Extension to time with unknown alignment. (d) Extension to multi-view learning without paired samples. The images represent data matrices, where rows are samples and columns are variables. The illustration represents the experimental design of each task, composed of a combination of standard covariates (disease, treatment), time-series information, and integration of multiple views.

The main message and contribution of this paper is that each of the three introduced new machine learning problems is too complicated to be analyzed with any existing method. We will show how to conceptualize each of these problems as an extended multi-way modelling task involving novel covariates. We then introduce a hierarchical generative model for each problem.

2 Model

We now present a unified framework to each of the novel tasks as an extended multi-way model. In each case, it turns out that the model can be formulated as a single hierarchical generative model, which guarantees that uncertainties are propagated properly between the model parts. We use Gibbs sampling for the computations.

The models need three components: (1) a regularized dimension reduction to transform the modelling into low-dimensional latent factor spaces, (2) ANOVA-type modelling of population priors acting on the low-dimensional latent factors, (3) a proper structuring of the analysis setup according to the task. The structure of the tasks and the methodological contributions of this paper are as follows: (i) in the multi-view learning with paired samples case the co-occurring sources are integrated with a generative model of CCA, (ii) in the time-covariate case the means of the emission distributions of an HMM act as one of the latent effects while HMM-alignment is done simultaneously, and (iii) in the case of multi-view learning in different domains without paired samples, the views only share common latent effects.

2.1 Multi-way Learning with Standard Covariates

Multi-way modelling [6] in a low-dimensional factor space requires two parts: regularized dimension reduction and an ANOVA-model formulated as population priors on the latent variables. In our model these parts are integrated into a single generative model, shown in Figure 2 (a). The dimension reduction is done by a factor analyzer that is regularized to find similarly behaving, correlated groups of variables and the ANOVA-effects act on the factors, each representing a cluster of variables.

Regularized Factor analyzer. The basis of the model is a Factor Analyzer (FA). The hierarchical model implementing the factor analyzer is [6]

$$\begin{aligned} \mathbf{x}_j^{lat} &\sim \mathcal{N}(0, \mathbf{I}) \\ \mathbf{x}_j &\sim \mathcal{N}(\boldsymbol{\mu} + \mathbf{V}\mathbf{x}_j^{lat}, \boldsymbol{\Lambda}) \,. \end{aligned} \tag{2}$$

Here $\mathbf{x}_j$ is a p-dimensional data vector, $\mathbf{V}$ is the projection matrix, and $\mathbf{x}_j^{lat}$ is the latent variable, $\boldsymbol{\Lambda}$ is a diagonal residual variance matrix with diagonal elements σ_i^2, $\boldsymbol{\mu}$ is the mean vector (parameters), $\mathbf{I}$ is the identity matrix and $\mathcal{N}$ denotes the normal distribution with mean being the first argument and covariance matrix being the second.

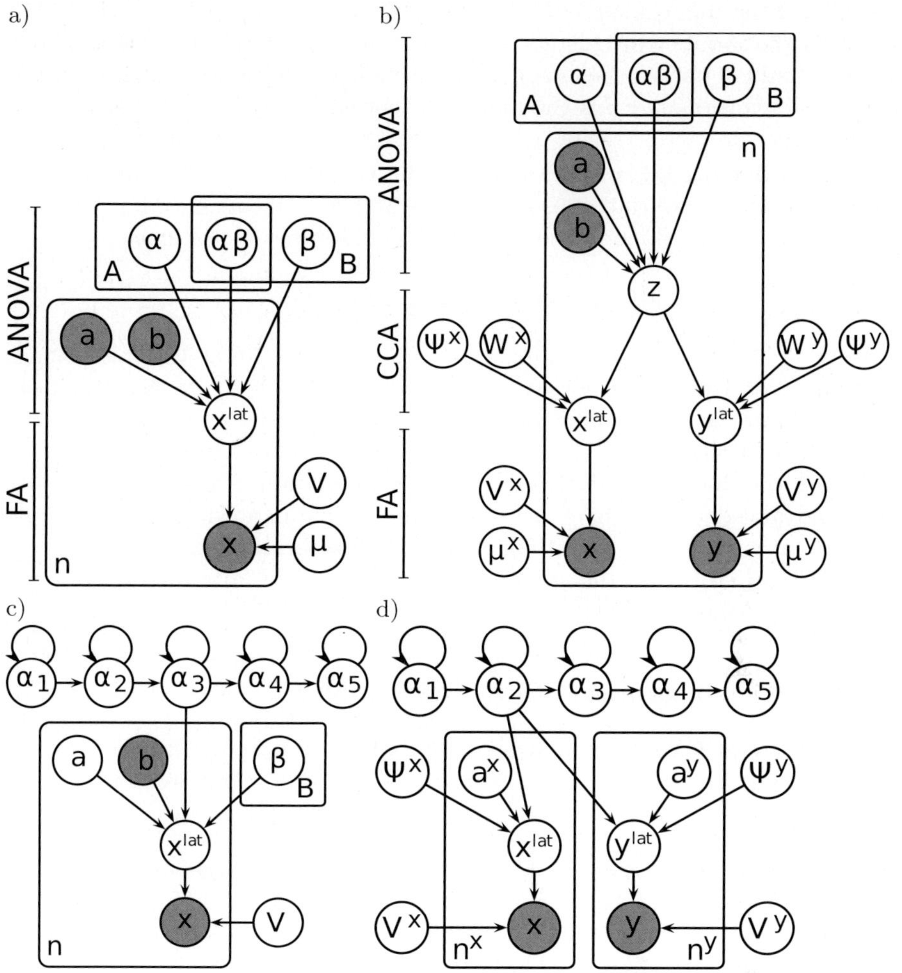

Fig. 2. The introduced model variants. (a) The hierarchical latent-variable model for standard multi-way learning with standard covariates, under "large p, small n" conditions, (b) model for multi-view learning with paired samples, (c) time with unknown alignment, (d) multi-view learning without paired samples, coupled only by shared time-course (with unknown alignment) and shared multi-way experimental design.

Since a standard factor analyzer cannot be used when $n \ll p$, we regularize projection matrix such that each variable comes from one factor only, implying a clustering assumption. The cluster indices are drawn from a multinomial distribution. The ANOVA-effects are then modelled for each cluster of correlated variables. Assuming the scales of the variables can be different, they need to be learned from data as well. For simplicity, we use a point-estimate in this paper for the scales, by scaling the variables to unit variance prior to the analysis. The number of clusters is selected by predictive likelihood [6]. The computational complexity of the models is $O(nKp + pK^2 + K^3 + nZK^2 + KZ^2 + nZ^2 + Z^3)$, where K is the number of clusters and Z is the number of CCA components. The most complex part is the clustering step, being $O(nKp)$. In small n large p conditions, dimensionality p is the main bottleneck.

ANOVA-model on latent factors. In the two-way case, the samples have two observed class covariates, $a = 0, \ldots, A$ and $b = 0, \ldots, B$. The ANOVA-modelling can now be done in the low-dimensional latent factor space,

$$\mathbf{x}_j^{lat}|_{(a,b)} = \boldsymbol{\alpha}_a + \boldsymbol{\beta}_b + (\boldsymbol{\alpha\beta})_{ab} + \text{noise}. \tag{3}$$

The ANOVA effects are set as population priors to the latent variables, which in turn are given Gaussian priors $\boldsymbol{\alpha}_a$, $\boldsymbol{\beta}_b$, $(\boldsymbol{\alpha\beta})_{ab} \sim \mathcal{N}(0, \mathbf{I})$. Note that the mean $\boldsymbol{\mu}$ is modelled in the actual data space (Equation (2)) and does not appear here.

We are now at the point where ANOVA-modelling is done in the latent factor space where the linear ANOVA-model acts as population priors. We now move into the advanced cases where "view", "time with unknown alignment" and "view without paired samples" are covariates. Gibbs-formulas have been derived analogously to [5,6,7] and the standard HMM-formalism, and are omitted due to space constraints.

2.2 Multi-view Learning with Paired Samples

We consider an ANOVA-type analysis when data comes from different views. If the data domains of the two views were the same, one might want to write a linear model

$$\mathbf{x}_d = \boldsymbol{\mu}_d + \boldsymbol{\alpha}_a + \boldsymbol{\beta}_b + (\boldsymbol{\alpha\beta})_{ab} + \boldsymbol{\gamma}_d + (\boldsymbol{\alpha\gamma})_{ad} + (\boldsymbol{\beta\gamma})_{bd} + (\boldsymbol{\alpha\beta\gamma})_{abd} + \text{noise},$$

where a and b are the two standard independent covariates, and d denotes the view. However, since the different views have different domains in general, a model cannot be written as such. It turns out that if the samples are paired (co-occur), it is possible to map the effects from latent effects to the actual data spaces $\mathbf{x}$ and $\mathbf{y}$ with unknown (estimated from the data) projections f^x and f^y as

$$\mathbf{x} = \boldsymbol{\mu}^x + f^x(\boldsymbol{\alpha}_a + \boldsymbol{\beta}_b + (\boldsymbol{\alpha\beta})_{ab}) + f^x(\boldsymbol{\alpha}_a^x + \boldsymbol{\beta}_b^x + (\boldsymbol{\alpha\beta})_{ab}^x) + \epsilon,$$
$$\mathbf{y} = \boldsymbol{\mu}^y + f^y(\boldsymbol{\alpha}_a + \boldsymbol{\beta}_b + (\boldsymbol{\alpha\beta})_{ab}) + f^y(\boldsymbol{\alpha}_a^y + \boldsymbol{\beta}_b^y + (\boldsymbol{\alpha\beta})_{ab}^y) + \epsilon\,.$$

Here the f^x and f^y represent a chain of projections from latent variables into the actual data spaces, shown in Figure 2 (b), for which the projection matrices are estimated from the data; we will define them implicitly in Equation (5) below.

This model now presents a desired decomposition into shared main and interaction effects $\boldsymbol{\alpha}_a$, $\boldsymbol{\beta}_b$, $(\boldsymbol{\alpha\beta})_{ab}$, and to view-specific main and interaction effects $\boldsymbol{\alpha}_a^x$, $\boldsymbol{\beta}_b^x$, $(\boldsymbol{\alpha\beta})_{ab}^x$. Equations are similar for $\mathbf{y}$. Note in particular that it is not meaningful to define a main effect for d since it is a view-effect (as discussed in the introduction), but the possibility to define interaction effects of a standard covariate and a view-covariate, such as $\boldsymbol{\alpha}_a^x$, allows ultimately the decomposition of effects into shared and view-specific ones. To our knowledge, there exist no methods capable of decomposing the covariate effects into shared and view-specific effects in a multi-way scenario.

We now fit the model into the extended multi-way modelling framework, depicted in Figure 2 (b). The integration of different domains takes place in the low-dimensional latent factor spaces $\mathbf{x}^{lat}$ and $\mathbf{y}^{lat}$. These factor spaces can be integrated by combining the factor analyzers into a generative model of Bayesian CCA [1,7]. This introduces a new hierarchy level where a latent variable $\mathbf{z}$ captures the shared variation between the views.

The generative model of BCCA has been formulated [1,7] for sample j as

$$\begin{aligned} \mathbf{z}_j &\sim \mathcal{N}(0, \mathbf{I}), \\ \mathbf{x}_j^{lat} &\sim \mathcal{N}(\mathbf{W}^x \mathbf{z}_j, \mathbf{\Psi}^x)\ , \end{aligned} \tag{4}$$

and likewise for $\mathbf{y}$. Note that here we have assumed no mean parameter since the mean of the data is estimated in the factor analysis part. The $\mathbf{W}^x$ is a projection matrix from the latent variables $\mathbf{z}_j$, and $\mathbf{\Psi}^x$ is a matrix of marginal variances modelling the source-specific effects not responding to external covariates. The prior distributions were chosen as in [7]; $\mathbf{W}^x$ has an Automatic Relevance Determination (ARD) prior [2]; $\mathbf{\Psi}^x$ has an inverse Wishart prior.

Decomposition into shared and view-specific effects. The decomposition into shared and view-specific effects is done by adding view-specific latent variables in addition to the shared ones, and the latent effects acting as population-specific priors on shared and specific latent variables identify the effects. The Bayesian CCA assumes that the data is generated by a sum of view-specific $\mathbf{z}^x$ and $\mathbf{z}^y$, and shared latent variables $\mathbf{z}$, as shown in Figure 3. In practice, the decomposition in Figure 3 can be implemented easily by restricting a column of $\mathbf{W}^x$ to be zero for the y-specific components and vice versa for x. As a summary the complete generative model is

$$\begin{aligned} \boldsymbol{\alpha}_0 = 0, \boldsymbol{\beta}_0 = 0, (\boldsymbol{\alpha\beta})_{a0} &= 0, (\boldsymbol{\alpha\beta})_{0b} = 0 \\ \boldsymbol{\alpha}_a, \boldsymbol{\beta}_b, (\boldsymbol{\alpha\beta})_{ab}, \boldsymbol{\alpha}_a^x, \boldsymbol{\beta}_b^x, (\boldsymbol{\alpha\beta})_{ab}^x &\sim \mathcal{N}(0, \mathbf{I}) \\ \mathbf{z}_j|_{j\in a,b} &\sim \mathcal{N}(\boldsymbol{\alpha}_a + \boldsymbol{\beta}_b + (\boldsymbol{\alpha\beta})_{ab}, \mathbf{I}) \\ \mathbf{z}_j^x|_{j\in a,b} &\sim \mathcal{N}(\boldsymbol{\alpha}_a^x + \boldsymbol{\beta}_b^x + (\boldsymbol{\alpha\beta})_{ab}^x, \mathbf{I}) \\ \mathbf{x}_j^{lat} &\sim \mathcal{N}(\mathbf{W}_{\text{shared}}^x \mathbf{z}_j + \mathbf{W}_{\text{specific}}^x \mathbf{z}_j^x, \mathbf{\Psi}^x) \end{aligned}$$

$$\mathbf{x}_j \sim \mathcal{N}(\boldsymbol{\mu}^x + \mathbf{V}^x \mathbf{x}_j^{lat}, \mathbf{\Lambda}^x). \tag{5}$$

2.3 Time with Unknown Alignment

We concentrate here on the case of a small number (~ 10) of replicate time-series from multiple populations, in the unfavorable conditions of short ($\sim$ 10-20) time-series and high dimensionality.

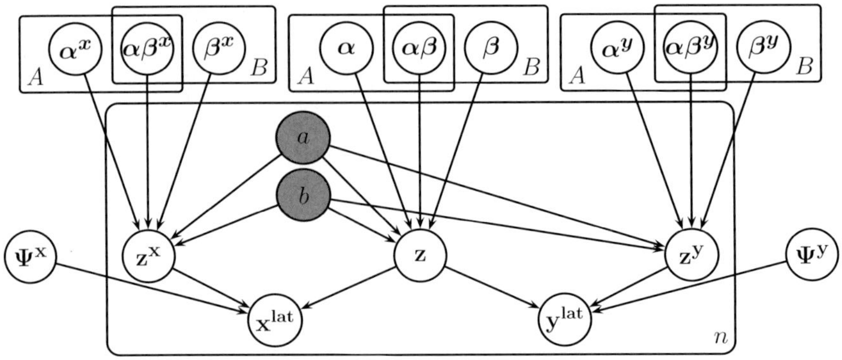

Fig. 3. The graphical model describing the decomposition of covariate effects into shared and view-specific ones. The figure expands the top part of Figure 2 (b).

We consider "time with unknown alignment" as one covariate in an extended multi-way model; a particular case is HMM-alignment. The extended multi-way model with HMM-time as a covariate, is shown in Figure 2 (c). We assume that the time operates on the latent variables as the other covariates, with the unknown alignment modelled by HMM. This can be accomplished by having the HMM emit values for the latent factors. In addition, there is another covariate effect $\boldsymbol{\beta}_b$. The model becomes

$$\mathbf{x}_j^{lat}|_{state(j,t)=s,b} \sim \mathcal{N}(\boldsymbol{\alpha}_s + \boldsymbol{\beta}_b + (\boldsymbol{\alpha\beta})_{sb}, \mathbf{I}). \tag{6}$$

Here $\boldsymbol{\alpha}_s$ is the effect of HMM-time in the multi-way model, that is the mean in the Gaussian emission distribution of HMM-state s. The $\boldsymbol{\beta}_b$ is the effect of the other, observed, covariate b, and $(\boldsymbol{\alpha\beta})_{sb}$ is the interaction effect. Here $state(j,t) = s$ means that time-point t of sample j belongs to state s.

Assignments to HMM states are sampled according to a standard Bayesian HMM formalism, the prior for the transition matrix of the linear HMM allowing only self-transitions and transitions to the next state.

In biological case studies where time-series measurements are taken from multiple populations, e.g. healthy and diseased, there is a need for HMM alignment when intervals between measurements are long and irregular within- and between patients. In addition, patients are assumed to develop to different biological states at individual times/ages. Previous works [3,11] have resorted to training a separate HMM for each population and comparing them afterwards, additionally restricting to strong feature selection, only allowing favorable $n > p$-conditions.

In our experiments, we will have 5 states, $b = 0, ..., 4$ of β_b-effects for the diseased, corresponding to the observed disease-development states in the time-series. For simplicity, we do not consider the interaction effect, restricting to $\mathbf{x}_j^{lat}|_{state(j,t)=s,b} \sim \mathcal{N}(\boldsymbol{\alpha}_s + \boldsymbol{\beta}_b, \mathrm{I})$.As a summary, "time with unknown alignment", such as "HMM-time", can be seen as one covariate in an extended multi-way model, where the covariate assignments (alignment to HMM-states), are inferred from the observed data. A main benefit of building a unified model is that after explaining away the effect of "aligned time", $\boldsymbol{\alpha}_s$, one can answer the following statistical question: is there a difference in the populations, that is; is $\boldsymbol{\beta}_b$ statistically significant for some b? Earlier HMM approaches training a separate model for each population cannot fully rigorously answer this question.

2.4 Multi-view Learning without Paired Samples

Finally, we consider integrating data sources in different domains, without paired samples, which is a much more difficult problem. In a similar case in [13], the underlying assumption was an unobserved pairing between the samples, and the pairing was found by iteratively alternating between searching for pairing and maximizing dependencies between the sources by CCA. However, the assumption of latent unknown pairing might be too restrictive in many cases, and the non-generative solution in [13] cannot easily be extended to the present tasks.

We propose an alternative assumption, allowing to integrate multiple unmatched data sources under the assumption of shared, underlying multi-way covariate-related behavior. For brevity, we concentrate in this article on one standard covariate, say, time with unknown alignment $\boldsymbol{\alpha}$, and the "view without paired samples" is the other covariate. Again, since "view without paired samples" is a **view-covariate**, it cannot be defined at all as a main effect due to data domains being different. However, we can define an interaction effect of time and "view without paired samples". The model becomes

$$\begin{aligned} \mathbf{x} &= \boldsymbol{\mu}^x + f^x(\boldsymbol{\alpha}_a) + f^x(\boldsymbol{\alpha}_a^x) + \epsilon, \\ \mathbf{y} &= \boldsymbol{\mu}^y + f^y(\boldsymbol{\alpha}_a) + f^y(\boldsymbol{\alpha}_a^y) + \epsilon \ , \end{aligned} \tag{7}$$

where $\boldsymbol{\alpha}_a$ is the shared effect of time, and $\boldsymbol{\alpha}_a^x$ and $\boldsymbol{\alpha}_a^y$ are the view-specific time-effects, and f^x and f^y are again functional mappings from latent states to actual data spaces. The possibility to decompose the time-related behavior in the two datasets into shared and view-specific covariate-related behaviors connects the views.

To make the translational problem (presented below) more realistic, we consider the time-covariate to be "HMM-time". This allows us to study more flexible translational cases with time-measurement having irregular intervals, and time-spans being different [9], important in cross-species biological applications.

The model can be formulated as a graphical multi-way model with the techniques presented in the previous sections. For simplicity, the view-specific time-behavior is integrated out with $\boldsymbol{\Psi}_x$ and $\boldsymbol{\Psi}_y$ following [7], and we only search for the shared effects in the simulations. The graphical model of the problem is shown in Figure 2 (d). The key difference to the multi-view learning with paired samples case in Figure 2 (b) is that since there is no known pairing of samples, there is no latent variable $\mathbf{z}_j$ shared by the samples from different views.

The learning algorithm faces a matching problem since a shared time behavior might be identified in, e.g., cluster 1 of $\mathbf{x}$ and cluster 3 of $\mathbf{y}$. For the model to identify the effect as a shared effect, it should be found for the same cluster identity. We include a Metropolis-Hastings step in our Gibbs sampler that proposes to switch identities of two clusters, attempting to maximize similar time-related behavior.

Related work in translational studies. A main application is translating biological findings between experiments on model organisms and actual human experiments [9,10]. The common setup is doing a similar experiment (time-series, same disease) to the two different organisms and comparing the results. High-dimensional biological measurements usually have different unmatched domains in different species. Most multi-species approaches [9] are restricted to the subset of variables that are a priori matched between the species. Since this assumption is restrictive, we have wanted to consider the more general case where the domains are different, making it possible to use all the data, and while doing so to actually search for the matching of variables.

3 Results

3.1 Multi-view Learning with Paired Samples

Generated data. We integrate three data-sources, $\mathbf{x}$, $\mathbf{y}$ and $\mathbf{u}$, with pairing between the samples, which have a two-way experimental setup, generated from the model of Figure 2 (b). The datasets are 200-dimensional, there are three clusters of variables in each dataset. The $\sigma_i = 1$ for each variable. The model is learned by Gibbs sampling, with 2000 samples and 2000 burn-in samples. The optimal number of clusters is found for each data source separately as explained in [6], and always correctly recovered. Unless otherwise stated, these parameters are the same throughout the results section. Effects $\alpha = +2$, $\beta^y = +2$ and $(\alpha\beta)^x = +2$ have been generated. We learned 4 components: one shared and three source-specific. Shared and source-specific α, β and $(\alpha\beta)$ are therefore to be estimated. The model always finds the correct clusterings (data not shown). The results in Figure 4 show how the model finds the generated effects as a function of number of samples. According to the results, the model finds the generated effects with relatively small sample-sizes, and the uncertainty decreases with increasing sample-size. The shared effect is found with considerably less uncertainty since there is evidence from both sources. In a typical biological dataset there may be 20-60 samples.

Lipidomic multi-tissue data. We now apply the method on an unpublished lipidomic lung cancer study, where lipidomic measurements have been taken from several tissues of mice. There are cancerous and healthy mice, and additionally half of both populations have been given a test anti-cancer drug. This is a typical two-way setup with healthy untreated (10 mice), diseased untreated (10), healthy treated (9), diseased treated (10) mice. The tissues have different lipids. We first integrated the lung tissue (68 lipids) with spleen tissue (44 lipids). We learned 3 components, one shared and one for each view. According to the results in Figure 5 (left), the model finds a shared disease effect α and a shared treatment effect β. The result shows that the treatment enhances, not

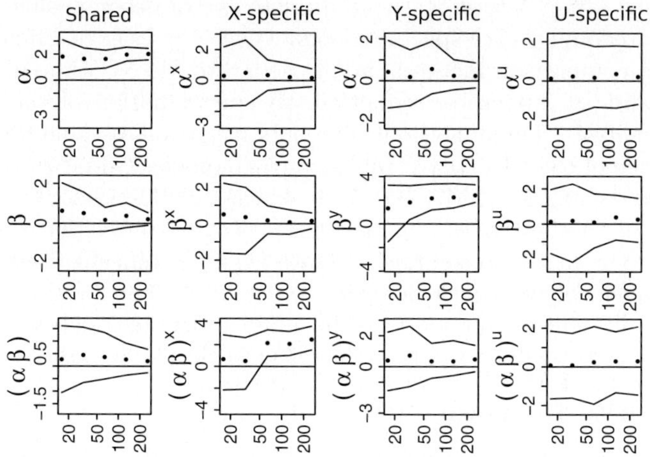

Fig. 4. The method finds the generated effects $\alpha = +2$, $\beta^y = +2$ and $(\alpha\beta)^x = +2$ in a three-view, two-way study. The points show posterior means, and lines the 95% posterior mass of the effects. The posterior distributions have been mirrored to have a positive mean. A consistently non-zero posterior of the effects indicates a statistically significant effect. This corresponds to a classical p-value being $p < 0.05$.

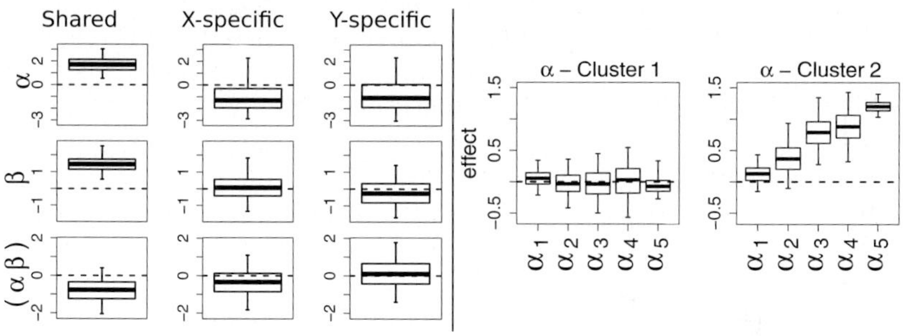

Fig. 5. In the experiment on multi-tissue data (left), the method finds a disease effect α and a treatment effect β shared between the two views, spleen ($\mathbf{x}$) and lung ($\mathbf{y}$) tissues. (right) The shared HMM finds shared effects in two generated datasets (Section 3.3) in different domain with no paired samples. A growing HMM-time effect was generated in cluster 2. A consistently non-zero posterior implies an effect found.

diminishes the effect of the disease, therefore not being effective. In lung, for instance, a cluster of 12 lipids containing ether lipids known to be co-regulated, was coherently up-regulated due to disease, and additionally up-regulated by the treatment. Another cluster of 13 lipids in lung was found down-regulated due to the disease and additionally down-regulated due to treatment. The lipids of the down-regulated cluster are thus negatively correlated with the up-regulated clusters. The effect can be traced back to the clusters of lipids by identifying the responsible elements in $\mathbf{W}^x$, and to the actual lipids from $\mathbf{V}^x$.

No existing ANOVA-type methods are capable of decomposing covariate effects into shared and source-specific effects, when sources have different domains. The possible comparison methods are 1) separate MANOVA-analysis for each source including a dimension reduction, 2) concatenation of the sources and MANOVA-analysis. These methods give only an overall p-value for the statistical significance of the effects. We compare the biological result to concatenating the sources and using 50-50 MANOVA [8], which includes a prior PCA-dimension reduction. The method gives p-values 0.01, 0.71 and 0.071, for α, β and $(\alpha\beta)$, respectively. The method only finds a statistically significant disease effect, not finding the effect of treatment, showing the superior behavior of an integrated dimension reduction in our model. The main difference is, however, that the method cannot distinguish whether the effect is shared or source-specific.

3.2 Time with Unknown Alignment

Generated data. We show results on data generated from the model in Figure 2 (c). There are 5 HMM-states α, 23 replicate time-series from healthy and 21 from diseased population for which there are 3 disease states β. Each time-series has a length of 5-15 time-points at random times (no matching of time-points), dimensionality is $p = 400$. Disease state-type covariates $b_{jt} = \{1, 2, 3\}$ are observed for the diseased patients, healthy patients only have HMM-states. Effects $\boldsymbol{\alpha} = 0, +0.5, +1, +1.5, +2$ have been generated in the consecutive HMM-states in the first cluster of α. In the disease states of β, effects $\boldsymbol{\beta} = -0.5, -1, -2$ have been generated in the consecutive disease states, equally in the first cluster. In the other clusters, there are no covariate-related effects, only structured noise from the model. The model is able to identify the clusters correctly. The results in Figure 6 (left) show that a proper HMM-alignment is achieved, and the model found the generated, growing time-behavior $\boldsymbol{\alpha}$ in cluster 1 and especially is able to separate the correct descending disease-state behavior $\boldsymbol{\beta}$ related to the known covariates.

Lipidomic time-series data. We then applied the model to a recently published lipidomic dataset [12]. There are 71 healthy patients and 53 patients that later developed into type 1 diabetes, there are 3-29 time-points in each time-series, measured at irregular intervals. In addition, for the patients that later developed into type 1 diabetes, the progress of the disease (disease state) is observed at each time point and used here as a covariate b, with 5 disease states. There are 53 lipids. We show results on 2 clusters in Figure 6 (right). The model was capable of identifying the normal aging effects for clusters of similarly behaving lipids (HMM-time effects), and it was able to separate disease state-related effects for each cluster. The model found consistent clusters of lipids known to be co-regulated. In Figure 3 of [12], the data analysis was done by univariate t-tests for each lipid, and time was separated to bins of length 1 year. Our multivariate modelling was able to take into account that different individuals enter age-related metabolic states at largely varying, individual times. In addition, the model could separate the disease-state related behavior from the normal aging effects, all done for similarly behaving groups of lipids. Our results were consistent

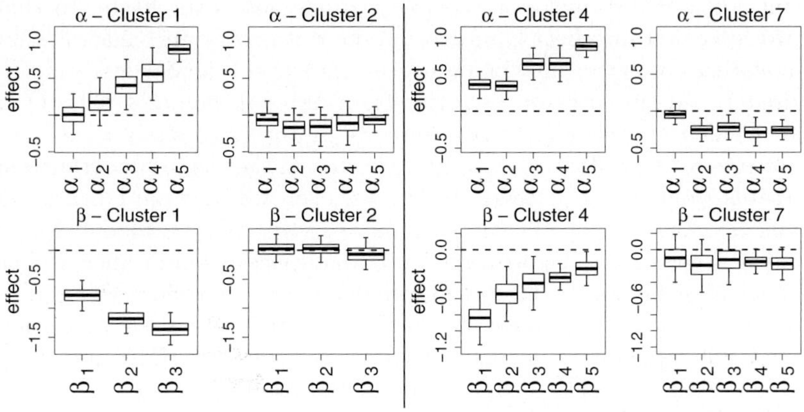

Fig. 6. (left) The HMM-model finds the effect of time with unknown alignment α_s in cluster 1 from two populations of generated data, and is able to separate a disease-progression type effets β generated in the other population. (right) The HMM-model separates normal aging α_s for clusters of similarly behaving lipids, from effects related to known disease progression-states in a real lipidomics type 1 diabetes study. A consistently non-zero posterior shown by box-plots implies an effect found.

with those in the paper. In addition, our model suggested that PC(14:0/18:2) and PC(18:2/16:1) in cluster 4 have a strong down-regulation in the early disease development states, and might act as early biomarkers of a developing disease. This was not revealed by the prior analysis. Existing methods are limited to training a separate HMM for each population, which is an appropriate approach for classification, but cannot be used to rigorously compare the effects of disease states in the data under the assumption that normal aging effects have to be modelled (away) by a HMM-alignment.

3.3 Multi-view Learning without Paired Samples

We now show results of multi-way learning when the domains of multiple data sources are different and samples have no pairing. We consider a non-trivial case where we have two generated time-series datasets, with irregular lengths and measurement times where we assume, however, that behavior of HMM-states is similar. We can now search for similar HMM-behavior in two unpaired datasets in different domains. We can make the assumption that in addition to view-specific effects, there is a shared HMM-chain that emits latent variables $\mathbf{x}^{lat}$ and $\mathbf{y}^{lat}$, which in turn generate the actual data to different domains.

We have generated two data sets from the model with 10 and 11 replicate time-series of irregular lengths between 8-12, and datasets have 100 and 110 variables, respectively. There are 3 clusters in each, and the corresponding $\mathbf{x}^{lat}$ and $\mathbf{y}^{lat}$ have been generated according to the shared HMM-chain where the effects 0,+0.5,+1,+1.5,+2 have been generated to the second factor (cluster) of $\boldsymbol{\alpha}$ in the five HMM-states of the shared HMM-chain, an x-specific time-effect in cluster 1, the third cluster (not shown) does not have effects. In this case study, the specific time effect is integrated out with covariance matrices $\mathbf{\Psi}_x$ and $\mathbf{\Psi}_y$. According to the results in Figure 5 (right), the model was able to find the shared HMM-time-related behavior from different domains without paired samples, while x-specific effects were integrated out successfully.

The results of the simple case study show that in the case of underlying shared HMM-states, connection between two views, even without paired samples, can be formed by formulating the analysis as an ANOVA-type model over views. This makes it possible to rigorously evaluate statistical significance of a similar covariate-related behavior. To our knowledge, such a possibility has not been proposed in any previous studies. To our knowledge, there exists no comparable method to this modelling task, except training a separate HMM for each view, which allows only qualitative comparison of the results.

4 Conclusions

We have extended multi-way learning to three novel cases: (i) multi-view learning with paired samples, where data comes from different domains, (ii) one of the covariates has an unknown structure (iii) data comes from different domains with no pairing of samples, but covariates are shared. In (i) we have shown how covariate-related behavior can be decomposed into shared and view-specific effects, when integrating data sources with paired samples. In (ii) we have presented a multi-way model where one of the covariates has an unknown structure which can be learned jointly. In (iii) we have shown that it is possible to integrate multiple data sources without paired samples, if the datasets have a similar covariate-structure. We have shown that unified hierarchical graphical models can be used to structure each case as a graphical multi-way model.

Each of the presented multi-way models has direct applications for biological experiments, but they also offer novel possibilities for other application domains such as brain signal analysis (multi-way time-series in fMRI), detection problems in sensor fusion and cold-start problem in content-based retrieval given second content descriptors etc. We showed that the models are capable of finding ANOVA-type effects from real and simulated high-dimensional data, even with small sample-sizes. The biological results were plausible, comparing to previous studies.

References

1. Bach, F.R., Jordan, M.I.: A probabilistic interpretation of canonical correlation analysis. Tech. Rep. 688, Department of Statistics, University of California, Berkeley (2005)
2. Bishop, C.M.: Bayesian PCA. In: Kearns, M.S., Solla, S., Cohn, D. (eds.) Advances in Neural Information Processing Systems, vol. 11, pp. 382–388. MIT Press, Cambridge (1999)

3. Costa, I.G., Schonhuth, A., Hafemeister, C., Schliep, A.: Constrained mixture estimation for analysis and robust classification of clinical time series. Bioinformatics 25(12), i6–i14 (2009)
4. Fisher, R.: The correlation between relatives on the supposition of mendelian inheritance. Royal Society of Edinburgh from Transactions of the Society 52, 399–433 (1918)
5. Huopaniemi, I., Suvitaival, T., Nikkilä, J., Orešič, M., Kaski, S.: Multivariate multiway analysis of multi-source data. Bioinformatics 26, i391–i398 (2010)
6. Huopaniemi, I., Suvitaival, T., Nikkilä, J., Orešič, M., Kaski, S.: Two-way analysis of high-dimensional collinear data. Data Mining and Knowledge Discovery 19(2), 261–276 (2009)
7. Klami, A., Kaski, S.: Local dependent components. In: Ghahramani, Z. (ed.) Proceedings of ICML 2007, the 24th International Conference on Machine Learning, pp. 425–432. Omni Press (2007)
8. Langsrud, O.: 50-50 multivariate analysis of variance for collinear responses. Journal of the Royal Statistical Society Series D-the Statistician 51, 305–317 (2002)
9. Lu, Y., Huggins, P., Bar-Joseph, Z.: Cross species analysis of microarray expression data. Bioinformatics 25(12), 1476–1483 (2009)
10. Lucas, J., Carvalho, C., West, M.: A bayesian analysis strategy for cross-study translation of gene expression biomarkers. Statistical Applications in Genetics and Molecular Biology 8(1), 11 (2009)
11. Nikkilä, J., Sysi-Aho, M., Ermolov, A., Seppänen-Laakso, T., Simell, O., Kaski, S., Orešič, M.: Gender dependent progression of systemic metabolic states in early childhood. Molecular Systems Biology 4, 197 (2008)
12. Orešič, M., et al.: Dysregulation of lipid and amino acid metabolism precedes islet autoimmunity in children who later progress to type 1 diabetes. Journal of Experimental Medicine 205(13), 2975–2984 (2008)
13. Tripathi, A., Klami, A., Kaski, S.: Using dependencies to pair samples for multiview learning. In: Proceedings of ICASSP 2009, the International Conference on Acoustics, Speech, and Signal Processing, pp. 1561–1564 (2009)
14. West, M.: Bayesian factor regression models in the large p, small n paradigm. Bayesian Statistics 7, 723–732 (2003)

Exploration-Exploitation of Eye Movement Enriched Multiple Feature Spaces for Content-Based Image Retrieval

Zakria Hussain[1], Alex P. Leung[2], Kitsuchart Pasupa[3], David R. Hardoon[4], Peter Auer[2], and John Shawe-Taylor[1]

[1] University College London, UK
{z.hussain,jst}@cs.ucl.ac.uk
[2] University of Leoben, Austria
{auer,alex.leung}@unileoben.ac.at
[3] University of Southampton, UK
k.pasupa@gmail.com
[4] Institute for Infocomm Research (I^{2}R), Singapore
drhardoon@i2r.a-star.edu.sg

Abstract. In content-based image retrieval (CBIR) with relevance feedback we would like to retrieve relevant images based on their content features and the feedback given by users. In this paper we view CBIR as an Exploration-Exploitation problem and apply a kernel version of the LinRel algorithm to solve it. By using multiple feature extraction methods and utilising the feedback given by users, we adopt a strategy of multiple kernel learning to find a relevant feature space for the kernel LinRel algorithm. We call this algorithm LinRelMKL. Furthermore, when we have access to eye movement data of users viewing images we can enrich our (multiple) feature spaces by using a tensor kernel SVM. When learning in this enriched space we show that we can significantly improve the search results over the LinRel and LinRelMKL algorithms. Our results suggest that the use of exploration-exploitation with multiple feature spaces is an efficient way of constructing CBIR systems, and that when eye movement features are available, they should be used to help improve CBIR.

Keywords: content-based image retrieval, LinRel, images, eye movements, multiple kernel learning, tensor kernel SVM.

1 Introduction

Assume we have a database $\mathcal{D}$ of images and would like to find an image $x \in \mathcal{D}$ without having to conduct an exhaustive search of $\mathcal{D}$. If we have not tagged the images in the database, then we would need to retrieve images using some content-based image retrieval (CBIR) system utilising relevance-feedback [9]; a tool designed for accessing *relevant* images from a database using their *content* such as colour, shape, texture etc., and receiving *feedback* from the user on

J.L. Balcázar et al. (Eds.): ECML PKDD 2010, Part I, LNAI 6321, pp. 554–569, 2010.

the relevance of images presented so far. We will refer to content-based image retrieval using relevance-feedback as CBIR throughout this paper. Using this relevance information the CBIR system should find relevant images quickly. The goal of the CBIR system is to produce the target image in the least number of presented images.

In this paper, we view the underlying mechanics of a CBIR system as an exploration-exploitation problem. We apply the LinRel algorithm [3], a linear regression algorithm that efficiently carries out exploration-exploitation. LinRel makes use of side information – typically various image features – to estimate the relevance of an image. Since these estimates are imperfect, when selecting possibly relevant images the algorithm needs to trade-off between high estimated relevance and possible gain of information to improve the estimates.

Each image can be represented using many different feature extraction methods such as SIFT, histogram, colour, *etc.*, and so its not clear which ones to use for the LinRel algorithm – as some may be good for some search tasks but not for others. A recent line of research, called multiple kernel learning (MKL) aims at finding a good linear combination of feature spaces [15]. The idea is to find the best weighting of feature spaces for a given classification task. By viewing the relevance feedback as our classification outputs we can learn a combination of feature spaces using MKL, and present this new kernel to the kernelized version of the LinRel algorithm. This removes the need of choosing explicitly which feature spaces to use and also tackles the problem of combining feature spaces with a non-uniform combination, unlike [8]. This algorithm will be called LinRelMKL.

Finally, we will also assume that our CBIR system has access to an eye tracking device, allowing us to record users' eye movements whilst they view images.[1] This gives us an *extra set of features* for each image *viewed*. However, we do not have access to the eye movement features for the images not shown by the CBIR system. Therefore, we use a recent approach [11] of mapping the combination of image features we have derived (using MKL) into the eye movement feature spaces, using the tensor kernel support vector machine [12]. When used with LinRel alone we call this algorithm LinRelTensor and when also combined with MKL we call it LinRelMKLTensor.

We start with our problem definition in Section 2, followed by the LinRel, MKL, and tensor learning algorithms in Section 3, 4 and 5, respectively. Section 6 describes our final CBIR system. The experiments are described in Section 7, with concluding remarks in Section 8. We begin with some preliminary definitions.

Notation: Let $\mathbf{x} \in \mathbb{R}^m$ be an m-dimensional row vector, with $\mathbf{x}^\top$ denoting its transpose. Let $\phi : \mathbf{x} \rightarrow \mathcal{H} \subset \mathbb{R}^M$ be a mapping into a high dimensional feature space $\mathcal{H}$. A kernel function will be defined as $\kappa(I, I) = \langle \phi(\mathbf{x}_I), \phi(\mathbf{x}_I) \rangle$ and accessed using an index set I. The cardinality of I will be made clear from context. We denote $\mathbf{X} = (\mathbf{x}_1, \ldots, \mathbf{x}_t)^\top$ as the matrix containing t row vectors and $\mathbf{I}$ as the identity matrix. The covariance matrix can be written as $\mathbf{X}^\top\mathbf{X}$,

[1] We will show later that having access to eye movement features can improve search results.

with its inverse as $\left(\mathbf{X}^\top\mathbf{X}\right)^{-1}$. The notation $\|\mathbf{x}\|_p$ will denote the p-norm of vector $\mathbf{x}$, where $p = 1, 2$.

2 Problem Definition

In content-based image retrieval (CBIR) with relevance feedback [24,8,19,18], one can consider a model motivated by a filtering task, where a user wants to find a set of images relevant to his query. When presented with a collage of images, the user marks[2] all images which are relevant to his query. The goal of the search algorithm is to present as many relevant images as possible in early iterations of the search.

Our work focuses on algorithms and experiments for this model. The online algorithms build upon the LinRel algorithm proposed in [3, Section 4]. Before discussing the LinRel algorithm we will describe the user model we consider, the model for relevance scores and also the different image feature extraction methods used throughout.

2.1 The User Model for the Filtering Task

We assume that the user is looking for a set of relevant images I in an image database $\mathcal{D}$. The relevance of an image is determined by the user query, about which the search engine is informed only by relevance feedback. The goal of the search engine is to present mostly relevant images to the user, and only a small number of irrelevant images. The feedback of the user is given by a relevance score $y \in [0, 1]$, with 0 meaning not relevant, 1 meaning relevant, and possible degrees of relevance for values in between. The relevance score can be given by an explicit binary feedback (e.g. mouse clicks), or implicitly (e.g. by recorded eye movements). The formal protocol for this model is outlined in Fig. 1.

- In each iteration $t = 1, 2, \ldots$
  - The search engine selects an image $I_t \in \mathcal{D}$ and presents it to the user.
  - The user's feedback is given by the relevance score $y_t \in \{0, 1\}$.

Fig. 1. Original protocol for LinRel

The performance of the search engine is determined by the number of relevant images returned in a certain number of iterations. We note that in the protocol of Figure 1 only a *single* image is presented in each round and the relevance score is *binary*. This is the original protocol used to develop the LinRel algorithm in [3]. In a realistic CBIR setting, the protocol should be extended to Fig. 2.

[2] This will be done explicitly using mouse clicks. An implicit score of relevance could be found using eye movements. However we will not address this extension in the current paper.

- In each iteration $t = 1, 2, \ldots$
  - The search engine selects n images $I_{t,1}, \ldots, I_{t,n} \in \mathcal{D}$ and presents them to the user.
  - For each presented image the search engine receives a relevance score $y_{t,i} \in [0, 1]$, $i = 1, \ldots, n$.

Fig. 2. A CBIR protocol for LinRel

Thus the search engine described in Figure 2 needs to select a fixed number n of images, with the relevance scores ranging between 0 and 1.

2.2 Modelling the Relevance Scores

To be able to learn about the user's query from the relevance scores, we are making assumptions about how the relevance scores are generated. We assume that an image I is represented by a normalised vector $\mathbf{x}_I \in \mathbb{R}^d$ of features (see subsection 2.3), with $\|\mathbf{x}_I\|_2 = 1$. Furthermore, we assume that the relevance score y_I of an image I is a random variable with expected value $\mathbb{E}[y_I] = \mathbf{x}_I \cdot \mathbf{w}$, $\mathbf{w} \in \mathbb{R}^m$, such that the expected relevance score is a linear function of the image features. The unknown weight vector $\mathbf{w}$ is essentially the representation of the user's query and determines the relevance of images.

2.3 Features Extracted from Images

The feature extraction methods can be found below in Table 1, with a detailed description of each method given in [14]. For each image I we will assume that each of the 11 feature extraction methods of Table 1 has been carried out. Hence, for each image I we will have feature vectors $\mathbf{x}_{I,1}, \ldots, \mathbf{x}_{I,11}$.

A typical method of using all of these feature vectors would be to concatenate them together and construct a single kernel matrix, and run this through

Table 1. Visual features extracted from images

Feature	dimensions
DCT coefficients of average colour in rectangular grid	12
CIE L*a*b* colour of two dominant colour clusters	6
Histogram of local edge statistics	80
Haar transform of quantised HSV colour histogram	256
Histogram of interest point SIFT features	256
Average CIE L*a*b* colour	15
Three central moments of CIE L*a*b* colour distribution	45
Histogram of four Sobel edge directions	20
Co-occurrence matrix of four Sobel edge directions	80
Magnitude of the 16 × 16 FFT of Sobel edge image	128
Histogram of relative brightness of neighbouring pixels	40

a kernelized CBIR system [8]. This would correspond to a uniform weighting of the feature vectors (*i.e.*, kernels). However, it may be the case for some search tasks that, for instance *"Histogram of four Sobel edge direction"* would be more important than *"CIE L*a*b* colour of two dominant colour clusters"* features. Using a uniform weighting only assumes that both features are equally important. Therefore we will apply kernels for each of the feature extraction methods outlined in Table 1, and apply MKL to find a linear combination of these feature spaces (*i.e.*, kernels). For certain search tasks, this linear combination results in a non-uniform weighting, giving rise to higher weightings for more useful feature extraction methods.

3 The LinRel Algorithm and Its "Kernelized" Counterpart

In [3] the LinRel algorithm was devised for a slight variant of the model described in the previous section. In this section we describe the LinRel algorithm with the necessary modifications to accommodate our model. We restrict ourselves to the case $n = 1$, *i.e.*, only one image is presented to the user in each iteration. The general case *i.e.*, image collages, will be discussed later.

The user model for the filtering tasks (in particular the model for the relevance scores) confronts the search engine with an exploration-exploitation trade-off. Typically the search engine will maintain an implicit or explicit representation of an estimate $\hat{\mathbf{w}}$ of the unknown weight vector $\mathbf{w}$. When selecting the next image for presentation to the user, the search engine might simply select the image with highest estimated relevance score based on $\hat{\mathbf{w}}$. But since the estimate $\hat{\mathbf{w}}$ might be inaccurate, this exploitative choice might be suboptimal. Alternatively, the search engine might exploratively select an image for which the user feedback improves the accuracy of the estimate $\hat{\mathbf{w}}$, enabling better image selections in subsequent iterations.

In each iteration t, LinRel obtains an estimate $\hat{\mathbf{w}}_t$ by solving the linear regression problem

$$\mathbf{y}_t \approx \mathbf{X}_t \cdot \hat{\mathbf{w}}_t,$$

where

$$\mathbf{y}_t = \begin{pmatrix} y_1 \\ \vdots \\ y_{t-1} \end{pmatrix}$$

is the column vector of relevance scores received so far, and

$$\mathbf{X}_t = \begin{pmatrix} \mathbf{x}_1 \\ \vdots \\ \mathbf{x}_{t-1} \end{pmatrix}$$

is the matrix of row feature vectors of the images presented so far. Based on the estimated weight vector $\hat{\mathbf{w}}$, LinRel calculates an estimated relevance score

$\hat{y}_I = \mathbf{x}_I \cdot \hat{\mathbf{w}}$ for each image I that has not already been presented to the user. To deal with the exploration-exploitation trade-off, LinRel selects for presentation not the image with largest estimated relevance score, but the image with the largest upper confidence bound for the relevance score. The upper confidence bound for an image I is calculated as $\hat{y}_I + c\hat{\sigma}_I$, where $\hat{\sigma}_I$ is an upper bound on the standard deviation of the relevance estimate $\hat{y}_I$. The constant c is used to adjust the confidence level of the upper confidence bound.

The rationale for using upper confidence bounds is that an image gets selected if (a) its relevance score is indeed large, or (b) the estimated relevance score is rather unreliable and the resulting confidence interval is large. Case (a) gives an exploitative choice, while case (b) improves the estimates and thus is explorative. It has been shown that upper confidence bounds are a versatile tool to balance exploration and exploitation in online selection problems [1,4,3]. In [3] rigorous bounds on the performance of LinRel are proven.

In each iteration t the regularised LinRel algorithm for $n = 1$ [5], calculates

$$\mathbf{a}_I = \mathbf{x}_I \cdot (\mathbf{X}_t^\top \mathbf{X}_t + \mu \mathbf{I})^{-1} \mathbf{X}_t^\top \tag{1}$$

for each image I and selects for presentation the image I_t which maximises

$$I_t = \arg\max_I \left\{ \mathbf{a}_I \cdot \mathbf{y}_t + \frac{c}{2} ||\mathbf{a}_I|| \right\} \tag{2}$$

for some specified constant $c > 0$.

3.1 Kernelization and Selection of Image Collages

Kernel learning [21] can be integrated into the LinRel algorithm. A kernel learning algorithm learns a suitable metric between images in respect to a user query, by finding a good kernel function. Since only in each iteration the kernelized LinRel algorithm relies on a fixed kernel function, the integration of kernel learning into LinRel is very simple: LinRel calls the kernel learning algorithm at the beginning of each iteration and then uses the kernel matrix returned by the kernel learning algorithm. To kernelize LinRel [5],

$$\mathbf{a}_I = \big(\kappa(I, I_1) \ \cdots \ \kappa(I, I_{t-1})\big) \cdot (\mathbf{K}_t + \mu \mathbf{I})^{-1},$$

where $I_1, \ldots, I_{t-1}$ are the images selected in iterations $i = 1, \ldots, t-1$ and $\mathbf{K}_t$ is the Gram matrix

$$\mathbf{K}_t = (k(I_i, I_j))_{1 \le i,j \le t-1}.$$

Thus $\mathbf{a}_I$ can be calculated by using only the kernel function $\kappa(\cdot,\cdot)$. Since the selection rule (2) remains unchanged, this gives the kernelized version of LinRel.

Furthermore, three methods for the selection of image collages are used and evaluated in this paper:

1. Select the images $I_{t,1}, \ldots, I_{t,n}$ with maximal upper confidence bounds $\mathbf{a}_I \cdot \mathbf{y}_t + \frac{c}{2}||\mathbf{a}_I||$, where $\mathbf{a}_I = \mathbf{x}_I \cdot (\mathbf{X}_t^\top \mathbf{X}_t + \mu \mathbf{I})^{-1} \mathbf{X}_t^\top$ with $\mathbf{X}_t$ and $\mathbf{y}_t$,

2. Select image $I_{t,1}$ with the maximal upper confidence bound $\mathbf{a}_I \cdot \mathbf{y}_t + \frac{c}{2}||\mathbf{a}_I||$. Select the images $I_{t,2}, \ldots, I_{t,n}$ with maximal estimated relevance score $\mathbf{a}_I \cdot \mathbf{y}_t$,
3. For $k = 1, \ldots, n$ select image $I_{t,k}$ which maximises $\mathbf{a}_I \cdot \mathbf{y}_{t,k} + \frac{c}{2}||\mathbf{a}_I||$, where $\mathbf{a}_I = \mathbf{x}_I \cdot (\mathbf{X}_{t,k}^\top \mathbf{X}_{t,k} + \mu \mathbf{I})^{-1} \mathbf{X}_{t,k}^\top$,[3]

where the matrix $\mathbf{X}_{t,k}$ augments $\mathbf{X}_t$ by the feature vectors of the already selected images $I_{t,1}, \ldots, I_{t,k-1}$,

$$\mathbf{X}_t = \begin{pmatrix} \phi(\mathbf{x}_{I_{1,1}}) \\ \vdots \\ \phi(\mathbf{x}_{I_{t-1,n}}) \end{pmatrix}, \quad \mathbf{X}_{t,k} = \begin{pmatrix} \mathbf{X}_t \\ \phi(\mathbf{x}_{I_{t,1}}) \\ \vdots \\ \phi(\mathbf{x}_{I_{t,k-1}}) \end{pmatrix}.$$

The vector $\mathbf{y}_{t,k}$ augments the vector of previous outcomes $\mathbf{y}_t$ by the *expected outcomes* for the already selected images,

$$\mathbf{y}_t = \begin{pmatrix} y_1 \\ \vdots \\ y_{t-1} \end{pmatrix}, \quad \mathbf{y}_{t,k} = \begin{pmatrix} \mathbf{y}_t \\ \hat{y}_{t,1} \\ \vdots \\ \hat{y}_{t,k-1} \end{pmatrix},$$

with $\hat{y}_{t,j} = \hat{y}_{I_{t,j}}$.

The first method presents those n images with the highest upper confidence bounds. However, a drawback of relying only on the upper confidence values is that the similarities of the selected images are not considered. A simple method to avoid this problem is to use just one image for exploration (the second method) — by selecting the image with the maximum upper confidence bound $\hat{y}_I + c\hat{\sigma}_I$ — and selecting the remaining $n-1$ images exploitatively just according to the maximum estimated relevance scores $\hat{y}_I$. Thus this second method does as little exploration as possible, while the first method does the maximal possible exploration. The third method we consider covers the middle ground between these two extreme methods. It selects the images $I_1, \ldots, I_n$ for presentation sequentially, still relying on the upper confidence bounds, but when selecting image I_k it takes into account the exploration already done for images $I_1, \ldots, I_{k-1}$.

4 Multiple Kernel Learning to Update the Feature Space

In multiple kernel learning (MKL) [15,7,2] we would like to solve a classification[4] problem using a combination of kernels to find (a) the weights of each kernel, and (b) the optimal dual weight vector of the combined kernel. The standard MKL framework has been proposed using the SVM and so we will also adopt this algorithm in our system. Given a set of kernels, the main goal of MKL is to

[3] Combining this third method with kernel learning is a bit more involved, since for each selection $I_{t,k}$, $k = 1, \ldots, n$, the kernel learning algorithm needs to be called.

[4] Regression problems can also be solved, but we restrict ourselves to classification.

find a non-uniform weighting of these kernels to help improve the classification task, over and above what could be achieved using a single kernel from the set.

More formally, assume we have a set of kernel functions $\mathcal{K} = \{\kappa_1, \ldots, \kappa_K\}$. Given a vector $\boldsymbol{\eta} = (\eta_1, \ldots, \eta_K)$ of coefficients and kernel functions $\kappa_i(\cdot, \cdot)$ for $i = 1, \ldots, K$, we can define the following linear combination of kernel functions:

$$\kappa_{\boldsymbol{\eta}}(\cdot, \cdot) = \sum_{i=1}^{K} \eta_i \kappa_i(\cdot, \cdot).$$

In our case the kernels correspond to the $K = 11$ feature extraction methods from Table 1 and each component of $\boldsymbol{\eta}$ is the weighting of each of these features. From the previous section we know that the kernelized LINREL requires a kernel function between images. Therefore, after the user is presented with a set of images and classifies them as relevant or not, we will have access to these viewed images and their classifications $\in \{0, 1\}$. Using this information we can have the LINRELMKL algorithm:

$$\mathbf{a}_I(\boldsymbol{\eta}) = \left(\kappa_{\boldsymbol{\eta}}(I, I_1) \; \cdots \; \kappa_{\boldsymbol{\eta}}(I, I_{t-1})\right) \cdot \left(\mathbf{K}_t^{\boldsymbol{\eta}} + \mu \mathbf{I}\right)^{-1},$$

where $I_1, \ldots, I_{t-1}$ are the images selected in iterations $i = 1, \ldots, t-1$ and $\mathbf{K}_t^{\boldsymbol{\eta}}$ is the Gram matrix

$$\mathbf{K}_t^{\boldsymbol{\eta}} = (\kappa_{\boldsymbol{\eta}}(I_i, I_j))_{1 \leq i,j \leq t-1}.$$

Thus $\mathbf{a}_I(\boldsymbol{\eta})$ can be calculated by using the MKL derived kernel function $\kappa_{\boldsymbol{\eta}}(\cdot, \cdot)$. It should be noted that the LINRELMKL algorithm can only learn after feedback has been given. In the CBIR protocol we follow, learning occurs after every page of images has been presented.

We follow an MKL approach recently proposed [13] that looks for a 1-norm 2-norm regularisation of the primal weight vectors in each feature space. The idea is to constrain the 2-norm of each weight vector $\mathbf{w}_k \in \mathbb{R}^m$ (note $\mathbf{w}_k$ is the weight vector in the kth feature space):

$$\min_{\mathbf{w}_k, \boldsymbol{\xi}} \; \lambda \underbrace{\left(\sum_{k=1}^{K} \|\mathbf{w}_k\|_2\right)^2}_{1-\text{norm}} + (1 - \lambda) \underbrace{\sum_{k=1}^{K} \|\mathbf{w}_k\|_2^2}_{2-\text{norm}} + C\|\boldsymbol{\xi}\|_1,$$

where $\lambda \in [0, 1]$ is a parameter, which gives us the standard 1-norm MKL regularisation when $\lambda = 1$ and the 2-norm when $\lambda = 0$, C is the SVM penalty parameter and $\boldsymbol{\xi}$ denotes the slack variables. Any values of λ between 1 and 0 will solve the MKL problem with a regularisation between a 1-norm and 2-norm. The justification of this form of regularisation is that in our CBIR system we expect that early on in the search, all features will need to be active, but as the search continues and more feedback is given we should be able to concentrate on a smaller number of relevant feature spaces (as $\lambda \rightarrow 1$).

We should point out that computing the kernels at each stage of a search page may become cumbersome for a large number of search pages. However, typically a CBIR search may only last 20 search pages, which, in our setting of presenting 15 images per search page (see Section 7) would correspond to kernel matrices of size 300×300. Therefore, the use of MKL together with LinRel in a CBIR system can be practical for the very reason that CBIR searches will never exceed a large number of search pages.

We have now described all of the principle components needed for our CBIR system when it has access to several different feature extraction methods. In the next section we discuss the integration of eye movement features.

5 Tensor Learning for Eye Movements as New Features

In this section we outline the approach taken when eye movement features are also available. We assume that the CBIR system has access to an eye tracking device, which tracks the eye movements of users whilst they view images. We follow the approach outlined in [11]. The idea is to create a tensor kernel between the image and eye movement features, and then to solve a tensor kernel SVM, in order to enrich our image feature space with eye movements. Note that we do not use eye movement features for relevance feedback but as an extra set of features. Our goal is to incorporate the information gained from the eye movements, into a new set of features and combine them with the content-based image features described in Table 1, using the MKL approach of the previous section.

The underlying idea for using tensors is for the creation of rich semantic representation that captures all possible relationships between features of two or more views (sources). This tensor representation creates an implicit correlation space in which we solve our learning problem – of course we must have an existing belief that there is some relationship between the different representations – which in our case is a valid assumption as we extract image and eye movement features of the *same* images.

An explicit mapping of the feature spaces and the tensor computation become computationally infeasible. However, recent papers have shown that the mapping can be made implicitly using the kernel trick [22,17] – by simply taking the dot product between each individual kernel. For instance, let $\phi(\mathbf{x}_I)$ be the vector $\mathbf{x}_I$ mapped using feature mapping ϕ (*i.e.*, image features), and let $\psi(\mathbf{x}_I)$ be the vector $\mathbf{x}_I$ mapped using feature mapping ψ (*i.e.*, eye movement features). Therefore we have:

$$\begin{aligned}\langle \phi(\mathbf{x}_{I_i}) \circ \psi(\mathbf{x}_{I_i}), \phi(\mathbf{x}_{I_j}) \circ \psi(\mathbf{x}_{I_j}) \rangle &= \langle \phi(\mathbf{x}_{I_i}), \phi(\mathbf{x}_{I_j}) \rangle \langle \psi(\mathbf{x}_{I_i}), \psi(\mathbf{x}_{I_j}) \rangle \\ &= \kappa_\phi(I_i, I_j) \kappa_\psi(I_i, I_j),\end{aligned}$$

where $\circ$ denotes the tensor product. Hence, the Gram matrix $\mathbf{K}_\phi$ and $\mathbf{K}_\psi$ of each view is multiplied together using a component-wise product:

$$\mathbf{K}_{\phi\circ\psi} = \mathbf{K}_\phi \cdot \mathbf{K}_\psi,$$

where $\cdot$ denotes the component-wise product. Further to this the kernel matrix $\mathbf{K}_{\phi\circ\psi}$ is used to train a tensor kernel SVM [12] to generate a weight matrix W

(see below). However, this weight matrix is composed of both views, and needs to be decomposed into one weight vector per view (*i.e.*, one for ϕ and one for ψ). This has been resolved by [12] who propose a novel singular value decomposition (SVD) like approach for decomposing the resulting tensor weight matrix into its two component parts, without needing to directly access the feature space. They demonstrate that the weight matrix can be decomposed into a sum of tensor products of corresponding weight components. Hence

$$W \approx W(D) = \sum_{d=1}^{D} \mathbf{w}_\phi^d {\mathbf{w}_\psi^d}^\top,$$

where D is similar to the number of components in SVD and where each $\mathbf{w}_\phi^d, \mathbf{w}_\psi^d$ is a projection vector (like an eigenvector computed in PCA) in the dimension d. Furthermore, from the Representer theorem we know that each $\mathbf{w}_\phi^d, \mathbf{w}_\psi^d$ can be rewritten as a weighted combination of observed examples, such that:

$$\mathbf{w}_\phi^d = \sum_{i=1}^{m} \beta_i^d \phi(\mathbf{x}_{I_i}),$$
$$\mathbf{w}_\psi^d = \sum_{i=1}^{m} \gamma_i^d \psi(\mathbf{x}_{I_i}),$$

where $\boldsymbol{\beta}^d, \boldsymbol{\gamma}^d$ are dual variables (see [12] for decomposition details).

In our CBIR system we do not have the eye movement features for images not yet displayed to the user. Despite this restriction we are still interested in improving the semantic space by using both eye movements and image features during the phase when users have access to them, but then switching to using image features when deciding on which images to present next. Recall that we have access to K feature extraction methods (see Table 1) corresponding to feature maps $\phi_1, \ldots, \phi_K$, and we construct a kernel $\kappa_{\boldsymbol{\eta}}$ using a convex combination of the K feature spaces (see section 4). Following [11] we use the proposed decomposition and create a new feature representation for image features

$$\hat{\phi}(\mathbf{x}_{I_j}) = \left[\sum_{i=1}^{m} \kappa_{\boldsymbol{\eta}}(I_i, I_j) \beta_i^d \right]_{d=1}^{D},$$

where we now have a kernel $\hat{\kappa}_{\boldsymbol{\eta}}(I, I) = \langle \hat{\phi}(\mathbf{x}_I), \hat{\phi}(\mathbf{x}_I) \rangle$ constructed from these new image features $\hat{\phi}(\mathbf{x})$. Hence, the LinRelMKLTensor algorithm is:

$$\hat{\mathbf{a}}_I(\boldsymbol{\eta}) = \left(\hat{\kappa}_{\boldsymbol{\eta}}(I, I_1) \ \cdots \ \hat{\kappa}_{\boldsymbol{\eta}}(I, I_{t-1}) \right) \cdot (\hat{\mathbf{K}}_t^{\boldsymbol{\eta}} + \mu \mathbf{I})^{-1},$$

where $I_1, \ldots, I_{t-1}$ are the images selected in iterations $i = 1, \ldots, t-1$ and $\hat{\mathbf{K}}_t^{\boldsymbol{\eta}}$ is the Gram matrix

$$\hat{\mathbf{K}}_t^{\boldsymbol{\eta}} = (\hat{\kappa}_{\boldsymbol{\eta}}(I_i, I_j))_{1 \leq i,j \leq t-1}.$$

6 The Final CBIR System

Figure 3 describes pictorially the components of our proposed CBIR system. We have the "DATABASE" which contains all images we have access to. The LinRel algorithm has direct access to the images from the database. Furthermore, after presenting some images to the "USER", the system receives feedback on the images that are relevant/non relevant by the user. This feedback is then passed to "MKL" and "Tensor" (when eye movements are available) in order to learn an enriched feature space utilising image and eye movement features as outlined in Sections 4 and 5 (*i.e.*, LinRelMKL and LinRelMKLTensor). This protocol is repeated until the user is satisfied with the search, and retrieved the image(s) they were looking for. In general, any number of images may be presented to the user but in our experiments we will present 15 images per search page.

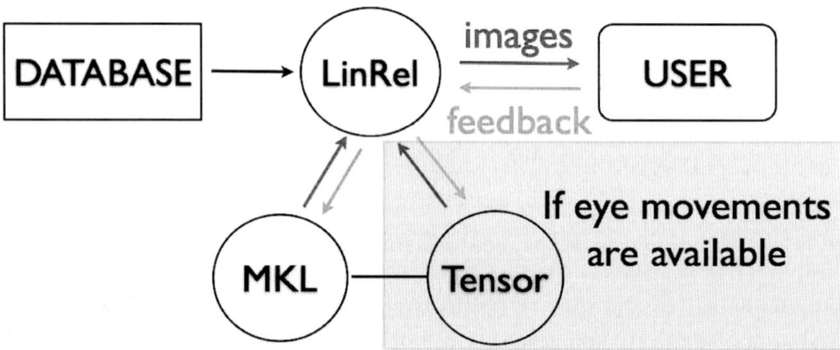

Fig. 3. Our CBIR system: the LinRelMKL algorithm and LinRelMKLTensor when eye movements are available. The red lines indicate features (inputs) and the green lines indicate relevance-feedback given by the user.

When "MKL" is not utilised, but eye movements are available for "Tensor" then we will assume that LinRel receives a sum of uniform weighted kernels *i.e.*, $\boldsymbol{\eta} = (1/K, \ldots, 1/K)$. We call this algorithm LinRelTensor.

7 Experiments

The database used for images was the PASCAL VOC 2007 challenge database [10] which contains over 9000 images, each of which contains at least 1 object from a possible list of 20 objects such as cars, trains, cows, people, *etc.*,. There were $T = 23$ users in the experiments, and participants were asked to view twenty pages, each of which contained fifteen images. Their eye movements were recorded by a Tobii X120 eye tracker [23] connected to a PC using a 19-inch

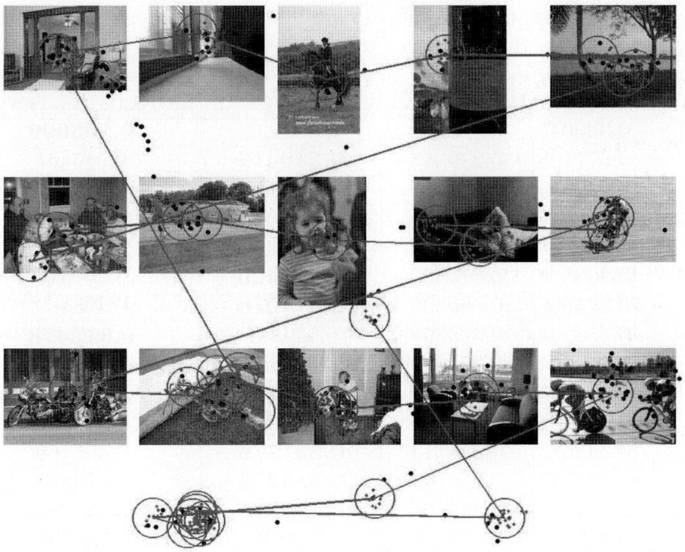

Fig. 4. An example page shown to a participant. Eye movements of the participant are shown in red. The red circles mark fixations and small red dots correspond to raw measurements that belong to those fixations. The black dots mark raw measurements that were not included in any of the fixations.

monitor (resolution of 1280x1024). The eye tracker has approximately 0.5 degrees of accuracy with a sample rate of 120 Hz and used infrared lens to detect pupil centres and corneal reflection. Figure 4 shows an example image presented to a user along with the eye movements (in red).

Each participant was asked to carry out one search task among a possible of three, which included looking for a Bicycle (8 users), a Horse (7 users) or some form of Transport (8 users). Any images within a search page that a user thought contained images from there search, they marked as relevant using mouse clicks.

The eye movement features extracted from the Tobii eye tracker can be found in Table 1 of [16] and was collected by [20]. Most of these are motivated by features considered in the literature for text retrieval studies – however, in addition, they also consider more image-based eye movement features. These are computed for each full image and based on the eye trajectory and locations of the images in the page. They cover the three main types of information typically considered in reading studies: fixations, regressions, and re-fixations. The number of features extracted were 33, computed from: (i) raw measurements from the eye tracker; (ii) fixations estimated from the raw measurements using the standard ClearView fixation filter provided with the Tobii eye tracking software, with settings of: "radius 50 pixels, minimum duration 100 ms".

For each subjects feedback and eye movement data, we train our system to find the most relevant images for the given search task (*i.e.*, Train, Horse or

Table 2. Average Precision results for 23 subjects (users) searching for Bicycle (user 1-8), Horse (user 9-15) and Transport (16-23)

Subject	LinRel	LinRelMKL	LinRelTensor	LinRelMKLTensor
1	0.320045	0.335052	0.498231	**0.505099**
2	0.377128	0.379353	**0.519415**	0.516407
3	0.240924	0.248338	0.409046	**0.415642**
4	0.358075	0.368579	**0.501031**	0.498201
5	0.375263	0.388169	**0.521616**	0.515576
6	0.368568	0.374660	**0.523426**	0.507518
7	0.373300	0.369395	**0.514916**	0.498389
8	0.358244	0.375362	0.514891	**0.512478**
9	0.369601	0.365979	0.614555	**0.624961**
10	0.322203	0.322497	**0.522267**	0.510010
11	0.349775	0.343399	**0.567960**	0.566198
12	0.345682	0.344114	**0.557776**	0.555769
13	0.348333	0.357742	**0.585215**	0.574832
14	0.326431	0.319506	**0.497876**	0.481860
15	0.327132	0.334904	0.525686	**0.530794**
16	0.393561	0.389895	0.616144	**0.624988**
17	0.330865	0.333592	0.578743	**0.592307**
18	0.424255	0.429772	0.649531	**0.658993**
19	0.372035	0.372421	0.602936	**0.621422**
20	0.307158	0.309784	0.560447	**0.580845**
21	0.383495	0.385278	0.611653	**0.627579**
22	0.321837	0.329154	0.531695	**0.563895**
23	0.313631	0.309275	0.568597	**0.577140**
Average	0.348154	0.351575	0.547550	**0.550474**

Transport). We randomly permute the images and choose 15 images initially to present to our CBIR system. Our system then trains the MKL (if used) and tensor kernel SVM (if used) to pass a new kernel to LinRel, which then chooses the next set of 15 images to be presented. This is repeated until all images are presented. We measure our success using Average Precision – in other words, by looking at the number of retrieved images with respect to their rank. For our results we used several random seeds to initialise the first set of images presented to the CBIR system, and averaged over them.

For our experiments we use linear kernels[5] constructed using the features presented in Table 1 and the eye movement features described above. Table 2 presents the Average Precision results, where we used a leave-one-subject-out strategy to optimise the parameters for the LinRel algorithm, for various values[6] of c and μ. Furthermore, we selected method 2 from Subsection 3.1 which seemed to generate the best performance out of the three methods outlined there.

[5] Nonlinear kernels can also be used, but we chose linear kernels for their simplicity and for proof of concept.

[6] $\mu = (10^{-3}, 10^{-2}, 10^{-1}, 10^{0}, 10^{1}, 10^{2}, 10^{3})$, $c = (10^{-3}, 10^{-2}, 10^{-1}, 10^{0}, 10^{1}, 10^{2}, 10^{3})$.

Parameters for all other algorithms were kept fixed. The LinRel column is the kernelized LinRel algorithm using a sum of uniform weighted linear kernels. The LinRelMKL column corresponds to a kernel being learnt for LinRel using MKL.[7] The LinRelTensor column corresponds to no MKL being learnt (hence, a default uniform weighting of kernels) but tensor kernel SVM[8] being trained to find a new feature space. Finally, the LinRelMKLTensor column corresponds to both the MKL and tensor kernel SVM being used to find an appropriate kernel for LinRel.

We can see by comparing the LinRel and LinRelMKL methods that applying MKL can improve the Average Precision. However, when eye movements are available then LinRelTensor and LinRelMKLTensor are clearly superior and help improve the search results, with LinRelMKLTensor being better than LinRelMKL with a statistical significance (Wilcoxon signed test) of $p = 0.05$.

8 Conclusions

We described a novel content-based image retrieval system (CBIR) using relevance-feedback, which views the problem as an exploration-exploitation problem – by using a kernelized version of the LinRel algorithm together with multiple kernel learning (when several different feature extraction methods are available) we demonstrated that we can combine these two algorithms together to learn different feature weightings (LinRelMKL). We also proposed a novel technique of combining the kernel constructed using MKL (on image features) and the kernel constructed using eye movements. This kernel was then trained using the tensor kernel SVM to yield a new feature space combining eye movement and image feature spaces. We called this algorithm LinRelMKLTensor and observed that the search results improved over all other methods (on average). Hence, when eye movement features are available then they should be used in order to improve CBIR systems.

In the current work we only used linear kernels constructed from the image and eye movement features. In future work we plan to use a parameterised family of kernels such as Gaussian kernels. Initial results on Gaussian kernels for the kernelized LinRel algorithm have reported good results. Furthermore, by constructing several different Gaussian kernels per feature extraction method our model would find a weighted combination of important kernels using the MKL component of the system. This would also give us a much larger set of kernels – shown to be effective at improving MKL [6].

Also, we used the discrete values of the feedback received from users to train the MKL and tensor kernel SVM. However, we could of course apply a regression version of MKL and a tensor kernel support vector regression (for instance), by using the real valued relevance estimates provided by users. This would also more

[7] MKL used fixed parameter $\lambda = 0.5$ throughout the experiments, and a penalty parameter $C = 1$.

[8] Penalty parameter $C = 1$.

closely resemble the optimisation problem solved by LinRel, which is itself a linear regression algorithm.

Another line of further research is to use the eye movements for (implicit) relevance feedback. The system we have outlined in this paper *only* uses explicit feedback from mouse clicks to determine the relevance of images. However, by utilising the eye movements we could predict the relevance of images based on what users look at. This would make the CBIR system much more automated and require less explicit interaction by the user. We have carried out preliminary work using our system with such a mechanism in place, and the results are encouraging. We plan to discuss these results in future studies.

Acknowledgements

We would like to thank the anonymous reviewers for their valuable comments. The research leading to these results has received funding from the European Community's Seventh Framework Programme (FP7/2007–2013) under *grant agreement* no. 216529, Personal Information Navigator Adapting Through Viewing, PinView. This work was supported in part by the IST Programme of the European Community, under the PASCAL2 Network of Excellence, IST-2007-216886. This publication only reflects the authors' views.

References

1. Agrawal, R.: Sample mean based index policies with O(log n) regret for the multi-armed bandit problem. Advances in Applied Probability 27(4), 1054–1078 (1995)
2. Argyriou, A., Micchelli, C.A., Pontil, M.: Learning convex combinations of continuously parameterized basic kernels. In: Auer, P., Meir, R. (eds.) COLT 2005. LNCS (LNAI), vol. 3559, pp. 338–352. Springer, Heidelberg (2005)
3. Auer, P.: Using confidence bounds for exploitation-exploration trade-offs. Journal of Machine Learning Research 3, 397–422 (2003)
4. Auer, P., Cesa-Bianchi, N., Fischer, P.: Finite-time analysis of the multiarmed bandit problem. Machine Learning 47(2-3), 235–256 (2002)
5. Auer, P., Leung, A., Hussain, Z., Shawe-Taylor, J.: Report on using side information for exploration-exploitation trade-offs. PinView FP7-216529 Project Deliverable Report D4.2.1 (December 2009)
6. Bach, F.R.: Exploring large feature spaces with hierarchical multiple kernel learning. In: Koller, D., Schuurmans, D., Bengio, Y., Bottou, L. (eds.) Advances in Neural Information Processing Systems, vol. 21, pp. 105–112 (2009)
7. Bach, F.R., Lanckriet, G.R.G., Jordan, M.I.: Multiple kernel learning, conic duality, and the smo algorithm. In: Proceedings of the Twenty-First International Conference on Machine Learning, vol. 6. ACM, New York (2004)
8. Chen, Y., Zhou, X.S., Huang, T.: One-class SVM for learning in image retrieval. In: Proceedings of International Conference on Image Processing 2001, vol. 1, pp. 34–37 (2001)
9. Datta, R., Joshi, D., Li, J., Wang, J.Z.: Image retrieval: Ideas, influences, and trends of the new age. ACM Computing Surveys 40, 5:1–5:60 (2008)

10. Everingham, M., Van Gool, L., Williams, C.K.I., Winn, J., Zisserman, A.: The PASCAL Visual Object Classes Challenge (VOC 2007) (2007), Results, http://www.pascal-network.org/challenges/VOC/voc2007/workshop/index.html
11. Hardoon, D.R., Pasupa, K.: Image ranking with implicit feedback from eye movements. In: Proceedings of ETRA 2010: ACM Symposium on Eye-Tracking Research & Applications, pp. 291–298. ACM, New York (2010)
12. Hardoon, D.R., Shawe-Taylor, J.: Decomposing the tensor kernel support vector machine for neuroscience data with structure labels. Machine Learning Journal: Special Issue on Learning From Multiple Sources 79(1-2), 29–46 (2010)
13. Hussain, Z., Pasupa, K., Saunders, C.J., Shawe-Taylor, J.: Basic metric learning. PinView FP7-216529 Project Deliverable Report D3.1 (December 2008)
14. Laaksonen, J., Viitaniemi, V.: Evaluation of pointer click relevance feedback in picsom. PinView FP7-216529 Project Deliverable Report D1.2 (August 2008)
15. Lanckriet, G.R.G., Cristianini, N., Bartlett, P., Ghaoui, L.E., Jordan, M.I.: Learning the kernel matrix with semidefinite programming. Journal of Machine Learning Research 5, 27–72 (2004)
16. Pasupa, K., Saunders, C., Szedmak, S., Klami, A., Kaski, S., Gunn, S.: Learning to rank images from eye movements. In: HCI 2009: Proceeding of the IEEE 12th International Conference on Computer Vision (ICCV 2009) Workshops on Human-Computer Interaction, pp. 2009–2016 (2009)
17. Pulmannová, S.: Tensor products of hilbert space effect algebras. Reports on Mathematical Physics 53(2), 301–316 (2004)
18. Rocchio, J.: Relevance Feedback in Information Retrieval, pp. 313–323 (1971)
19. Rui, Y., Huang, T.: Optimizing learning in image retrieval. In: Proceedings of the IEEE Conference on Computer Vision and Pattern Recognition, pp. 236–243 (2000)
20. Saunders, C., Klami, A.: Database of eye-movement recordings. Technical Report Project Deliverable Report D8.3, PinView FP7-216529 (2008), http://www.pinview.eu/deliverables.php
21. Shawe-Taylor, J., Cristianini, N.: Kernel Methods for Pattern Analysis. Cambridge University Press, Cambridge (2004)
22. Szedmak, S., Shawe-Taylor, J., Parado-Hernandez, E.: Learning via linear operators: Maximum margin regression; multiclass and multiview learning at one-class complexity. Technical report, University of Southampton (2005)
23. L. Tobii Technology. Tobii Studio Help, http://studiohelp.tobii.com/StudioHelp_1.2/
24. Tong, S., Chang, E.: Support vector machine active learning for image retrieval. In: MULTIMEDIA 2001: Proceedings of the Ninth ACM International Conference on Multimedia, pp. 107–118. ACM, New York (2001)

Graph Regularized Transductive Classification on Heterogeneous Information Networks*

Ming Ji, Yizhou Sun, Marina Danilevsky, Jiawei Han, and Jing Gao

Dept. of Computer Science, University of Illinois at Urbana-Champaign, Urbana, IL
{mingji1,sun22,danilev1,hanj,jinggao3}@illinois.edu

Abstract. A heterogeneous information network is a network composed of multiple types of objects and links. Recently, it has been recognized that strongly-typed heterogeneous information networks are prevalent in the real world. Sometimes, label information is available for some objects. Learning from such labeled and unlabeled data via transductive classification can lead to good knowledge extraction of the hidden network structure. However, although classification on homogeneous networks has been studied for decades, classification on heterogeneous networks has not been explored until recently.

In this paper, we consider the transductive classification problem on heterogeneous networked data which share a common topic. Only some objects in the given network are labeled, and we aim to predict labels for all types of the remaining objects. A novel graph-based regularization framework, GNetMine, is proposed to model the link structure in information networks with arbitrary network schema and arbitrary number of object/link types. Specifically, we explicitly respect the type differences by preserving consistency over each relation graph corresponding to each type of links separately. Efficient computational schemes are then introduced to solve the corresponding optimization problem. Experiments on the DBLP data set show that our algorithm significantly improves the classification accuracy over existing state-of-the-art methods.

1 Introduction

Information networks, composed of large numbers of data objects linking to each other, are ubiquitous in real life. Examples include co-author networks and paper citation networks extracted from bibliographic data, and webpage networks interconnected by hyperlinks in the World Wide Web. Extracting knowledge from such gigantic sets of networked data has recently attracted substantial interest

* Research was sponsored in part by the U.S. National Science Foundation under grant IIS-09-05215, and by the Army Research Laboratory under Cooperative Agreement Number W911NF-09-2-0053 (NS-CTA). The views and conclusions contained in this document are those of the authors and should not be interpreted as representing the official policies, either expressed or implied, of the Army Research Laboratory or the U.S. Government. The U.S. Government is authorized to reproduce and distribute reprints for Government purposes notwithstanding any copyright notation here on.

J.L. Balcázar et al. (Eds.): ECML PKDD 2010, Part I, LNAI 6321, pp. 570–586, 2010.

[11] [15] [16] [19]. Sometimes, label information is available for some data objects. Learning from labeled and unlabeled data is often called semi-supervised learning [22] [21] [3], which aims to classify the unlabeled data based on known information. Classification can help discover the hidden structure of the information network, and give deep insight into understanding different roles played by each object. In fact, applications like research community discovery, fraud detection and product recommendation can all be cast as a classification problem [11] [15]. Generally, classification can be categorized into two groups: (1) transductive classification [10] [11] [22] [21] [19]: to predict labels for the given unlabeled data; and (2) inductive classification [9] [15] [12] [17] [3]: to construct a decision function in the whole data space. In this paper, we focus on transductive classification, which is a common scenario in networked data.

Current studies about transductive classification on networked data [9] [10] [11] [15] mainly focus on homogeneous information networks, i.e., networks composed of a single type of objects, as mentioned above. But in real life, there could be multiple types of objects which form heterogeneous information networks. Beyond co-author networks and citation networks, bibliographic data naturally forms a network among papers, authors, conferences, terms, etc. It has been recognized that heterogeneous information networks, where interconnected links can occur between any two types of objects, are prevalent.

Example 1. Bibliographic Information Network. A bibliographic information network generally contains four types of data objects: *papers*, *authors*, *venues* (conferences and journals) and *terms*. *Papers* and *authors* are linked by the relation of "written by" and "write". *Papers* and *venues* are linked by the relation of "published in" and "publish". *Papers* and *terms* are linked by the relation of "contain" and "contained in". ■

As a natural generalization of classification on homogeneous networked data, we consider the problem of classifying heterogeneous networked data into classes, each of which is composed of multi-typed data sharing a common topic. For instance, a research community in a bibliographic information network contains not only authors, but also papers, venues and terms all belonging to the same research area. Other examples include movie networks in which movies, directors, actors and keywords relate to the same genre, and E-commerce networks where sellers, customers, items and tags belong to the same shopping category.

The general problem of classification has been well studied in the literature. Transductive classification on strongly-typed heterogeneous information networks, however, is much more challenging due to the following characteristics of data:

1. *Complexity of the network structure.* When dealing with the multi-typed network structure in a heterogeneous information network, one common solution is to transform it into a homogenous network and apply traditional classification methods [11] [15]. However, this simple transformation has several drawbacks. For instance, suppose we want to classify *papers* into different research areas. Existing methods would most likely extract a *citation*

network from the whole bibliographic network. Then some valuable discriminative information is likely to be lost (e.g., *authors* of the paper, and the *venue* the paper is published in.) Another solution to make use of the whole network is to ignore the type differences between objects and links. Nevertheless, different types of objects naturally have different data distributions, and different types of links have different semantic meanings, therefore treating them equally is likely to be suboptimal. It has been recognized [8] [16] that while mining heterogeneous information networks, the type differences among links and objects should be respected in order to generate more meaningful results.
2. *Lack of features.* Traditional classification methods usually learn from local features or attributes of the data. However, there is no natural feature representation for all types of networked data. If we transform the link information into features, we will likely generate very high dimensional and sparse data as the number of objects increases. Moreover, even if we have feature representation for some objects in a heterogeneous information network, the features of different types of objects are in different spaces and are hardly comparable. This is another reason why traditional feature-based methods including Support Vector Machines, Naïve Bayes and logistic regression are difficult to apply in heterogeneous information networks.
3. *Lack of labels.* Many classification approaches need a reasonable amount of training examples. However, labels are expensive in many real applications. In a heterogeneous information network, we may even not be able to have a fully labeled subset of all types of objects for training. Label information for some types of objects are easy to obtain while labels for some other types are not. Therefore, a flexible transductive classifier should allow label propagation among different types of objects.

In this paper, we propose a novel graph-based regularization framework to address all three challenges, which simultaneously classifies all of the non-attributed, network-only data with an arbitrary network topology and number of object/link types, just based on the label information of any type(s) of objects and the link structure. By preserving consistency over each relation graph corresponding to each type of links separately, we explicitly respect the type differences in links and objects, thus encoding the typed information in a more organized way than traditional graph-based transductive classification on homogeneous networks.

The rest of the paper is structured as follows. In Section 2, we briefly review the existing work about classification on networked data and graph-based learning. In Section 3, we formally define the problem of transductive classification on heterogeneous information networks. Our graph-based regularization framework (denoted by GNetMine) is introduced in Section 4. Section 5 provides the experimental results. Finally, we conclude this work in Section 6.

2 Related Work

We summarize various transductive classification methods in Table 1, where one dimension represents whether the data has features/attributes or not, and

Table 1. Summary of related work about transductive classification

	Non-networked data	Homogenous networked data	Heterogeneous networked data
Attributed data	SVM, Logistic Regression, etc.	Statistical Relational Learning (Relational Dependency Networks, etc.)	
Non-attributed data	/	Network-only Link-based classifier, Relational Neighbor, etc.	***GNetMine***

the other dimension represents different kinds of network structure: from non-networked data to heterogeneous networked data. Our proposed method works on heterogeneous, non-attributed network-only data, which is the most general case requiring the least amount of information.

Classifying networked data has received substantial attention in recent years. The central idea is to infer the class label from the network structure together with local attributes, if there are any. When classifying webpages or documents, local text features and link information can be combined by using Naïve Bayes [4], logistic regression [9], graph regularization [20], etc. All of these methods assume that the network is homogeneous. Relational dependency networks [12] respect the type differences among relational data when learning the dependency structure by building a conditional model for each variable of interest, but still rely on local features just like other relational learning methods do. Moreover, statistical relational learning usually requires a fully labeled data set for training, which might be difficult to obtain in real applications.

Macskassy et al. [10] propose a relational neighbor classifier on network-only data. Through iteratively classifying an object by the majority class of its neighbors, this method performs very well compared to more complex models including Probabilistic Relational Models [6] [18], Relational Probability Trees [13] and Relational Bayesian Classifiers [14]. Macskassy et al. [11] further emphasize that homogeneousness is very important for their methods to perform within-network classification well.

Recently, there has been a surge of interest in mining heterogeneous information networks [7] [2] [8] [1] . NetClus [16] uses a ranking-clustering mutual enhancement method to generate clusters composed of multi-typed objects. However, clustering does not effectively make use of prior knowledge when it is available. Yin et al. [19] explore social tagging graphs for heterogeneous web object classification. They construct a bipartite graph between tags and web objects to boost classification performance. Nevertheless, they fail to distinguish between different types of links. And their method is confined to the specific network schema between tags and web data, thus cannot be applied to an arbitrary link structure.

Meanwhile, graph-based learning has enjoyed long-lasting popularity in transductive classification. Most of the methods construct an affinity graph over both labeled and unlabeled examples based on local features to encode the similarity between instances. They then design a learner which preserves the smoothness and consistency over the geometrical structure of the data. Zhu et al. [22] formulate the problem using a Gaussian random field model defined with respect to the graph. Zhou et al. [21] propose to let each point iteratively spread its label information to neighbors so as to ensure both local and global consistency. When local features are not available in information networks, graph-based methods can sometimes use the inherent network structure to play the role of the affinity graph. However, traditional graph-based learning mainly works on homogeneous graphs covering all the examples as a whole, and thus cannot distinguish the different semantic meaning of multi-typed links and objects very well. In this paper, we extend the graph-based learning framework to fit the special characteristics of heterogeneous networked data.

3 Problem Definition

In this section, we introduce several related concepts and notations, and then formally define the problem.

Definition 1. Heterogeneous information network. Given m types of data objects, denoted by $\mathcal{X}_1 = \{x_{11}, \ldots, x_{1n_1}\}, \ldots, \mathcal{X}_m = \{x_{m1}, \ldots, x_{mn_m}\}$, a graph $G = \langle V, E, W \rangle$ is called a heterogeneous information network if $V = \bigcup_{i=1}^{m} \mathcal{X}_i$ and $m \geq 2$, E is the set of links between any two data objects of V, and W is the set of weight values on the links. When $m = 1$, G reduces to a homogeneous information network.

Definition 2. Class. Given a heterogeneous information network $G = \langle V, E, W \rangle$, $V = \bigcup_{i=1}^{m} \mathcal{X}_i$, a class is defined as $G' = \langle V', E', W' \rangle$, where $V' \subseteq V$, $E' \subseteq E$. $\forall e = \langle x_{ip}, x_{jq} \rangle \in E'$, $W'_{x_{ip}x_{jq}} = W_{x_{ip}x_{jq}}$. Note here, V' also consists of multiple types of objects from $\mathcal{X}_1$ to $\mathcal{X}_m$.

Definition 2 follows [16]. Notice that a class in a heterogeneous information network is actually a sub-network containing multi-typed objects that are closely related to each other. Now our problem can be formalized as follows.

Definition 3. Transductive classification on heterogeneous information networks. Given a heterogeneous information network $G = \langle V, E, W \rangle$, a subset of data objects $V' \subseteq V = \bigcup_{i=1}^{m} \mathcal{X}_i$, which are labeled with values $\mathcal{Y}$ denoting which class each object belongs to, predict the class labels for all the unlabeled objects $V - V'$.

We design a set of one-versus-all soft classifiers in the multi-class classification task. Suppose the number of classes is K. For any object type $\mathcal{X}_i$, $i \in \{1, \ldots, m\}$, we try to compute a class indicator matrix $\mathbf{F}_i = [\boldsymbol{f}_i^{(1)}, \ldots, \boldsymbol{f}_i^{(K)}] \in \mathbb{R}^{n_i \times K}$, where each $\boldsymbol{f}_i^{(k)} = [f_{i1}^{(k)}, \ldots, f_{in_i}^{(k)}]^T$ measures the confidence that each object $x_{ip} \in \mathcal{X}_i$ belongs to class k. Then we can assign the p-th object in type $\mathcal{X}_i$ to class c_{ip} by finding the maximum value in the p-th row of $\mathbf{F}_i$: $c_{ip} = \arg\max_{1 \leq k \leq K} f_{ip}^{(k)}$.

In a heterogeneous information network, a relation graph $\mathcal{G}_{ij}$ can be built corresponding to each type of link relationships between two types of data objects $\mathcal{X}_i$ and $\mathcal{X}_j$, $i, j \in \{1, \ldots, m\}$. Note that it is possible for $i = j$. Let $\mathbf{R}_{ij}$ be an

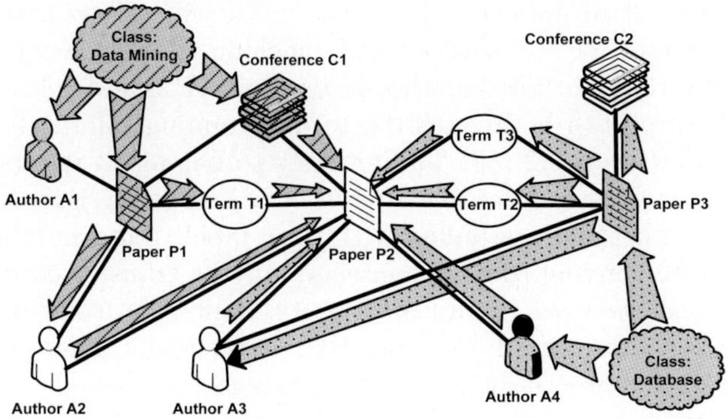

Fig. 1. Knowledge propagation in a bibliographic information network

$n_i \times n_j$ relation matrix corresponding to graph $\mathcal{G}_{ij}$. The element at the p-th row and q-th column of $\mathbf{R}_{ij}$ is denoted as $R_{ij,pq}$, representing the weight on link $\langle x_{ip}, x_{jq} \rangle$. There are many ways to define the weights on the links, which can also incorporate domain knowledge. A simple definition is as follows:

$$R_{ij,pq} = \begin{cases} 1 & \text{if data objects } x_{ip} \text{ and } x_{jq} \text{ are linked together} \\ 0 & \text{otherwise.} \end{cases}$$

Here we consider undirected graphs such that $\mathbf{R}_{ij} = \mathbf{R}_{ji}^T$.

In order to encode label information, we basically set a vector $\mathbf{y}_i^{(k)} = [y_{i1}^{(k)}, \ldots, y_{in_i}^{(k)}]^T \in \mathbb{R}^{n_i}$ for each data object type $\mathcal{X}_i$ such that:

$$y_{ip}^{(k)} = \begin{cases} 1 & \text{if } x_{ip} \text{ is labeled to the } k\text{-th class} \\ 0 & \text{otherwise.} \end{cases}$$

Then for each class $k \in \{1, \ldots, K\}$, our goal is to infer a set of $\boldsymbol{f}_i^{(k)}$ from $\mathbf{R}_{ij}$ and $\mathbf{y}_i^{(k)}$, $i, j \in \{1, \ldots, m\}$.

4 Graph-Based Regularization Framework

In this section, we begin by describing the intuition of our method. Then we formulate the problem using a graph-based regularization framework. Finally, efficient computational schemes are proposed to solve the optimization problem.

4.1 Intuition

Consider a simple bibliographic information network in Figure 1. Four types of objects (*paper*, *author*, *conference* and *term*) are interconnected by multi-typed links (denoted by solid black lines) as described in Example 1. Suppose we want to classify them into research communities. Labeled objects are shaded, whereas the labels of unshaded objects are unknown. Given prior knowledge that author A1, paper P1 and conference C1 belong to the area of data mining, it is easy to infer that author A2 who *wrote* paper P1, and term T1 which is *contained* in P1, are both highly related to data mining. Similarly, author A3, conference C2, and terms T2 and T3 are likely to belong to the database area, since they link directly to a database paper P3. For paper P2, things become more complicated because it is linked with both labeled and unlabeled objects. The confidence of belonging to a certain class may be transferred not only from labeled objects (conference C1 and author A4), but also from unlabeled ones (authors A2 and A3, terms T1, T2 and T3). The classification process can be intuitively viewed as a process of knowledge propagation throughout the network as shown in Figure 1, where the thick shaded arrows indicate possible knowledge flow. The more links between an object x and other objects of class k, the higher the confidence that x belongs to class k. Accordingly, labeled objects serve as the source of prior knowledge. Although this intuition is essentially consistency preserving over the network, which is similar to [10] and [21], the interconnected relationships in heterogeneous information networks are more complex due to the typed information. Knowledge propagation through different types of links contains different semantic meaning, and thus should be considered separately.

In this way, our framework is based on the consistency assumption that the class assignments of two linked objects are likely to be similar. And the class prediction on labeled objects should be similar to their pre-assigned labels. In order to respect the type differences between links and objects, we ensure that such consistency is preserved over each relation graph corresponding to each type of links separately. We formulate our intuition as follows:

1. The estimated confidence measure of two objects x_{ip} and x_{jq} belonging to class k, $f_{ip}^{(k)}$ and $f_{jq}^{(k)}$, should be similar if x_{ip} and x_{jq} are linked together, i.e., the weight value $R_{ij,pq} > 0$.
2. The confidence estimation $\boldsymbol{f}_i^{(k)}$ should be similar to the ground truth, $\mathbf{y}_i^{(k)}$.

4.2 The Algorithm

For each relation matrix $\mathbf{R}_{ij}$, we define a diagonal matrix $\mathbf{D}_{ij}$ of size $n_i \times n_i$. The (p,p)-th element of $\mathbf{D}_{ij}$ is the sum of the p-th row of $\mathbf{R}_{ij}$. Following the above discussion, $\boldsymbol{f}_i^{(k)}$ should be as consistent as possible with the link information and prior knowledge within each relation graph, so we try to minimize the following objective function:

$$J(\boldsymbol{f}_1^{(k)},\ldots,\boldsymbol{f}_m^{(k)}) = \sum_{i,j=1}^{m} \lambda_{ij} \sum_{p=1}^{n_i}\sum_{q=1}^{n_j} R_{ij,pq}\Big(\frac{1}{\sqrt{D_{ij,pp}}} f_{ip}^{(k)} - \frac{1}{\sqrt{D_{ji,qq}}} f_{jq}^{(k)}\Big)^2 + \sum_{i=1}^{m} \alpha_i (\boldsymbol{f}_i^{(k)} - \mathbf{y}_i^{(k)})^T (\boldsymbol{f}_i^{(k)} - \mathbf{y}_i^{(k)}). \quad (1)$$

where $D_{ij,pp}$ is the (p,p)-th element of $\mathbf{D}_{ij}$, and $D_{ji,qq}$ is the (q,q)-th element of $\mathbf{D}_{ji}$. The first term in the objective function (1) is the *smoothness* constraints formulating the first intuition. This term is normalized by $\sqrt{D_{ij,pp}}$ and $\sqrt{D_{ji,qq}}$ in order to reduce the impact of popularity of nodes. In other words, we can, to some extent, suppress popular nodes from dominating the confidence estimations. The normalization technique is adopted in traditional graph-based learning and its effectiveness is well proved [21]. The second term minimizes the difference between the prediction results and the labels, reflecting the second intuition.

The trade-off among different terms is controlled by regularization parameters λ_{ij} and α_i, where $0 \leq \lambda_{ij} < 1$, $0 < \alpha_i < 1$. For $\forall i,j \in \{1,\ldots,m\}$, $\lambda_{ij} > 0$ indicates that object types $\mathcal{X}_i$ and $\mathcal{X}_j$ are linked together and this relationship is taken into consideration. The larger λ_{ij}, the more value is placed on the relationship between object types $\mathcal{X}_i$ and $\mathcal{X}_j$. For example, in a bibliographic information network, if a user believes that the links between *authors* and *papers* are more trustworthy and influential than the links between *conferences* and *papers*, then the λ_{ij} corresponding to the *author-paper* relationship should be set larger than that of *conference-paper*, and the classification results will rely more on the *author-paper* relationship. Similarly, the value of α_i, to some extent, measures how much the user trusts the labels of object type $\mathcal{X}_i$. Similar strategy has been adopted in [8] to control the weights between different types of relations and objects. However, we will show in Section 5 that the parameter setting will not influence the performance of our algorithm dramatically.

To facilitate algorithm derivation, we define the normalized form of $\mathbf{R}_{ij}$:

$$\mathbf{S}_{ij} = \mathbf{D}_{ij}^{(-1/2)}\mathbf{R}_{ij}\mathbf{D}_{ji}^{(-1/2)}, i,j \in \{1,\ldots,m\} \tag{2}$$

With simple algebraic formulations, the first term of (1) can be rewritten as:

$$\begin{aligned}
&\sum_{i,j=1}^{m} \lambda_{ij} \sum_{p=1}^{n_i}\sum_{q=1}^{n_j} R_{ij,pq}\Big(\frac{1}{\sqrt{D_{ij,pp}}}f_{ip}^{(k)} - \frac{1}{\sqrt{D_{ji,qq}}}f_{jq}^{(k)}\Big)^2 \\
&= \sum_{i,j=1}^{m} \lambda_{ij} \sum_{p=1}^{n_i}\sum_{q=1}^{n_j} R_{ij,pq}\Big(\frac{(f_{ip}^{(k)})^2}{D_{ij,pp}} - 2\frac{f_{ip}^{(k)}f_{jq}^{(k)}}{\sqrt{D_{ij,pp}D_{ji,qq}}} + \frac{(f_{jq}^{(k)})^2}{D_{ji,qq}}\Big) \\
&= \sum_{i,j=1}^{m} \lambda_{ij}\Big(\sum_{p=1}^{n_i}(f_{ip}^{(k)})^2 + \sum_{q=1}^{n_j}(f_{jq}^{(k)})^2 - 2\sum_{p=1}^{n_i}\sum_{q=1}^{n_j}(f_{ip}^{(k)}S_{ij,pq}f_{jq}^{(k)})\Big) \\
&= \sum_{i,j=1}^{m} \lambda_{ij}\big((\boldsymbol{f}_i^{(k)})^T\boldsymbol{f}_i^{(k)} + (\boldsymbol{f}_j^{(k)})^T\boldsymbol{f}_j^{(k)} - 2(\boldsymbol{f}_i^{(k)})^T\mathbf{S}_{ij}\boldsymbol{f}_j^{(k)}\big)
\end{aligned} \tag{3}$$

Then we can rewrite (1) in the following form:

$$\begin{aligned}
J(\boldsymbol{f}_1^{(k)},\ldots,\boldsymbol{f}_m^{(k)}) = &\sum_{i,j=1}^{m} \lambda_{ij}\big((\boldsymbol{f}_i^{(k)})^T\boldsymbol{f}_i^{(k)} + (\boldsymbol{f}_j^{(k)})^T\boldsymbol{f}_j^{(k)} - 2(\boldsymbol{f}_i^{(k)})^T\mathbf{S}_{ij}\boldsymbol{f}_j^{(k)}\big) \\
&+\sum_{i=1}^{m}\alpha_i(\boldsymbol{f}_i^{(k)} - \mathbf{y}_i^{(k)})^T(\boldsymbol{f}_i^{(k)} - \mathbf{y}_i^{(k)})
\end{aligned} \tag{4}$$

Connection to homogeneous graph-based learning. Here we first show that the homogenous version of our algorithm is equivalent to the graph-based learning method [21]. Then we show the connection and difference between our algorithm and [21] on heterogeneous information networks.

We first define $\mathbf{L}_{ii} = \mathbf{I}_i - \mathbf{S}_{ii}$, where $\mathbf{I}_i$ is the identity matrix of size $n_i \times n_i$. Note that $\mathbf{L}_{ii}$ is the *normalized graph Laplacian* [5] of the homogeneous sub-network on object type $\mathcal{X}_i$.

Lemma 1. *In homogeneous information networks, the objective function (4) reduces to:*

$$J(\boldsymbol{f}_1^{(k)}) = 2\lambda_{11}(\boldsymbol{f}_1^{(k)})^T\mathbf{L}_{11}\boldsymbol{f}_1^{(k)} + \alpha_1(\boldsymbol{f}_1^{(k)} - \mathbf{y}_1^{(k)})^T(\boldsymbol{f}_1^{(k)} - \mathbf{y}_1^{(k)}) \qquad \blacksquare$$

The proof can be done by simply setting $m = 1$ in function (4). It is easy to see that the homogeneous version of our algorithm is equivalent to the objective function of [21].

When the information network is heterogeneous, we can consider all types of objects as a whole set. We define:

$$\boldsymbol{f}^{(k)} = [(\boldsymbol{f}_1^{(k)})^T, \ldots, (\boldsymbol{f}_m^{(k)})^T]^T \; ; \mathbf{y}^{(k)} = [(\mathbf{y}_1^{(k)})^T, \ldots, (\mathbf{y}_m^{(k)})^T]^T$$
$$\boldsymbol{\alpha}_i = \alpha_i \mathbf{1}_{n_i}, \; i = 1, \ldots, m \; ; \boldsymbol{\alpha} = \text{diag}\{[\boldsymbol{\alpha}_1^T, \ldots, \boldsymbol{\alpha}_m^T]\}$$

where $\mathbf{1}_{n_i}$ is an n_i-dimensional column vector of all ones. We further construct a matrix corresponding to each type of relationship between two different object types $\mathcal{X}_i$ and $\mathcal{X}_j$ as follows:

$$\mathbf{L}_{ij} = \begin{bmatrix} \mathbf{I}_i & -\mathbf{S}_{ij} \\ -\mathbf{S}_{ji} & \mathbf{I}_j \end{bmatrix}, \text{ where } i \neq j$$

Suppose $\sum_{i=1}^m n_i = n$, let $\mathbf{H}_{ij}$ be the $n \times n$ symmetric matrix where each row/column corresponds to an object, with the order the same as that in $\boldsymbol{f}^{(k)}$. The elements of $\mathbf{H}_{ij}$ at rows and columns corresponding to object types $\mathcal{X}_i$ and $\mathcal{X}_j$ are equal to $\mathbf{L}_{ij}$, and all the other elements are 0. This also holds for $i = j$.

Lemma 2. *On heterogeneous information networks, the objective function (4) is equivalent to the following:*

$$J(\boldsymbol{f}_1^{(k)}, \ldots, \boldsymbol{f}_m^{(k)}) = (\boldsymbol{f}^{(k)})^T\mathbf{H}\boldsymbol{f}^{(k)} + \left(\boldsymbol{f}^{(k)} - \mathbf{y}^{(k)}\right)^T\boldsymbol{\alpha}\left(\boldsymbol{f}^{(k)} - \mathbf{y}^{(k)}\right) \qquad (5)$$

where $\mathbf{H} = \sum_{i \neq j} \lambda_{ij}\mathbf{H}_{ij} + 2\sum_{i=1}^m \lambda_{ii}\mathbf{H}_{ii}$. $\blacksquare$

The proof can be done by considering each term in the objective function (4) separately for $i \neq j$ and $i = j$, respectively, and then summing them up. Lemma 2 shows that our proposed GNetMine algorithm has a consistent form with the graph-based learning framework on homogeneous data [21], in which $\mathbf{H}$ is replaced by the *normalized graph Laplacian* $\mathbf{L}$ [5]. Moreover, we respect the different semantic meanings of the multi-typed links by applying graph regularization on each relation graph corresponding to each type of links separately rather than on the whole network. Different regularization parameters λ_{ij} also provide more flexibility in incorporating user preference on how much the relationship between object types $\mathcal{X}_i$ and $\mathcal{X}_j$ is valued among all types of relationships. However, even if all the λ_{ij} are set the same, we can see that $\mathbf{H}$ is different from the *normalized graph Laplacian* $\mathbf{L}$ [5] on the whole network as long as there is at least one type of objects linking to other objects via multiple types of relationships.[1]

Closed form solution. It is easy to check that $\mathbf{L}_{ii}$ is positive semi-definite, and so is $\mathbf{H}_{ii}$. We now show that $\mathbf{L}_{ij}$ is also positive semi-definite.

Proof. Recall that $D_{ij,pp} = \sum_{q=1}^{n_j} R_{ij,pq}$ and $\mathbf{R}_{ij} = \mathbf{R}_{ji}^T$, we define:

$$\widehat{\mathbf{L}}_{ij} = \begin{bmatrix} \mathbf{D}_{ij} & -\mathbf{R}_{ij} \\ -\mathbf{R}_{ji} & \mathbf{D}_{ji} \end{bmatrix} = \begin{bmatrix} \mathbf{D}_{ij} & \mathbf{0} \\ \mathbf{0} & \mathbf{D}_{ji} \end{bmatrix} - \begin{bmatrix} \mathbf{0} & \mathbf{R}_{ij} \\ \mathbf{R}_{ji} & \mathbf{0} \end{bmatrix} = \widehat{\mathbf{D}} - \widehat{\mathbf{W}}$$

It can be observed that $\widehat{\mathbf{L}}_{ij}$ has the same form as the *graph Laplacian* [5], where $\widehat{\mathbf{D}}$ is a diagonal matrix whose entries are column (or row, since $\widehat{\mathbf{W}}$ is symmetric) sums of $\widehat{\mathbf{W}}$. So $\widehat{\mathbf{L}}_{ij}$ is positive semi-definite. Hence

$$\mathbf{L}_{ij} = \begin{bmatrix} \mathbf{D}_{ij} & \mathbf{0} \\ \mathbf{0} & \mathbf{D}_{ji} \end{bmatrix}^{-1/2} \widehat{\mathbf{L}}_{ij} \begin{bmatrix} \mathbf{D}_{ij} & \mathbf{0} \\ \mathbf{0} & \mathbf{D}_{ji} \end{bmatrix}^{-1/2}$$

is positive semi-definite.

In this way, $\mathbf{H}_{ij}$ is positive semi-definite. We further check the Hessian matrix of the objective function (4), which is easy to derive from equation (5):

$$H\big(J(\boldsymbol{f}_1^{(k)}, \ldots, \boldsymbol{f}_m^{(k)})\big) = 2\mathbf{H} + 2\boldsymbol{\alpha}$$

$\mathbf{H}$ is the weighted summation of $\mathbf{H}_{ii}$ and $\mathbf{H}_{ij}$, which is also positive semi-definite. Since $\alpha_i > 0$ for all i, we conclude that $H\big(J(\boldsymbol{f}_1^{(k)}, \ldots, \boldsymbol{f}_m^{(k)})\big)$ is positive definite. Therefore, the objective function (4) is strictly convex. The unique global minimum is obtained by differentiating (4) with respect to each $(\boldsymbol{f}_i^{(k)})^T$:

$$\frac{\partial J}{\partial (\boldsymbol{f}_i^{(k)})^T} = \sum_{j=1, j\neq i}^{m} \lambda_{ij}(2\boldsymbol{f}_i^{(k)} - 2\mathbf{S}_{ij}\boldsymbol{f}_j^{(k)}) + 4\lambda_{ii}\mathbf{L}_{ii}\boldsymbol{f}_i^{(k)} + 2\alpha_i(\boldsymbol{f}_i^{(k)} - \mathbf{y}_i^{(k)}) \quad (6)$$

and letting $\frac{\partial J}{\partial (\boldsymbol{f}_i^{(k)})^T} = 0$ for all i.

Finally, we give the closed form solution by solving the following linear equation system:

$$\boldsymbol{f}_i^{(k)} = \Big(\big(\sum_{j=1, j\neq i}^{m} \lambda_{ij} + \alpha_i\big)\mathbf{I}_i + 2\lambda_{ii}\mathbf{L}_{ii}\Big)^{-1}\Big(\alpha_i \mathbf{y}_i^{(k)} + \sum_{j=1, j\neq i}^{m} \lambda_{ij}\mathbf{S}_{ij}\boldsymbol{f}_j^{(k)}\Big), \ i \in \{1, \ldots, m\}$$

It can be proven that $\Big((\sum_{j=1, j\neq i}^{m} \lambda_{ij} + \alpha_i)\mathbf{I}_i + 2\lambda_{ii}\mathbf{L}_{ii}\Big)$ is positive definite and invertible.

[1] If a network has only two types of objects $\mathcal{X}_1$ and $\mathcal{X}_2$, and only one type of relationship $\mathbf{R}_{12}$, then $\mathbf{H}$ reduces to $\lambda_{12}\mathbf{L}$.

Iterative solution. Though the closed form solution is obtained, sometimes the iterative solution is preferable. Based on equation (6), we derive the iterative form of our algorithm as follows:

- Step 0: For $\forall k \in \{1, \ldots, K\}$, $\forall i \in \{1, \ldots, m\}$, initialize confidence estimates $\boldsymbol{f}_i^{(k)}(0) = \mathbf{y}_i^{(k)}$ and $t = 0$.
- Step 1: Based on the current $\boldsymbol{f}_i^{(k)}(t)$, compute:

$$\boldsymbol{f}_i^{(k)}(t+1) = \frac{\sum_{j=1, j\neq i}^{m} \lambda_{ij} \mathbf{S}_{ij} \boldsymbol{f}_j^{(k)}(t) + 2\lambda_{ii} \mathbf{S}_{ii} \boldsymbol{f}_i^{(k)}(t) + \alpha_i \mathbf{y}_i^{(k)}}{\sum_{j=1, j\neq i}^{m} \lambda_{ij} + 2\lambda_{ii} + \alpha_i}$$

  for $\forall k \in \{1, \ldots, K\}$, $\forall i \in \{1, \ldots, m\}$.
- Step 2: Repeat step 1 with $t = t + 1$ until convergence, i.e., until $\boldsymbol{f}_i^{(k)*} = \boldsymbol{f}_i^{(k)}(t)$ do not change much for all i.
- Step 3: For each $i \in \{1, \ldots, m\}$, assign the class label to the p-th object of type $\mathcal{X}_i$ as $c_{ip} = \arg\max_{1\leq k\leq K} f_{ip}^{(k)*}$, where $\boldsymbol{f}_i^{(k)*} = [f_{i1}^{(k)*}, \ldots, f_{in_i}^{(k)*}]^T$.

Following an analysis similar to [21], the iterative algorithm can be proven to converge to the closed form solution. The iterative solution can be viewed as a natural extension of [21], where each object iteratively spreads label information to its neighbors until a global stable state is achieved. At the same time, we explicitly distinguish the semantic differences between the multi-typed links and objects by employing different normalized relation graphs corresponding to each type of links separately rather than a single graph covering all the instances.

4.3 Time Complexity Analysis

We analyze the computational complexity of the iterative solution here. Step 0 takes $O(K|V|)$ time for initialization, where K is the number of classes and $|V|$ the total number of objects. At each iteration of step 1, we need to process each link twice, once for the object at each end of the link. And we need $O(K|V|)$ time to incorporate label information in $\alpha_i \mathbf{y}_i^{(k)}$. So the time for each iteration is $O(K(|E| + |V|))$, where $|E|$ is the total number of links in the information network. Finally, it takes $O(K|V|)$ time to compute the class prediction result in step 3. Hence the total time complexity of the iterative algorithm is $O\big(NK(|E|+|V|)\big)$, where N is the number of iterations.

The time complexity of the closed form solution is dependent on the particular network structure. We omit the analysis due to space limitation. In general, the iterative solution is more computationally efficient because it bypasses the matrix inversion operation.

After all, the classification task is done offline, where all the objects can be classified once and the results stored for future querying.

5 Experimental Results

In this section, we present an empirical study of the effectiveness of our graph-based regularization framework for transductive classification (denoted by

GNetMine) on the real heterogeneous information network of DBLP[2]. As discussed before, we try to classify the bibliographic data into research communities, each of which contains multi-typed objects all closely related to the same area.

5.1 Data Set

We extract a sub-network of the DBLP data set on four areas: database, data mining, information retrieval and artificial intelligence, which naturally form four classes. By selecting five representative conferences in each area, papers published in these conferences, the authors of these papers and the terms that appeared in the titles of these papers, we obtain a heterogeneous information network that consists of four types of objects: *paper, conference, author* and *term*. Within that heterogeneous information network, we have three types of link relationships: *paper-conference, paper-author*, and *paper-term*. The data set we used contains 14376 papers, 20 conferences, 14475 authors and 8920 terms, with a total number of 170794 links[3]. By using our GNetMine algorithm, we can simultaneously classify all types of objects regardless of how many types of objects we labeled.

For accuracy evaluation, we use a labeled data set of 4057 authors, 100 papers and all 20 conferences. For more details about the labeled data set, please refer to [7] [16]. In the following sections, we randomly choose a subset of labeled objects and use their label information as prior knowledge. The classification accuracy is evaluated by comparing with manually labeled results on the rest of the labeled objects. Since terms are difficult to label even manually, i.e., many terms are closely related to multiple areas, we did not evaluate the accuracy on terms here.

5.2 Algorithms for Comparison

We compare GNetMine with the following state-of-the-art algorithms:

- Learning with Local and Global Consistency (LLGC) [21]
- Weighted-vote Relational Neighbor classifier (wvRN) [10] [11]
- Network-only Link-based classification (nLB) [9] [11]

LLGC is a graph-based transductive classification algorithm, which is also the homogenous reduction of GNetMine if we use the intrinsic network structure to play the role of the affinity graph. Weighted-vote relational neighbor classifier and link-based classification are two popular classification algorithms on networked data. Since local attributes/features are not available in our problem, we use the network-only derivative of the link-based classifier (nLB). Following [11], nLB creates a feature vector for each node based on neighboring information.

---

[2] http://www.informatik.uni-trier.de/~ley/db/

[3] The data set is available at www.cs.illinois.edu/homes/mingji1/DBLP_four_area.zip for sharing and experiment repeatability.

Note that none of the algorithms above can be directly applied to heterogeneous information networks. In order to make all the algorithms comparable, we can transform a heterogenous information network into a homogeneous one in two ways: (1) disregard the type differences between objects and treat all of them as the same type; or (2) extract a homogeneous sub-network on one single type of objects, if that object type is partially labeled. We try both approaches in the accuracy study. The open-source implementation of NetKit-SRL[4] [11] is employed in our experiments.

5.3 Accuracy Study

In this experiment, we choose labels on both *authors* and *papers* to test the classification accuracy. In order to address the label scarcity problem, we randomly choose $(a\%, p\%) = [(0.1\%, 0.1\%), (0.2\%, 0.2\%), \ldots, (0.5\%, 0.5\%)]$ of authors and papers, and use their label information for transductive classification. For each given $(a\%, p\%)$, we average the results over 10 random selections. Note that the very small percentage of labeled objects here are likely to be disconnected, so we may not even be able to extract a fully labeled sub-network for training, making many state-of-the-art algorithms inapplicable.

Since the homogeneous LLGC algorithm just has one α and one λ, only the ratio $\frac{\alpha}{\lambda}$ matters in the model selection. The $\frac{\alpha}{\lambda}$ is set by searching the grid $\{0.01, 0.05, 0.1, 0.5, 1, 5, 10\}$, where the best results are obtained by $\frac{\alpha}{\lambda} = 0.5$. For GNetMine, we do not treat any object/link type as particularly important here and use the same set of parameters as LLGC, i.e., $\alpha_i = 0.1$, $\lambda_{ij} = 0.2$, $\forall i, j \in \{1, \ldots, m\}$. This may not be the best choice, but it is good enough to show the effectiveness of GNetMine. As label information is given on authors and papers, the results on conferences of wvRN, nLB and LLGC can only be obtained by disregarding the type differences between objects and links, denoted by (A-C-P-T). While classifying authors and papers, we also tried constructing homogeneous *author-author* (A-A) and *paper-paper* (P-P) sub-networks in different ways, where the best results presented for authors are given by the *co-author* network, and the best results for papers are generated by linking two papers if they are published in the same conference. We show the classification accuracy on authors, papers and conferences in Tables 2, 3 and 4, respectively.

When classifying authors and papers, it is interesting to notice that the performances of wvRN and nLB on the *author-author* and *paper-paper* sub-networks are better than working on the whole heterogeneous information network, verifying the importance of working with homogeneous data for such homogeneous relational classifiers. However, the transformation from the original heterogeneous network to the homogeneous sub-network causes some information loss, as discussed before. And only one type of label information can be used in the homogeneous sub-network, even if the prior knowledge of another type of objects is available.

When the entire heterogeneous information network (A-C-P-T) is taken into consideration, the task actually becomes more challenging, since the total

[4] http://www.research.rutgers.edu/~sofmac/NetKit.html

Table 2. Comparison of classification accuracy on authors (%)

$(a\%, p\%)$ of authors and papers labeled	nLB (A-A)	nLB (A-C-P-T)	wvRN (A-A)	wvRN (A-C-P-T)	LLGC (A-A)	LLGC (A-C-P-T)	GNetMine (A-C-P-T)
(0.1%, 0.1%)	25.4	26.0	40.8	34.1	41.4	61.3	**82.9**
(0.2%, 0.2%)	28.3	26.0	46.0	41.2	44.7	62.2	**83.4**
(0.3%, 0.3%)	28.4	27.4	48.6	42.5	48.8	65.7	**86.7**
(0.4%, 0.4%)	30.7	26.7	46.3	45.6	48.7	66.0	**87.2**
(0.5%, 0.5%)	29.8	27.3	49.0	51.4	50.6	68.9	**87.5**

Table 3. Comparison of classification accuracy on papers (%)

$(a\%, p\%)$ of authors and papers labeled	nLB (P-P)	nLB (A-C-P-T)	wvRN (P-P)	wvRN (A-C-P-T)	LLGC (P-P)	LLGC (A-C-P-T)	GNetMine (A-C-P-T)
(0.1%, 0.1%)	49.8	31.5	62.0	42.0	67.2	62.7	**79.2**
(0.2%, 0.2%)	73.1	40.3	71.7	49.7	72.8	65.5	**83.5**
(0.3%, 0.3%)	77.9	35.4	77.9	54.3	76.8	66.6	**83.2**
(0.4%, 0.4%)	79.1	38.6	78.1	54.4	77.9	70.5	**83.7**
(0.5%, 0.5%)	80.7	39.3	77.9	53.5	79.0	73.5	**84.1**

Table 4. Comparison of classification accuracy on conferences (%)

$(a\%, p\%)$ of authors and papers labeled	nLB (A-C-P-T)	wvRN (A-C-P-T)	LLGC (A-C-P-T)	GNetMine (A-C-P-T)
(0.1%, 0.1%)	25.5	43.5	79.0	**81.0**
(0.2%, 0.2%)	22.5	56.0	83.5	**85.0**
(0.3%, 0.3%)	25.0	59.0	**87.0**	**87.0**
(0.4%, 0.4%)	25.0	57.0	86.5	**89.5**
(0.5%, 0.5%)	25.0	68.0	90.0	**94.0**

number of objects rises to 14376 $(papers)$+20 $(conferences)$+14475 $(authors)$+ 8920 $(terms)$ = 37791, out of which at most (14376 $(papers)$+14475 $(authors)$)× 0.5%/37791 = 0.4% objects are labeled. Similar results have been reported [11] that when the percentage of labeled objects is less than 20%, the classification accuracy can drop below random guess (here 25%). Therefore, wvRN and nLB perform less well due to the lack of labels. And increasing the label ratio from 0.1% to 0.5% does not make a big difference in improving the accuracy of nLB.

Overall, GNetMine performs the best on all types of objects via learning from labeled authors and papers. Even though the parameters for all types of objects and links are set to the same values, GNetMine still outperforms its homogeneous reduction, LLGC, by preserving consistency on each subgraph corresponding to each type of links separately and minimizing the aggregated error, thus modeling the heterogenous network structure in a more organized way.

5.4 Model Selection

The α_i's and λ_{ij}'s are essential parameters in GNetMine which control the relative importance of different terms. We empirically set all the α_i's as 0.1, and all the λ_{ij}'s as 0.2 in the previous experiment. In this subsection, we try to study the impact of parameters on the performance of GNetMine. Since labels are given on authors and papers, the α_i associated with authors (denoted by α_a) and papers (denoted by α_p), as well as the λ_{ij} associated with the *author-paper* relationship (denoted by λ_{pa}) are empirically more important than other parameters. So we fix all the other parameters and let α_a, α_p and λ_{pa} vary. We also change α and λ in LLGC accordingly. Figure 2 shows the average classification accuracy on three types of objects (author, paper, conference) as a function of the parameters, with $(a\%, p\%) = (0.5\%, 0.5\%)$ authors and papers labeled.

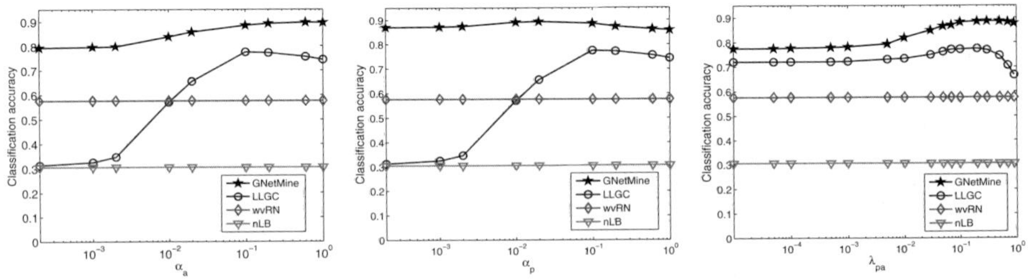

(a) Varying α_a (b) Varying α_p (c) Varying λ_{pa}

Fig. 2. Model Selection when (0.5%, 0.5%) of authors and papers are labeled

It can be observed that over a large range of parameters, GNetMine achieves significantly better performance than all the other algorithms, including its homogeneous reduction, LLGC, with the parameters varying the same way. It is interesting to note that the accuracy curve of α_a is different from that of α_p, indicating that authors and papers do play different roles in the classification process. Generally, the performance of GNetMine with varying α_p is more stable than that with varying α_a. From the accuracy curve of λ_{pa}, it can be seen that setting λ_{pa} larger than all other λ_{ij}'s (which are set to 0.2) improves the accuracy. This is because that increasing λ_{pa} enhances the knowledge propagation between the two types of labeled data, which is beneficial.

Overall, the parameter selection will not critically affect the performance of GNetMine. And if the user has some knowledge about the importance of certain types of links, the parameters can be adjusted accordingly to model the special characteristics of the network.

6 Conclusions

In this paper, we develop a novel graph-based regularization framework to address the transductive classification problem on heterogeneous information networks. We propose that different types of objects and links should be treated separately due to different semantic meanings, which is then proved by both theory and practice. By applying graph regularization to preserve consistency over each relation graph corresponding to each type of links separately and minimizing the aggregated error, we make full use of the multi-typed link information to predict the class label for each object. In this way, our framework can be generally applied to heterogeneous information networks with an arbitrary schema consisting of a number of object/link types. Experiments on the real DBLP data set illustrate the superiority of our method over existing algorithms.

The presented framework classifies the unlabeled data by labeling some randomly selected objects. However, the quality of labels can significantly influence the classification results, as observed in many past studies. In the future, we plan to automatically detect the most informative objects, which can lead to better classification quality if they are labeled. Objects that will potentially have high ranks or lie in the centrality of sub-networks might be good candidates.

References

1. Banerjee, A., Basu, S., Merugu, S.: Multi-way clustering on relation graphs. In: SDM 2007 (2007)
2. Bekkerman, R., El-Yaniv, R., McCallum, A.: Multi-way distributional clustering via pairwise interactions. In: ICML 2005, pp. 41–48 (2005)
3. Belkin, M., Niyogi, P., Sindhwani, V.: Manifold regularization: A geometric framework for learning from examples. J. Mach. Learn. Res. 7, 2399–2434 (2006)
4. Chakrabarti, S., Dom, B., Indyk, P.: Enhanced hypertext categorization using hyperlinks. In: SIGMOD 1998, pp. 307–318. ACM, New York (1998)
5. Chung, F.R.K.: Spectral Graph Theory. Regional Conference Series in Mathematics, vol. 92. AMS, Providence (1997)
6. Friedman, N., Getoor, L., Koller, D., Pfeffer, A.: Learning probabilistic relational models. In: IJCAI 1999 (1999)
7. Gao, J., Liang, F., Fan, W., Sun, Y., Han, J.: Graph-based consensus maximization among multiple supervised and unsupervised models. In: Advances in Neural Information Processing Systems (NIPS), vol. 22, pp. 585–593 (2009)
8. Long, B., Zhang, Z.M., Wu, X., Yu, P.S.: Spectral clustering for multi-type relational data. In: ICML 2006, pp. 585–592 (2006)
9. Lu, Q., Getoor, L.: Link-based classification. In: ICML 2003 (2003)
10. Macskassy, S.A., Provost, F.: A simple relational classifier. In: Proc. of MRDM-2003 at KDD-2003, pp. 64–76 (2003)
11. Macskassy, S.A., Provost, F.: Classification in networked data: A toolkit and a univariate case study. J. Mach. Learn. Res. 8, 935–983 (2007)
12. Neville, J., Jensen, D.: Relational dependency networks. J. Mach. Learn. Res. 8, 653–692 (2007)
13. Neville, J., Jensen, D., Friedland, L., Hay, M.: Learning relational probability trees. In: KDD 2003, pp. 625–630 (2003)
14. Neville, J., Jensen, D., Gallagher, B.: Simple estimators for relational bayesian classifiers. In: ICDM 2003, p. 609 (2003)

15. Sen, P., Getoor, L.: Link-based classification. Technical Report CS-TR-4858, University of Maryland (February 2007)
16. Sun, Y., Yu, Y., Han, J.: Ranking-based clustering of heterogeneous information networks with star network schema. In: KDD 2009, pp. 797–806 (2009)
17. Taskar, B., Abbeel, P., Koller, D.: Discriminative probabilistic models for relational data. In: UAI, pp. 485–492 (2002)
18. Taskar, B., Segal, E., Koller, D.: Probabilistic classification and clustering in relational data. In: IJCAI 2001, pp. 870–876 (2001)
19. Yin, Z., Li, R., Mei, Q., Han, J.: Exploring social tagging graph for web object classification. In: KDD 2009, pp. 957–966 (2009)
20. Zhang, T., Popescul, A., Dom, B.: Linear prediction models with graph regularization for web-page categorization. In: KDD 2006, pp. 821–826 (2006)
21. Zhou, D., Bousquet, O., Lal, T.N., Weston, J., Schölkopf, B.: Learning with local and global consistency. In: NIPS 16 (2003)
22. Zhu, X., Ghahramani, Z., Lafferty, J.D.: Semi-supervised learning using gaussian fields and harmonic functions. In: ICML 2003, pp. 912–919 (2003)

Temporal Maximum Margin Markov Network

Xiaoqian Jiang[1], Bing Dong[2], and Latanya Sweeney[1]

[1] Data Privacy Lab, School of Computer Science
[2] Center for Building Performance and Diagnostics, School of Architecture
Carnegie Mellon University

Abstract. Typical structured learning models consist of a regression component of the explanatory variables (observations) and another regression component that accounts for the neighboring states. Such models, including Conditional Random Fields (CRFs) and Maximum Margin Markov Network (M3N), are essentially Markov random fields with the pairwise spatial dependence. They are effective tools for modeling spatial correlated responses; however, ignoring the temporal correlation often limits their performance to model the more complex scenarios. In this paper, we introduce a novel Temporal Maximum Margin Markov Network (TM3N) model to learn the spatial-temporal correlated hidden states, simultaneously. For learning, we estimate the model's parameters by leveraging on loopy belief propagation (LBP); for predicting, we forecast hidden states use linear integer programming (LIP); for evaluation, we apply TM3N to the simulated datasets and the real world challenge for occupancy estimation. The results are compared with other state-of-the-art models and demonstrate superior performance.

1 Introduction

Traditional Markov random field models concentrate on either the spatial dependence or the temporal correlation. Lafferty et al. [6] developed a statistical framework, Conditional Random Fields (CRFs), which accounts for spatial dependence, in addition to the explanatory variables (observations). Later, Taskar [14] extended the Support Vector Machine (SVM) to the Maximum Margin Markov Network (M3N), which has the same modeling capacity of the CRFs but can be computed more efficiently. Similar models considering spatial dependence include, the Structured SVM [15] and the Maximum Margin Training [12]. All of these models aim to combine spatial dependence and the information from observations for a single end task, multivariate classification. They have been successfully applied to problems like optical character recognition [10], object detection [1] and scene understanding [4]. However, these models overlook the state correlations over time, hence, are insufficient to handle data with strong temporal pattern.

On the other hand, temporal correlated models has been developed over the decades, models including Kalman filter [5], HMM [17] have been carefully studied by the optimization and control community. Successful applications including time series forecasting [3], speech recognition [11] and behavior classification

J.L. Balcázar et al. (Eds.): ECML PKDD 2010, Part I, LNAI 6321, pp. 587–600, 2010.

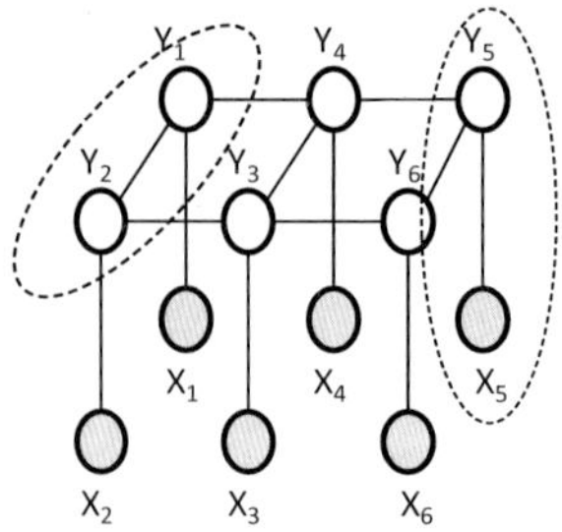

Fig. 1. Graphical model of CRFs and M3N. X_i and Y_i correspond to the local observations and their labels. Two dashed ovals encompass $[X_5, Y_5]$ and $[Y_1, Y_2]$, which correspond to a unary feature and a pairwise Markovian spatial feature, respectively.

[16]. These models are well known for their capability of capturing hidden temporal correlations; modeling the unknown state process from observations made in noisy environments. However, they ignore the structural correlations in the environment, which oftentimes hurt their performance.

Clearly, both temporal correlated models and spatial dependent models have limitations. [9] thus advocated a variational inference method for switching Linear Dynamical system (SLDS) that learns different dynamical processes at various time ticks; [7] extended this work to combine HMM and LDS with tractable computation. However, these methods treat temporal and spatial (structural) information once at a time; they fail to provide a comprehensive interface to model the temporal and spatial correlated real-world scenarios.

To close the gap, we propose a novel model that considers spatial correlations aggregated over time for tractable inference. The proposed model has advantages over models concentrating on either aspect, as the temporal and structural information are oftentimes complementary. We intend to provide a principled approach which accounts for spatial dependence and temporal correlations, simultaneously.

The remaining of the paper is organized as follows. In Section 2, we review the spatial-dependent structured learning framework. In Section 3, we suggest the spatial-temporal correlated framework extending the existing works. In Section 4, we instantiate the framework to propose a novel model: Temporal Maximum Margin Markov Network (TM3N). In Section 5, we introduce algorithms for estimating model parameters, leveraging on loopy belief propagation. In Section 6, we propose a linear integer programming interface for predicting hidden states. In Section 7, the TM3N model is applied to both the simulated datasets and a real world building occupancy estimation challenge. We compare the results with other state of the arts methods. Finally, in Section 8, we conclude the paper.

2 Overview

2.1 Notation

We summarize the basic notation of this paper in the following table.

Table 1. Summary of the notation

Variables	Summary
$X_{k,i,t}$	The k dimensional feature at site i in time tick t.
$Y_{i,t}$	The discrete valued state at site i in time tick t.
$\mathbf{Y}_t = (Y_{1,t}, \ldots, Y_{n,t})'$	The estimated states at time tick t.
$\hat{\mathbf{Y}}_t = (\hat{Y}_{1,t}, \ldots, \hat{Y}_{n,t})'$	The ground-truth states at time tick t.
$\theta_k^1, \theta_{ij,t}^2$ and $\theta_{it,t-1}^3$	The unary, spatial and temporal regression coefficients.
$\varphi(\cdot)$	The feature function.
$\ell_t(\mathbf{Y}_t)$	The loss at time t.
$h_\theta(\mathbf{X}_t)$	The state estimation function.

2.2 Backgrounds

We first summarize the framework that encompasses spatial dependent models on a regular or irregular lattice [6, 13]. Define $s_1, \ldots, s_n$ to be the sites on a spatial lattice. For notational convenience, let $j \sim i$ denote $j \in N_i$, where N_i=$\{j$: s_j is a neighbor of $s_i\}$ defines the neighbors of the site s_i. Let $Y_1, \ldots, Y_n$ denote hidden states on the lattice, where $Y_i = Y(s_i) \in (1, \ldots, C)$, and C is the number of classes. The joint distribution of $\mathbf{Y} = (Y_1, \ldots, Y_n)'$ can be formulated as:

$$p_\theta(\mathbf{X}, \mathbf{Y}) \propto \exp\left\{\sum_{i=1}^{n}\sum_{k=1}^{p} \theta_k \varphi(X_{k,i}, Y_i) + \sum_{j \sim i} \theta_{ij} \varphi(Y_i, Y_j)\right\}, \tag{1}$$

$$p_\theta(\mathbf{Y}|\mathbf{X}) = \frac{1}{Z_\theta(\mathbf{X})} \exp\left\{\sum_{i=1}^{n}\sum_{k=1}^{p} \theta_k \varphi(X_{k,i}, Y_i) + \sum_{j \sim i} \theta_{ij} \varphi(Y_i, Y_j)\right\}, \tag{2}$$

where

$$Z_\theta(\mathbf{X}) = \sum_{Y} \exp\{\sum_{i=1}^{n}\sum_{k=1}^{p} \theta_k \varphi(X_{k,i}, Y_i) + \sum_{j \sim i} \theta_{ij} \varphi(Y_i, Y_j)\} \tag{3}$$

is called the partition function; $X_{k,i} = X_k(s_i)$ denotes the k-th explanatory variable at site s_i; θ_k denotes the k-th regression coefficients correspond to the feature function $\varphi(X_{k,i}, Y_i)$, with $k = 1, \ldots, p$; θ_{ij} denotes the spatial-dependent regression coefficients for the i-th and j-th sites so that $\theta_{ij} = \theta_{ji}$ and $\theta_{ij} \geq 0$ if $j \sim i$.

Such model relates a discrete valued response variable to a hidden state by two regression components; and it is capable of estimating the probability of hidden states at a given site; and predicting a certain outcome at an unsampled site. However, this formulation ignores the fact that observations are oftentimes made repeatedly over time and past states on the same spatial lattice may contribute to the states in a future time tick. That is, for a given location s_i at a given time tick t, the state is $Y(s_i,t) = Y_{i,t} \perp (Y_{i,t-1} \cup \{Y_{j,t}\}_{j\sim N_i})$, where $i = 1,\ldots,n$ and $t = 1, 2, \ldots$. To close the gap, we will extend the model to include temporal correlations.

3 Spatial-Temporal Structured Model

We generalize the previous framework to include an additional temporal component. With the additional regression term, the new framework is capable of modeling: information carried by observations, spatial dependence at fixed time tick, and temporal correlations of the hidden states.

Consider a discrete valued spatial-temporal process $\{Y_{i,t} : i = 1,\ldots,n, t = 1, 2, \ldots\}$, where $Y_{i,t} = Y(s_i,t) \in (1,\ldots,C)$ corresponds to the i-th site s_i at the time tick t; $i = 1,\ldots,n$ and $t = 1, 2, \ldots$. For a given time tick t, let $\mathbf{Y}_t = (Y_{1,t},\ldots,Y_{n,t})'$ denote the discrete valued hidden states on a graph structure $\{(s_i),(s_i \times s_j)\}\,|_{i,j=1}^{n}$. We model $\{\mathbf{Y}_t : t = 1, 2, \ldots\}$ by a n-dimensional Markov chain with the following transition probability:

$$p_\theta(\mathbf{Y}_t|\mathbf{Y}_{t-1}) = q(\mathbf{Y}_t|\mathbf{Y}_{t-1})/G_{\mathbf{X}}. \tag{4}$$

Here $G_{\mathbf{X}}$ is a normalization constant and,

$$q(\mathbf{Y}_t|\mathbf{Y}_{t-1}) = \exp\left\{\sum_{i=1}^{n}\sum_{k=1}^{p}\theta_k^1\varphi^1(X_{k,i,t},Y_{i,t}) + \sum_{j\sim i}\theta_{ij,t}^2\varphi^2(Y_{i,t},Y_{j,t}) + \sum_{i=1}^{n}\theta_{it,t-1}^3\varphi^3(Y_{i,t},Y_{i,t-1})\right\}, \tag{5}$$

where $X_{k,i,t} = X_k(s_i,t)$ denotes the k-th explanatory variable at the site s_i and the time tick t; θ_k is the linear regression coefficients corresponding to explanatory feature $\varphi^1(X_{k,i,t},Y_{i,t}), k = 1,\ldots,p$; $\theta_{ij,t}^2$ represent the spatial regression coefficients. The difference between the Equation 5 and the Equation 2 is the additional parameters $\theta_{it,t-1}^3$ that represent the temporal coefficients. When $\theta_{it,t-1} = 0$, there is no correlation over time and the Markov network of $\{\mathbf{Y}_t\}$ reduces to a sequence of independent random vectors, each represents a set of spatially dependent observations at a given time tick. Clearly, the new framework incorporates the previous one described in Section 2 as $p_\theta(\mathbf{Y}_t|\mathbf{Y}_{t-1})$ reduces to $p_\theta(\mathbf{Y}_t)$.

On the other hand, when $\theta_{it,t-1} \neq 0$, the framework considers state correlations over time; the magnitude of $\theta_{it,t-1}$ is related to the mean difference between two consecutive time ticks of the same site. To simplify the representations, we abbreviate the model parameters by $\theta = \left(\{\theta^1\}, \{\theta^2\}, \{\theta^3\}\right)'$; model features by $\psi(\mathbf{X}_t, \mathbf{Y}_t, \mathbf{Y}_{t-1}) = \left(\{\varphi^1(X_{k,i,t}, Y_{i,t})\}, \{\varphi^2(Y_{i,t}, Y_{j,t})\}, \{\varphi^3(Y_{i,t}, Y_{i,t-1})\}\right)'$; and observations from T time points by $\mathbf{Y}_1, \ldots, \mathbf{Y}_T$, where $\mathbf{Y}_t = (Y_{1,t}, \ldots, Y_{n,t})'$, $t = 1, \ldots, T$.

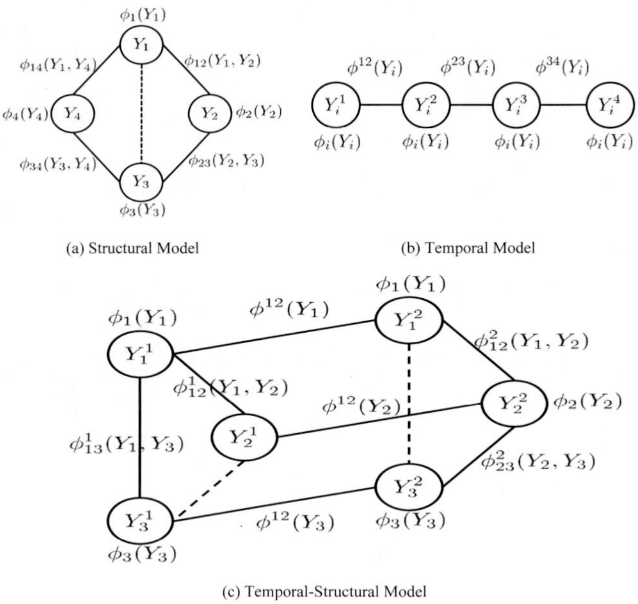

(a) Structural Model (b) Temporal Model (c) Temporal-Structural Model

Fig. 2. (a) Typical spatial dependent model - first order Markov network: $\phi_i(Y_i) = \exp\left\{\sum_{i=1}^{n}\sum_{k=1}^{p}\theta_k\varphi(X_{k,i}, Y_i)\right\}$ correspond to node potentials, $\phi_{i,i+1}(Y_i, Y_{i+1}) = \exp\left\{\sum_{j\sim i}\theta_{ij}\varphi(Y_i, Y_j)\right\}$ correspond to spatial edge potentials. (b) Typical temporal correlated model - first order Markov chain: $\phi_i(Y_i) = \exp\left\{\sum_{i=1}^{n}\sum_{k=1}^{p}\theta_k\varphi(X_{k,i}, Y_i)\right\}$ correspond to node potentials, $\phi^{t-1,t}(Y_i) = \exp\left\{\sum_{i=1}^{n}\theta^3_{it,t-1}\varphi^3(Y_{i,t}, Y_{i,t-1})\right\}$ correspond to temporal edge potentials. (c) We propose a new framework that generalizes both spatial dependent models and temporal correlated models. For illustration purpose, we only show correlated states of two consequent time ticks but the framework indeed depicts a gigantic network over time. Thus, traditional approaches such as CRFs and M3N fail to solve it with tractable computation.

The equation 5 represents a general framework considering spatial-temporal correlations, which generalizes both temporal correlated models and spatial dependent models, as indicated by Figure 2. However, there are more states to be considered together in the new framework due to the spatial and temporal coupling. Thus, traditional solutions such as constructing a gigantic CRFs network would be computationally intractable.

4 Temporal Maximum Margin Markov Network

There are two typical tasks in a machine learning problem like Equation 5: learning and predicting. For learning, we want to estimate parameters θ so that

$$h_\theta(\mathbf{X}_t) = \text{argmax}_{\mathbf{Y}} \theta' \psi(\mathbf{X}_t, \mathbf{Y}, \hat{\mathbf{Y}}_{t-1}) \approx \hat{\mathbf{Y}}_t, \forall t, \tag{6}$$

where $\hat{\mathbf{Y}}_t$ is the ground-truth states. For predicting, we would like to infer the most likely states

$$\mathbf{Y}^*_{t+1} = \text{argmax}_{\mathbf{Y}} \theta' \psi(\mathbf{X}_{t+1}, \mathbf{Y}, \hat{\mathbf{Y}}_t), \tag{7}$$

given the parameter θ and the novel observation $\mathbf{X}_{t+1}$ and past states $\hat{\mathbf{Y}}_t$. We will now describe a convex instantiation of the spatial-temporal correlated framework to handle both tasks.

First, we need to measure the error of the approximation $h(\cdot)$ using a loss function ℓ. Here we use a Hamming distance error measurement $\ell_t(\mathbf{Y}_t)$ to indicate the number of variables predicted incorrectlys, which essentially measure the loss on the label sequences,

$$\ell_t(\mathbf{Y}_t) = \sum_i \Delta(Y_{i,t}, \hat{Y}_{i,t}) \text{ and } \Delta(Y_{i,t}, \hat{Y}_{i,t}) = \begin{cases} 1 & Y_{i,t} \neq \hat{Y}_{i,t} \\ 0 & Y_{i,t} = \hat{Y}_{i,t} \end{cases}.$$

We adapt the hinge upper bound $\bar{\ell}(h_\theta(\mathbf{X}_t))$ on the loss function for structured classification inspired by max-margin criterion:

$$\bar{\ell}_t(h_\theta(\mathbf{X}_t)) = \max_{\mathbf{Y}_t} \Big[\theta' \psi(\mathbf{X}_t, \mathbf{Y}_t, \hat{\mathbf{Y}}_{t-1})) + \ell_t(\mathbf{Y}_t)\Big] - \theta' \psi(\mathbf{X}_t, \hat{\mathbf{Y}}_t, \hat{\mathbf{Y}}_{t-1})) \tag{8}$$

$$\geq \ell_t(h_\theta(\mathbf{X}_t)), \tag{9}$$

where $\bar{\ell}_t(h_\theta(\mathbf{X}_t)) = \bar{\ell}(h_\theta(\mathbf{X}_t), \hat{\mathbf{Y}}_t)$ and $\ell_t(h_\theta(\mathbf{X}_t)) = \ell(h_\theta(\mathbf{X}_t), \hat{\mathbf{Y}}_t)$. With this upper bound, the min-max formulation for structured classification problem is analogous to SVM,

$$\min_{\theta, \mathbf{Y}_t} \frac{\lambda}{2} ||\theta||^2 + \frac{1}{T} \sum_{t=1}^{T} \xi_t \tag{10}$$

$$s.t. \Big\langle \theta, \Phi(\mathbf{X}_t, \mathbf{Y}_t, \hat{\mathbf{Y}}_{t-1}, \hat{\mathbf{Y}}_t) \Big\rangle \geq \bar{\ell}(\mathbf{Y}_t, \hat{\mathbf{Y}}_t) - \xi_t, \forall t, \forall \mathbf{Y}_t, \tag{11}$$

where $\Phi(\mathbf{X}_t, \mathbf{Y}_t, \hat{\mathbf{Y}}_{t-1}, \hat{\mathbf{Y}}_t) = \psi(\mathbf{X}_t, \hat{\mathbf{Y}}_t, \hat{\mathbf{Y}}_{t-1}) - \psi(\mathbf{X}_t, \mathbf{Y}_t, \hat{\mathbf{Y}}_{t-1})$. This formulation incorporates the "maximum margin" criteria. We can interpret

$$\mathrm{M} = \frac{1}{||\theta||} \Big\langle \theta, \Phi(\mathbf{X}_t, \mathbf{Y}_t, \hat{\mathbf{Y}}_{t-1}, \hat{\mathbf{Y}}_t) \Big\rangle \tag{12}$$

as the margin of the state configuration $\hat{\mathbf{Y}}_t$ over another state configuration $\mathbf{Y}_t$. Assuming ξ_i are all zeros (because λ is very small), the constraints enforce,

$$\theta' \Big(\psi(\mathbf{X}_t, \hat{\mathbf{Y}}_t, \hat{\mathbf{Y}}_{t-1}) - \psi(\mathbf{X}_t, \mathbf{Y}_t, \hat{\mathbf{Y}}_{t-1})\Big) \geq \bar{\ell}(\mathbf{Y}_t, \hat{\mathbf{Y}}_t), \tag{13}$$

so minimizing $||\theta||^2$ essentially maximizes the smallest of such margins, scaled by the loss $\ell_i(\mathbf{Y}_t, \hat{\mathbf{Y}}_t)$. The above formulation is a standard QP and can be solved use optimization packages, but it is exponential in the size and computation is generally prohibitive. Another way to express this problem is the following representation,

$$\min_{\theta, \mathbf{Y}_t} \frac{\lambda}{2}||\theta||^2 + \frac{1}{T}\sum_{t=1}^{T} \xi_t \tag{14}$$

$$s.t. \theta' \psi(\mathbf{X}_t, \hat{\mathbf{Y}}_t, \hat{\mathbf{Y}}_{t-1}) + \xi_t \geq \max_{\mathbf{Y}_t} \left[\theta' \psi(\mathbf{X}_t, \mathbf{Y}_t, \hat{\mathbf{Y}}_{t-1})) + \ell_t(\mathbf{Y}_t)\right], \forall t, \tag{15}$$

which is a convex quadratic program in θ, since

$$\max_{\mathbf{Y}_t} \left[\theta' \psi(\mathbf{X}_t, \mathbf{Y}_t, \hat{\mathbf{Y}}_{t-1})) + \ell_t(\mathbf{Y}_t)\right], \tag{16}$$

is convex in θ. It might be easier to interpret Equation 14 in its alternative representation Equation 17 by eliminating the constraints,

$$\min_{\theta, \mathbf{Y}_t} \frac{\lambda}{2}||\theta||^2 + \frac{1}{T}\sum_{i=1}^{T} \left\{\max_{\mathbf{Y}_t} \left[\theta' \psi(\mathbf{X}_t, \mathbf{Y}_t, \hat{\mathbf{Y}}_{t-1})) + \ell_t(\mathbf{Y}_t)\right] - \theta' \psi(\mathbf{X}_t, \hat{\mathbf{Y}}_t, \hat{\mathbf{Y}}_{t-1})\right\}, \tag{17}$$

careful readers might notice that $\theta' \psi(\mathbf{X}_t, \hat{\mathbf{Y}}_t, \hat{\mathbf{Y}}_{t-1})$ is invariant to $\mathbf{Y}_t$ and we can run the algorithm in two separate steps: first, fix θ and optimize $\max_{\mathbf{Y}_t} \left[\theta' \psi(\mathbf{X}_t, \mathbf{Y}_t, \hat{\mathbf{Y}}_{t-1})) + \ell_t(\mathbf{Y}_t)\right]$; second, fix $\mathbf{Y}_t$ obtained in the first step to calculate θ that minimize Equation 17. The procedure is similar to the Expectation-Maximization algorithm and we are guaranteed not to increase the objective function at each step.

5 Learning

Recall the objective in Equation 17 is a convex function, an intuitive way to estimate its parameters θ is to use a gradient descent approach. In this case, the gradients only depends on the most violated state configuration,

$$\mathbf{Y}_t^* = \text{argmax}_{\mathbf{Y_t}} \left(\theta' \psi(\mathbf{X}_t, \mathbf{Y}_t, \hat{\mathbf{Y}}_{t-1})) + \ell_t(\mathbf{Y}_t)\right), \tag{18}$$

which can be computed as:

$$g(\theta) = \lambda\theta + \frac{1}{T}\sum_{i=1}^{T} \left(\psi(\mathbf{X}_t, \mathbf{Y}_t^*, \hat{\mathbf{Y}}_{t-1}) - \psi(\mathbf{X}_t, \hat{\mathbf{Y}}_t, \hat{\mathbf{Y}}_{t-1})\right), \tag{19}$$

the following algorithm thus summarizes the procedure of gradient optimization,

Algorithm 1. Subgraident Optimization

Input: training data $\mathcal{D} = \{(\mathbf{X}_t, \mathbf{Y}_t)\}|_{t=1}^{T}$, regularization parameter λ, step size σ, tolerance ϵ, number of iterations T

Output: parameter vector θ

```
Initialize θ ← 0, t ← 1
repeat
    for t = 1 to T do
        Set violation function
        H(Y_t) = θ'ψ(X_t, Y_t, Ŷ_{t-1})) + ℓ_t(Y_t) − θ'ψ(X_t, Ŷ_t, Ŷ_{t-1})
        Find most violated label for (X_t, Y_t) : Y*_t = arg max_{Y_t} H(Y_t)
    end for
    Compute g(θ), update θ^(t) ← θ^(t-1) − σg(θ).
    Update t ← t + 1
until t ≥ T or MSE(||θ^(t)|| − ||θ^(t-1)||) ≤ ε
```

A critical part of Algorithm 1 is to compute the most violated constraint at each time step efficiently. The exact inference of this step is usually intractable as irregular lattices often involve loops that cannot be handled by deterministic algorithms in polynomial time.To this end, we leverage on a well established approximation algorithm, loopy belief propagation (LBP) [8] to solve this. To use LBP, we define the following potentials:

- **Unary potentials** represent the impact of local observation in $\mathbf{X}_t$ to the states $\mathbf{Y}_t$, this potential function at each site s_i takes the form,

$$\exp\left(\sum_{k=1}^{p} \theta_k^1 \varphi^1(X_{k,i,t}, Y_{i,t}) + \ell_t(Y_{i,t})\right), \forall i, \tag{20}$$

- **Environmental potentials** represent the influence between states and over time, these potential functions take the form,

$$\exp\left(\theta_{ij,t}^2 \varphi^2(Y_{i,t}, Y_{j,t})\right), \forall i, j \sim i, \ \text{Structural Potential}, \tag{21}$$

$$\exp\left(\theta_{it,t-1}^3 \varphi^3(Y_{i,t}, Y_{i,t-1})\right), \forall i, \quad \text{Temporal Potential}. \tag{22}$$

6 Predicting

Now we will introduce our linear integer programming interface for predicting. The goal is to predict a hidden state as the most likely configuration:

$$\mathbf{Y}_{T+1}^* = \text{argmax}_{\mathbf{Y_T}} \left(\theta' \psi(\mathbf{X}_{T+1}, \mathbf{Y}_T, \hat{\mathbf{Y}}_T)\right). \tag{23}$$

Denote $Z^t = (\{z_i^t\}|_{i=1}^{n}, \{z_{ij}^t\}|_{i=1}^{n,j\sim i}, \{z_i^{t,t-1}\}|_{i=1}^{n})$ as indicator variables at time t so that: $z_i(m) = 1$ indicates i-th site takes state m, $z_{ij}(m,n)$ indicates i and j-th sites take states m and n, and $z_i^{t,t-1}(m,n) = 1$ indicates i-th site take states m and n at time t and $t-1$, respectively. If we factorize Equation 23, the following linear integer programming interface defines an exact mapping,

$$\max_{Z^t} \sum_{i,m} z_i^t(m) \left[\theta^1_{(\cdot)}\varphi^1(X_{(\cdot),i,t}, m)\right] + \tag{24}$$

$$\sum_{i,j,m,n} z_{ij}^t(m,n) \left[\theta^2_{ij,t}\varphi^2(m,n)\right] + \sum_i \left[\theta^3_{it,t-1}\varphi^3(m, \hat{Y}_{i,t-1})\right],$$

$$s.t. \; z_i^t(m) \in \{0,1\}, \; z_{ij}^t(m,n) \in \{0,1\}, \tag{25}$$

$$\sum_m z_i^t(m) = 1, \tag{26}$$

$$\sum_n z_{ij}^t(m,n) = z_i^t(m), \tag{27}$$

$$\sum_m z_{ij}^t(m,n) = z_j^t(n). \tag{28}$$

The constraint Equation 26 enforces only one state is allocated for each site s_i; the constraint equation 27 enforces the structural consistency. Note we assign $z_i^{t,t-1}(m, \hat{Y}_{i,t-1}) = 1, \forall i$ so that $Y_{i,t}$ is influenced by its previous state $\hat{Y}_{i,t-1}$of the same site s_i. The above linear integer programming is an intractable combinatorial problem but we can obtain an approximated solution by relaxing the binary constraint in Equation 25 to be $z_i^t(m) \geq 0, \; z_{ij}^t(m,n) \geq 0$. A threshold χ, usually equals to 0.5, is used to discretize the final outputs Z^t for predicting the states.

7 Experiments

7.1 Simulation Results

We use the following temporal-spatial correlated Linear Dynamic System (LDS) to generate the simulation. This system specifies the hidden state Y_t^i, which depends temporally on the previous state Y_{t-1}^i and correlates spatially with the states of the neighboring sites Y_t^j, $j \in N_{-i}$.

$$Y_t^i = \alpha Y_{t-1}^i + (1-\alpha) \sum_{j \in N_{-i}} \beta^j Y_t^j + e_1, \; , e_1 \sim N(0, \sigma_{e_1}^2), \tag{29}$$

$$X_t^i = A Y_t^i + e_2, \; , e_2 \sim N(0, \sigma_{e_2}^2), \tag{30}$$

where N_{-i} corresponds to the neighboring sites of i; A is a projection vector that maps hidden states to the observations; X_t^i corresponds to the observations at site i, time tick t; e_1 and e_2 are the environmental Gaussian noises;α represents the temporal/spatial trade-off parameter. If α is set to be zero, the simulation considers no time dependence. Otherwise, if α is set to be one, the simulation ignores spatial correlations.

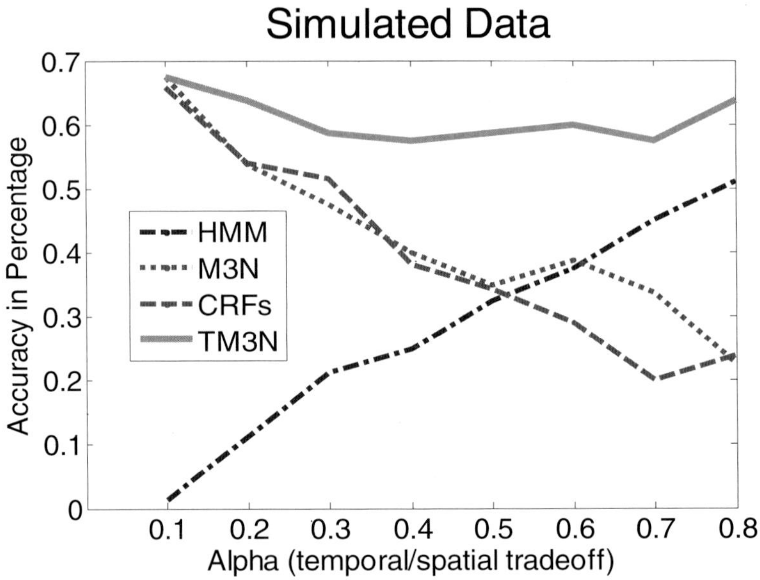

Fig. 3. Model comparison on simulated temporal-spatial correlated data. The X axis corresponds to the Alpha (temporal/spatial trade-off parameter) value and Y axis represents the accuracy in percentage. HMM's performance increases as the temporal influence becomes larger while CRFs/M3N's accuracy decreases at the meantime. TM3N outperforms all these three models and demonstrates its efficacy.

We initialize $Y_t^i \sim \text{Uniform}(0,1)$, $\beta^j \sim \text{Uniform}(0,1)$; set total sites number N equals to four; specify the error term $e_1 \sim N(0, 0.05)$ and $e_l \sim N(0,1)$; and let the projection matrix $A = [10; 20]$. To simulate the hidden states, we use an approach similar to Gibbs sampling that iteratively samples Y_t^i until the system converges. These simulated states are rounded to be real valued states and the simulated observations are calculated use Equation 30.

In the experiment, we vary the temporal/spatial trade-off parameter α from 0.1 to 0.8 at an interval of 0.1 to evaluate the performances of four different models: HMM, M3N, CRF and TM3N. For every α value, we run the simulation for 50 times and calculate the averaged accuracy. The results are demonstrated in Figure 3, where the blue curve corresponds to the accuracy of TM3N model at various α values. Obviously, TM3N shows superior performance comparing to HMM, CRF and M3N.

7.2 Real World Applications

Recently, buildings began to have sensor networks installed for energy and comfort management. The control strategy for lighting, heating ventilation and air-conditioning (HVAC) can be updated adaptively as needed [2]. For the cost-efficient operation, understanding occupancy behavior in buildings is becoming

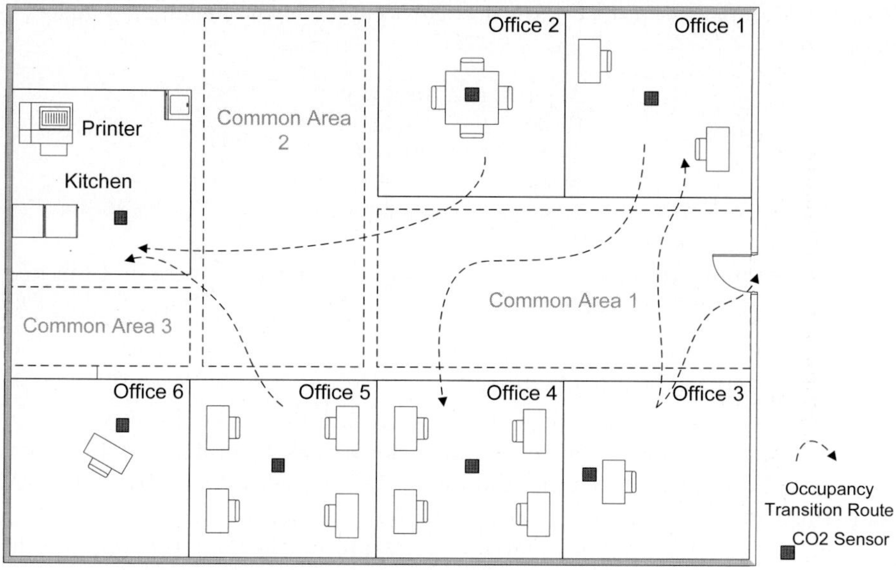

Fig. 4. Geometric flat view of the office area testbed

crucial problems to success. One specific question is to estimate office occupant number over the time, which naturally fits our proposed model.

7.2.1 Data Collection

The sensor network is setup in an open plan office space with six rooms and one kitchen/printer room. It provides offices for two faculty members and ten Ph.D. students. Since it is an open plan office, the faculties and students have discussions frequently. The entire indoor environment can be considered very dynamic. Occupants have different activities such as reading, talking on the phone, drop-by and discussion.An occupant may leave his own area and go to other areas, such as printer room, kitchen, and restroom. The physical sensor network includes a wired CO_2 network and a data server. One CO_2 sensor is installed in the center of each office at the nose level (1.1m) above the ground. To establish ground truth about occupancy information, we use a network of commercial cameras. Figure 4 shows the geometric view of the test-bed. Note the CO_2 sensors are preferred over the vision sensors because of the privacy reasons, e.g., we cannot easily distinguish the occupants by the CO_2 measurements.

Data collection for this paper was for one continuous period, with a sampling rate of every one minutes, capturing CO_2 measurements and the number of occupants in four offices. The time period is three weeks from March 17th, 2008 to April 4th, 2008 excluding weekends. Occupancy data was recorded from 8:00am to 8:00pm from the four offices (2, 3, 4 and 5). Office 2 and 5 have four Ph.D. students; office 4 has two graduate students ; and office office 3 have 1 faculty. We synchronize the measurements from all sensors; and aggregate measurements

Table 2. Comparison results of the average accuracy in the building occupancy estimation task

Algorithm	Accuracy
HMM	36.5%
CRFs	49.81%
M3N	49.05%
TM3N	**69.76%**

for every 10 minutes to predict the averaged occupant numbers in a ten minute window.

7.2.1 Results and Discussions

We split the data into training and testing: the first week from 3/17 to 3/21 is used as training data; the second and third week from 3/24 to 4/4 are used as testing. Along with our proposed model, we implement and test CRFs, M3N and HMM models.

Table 2 summarizes the comparison results of the four different methods. If we consider only temporal correlations and assume structural independence, HMM model gives an average accuracy of 36.5% for four offices. Clearly, first order Markov model is not suffice to capture the dynamics in the environment. On the other hand, structural models such as CRFs and M3N shows similar results, although slightly better than HMM, are still unsatisfactory. A significant improvement in average accuracy is observed when we combine both temporal and structural influence into a unified model, TM3N. This improvement owns a big part of its success to the joint modeling of both temporal and structural information, which oftentimes complement to each other.

To make the figures uncluttered, we only show the prediction results of proposed TM3N model on 3/25/2008. Figure 5 illustrates the results in four offices from 8am to 8pm. The solid blue curve corresponds to the prediction results of TM3N approach and the dashed magenta curve corresponds to the ground-truth value. The X axis show the time tick for every 50 minutes from 8am to 8pm. The Y axis shows the number of occupants. The estimation accuracies are 0.76, 0.86, 0.74 and 0.67 for office 2, 3, 4 and 5, respectively. As indicated by these figures, the TM3N occupant estimation results are close to ground-truth, which shows our method captures the occupancy dynamics quickly.

For the faculty office 3, the occupancy number is usually between zero and one during the day. For student's offices, there are much faster changes of occupant numbers, for example, the occupant number in office 2 and office 5 changes hourly. These changes are due to the "drop-by" activities of visitors.

In some of those cases, the estimation does not capture it well as the length of our estimation window is ten minutes while the visitors stay for a few minutes. However, such abrupt changes usually will not affect the operation of building energy management systems because these systems such as HVAC cannot response in high frequencies. Hence, in the practical application, this abrupt change will be ignored.

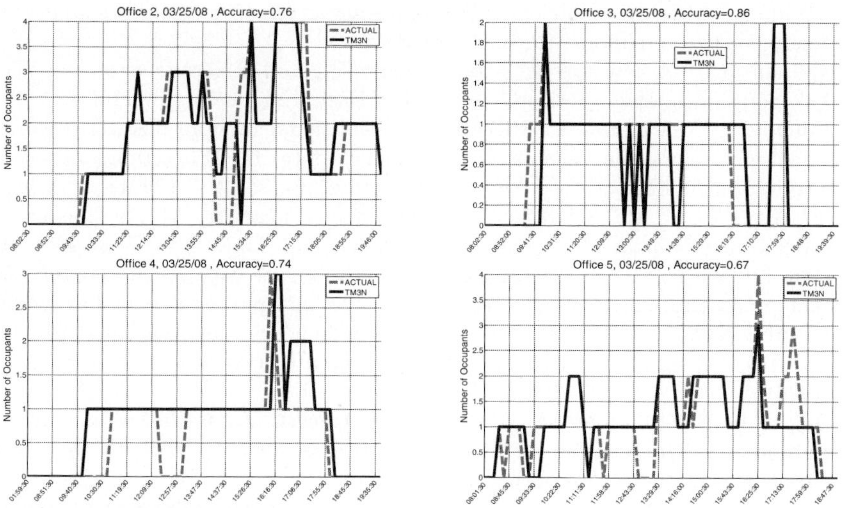

Fig. 5. Occupancy Estimation of 4 offices on March 25. The dashed magenta curve corresponds to the ground-truth; the solid blue curve corresponds to the estimation results of TM3N. The states are discrete valued variables, e.g. 0, 1, 2, 3 and 4. We connect these discrete states for visualization purpose; thus, a sharp jump mismatch does not mean a large deviation from the ground-truth.

8 Conclusion

This paper presents a maximum margin structured learning model, TM3N to model temporal-spatial correlated environments. The main goal of this work is to synthesize information from different perspectives to model real world systems more faithfully. We demonstrate how the proposed framework incorporates, generalizes, and extends existing approaches presented in the literature. The experiments show superior performance of proposed model against other state of the arts approaches in the building occupancy estimation task.

References

1. Desai, C., Ramanan, D., Fowlkes., C.: Discriminative models for multi-class object layout. In: ICCV (2009)
2. Dong, B., Andrews, B., Lam, K., Hoynck, M., Chiou, Y., Zhang, R., Benitez, D.: An Information Technology Enabled Sustainability Test-Bed (ITEST) For Occupancy Detection Through An Environmental Sensing Network. Energy and Buildings. Energy and Buildings 42(7) (2010)
3. Fildes, R.: Forecasting structural time series models and the kalman filter: Andrew harvey, p. 554. Cambridge University Press, Cambridge (1989); International Journal of Forecasting 8(4), 635–635 (December 1992), `http://ideas.repec.org/a/eee/intfor/v8y1992i4p635-635.html`, ISBN: 0-521-32196-4

[4] Gould, S., Fulton, R., Koller, D.: Decomposing a scene into geometric and semantically consistent regions. In: ICCV (2009)

[5] Kalman, R.E.: A new approach to linear filtering and prediction problems. Trans. Am. Soc. Mech. Eng., Series D. Journal of Basic Engineering 82 (1960)

[6] Lafferty, J., McCallum, A., Pereira, F.: Conditional random fields: Probabilistic models for segmenting and labeling sequence data. In: Proceedings of the 18th International Conference on Machine Learning, pp. 282–289 (2001)

[7] Li, R., Tian, T.P., Sclaroff, S.: Simultaneous Learning of Nonlinear Manifold and Dynamical Models for High-dimensional Time Series. In: 2007 IEEE 11th International Conference on Computer Vision, pp. 1–8 (October 2007), http://ieeexplore.ieee.org/lpdocs/epic03/wrapper.htm?arnumber=4409044

[8] Murphy, K.P., Weiss, Y., Jordan, M.I.: Loopy belief propagation for approximate inference: An empirical study. In: Proceedings of Uncertainty in AI, pp. 467–475 (1999)

[9] Oh, S.M., Ranganathan, A., Rehg, J.M., Dellaert, F.: A Variational inference method for Switching Linear Dynamic Systems Need for Variational methods Variational method. Computing (2005)

[10] Qian, X., Jiang, X., Zhang, Q., Huang, X., Wu, L.: Sparse higher order conditional random fields for improved sequence labeling. In: ICML, p. 107 (2009)

[11] Rabiner, L.R.: A tutorial on hidden markov models and selected applications in speech recognition. Proceedings of the IEEE, 257–286 (1989)

[12] Sarawagi, S., Gupta, R.: Accurate max-margin training for structured output spaces. In: ICML 2008: Proceedings of the 25th International Conference on Machine Learning, pp. 888–895. ACM, New York (2008)

[13] Taskar, B., Guestrin, C., Koller, D.: Max-margin markov networks. Advances in Neural Information Processing Systems 16, 25–32 (2003)

[14] Taskar, B.: Learning structured prediction models: A large margin approach. Stanford University, Ph.D. thesis (2004)

[15] Tsochantaridis, I., Joachims, T., Hofmann, T., Altun, Y.: Large margin methods for structured and interdependent output variables. J. Mach. Learn. Res. 6, 1453–1484 (2005)

[16] Duong, T.V., Phung, D.Q., Bui, H.H., Venkatesh, S.: Behavior recognition with generic exponential family duration modeling in the hidden semi-markov model. In: Proceedings of the 18th International Conference on Pattern Recognition, vol. 3 (2006)

[17] Welch, L.R.: Hidden markov models and the baum-welch algorithm. IEEE Information Theory Society Newsletter 53(4) (December 2003), http://www.itsoc.org/publications/nltr/it_dec_03final.pdf

Gaussian Processes for Sample Efficient Reinforcement Learning with RMAX-Like Exploration

Tobias Jung and Peter Stone

Department of Computer Science
University of Texas at Austin
{tjung,pstone}@cs.utexas.edu

Abstract. We present an implementation of model-based online reinforcement learning (RL) for continuous domains with deterministic transitions that is specifically designed to achieve low sample complexity. To achieve low sample complexity, since the environment is unknown, an agent must intelligently balance exploration and exploitation, and must be able to rapidly generalize from observations. While in the past a number of related sample efficient RL algorithms have been proposed, to allow theoretical analysis, mainly model-learners with weak generalization capabilities were considered. Here, we separate function approximation in the model learner (which does require samples) from the interpolation in the planner (which does not require samples). For model-learning we apply Gaussian processes regression (GP) which is able to automatically adjust itself to the complexity of the problem (via Bayesian hyperparameter selection) and, in practice, often able to learn a highly accurate model from very little data. In addition, a GP provides a natural way to determine the uncertainty of its predictions, which allows us to implement the "optimism in the face of uncertainty" principle used to efficiently control exploration. Our method is evaluated on four common benchmark domains.

1 Introduction

In reinforcement learning (RL), an agent interacts with an environment and attempts to choose its actions such that an externally defined performance measure, the accumulated per-step reward, is maximized over time. One defining characteristic of RL is that the environment is *unknown* and that the agent has to *learn* how to act directly from experience. In practical applications, e.g., in robotics, obtaining this experience means having a physical system interact with the physical environment in real time. Therefore, RL methods that are able to learn quickly and minimize the amount of time the robot needs to interact with the environment until good or optimal behavior is learned, are highly desirable.

In this paper we are interested in online RL for tasks with continuous state spaces and smooth transition dynamics that are typical for robotic control domains. Our primary goal is to have an algorithm which keeps sample complexity as low as possible.

J.L. Balcázar et al. (Eds.): ECML PKDD 2010, Part I, LNAI 6321, pp. 601–616, 2010.

1.1 Overview of the Contribution

To maximize sample efficiency, we consider online RL that is *model-based* in the spirit of RMAX [3], but extended to continuous state spaces similar to [1,10,5]. As in RMAX and related methods, our algorithm, GP-RMAX, consists of two parts: a *model-learner* and a *planner*. The model-learner estimates the dynamics of the environment from the sample transitions the agent experiences while interacting with the environment. The planner is used to find the best possible action, given the current model. As the predictions of the model-learner become increasingly more accurate, the actions derived become increasingly closer to optimal. To control the amount of exploration, the "optimism in the face of uncertainty" principle is employed which makes the agent visit unexplored states first. In our algorithm, the model-learner is implemented by Gaussian process (GP) regression; being non-parametric, GPs give us enhanced modeling flexibility. GPs allow Bayesian model selection and automatic relevance determination. In addition, GPs provide a natural way to determine the uncertainty of predictions, which allows us to implement the "optimism in the face of uncertainty" exploration of RMAX in a principled way. The planner uses the estimated transition function (as estimated by the model) to solve the Bellman equation via value iteration on a uniform grid.[1]

The key point of our algorithm is that we separate the steps estimating a function from samples in the model-learner from solving the Bellman equation in the planner. The rationale behind this is that, if the transition function is relatively simple, it can be estimated accurately from only few sample transitions. On the other hand, the optimal value function, due to the inclusion of the max operator, often is a complex function with sharp discontinuities. Solving the Bellman equation, however, does not require actual "samples"; instead, we must only be able to evaluate the Bellman operator in arbitrary points of the state space. This way, when the transition function can be learned from only a few samples, large gains in sample efficiency are possible. Competing model-free methods, such as fitted Q-iteration [18,8,15] or policy iteration based LSPI/LSTD/LSPE [12,4,13,11], do not have this advantage, as they need the actual sample transitions to estimate and represent the value function.

Conceptually, our approach is closely related to Fitted R-MAX, which was proposed in [10] and uses an instance-based approach in the model-learner, and related work in [5,1], which uses grid-based interpolation in the model-learner. The primary contribution of this paper is to use GPs instead. Doings this means we are willing to trade off theoretical analysis with practical performance. For example, unlike the recent ARL [1], for which PAC-style performance bounds could be derived (because of its grid-based implementation of model-learning), a GP is much better able to handle generalization and as a consequence can achieve much lower sample complexity.

[1] While certainly more advanced methods exist, e.g., [9,14], for our purpose here, a uniform grid is sufficient as proof of concept.

1.2 Assumptions and Limitations

Our approach makes the following assumptions (most of which are also made in related work, even if it is not always explicitly stated):

- *Low dimensionality of the state space.* With a uniform grid, the number of grid points for solving the Bellman equation scales exponentially with the dimensionality. While more advanced methods, such as sparse grids or adaptive grids, may allow us to somewhat reduce this exponential increase, at the end they do not break the curse of dimensionality. Alternatively, one can use nonlinear function approximation; however, despite some encouraging results, it is unclear as to whether this approach would really do any better in general applications. Today, breaking the curse of dimensionality is still an open research problem.
- *Discrete actions.* While continuous actions may be discretized, in practice, for higher dimensional action spaces this becomes infeasible.
- *Smooth transition function.* Performing an action from states that are "close" must lead to successor states that are "close". (Otherwise both the generalization in the model learner and the interpolation in the value function approximation would not work).
- *Deterministic transitions.* This is not a fundamental requirement of our approach, since GPs can also learn noisy functions (either due to observation noise or random disturbances with small magnitude), and the Bellman operator can be evaluated in the resulting predictive distribution. Rather it is one taken for convenience.
- *Known reward function.* Assuming that the reward function is known and only the transition function needs to be learned is what is different from most comparable work. While it is not a fundamental requirement of our approach (since we could learn the reward function as well), it is an assumption that we think is well justified: for one, reward is the performance criterion and specifies the goal. For the type of control problems we consider here, reward is always externally defined and never something that is "generated" from within the environment. Two, reward sometimes is a discontinuous function, e.g., +1 at the goal state and 0 elsewhere. Which makes it not very amenable for function approximation.

2 Background: Planning When the Model Is Exact

Consider the reinforcement learning problem for MDPs with continuous state space, finite action space, discounted reward criterion and *deterministic* dynamics [19]. In this section we assume that dynamics and rewards are available to the learning agent. Let state space $\mathcal{X}$ be a hyperrectangle in $\mathbb{R}^d$ (this assumption is justified if, for example, the system is a motor control task), $\mathcal{A}$ be the finite action space (assuming continuous controls are discretized), $x_{t+1} = f(x_t, a_t)$ be the transition function (assuming that continuous time problems are discretized in time), and $r(x, a)$ be the reward function. For the following theoretical argument

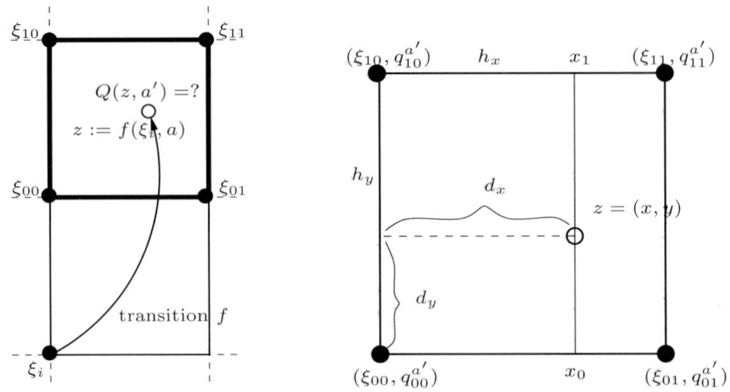

Fig. 1. Bilinear interpolation to determine $Q(f(\xi_i, a), a')$ in $\mathbb{R}^2$

we require that both transition and reward function are Lipschitz continuous in the actions; i.e., there exist constants L_f, L_r such that $\|f(x,a) - f(x',a)\| \leq L_f \|x - x'\|$, and $|r(x,a) - r(x',a)| \leq L_r \|x - x'\|$, $\forall x, x' \in \mathcal{X}, a \in \mathcal{A}$. In addition, we assume that the reward is bounded, $|r(x,a)| \leq R_{\text{MAX}}$, $\forall x, a$. Note that in practice, while the first condition, continuity in the transition function, is usually fulfilled for domains derived from physical systems, the second condition, continuity in the rewards, is often violated (e.g. in the mountain car domain, reward is 0 in the goal and -1 everywhere else). Despite that we find that in many of these cases the outlined procedure may still work well enough.

For any state x, we are interested in determining a sequence of actions $a_0, a_1, a_2, \ldots$ such that the accumulated reward is maximized,

$$V^*(x) := \max_{a_0, a_1, \ldots} \Big\{ \sum_{t=0}^{\infty} \gamma^t r(x_t, a_t) \mid x_0 = x, x_{t+1} = f(x_t, a_t) \Big\},$$

where $0 < \gamma < 1$. Using the Q-notation, where $Q^*(x,a) := r(x,a) + \gamma V^*(f(x,a))$, the optimal decision policy π^* is found by first solving the Bellman equation in the unknown function Q,

$$Q(x,a) = r(x,a) + \gamma \max_{a'} Q(f(x,a), a') \qquad \forall x \in \mathcal{X}, a \in \mathcal{A} \tag{1}$$

to yield Q^*, and then choosing the action with the highest Q-value,

$$\pi^*(x) = \operatorname*{argmax}_{a'} Q^*(x, a').$$

The Bellman operator T related to (1) is defined by

$$\big(TQ\big)(x,a) := r(x,a) + \gamma \max_{a'} Q(f(x,a), a'). \tag{2}$$

It is well known that T is a contraction and Q^* the unique bounded solution to the fixed point problem $Q(x,a) = \big(TQ\big)(x,a)$, $\forall x, a$.

In order to solve the infinite dimensional problem in (1) numerically, we have to reduce it to a finite dimensional problem. This is done by introducing a discretization Γ of $\mathcal{X}$ into a finite number of elements, applying the Bellman operator to only the nodes and interpolating in between.

In the following we will consider a uniform grid Γ_h with N vertices ξ_i and d-dimensional tensor B-spline interpolation of order 1. The solution of (1) is then obtained in the space of piecewise affine functions.

For a fixed action a', the value $Q^{\Gamma_h}(z, a')$ of any state z with respect to grid Γ_h can be written as a convex combination of the vertices ξ_j of the grid cell enclosing z with coefficients w_{ij} (see Figure 1a). For example, consider the 2-dimensional case (bilinear interpolation) in Figure 1b. Let $z = (x, y) \in \mathbb{R}^2$. To determine $Q^{\Gamma_h}(z, a')$, we find the four vertices $\xi_{00}, \xi_{01}, \xi_{10}, \xi_{11} \in \mathbb{R}^2$ of the enclosing cell with known function values $q_{00}^{a'} := Q^{\Gamma_h}(\xi_{00}, a'), \ldots$ etc. We then perform two linear interpolations along the x-coordinate (order invariant) in the auxilary points x_0, x_1 to obtain

$$Q^{\Gamma_h}(x_0, a') = (1 - \lambda_0)q_{00}^{a'} + \lambda_0 q_{01}^{a'}$$
$$Q^{\Gamma_h}(x_1, a') = (1 - \lambda_0)q_{10}^{a'} + \lambda_0 q_{11}^{a'}$$

where $\lambda_0 := d_x/h_x$ (see Figure 1b for a definition of d_x, h_x, x_0, x_1). We then perform another linear interpolation in x_0, x_1 along the y-coordinate to obtain

$$Q^{\Gamma_h}(z, a') = (1 - \lambda_1)(1 - \lambda_0)q_{00}^{a'} + (1 - \lambda_1)\lambda_0 q_{01}^{a'} + \lambda_1(1 - \lambda_0)q_{10}^{a'} + \lambda_1\lambda_0 q_{11}^{a'} \quad (3)$$

where $\lambda_1 := d_y/h_y$. Weights w_{ij} now correspond to the coefficients in (3). An analogous procedure applies to higher dimensions.

Let $Q^{a'}$ be the $N \times 1$ vector with entries $[Q^{a'}]_i = Q^{\Gamma_h}(\xi_i, a')$. Let $z_1^a, \ldots, z_N^a \in \mathbb{R}^d$ denote the successor state we obtain when we apply the transition function f to vertices ξ_i using action a, i.e., $z_i^a := f(\xi_i, a)$. Let $[w_i^a]_j = w_{ij}^a$ denote the $1 \times N$ vector of coefficients for z_i^a from (3). The Q-value of z_i^a for any action a' with respect to grid Γ_h can thus be written as $Q^{\Gamma_h}(z_i^a, a') = \sum_{j=1}^N [w_i^a]_j [Q^{a'}]_j$. Let W^a with rows $[w_i^a]$ be the $N \times N$ matrix of all coefficients. (Note that this matrix is sparse: each row contains only 2^d nonzero entries).

Let R^a be the $N \times 1$ vector of associated rewards, $[R^a]_i := r(\xi_i, a)$. Now we can use (2) to obtain a fixed point equation in the vertices of the grid Γ_h,

$$Q^{\Gamma_h}(\xi_i, a) = \left(T^{\Gamma_h} Q^{\Gamma_h}\right)(\xi_i, a) \quad i = 1, \ldots, N, \ a = 1, \ldots, |\mathcal{A}|, \quad (4)$$

where

$$\left(T^{\Gamma_h} Q^{\Gamma_h}\right)(\xi_i, a) := r(\xi_i, a) + \gamma \max_{a'} Q^{\Gamma_h}(f(\xi_i, a), a').$$

Slightly abusing the notation, we can write this more compactly in terms of matrices and vectors,

$$T^{\Gamma_h} Q^{\Gamma_h} := R^a + \gamma \max_{a'} \left\{W^a Q^{a'}\right\} \quad \forall a. \quad (5)$$

The Q-function is now represented by $|\mathcal{A}|$ N-dimensional vectors $Q^{a'}$, each containing the values for the vertices ξ_i. The discretized Bellman operator T^{Γ_h} is

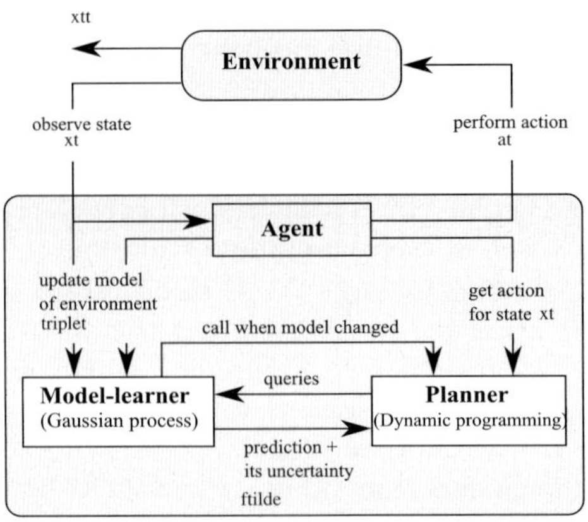

Fig. 2. High-level overview of the GP-RMAX framework

a contraction in $\mathbb{R}^d \times \mathcal{A}$ and therefore has a unique fixed point $Q^* \in \mathbb{R}^d \times \mathcal{A}$. Let function $Q^{*,\Gamma_h} : (\mathbb{R}^d \times \mathcal{A}) \to \mathbb{R}$ be the Q-function obtained by linear interpolation of vector Q^* along states. The function Q^{*,Γ_h} can now be used to determine (approximately) optimal control actions: for any state $x \in \mathcal{X}$, we simply determine

$$\pi^{*,\Gamma_h}(x) = \underset{a'}{\operatorname{argmax}}\ Q^{*,\Gamma_h}(x, a').$$

In order to estimate how well function Q^{*,Γ_h} approximates the true Q^*, *a posteriori* estimates can be defined that are based on local errors, i.e. the maximum of residual in each grid cell. The local error in a grid cell in turn depends on the granularity of the grid, h, and the modulus of continuity L_f, L_g (e.g., see [9,14] for details).

3 Our Algorithm: GP-RMAX

In the last section we have seen how, for a continuous state space, optimal behavior of an agent can be obtained in a numerically robust way, given that the transition function $x_{t+1} = f(x_t, a_t)$ is known.[2]

For model-based RL we are now interested in solving the same problem for the case that the transition function is not known. Instead, the agent has to *interact* with the environment, and only use the samples it observes to compute optimal behavior. Our goal in this paper is to develop a learning framework where this

[2] Remember our working assumption: reward as a performance criterion is externally given and does not need to be estimated by the agent. Also note that discretization (even with more advanced methods like adaptive or sparse grids) is likely to be feasible only in state spaces with low to medium dimensionality. Breaking the curse of dimensionality is an open research problem.

number is kept as small as possible. This will be done by using the samples to learn an estimate $\tilde{f}(x,a)$ of $f(x,a)$ and then use this estimate $\tilde{f}$ in place of f in the numerical procedure outlined in the previous section.

3.1 Overview

A sketch of our architecture is shown in Figure 2. GP-RMAX consists of the two parts model learning and planning which are interwoven for online learning. The model-learner estimates the dynamics of the environment from the sample transitions the agent experiences while interacting with the environment. The planner is used to find the best possible action, given the current model. As the predictions of the model-learner become increasingly more accurate, the actions derived from the planner become increasingly closer to optimal. Below is a high-level overview of the algorithm:

- **Input:**
  - Reward function $r(x,a)$
  - Discount factor γ
  - Performance parameters:
    - planning and model-update frequency K
    - model accuracy δ_1^M, δ_2^M (stopping criterion for model-learning)
    - discretization of planner N
- **Initialize:**
  - Model $\mathcal{M}_1$, Q-function $\mathcal{Q}_1$, observed transitions $\mathcal{D}_1$
- **Loop:** $t = 1, 2, \ldots$
  - **Interact with system:**
    - observe current state x_t
    - choose action a_t greedy with respect to $\mathcal{Q}_t$

$$a_t = \underset{a'}{\operatorname{argmax}}\, \mathcal{Q}_t(x_t, a')$$

    - execute action a_t, observe next state x_{t+1}, store transition $\mathcal{D}_{t+1} = \mathcal{D}_t \cup \{x_t, a_t, x_{t+1}\}$
  - **Model learning:** (see Section 3.2)
    - only every K steps, and only if $\mathcal{M}_t$ is not sufficiently exact (as determined by evaluating the stopping criterion)
      - $\mathcal{M}_{t+1} = \text{update\_model}\,(\mathcal{M}_t, \mathcal{D}_{t+1})$
      - evaluate_stopping_criterion $(\mathcal{M}_{t+1}, \mathcal{M}_t, \delta_1^M, \delta_2^M)$
    - else
      - $\mathcal{M}_{t+1} = \mathcal{M}_t$
  - **Planning with model:** (see Section 3.3)
    - only every K steps, and only if $\mathcal{M}_t$ is not sufficiently exact (as determined by evaluating the stopping criterion)
      - $\mathcal{Q}_{t+1} = \text{augmented\_value\_iteration}\,(\mathcal{Q}_{\sqcup}, \mathcal{M}_{t+1}, @r(x,u), \gamma, N)$
    - else
      - $\mathcal{Q}_{t+1} = \mathcal{Q}_t$

Next, we will explain in more detail how each of the two functional modules "model-learner" and "planner" is implemented.

3.2 Model Learning with GPs

In essence, estimating $\tilde{f}$ from samples is a regression problem. While in theory any nonlinear regression algorithm could serve this purpose, we believe that GPs are particularly well-suited: (1) being non-parametric means great modeling flexibility; (2) setting the hyperparameters can be done automatically (and in a principled way) via optimization of the marginal likelihood and allows automatic determination of relevant inputs; and (3) GPs provide a natural way to determine the uncertainty of its predictions which will be used to guide exploration. Furthermore, uncertainty in GPs is *supervised* in that it depends on the target function that is estimated (because of (2)); other methods only consider the density of the data (*unsupervised*) and will tend to overexplore if the target function is simple.

Assume we have observed a number of transitions, given as triplets of state, performed action, and resulting successor state, e.g., $\mathcal{D} = \{x_t, a_t, x_{t+1}\}_{t=1,2,\ldots}$ where $x_{t+1} = f(x_t, a_t)$. Note that f is a d-dimensional function, $f(x_t, a_t) = \left[f_1(x_t, a_t), \ldots, f_d(x_t, a_t)\right]^T$. Instead of trying to estimate f directly (which corresponds to absolute transitions), we try to estimate the relative change $x_{t+1} - x_t$ as in [10]. The effect of each action on each state variable will be treated independently: we train multiple univariate GPs and combine the individual predictions afterwards. Each individual $\mathcal{GP}_{ij}$ is trained in the respective subset of data in $\mathcal{D}$, e.g., $\mathcal{GP}_{ij}$ is trained on all x_t as input, and $x_{t+1}^{(i)} - x_t^{(i)}$ as output, where $a_t = j$. Each individual $\mathcal{GP}_{ij}$ has its own set of hyperparameters obtained from optimizing the marginal likelihood.

The details of working[3] with GPs can be found in [17]; using GPs to learn a model for RL was previously also studied in [6] (for offline RL and without uncertainty-guided exploration). One characteristic of GPs is that their functional form is given in terms of a parameterized covariance function. Here we use the squared exponential,

$$k(x, x'; v_0, b, \boldsymbol{\theta}) = v_0 \exp\Big\{-0.5(x - x')^T \Omega (x - x')\Big\} + b,$$

where matrix Ω is either one of the following: (1) $\Omega = \theta I$ (uniform), (2) $\Omega = \mathrm{diag}(\theta_1, \ldots, \theta_d)$ (axis aligned ARD), (3) $\Omega = M_k M_k^T$ (factor analysis). Scalars v_0, b and the (Ω-dependent number of) entries of $\boldsymbol{\theta}$ constitute the hyperparameters of the GP and are adapted from the training data (likelihood optimization). Note that variant (2) and (3) implement *automatic relevance determination*: relevant inputs or linear projections of inputs are automatically identified, whereby model complexity is reduced and generalization sped up.

[3] There is also the problem of implementing GPs *efficiently* when dealing with a possible large number of data points. For the lack of space we can only sketch our particular implementation, see [16] for more detailed information. Our GP implementation is based on the *subset of regressors* approximation. The elements of the subset are chosen by a stepwise greedy procedure aimed at minimizing the error incurred from using a low rank approximation (incomplete Cholesky decomposition). Optimization of the likelihood is done on random subsets of the data of fixed size. To avoid a degenerate predictive variance, the *projected process* approximation was used.

Once trained, for any testpoint x, $\mathcal{GP}_{ij}$ provides a distribution over target values, $\mathcal{N}(\mu_{ij}(x), \sigma^2_{ij}(x))$, with mean $\mu_{ij}(x)$ and variance $\sigma^2_{ij}(x)$ (exact formulas for μ and σ can be found in [17]). Each individual mean μ_{ij} predicts the change in the i-th coordinate of the state under the j-th action. Each individual variance σ^2_{ij} can be interpreted as the associated uncertainty; it will be close to 0 if $\mathcal{GP}_{ij}$ is certain, and close to $k(x,x)$ if it is uncertain (the value of $k(x,x)$ depends on the hyperparameters of $\mathcal{GP}_{ij}$). Stacking the individual predictions together, our model-learner produces in summary

$$\tilde{f}(x,a) := \begin{bmatrix} x^{(1)} \\ \vdots \\ x^{(d)} \end{bmatrix} + \begin{bmatrix} \mu_{1a}(x) \\ \vdots \\ \mu_{da}(x) \end{bmatrix}, \qquad c(x,a) := \max_{i=1,\ldots,d} \left(\text{normalize}_{ia}(\sigma^2_{ia})\right), \quad (6)$$

where $\tilde{f}(x,a)$ is the predicted successor state and $c(x,a)$ the associated uncertainty (taken as maximum over the normalized per-coordinate uncertainties, where normalization ensures that the values lie between 0 and 1).

3.3 Planning with a Model

At any time t, the planner receives as input model $\mathcal{M}_t$. For any state x and action a, model $\mathcal{M}_t$ can be evaluated to "produce" the transition $\tilde{f}(x,a)$ along with normalized scalar uncertainty $c(x,a) \in [0,1]$, where 0 means maximally certain and 1 maximally uncertain (see Section 3.2)

Let Γ_h be the discretization of the state space $\mathcal{X}$ with nodes ξ_i, $i = 1,\ldots,N$. We now solve the planning stage by plugging $\tilde{f}$ into the procedure described in Section 2. First, we compute $\tilde{z}^a_i = \tilde{f}(\xi_i, a)$, $c(\xi_i, a)$ from (6) and the associated interpolation coefficients w^a_{ij} from (3) for each node ξ_i and action a.

Let C^a denote the $N \times 1$ vector corresponding to the uncertainties, $[C^a]_i = c(\xi_i, a)$; and R^a be the $N \times 1$ vector corresponding to the rewards, $[R^a]_i = r(\xi_i, a)$. To solve the discretized Bellman equation in Eq. (4), we perform basic Jacobi iteration:

- **Initialize** $[Q^a_0]_i$, $i = 1,\ldots,N$, $a = 1,\ldots,|A|$
- **Repeat** for $k = 0,1,2,\ldots$

$$[Q^a_{k+1}]_i = [R^a]_i + \gamma \max_{a'} \left\{ \sum_{j=1}^{N} w^a_{ij} [Q^{a'}_k]_j \right\} \qquad \forall i, a \quad (7)$$

until $|Q^a_{k+1} - Q^a_k|_\infty < tol$, $\forall a$, or a maximum number of iterations is reached.

To reduce the number of iterations necessary, we adapt Grüne's *increasing coordinate algorithm* [9] to the case of Q-functions: instead of Eq. (7), we perform updates of the form

$$[Q^a_{k+1}]_i = [1 - \gamma w^a_{ii}]^{-1} \left([R^a]_i + \gamma \max_{a'} \left\{ \sum_{j=1, j\neq i}^{N} w^a_{ij} [Q^{a'}_k]_j \right\} \right). \quad (7')$$

In [9] it was proved that Eq. (7') converges to the same fixed point as Eq. (7), and it was empirically demonstrated that convergence can occur in significantly fewer iterations. The exact reduction is problem-dependent, savings will be greater for small γ and large cells where self-transitions occur (i.e., ξ_i is among the vertices of the cell enclosing $\tilde{z}_i^a$).

To implement the "optimism in the face of uncertainty" principle, that is, to make the agent explore regions of the state space where the model predictions are uncertain, we employ the heuristic modification of the Bellman operator which was suggested in [15] and shown to perform well. Instead of Eq. (7'), the update rule becomes

$$[Q_{k+1}^a]_i = (1 - [C^a]_i)[1 - \gamma w_{ii}^a]^{-1} \left([R^a]_i + \gamma \max_{a'} \left\{ \sum_{j=1, j \neq i}^{N} w_{ij}^a [Q_k^{a'}]_j \right\} \right) + \\ + [C^a]_i V_{\mathrm{MAX}} \quad (7")$$

where $V_{\mathrm{MAX}} := R_{\mathrm{MAX}}/(1-\gamma)$. Eq. (7") can be seen as a generalization of the binary uncertainty in the original RMAX paper to continuous uncertainty; whereas in RMAX a state was either "known" (sufficiently explored), in which case the unmodified update was used, or "unknown" (not sufficiently explored), in which case the value V_{MAX} was assigned, here the shift from exploration to exploitation is more gradual.

Finally we can take advantage of the fact that the planning function will be called many times during the process of learning. Since the discretization Γ_h is kept fixed, we can reuse the final Q-values obtained in one call to plan as initial values for the next call to plan. Since updates to the model often affect only states in some local neighborhood (in particular in later stages), the number of necessary iterations in each call to planning will be further reduced.

A summary of our model-based planning function is shown below.

- **Input:**
  - Model $\mathcal{M}_t$, initial $[Q_0^a]_i$, $i = 1, \ldots, N$, $a = 1, \ldots, |A|$
- **Static inputs:**
  - Grid Γ_h with nodes $\xi_1, \ldots, \xi_N$, discount factor γ, reward function $r(x, a)$ evaluated in nodes giving $[R^a]_i$
- **Initialize:**
  - Compute $\tilde{z}_i^a = \tilde{f}(\xi_i, a)$ and $[C^a]_i$ from $\mathcal{M}_t$ (see Eq. (6))
  - Compute weights w_{ij}^a for each $\tilde{z}_i^a$ (see Eq. (3))
- **Loop:**
  - Repeat update Eq. (7") until $|Q_{k+1}^a - Q_k^a|_\infty < tol$, $\forall a$, or the maximum number of iterations is reached.

4 Experiments

We now examine the online learning performance of GP-RMAX in various well-known RL benchmark domains.

4.1 Description of Domains

In particular, we choose the following domains (where a large number of comparative results is available in the literature):

Mountain car: In mountain car, the goal is to drive an underpowered car from the bottom of a valley to the top of one hill. The car is not powerful enough to climb the hill directly, instead it has to build up the necessary momentum by reversing throttle and going up the hill on the opposite side first. The problem is 2-dimensional, state variable $x_1 \in [-1.2, 0.5]$ describes the position of the car, $x_2 \in [-0.07, 0.07]$ its velocity. Possible actions are $a \in \{-1, 0, +1\}$. Learning is episodic: every step gives a reward of -1 until the top of the hill at $x_1 \geq 0.5$ is reached. Our experimental setup (dynamics and domain specific constants) is the same as in [19], with the following exceptions: maximal episode length is 500 steps, discount factor $\gamma = 0.99$ and every episode starts with the agent being at the bottom of the valley with zero velocity, $x_{\text{start}} = (-\pi/6, 0)$.

Inverted pendulum: The next task is to swing up and stabilize a single-link inverted pendulum. As in mountain car, the motor does not provide enough torque to push the pendulum up in a single rotation. Instead, the pendulum needs to be swung back and forth to gather energy, before being pushed up and balanced. This creates a more difficult, nonlinear control problem. The state space is 2-dimensional, $\theta \in [-\pi, \pi]$ being the angle, $\dot{\theta} \in [-10, 10]$ the angular velocity. Control force is discretized to $a \in \{-5, -2.5, 0, +2.5, +5\}$ and held constant for 0.2sec. Reward is defined as $r(x, a) := -0.1x_1^2 - 0.01x_2^2 - 0.01a^2$. The remaining experimental setup (equations of motion and domain specific constants) is the same as in [6]. The task is made episodic by resetting the system every 500 steps to the initial state $x_{\text{start}} = (0, 0)$. Discount factor $\gamma = 0.99$.

Bicycle: Next we consider the problem of balancing a bicycle that rides at a constant speed [8],[12]. The problem is 4-dimensional: state variables are the roll angle $\omega \in [-12\pi/180, 12\pi/180]$, roll rate $\dot{\omega} \in [-2\pi, 2\pi]$, angle of the handle bar $\alpha \in [-80\pi/180, 80\pi/180]$, and the angular velocity $\dot{\alpha} \in [-2\pi, 2\pi]$. The action space is inherently 2-dimensional (displacement of rider from the vertical and turning the handlebar); in RL it is usually discretized into 5 actions. Our experimental setup so far is similar to [8]. To allow a more conclusive comparison of performance, instead of just being able to keep the bicycle from falling, we define a more discriminating reward $r(x, a) = -x_1^2$, and $r(x, a) = -10$ for $|x_1| < 12\pi/180$ (bicycle has fallen). Learning is episodic: every episode starts in one of two (symmetric) states close to the boundary from where recovery is impossible: $x_{\text{start}} = (10\pi/180, 0, 0, 0)$ or $x_{\text{start}} = (-10\pi/180, 0, 0, 0)$, and proceeds for 500 steps or until the bicycle has fallen. Discount factor $\gamma = 0.98$.

Acrobot: Our final problem is the acrobot swing-up task [19]. The goal is to swing up the tip of the lower link of an underactuated two-link robot over a given height (length of first link). Since only the lower link is actuated, this is a rather challenging problem. The state space is 4-dimensional: $\theta_1 \in [-\pi, \pi]$, $\dot{\theta}_1 \in [-4\pi, 4\pi]$, $\theta_2 \in [-\pi, \pi]$, $\dot{\theta}_2 \in [-9\pi, 9\pi]$. Possible actions are $a \in \{-1, +1\}$. Our experimental setup and implementation of state transition dynamics is similar to [19]. The objective of learning is to reach a goal state as quickly as possible, thus $r(x, a) = -1$ for every step. The initial state for every episode is $x_{\text{start}} = (0, 0, 0, 0)$. An episode ends if either a goal state is reached or 500 steps have passed. The discount factor was set to $\gamma = 1$, as in [19].

4.2 Results

We now apply our algorithm GP-RMAX to each of the four problems. The granularity of the discretization Γ_h in the planner is chosen such that for the 2-dimensional problems, the loss in performance due to discretization is negligible. For the 4-dimensional problems, we ran offline trials with the true transition function to find the best compromise of granularity and computational efficiency. As result, we use a 100×100 grid for mountain car and inverted pendulum, a $20 \times 20 \times 20 \times 20$ grid for the bicycle balancing task, and a $25 \times 25 \times 25 \times 25$ grid for the acrobot. The maximum number of value iterations was set to 500, tolerance was $< 10^{-2}$. In practice, running the full planning step took between 0.1-10 seconds for the small problems, and less than 5 min for the large problems (where often more than 50% of the CPU time was spent on computing the GP predictions in all the nodes of the grid). Using the planning module offline with the true transition function, we computed the best possible performance for each domain in advance. We obtained: mountain car (103 steps), inverted pendulum (-18.41 total cost), bicycle balancing (-3.49 total cost), and acrobot (64 steps).[4]

For the GP-based model-learner, we set the maximum size of the subset to 1000, and ICD tolerance to 10^{-2}. The hyperparameters of the covariance were not manually tuned, but found from the data by likelihood optimiziation.

Since it would be computationally too expensive to update the model and perform the full planning step after every single observation, we set the planning frequency K to 50 steps. To gauge if optimal behavior is reached and further learning becomes unnessecary, we monitor the change in the model predictions and uncertainties between successive updates and stop if both fall below a threshold (test points in a fixed coarse grid).

We consider the following variations of the base algorithm: (1) GP-RMAXEXP, which actively explores by adjusting the Bellman updates in Eq. (7”) according to the uncertainties produced by the GP prediction; (2) GP-RMAXGRID, which does the same but uses binary uncertainty by overlaying a uniform grid on top of the state-action space and keeping track which cells are visited; and (3) GP-RMAXNOEXP, which does not actively explore (see Eq. (7’)). For comparison, we repeat the experiments using the standard online model-free RL algorithm Sarsa(λ) with tile coding [19], where we consider two different setup of the tilings (one finer and one coarser).

[4] Note that 64 steps is not the optimal solution, [2] demonstrated swing-up with 61 steps.

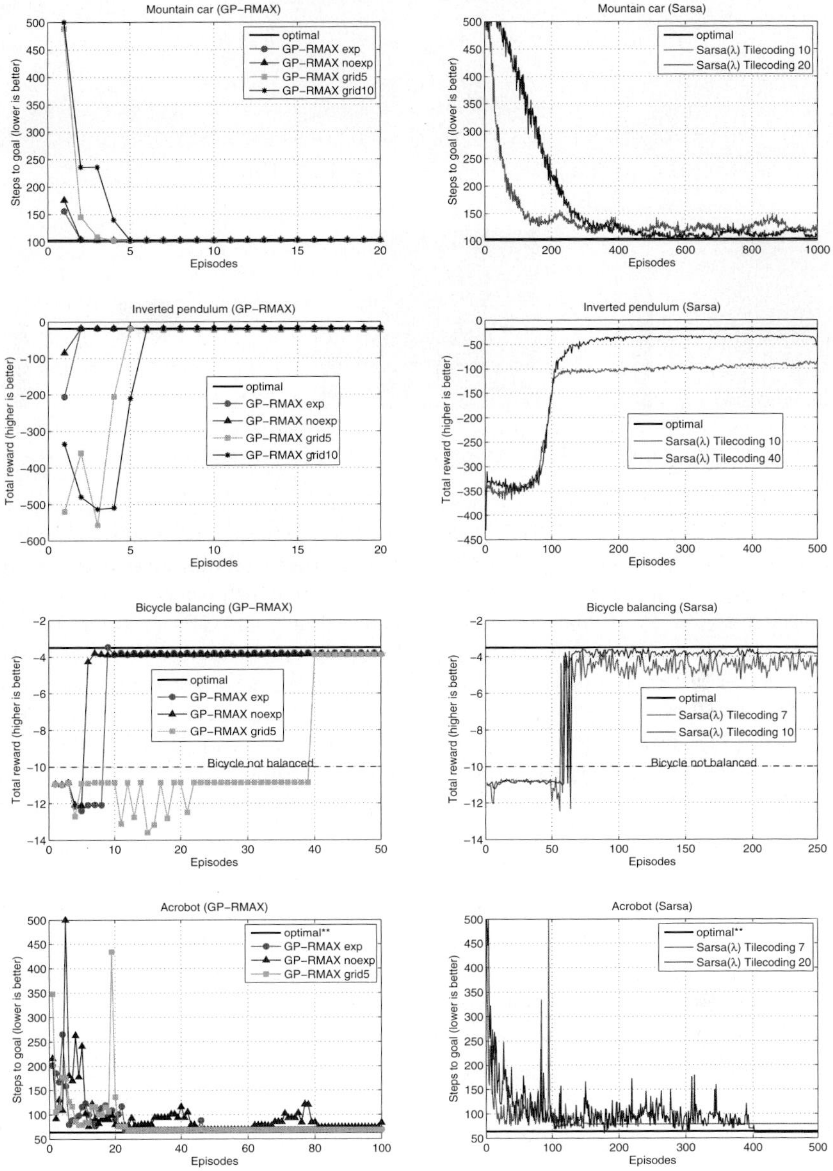

Fig. 3. Learning curves of our algorithm GP-RMAX (left column) and the standard method Sarsa(λ) with tile coding (right column) in the four benchmark domains. Each curve shows the online learning performance and plots the total reward as a function of the episode (and thus sample complexity). The black horizontal line denotes the best possible performance computed offline. Note the different scale of the x-axis between GP-RMAX and Sarsa.

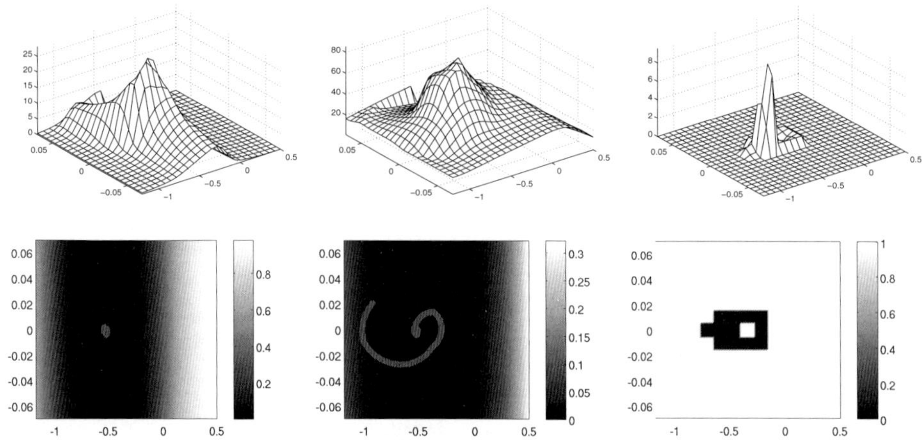

(a) GP after 40 samples (b) GP after 120 samples (c) Grid after 120 samples

Fig. 4. Model-learning and propagation of "knownness" of state-action pairs with GPs. The top row shows the value function that results from applying value iteration with the update modified for uncertainty, see Eq. (7"). The bottom row shows the actual samples (red circles) and the induced uncertainty of all states: black is perfectly "known", white is perfectly "unknown". Panels (a) and (b) show that with GPs certainty of model predictions is rapidly propagated through the whole state space, leading to strong generalization and targeted exploration. This in turn allows the optimal value function to be learned from very few sample transitions: panel (b) shows that after only 120 transitions (still in the middle of the very first episode) the approximated value function already resembles the true one [19]. Panel (c) shows the same for a counter-based binary uncertainty; most of the grid cells are unvisited and the thus the approximate value function is zero in most parts of the state space.

Figure 3 shows the result of online learning with GP-RMAX and Sarsa. In short, the graphs show us two things in particular: (1) GP-RMAX learns very quickly; and (2) GP-RMAX learns a behavior that is very close to optimal. In comparison, Sarsa(λ) has a much higher sample complexity and does not always learn the optimal behavior (exception is the acrobot). While direct comparison with other high performance RL algorithms, such as fitted value iteration [18,8,15], policy iteration based LSPI/LSTD/LSPE [12,4,13,11], or other kernel-based methods [7,6] is difficult, because they are either batch methods or handle exploration in a more ad-hoc way, from the respective results given in the literature it is clear that for the domains we examined GP-RMAX performs relatively well.

Examining the plots in more detail, we find that, while GP-RMAXGRID is somewhat less sample efficient (explores more), GP-RMAXEXP and GP-RMAXNOEXP perform nearly the same. Initially, this appears to be in contrast with the whole point of RMAX, which is efficient exploration guided by the uncertainty of the predictions. Here, we believe that this behavior can be explained by the good generalization capabilities of GPs. Figure 4 illustrates model learning and certainty propagation with GPs in the mountain car domain (predicting acceleration as function of state). The state of the model-learner is shown for two snapshots: after 40 transitions and after 120 transitions. The top row shows the value function that results from applying value iteration with the update modified for uncertainty, see Eq. (7”). The bottom row shows the observed samples and the associated certainty of the predictions. As expected, certainty is high in regions where data was observed. However, due to the generalization of GPs and data-dependent hyperparameter selection, certainty is also high in unexplored regions; and in particular it is constant along the y-coordinate. To understand this, we have to look at the state transition function of the mountain car: acceleration of the car indeed only depends on the position, but not on velocity. This shows that certainty estimates of GPs are *supervised* and take the properties of the target function into account, whereas prior RMAX treatments of uncertainty are *unsupervised* and only consider the density of samples to decide if a state is “known”. For comparison, we also show what GP-RMAX with grid-based uncertainty would produce in the same situation.

5 Summary

We presented an implementation of model-based online reinforcement learning similar to RMAX for continuous domains by combining GP-based model learning and value iteration on a grid. Doing so, our algorithm separates the problem function approximation in the model-learner from the problem function approximation/interpolation in the planner. If the transition function is easier to learn, i.e., requires only few samples relative to the representation of the optimal value function, then large savings in sample-complexity can be gained. Related model-free methods, such as fitted Q-iteration, can not take advantage of this situation. The fundamental limitation of our approach is that it relies on solving the Bellman equation globally over the state space. Even with more advanced discretization methods, such as adaptive grids, or sparse grids, the curse of dimensionality limits the applicability to problems with low or moderate dimensionality. Other, more minor limitations, concern the simplifying assumptions we made: deterministic state transitions and known reward function. However, these are not conceptual limitations but rather simplifying assumptions made for the present paper; they could be easily addressed in future work.

Acknowledgments

This work has taken place in the Learning Agents Research Group (LARG) at the Artificial Intelligence Laboratory, The University of Texas at Austin. LARG research is supported in part by grants from the National Science Foundation (IIS-0917122), ONR (N00014-09-1-0658), DARPA (FA8650-08-C-7812), and the Federal Highway Administration (DTFH61-07-H-00030).

References

1. Bernstein, A., Shimkin, N.: Adaptive-resolution reinforcement learning with efficient exploration. Machine Learning (2010), doi:10.1007/s10994-010-5186-7 (published online: May 5, 2010)
2. Boone, G.: Minimum-time control of the acrobot. In: Proc. of IEEE International Conference on Robotics and Automation, vol. 4, pp. 3281–3287 (1997)
3. Brafman, R., Tennenholtz, M.: R-MAX, a general polynomial time algorithm for near-optimal reinforcement learning. JMLR 3, 213–231 (2002)
4. Busoniu, L., Ernst, D., De Schutter, B., Babuska, R.: Online least-squares policy iteration for reinforcement learning control. In: American Control Conference, ACC 2010 (2010)
5. Davies, S.: Multidimensional triangulation and interpolation for reinforcement learning. In: NIPS 9. Morgan, San Francisco (1996)
6. Deisenroth, M.P., Rasmussen, C.E., Peters, J.: Gaussian process dynamic programming. Neurocomputing 72(7-9), 1508–1524 (2009)
7. Engel, Y., Mannor, S., Meir, R.: Bayes meets Bellman: The Gaussian process approach to temporal difference learning. In: Proc. of ICML 20, pp. 154–161 (2003)
8. Ernst, D., Geurts, P., Wehenkel, L.: Tree-based batch mode reinforcement learning. JMLR 6, 503–556 (2005)
9. Grüne, L.: An adaptive grid scheme for the discrete Hamilton-Jacobi-Bellman equation. Numerische Mathematik 75, 319–337 (1997)
10. Jong, N.K., Stone, P.: Model-based exploration in continuous state spaces. In: The 7th Symposium on Abstraction, Reformulation and Approximation (2007)
11. Jung, T., Polani, D.: Learning robocup-keepaway with kernels. JMLR: Workshop and Conference Proceedings (Gaussian Processes in Practice) 1, 33–57 (2007)
12. Lagoudakis, M.G., Parr, R.: Least-squares policy iteration. JMLR 4, 1107–1149 (2003)
13. Li, L., Littman, M.L., Mansley, C.R.: Online exploration in least-squares policy iteration. In: Proc. of 8th AAMAS (2009)
14. Munos, R., Moore, A.: Variable resolution discretization in optimal control. Machine Learning 49, 291–323 (2002)
15. Nouri, A., Littman, M.L.: Multi-resolution exploration in continuous spaces. In: NIPS 21 (2008)
16. Quiñonero-Candela, J., Rasmussen, C.E., Williams, C.K.I.: Approximation methods for gaussian process regression. In: Bottou, L., Chapelle, O., DeCoste, D., Weston, J. (eds.) Large Scale Learning Machines, pp. 203–223. MIT Press, Cambridge (2007)
17. Rasmussen, C.E., Williams, C.K.I.: Gaussian Processes for Machine Learning. MIT Press, Cambridge (2006)
18. Riedmiller, M.: Neural fitted q-iteration. In: Proc. of 16th ECML (2005)
19. Sutton, R., Barto, A.: Reinforcement Learning: An Introduction. MIT Press, Cambridge (1998)

Author Index

Printing: Mercedes-Druck, Berlin
Binding: Stein+Lehmann, Berlin